3.3 Function A function from a set X to a set Y is a correspondence that a[ssigns] one element of Y.

3.3 Domain, range The domain of a function defined by $y = f(x)$ is the set of all possible values of x; the range is the set of all possible values of y.

3.4 Slope The slope m of the line through (x_1, y_1) and (x_2, y_2), with $x^2 - x_1 \neq 0$, is $\dfrac{y_2 - y_1}{x_2 - x_1}$.

The slope of a vertical line is undefined.

3.5 Point-slope form The line with slope m passing through the point (x_1, y_1) has an equation $y - y_1 = m(x - x_1)$.

3.5 Slope-intercept form The line with slope m and y-intercept b has an equation $y = mx + b$.

3.8 Composition of functions $(g \circ f)(x) = g[f(x)]$

3.9 One-to-one function A function f is one to one if $f(a) = f(b)$ implies that $a = b$.

3.9 Inverse functions Let f be a one-to-one function with domain X and range Y. Let g be a function with domain Y and range X. Then g is the inverse function of f if

$$(f \circ g)(x) = x \text{ for every } x \text{ in } Y,$$

and $(g \circ f)(x) = x$ for every x in X.

5.2 Logarithm For $a > 0$, $a \neq 1$, and $x > 0$, $y = \log_a x$ if and only if $x = a^y$.

5.2 Properties of logarithms For $a > 0$, $a \neq 1$, $x > 0$, and $y > 0$,

$$\log_a xy = \log_a x + \log_a y \qquad \log_a \frac{x}{y} = \log_a x - \log_a y$$

$$\log_a x^r = r \cdot \log_a x \qquad \log_a a = 1$$

$$\log_a 1 = 0$$

11.1 Binomial theorem For any positive integer n and any complex numbers x and y,

$$(x + y)^n = x^n + \binom{n}{n-1} x^{n-1}y + \ldots + \binom{n}{n-r} x^{n-r}y^r + \ldots + \binom{n}{1} xy^{n-1} + y^n.$$

11.2 Arithmetic sequences nth term: $a_n = a_1 + (n-1)d \qquad S_n = \dfrac{n}{2}(a_1 + a_n) \quad \text{or} \quad S_n = \dfrac{n}{2}[2a_1 + (n-1)d]$

11.3 Geometric sequences nth term: $a_n = a_1 r^{n-1} \qquad S_n = \dfrac{a_1(1 - r^n)}{1 - r} \quad (r \neq 1) \qquad S_\infty = \dfrac{a_1}{1 - r} \quad (-1 < r < 1)$

11.6 Permutations The number of arrangements of n things taken r at a time is $P(n, r) = \dfrac{n!}{(n-r)!}$.

11.6 Combinations The number of ways to choose r things from a group of n things is $\binom{n}{r} = \dfrac{n!}{(n-r)!r!}$.

11.7 Properties of probability For any events E and F:

$P(\text{a certain event}) = 1$ $\qquad\qquad\qquad\qquad 0 \leq P(E) \leq 1$

$P(\text{an impossible event}) = 0$ $\qquad\qquad\qquad P(E') = 1 - P(E)$

$P(E \text{ or } F) = P(E \cup F) = P(E) + P(F) - P(E \cap F)$.

FIFTH EDITION

ALGEBRA AND TRIGONOMETRY

FIFTH EDITION

ALGEBRA AND TRIGONOMETRY

Margaret L. Lial
American River College

Charles D. Miller

David I. Schneider
University of Maryland

SCOTT, FORESMAN/LITTLE, BROWN HIGHER EDUCATION
A Division of Scott, Foresman and Company
Glenview, Illinois   London, England

To the Student

If you need further help with this course, you may want to obtain a copy of the *Student's Solutions Manual and Study Guide* that accompanies this textbook. This manual provides detailed step-by-step solutions to the odd-numbered exercises in the textbook as well as a practice test for every chapter. Your college bookstore either has this manual or can order it for you.

Cover photo: © William J. Warren/West Light

Library of Congress Cataloging-in-Publication Data

Lial, Margaret L.
    Algebra and trigonometry/Margaret L. Lial, Charles D. Miller,
David I. Schneider.—5th ed.
      p.   cm.
    ISBN 0-673-38576-0
    1. Algebra.   2. Trigonometry.   I. Miller, Charles David.
    II. Schneider, David I.   III. Title.
QA154.2.L49   1990
512'.13—dc20   89-10738

Artwork, illustrations, and other materials supplied by the publisher. Copyright © 1990 Scott, Foresman and Company.

23456-VHJ-94 93 92 91 90

PREFACE

The fifth edition of *Algebra and Trigonometry* gives a mathematically sound development of the topics in algebra and trigonometry needed for success in later courses. In particular, it is intended to prepare students for calculus by providing them with thorough explanations that are precise, yet written for student comprehension. To answer the question "How will I use this?" we have included many examples and exercises designed to show students how the topics they are studying will be used in calculus as well as in many applications in various fields.

The book can be used for courses given in one semester, two semesters, or two quarters. The prerequisites are a course in intermediate algebra and a course in geometry. For students whose background is not solid or who have not studied the prerequisites recently, we have included a review of the basics of algebra in Chapter 1 and a review of equations and inequalities in Chapter 2. (A special set of cumulative review exercises at the end of Chapter 2 can be used to decide which portions of the first two chapters need to be covered or reviewed.)

Key Features

New historical comments are included throughout the text to give insights into how mathematics developed. These provide interest and background information.

New exercises that test understanding are included in most exercise sets. These are not difficult and can be answered quickly if the topic is understood. This provides both the student and the instructor with a way to determine whether real comprehension is occurring.

Challenging problems of the sort that students will encounter in calculus are included in most exercise sets, usually at the end of the set. Many of these have references to their use in calculus. These exercises will prepare students for the level of algebraic competence they will need in calculus and further mathematics.

Cautions to help students recognize and avoid common pitfalls and errors and **Notes** to emphasize unique situations are included throughout and highlighted with a striped bar.

Numerous carefully detailed **examples** and extensive **exercise sets,** nearly 6000 problems in all, are hallmarks of this book.

Many graphs and illustrations are given in the exercise sets as well as in the exposition to enhance understanding and to aid students in picturing a situation that has been described in words.

A summary of key words referenced to the sections in which they are introduced is given at the end of each chapter. These provide a quick check for coverage of concepts.

Review exercises are included at the end of each chapter to help students review and assimilate the material studied in the chapter.

The **format** of the textbook is designed to be helpful to learning as well as appealing. Color is used to highlight explanations both in text and in the figures. Important rules and definitions are set off with colored screens and titles. The latter feature helps students find definitions and rules easily and helps instructors see what must be covered in the lecture.

Content Highlights

Linear and quadratic functions are presented right before the general topics of symmetry, translation, algebra of functions, and inverse functions. In this way, material on functions that the student has just studied and is familiar with can be used to discuss the new function concepts.

Zeros of polynomials are included in the same chapter as polynomial functions. This allows the use of synthetic division and the theorems on zeros of polynomial functions for graphing polynomials.

Calculator use is encouraged throughout the text and explained as needed. Since calculator use is assumed, all angle measures in the chapters on trigonometry are given in decimal degrees. Discussions of the use of tables to find trigonometric values and common logarithms are given in the appendices.

The trigonometric functions are introduced in Chapter 6 with the unit circle. Trigonometry is treated thoroughly in order to prepare students for what can be a troublesome area in the study of calculus.

The material on graphing trigonometric functions has been expanded. Students are encouraged to use key points for each function as an aid in graphing the function.

The chapter on systems, Chapter 9, has been extensively reorganized to present first those topics most frequently taught: linear systems solved by elimination and substitution, nonlinear systems, and systems of inequalities.

Ellipses and hyperbolas have been combined with a complete discussion of the conic sections, including rotation of axes, in Chapter 10.

Supplements

The **Instructor's Answer Manual** gives answers to each text exercise, both even-numbered and odd-numbered.

The **Instructor's Guide** includes an extensive set of test questions for each chapter, organized by section, plus answers to all the questions. It also includes six ready-to-duplicate test forms for each chapter, two of which are multiple-choice, plus answers.

The **Instructor's Solutions Manual** provides detailed, worked-out solutions to all of the even-numbered text exercises.

The **Scott, Foresman Test Generator for Mathematics** enables instructors to select questions for any section in the text, or to use a ready-made test for each chapter. Instructors may generate tests in multiple-choice or open-response formats, scramble the order of questions while printing, and produce multiple versions of each test (up to 9 with Apple, up to 25 with IBM and Macintosh). The system features printed graphics and accurate mathematics symbols. It also features a preview option that allows instructors to view questions before printing, to regenerate variables, and to replace or skip questions if desired. The IBM version includes an editor that allows instructors to add their own problems to existing data disks.

The **Student's Solutions Manual and Study Guide** gives detailed, worked-out solutions to all of the odd-numbered text exercises, plus a practice chapter test for each chapter. Solutions to the practice test questions are provided.

A set of approximately 75 two-color **overhead transparencies** accompanies the text. These mathematical figures from the text can enhance classroom presentations.

Acknowledgments

Many people helped in the preparation of this revision. Special thanks are due to Florence Fasanelli, National Science Foundation, for serving as our consultant on the history of mathematics, and to David Deutsch, University of Maryland, for making many valued contributions to the writing process as well as helping check the answers.

We also thank the following people who helped check the answers: Hilde Achepohl; Patricia Bradley; Brian Hayes, Triton College; Karen Livengood, College of DuPage; and Abby Tanenbaum, College of DuPage.

The following people reviewed the manuscript and made many helpful suggestions for improving this edition: Evan Alderfer, Ocean County College; Ruth E. Briggs, Detroit College of Business; Kathryn Conroy, Anoka-Ramsey Community College; Roger L. Creech, East Carolina University; Roger E. Davis, Williamsport Area Community College; Joe P. Everett, Chattanooga State Technical Community College; Peter U. Georgakis, Santa Barbara City College; Walter P. Gerlach, Jacksonville State University; Joel Greenstein, New York City Technical College; Kenneth R. Gurganus, University of North Carolina at Wilmington; Edwin M. Klein, University of Wisconsin-Whitewater; Judith E. Land, Boston University; Rudy Maglio, Oakton Community College; Peter L.A. Oakes, Muskegon Community College; Glen Power, Indian River Community College; George W. Schultz, Saint Petersburg Junior College; Ralph Selig, Fairleigh Dickinson University; Mark Serebransky, Camden County College; Charles W. Sinclair, Portland State University; Donna M. Szott, Community College of Allegheny County, South Campus; Frank Ward, Indian River Community College; Karen E. Zak, Washington College.

As always, the excellent staff at Scott, Foresman has been extremely helpful to us. In particular, we wish to thank Bill Poole, Linda Youngman, and Sarah Joseph.

Margaret L. Lial

David I. Schneider

CONTENTS

APPENDICES

TABLES

ANSWERS TO SELECTED EXERCISES   687

INDEX   729

USING A CALCULATOR

Buying a Calculator   A scientific calculator is very helpful in doing mathematics, such as trigonometry. These calculators once cost hundreds of dollars, but now can be purchased for less than the cost of a pizza. Scientific calculators can be recognized by the following keys (among others):

The first of these keys is used in trigonometry, the second to find the logarithm of a number, and the third to raise a number to a power. A scientific calculator has the advantage of doing away with the need for most tables; with a scientific calculator you seldom if ever need to use a table in a trigonometry course. (The general skill of table reading is still useful, however, so we include it in the book and many instructors discuss it in class.)

Some advanced calculators are *programmable:* instead of starting each new problem of a given type from scratch, only the necessary keystrokes are entered. Then the data for a new problem can be entered, with only one or two keys needed to get the result. A programmable calculator is not really necessary for this course. However, such calculators do offer two advantages: first, since you cannot program a calculator for a group of problems unless you completely understand the basic ideas of the problems, your understanding would be enhanced; and second, the programming skills taught in using such calculators are useful in further course work in science or mathematics.

There are two types of logic in common use in calculators today. Both algebraic and Reverse Polish Notation (RPN) have advantages and disadvantages. Algebraic logic is the easiest to learn. For example, the problem 8 + 17 is entered into an algebraic machine by pressing

$$8 + 17 = .$$

On a machine with Reverse Polish Notation (named for the eminent Polish mathematicians who developed the system), this same problem would be entered as

$$8 \text{ ENTER } 17 + .$$

Some people claim that Reverse Polish machines work advanced problems more easily than algebraic machines. Others claim that algebraic machines are easier to use for the great bulk of ordinary, common problems. It is up to you to decide which to buy. You may wish to discuss your calculator purchase with your instructor for further guidance.

Calculator Errors   A calculator can store only so many digits in its memory. Because of this, numbers that have more digits than can be stored must be rounded. For example, 1/3 is not stored as the exact fraction 1/3, but rather as a decimal, perhaps .3333333333333. Since this rounded form of 1/3 is used, errors can occur in calculations. To see how this happens, use a calculator to divide 1 by 3, and then multiply the result by 3. You should get 1 (exactly), but many machines produce

$$(1 \div 3) \times 3 = \left(\frac{1}{3}\right) \times 3 = .9999999999.$$

Some machines round this result to 1; however, the machine does not treat the number internally as 1. To see this, subtract 1 from the result above; you should get 0 but probably will not.

Another calculator error results when numbers of greatly different size are used in addition. For example,

$$10^9 + 10^{-5} - 10^9 = 10^{-5}.$$

However, most calculators would give

$$10^9 + 10^{-5} - 10^9 = 0.$$

These calculator errors seldom occur in realistic problems, but if they do occur you should know what is happening.

Using a Calculator   While this introduction is not designed to replace your calculator instruction manual, we do list a few things to keep in mind as you use your calculator.

Parenthesis   Many calculators have parentheses keys, $\boxed{(}$ and $\boxed{)}$ These are used as in algebra. For example, $(3 \cdot 5 + 8 \cdot 2) \cdot 4$ could be found as follows.

$$\boxed{(} \quad 3 \quad \boxed{\times} \quad 5 \quad \boxed{+} \quad 8 \quad \boxed{\times} \quad 2 \quad \boxed{)} \quad \boxed{\times} \quad 4 \quad \boxed{=} \quad 124$$

Memory   A memory key is like an electronic piece of scratch paper. Pressing $\boxed{M}$ or $\boxed{STO}$ will cause the number in the display to be stored, and pressing $\boxed{MR}$ or $\boxed{RCL}$ will cause it to be recalled.

Scientific Notation   A key labeled $\boxed{\text{EE}}$ permits numbers to be entered in scientific notation. For example, entering 9.68, pressing $\boxed{\text{EE}}$, and then entering 5, followed by $\boxed{+/-}$, results in the display

$$9.68 - 05,$$

which represents 9.68×10^{-5}. With some calculators, pressing $\boxed{\text{INV}}$ and then $\boxed{\text{EE}}$ causes a number displayed in scientific notation to be written in regular notation.

Significant Digits   A common mistake with calculator use is to quote the displayed answer to more accuracy than is warranted by the original data. For example, if we measure a wall to the nearest meter and say that it is 18 meters long, then we are really saying that the wall has a length from 17.5 meters to 18.5 meters. If we measure the wall more accurately, and say that it is 18.3 meters long, then we know that its length is really in the range 18.25 meters to 18.35 meters. A measurement of 18.00 meters would indicate that the wall's length is in the range 17.995 to 18.005 meters. The measurement 18 meters is said to have 2 significant digits of accuracy; 18.3 has 3 significant digits and 18.00 has 4.

The following chart shows some numbers, the number of significant digits in each number, and the range represented by each number.

Number	Number of significant digits	Range represented by number
29.6	3	29.55 to 29.65
1.39	3	1.385 to 1.395
.000096	2	.0000955 to .0000965
.03	1	.025 to .035
100.2	4	100.15 to 100.25

1

FUNDAMENTALS OF ALGEBRA

Today, algebra is required in a great many fields, ranging from accounting to ecology. This is not surprising, since most topics in algebra were developed to help solve applied problems. In this text, algebra is used to predict population growth, to determine the path of objects orbiting in space, and to investigate the costs versus the benefits of removing pollutants from a substance. To prepare for this study, the book begins with a review of the basics of algebra.

The name *algebra* is a Latin translation of the Arabic *al-jabr*. This word is part of the title of a book written by āl-Khwarizmī (*ca.* 800–847), a mathematician and astronomer who lived in Baghdad. The book is a collection of rules for solving equations. Since some techniques had been developed to solve geometric problems, negative numbers were not accepted as solutions. Part of the development of algebra to its current usefulness involves the movement mathematicians made toward the recognition of negative and imaginary numbers. These enable us to write down *all* possible solutions of equations. We begin our study with a review of the real number system.

1.1 THE REAL NUMBERS

Numbers are the foundation of mathematics. The most common numbers in mathematics are the **real numbers.** These numbers can be written as a decimal, either repeating, such as

$$\frac{1}{3} = .333\overline{3}, \quad \frac{3}{4} = .7500\overline{0}, \quad \text{or} \quad 2\frac{4}{7} = 2.571428\overline{571428},$$

or nonrepeating, such as

$$\sqrt{2} = 1.4142135 \ldots \quad \text{or} \quad \pi = 3.14159. \ldots$$

The real numbers are said to be **closed** under the operations of addition and multiplication. That is, for any two real numbers a and b, the **sum** $a + b$, and the **product** $a \cdot b$ are unique real numbers. Multiplication is written in a variety of ways. The symbols 2×8, $2 \cdot 8$, $2(8)$, and $(2)(8)$ all represent the product of 2 and 8, or 16. For writing products involving **variables** (letters used to represent numbers), no operation symbols may be necessary: $2x$ represents the product of 2 and x, while xy indicates the product of x and y.

The set of real numbers, together with the operations addition and multiplication, form the **real number system.** (Informally, a **set** is a collection of objects.) The key properties of the real number system are given below, where a, b, and c are letters used to represent any real number.

Properties of the Real Numbers

Closure properties	$a + b$ is a real number.
	ab is a real number.
Commutative properties	$a + b = b + a$
	$ab = ba$
Associative properties	$(a + b) + c = a + (b + c)$
	$(ab)c = a(bc)$
Identity properties	There exists a unique real number 0 such that
	$$a + 0 = a \quad \text{and} \quad 0 + a = a.$$
	There exists a unique real number 1 such that
	$$a \cdot 1 = a \quad \text{and} \quad 1 \cdot a = a.$$
Inverse properties	There exists a unique real number $-a$ such that
	$$a + (-a) = 0 \quad \text{and} \quad (-a) + a = 0.$$
	If $a \neq 0$, there exists a unique real number $1/a$ such that
	$$a \cdot \frac{1}{a} = 1 \quad \text{and} \quad \frac{1}{a} \cdot a = 1.$$
Distributive property	$a(b + c) = ab + ac$

Let's consider some consequences of the last four properties.

The associative properties are used to add or multiply three or more numbers. For example, the associative property for addition says that the sum $a + b + c$ of the real numbers a, b, and c can be found either by first adding a and b, and then adding c to the result, indicated by the association

$$(a + b) + c,$$

or by first adding b and c, and then adding a to the result, indicated by

$$a + (b + c),$$

since, by the associative property, either method gives the same result.

The identity properties show that 0 and 1 are special numbers: the sum of 0 and any real number a is the number a, so that 0 preserves the identity of a real number under addition. For this reason, 0 is the **identity element for addition.** In the same way, 1 preserves the identity of a real number under multiplication, making 1 the **identity element for multiplication.**

According to the additive inverse property, for any real number a there is a real number, written $-a$, such that the sum of a and $-a$ is 0, or $a + (-a) = 0$. The number $-a$ is called the **additive inverse** or **negative** of a. The additive inverse property also says that this number $-a$ is *unique;* that is, a given number has only one additive inverse.

 CAUTION   Don't confuse the *negative of a number* with a *negative number*. Since a is a variable, it can represent a positive or a negative number (as well as zero). The negative of a, written $-a$, can also be either a negative or a positive number (or zero). It is a common mistake to think that $-a$ *must* represent a negative number, although, for example, if a is -3, then $-a$ is $-(-3)$ or 3.

For each real number a except 0, there is a real number $1/a$ such that the product of a and $1/a$ is 1, or

$$a \cdot \frac{1}{a} = 1, \quad a \neq 0.$$

The symbol $1/a$ is often written a^{-1}.

Definition of a^{-1}

For every nonzero real number a,

$$a^{-1} = \frac{1}{a}.$$

For example, $2^{-1} = 1/2$, $5^{-1} = 1/5$, and $-3^{-1} = -1/3$.

The number $1/a$ or a^{-1} is called the **multiplicative inverse** or **reciprocal** of the number a. Every real number except 0 has a reciprocal. As with the additive inverse, the multiplicative inverse is unique—a given nonzero real number has only one multiplicative inverse.

The distributive property is particularly useful; we use it to rewrite certain sums as products or to rewrite some products as sums. Using the commutative property, we can rewrite the distributive property in an alternative form. Since $a(b + c) = (b + c)a$,

$$(b + c)a = ba + ca$$

The distributive property can be extended to include more than two numbers in the sum, as follows.

$$a(b + c + d + e + \ldots + n) = ab + ac + ad + ae + \ldots + an$$

This form is called the **extended distributive property.**

. . . .

EXAMPLE 1   The following statements illustrate some properties of the real numbers.

(a) $-2 + 3 = 3 + (-2)$ Commutative property of addition

(b) $\sqrt{2}m = m\sqrt{2}$ Commutative property of multiplication
    (The product of $\sqrt{2}$ and m is often written $m\sqrt{2}$, since $\sqrt{2}m$ is too easily confused with $\sqrt{2m}$.)

(c) $-\pi + (\pi + 3) = (-\pi + \pi) + 3$ Associative property of addition

(d) $3(9x) = (3 \cdot 9)x$ Associative property of multiplication

(e) $9 + (-9) = 0$ Inverse property of addition

(f) $6 \cdot 6^{-1} = 6 \cdot \dfrac{1}{6} = 1$ Inverse property of multiplication

(g) $6k^2 + 3k^3 = 2 \cdot 3k^2 + k \cdot 3k^2 = (2 + k)3k^2$ Distributive property
    (This example shows that the distributive property can also be used in reverse to factor an expression.) ●

Other properties useful for solving equations are given below.

. . . .

Further Properties of Real Numbers

For all real numbers a, b, and c:

Substitution property
    If $a = b$, then a may replace b or b replace a in any expression without affecting the truth or falsity of the statement.

Addition property   If $a = b$, then $a + c = b + c$.

Multiplication property   If $a = b$, then $ac = bc$.

Properties of negatives $-(-a) = a$
                $-a(b) = -(ab)$
                $a(-b) = -(ab)$
                $(-a)(-b) = ab$

Properties of zero $a \cdot 0 = 0$
                $ab = 0$ if and only if $a = 0$ or $b = 0$

The addition and multiplication properties say that the same number may be added or multiplied on both sides of a statement of equality. The second property of zero contains the phrase "if and only if." This phrase implies that "if $ab = 0$, then $a = 0$ or $b = 0$" and, conversely, "if $a = 0$ or $b = 0$, then $ab = 0$."

Definition of
If and Only If

For any statements p and q,

$$p \text{ if and only if } q$$

means

if p is true, then q is true   and
if q is true, then p is true.

The properties of real numbers given above apply to addition or multiplication. The two other common operations for the real numbers, subtraction and division, are defined in terms of the operations of addition and multiplication, respectively.

Subtraction is defined by saying that the difference of the numbers a and b, written $a - b$, is found by adding a and the *negative* of b.

Definition of
Subtraction

For all real numbers a and b,

$$a - b = a + (-b).$$

Division of a real number a by a nonzero real number b is defined in terms of multiplication as follows.

Definition of
Division

For all real numbers a and b, with $b \neq 0$,

$$\frac{a}{b} = a \cdot \frac{1}{b} = ab^{-1}$$

That is, to divide a by b, multiply a by the reciprocal of b. The symbol $\div$ is also used to indicate division, as in $12 \div 3 = 4$.

Several useful properties of quotients are listed below.

Properties of
Quotients

For all real numbers a, b, c, and d, with all denominators nonzero,

$$\frac{a}{b} = \frac{c}{d} \text{ if and only if } ad = bc \qquad \frac{a}{b} + \frac{c}{d} = \frac{ad + bc}{bd}$$

$$\frac{ac}{bc} = \frac{a}{b} \qquad\qquad\qquad \frac{a}{b} - \frac{c}{d} = \frac{ad - bc}{bd}$$

$$\frac{a}{-b} = \frac{-a}{b} = -\frac{a}{b} \qquad\qquad \frac{a}{b} \cdot \frac{c}{d} = \frac{ac}{bd}$$

$$\frac{-a}{-b} = \frac{a}{b} \qquad\qquad\qquad \frac{a}{b} \div \frac{c}{d} = \frac{a}{b} \cdot \frac{d}{c}.$$

EXAMPLE 2

(a) $6 - (-15) = 6 + (-(-15))$
$$= 6 + 15$$
$$= 21$$

(b) $\dfrac{-8}{4} = -8 \cdot \dfrac{1}{4} = -2$

(c) $\dfrac{-9}{0}$ is undefined since 0 has no multiplicative inverse; also, $\dfrac{0}{0}$ is undefined.

(d) $\dfrac{\frac{2}{3}}{\frac{5}{7}} = \dfrac{2}{3} \div \dfrac{5}{7}$

$$= \dfrac{2}{3} \cdot \dfrac{7}{5} = \dfrac{14}{15} \quad \bullet$$

To avoid ambiguity when working problems, use the following **order of operations,** which has been generally agreed upon. (By the way, this order of operations is used by computers and many calculators.)

Order of Operations

1. Work separately above and below any fraction bar.
2. Use the rules below within each set of parentheses or square brackets. Start with the innermost and work outward.
3. Evaluate all exponents and roots.
4. Evaluate negations.
5. Do any multiplications or divisions in the order in which they occur, working from left to right.
6. Do any additions or subtractions in the order in which they occur, working from left to right.

EXAMPLE 3   Use the order of operations given above to simplify the following:

(a) $9 + 6 \div 3 = 9 + 2 = 11$

(b) $\dfrac{1 - 2(3 + 4)}{5 + 6 \cdot 2} = \dfrac{1 - 2(7)}{5 + 12}$

$$= \dfrac{1 - 14}{17} = \dfrac{-13}{17} = -\dfrac{13}{17}$$

(c) $4 \cdot 2^3 = 4 \cdot 8 = 32$

(d) $-2^2 = -(2^2) = -4$ •

CAUTION   A common error is evaluating negations before exponents. For instance, in Example 3(d) this type of error would give the incorrect calculation $-2^2 = (-2)^2 = 4$.

There are several subsets* of the set of real numbers that come up so often they are given special names, as listed below. Some of the subsets are written with **set-builder notation;** with this notation,

$$\{x \mid x \text{ has property } P\},$$

read ''the set of all elements x such that x has property P,'' represents the set of all elements having some specified property P.

Subsets of the Real Numbers

Natural numbers	$\{1, 2, 3, 4, \ldots\}$
Whole numbers	$\{0, 1, 2, 3, 4, \ldots\}$
Integers	$\{\ldots, -3, -2, -1, 0, 1, 2, 3, \ldots\}$
Rational numbers	$\left\{\dfrac{p}{q} \,\middle|\, p \text{ and } q \text{ are integers, } q \neq 0\right\}$
Irrational numbers	$\{x \mid x \text{ is a real number that is not rational}\}$

EXAMPLE 4   Let set $A = \{-8, -6, -3/4, 0, 3/8, 1/2, 1, \sqrt{2}, \sqrt{5}, 6, \sqrt{-1}\}$.

(a) The *natural numbers* in set A are 1 and 6.

(b) The *whole numbers* are 0, 1, and 6.

(c) The *integers* are $-8, -6, 0, 1$, and 6.

(d) The *rational numbers* are $-8, -6, -3/4, 0, 3/8, 1/2, 1$, and 6. (The number -8 is rational since -8 can be written as the quotient $-8/1$. Also, 6 is rational since $6 = 6/1$.)

(e) The *irrational numbers* are $\sqrt{2}$ and $\sqrt{5}$. (In further mathematics courses you will see that these numbers do not have repeating or terminating decimal expansions.)

(f) All elements of A are *real numbers* except $\sqrt{-1}$. Square roots of negative numbers are discussed later in this chapter.   •

*Set A is a **subset** of set B if and only if every element of set A is also an element of set B.

The relationships among the various sets of numbers are shown in Figure 1. All the numbers shown are real numbers. (The number π shown in Figure 1 is the ratio of the circumference of a circle to its diameter; π is approximately 3.14159. Also, e is an irrational number discussed later in this text in connection with exponential and logarithmic functions; e is approximately 2.7182818.)

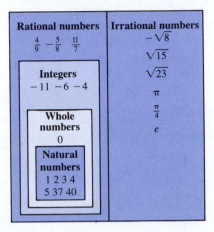

Figure 1

We often need to know which of two given real numbers is the smaller. Deciding which is smaller is sometimes easier with a **number line,** a geometric representation of the set of real numbers. Figure 2 shows a number line with the points corresponding to several different numbers marked on it. A number that corresponds to a particular point on a line is called the **coordinate** of the point. For example, the leftmost marked point in Figure 2 has coordinate -4. The correspondence between points on a line and the real numbers is called a **coordinate system** for the line. (From now on, the phrase "the point on a number line with coordinate a" will be abbreviated as "the point with coordinate a," or simply "the point a.")

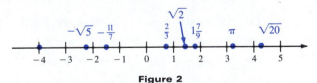

Figure 2

EXAMPLE 5   Locate the elements of the set $\{-2/3, 0, \sqrt{2}, \sqrt{5}, \pi, 4\}$ on a number line.

The number π is irrational, with $\pi \approx 3.14159$ ($\approx$ means "approximately equal to"). From a calculator, $\sqrt{2} \approx 1.414$ and $\sqrt{5} \approx 2.236$. Using this information, place the given points on a number line as shown in Figure 3.   ●

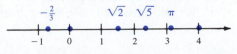

Figure 3

Suppose a and b are two real numbers. If the difference $a - b$ is positive, then a is greater than b, written $a > b$. If the difference $a - b$ is negative, then a is less than b, written $a < b$. The signs $<$ and $>$ were invented by the English mathematician Thomas Harriot (1560–1621), to simplify cumbersome notations used earlier.

These algebraic statements can be given a geometric interpretation. If $a - b$ is positive, so that $a > b$, then a would be to the *right* of b on a number line. Also, if $a < b$, then a would have to be to the *left* of b. Both the algebraic and geometric statements are summarized below.

Inequality Statements

Statement	Algebraic form	Geometric form
$a > b$	$a - b$ is positive	a is to the right of b
$a < b$	$a - b$ is negative	a is to the left of b

EXAMPLE 6

(a) In Figure 3, $-2/3$ is to the left of $\sqrt{2}$, so $-2/3 < \sqrt{2}$. Also, $\sqrt{2}$ is to the right of $-2/3$, giving $\sqrt{2} > -2/3$.

(b) The difference $-3 - (-8)$ is positive, showing that $-3 > -8$. The difference $-8 - (-3)$ is negative, so that $-8 < -3$.   ●

The following variations on $<$ and $>$ are often used.

Definitions of Other Inequality Symbols

Symbol	Meaning
$\leq$	is less than or equal to
$\geq$	is greater than or equal to
$\nless$	is not less than
$\ngtr$	is not greater than

Statements involving these symbols, as well as $<$ and $>$, are called **inequalities.**

EXAMPLE 7

(a) $8 \leq 10$ (since $8 < 10$)

(b) $8 \leq 8$ (since $8 = 8$)

(c) $-9 \geq -14$ (since $-9 > -14$)

(d) $-8 \ngtr -2$ (since $-8 < -2$)

(e) $4 \nless 2$ (since $4 > 2$)   ●

CAUTION   The expression $a < b < c$ says that b is *between* a and c, since $a < b < c$ means $a < b$ and $b < c$. Also, $a \le b \le c$ means $a \le b$ and $b \le c$. When writing these "between" statements, make sure that both inequality symbols point in the same direction. For example, both $2 < 7 < 11$ and $5 > -1 > -6$ are true statements, but a statement such as $3 < 5 > -1$ is meaningless.

The following **properties of order** give the basic properties of $<$ and $>$.

Properties of Order

For all real numbers a, b, and c,

Transitive property	If $a < b$ and $b < c$, then $a < c$.
Addition property	If $a < b$, then $a + c < b + c$.
Multiplication property	If $a < b$, and $c > 0$, then $ac < bc$.
	If $a < b$, and $c < 0$, then $ac > bc$.
Trichotomy property	Given the real numbers a and b, either $a < b$, $a > b$, or $a = b$.

The distance on the number line from a number to 0 is called the **absolute value** of that number. The absolute value of the number a is written $|a|$. For example, the distance on the number line from 9 to 0 is 9, as is the distance from -9 to 0 (see Figure 4), so $|9| = 9$ and $|-9| = 9$.

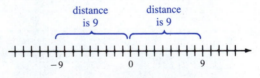

Figure 4

EXAMPLE 8

(a) $|-4| = 4$

(b) $|2\pi| = 2\pi$

(c) $-|8| = -(8) = -8$

(d) $-|-2| = -(2) = -2$ •

The definition of absolute value can be stated as follows.

Definition of Absolute Value

For every real number a,

$$|a| = \begin{cases} a \text{ if } a \ge 0 \\ -a \text{ if } a < 0. \end{cases}$$

The second part of this definition requires some thought. If a is a negative number, that is, if $a < 0$, then $-a$ is positive. Thus, for a *negative a*,

$$|a| = -a.$$

For example, if $a = -5$, then $|a| = |-5| = -(-5) = 5$.

. . . .

EXAMPLE 9   Write each of the following without absolute value bars.

(a) $|\sqrt{5} - 2|$

  For positive numbers we can reason as follows. Since $(\sqrt{5})^2 > 2^2$, $\sqrt{5} > 2$, making $\sqrt{5} - 2 > 0$, and $|\sqrt{5} - 2| = \sqrt{5} - 2$.

(b) $|\pi - 4|$

  From Figure 3, $\pi < 4$, so $\pi - 4 < 0$, and $|\pi - 4| = -(\pi - 4) = -\pi + 4 = 4 - \pi$.

(c) $|m - 2|$ if $m < 2$

  If $m < 2$, then $m - 2 < 0$, so $|m - 2| = -(m - 2) = 2 - m$.   ●

The definition of absolute value can be used to prove the following properties of absolute value.

. . . .

**Properties of
Absolute
Value**

. . . .

For all real numbers a and b,

$$|a| \geq 0 \qquad\qquad |a| \cdot |b| = |ab|$$

$$|-a| = |a| \qquad\qquad \left|\frac{a}{b}\right| = \frac{|a|}{|b|}, \quad b \neq 0$$

$$|a + b| \leq |a| + |b| \quad \text{(the triangle inequality)}.$$

The number line in Figure 5 shows the point A, with coordinate -3, and the point B, with coordinate 5. The distance between points A and B is 8 units, which can be found by subtracting the smaller coordinate from the larger. If $d(A, B)$ represents the distance between points A and B, then

$$d(A, B) = 5 - (-3) = 8.$$

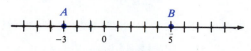

Figure 5

To avoid worrying about which coordinate is smaller, use the absolute value as in the definition on the following page.

Definition of Distance

Suppose points A and B have coordinates a and b respectively. The distance between A and B, written $d(A, B)$, is

$$d(A, B) = |a - b|.$$

EXAMPLE 10   Let points A, B, C, D, and E have coordinates as shown on the number line in Figure 6. Find the indicated distances.

(a) $d(B, E)$

Since B has coordinate -1 and E has coordinate 5,

$$d(B, E) = d(-1, 5)$$
$$= |-1 - 5| = 6.$$

(b) $d(D, A) = \left| 2\dfrac{1}{2} - (-3) \right| = 5\dfrac{1}{2}$

(c) $d(B, C) = |-1 - 0| = 1$

(d) $d(E, E) = |5 - 5| = 0$ ●

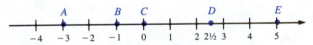

Figure 6

1.1 Exercises

Identify the properties illustrated in each of the following. Some illustrate more than one property. Assume all variables represent real numbers.

1. $8 \cdot 9 = 9 \cdot 8$

2. $3 + (-3) = 0$

3. $3 + (-3) = (-3) + 3$

4. $0 + (-7) = (-7) + 0$

5. $-7 + 0 = -7$

6. $8 + (12 + 6) = (8 + 12) + 6$

7. $[9(-3)] \cdot 2 = 9[(-3) \cdot 2]$

8. $8(m + 4) = 8m + 8 \cdot 4$

9. $(7 - y) + 0 = 7 - y$

10. $8(4 + 2) = (2 + 4)8$

Simplify each of the following expressions using the order of operations given in the text.

11. $9 \div 3 \cdot 4 \cdot 2$

12. $18 \cdot 3 \div 9 \div 2$

13. $8 + 7 \cdot 2 + (-5)$

14. $-9 + 6 \cdot 5 + (-8)$

15. $(-9 + 4 \cdot 3)(-7)$

16. $-15(-8 - 4 \div 2)$

17. $\dfrac{-8 + (-4)(-6) \div 12}{4 - (-3)}$

18. $\dfrac{15 \div 5 \cdot 4 \div 6 - 8}{-6 - (-5) - 8 \div 2}$

19. $-3^2 + 2 \cdot 3$

20. $1 + 2(-3)^2$

For Exercises 21–32 choose all words from the following list that apply.

(a) natural number   (b) whole number   (c) integer   (d) rational number

(e) irrational number   (f) real number   (g) undefined

21. 12

22. 0

23. -9

24. 3/4

25. $-5/9$

26. $\sqrt{8}$

27. $-\sqrt{2}$

28. $\sqrt{25}$

29. $-\sqrt{36}$

30. 8/0

31. $-9/(0-0)$

32. $0/(3+3)$

Use the distributive property to calculate the following values in your head.

33. $86 \cdot 19 + 14 \cdot 19$

34. $106\frac{5}{6} \cdot 1\frac{1}{2} - 6\frac{5}{6} \cdot 1\frac{1}{2}$

35. $15\frac{2}{5} \cdot 12\frac{3}{4} - 15\frac{2}{5} \cdot 2\frac{3}{4}$

36. $23 \cdot 80 + 23 \cdot 20$

Write the following numbers in numerical order, from smallest to largest. Use a calculator as necessary.

37. $\sqrt{8}, -4, -\sqrt{3}, -2, -5, \sqrt{6}, 3$

38. $\sqrt{2}, -1, 4, 3, \sqrt{8}, -\sqrt{6}, \sqrt{7}$

39. $3/4, \sqrt{2}, \pi/2, 1.2, 8/5, 22/15$

40. $-9/8, -3, \pi/3, 1, -\sqrt{5}, -9/5, -8/5$

Write an equivalent expression for each of the following without using absolute value bars.

41. $-|-8| + |-2|$

42. $3 - |-4|$

43. $|2 - \sqrt{3}|$

44. $|\sqrt{7} - 5|$

45. $|\sqrt{2} - 3|$

46. $|\pi - 3|$

47. $|x - 4|$, if $x > 4$

48. $|y - 3|$, if $y < 3$

49. $|2k - 8|$, if $k < 4$

50. $|3r - 15|$, if $r > 5$

51. $|7m - 56|$, if $m < 4$

52. $|2k - 7|$, if $k > 4$

53. $|x - y|$, if $x < y$

54. $|x - y|$, if $x > y$

55. $|3 + x^2|$

56. $|x^2 + 4|$

57. $|-1 - p^2|$

58. $|-r^4 - 16|$

59. $|\pi - 5| + 1$

60. $|3 - \sqrt{11}| + 2$

61. $|m - 3| + |m - 4|$, if $3 < m < 4$

62. $|z - 6| - |z - 5|$, if $5 < z < 6$

In Exercises 63–66, the coordinates of four points are given. Find (a) d(A, B), (b) d(B, C), (c) d(D, A), (d) d(A, C), (e) d(A, B) + d(B, C).

63. $A, -4; B, -3; C, -2; D, 10$

64. $A, -8; B, -7; C, 11; D, 5$

65. $A, -3; B, -5; C, -12; D, -3$

66. $A, 0; B, 6; C, 9; D, -1$

67. Suppose that *A, B,* and *C* are three points on the number line with $d(A, B) = 3$ and $d(B, C) = 4$. What can you say about $d(A, C)$?

68. Under what conditions will the following be true? $a < b$ but $|b| > |a|$

Justify each of the following statements by giving the correct property from this section. Assume that all variables represent real numbers.

69. If $2k < 8$, then $k < 4$.

70. If $x + 8 < 15$, then $x < 7$.

71. If $-4x < 24$, then $x > -6$.

72. If $x < 5$ and $5 < m$, then $x < m$.

73. If $m > 0$, then $9m > 0$.

74. If $k > 0$, then $8 + k > 8$.

75. $|8 + m| \leq |8| + |m|$ **76.** $|k - m| \leq |k| + |-m|$ **77.** $|8| \cdot |-4| = |-32|$

78. $|12 + 11r| \geq 0$ **79.** $\left|\dfrac{-12}{5}\right| = \dfrac{|-12|}{|5|}$ **80.** $\left|\dfrac{6}{5}\right| = \dfrac{|6|}{|5|}$

81. If p is a real number, then $p < 5$, $p > 5$, or $p = 5$.

82. If z is a real number, then $z > -2$, $z < -2$, or $z = -2$.

For each of the following inequalities, what can be said about the signs of x and y?

83. $xy > 0$ **84.** $x^2 y > 0$ **85.** $\dfrac{x^3}{y} < 0$ **86.** $\dfrac{y^2}{x} < 0$

Under what conditions are the following statements true?

87. $|x| = |y|$ **88.** $|x + y| = |x| + |y|$ **89.** $|x| \leq 0$ **90.** $|x - y| = |x| - |y|$

Evaluate the expressions in Exercises 91–94 for real numbers x and y, if no denominators are equal to 0.

91. $\dfrac{|x|}{x}$ **92.** $\left|\dfrac{x}{|x|}\right|$ **93.** $\left|\dfrac{x - y}{y - x}\right|$ **94.** $\left|\dfrac{x + y}{-x - y}\right|$

95. Suppose $x^2 \leq 81$. Must it then be true that $x \leq 9$?   **96.** Suppose $x^2 \geq 81$. Must it then be true that $x \geq 9$?

1.2 EXPONENTS

Exponents are used to write the products of repeated factors. For example, the product $2 \cdot 2 \cdot 2$ can be written as 2^3, where the 3 shows that three factors of 2 appear in the product. The French philosopher and mathematician René Descartes (1596–1650) first used exponents extensively as an efficient way of writing $a \cdot a$ as a^2 and $a^2 \cdot a$ as a^3 and so on. Much progress in algebra has depended on such improved notations.

The symbol a^n is defined as follows.

Definition of a^n

If n is any positive integer and a is any real number,

$$a^n = a \cdot a \cdot a \cdots a,$$

where a appears as a factor n times.

The integer n is called the **exponent,** and a is the **base.** (Read a^n as "a to the nth power," or just "a to the nth.") For example,

$$(-6)^2 = (-6)(-6) = 36, \qquad 4^3 = 4 \cdot 4 \cdot 4 = 64,$$

$$\text{and} \qquad \left(\frac{2}{3}\right)^4 = \frac{2}{3} \cdot \frac{2}{3} \cdot \frac{2}{3} \cdot \frac{2}{3} = \frac{16}{81}.$$

Expressions with exponents can be simplified by using various properties of exponents. By definition, a^m (where m is a positive integer and a is a real number) means a appears as a factor m times. In the same way, a^n (where n is a positive integer) means that a appears as a factor n times. In the product $a^m \cdot a^n$, a will therefore appear as a factor $m + n$ times, so that for any positive integers m and n and any real number a,

$$a^m \cdot a^n = a^{m+n}.$$

Similar arguments can be given for the other **properties of exponents** listed below.

Properties of Exponents

For any positive integers m and n and for any real numbers a and b,

(a) $a^m \cdot a^n = a^{m+n}$

(b) $\dfrac{a^m}{a^n} = \begin{cases} a^{m-n} & \text{if } m > n \\ 1 & \text{if } m = n \\ \dfrac{1}{a^{n-m}} & \text{if } m < n \end{cases}$ $a \neq 0$

(c) $(a^m)^n = a^{mn}$

(d) $(ab)^m = a^m b^m$

(e) $\left(\dfrac{a}{b}\right)^m = \dfrac{a^m}{b^m}, \quad b \neq 0$

Complete proofs of the various properties of exponents require the method of mathematical induction, discussed in the last chapter of this book. Properties (a) and (d) can be generalized: for example,

$$a^m \cdot a^n \cdot a^p = a^{m+n+p} \quad \text{and} \quad (abc)^n = a^n \cdot b^n \cdot c^n.$$

EXAMPLE 1   Each of the following examples is justified by one or more of the properties of exponents and by the properties of real numbers.

(a) $(4x^3)(-3x^5) = (4)(-3)x^3 \cdot x^5 = -12x^{3+5} = -12x^8$

(b) $\dfrac{m^4 p^2}{m^3 p^5} = \dfrac{m^4}{m^3} \cdot \dfrac{p^2}{p^5} = m \cdot \dfrac{1}{p^3} = \dfrac{m}{p^3}$ $(m \neq 0, p \neq 0)$

(c) $\left(\dfrac{2x^5}{7}\right)^3 = \dfrac{2^3(x^5)^3}{7^3} = \dfrac{8x^{15}}{343}$

(d) $(2z^{2r})(z^{r+1}) = 2z^{2r+(r+1)} = 2z^{3r+1}$ (r is a positive integer)

(e) $\dfrac{k^{r+1}}{k^r} = k^{(r+1)-r} = k$ (r is a positive integer and k is nonzero) ●

By part (a) of the properties of exponents, $a^m \cdot a^n = a^{m+n}$. If $n = 0$, then

$$a^m \cdot a^n = a^{m+n} \quad \text{becomes} \quad a^m \cdot a^0 = a^{m+0} = a^m.$$

The only way that $a^m \cdot a^0$ can equal a^m is if $a^0 = 1$. Therefore, a^0 is defined as on the following page.

**Definition
of Zero
Exponent**

If a is a nonzero real number, then

$$a^0 = 1.$$

The symbol 0^0 is not defined.

EXAMPLE 2

(a) $3^0 = 1$ 　　　　　　　　　　　　　　**(b)** $(-4)^0 = 1$

(c) $\left(\dfrac{1}{7}\right)^0 = 1$ 　　　　　　　　**(d)** $(-\sqrt{11})^0 = 1$

(e) $-4^0 = -1$ 　　　　　　　　　　**(f)** $-(-4)^0 = -1$ 　●

　　In Section 1.1 the symbol a^{-1} was defined for any nonzero real number a as $a^{-1} = 1/a$. This definition can be extended so as to give meaning to an expression of the form a^{-n}, where n is any positive integer, not just 1. As with zero exponents, any definition for a^{-n} must be consistent with the properties of exponents given earlier. Property (c) from above can be valid only if

$$a^{-n} = a^{(-1)n} = (a^{-1})^n.$$

Since $a^{-1} = \dfrac{1}{a}$,

$$a^{-n} = (a^{-1})^n = \underbrace{\frac{1}{a} \cdot \frac{1}{a} \cdot \frac{1}{a} \cdots \frac{1}{a}}_{n \text{ factors}} = \frac{1}{a^n}.$$

Based on this, for the listed properties of exponents to remain valid, a^{-n} must be defined as $1/a^n$.

**Definition of
Negative
Exponent**

For all positive integers n and all nonzero real numbers a,

$$a^{-n} = \frac{1}{a^n}.$$

EXAMPLE 3

(a) $5^{-3} = \dfrac{1}{5^3} = \dfrac{1}{125}$ 　　　　　　**(b)** $\left(\dfrac{2}{3}\right)^{-3} = \dfrac{1}{\left(\dfrac{2}{3}\right)^3} = \dfrac{1}{\dfrac{8}{27}} = \dfrac{27}{8}$

(c) $-4^{-2} = -\dfrac{1}{4^2} = -\dfrac{1}{16}$ 　●

Part (b) of Example 3 involves work with fractions of a type that can lead to errors. For a shortcut useful with such fractions, assume that neither a nor b is zero, so that $(a/b) \cdot (b/a) = 1$, showing that $(a/b)^{-1} = b/a$. Use this, along with properties of exponents, to get

$$\left(\frac{a}{b}\right)^{-n} = \left[\left(\frac{a}{b}\right)^{-1}\right]^n = \left(\frac{b}{a}\right)^n.$$

Reciprocal Property of Negative Exponents

For all positive integers n and nonzero real numbers a and b,

$$\left(\frac{a}{b}\right)^{-n} = \left(\frac{b}{a}\right)^n.$$

EXAMPLE 4

(a) $\left(\frac{1}{3}\right)^{-2} = \left(\frac{3}{1}\right)^2 = 9$ **(b)** $\left(\frac{2}{3}\right)^{-3} = \left(\frac{3}{2}\right)^3 = \frac{27}{8}$ ●

It can be shown that all the properties of exponents given above are valid for *all* integer exponents, and not just positive integer exponents. In particular, property (b) from above can now be simplified to just one case:

$$\frac{a^m}{a^n} = a^{m-n}, \quad \text{if } a \neq 0.$$

The meaning of a^n can be extended to rational values of the exponent. To start, let us define $a^{1/n}$ for positive integers n. Any definition of $a^{1/n}$ should be consistent with past rules of exponents. In particular, $(b^m)^n = b^{mn}$ should still be valid. Replacing b^m with $a^{1/n}$ gives

$$(a^{1/n})^n = a^{n/n} = a^1 = a.$$

This means that $a^{1/n}$ should be a real number whose nth power is a. Such a number is called an **nth root of a**. When n is 2, an nth root is called a **square root** and when n is 3, it is called a **cube root**. Depending on the values of a and n, there are one, two, or no nth roots. For instance, -5 has one cube root, 5 has two square roots (one positive and one negative), and -5 has no square root. Two nth roots occur when a is positive and n is even. The value of $a^{1/n}$ is defined as follows.

Definition of $a^{1/n}$

Let n be a positive integer and a a real number. Then

$$a^{1/n} = \begin{cases} \text{the } n\text{th root of } a \text{ if } n \text{ is odd} \\ \text{the nonnegative } n\text{th root of } a \text{ if } n \text{ is even and } a \geq 0. \end{cases}$$

EXAMPLE 5

(a) $(-32)^{1/5} = -2$ since $(-2)^5 = -32$

(b) $\left(\dfrac{16}{9}\right)^{1/2} = \dfrac{4}{3}$ since $\left(\dfrac{4}{3}\right)^2 = \dfrac{16}{9}$

(c) $(-9)^{1/2}$ is not defined. ●

For rational exponents in general, $a^{m/n}$ must be defined so that all the rules for integer exponents still hold. Isaac Newton, the greatest of the English mathematicians, first did this in 1676. For the rule $(b^m)^n = b^{mn}$ to hold, $(a^{1/n})^m$ must equal $a^{m/n}$. Thus $a^{m/n}$ is defined as follows.

Definition of $a^{m/n}$

> For all integers m and all positive integers n such that m/n is in lowest terms, and all nonzero real numbers a for which $a^{1/n}$ is a real number,
>
> $$a^{m/n} = (a^{1/n})^m.$$

EXAMPLE 6

(a) $8^{5/3} = (8^{1/3})^5 = 2^5 = 32$

(b) $25^{1.5} = 25^{3/2} = (25^{1/2})^3 = 5^3 = 125$ ●

We shall now show that $(a^{1/n})^m = (a^m)^{1/n}$, so that $a^{m/n}$ is also equal to $(a^m)^{1/n}$. To prove that $(a^{1/n})^m = (a^m)^{1/n}$, we shall use some of the properties of exponents. Start with $(a^{1/n})^m$, and raise it to the nth power, to get

$$[(a^{1/n})^m]^n.$$

Now use properties of exponents, getting

$$[(a^{1/n})^m]^n = (a^{1/n})^{mn} = [(a^{1/n})^n]^m = a^m.$$

Because of this result, $(a^{1/n})^m$ must be an nth root of a^m, so that

$$(a^{1/n})^m = (a^m)^{1/n}.$$

Since $a^{m/n} = (a^{1/n})^m$ by definition, then

$$a^{m/n} = (a^m)^{1/n}.$$

This result is summarized in the next theorem.

Theorem on $a^{m/n}$

> For all integers m, and all positive integers n, such that m/n is in lowest terms, and all nonzero real numbers a for which all indicated powers exist,
>
> $$a^{m/n} = (a^{1/n})^m \quad \text{or} \quad a^{m/n} = (a^m)^{1/n}.$$

This result gives two ways to evaluate $a^{m/n}$. Find $a^{1/n}$ and raise the result to the mth power, or find the nth root of a^m. In practice, it is usually easier to find $(a^{1/n})^m$. For example, $27^{4/3}$ can be evaluated in either of two ways:

$$27^{4/3} = (27^{1/3})^4 = 3^4 = 81$$

or $$27^{4/3} = (27^4)^{1/3} = 531{,}441^{1/3} = 81.$$

The form $(27^{1/3})^4$ is easier to evaluate.

It can be shown that all the earlier results concerning integer exponents also apply to rational exponents. These rules for exponents are listed below.

Definitions and Properties of Exponents

Let r and s be rational numbers. The results below are valid for all real numbers a and b for which all indicated expressions exist.

(a) $a^r \cdot a^s = a^{r+s}$ (b) $\dfrac{a^r}{a^s} = a^{r-s}$

(c) $(a^r)^s = a^{rs}$ (d) $(ab)^r = a^r b^r$

(e) $\left(\dfrac{a}{b}\right)^r = \dfrac{a^r}{b^r}$ (f) $a^0 = 1$

(g) $a^{-r} = \dfrac{1}{a^r}$ (h) $\left(\dfrac{a}{b}\right)^{-r} = \left(\dfrac{b}{a}\right)^r$

EXAMPLE 7   Rewrite each of the following using only positive exponents. Assume that all variables represent nonzero real numbers.

(a) $6y^{2/3} \cdot 2y^{1/2} = 12y^{2/3+1/2}$

$$= 12y^{7/6}$$

(b) $3x^{-2}(4^{-1}x^{-5})^2 = 3x^{-2}(4^{-2}x^{-10})$

$$= 3 \cdot 4^{-2} \cdot x^{-2+(-10)}$$

$$= 3 \cdot 4^{-2} \cdot x^{-12} = 3 \cdot \frac{1}{4^2} \cdot \frac{1}{x^{12}} = \frac{3}{16x^{12}}$$

(c) $(x + 2)(x + 2)^{-1/3}(x + 2)^{-2} = (x + 2)^{1+(-1/3)+(-2)}$

$$= (x + 2)^{-4/3} = \frac{1}{(x + 2)^{4/3}}$$

(d) $\dfrac{5k^{-3}}{10k^{-5}} = \dfrac{5}{10}k^{-3-(-5)} = \dfrac{1}{2}k^2$, or $\dfrac{k^2}{2}$

(e) $(8ab)^{-2/3}4a^{-1} = 8^{-2/3}a^{-2/3}b^{-2/3}4a^{-1} = \dfrac{1}{4} \cdot 4a^{-2/3+(-1)}b^{-2/3}$

$$= a^{-5/3}b^{-2/3} = \frac{1}{a^{5/3}b^{2/3}}$$

(f) $\dfrac{(3m^3)^2(my)^{-1}}{(25y^{-4}m^{14})^{1/2}} = \dfrac{3^2m^6m^{-1}y^{-1}}{25^{1/2}y^{-2}m^7}$

$\qquad\qquad\qquad = \dfrac{9}{5} \cdot \dfrac{m^{6+(-1)}}{m^7} \cdot \dfrac{y^{-1}}{y^{-2}}$

$\qquad\qquad\qquad = \dfrac{9}{5} \cdot m^{5-7}y^{-1-(-2)}$

$\qquad\qquad\qquad = \dfrac{9}{5}\,m^{-2}y = \dfrac{9y}{5m^2}$ ●

1.2 Exercises ···

Simplify each of the following. Assume all variables represent nonnegative real numbers. Write answers without exponents when possible.

1. 7^3

2. 6^4

3. $4^{1/2}$

4. $25^{1/2}$

5. $8^{2/3}$

6. $81^{3/4}$

7. -2^{-3}

8. -3^{-2}

9. $27^{-2/3}$

10. $32^{-4/5}$

11. $(-5)^{-2}$

12. $\left(\dfrac{2}{7}\right)^{-3}$

13. $\left(\dfrac{4}{9}\right)^{-3/2}$

14. $(121)^{-3/2}$

15. $\dfrac{12}{10^{-2}}$

16. $\dfrac{5}{2^{-4}}$

17. $(27x^6)^{2/3}$

18. $(64a^{12})^{5/6}$

19. $(9.864)^{-3}$

20. $(14.259)^{-2}$

21. $(.0001)^{-3/2}$

22. $(.001)^{-5/3}$

Simplify the expressions in Exercises 23–46 by writing each with only positive exponents. Assume that all variables represent positive real numbers and that variables used as exponents represent rational numbers.

23. $(3m)^2(-2m)^3$

24. $(-2a^4b^7)(3b)^2$

25. $(m^{2/3})(m^{5/3})$

26. $(x^{4/5})(x^{2/5})$

27. $(6^2)(6^{-5})(6^3)$

28. $(4^8)(4^{-10})(4^2)$

29. $(1+n)^{1/2}(1+n)^{3/4}$

30. $(m+7)^{-1/6}(m+7)^{-2/3}$

31. $(2y^{3/4}z)(3y^{1/4}z^{-1/3})$

32. $(4a^{-1/2}b^{2/3})(2a^{3/2}b^{-1/3})$

33. $\dfrac{(d^{-1})(d^{-2})}{(d^8)(d^{-3})}$

34. $\dfrac{(t^5)(t^{-3})}{(t^4)(t^{-7})}$

35. $\dfrac{a^{4/3} \cdot b^{1/2}}{a^{2/3} \cdot b^{-3/2}}$

36. $\dfrac{x^{1/3}y^{2/3}}{x^{5/3}y^{-1/3}}$

37. $\left(\dfrac{a^{-1}}{b^2}\right)^{-3}$

38. $\left(\dfrac{2c^2}{d^3}\right)^{-2}$

39. $\dfrac{(5x)^{-2}(x^{-3})^{-4}}{(25^{-1}x^{-3})^{-1}}$

40. $\dfrac{(2k)^{-3}(k^{-5})^{-1}}{(6k^{-2})^{-1}(k^3)^{-6}}$

41. $\left(\dfrac{x^4y^3z}{16x^{-6}yz^5}\right)^{-1/2}$

42. $\left(\dfrac{p^3r^9}{27p^{-3}r^{-6}}\right)^{-1/3}$

43. $5p^r(6p^{3-2r})$, $r < 3$

44. $(-z^{2r})(4z^{r+3})$

45. $\dfrac{(b^2)^{y+1}}{(2b^y)^3}$, $y > 2$

46. $\dfrac{(3x^n)^3}{(x^2)^{n-1}}$

47. Show that $(1 + 2^3)^{-1} + (1 + 2^{-3})^{-1} = 1$.

48. Show that $(1 + x^m)^{-1} + (1 + x^{-m})^{-1} = 1$.

49. If $a^7 = 30$, what is a^{21}?

50. If $a^{-3} = .2$, what is a^6?

51. If the lengths of the sides of a cube are tripled, by what factor will the volume change?

52. If the radius of a circle is doubled, by what factor will the area change?

One important application of mathematics to business and management concerns supply and demand. Usually, as the price of an item increases, the supply increases and the demand decreases. By studying past records of supply and demand at different prices, economists can construct an equation which describes (approximately) supply and demand for a given item. Exercises 53 and 54 show examples of this.

53. The price of a certain solar heater (in hundreds of dollars) is approximated by p, where $p = 2x^{1/2} + 3x^{2/3}$ and x is the number of units supplied. Find the price when the supply is 64 units.

54. The demand for a certain commodity and the price in dollars are related by $p = 1000 - 200x^{-2/3}$, where x is the number ($x > 0$) of units of the product demanded. Find the price when the demand is 27.

In our system of government, the president is elected by the electoral college and not by individual voters. Because of this, smaller states have a greater voice in the selection of a president than they would otherwise. Two political scientists have studied the problems of campaigning for president under the current system and have concluded that candidates should allot their money according to the formula

$$\frac{\text{amount for}}{\text{large state}} = \left(\frac{E_{\text{large}}}{E_{\text{small}}}\right)^{3/2} \times \frac{\text{amount for}}{\text{small state}}.$$

Here E_{large} represents the electoral vote of the large state, and E_{small} represents the electoral vote of the small state. Find the amount that should be spent in the larger states in Exercises 55–58 if $1,000,000 is spent in the small state and the following statements are true.

55. The large state has 48 electoral votes, and the small state has 3.

56. The large state has 36 electoral votes, and the small state has 4.

57. There are six votes in the small state; 28 in the large.

58. There are nine votes in the small state; 32 in the large.

A Delta Airlines map gives a formula for calculating the visible distance from a jet plane to the horizon. On a clear day, this distance is approximated by $D = 1.22x^{1/2}$, where x is altitude in feet and D is distance to the horizon in miles. Find D for each of the following altitudes.

59. 5000 feet

60. 10,000 feet

61. 30,000 feet

62. 40,000 feet

The Galápagos Islands are a chain of islands ranging in size from 2 to 2249 square miles. A biologist has shown that the number of different land-plant species on an island in this chain is related to the size of the island by $S = 28.6A^{.32}$, where A is the area of an island in square miles and S is the number of different plant species on that island. Estimate S (rounding to the nearest whole number) for islands with the following areas.

63. 1 square mile

64. 25 square miles

65. 300 square miles

66. 2000 square miles

1.3 POLYNOMIALS

A variable is a letter used to represent an element from a given set. Unless otherwise specified, in this book variables will represent real numbers. An **algebraic expression** is the result of performing the basic operations of addition, subtraction, multiplication, division (except by 0), or extraction of roots, on any collection of variables and numbers. The simplest algebraic expression, a polynomial, is discussed in this section.

To begin, a **term** is the product of a real number and one or more variables raised to powers. The real number is the **numerical coefficient,** or just the **coefficient.** For example, -3 is the coefficient in $-3m^4$, while -1 is the coefficient in $-p^2$.

A **polynomial** is defined as a finite sum of terms, with only nonnegative integer exponents permitted on the variables. If the terms of a polynomial contain only the variable x, then the polynomial is called a *polynomial in x.* (Polynomials in other variables are defined similarly.)

Definition of a Polynomial in x

A **polynomial in x** is an expression of the form

$$a_nx^n + a_{n-1}x^{n-1} + \ldots + a_1x + a_0,$$

where a_n (read "a-sub-n"), $a_{n-1}, \ldots, a_1,$ and a_0 are real numbers and n is a nonnegative integer. If $a_n \neq 0$, then n is the **degree** of the polynomial, and a_n is called the **leading coefficient.**

By this definition,

$$3x^4 - 5x^2 + 2$$

is a polynomial of degree 4 with leading coefficient 3. A nonzero constant polynomial is said to have degree 0, but no degree is assigned to the polynomial 0. If all the coefficients of a polynomial are 0, the polynomial is called the **zero polynomial.** Polynomials are often denoted by names such as $f(x)$ and $g(x)$.

A polynomial can have more than one variable. A term containing more than one variable is said to have degree equal to the sum of all the exponents appearing on the variables in the term. For example, $-3x^4y^3z^5$ is of degree 12 because $4 + 3 + 5 = 12$. The degree of a polynomial in more than one variable is the highest degree of any term appearing in the polynomial. With this definition, the polynomial

$$2x^4y^3 - 3x^5y + x^6y^2$$

is of degree 8 because of the x^6y^2 term.

A polynomial containing exactly three terms is called a **trinomial;** one containing exactly two terms is a **binomial;** and a single-term polynomial is called a **monomial.** For example, $7x^9 - \sqrt{2}x^4 + 1$ is a trinomial of degree 9.

Since the variables used in polynomials represent real numbers, a polynomial represents a real number. This means that all the properties of the real numbers mentioned at the beginning of this chapter hold for polynomials. In particular, the distributive property holds, so that

$$3m^5 - 7m^5 = (3 - 7)m^5 = -4m^5.$$

Addition and Subtraction   To *add* polynomials of one variable, use the distributive property to add coefficients of variables having the same powers; to *subtract* polynomials, subtract coefficients of variables having the same powers, again with the distributive property. That is, polynomials are added by adding like terms and subtracted by subtracting like terms.

· · · · · · · · ·

EXAMPLE 1   Add or subtract, as indicated.

(a) $(2y^4 - 3y^2 + y) + (4y^4 + 7y^2 + 6y)$

$$= (2 + 4)y^4 + (-3 + 7)y^2 + (1 + 6)y$$

$$= 6y^4 + 4y^2 + 7y$$

(b) $(6r^4 + 2r^2) + (3r^3 + 9r) = 6r^4 + 3r^3 + 2r^2 + 9r$

(c) $(-3m^3 - 8m^2 + 4) - (m^3 + 7m^2 - 3)$

$$= (-3 - 1)m^3 + (-8 - 7)m^2 + [4 - (-3)]$$

$$= -4m^3 - 15m^2 + 7 \quad \bullet$$

Multiplication   The distributive property, together with the properties of exponents, can also be used to find the product of two polynomials. For example, to find the product of $3x - 4$ and $2x^2 - 3x + 5$, treat $3x - 4$ as a single expression and use the distributive property as follows.

$$(3x - 4)(2x^2 - 3x + 5) = (3x - 4)(2x^2) - (3x - 4)(3x) + (3x - 4)(5)$$

Now use the distributive property three separate times on the right of the equals sign to get

$$(3x - 4)(2x^2 - 3x + 5) = (3x)(2x^2) - 4(2x^2) - (3x)(3x) - (-4)(3x)$$
$$+ (3x)5 - 4(5)$$
$$= 6x^3 - 8x^2 - 9x^2 + 12x + 15x - 20$$
$$= 6x^3 - 17x^2 + 27x - 20.$$

It is sometimes more convenient to find such a product as follows.

$$
\begin{array}{r}
2x^2 - 3x + 5 \\
3x - 4 \\
\hline
-8x^2 + 12x - 20 \quad \leftarrow -4(2x^2 - 3x + 5) \\
6x^3 - 9x^2 + 15x \quad\quad\ \longleftarrow 3x(2x^2 - 3x + 5) \\
\hline
6x^3 - 17x^2 + 27x - 20 \quad \text{Add in columns.}
\end{array}
$$

· · · · · · · · · ·

EXAMPLE 2   Find each product.

(a) $(6m + 1)(4m - 3) = (6m)(4m) - (6m)(3) + 1(4m) - 1(3)$
$$= 24m^2 - 14m - 3$$

(b) $(2k^n - 5)(k^n + 3) = 2k^{2n} + 6k^n - 5k^n - 15$
$$= 2k^{2n} + k^n - 15$$

(c) $(2x + 7)(2x - 7) = 4x^2 - 14x + 14x - 49$
$$= 4x^2 - 49 \quad \bullet$$

The products in Example 2 were found by multiplying the first terms, the outer terms, the inner terms, the last terms, and then adding the four products. This method, which is used to multiply any two binomials, is often called the **FOIL method** (for first, outer, inner, last).

Special Products   Certain products occur so frequently that they should be memorized.

· · · · · · · · · · · · ·

Special Products

Difference of Two Squares	$(x + y)(x - y) = x^2 - y^2$
Square of a Binomial	$(x + y)^2 = x^2 + 2xy + y^2$ $(x - y)^2 = x^2 - 2xy + y^2$
Cube of a Binomial	$(x + y)^3 = x^3 + 3x^2y + 3xy^2 + y^3$ $(x - y)^3 = x^3 - 3x^2y + 3xy^2 - y^3$

· · · · · · · · ·

EXAMPLE 3   Find each product.

(a) $(3p + 11)(3p - 11)$

Using the result above, replace x with $3p$ and y with 11. This gives
$$(3p + 11)(3p - 11) = (3p)^2 - 11^2 = 9p^2 - 121.$$

(b) $(2m + 5)^2 = (2m)^2 + 2(2m)(5) + (5)^2$
$$= 4m^2 + 20m + 25$$

(c) $(5k - 2z^5)^3 = (5k)^3 - 3(5k)^2(2z^5) + 3(5k)(2z^5)^2 - (2z^5)^3$
$$= 125k^3 - 150k^2z^5 + 60kz^{10} - 8z^{15} \quad \bullet$$

It is useful to memorize and be able to apply these special products. (Both the square of a binomial and the cube of a binomial are special cases of the binomial theorem, discussed in the last chapter of this book.)

Division   To divide a polynomial by a monomial, divide each term of the polynomial by the monomial.

• • • • • • • • •

EXAMPLE 4   Divide.

(a) $\dfrac{2m^5 - 6m^3}{2m^3} = \dfrac{2m^5}{2m^3} - \dfrac{6m^3}{2m^3} = m^2 - 3$

The polynomial $m^2 - 3$ is the quotient of $2m^5 - 6m^3$ and $2m^3$.

(b) $\dfrac{3y^6x^3 - 6y^3x^6 + 8y^5x}{3y^3x^3} = \dfrac{3y^6x^3}{3y^3x^3} - \dfrac{6y^3x^6}{3y^3x^3} + \dfrac{8y^5x}{3y^3x^3}$

$$= y^3 - 2x^3 + \dfrac{8y^2}{3x^2}$$

This result is not a polynomial. ●

The quotient of two polynomials can be found with a **division algorithm** very similar to that used for dividing whole numbers. (An *algorithm* is a step-by-step procedure for working a problem.) This algorithm requires that both polynomials be written in descending order.

• • • • • • • • •

EXAMPLE 5   Divide $4m^3 - 8m^2 + 4m + 6$ by $2m - 1$.

Work as follows.

$4m^3$ divided by $2m$ is $2m^2$.

$- 6m^2$ divided by $2m$ is $- 3m$.

m divided by $2m$ is $\frac{1}{2}$.

$$
\begin{array}{r}
2m^2 - 3m + \frac{1}{2} \\
2m - 1 \overline{)\,4m^3 - 8m^2 + 4m + 6} \\
\underline{4m^3 - 2m^2} \\
- 6m^2 + 4m \\
\underline{- 6m^2 + 3m} \\
m + 6 \\
\underline{m - \frac{1}{2}} \\
\frac{13}{2}
\end{array}
$$

$2m^2(2m - 1) = 4m^3 - 2m^2$
Subtract; bring down the next term.
$-3m(2m - 1) = -6m^2 + 3m$
Subtract; bring down the next term.
$(1/2)(2m - 1) = m - (1/2)$
Subtract. The remainder is 13/2.

In dividing these polynomials, $4m^3 - 2m^2$ is subtracted from $4m^3 - 8m^2 + 4m + 6$. The complete result, $- 6m^2 + 4m + 6$, should be written under the line. However, it is customary to save work and "bring down" just the $4m$, the only term needed for the next step. By this work,

$$\frac{4m^3 - 8m^2 + 4m + 6}{2m - 1} = 2m^2 - 3m + \frac{1}{2} + \frac{13/2}{2m - 1}. \quad ●$$

The polynomial $3x^3 - 2x^2 - 150$ has a missing term, the term in which the power of x is 1. When a polynomial with a missing term is divided, it is useful to allow for that term by inserting a zero coefficient for the missing term as shown in the next example.

EXAMPLE 6   Divide $3x^3 - 2x^2 - 150$ by $x^2 - 4$.

Both polynomials have missing terms. Insert each missing term with a 0 coefficient.

$$
\begin{array}{r}
3x - 2 \\
x^2 + 0x - 4 \overline{)3x^3 - 2x^2 + 0x - 150} \\
\underline{3x^3 + 0x^2 - 12x} \\
-2x^2 + 12x - 150 \\
\underline{-2x^2 + 0x + 8} \\
12x - 158
\end{array}
$$

Since $12x - 158$ has lower degree than the divisor, it is the remainder, and the result of the division is written

$$\frac{3x^3 - 2x^2 - 150}{x^2 - 4} = 3x - 2 + \frac{12x - 158}{x^2 - 4}. \quad \bullet$$

The examples suggest the following generalization.

$$
\begin{array}{r}
q(x) + \dfrac{r(x)}{g(x)} \\
\hline
g(x) \overline{)f(x)}
\end{array}
$$

The *division algorithm* below expresses this process another way, as follows.

Division Algorithm

Let $f(x)$ and $g(x)$ be polynomials with $g(x)$ of lower degree than $f(x)$ and $g(x) \neq 0$. There exist unique polynomials $q(x)$ and $r(x)$ such that

$$f(x) = g(x) \cdot q(x) + r(x),$$

where $r(x) = 0$ or the degree of $r(x)$ is less than the degree of $g(x)$.

The polynomial $f(x)$ in the division algorithm is the **dividend** and $g(x)$ is the **divisor**. The polynomial $q(x)$ is the **quotient polynomial** or the **quotient**, while $r(x)$ is the **remainder polynomial** or the **remainder**.

The division algorithm applies to the polynomials in Examples 5 and 6. In Example 5,

$$\frac{4m^3 - 8m^2 + 4m + 6}{2m - 1} = 2m^2 - 3m + \frac{1}{2} + \frac{13/2}{2m - 1}$$

from which

$$4m^3 - 8m^2 + 4m + 6 = \underbrace{(2m - 1)}_{} \underbrace{\left(2m^2 - 3m + \frac{1}{2}\right)}_{} + \underbrace{\frac{13}{2}}_{}.$$

$$\underbrace{}_{f(x)} = \underbrace{}_{g(x)} \cdot \underbrace{}_{q(x)} + \underbrace{}_{r(x)}$$

Identify $q(x)$ and $r(x)$ in Example 6.

1.3 Exercises ··

Perform the indicated operations.

1. $(x^2 + 4x) + (3x^3 - 4x^2) + (2x + 2)$

2. $(r^5 - r^3 + r) + (3r^5 - 4r^4) + (r^3 + 2r)$

3. $(12y^2 - 8y + 6) - (3y^2 - 4y) + 2$

4. $(8p^2 - 5p) - (3p^2 - 2p) + 4$

5. $(5b^2 - 4b + 3) - [(2b^2 + b) - (3b + 4)]$

6. $-[(8x^3 + x - 3) + (2x^3 + x^2)] - (4x^2 + 3x - 1)$

7. $(x - 2)(3x + 1)$

8. $(2x - y)(x - 2y)$

9. $(4r - 1)(7r + 2)$

10. $(5m - 6)(3m + 4)$

11. $(6p + 5q)(3p - 7q)$

12. $(2z + y)(3z - 4y)$

13. $\left(3x - \dfrac{2}{3}\right)\left(5x + \dfrac{1}{3}\right)$

14. $\left(2m - \dfrac{1}{4}\right)\left(3m + \dfrac{1}{2}\right)$

15. $\left(\dfrac{2}{5}y + \dfrac{1}{8}z\right)\left(\dfrac{3}{5}y + \dfrac{1}{2}z\right)$

16. $\left(\dfrac{3}{4}r - \dfrac{2}{3}s\right)\left(\dfrac{5}{4}r + \dfrac{1}{3}s\right)$

17. $(5r + 2)(5r - 2)$

18. $(6z + 5)(6z - 5)$

19. $(4x + 3y)(4x - 3y)$

20. $(7m + 2n)(7m - 2n)$

21. $(6k - 3)^2$

22. $(3p + 5)^2$

23. $(4m + 2n)^2$

24. $(a - 6b)^2$

25. $(2z - 1)^3$

26. $(3m + 2)^3$

27. $(2x^3 + y^2)^2$

28. $(x^2 - y^2)^2$

29. $(x - 2y^2)^3$

30. $4x^2(3x^3 + 2x^2 - 5x + 1)$

31. $2b^3(b^2 - 4b + 3)$

32. $5m(3m^3 - 2m^2 + m - 1)$

33. $4y^3(y^3 + 2y^2 - 6y + 3)$

34. $(2z - 1)(-z^2 + 3z - 4)$

35. $(x - 1)(x^2 - 1)$

36. $(3p - 1)(9p^2 + 3p + 1)$

37. $(2p - 1)(3p^2 - 4p + 5)$

38. $(2m + 1)(4m^2 - 2m + 1)$

39. $(k + 2)(12k^3 - 3k^2 + k + 1)$

40. $(m - n + k)(m + 2n - 3k)$

41. $(a - b + 2c)^2$

42. $(k - y + 3m)^2$

Find each of the following products. Assume all variables used as exponents represent integers.

43. $(k^m + 2)(k^m - 2)$

44. $(y^x - 4)(y^x + 4)$

45. $(3p^x + 1)(p^x - 2)$

46. $(2^a + 5)(2^a + 3)$

47. $(m^x - 2)^2$

48. $(z^r + 5)^2$

49. $(3k^a - 2)^3$

50. $(r^x - 4)^3$

Find each of the following products. Assume that all variables represent positive real numbers.

51. $y^{5/8}(y^{3/8} - 10y^{11/8})$

52. $p^{11/5}(3p^{4/5} + 9p^{19/5})$

53. $-4k(k^{7/3} - 6k^{1/3})$

54. $-5y(3y^{9/10} + 4y^{3/10})$

55. $(x + x^{1/2})(x - x^{1/2})$

56. $(2z^{1/2} + z)(z^{1/2} - z)$

57. $(r^{1/2} - r^{-1/2})^2$

58. $(p^{1/2} - p^{-1/2})(p^{1/2} + p^{-1/2})$

Perform each of the following divisions. Assume that all variables appearing in denominators represent nonzero real numbers.

59. $\dfrac{15x^4 + 30x^3 + 12x^2 - 9}{3x}$

60. $\dfrac{16a^6 + 24a^5 - 48a^4 + 12a}{8a^2}$

61. $\dfrac{25x^2y^4 - 15x^3y^3 + 40x^4y^2}{5x^2y^2}$

62. $\dfrac{-8r^3s - 12r^2s^2 + 20rs^3}{4rs}$

63. $\dfrac{4p^4 - 6p^3 + 4p^2 + 3p + 8}{2p - 1}$

64. $\dfrac{6m^3 + 7m^2 - 4m + 2}{3m + 2}$

65. $\dfrac{k^4 - 4k^2 + 2k + 5}{k^2 + 1}$

66. $\dfrac{3x^4 + 2x^2 + 6x - 1}{3x^2 - x}$

67. $\dfrac{5y^5 + 10y^4 - 5y^2 + 15y}{5y^3 - 2y + 1}$

68. $\dfrac{8z^5 + 4z^4 + 2z^2 - 5z + 16}{4z^2 - z + 2}$

In Exercises 69–74, find the coefficient of x^3 without finding the entire product.

69. $(x^2 + 4x)(-3x^2 + 4x - 1)$

70. $(4x^3 - 2x + 5)(x^3 + 2)$

71. $(1 + x^2)(1 + x)$

72. $(3 - x)(2 - x^2)$

73. $x^2(4 - 3x)^2$

74. $-4x^2(2 - x)(2 + x)$

75. Show that $(y - x)^3 = -(x - y)^3$.

76. Show that $(y - x)^2 = (x - y)^2$.

77. Show that $\left(x + \dfrac{1}{2}\right)^2 = x(x + 1) + \dfrac{1}{4}$.

78. Use the result of Exercise 77 to calculate the square of 9 1/2 in your head.

79. Suppose one polynomial has degree 3 and another also has degree 3. Find all possible values for the degree of their **(a)** sum, **(b)** difference, **(c)** product.

80. If one polynomial has degree 3 and another has degree 4, find all possible values for the degree of their **(a)** sum, **(b)** difference, **(c)** product.

81. Generalize the results of Exercises 79 and 80: Suppose one polynomial has degree m and another has degree n, where m and n are natural numbers with $n < m$. Find all possible values for the degree of their **(a)** sum, **(b)** difference, **(c)** product.

82. Derive a formula for $(x + y + z)^2$.

83. Derive a formula for $(x + y + z)(x + y - z)$.

1.4 FACTORING

The process of finding polynomials whose product equals a given polynomial is called **factoring**. For example, since $4x + 12 = 4(x + 3)$, both 4 and $x + 3$ are **factors** of $4x + 12$. Also, $4(x + 3)$ is the **factored form** of $4x + 12$. A polynomial is **factored completely** when it is written as a product of polynomials, none of which can be written as the product of polynomials of positive degree. A polynomial that cannot be written as a product of two polynomials of positive degree is a **prime** or **irreducible** polynomial.

The first step in factoring a polynomial is to look for any **common factors;** that is, expressions that are factors of each term of the given polynomial.

· · · · · · · · ·

EXAMPLE 1   Factor a common factor from each of the following polynomials.

(a) $6x^2y^3 + 9xy^4 + 18y^5$

Each term of this polynomial can be written with a factor of $3y^3$, so that $3y^3$ is a common factor. Use the reverse of the distributive property to get

$$6x^2y^3 + 9xy^4 + 18y^5 = (3y^3)(2x^2) + (3y^3)(3xy) + (3y^3)(6y^2)$$
$$= 3y^3(2x^2 + 3xy + 6y^2).$$

(b) $6x^2t + 8xt + 12t = 2t(3x^2 + 4x + 6)$ ●

Each of the special patterns of multiplication from the previous section can be used in reverse to get a pattern for factoring. One of the most common of these is the difference of two squares.

Difference of Two Squares

$$x^2 - y^2 = (x + y)(x - y)$$

EXAMPLE 2   Factor each of the following polynomials.

(a) $4m^2 - 9$

First, recognize that $4m^2 - 9$ is the difference of two squares, since $4m^2 = (2m)^2$ and $9 = 3^2$. Use the pattern for the difference of two squares with $2m$ replacing x and 3 replacing y.

$$4m^2 - 9 = (2m)^2 - 3^2$$
$$= (2m + 3)(2m - 3)$$

(b) $144r^2 - 25s^2 = (12r + 5s)(12r - 5s)$

(c) $256k^4 - 625m^4$

Use the difference of two squares pattern twice, as follows.

$$256k^4 - 625m^4 = (16k^2)^2 - (25m^2)^2$$
$$= (16k^2 + 25m^2)(16k^2 - 25m^2)$$
$$= (16k^2 + 25m^2)(4k + 5m)(4k - 5m)$$

(d) $(a + 2b)^2 - 4c^2 = (a + 2b)^2 - (2c)^2$
$$= [(a + 2b) + 2c][(a + 2b) - 2c]$$
$$= (a + 2b + 2c)(a + 2b - 2c) \quad \bullet$$

In this chapter, a polynomial with only integer coefficients will be factored so that all factors have only integer coefficients. This assumption is sometimes summarized by saying that only **factoring over the integers** is permitted. With factoring over the integers, the polynomial $x^2 - 5$, for example, cannot be factored. While it is true that

$$x^2 - 5 = (x + \sqrt{5})(x - \sqrt{5}),$$

the two factors $x + \sqrt{5}$ and $x - \sqrt{5}$ have noninteger coefficients. When factored over the integers, $x^2 - 5$ is prime; when factored over the reals, it is not.

Now we need to look at methods of factoring a trinomial of degree 2, such as $kx^2 + mx + n$, where k, m, and n are integers. Any factorization will be of the form $(ax + b)(cx + d)$ with a, b, c, and d integers. Multiplying out the product $(ax + b)(cx + d)$ gives

$$(ax + b)(cx + d) = acx^2 + (ad + bc)x + bd,$$

which equals $kx^2 + mx + n$ if

$$ac = k, \quad ad + bc = m, \quad \text{and} \quad bd = n. \qquad (*)$$

In summary, to factor a trinomial $kx^2 + mx + n$, look for four integers a, b, c, and d satisfying the conditions given in (*). If no such integers exist, the trinomial is prime.

. . . .

EXAMPLE 3 Factor each of the following polynomials.

(a) $6p^2 - 7p - 5$

To find integers a, b, c, and d so that

$$6p^2 - 7p - 5 = (ap + b)(cp + d),$$

use the results given in (*) above and try to find integers satisfying $ac = 6$, $ad + bc = -7$, and $bd = -5$. Look for these integers by trying various possibilities. Since $ac = 6$, we might let $a = 2$ and $c = 3$. Since $bd = -5$, we might let $b = -5$ and $d = 1$, giving

$$(2p - 5)(3p + 1) = 6p^2 - 13p - 5. \qquad \text{Incorrect}$$

To make another attempt, try

$$(3p - 5)(2p + 1) = 6p^2 - 7p - 5. \qquad \text{Correct}$$

Finally, $6p^2 - 7p - 5$ is factored as $(3p - 5)(2p + 1)$.

(b) $4x^3 + 6x^2r - 10xr^2$

There is a common factor of $2x$.

$$4x^3 + 6x^2r - 10xr^2 = 2x(2x^2 + 3xr - 5r^2).$$

Factoring $2x^2 + 3xr - 5r^2$ requires factors of $2x^2$ and of $-5r^2$ that will yield the correct middle term of $3xr$. By inspection,

$$4x^3 + 6x^2r - 10xr^2 = 2x(2x + 5r)(x - r).$$

(c) $r^2 + 6r + 7$

It is not possible to find integers a, b, c, and d so that

$$r^2 + 6r + 7 = (ar + b)(cr + d);$$

therefore, $r^2 + 6r + 7$ is a prime polynomial. ●

Two other special types of factoring are listed below.

. . . .

Difference and Sum of Cubes

$$x^3 - y^3 = (x - y)(x^2 + xy + y^2) \qquad \textbf{Difference of two cubes}$$
$$x^3 + y^3 = (x + y)(x^2 - xy + y^2) \qquad \textbf{Sum of two cubes}$$

· · · · · · · ·

EXAMPLE 4   Factor each polynomial.

(a) $m^3 - 64n^3$

Since $64n^3 = (4n)^3$, the given binomial is a difference of two cubes. To factor, use the first pattern above, replacing x with m and y with $4n$, to get

$$m^3 - 64n^3 = m^3 - (4n)^3$$
$$= (m - 4n)[m^2 + m(4n) + (4n)^2]$$
$$= (m - 4n)(m^2 + 4mn + 16n^2).$$

(b) $8q^6 + 125p^9$

Write $8q^6$ as $(2q^2)^3$ and $125p^9$ as $(5p^3)^3$, showing that the given polynomial is a sum of two cubes. Factor as

$$8q^6 + 125p^9 = (2q^2)^3 + (5p^3)^3$$
$$= (2q^2 + 5p^3)[(2q^2)^2 - (2q^2)(5p^3) + (5p^3)^2]$$
$$= (2q^2 + 5p^3)(4q^4 - 10q^2p^3 + 25p^6).$$

(c) $(2a - 1)^3 + 8$

Use the pattern for the sum of two cubes, with $x = 2a - 1$ and $y = 2$. Doing so gives

$$(2a - 1)^3 + 8 = [(2a - 1) + 2][(2a - 1)^2 - (2a - 1)2 + 2^2]$$
$$= (2a - 1 + 2)(4a^2 - 4a + 1 - 4a + 2 + 4)$$
$$= (2a + 1)(4a^2 - 8a + 7). \quad \bullet$$

When a polynomial has more than three terms, it can often be factored by grouping. For example, to factor

$$ax + ay + 6x + 6y,$$

collect the terms into two groups,

$$ax + ay + 6x + 6y = (ax + ay) + (6x + 6y),$$

and then factor each group, getting

$$ax + ay + 6x + 6y = a(x + y) + 6(x + y).$$

The quantity $(x + y)$ is now a common factor, which can be factored out, producing

$$ax + ay + 6x + 6y = (x + y)(a + 6).$$

It is not always obvious which terms should be grouped. Experience is the best teacher for techniques of factoring.

· · · · · · · ·

EXAMPLE 5   Factor by grouping.

(a) $mp^2 + 7m + 3p^2 + 21$

Group the terms as follows.

$$mp^2 + 7m + 3p^2 + 21 = (mp^2 + 7m) + (3p^2 + 21)$$

Find the common factor for each part.

$$(mp^2 + 7m) + (3p^2 + 21) = m(p^2 + 7) + 3(p^2 + 7)$$
$$= (p^2 + 7)(m + 3)$$

(b) $2y^2 - 2z - ay^2 + az$

Grouping terms as above gives

$$2y^2 - 2z - ay^2 + az = (2y^2 - 2z) + (-ay^2 + az)$$
$$= 2(y^2 - z) + a(-y^2 + z).$$

The expression $-y^2 + z$ is the negative of $y^2 - z$, so the terms should be grouped as follows.

$$2y^2 - 2z - ay^2 + az = (2y^2 - 2z) - (ay^2 - az)$$
$$= 2(y^2 - z) - a(y^2 - z)$$
$$= (y^2 - z)(2 - a)$$

(c) $x^2 - 6x + 9 - y^2 = (x^2 - 6x + 9) - y^2$
$$= (x - 3)^2 - y^2$$
$$= (x - 3 + y)(x - 3 - y) \quad \bullet$$

Expressions that are not polynomials can also be factored.

· · · · · · · · ·

EXAMPLE 6 Factor out the specified common factor from the following expressions. Assume all variables represent positive numbers.

(a) $5x^{4/3} + 2x^{1/3}; \ x^{1/3}$

$$5x^{4/3} + 2x^{1/3} = x^{1/3}(5x + 2)$$

(b) $4y^{-1/2} - 3y^{1/2}; \ y^{-1/2}$

$$4y^{-1/2} - 3y^{1/2} = y^{-1/2}(4 - 3y) \quad \bullet$$

· · · · · · · · ·

EXAMPLE 7 Simplify $\dfrac{3(m - 1)^{1/2} + (m - 1)^{-1/2}}{m - 1}$.

Factor the numerator and then simplify.

$$\frac{3(m - 1)^{1/2} + (m - 1)^{-1/2}}{m - 1} = \frac{(m - 1)^{-1/2}[3(m - 1) + 1]}{m - 1}$$

$$= \frac{3m - 2}{(m - 1)^{3/2}} \quad \bullet$$

1.4 Exercises ···

Factor as completely as possible. Assume all variables appearing as exponents represent integers.

1. $12mn - 8m$

2. $3pq - 18pqr$

3. $12r^3 + 6r^2 - 3r$

4. $-5k^2g - 25kg - 30kg^2$

5. $6px^2 - 8px^3 - 12px$

6. $9m^2n^3 - 18m^3n^2 + 27m^2n^4$

7. $2(a + b) + 4m(a + b)$

8. $4(y - 2)^2 + 3(y - 2)$

9. $x^2 - 11x + 24$

10. $y^2 - 2y - 35$

11. $4p^2 + 3p - 1$

12. $6x^2 + 7x - 3$

13. $12r^2 + 24r - 15$

14. $12p^2 + p - 20$

15. $18r^2 - 3rs - 10s^2$

16. $12m^2 + 16mn - 35n^2$

17. $9x^2 - 6x^3 + x^4$

18. $30a + am - am^2$

19. $4m^2 - 25$

20. $25a^2 - 16$

21. $144r^2 - 81s^2$

22. $81m^2 - 16n^2$

23. $121p^4 - 9q^4$

24. $81q^4 - 256m^4$

25. $p^8 - 1$

26. $y^{16} - 1$

27. $8m^3 - 27n^3$

28. $125x^3 - 1$

29. $4x + 4y + mx + my$

30. $x^2 + xy + 5x + 5y$

31. $q^2 + 6q + 9 - p^2$

32. $4b^2 + 4bc + c^2 - 16$

33. $a^2 + 2ab + b^2 - x^2 - 2xy - y^2$

34. $d^2 - 10d + 25 - c^2 + 4c - 4$

35. $x^2 - (x - y)^2$

36. $16m^2 - 25(m - n)^2$

37. $49(m - 1)^2 - 16m^2$

38. $36(a + 2)^2 - 25a^2$

39. $(x + y)^2 + 2(x + y)z - 15z^2$

40. $(m + n)^2 + 3(m + n)p - 10p^2$

41. $(p + q)^2 - (p - q)^2$

42. $(p - q)^2 - (p + q)^2$

43. $(r + 6)^3 - 216$

44. $(b + 3)^3 - 27$

45. $27 - (m + 2n)^3$

46. $125 - (4a - b)^3$

47. $27k^3 + 1$

48. $(2x + 1)^3 + 64$

49. $3mx - 6my - nx + 2ny$

50. $2p - 6q - xp + 3xq$

51. $x^3 - 3x^2y + 3xy^2 - y^3$

52. $p^3 + 3p^2q + 3pq^2 + q^3$

53. $m^{2n} - 16$

54. $p^{4n} - 49$

55. $x^{3n} - y^{6n}$

56. $a^{4p} - b^{12p}$

57. $2x^{2n} - 23x^ny^n - 39y^{2n}$

58. $3a^{2x} + 7a^xb^x + 2b^{2x}$

59. $25q^{2r} - 30q^rt^p + 9t^{2p}$

60. $16m^{2p} + 56m^pn^q + 49n^{2q}$

61. $6(m + p)^{2k} + (m + p)^k - 15$

Factor each expression. (The expressions in Exercises 62–71 arise in calculus from techniques called the product and quotient rules.)

62. $2(3x - 4)^2 + (x - 5)(2)(3x - 4)(3)$

63. $(5 - 2x)(3)(7x - 8)^2(7) + (7x - 8)^3(-2)$

64. $\dfrac{(p + 1)^{1/2} - p(\frac{1}{2})(p + 1)^{-1/2}}{p + 1}$

65. $\dfrac{(r - 2)^{2/3} - r(\frac{2}{3})(r - 2)^{-1/3}}{(r - 2)^{4/3}}$

66. $\dfrac{3(2x^2 + 5)^{1/3} - x(2x^2 + 5)^{-2/3}(4x)}{(2x^2 + 5)^{2/3}}$

67. $\dfrac{-(m^3 + m)^{2/3} + m(\frac{2}{3})(m^3 + m)^{-1/3}(3m^2 + 1)}{(m^3 + m)^{4/3}}$

68. $(x^{-1} - 5)^3(2)(2 - x^{-2})(2x^{-3}) + (2 - x^{-2})^2(3)(x^{-1} - 5)^2(-x^{-2})$

69. $(6 + x^{-4})^2(3)(3x - 2x^{-1})^2(3 + 2x^{-2}) + (3x - 2x^{-1})^3(2)(6 + x^{-4})(-4x^{-5})$

70. Factor $4(x^2 + 3)^{-3}(4x + 7)^{-4}$ from $(x^2 + 3)^{-2}(-3)(4x + 7)^{-4}(4) + (4x + 7)^{-3}(-2)(x^2 + 3)^{-3}(2x)$.

71. Factor $(7x - 8)^{-6}(3x + 2)^{-3}$ from $(7x - 8)^{-5}(-2)(3x + 2)^{-3}(3) + (3x + 2)^{-2}(-5)(7x - 8)^{-6}(7)$.

Factor the variable of smallest exponent, together with any numerical common factor, from each of the following expressions. (For example, factor $9x^{-2} - 6x^{-3}$ as $3x^{-3}(3x - 2)$.)

72. $p^{-4} - p^{-2}$

73. $m^{-1} + 3m^{-5}$

74. $12k^{-3} + 4k^{-2} - 8k^{-1}$

75. $6a^{-5} - 10a^{-4} + 18a^{-2}$

76. $100p^{-6} - 50p^{-2} + 75p^2$

77. $32y^{-3} + 48y - 64y^2$

78. $4k^{7/4} + k^{3/4}$

79. $y^{9/2} - 3y^{5/2}$

80. $9z^{-1/2} + 2z^{1/2}$

81. $3m^{2/3} - 4m^{-1/3}$

82. $-(2a - 5)^{-3/2} + 3(2a - 5)^{-1/2}$

83. $(3k - 2)^{-1/2} + 4(3k - 2)^{-3/2}$

Factor $x^4 + 4x^2 + 16$ as follows.

$$x^4 + 4x^2 + 16 = (x^4 + 8x^2 + 16) - 4x^2$$
$$= (x^2 + 4)^2 - (2x)^2$$
$$= (x^2 + 4 + 2x)(x^2 + 4 - 2x)$$

Use this procedure to factor each of the following polynomials.

84. $x^4 + 64$

85. $r^4 - 6r^2 + 1$

86. $p^4 + 9p^2 + 81$

87. $x^4 - 18x^2 + 1$

88. $z^4 - 11z^2 + 25$

89. $m^4 - 22m^2 + 9$

1.5 RATIONAL EXPRESSIONS

The quotient of two algebraic expressions (with denominator not 0) is a **fractional expression.** The most common fractional expressions are the quotients of two polynomials; these quotients are called **rational expressions.** Since fractional expressions involve quotients, it is important to note the values of the variables that would make the denominator zero. For example, -2 cannot replace x in the rational expression

$$\frac{x + 6}{x + 2}$$

since -2 (when substituted for x) makes the denominator equal 0. This restriction on x can be written $x \neq -2$. In a similar way, the restriction on x for

$$\frac{(x + 6)(x + 4)}{(x + 2)(x + 4)}$$

can be written $x \neq -2$ and $x \neq -4$.

Just as the fraction 6/8 is written in lowest terms as 3/4, rational expressions can also be written in lowest terms. Do this with the *fundamental principle*.

Fundamental Principle

For any rational number a/b and nonzero real number c,

$$\frac{ac}{bc} = \frac{a}{b}.$$

. . . .

EXAMPLE 1   Write each expression in lowest terms.

(a) $\dfrac{2p^2 + 7p - 4}{5p^2 + 20p}$

Factor the numerator and denominator to get

$$\frac{2p^2 + 7p - 4}{5p^2 + 20p} = \frac{(2p - 1)(p + 4)}{5p(p + 4)}.$$

By the fundamental principle,

$$\frac{2p^2 + 7p - 4}{5p^2 + 20p} = \frac{2p - 1}{5p}.$$

The restriction on p for the original expression is $p \neq 0$ and $p \neq -4$, so this result is valid only for values of p other than 0 and -4. From now on, we shall always assume such restrictions when writing rational expressions in lowest terms.

(b) $\dfrac{6 - 3k}{k^2 - 4}$

Factor to get $\dfrac{6 - 3k}{k^2 - 4} = \dfrac{3(2 - k)}{(k + 2)(k - 2)}.$

The factors $2 - k$ and $k - 2$ have exactly opposite signs. Because of this, multiply numerator and denominator by -1, as follows.

$$\frac{6 - 3k}{k^2 - 4} = \frac{3(2 - k)(-1)}{(k + 2)(k - 2)(-1)}$$

Since $(k - 2)(-1) = -k + 2$, or $2 - k$, the fraction becomes

$$\frac{6 - 3k}{k^2 - 4} = \frac{3(2 - k)(-1)}{(k + 2)(2 - k)},$$

finally giving $\dfrac{6 - 3k}{k^2 - 4} = \dfrac{-3}{k + 2}.$

Working in an alternate way would give the equivalent result $3/(-k - 2)$.   ●

To multiply or divide rational expressions, again use properties and definitions from Section 1.1.

. . . .
Multiplication and Division
. . . .

For rational numbers a/b and c/d,

$$\frac{a}{b} \cdot \frac{c}{d} = \frac{ac}{bd},$$

$$\frac{a}{b} \div \frac{c}{d} = \frac{a}{b} \cdot \frac{d}{c}, \quad c \neq 0.$$

· · · · · · · · · ·

EXAMPLE 2   Multiply or divide, as indicated.

(a) $\dfrac{3m^2 - 2m - 8}{3m^2 + 14m + 8} \cdot \dfrac{3m + 2}{3m + 4} = \dfrac{(m - 2)(3m + 4)}{(m + 4)(3m + 2)} \cdot \dfrac{3m + 2}{3m + 4}$

$$= \dfrac{(m - 2)(3m + 4)(3m + 2)}{(m + 4)(3m + 2)(3m + 4)} = \dfrac{m - 2}{m + 4}$$

(b) $\dfrac{5}{8m + 16} \div \dfrac{7}{12m + 24} = \dfrac{5}{8(m + 2)} \div \dfrac{7}{12(m + 2)}$

$$= \dfrac{5}{8(m + 2)} \cdot \dfrac{12(m + 2)}{7}$$

$$= \dfrac{5 \cdot 12(m + 2)}{8 \cdot 7(m + 2)}$$

$$= \dfrac{15}{14}$$

(c) $\dfrac{3p^2 + 11p - 4}{24p^3 - 8p^2} \div \dfrac{9p + 36}{24p^4 - 36p^3} = \dfrac{(p + 4)(3p - 1)}{8p^2(3p - 1)} \div \dfrac{9(p + 4)}{12p^3(2p - 3)}$

$$= \dfrac{(p + 4)(3p - 1)(12p^3)(2p - 3)}{8p^2(3p - 1)(9)(p + 4)}$$

$$= \dfrac{12p^3(2p - 3)}{9 \cdot 8p^2} = \dfrac{p(2p - 3)}{6} \quad \bullet$$

Add or subtract rational expressions with properties given earlier.

· · · · · · · · · · · · · · ·

Addition and
Subtraction

For rational numbers a/b, c/b, and c/d,

$$\dfrac{a}{b} + \dfrac{c}{b} = \dfrac{a + c}{b} \qquad \text{or} \qquad \dfrac{a}{b} + \dfrac{c}{d} = \dfrac{ad + bc}{bd}$$

$$\dfrac{a}{b} - \dfrac{c}{b} = \dfrac{a - c}{b} \qquad \text{or} \qquad \dfrac{a}{b} - \dfrac{c}{d} = \dfrac{ad - bc}{bd}.$$

The left hand results are for rational expressions with the same denominators. The results on the right come from the fundamental principle. For example,

$$\dfrac{a}{b} + \dfrac{c}{d} = \dfrac{a \cdot d}{b \cdot d} + \dfrac{c \cdot b}{d \cdot b} = \dfrac{ad}{bd} + \dfrac{bc}{bd} = \dfrac{ad + bc}{bd}.$$

In practice, rational expressions are normally added or subtracted after rewriting all the rational expressions with the same denominator. A **common denominator** of a set of rational expressions is an expression that can be divided evenly (without a remainder) by each denominator in the set.

· · · · · · · · ·

EXAMPLE 3   Add or subtract, as indicated.

(a) $\dfrac{5}{9x^2} + \dfrac{1}{6x}$

Both $9x^2$ and $6x$ can be divided into $18x^2$ evenly, so a common denominator is $18x^2$. Now write both of the given rational expressions with this denominator, giving

$$\frac{5}{9x^2} + \frac{1}{6x} = \frac{5 \cdot 2}{9x^2 \cdot 2} + \frac{1 \cdot 3x}{6x \cdot 3x} = \frac{10}{18x^2} + \frac{3x}{18x^2} = \frac{10 + 3x}{18x^2}.$$

(b) $\dfrac{y + 2}{y^2 - y} - \dfrac{3y}{2y^2 - 4y + 2}$

To find the common denominator, first factor each denominator.

$$\frac{y + 2}{y^2 - y} - \frac{3y}{2y^2 - 4y + 2} = \frac{y + 2}{y(y - 1)} - \frac{3y}{2(y - 1)^2}$$

The common denominator is $2y(y - 1)^2$. Write each rational expression with this denominator, as follows.

$$\frac{y + 2}{y(y - 1)} - \frac{3y}{2(y - 1)^2} = \frac{2(y - 1)(y + 2)}{2y(y - 1)^2} - \frac{y \cdot 3y}{2y(y - 1)^2}$$

$$= \frac{2(y^2 + y - 2)}{2y(y - 1)^2} - \frac{3y^2}{2y(y - 1)^2}$$

$$= \frac{2y^2 + 2y - 4 - 3y^2}{2y(y - 1)^2} = \frac{-y^2 + 2y - 4}{2y(y - 1)^2} \quad \bullet$$

Complex Fractions   Any quotient of two rational expressions is called a **complex fraction**. Complex fractions can often be simplified by the methods shown in the following examples.

· · · · · · · · ·

EXAMPLE 4   Simplify each complex fraction.

(a) $\dfrac{6 - \dfrac{5}{k}}{1 + \dfrac{5}{k}}$

Multiply both numerator and denominator by the common denominator k.

$$\frac{k\left(6 - \dfrac{5}{k}\right)}{k\left(1 + \dfrac{5}{k}\right)} = \frac{6k - k\left(\dfrac{5}{k}\right)}{k + k\left(\dfrac{5}{k}\right)} = \frac{6k - 5}{k + 5}$$

(b) $\dfrac{\dfrac{a}{a + 1} + \dfrac{1}{a}}{\dfrac{1}{a} + \dfrac{1}{a + 1}}$

Multiply both numerator and denominator by the common denominator of all the fractions, in this case $a(a + 1)$. Doing so gives

$$\frac{\dfrac{a}{a + 1} + \dfrac{1}{a}}{\dfrac{1}{a} + \dfrac{1}{a + 1}} = \frac{\left(\dfrac{a}{a + 1} + \dfrac{1}{a}\right) a(a + 1)}{\left(\dfrac{1}{a} + \dfrac{1}{a + 1}\right) a(a + 1)} = \frac{a^2 + (a + 1)}{(a + 1) + a} = \frac{a^2 + a + 1}{2a + 1}.$$

As an alternate method of solution, first perform the indicated additions in the numerator and denominator, and then divide.

$$\frac{\dfrac{a}{a + 1} + \dfrac{1}{a}}{\dfrac{1}{a} + \dfrac{1}{a + 1}} = \frac{\dfrac{a^2 + 1(a + 1)}{a(a + 1)}}{\dfrac{1(a + 1) + 1(a)}{a(a + 1)}} = \frac{\dfrac{a^2 + a + 1}{a(a + 1)}}{\dfrac{2a + 1}{a(a + 1)}}$$

$$= \frac{a^2 + a + 1}{a(a + 1)} \cdot \frac{a(a + 1)}{2a + 1} = \frac{a^2 + a + 1}{2a + 1}. \quad \bullet$$

The next example shows how negative exponents can lead to rational expressions.

· · · · · · · · ·

EXAMPLE 5 Simplify $\dfrac{(x + y)^{-1}}{x^{-1} + y^{-1}}$. Write the result so that it has only positive exponents.

Use the definition of negative integer exponent to get

$$\frac{(x + y)^{-1}}{x^{-1} + y^{-1}} = \frac{\dfrac{1}{x + y}}{\dfrac{1}{x} + \dfrac{1}{y}} = \frac{\dfrac{1}{x + y}}{\dfrac{y + x}{xy}} = \frac{1}{x + y} \cdot \frac{xy}{x + y} = \frac{xy}{(x + y)^2}. \quad \bullet$$

1.5 Exercises ·

Give the restrictions on x for each of the following.

1. $\dfrac{x - 2}{x + 6}$

2. $\dfrac{x + 5}{x - 3}$

3. $\dfrac{2x}{5x^2 + 2x - 3}$

4. $\dfrac{6x}{2x^2 - x - 1}$

5. $\dfrac{-8}{x^2 + 1}$

6. $\dfrac{3x}{3x^2 + 7}$

Write each of the following in lowest terms.

7. $\dfrac{25p^3}{10p^2}$

8. $\dfrac{14z^3}{6z^2}$

9. $\dfrac{8k + 16}{9k + 18}$

10. $\dfrac{20r + 10}{30r + 15}$

11. $\dfrac{3(t + 5)}{(t + 5)(t - 3)}$

12. $\dfrac{-8(y - 4)}{(y + 2)(y - 4)}$

13. $\dfrac{8x^2 + 16x}{4x^2}$

14. $\dfrac{36y^2 + 72y}{9y}$

15. $\dfrac{m^2 - 4m + 4}{m^2 + m - 6}$

16. $\dfrac{r^2 - r - 6}{r^2 + r - 12}$

17. $\dfrac{8m^2 + 6m - 9}{16m^2 - 9}$

18. $\dfrac{6y^2 + 11y + 4}{3y^2 + 7y + 4}$

Perform each operation.

19. $\dfrac{x(y + 2)}{4} \div \dfrac{x}{2}$

20. $\dfrac{mn}{m + n} \div \dfrac{p}{m + n}$

21. $\dfrac{15p^3}{9p^2} \div \dfrac{6p}{10p^2}$

22. $\dfrac{3r^2}{9r^3} \div \dfrac{8r^3}{6r}$

23. $\dfrac{2k + 8}{6} \div \dfrac{3k + 12}{2}$

24. $\dfrac{5m + 25}{10} \cdot \dfrac{12}{6m + 30}$

25. $\dfrac{9y - 18}{6y + 12} \cdot \dfrac{3y + 6}{15y - 30}$

26. $\dfrac{12r + 24}{36r - 36} \div \dfrac{6r + 12}{8r - 8}$

27. $\dfrac{x^2 + x}{5} \cdot \dfrac{25}{xy + y}$

28. $\dfrac{3m - 15}{4m - 20} \cdot \dfrac{m^2 - 10m + 25}{12m - 60}$

29. $\dfrac{4a + 12}{2a - 10} \div \dfrac{a^2 - 9}{a^2 - a - 20}$

30. $\dfrac{6r - 18}{9r^2 + 6r - 24} \cdot \dfrac{12r - 16}{4r - 12}$

31. $\dfrac{p^2 - p - 12}{p^2 - 2p - 15} \cdot \dfrac{p^2 - 9p + 20}{p^2 - 8p + 16}$

32. $\dfrac{x^2 + 2x - 15}{x^2 + 11x + 30} \cdot \dfrac{x^2 + 2x - 24}{x^2 - 8x + 15}$

33. $\dfrac{n^2 - n - 6}{n^2 - 2n - 8} \div \dfrac{n^2 - 9}{n^2 + 7n + 12}$

34. $\dfrac{2m^2 - 5m - 12}{m^2 - 10m + 24} \div \dfrac{4m^2 - 9}{m^2 - 9m + 18}$

35. $\left(1 + \dfrac{1}{x}\right)\left(1 - \dfrac{1}{x}\right)$

36. $\left(3 + \dfrac{2}{y}\right)\left(3 - \dfrac{2}{y}\right)$

37. $\dfrac{x^3 + y^3}{x^2 - y^2} \cdot \dfrac{x + y}{x^2 - xy + y^2}$

38. $\dfrac{8y^3 - 125}{4y^2 - 20y + 25} \cdot \dfrac{2y - 5}{y}$

39. $\dfrac{x^3 + y^3}{x^3 - y^3} \cdot \dfrac{x^2 - y^2}{x^2 + 2xy + y^2}$

40. $\dfrac{8}{r} + \dfrac{6}{r}$

41. $\dfrac{3}{y} + \dfrac{4}{y}$

42. $\dfrac{8}{5p} + \dfrac{3}{4p}$

43. $\dfrac{2}{3y} - \dfrac{1}{4y}$

44. $\dfrac{6}{11z} - \dfrac{5}{2z}$

45. $\dfrac{a + 1}{2} - \dfrac{a - 1}{2}$

46. $\dfrac{y + 6}{5} - \dfrac{y - 6}{5}$

47. $\dfrac{3}{p} + \dfrac{1}{2}$

48. $\dfrac{9}{r} - \dfrac{2}{3}$

49. $\dfrac{2}{y} - \dfrac{1}{4}$

50. $\dfrac{6}{11} + \dfrac{3}{a}$

51. $\dfrac{1}{6m} + \dfrac{2}{5m} + \dfrac{4}{m}$

52. $\dfrac{8}{3p} + \dfrac{5}{4p} + \dfrac{9}{2p}$

53. $\dfrac{1}{y} + \dfrac{1}{y + 1}$

54. $\dfrac{2}{3(x - 1)} + \dfrac{1}{4(x - 1)}$

55. $\dfrac{2}{a + b} - \dfrac{1}{2(a + b)}$

56. $\dfrac{3}{m} - \dfrac{1}{m - 1}$

57. $\dfrac{1}{a + 1} - \dfrac{1}{a - 1}$

58. $\dfrac{1}{x + z} + \dfrac{1}{x - z}$

59. $\dfrac{m + 1}{m - 1} + \dfrac{m - 1}{m + 1}$

60. $\dfrac{2}{x - 1} + \dfrac{1}{1 - x}$

61. $\dfrac{3}{a - 2} - \dfrac{1}{2 - a}$

62. $\dfrac{m - 4}{3m - 4} + \dfrac{3m + 2}{4 - 3m}$

63. $\dfrac{1}{a^2 - 5a + 6} - \dfrac{1}{a^2 - 4}$

64. $\dfrac{-3}{m^2 - m - 2} - \dfrac{1}{m^2 + 3m + 2}$

65. $\dfrac{4x - 1}{x^2 + 3x - 10} + \dfrac{2x + 3}{x^2 + 4x - 5}$

66. $\dfrac{3y + 5}{y^2 - 9y + 20} + \dfrac{2y - 7}{y^2 - 2y - 8}$

67. $\left(\dfrac{3}{p - 1} - \dfrac{2}{p + 1}\right)\left(\dfrac{p - 1}{p}\right)$

68. $\left(\dfrac{y}{y^2 - 1} - \dfrac{y}{y^2 - 2y + 1}\right)\left(\dfrac{y - 1}{y + 1}\right)$

69. $\dfrac{\dfrac{1}{x + h} - \dfrac{1}{x}}{h}$

70. $\dfrac{1}{h}\left(\dfrac{1}{(x + h)^2 + 9} - \dfrac{1}{x^2 + 9}\right)$

71. $\dfrac{1 + \dfrac{1}{x}}{1 - \dfrac{1}{x}}$

72. $\dfrac{2 - \dfrac{2}{y}}{2 + \dfrac{2}{y}}$

73. $\dfrac{\dfrac{1}{x + 1} - \dfrac{1}{x}}{\dfrac{1}{x}}$

74. $\dfrac{\dfrac{1}{y + 3} - \dfrac{1}{y}}{\dfrac{1}{y}}$

75. $\dfrac{1 + \dfrac{1}{1 - b}}{1 - \dfrac{1}{1 + b}}$

76. $m - \dfrac{m}{m + \dfrac{1}{2}}$

Perform all indicated operations and write all answers with positive integer exponents.

77. $\dfrac{3^{-1} - 4^{-1}}{4^{-1}}$

78. $\dfrac{6^{-1} + 5^{-1}}{6^{-1}}$

79. $\dfrac{a^{-1} + b^{-1}}{(ab)^{-1}}$

80. $\dfrac{p^{-1} - q^{-1}}{(pq)^{-1}}$

81. $\dfrac{r^{-1} + q^{-1}}{r^{-1} - q^{-1}} \cdot \dfrac{r - q}{r + q}$

82. $\dfrac{xy^{-1} + yx^{-1}}{x^2 + y^2}$

83. $(a + b)^{-1}(a^{-1} + b^{-1})$

84. $(m^{-1} + n^{-1})^{-1}$

Simplify each of the following expressions. (The expressions in Exercises 85–88 arise in calculus from techniques called the chain rule *and* quotient rule *that are used to determine the shape of a curve.)*

85. $\left(\dfrac{x^3 + 2}{x - 5}\right)^3 \left(\dfrac{(x - 5)(3x^2) - (x^3 + 2)}{(x - 5)^2}\right)$

86. $\left(\dfrac{2x^2 - 9}{x^2 + 1}\right)^2 \left(\dfrac{(x^2 + 1)(4x) - (2x^2 - 9)}{(x^2 + 1)^2}\right)$

87. $5\left(\dfrac{x + 1}{3x^2 - 4}\right)^4 \left(\dfrac{(3x^2 - 4) - (x + 1)(6x)}{(3x^2 - 4)^2}\right)$

88. $7\left(\dfrac{x^2 + 2}{5x^2 + 3}\right)^6 \left(\dfrac{(5x^2 + 3)(2x) - (x^2 + 2)(10x)}{(5x^2 + 3)^2}\right)$

1.6 RADICALS

Radicals provide a convenient notation for *n*th roots. In this section we explore this notation and develop techniques for simplifying expressions involving radicals.

Definition of $\sqrt[n]{a}$

If a is a real number, n is a positive integer, and $a^{1/n}$ is a real number, then

$$\sqrt[n]{a} = a^{1/n}.$$

As we saw in Section 1.2, $a^{1/n}$ is defined only for nonnegative values of a when n is even, but is defined for all values of a when n is odd. Combining the definition of $\sqrt[n]{a}$ with the definition of $a^{1/n}$ from Section 1.2, we obtain the following characterization of $\sqrt[n]{a}$.

$\sqrt[n]{a}$

If a and b are nonnegative real numbers and n is a positive integer, or if both a and b are negative and n is an odd positive integer, then

$$\sqrt[n]{a} = b \quad \text{if and only if} \quad a = b^n.$$

The expression $\sqrt[n]{a}$ is called the **principal nth root** of a (abbreviated as the **nth root of a**); n is the **index** of the radical expression $\sqrt[n]{a}$. The number a is the **radicand**, and $\sqrt[n]{}$ is a **radical.** Abbreviate $\sqrt[2]{a}$ as just $\sqrt{a}$.

If n is a positive integer and a is a real number, then the following chart summarizes the conditions necessary for $\sqrt[n]{a}$ to exist.

n is	$a > 0$	$a < 0$	$a = 0$
even	$\sqrt[n]{a}$ is the positive real number such that $(\sqrt[n]{a})^n = a$	$\sqrt[n]{a}$ is not a real number	$\sqrt[n]{a}$ is 0
odd	$\sqrt[n]{a}$ is the real number such that $(\sqrt[n]{a})^n = a$		$\sqrt[n]{a}$ is 0

If n is even there are two nth roots of any positive number a; one root is positive and the other root is negative. In radical notation, $\sqrt[n]{a}$ denotes the positive nth root of a and $-\sqrt[n]{a}$ denotes the negative nth root of a.

EXAMPLE 1

(a) $\sqrt{\dfrac{4}{9}} = \dfrac{2}{3}$

(b) $-\sqrt[4]{16} = -2$

(c) $\sqrt{-4}$ is undefined.

(d) $\sqrt[5]{-32} = -2$

(e) $\sqrt[3]{1000} = 10$ •

By the definition of $\sqrt[n]{a}$, for any positive integer n, if $\sqrt[n]{a}$ exists, then

$$(\sqrt[n]{a})^n = a.$$

If a is positive, or if a is negative and n is an odd positive integer, then

$$\sqrt[n]{a^n} = a.$$

Because of the conditions just given, it is *not* necessarily true that $\sqrt{x^2} = x$. For example, if $x = -5$,

$$\sqrt{x^2} = \sqrt{(-5)^2} = \sqrt{25} = 5 \neq x.$$

To take care of the fact that a negative value of x can produce a positive result for the square root, use the following rule, which involves absolute value.

For any real number a,

$$\sqrt{a^2} = |a|.$$

Also, if n is any *even* positive integer, then

$$\sqrt[n]{a^n} = |a|.$$

We shall prove only the first part of this statement, when $n = 2$. The statement is certainly true if a is positive. If a is negative, then $-a$ is positive. Since $(-a)^2 = a^2$,

$$\sqrt{(-a)^2} = -a.$$

Since $|a| = -a$ if a is negative, $\sqrt{a^2} = |a|$ for *all* real numbers a.

EXAMPLE 2

(a) $\sqrt{(-9)^2} = |-9| = 9$ (b) $\sqrt{13^2} = |13| = 13$

(c) $\sqrt{x^6} = |x^3|$ (d) $\sqrt[3]{x^6} = x^2$

(e) $\sqrt[3]{x^9} = x^3$ (f) $\sqrt[4]{x^4} = |x|$

(g) $\sqrt[4]{x^8} = x^2$ •

Rational exponents can be expressed in terms of radicals in two ways.

Radical Definition of Rational Exponents

For all integers m and all positive integers n such that m/n is in lowest terms, and all nonzero numbers a for which $\sqrt[n]{a}$ is a real number.

$$a^{m/n} = (\sqrt[n]{a})^m = \sqrt[n]{a^m}.$$

EXAMPLE 3   Write each of the following with a single radical. Assume all variables represent nonnegative real numbers.

(a) $\sqrt[5]{y^4} \cdot \sqrt[3]{y} = y^{4/5}y^{1/3} = y^{17/15} = y^{1+2/15} = y \cdot y^{2/15} = y\sqrt[15]{y^2}$

(b) $\sqrt[4]{p^3q} \cdot \sqrt[5]{p^2q^7} = (p^3q)^{1/4}(p^2q^7)^{1/5} = p^{3/4} \, p^{2/5} \, q^{1/4} \, q^{7/5}$

$$= p^{23/20} \, q^{33/20} = pq\sqrt[20]{p^3q^{13}} \quad \bullet$$

To avoid difficulties when we are working with variable radicands, we will usually assume that all variables in radicands represent only nonnegative real numbers.

Three key rules for working with radicals are given below.

Rules for Radicals

For all real numbers a and b and positive integers m and n for which the indicated radicals exist,

$$\sqrt[n]{a} \cdot \sqrt[n]{b} = \sqrt[n]{ab}$$

$$\sqrt[n]{\frac{a}{b}} = \frac{\sqrt[n]{a}}{\sqrt[n]{b}}, \quad b \neq 0$$

$$\sqrt[m]{\sqrt[n]{a}} = \sqrt[mn]{a}.$$

EXAMPLE 4   Use the rules of radicals to simplify each of the following. Assume that all variables represent nonnegative real numbers.

(a) $\sqrt{64 \cdot 5} = \sqrt{64} \cdot \sqrt{5} = 8\sqrt{5}$

(b) $\sqrt{7x^5yz^2} = \sqrt{x^4z^2} \cdot \sqrt{7xy} = x^2z\sqrt{7xy}$

(c) $\sqrt[3]{\frac{-11x^6}{64}} = \frac{\sqrt[3]{-11x^6}}{\sqrt[3]{64}} = \frac{x^2\sqrt[3]{-11}}{4} = \frac{-x^2\sqrt[3]{11}}{4}$

(d) $\sqrt[7]{\sqrt[3]{2}} = \sqrt[21]{2}$ (e) $\sqrt[4]{\sqrt{3}} = \sqrt[8]{3}$ $\bullet$

By definition, an expression containing a radical is **simplified** when the following four conditions are satisfied.

Simplified Radicals

1. All possible factors have been removed from under the radical sign.

2. The index on the radical is as small as possible.

3. All radicals are removed from any denominators (a process called **rationalizing** the denominator).

4. All indicated operations have been performed (if possible).

· · · · · · · ·

EXAMPLE 5   Simplify each of the following. Assume that all variables represent nonnegative real numbers.

(a) $\sqrt{98x^3y} + 3x\sqrt{32xy}$

First remove all perfect square factors from under the radical. Then use the distributive property, as follows.

$$\sqrt{98x^3y} + 3x\sqrt{32xy} = \sqrt{49 \cdot 2 \cdot x^2 \cdot x \cdot y} + 3x\sqrt{16 \cdot 2 \cdot x \cdot y}$$
$$= 7x\sqrt{2xy} + (3x)(4)\sqrt{2xy}$$
$$= 7x\sqrt{2xy} + 12x\sqrt{2xy}$$
$$= 19x\sqrt{2xy}$$

(b) $\sqrt[3]{64m^4p^5} - \sqrt[3]{-27m^{10}p^{14}} = \sqrt[3]{(64m^3p^3)(mp^2)} - \sqrt[3]{(-27m^9p^{12})(mp^2)}$
$$= 4mp\sqrt[3]{mp^2} - (-3m^3p^4)\sqrt[3]{mp^2}$$
$$= 4mp\sqrt[3]{mp^2} + 3m^3p^4\sqrt[3]{mp^2}$$
$$= (4 + 3m^2p^3)mp\sqrt[3]{mp^2} \quad \bullet$$

Multiplying radical expressions is much like multiplying polynomials.

· · · · · · · ·

EXAMPLE 6 $(\sqrt{2} + 3)(\sqrt{8} - 5) = \sqrt{2}(\sqrt{8}) - \sqrt{2}(5) + 3\sqrt{8} - 3(5)$
$$= \sqrt{16} - 5\sqrt{2} + 3(2\sqrt{2}) - 15$$
$$= 4 - 5\sqrt{2} + 6\sqrt{2} - 15$$
$$= -11 + \sqrt{2} \quad \bullet$$

The next example shows how to rationalize the denominator (remove any radicals from the denominator) in an expression containing radicals.

· · · · · · · ·

EXAMPLE 7   Simplify each of the following expressions.

(a) $\dfrac{4}{\sqrt{3}}$

To rationalize the denominator, multiply by $\sqrt{3}/\sqrt{3}$ (which equals 1) so that the denominator of the product is a rational number. Work as follows.

$$\frac{4}{\sqrt{3}} \cdot \frac{\sqrt{3}}{\sqrt{3}} = \frac{4\sqrt{3}}{3}$$

(b) $\sqrt[4]{\dfrac{3}{5}}$

Start by using the fact that the radical of a quotient can be written as the quotient of radicals. To rationalize the denominator, multiply numerator and denominator by $\sqrt[4]{5^3}$. Use this number so that the denominator will be a rational number.

$$\sqrt[4]{\frac{3}{5}} = \frac{\sqrt[4]{3}}{\sqrt[4]{5}} = \frac{\sqrt[4]{3} \cdot \sqrt[4]{5^3}}{\sqrt[4]{5} \cdot \sqrt[4]{5^3}}$$

$$= \frac{\sqrt[4]{3 \cdot 5^3}}{\sqrt[4]{5^4}} = \frac{\sqrt[4]{375}}{5}$$

(c) $\sqrt[3]{\dfrac{9m^3n^5}{k^2p}}$

Write the radical as the quotient of two radicals. Then multiply by $\sqrt[3]{kp^2}/\sqrt[3]{kp^2}$, which will make the denominator $\sqrt[3]{k^3p^3} = kp$, so the denominator has no radical.

$$\sqrt[3]{\frac{9m^3n^5}{k^2p}} = \frac{\sqrt[3]{9m^3n^5}}{\sqrt[3]{k^2p}} \cdot \frac{\sqrt[3]{kp^2}}{\sqrt[3]{kp^2}}$$

$$= \frac{\sqrt[3]{9m^3n^5kp^2}}{\sqrt[3]{k^3p^3}}$$

$$= \frac{mn\sqrt[3]{9n^2kp^2}}{kp} \quad \bullet$$

In Example 8 below, the denominator is $1 - \sqrt{2}$. To rationalize this denominator, multiply numerator and denominator by $1 + \sqrt{2}$. This number is chosen because if a is rational and b is a nonnegative rational number, then

$$(a - \sqrt{b})(a + \sqrt{b}) = a^2 - (\sqrt{b})^2 = a^2 - b,$$

a rational number.

EXAMPLE 8   Rationalize the denominator of $\dfrac{1}{1 - \sqrt{2}}$.

As just mentioned, multiply numerator and denominator by $1 + \sqrt{2}$.

$$\frac{1}{1 - \sqrt{2}} = \frac{1(1 + \sqrt{2})}{(1 - \sqrt{2})(1 + \sqrt{2})} = \frac{1 + \sqrt{2}}{-1} = -1 - \sqrt{2} \quad \bullet$$

1.6 Exercises

Simplify each of the following. Assume that all variables represent nonnegative real numbers, and that no denominators are zero.

1. $\sqrt{50}$

2. $\sqrt{12}$

3. $\sqrt[3]{250}$

4. $\sqrt[3]{128}$

5. $-\sqrt{\dfrac{9}{5}}$

6. $\sqrt{\dfrac{3}{8}}$

7. $\sqrt{5} + \sqrt{45}$

8. $\sqrt{6} + \sqrt{54}$

9. $4\sqrt{3} - 5\sqrt{12} + 3\sqrt{75}$

10. $2\sqrt{5} - 3\sqrt{20} + 2\sqrt{45}$

11. $-\sqrt[3]{\dfrac{3}{2}}$

12. $-\sqrt[3]{\dfrac{4}{5}}$

13. $\sqrt[4]{\dfrac{3}{2}}$

14. $\sqrt[4]{\dfrac{32}{81}}$

15. $\sqrt[4]{5\dfrac{1}{16}}$

16. $\sqrt[3]{3\dfrac{3}{8}}$

17. $\sqrt[3]{2} - \sqrt[3]{16} + 2\sqrt[3]{54}$

18. $2\sqrt[3]{3} + 4\sqrt[3]{24} - \sqrt[3]{81}$

19. $\dfrac{1}{\sqrt{3}} - \dfrac{2}{\sqrt{12}} + 2\sqrt{3}$

20. $\dfrac{1}{\sqrt{2}} + \dfrac{3}{\sqrt{8}} + \dfrac{3}{\sqrt{32}}$

21. $\dfrac{5}{\sqrt[3]{2}} - \dfrac{2}{\sqrt[3]{16}} + \dfrac{1}{\sqrt[3]{54}}$

22. $\dfrac{-4}{\sqrt[3]{3}} + \dfrac{1}{\sqrt[3]{24}} - \dfrac{2}{\sqrt[3]{81}}$

23. $\sqrt{98r^3s^4t^{10}}$

24. $\sqrt[3]{16z^5x^8y^4}$

25. $\sqrt[4]{x^8y^6z^{10}}$

26. $\sqrt{a^3b^5} - 2\sqrt{a^7b^3} + \sqrt{a^3b^9}$

27. $\sqrt{p^7q^3} - \sqrt{p^5q^9} + \sqrt{p^9q}$

28. $(\sqrt{2} + 3)(\sqrt{2} - 3)$

29. $(\sqrt{5} + \sqrt{2})(\sqrt{5} - \sqrt{2})$

30. $(\sqrt{3} + \sqrt{8})^2$

31. $(\sqrt{2} + 1)^2$

32. $(2\sqrt[3]{3} + 1)(\sqrt[3]{3} - 4)$

33. $(\sqrt[3]{4} + 3)(5\sqrt[3]{4} + 1)$

34. $\sqrt{\dfrac{2}{3x}}$

35. $\sqrt{\dfrac{5}{3p}}$

36. $\sqrt{\dfrac{x^5y^3}{z^2}}$

37. $\sqrt{\dfrac{g^3h^5}{r^3}}$

38. $-\sqrt[3]{\dfrac{k^5m^3r^2}{r^8}}$

39. $-\sqrt[3]{\dfrac{9x^5y^6}{z^5w^2}}$

40. $\sqrt[4]{\dfrac{g^3h^5}{9r^6}}$

41. $\sqrt[4]{\dfrac{32x^6}{y^5}}$

42. $\dfrac{\sqrt[3]{mn} \cdot \sqrt[3]{m^2}}{\sqrt[3]{n^2}}$

43. $\dfrac{\sqrt[3]{8m^2n^3} \cdot \sqrt[3]{2m^2}}{\sqrt[3]{32m^4n^3}}$

44. $\sqrt[3]{\sqrt{4}}$

45. $\sqrt[4]{\sqrt[3]{2}}$

46. $\sqrt[6]{\sqrt[3]{x}}$

47. $\sqrt[8]{\sqrt[4]{y}}$

48. $\sqrt{2(a - b)^2 + 2(a + b)^2}$

49. $\sqrt{9a^2 + 9b^2}$

50. $\sqrt{a^4 + a^4b^2}$

51. $\sqrt{-x^2 + (x + 4)^2}$

52. $\dfrac{3}{1 - \sqrt{2}}$

53. $\dfrac{2}{1 + \sqrt{5}}$

54. $\dfrac{\sqrt{3}}{4 + \sqrt{3}}$

55. $\dfrac{2\sqrt{7}}{3 - \sqrt{7}}$

56. $\dfrac{1}{\sqrt{m} - \sqrt{p}}$

57. $\dfrac{\sqrt{z}}{1 - \sqrt{z}}$

58. $\dfrac{2}{3 + \sqrt{1 + k}}$

59. $\dfrac{-5}{1 - \sqrt{3 - p}}$

60. $\dfrac{m}{\sqrt{p}} + \dfrac{p}{\sqrt{m}}$

61. $\dfrac{a}{\sqrt{b}} - \dfrac{b}{\sqrt{a}}$

62. $\dfrac{\sqrt{x} + \sqrt{x + 1}}{\sqrt{x} - \sqrt{x + 1}}$

63. $\dfrac{\sqrt{p} + \sqrt{p^2 - 1}}{\sqrt{p} - \sqrt{p^2 - 1}}$

64. $\dfrac{5}{\sqrt[3]{a} + \sqrt[3]{b}}$

65. $\dfrac{1}{\sqrt[3]{m} - \sqrt[3]{n}}$

66. $\dfrac{5}{(2 - \sqrt{3})(1 + \sqrt{2})}$

67. $\dfrac{-2}{(4 + \sqrt{3})(3 - \sqrt{2})}$

68. $\dfrac{6 - \sqrt{3}}{(5 - \sqrt{2})(3 + \sqrt{5})}$

69. $\dfrac{1 - \sqrt{7}}{(2 + \sqrt{10})(1 - \sqrt{5})}$

Take the squared terms outside of the following radicals.

70. $\sqrt{5(3 - \sqrt{10})^2}$

71. $\sqrt{(4 - \sqrt{17})^2}$

Simplify Exercises 72 and 73, where $x \le 5$.

72. $\sqrt{(x - 5)^2}$

73. $\sqrt{(5 - x)^2(x - 3)^4}$

74. Use a calculator to find an approximate value for $\sqrt{5 + 2\sqrt{6}}$.

75. Show that $\sqrt{5 + 2\sqrt{6}} = \sqrt{2} + \sqrt{3}$.

1.7 COMPLEX NUMBERS

So far, we have worked only with real numbers in this book. The set of real numbers, however, does not include all the numbers needed in algebra. For example, with real numbers alone, it is not possible to find a number whose square is -1. Although as early as 50 B.C. square roots of negative numbers were known, they were not incorporated into an integrated number system until much later. The Italian Girolamo Cardano (1501–76) and others computed with them. As algebra became necessary in many applications, mathematicians distinguished between two types of solutions. Eventually these were named "real" and "imaginary" by Descartes (1596–1650).

The German mathematician Gottfried Leibniz, one of the founders of calculus, wrote to his Dutch colleague Christian Huygens in 1679 of the need for an expanded number system: "I have no hope that we can get very far in physics until we have found some method of abridgment." The void was filled by the complex numbers, which combine real and imaginary numbers. It was the renowned Leonard Euler in 1748 who first wrote $\sqrt{-1} = i$, and in 1832 Gauss stated that numbers of the form $a + bi$ are "complex." The first application of the imaginary numbers to geometry came in 1796 when the Norwegian Caspar Wessel used them in surveying. Charles Steinmetz (1865–1923), an electrical engineer, is said to have "generated electricity with the square root of minus one" when he used complex numbers to develop a theory of alternating currents. Today complex numbers are used extensively in science and engineering.

To achieve the extension of the real number system that allows such numbers as the one whose square is -1, the new number i is defined as follows.

Definition of i

$$i^2 = -1$$

Numbers of the form $a + bi$, where a and b are real numbers, are called **complex numbers.** Each real number is a complex number, since a real number a may be thought of as the complex number $a + 0i$. A complex number of the form $0 + bi$, where b is nonzero, is called an **imaginary number** (sometimes a *pure* imaginary number). Both the set of real numbers and the set of imaginary numbers are subsets of the set of complex numbers. (See Figure 7, which is an extension of Figure 1 in Section 1.1.) A complex number that is written in the form $a + bi$ or $a + ib$ is in **standard form.** (The form $a + ib$ is used to simplify certain symbols such as $i\sqrt{5}$, since $\sqrt{5}i$ could be too easily mistaken for $\sqrt{5i}$.)

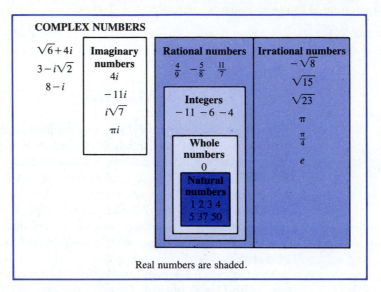

COMPLEX NUMBERS

Real numbers are shaded.

Figure 7

. . . .

EXAMPLE 1   The following statements identify different kinds of complex numbers.

(a) -8 and $\sqrt{7}$ and π are real numbers and complex numbers.

(b) $3i$ and $-11i$ and $i\sqrt{14}$ are imaginary numbers and complex numbers.

(c) $1 - 2i$ and $8 - 8i\sqrt{3}$ are complex numbers.   ●

. . . .

EXAMPLE 2   The list below shows several numbers, along with the standard form of the number.

Number	Standard form
$6i$	$0 + 6i$
-9	$-9 + 0i$
0	$0 + 0i$
$9 - i$	$9 - i$
$i - 1$	$-1 + i$ ●

Equality for complex numbers is defined as follows.

**Definition
of Equality**

For real numbers a, b, c, and d,

$$a + bi = c + di \quad \text{if and only if} \quad a = c \text{ and } b = d.$$

EXAMPLE 3 Solve $2 + mi = k + 3i$ for real numbers m and k.
By the definition of equality, $2 + mi = k + 3i$ if and only if $2 = k$ and $m = 3$. ●

The use of complex numbers allows the square root of a negative number to be defined as follows.

**Definition of
$\sqrt{-a}$**

If $a > 0$, then $\sqrt{-a} = i\sqrt{a}$.

EXAMPLE 4
(a) $\sqrt{-16} = i\sqrt{16} = 4i$ **(b)** $\sqrt{-70} = i\sqrt{70}$ ●

Products or quotients with square roots of negative numbers may be simplified using the definitions $\sqrt{-a} = i\sqrt{a}$ for positive numbers a and $i^2 = -1$. The next example shows how to do this.

EXAMPLE 5

(a) $\sqrt{-7} \cdot \sqrt{-7} = i\sqrt{7} \cdot i\sqrt{7}$
$\qquad\qquad\quad = i^2 \cdot (\sqrt{7})^2$
$\qquad\qquad\quad = (-1) \cdot 7$
$\qquad\qquad\quad = -7$

(b) $\sqrt{-6} \cdot \sqrt{-10} = i\sqrt{6} \cdot i\sqrt{10}$
$\qquad\qquad\qquad\quad = i^2 \cdot \sqrt{6 \cdot 10}$
$\qquad\qquad\qquad\quad = -1 \cdot 2\sqrt{15}$
$\qquad\qquad\qquad\quad = -2\sqrt{15}$

(c) $\dfrac{\sqrt{-20}}{\sqrt{-2}} = \dfrac{i\sqrt{20}}{i\sqrt{2}} = \sqrt{10}$ **(d)** $\dfrac{\sqrt{-48}}{\sqrt{24}} = \dfrac{i\sqrt{48}}{\sqrt{24}} = i\sqrt{2}$ ●

CAUTION When working with negative radicands, *use the definition $\sqrt{-a} = i\sqrt{a}$ before using any of the other rules for radicals*. In particular, the rule $\sqrt{c} \cdot \sqrt{d} = \sqrt{cd}$ is valid only when c and d are *not* both negative. For example,

$$\sqrt{(-4)(-9)} = \sqrt{36} = 6,$$

while
$$\sqrt{-4} \cdot \sqrt{-9} = 2i(3i) = 6i^2 = -6,$$

so
$$\sqrt{-4} \cdot \sqrt{-9} \neq \sqrt{(-4)(-9)}.$$

Operations on Complex Numbers With the definitions of $i^2 = -1$ and $\sqrt{-a} = i\sqrt{a}$, if $a > 0$, all the properties of real numbers can be extended to the complex numbers. As a result, complex numbers are added, subtracted, multiplied, and divided as shown by the following definitions and examples.

The *sum* of two complex numbers $a + bi$ and $c + di$ is defined as follows.

Definition of Addition

$$(a + bi) + (c + di) = (a + c) + (b + d)i$$

EXAMPLE 6

(a) $(3 - 4i) + (-2 + 6i) = [3 + (-2)] + [-4 + 6]i$
$$= 1 + 2i$$

(b) $(-9 + 7i) + (3 - 15i) = -6 - 8i$ ●

Since $(a + bi) + (0 + 0i) = a + bi$ for all complex numbers $a + bi$, the number $0 + 0i$ is called the **additive identity** for complex numbers. The sum of $a + bi$ and $-a - bi$ is $0 + 0i$, so the number $-a - bi$ is called the **negative** or **additive inverse** of $a + bi$.

Using this definition of additive inverse, *subtraction* of the complex numbers $a + bi$ and $c + di$ is defined as follows.

$$(a + bi) - (c + di) = (a + bi) + (-c - di)$$

This definition is usually written in the following form.

Definition of Subtraction

$$(a + bi) - (c + di) = (a - c) + (b - d)i$$

EXAMPLE 7 Subtract as indicated.

(a) $(-4 + 3i) - (6 - 7i) = (-4 - 6) + [3 - (-7)]i$
$$= -10 + 10i$$

(b) $(12 - 5i) - (8 - 3i) = 4 - 2i$ ●

The product of two complex numbers can be found by multiplying as if the numbers were binomials and using the fact that $i^2 = -1$, as follows.

$$\begin{aligned}
(a + bi)(c + di) &= ac + adi + bic + bidi \\
&= ac + adi + bci + bdi^2 \\
&= ac + (ad + bc)i + bd(-1) \\
(a + bi)(c + di) &= (ac - bd) + (ad + bc)i
\end{aligned}$$

Based on this, the *product* of the complex numbers $a + bi$ and $c + di$ is defined in the following way.

Definition of Product

$$(a + bi)(c + di) = (ac - bd) + (ad + bc)i$$

This definition is hard to remember. To find a given product, it is usually better to multiply as with binomials. The next example shows this.

EXAMPLE 8   Find each of the following products.

(a) $(2 - 3i)(3 + 4i) = 2(3) + 2(4i) - 3i(3) - 3i(4i)$

$$= 6 + 8i - 9i - 12i^2$$

$$= 6 - i - 12(-1)$$

$$= 18 - i$$

(b) $(6 + 5i)(6 - 5i) = 6 \cdot 6 - 6 \cdot 5i + 6 \cdot 5i - (5i)^2$

$$= 36 - 25i^2$$

$$= 36 - 25(-1)$$

$$= 36 + 25$$

$$= 61$$

(c) i^{15}

Since $i^2 = -1$, the value of a power of i is found by writing the given power as a product involving i^2. For example $i^3 = i^2 \cdot i = (-1) \cdot i = -i$. Also, $i^4 = i^2 \cdot i^2 = (-1)(-1) = 1$. Using i^4 to rewrite i^{15} gives

$$i^{15} = i^{12} \cdot i^3 = (i^4)^3 \cdot i^3 = (1)^3(-i) = -i. \quad \bullet$$

Methods similar to those in Example 8(c) give the following list of *powers of i*. Values of higher (or lower) powers continue in the same fashion.

Powers of *i*

$i^1 = i$	$i^5 = i$	$i^9 = i$
$i^2 = -1$	$i^6 = -1$	$i^{10} = -1$
$i^3 = -i$	$i^7 = -i$	$i^{11} = -i$
$i^4 = 1$	$i^8 = 1$	$i^{12} = 1$

Example 8(b) showed that $(6 + 5i)(6 - 5i) = 61$. The numbers $6 + 5i$ and $6 - 5i$ differ only in their middle signs; these numbers are **conjugates** of each other. Conjugates are useful because the product of a complex number and its conjugate is always a real number (see Exercise 72).

. . . .

EXAMPLE 9   The following list shows several pairs of conjugates, along with the product of the conjugates.

Number	Conjugate	Product
$3 - i$	$3 + i$	$(3 - i)(3 + i) = 10$
$2 + 7i$	$2 - 7i$	$(2 + 7i)(2 - 7i) = 53$
$-6i$	$6i$	$(-6i)(6i) = 36$ ●

The fact that the product of a complex number and its conjugate is always a real number is used to find the *quotient* of two complex numbers, as shown in the next example.

. . . .

EXAMPLE 10

(a) Find $\dfrac{3 + 2i}{5 - i}$.

Multiply numerator and denominator by the conjugate of $5 - i$.

$$\frac{3 + 2i}{5 - i} = \frac{(3 + 2i)(5 + i)}{(5 - i)(5 + i)}$$

$$= \frac{15 + 3i + 10i + 2i^2}{25 - i^2}$$

$$= \frac{13 + 13i}{26} = \frac{1}{2} + \frac{1}{2}i$$

To check this answer, show that

$$(5 - i)\left(\frac{1}{2} + \frac{1}{2}i\right) = 3 + 2i.$$

(b) Find $\dfrac{7 - 3i}{1 + 2i} \cdot \dfrac{1 - i}{4 - 3i}$.

$$\frac{7 - 3i}{1 + 2i} \cdot \frac{1 - i}{4 - 3i} = \frac{7 - 7i - 3i + 3i^2}{4 - 3i + 8i - 6i^2} = \frac{4 - 10i}{10 + 5i}$$

Multiply numerator and denominator by the conjugate of $10 + 5i$.

$$= \frac{4 - 10i}{10 + 5i} \cdot \frac{10 - 5i}{10 - 5i} = \frac{40 - 20i - 100i + 50i^2}{100 - 25i^2}$$

$$= \frac{-10 - 120i}{125} = \frac{-10}{125} - \frac{120}{125}i$$

$$= \frac{-2}{25} - \frac{24}{25}i \quad ●$$

1.7 Exercises ··

Identify each number as real, imaginary, or complex.

1. $-9i$

2. 6

3. π

4. $-\sqrt{7}$

5. $i\sqrt{6}$

6. $-3i$

7. $2 + 5i$

8. $-7 - 6i$

Write each of the following in standard form.

9. $\sqrt{-100}$

10. $\sqrt{-169}$

11. $-\sqrt{-400}$

12. $-\sqrt{-225}$

13. $5 + \sqrt{-4}$

14. $-7 + \sqrt{-100}$

15. $-6 - \sqrt{-196}$

16. $13 + \sqrt{-16}$

17. $\sqrt{-5} \cdot \sqrt{-5}$

18. $\sqrt{-20} \cdot \sqrt{-20}$

19. $\sqrt{-8} \cdot \sqrt{-2}$

20. $\sqrt{-27} \cdot \sqrt{-3}$

21. $\dfrac{\sqrt{-40}}{\sqrt{-10}}$

22. $\dfrac{\sqrt{-190}}{\sqrt{-19}}$

Add or subtract. Write each result in standard form.

23. $(3 + 2i) + (4 - 3i)$

24. $(4 - i) + (2 + 5i)$

25. $(-2 + 3i) - (-4 + 3i)$

26. $(-3 + 5i) - (-4 + 3i)$

27. $(2 - 5i) - (3 + 4i) - (-2 + i)$

28. $(-4 - i) - (2 + 3i) + (-4 + 5i)$

Multiply. Write each result in standard form.

29. $(2 + 4i)(-1 + 3i)$

30. $(1 + 3i)(2 - 5i)$

31. $(-3 + 2i)^2$

32. $(2 + i)^2$

33. $(2 + 3i)(2 - 3i)$

34. $(6 - 4i)(6 + 4i)$

35. $(\sqrt{6} + i)(\sqrt{6} - i)$

36. $(\sqrt{2} - 4i)(\sqrt{2} - 4i)$

37. $i(3 - 4i)(3 + 4i)$

38. $i(2 + 7i)(2 - 7i)$

Divide. Write each result in standard form.

39. $\dfrac{4 - 3i}{4 + 3i}$

40. $(3 - 4i) \div (2 - 5i)$

41. $(1 - 3i) \div (1 + i)$

42. $\dfrac{5 + 6i}{5 - 6i}$

43. $\dfrac{2}{i}$

44. $\dfrac{-7}{3i}$

45. $\dfrac{1 - \sqrt{-5}}{3 + \sqrt{-4}}$

46. $\dfrac{2 + \sqrt{-3}}{1 - \sqrt{-9}}$

Find each of the following powers of i.

47. i^5

48. i^{25}

49. i^{43}

50. $1/i^9$

51. $1/i^{12}$

52. i^{-6}

53. i^{-15}

54. i^{-49}

Perform the indicated operations and write your answers in standard form.

55. $\dfrac{2 + i}{3 - i} \cdot \dfrac{5 + 2i}{1 + i}$

56. $\dfrac{1 - i}{2 + i} \cdot \dfrac{4 + 3i}{1 + i}$

57. $\dfrac{6 + 2i}{5 - i} \cdot \dfrac{1 - 3i}{2 + 6i}$

58. $\dfrac{5 - 3i}{1 + 2i} \cdot \dfrac{2 - 4i}{1 + i}$

59. $\dfrac{5 - i}{3 + i} + \dfrac{2 + 7i}{3 + i}$

60. $\dfrac{4 - 3i}{2 + 5i} + \dfrac{8 - i}{2 + 5i}$

61. $\dfrac{6 + 2i}{1 + 3i} + \dfrac{2 - i}{1 - 3i}$

62. $\dfrac{4 - i}{3 + 4i} - \dfrac{3 + 2i}{3 - 4i}$

Use the definition of equality for complex numbers to solve the following equations for real numbers a and b.

63. $a + bi = 18 - 3i$

64. $2 + bi = a - 4i$

65. $a + 3i = 5 + 3bi + 2a$

66. $4a - 2bi + 7 = 3i + 3a + 5$

67. $i(2b + 6) - 3 = 4(bi + a)$

68. $3i + 2(a - 1) = 4 + 2i(b + 3)$

Let z = a + bi for real numbers a and b, and let $\bar{z}$ = a − bi, the conjugate of z. For example, if z = 8 − 9i, then $\bar{z}$ = 8 + 9i. Prove each of the following properties of conjugates.

69. $\bar{\bar{z}} = z$

70. $\bar{z} = z$ if and only if $b = 0$

71. $\overline{-z} = -\bar{z}$

72. $z \cdot \bar{z}$ is a real number

Evaluate $8z - z^2$ by replacing z with the indicated complex number in Exercises 73 and 74.

73. $2 + i$

74. $4 - 3i$

75. Find any restrictions on a and b so that the square $(a + bi)^2$ is real.

76. Find any restrictions on a and b so that the square $(a + bi)^2$ is imaginary.

77. Show that $\dfrac{\sqrt{2}}{2} + \dfrac{\sqrt{2}}{2} i$ is a square root of i.

78. Show that $\dfrac{\sqrt{3}}{2} - \dfrac{1}{2} i$ is a cube root of $-i$.

1 | CHAPTER SUMMARY

Key Words

To understand the concepts presented in this chapter, you should know the meaning and use of the following words and expressions. For easy reference, the section in the chapter in which a word or expression was first used is given with each item.

1.1 real number
variable
real number system
identity element
additive inverse
negative
multiplicative inverse
reciprocal
order of operations
natural number
whole number
integer
rational number
irrational number
coordinate

inequality
absolute value
1.2 exponent
base
nth root
square root
cube root
1.3 algebraic expression
term
coefficient
polynomial
degree of a polynomial
trinomial
binomial

monomial
1.4 factor
prime polynomial
common factor
1.5 rational expression
common denominator
complex fraction
1.6 principal nth root
index
radicand
radical
1.7 complex number
imaginary number
conjugates

Review Exercises

Identify the property which tells why the following are true.

1. $6 \cdot 4 = 4 \cdot 6$

2. $8(5 + 9) = (5 + 9)8$

3. $3 + (-3) = 0$

4. If $x \neq 0$, then $x \cdot \dfrac{1}{x} = 1$

5. $4 \cdot 6 + 4 \cdot 12 = 4(6 + 12)$

6. $3(4 \cdot 2) = (3 \cdot 4)2$

Simplify each of the following.

7. $(4 + 2 \cdot 8) \div 3$

8. $-7 + (-6)(-2) - 8$

9. $\dfrac{-9 + (2)(-8) + 7}{3(-4) - (-1 - 3)}$

10. $\dfrac{4 - (-8)(-2) - 7}{(-2)(-9) - 4(-4)}$

Let set $K = \{-12, -6, -9/10, -\sqrt{7}, 0, 1/8, \pi/4, 6, \sqrt{11}\}$. List the elements of K that belong to each of the following sets.

11. Natural numbers

12. Whole numbers

13. Integers

14. Rational numbers

15. Irrational numbers

16. Real numbers

Write without absolute value bars.

17. $|3 - \sqrt{7}|$

18. $|\sqrt{8} - 3|$

19. $|m - 3|$, if $m < 3$

20. $|-6 - x^2|$

In each of the following exercises, the coordinates of three points are given. Find (a) d(A, B), (b) d(A, B) + d(B, C).

21. $A, -2; B, -1; C, 10$

22. $A, -8; B, -12; C, -15$

Under what conditions are the following statements true?

23. $|x| = x$

24. $|x| \leq 0$

25. $d(A, B) = 0$

26. $d(A, B) + d(B, C) = d(A, C)$

Simplify each of the following. Assume all variables represent nonzero real numbers and variables used as exponents represent rational numbers.

27. $(-6x^2 - 4x + 11) + (-2x^2 - 11x + 5)$

28. $(3x^3 - 9x^2 - 5) - (-4x^3 + 6x^2) + (2x^3 - 9)$

29. $(8k - 7)(3k + 2)$

30. $(3r - 2)(r^2 + 4r - 8)$

31. $(x + 2y - z)^2$

32. $(r^z - 8)(r^z + 8)$

33. $\dfrac{2x^5t^2 \cdot 4x^3t}{16x^2t^3}$

34. $\dfrac{(2k^2)^2(3k^3)}{(4k^4)^2}$

Factor as completely as possible.

35. $7z^2 - 9z^3 + z$

36. $12p^5 - 8p^4 + 20p^3$

37. $6m^2 - 13m - 5$

38. $15a^2 + 7ab - 2b^2$

39. $30m^5 - 35m^4n - 25m^3n^2$

40. $2x^2p^3 - 8xp^4 + 6p^5$

41. $169y^4 - 1$

42. $49m^8 - 9n^2$

43. $(x - 1)^2 - 4$

44. $a^3 - 27b^3$

45. $r^9 - 8(r^3 - 1)^3$

46. $(2p - 1)^3 + 27p^3$

47. $3(m - n) + 4k(m - n)$

48. $6(z - 4)^2 + 9(z - 4)^3$

49. $2bx - b + 6x - 3$

50. $z^{6n} - 100$

Perform the indicated operation.

51. $\dfrac{3m - 9}{8m} \cdot \dfrac{16m + 24}{15}$

52. $\dfrac{5x^2y}{x + y} \cdot \dfrac{3x + 3y}{30xy^2}$

53. $\dfrac{x^2 + x - 2}{x^2 + 5x + 6} \div \dfrac{x^2 + 3x - 4}{x^2 + 4x + 3}$

54. $\dfrac{27m^3 - n^3}{3m - n} \div \dfrac{9m^2 + 3mn + n^2}{9m^2 - n^2}$

55. $\dfrac{1}{4y} + \dfrac{8}{5y}$

56. $\dfrac{m}{4 - m} + \dfrac{3m}{m - 4}$

57. $\dfrac{3}{x^2 - 4x + 3} - \dfrac{2}{x^2 - 1}$

58. $\left(\dfrac{1}{(x + h)^2 + 16} - \dfrac{1}{x^2 + 16} \right) \div h$

59. $\dfrac{3 + \dfrac{2m}{m^2 - 4}}{\dfrac{5}{m - 2}}$

60. $\dfrac{x^{-1} - 2y^{-1}}{y^{-1} - x^{-1}}$

Simplify each of the following. Assume all variables represent positive real numbers.

61. $\dfrac{(p^4)(p^{-2})}{p^{-6}}$

62. $(-6x^2y^{-3}z^2)^{-2}$

63. $\dfrac{6^{-1}r^3s^{-2}}{6r^4s^{-3}}$

64. $\dfrac{(3m^{-2})^{-2}(m^2n^{-4})^3}{9m^{-3}n^{-5}}$

Simplify. Assume that all variables represent positive real numbers.

65. $\sqrt{200}$

66. $\sqrt[4]{1250}$

67. $\sqrt{\dfrac{2^7y^8}{m^3}}$

68. $-\sqrt[3]{\dfrac{r^6m^5}{z^2}}$

69. $\sqrt[4]{\sqrt[3]{m}}$

70. $(\sqrt[3]{2} + 4)(\sqrt[3]{2^2} - 4\sqrt[3]{2} + 16)$

71. $\dfrac{\sqrt[4]{8p^2q^5} \cdot \sqrt[4]{2p^3q}}{\sqrt[4]{p^5q^2}}$

72. $\sqrt{18m^3} - 3m\sqrt{32m} + 5\sqrt{m^3}$

73. $\sqrt{75y^5} - y^2\sqrt{108y} + 2y\sqrt{27y^3}$

74. $\dfrac{-12}{\sqrt[3]{4}}$

75. $\dfrac{z}{\sqrt{z - 1}}$

76. $\dfrac{6}{3 - \sqrt{2}}$

77. $\dfrac{1}{\sqrt{5} + 2}$

78. $\dfrac{\sqrt{x} - \sqrt{x - 2}}{\sqrt{x} + \sqrt{x - 2}}$

Simplify each of the following. Assume that all variables represent positive real numbers, and variables used as exponents are rational numbers.

79. $36^{-3/2}$

80. $(125m^6)^{-2/3}$

81. $(8r^{3/4}s^{2/3})(2r^{3/2}s^{5/3})$

82. $(7r^{1/2})(2r^{3/4})(-r^{1/6})$

83. $\dfrac{p^{-3/4} \cdot p^{5/4} \cdot p^{-1/4}}{p \cdot p^{3/4}}$

84. $\left(\dfrac{y^6x^3z^{-2}}{16x^5z^4} \right)^{-1/2}$

85. $\dfrac{m^{2+p} \cdot m^{-2}}{m^{3p}}$

86. $\dfrac{z^{-p+1} \cdot z^{-8p}}{z^{-9p}}$

Find each product. Assume that all variables represent positive real numbers.

87. $2z^{1/3}(5z^2 - 2)$

88. $-m^{3/4}(8m^{1/2} + 4m^{-3/2})$

89. $(p + p^{1/2})(3p - 5)$

90. $(m^{1/2} - 4m^{-1/2})^2$

Write in standard form.

91. $(6 - 5i) + (2 + 7i) - (3 - 2i)$ **92.** $(4 - 2i) - (6 + 5i) - (3 - i)$

93. $(3 + 5i)(8 - i)$ **94.** $(4 - i)(5 + 2i)$ **95.** $(2 + 6i)^2$ **96.** $(6 - 3i)^2$

97. $(1 - i)^3$ **98.** $(2 + i)^3$ **99.** i^{17} **100.** i^{52}

101. $\dfrac{6 + 2i}{3 - i}$ **102.** $\dfrac{2 - 5i}{1 + i}$ **103.** $\dfrac{2 + i}{1 - 5i} \cdot \dfrac{1 + i}{3 - i}$ **104.** $\dfrac{4 + 3i}{1 - i} \cdot \dfrac{2 - 3i}{2 + i}$

105. $\sqrt{-12}$ **106.** $\sqrt{-18}$

Correct each incorrect statement in Exercises 107–118 by changing the right hand side.

107. $x(x^2 + 5) = x^3 + 5$ **108.** $-3^2 = 9$ **109.** $(m^2)^3 = m^5$

110. $(3x)(3y) = 3xy$ **111.** $\dfrac{\frac{a}{b}}{2} = \dfrac{2a}{b}$ **112.** $\dfrac{m}{r} \cdot \dfrac{n}{r} = \dfrac{mn}{r}$

113. $\dfrac{1}{\sqrt{a} + \sqrt{b}} = \dfrac{1}{\sqrt{a}} + \dfrac{1}{\sqrt{b}}$ **114.** $\dfrac{(2x)^3}{2y} = \dfrac{x^3}{y}$ **115.** $4 - (t + 1) = 4 - t + 1$

116. $\dfrac{1}{(-2)^3} = 2^{-3}$ **117.** $(-5)^2 = -5^2$ **118.** $\left(\dfrac{8}{7} + \dfrac{a}{b}\right)^{-1} = \dfrac{7}{8} + \dfrac{b}{a}$

119. For which of the following cases does $\sqrt{ab} = \sqrt{a} \cdot \sqrt{b}$? **(a)** a and b positive **(b)** a positive, b negative **(c)** a and b negative

120. For what integer values of n does $\sqrt[n]{a^n} = a$?

121. For what values of x does $\sqrt{9ax^2} = 3x\sqrt{a}$?

2

EQUATIONS AND INEQUALITIES

In many cases, the study of algebra is really the study of equations. Applications of mathematics frequently require the solution of one or more equations. The study of inequalities has also become important as more and more applications—especially in fields such as business—utilize inequalities.

An **equation** is a statement that two expressions are equal, such as $14x^2 + 2 = 8x - 4$. An **inequality** is a statement that two expressions are unequal, such as $11y > 5y + 3$. In this chapter we discuss the solution of several different kinds of equations and inequalities.

2.1 LINEAR EQUATIONS

To **solve** an equation means to find the number or numbers that make the equation a true statement. For example, the number 7 makes $x + 2 = 9$ a true statement, so 7 is the **solution** or the **root** of the equation and is said to **satisfy** the equation. The set of all solutions for an equation is the **solution set** of the equation. The solution set of $x^2 - 1 = 0$ is $\{-1, 1\}$.

Any two equations with the same solution set are **equivalent equations.** For example, $x + 1 = 5$ and $6x + 3 = 27$ are equivalent equations since they both have the solution set $\{4\}$. On the other hand, the equations

$$x + 1 = 5 \quad \text{and} \quad (x - 4)(x + 2) = 0$$

are not equivalent. The number 4 is a solution of both equations, but the equation $(x - 4)(x + 2) = 0$ also has -2 as a solution.

One way to solve an equation is to rewrite it as successively simpler equivalent equations. These simpler equivalent equations are derived using the properties of

Chapter 1 that allow the same number to be added to each side of an equation or to be multiplied on each side of an equation.

This section discusses methods of solving equations that are equivalent to linear equations.

Linear Equation

> An equation that can be written in the form
>
> $$ax + b = 0,$$
>
> where a and b are real numbers, with $a \neq 0$, is a **linear equation.**

The equation $ax + b = 0$ can be solved by writing the following sequence of equivalent equations

$$ax + b = 0 \qquad \text{Given equation}$$
$$ax + b + (-b) = 0 + (-b) \qquad \text{Add } -b \text{ on each side.}$$
$$ax = -b \qquad \text{Inverse and identity properties}$$
$$\frac{1}{a} \cdot ax = \frac{1}{a} \cdot (-b) \qquad \text{Multiply by } 1/a \text{ on each side.}$$
$$x = -\frac{b}{a} \qquad \text{Inverse and identity properties}$$

At which point in this solution was the fact that $a \neq 0$ used?

It is often necessary to first use the properties from Chapter 1 to change a linear equation into the form $ax + b = 0$ or $ax = -b$ so that it can be solved as shown above.

EXAMPLE 1   Solve $3(2x - 4) = 7 - (x + 5)$.

Using the distributive property and then collecting like terms gives the following sequence of simpler equivalent equations.

$$3(2x - 4) = 7 - (x + 5)$$
$$6x - 12 = 7 - x - 5 \qquad \text{Distributive property}$$
$$6x - 12 = 2 - x \qquad \text{Collect like terms.}$$

Now add the same expressions to each side of the equation.

$$x + 6x - 12 = x + 2 - x \qquad \text{Add } x \text{ to each side.}$$
$$7x - 12 = 2$$
$$12 + 7x - 12 = 12 + 2 \qquad \text{Add 12 to each side.}$$
$$7x = 14$$

Finally, multiply each side by the same number, $\frac{1}{7}$.

$$\frac{1}{7} \cdot 7x = \frac{1}{7} \cdot 14$$

$$x = 2$$

To check this proposed solution, replace x with 2 in the original equation.

$$3(2x - 4) = 7 - (x + 5) \quad \text{Original equation}$$
$$3(2 \cdot 2 - 4) = 7 - (2 + 5) \quad \text{Let } x = 2.$$
$$3(4 - 4) = 7 - (7)$$
$$0 = 0 \qquad\qquad\qquad \text{True}$$

Since replacing x with 2 results in a true statement, 2 is the solution of the given equation. The solution set is therefore {2}. ●

· · · · · · · · · ·

EXAMPLE 2 Solve $\dfrac{3p - 1}{3} - \dfrac{2p}{p - 1} = p$.

At first glance, this equation does not satisfy the definition of a linear equation given above. However, the equation does appear in proper form after algebraic simplification. To obtain a simpler equivalent equation, first multiply both sides by the common denominator, $3(p - 1)$, where we must assume $p \neq 1$.

$$3(p - 1)\left(\frac{3p - 1}{3}\right) - 3(p - 1)\left(\frac{2p}{p - 1}\right) = 3(p - 1)p$$
$$(p - 1)(3p - 1) - 3(2p) = 3p(p - 1)$$
$$3p^2 - 4p + 1 - 6p = 3p^2 - 3p$$

An even simpler equivalent equation comes from combining terms and adding $-3p^2$ to both sides.

$$-10p + 1 = -3p$$
$$1 = 7p \quad \text{Add } 10p \text{ to each side.}$$
$$\frac{1}{7} = p \quad \text{Multiply each side by } \frac{1}{7}.$$

Substitute 1/7 for p in the given equation to verify that 1/7 is the root of the equation. The restriction $p \neq 1$ does not affect the solution set here, since $1/7 \neq 1$. ●

· · · · · · · · ·

EXAMPLE 3 Solve $\dfrac{x}{x - 2} = \dfrac{2}{x - 2} + 2$.

Multiply both sides of the equation by $x - 2$, assuming that $x - 2 \neq 0$ (or $x \neq 2$).

$$x = 2 + 2(x - 2)$$
$$x = 2 + 2x - 4$$
$$x = 2$$

It was necessary to assume $x - 2 \neq 0$ in order to multiply both sides of the equation by $x - 2$. The proposed solution of 2, however, makes $x - 2 = 0$, meaning that the given equation has no solution. (To see that 2 is not a solution, substitute 2 for x in the given equation.) The solution set is $\emptyset$, the set containing no elements. (The set $\emptyset$ is called the **empty set**.)   •

EXAMPLE 4   Solve $5(2x + 1) - 2x = 8x + 5$.
   Use the distributive property on the left side to get

$$10x + 5 - 2x = 8x + 5$$
$$8x + 5 = 8x + 5$$
$$0 = 0.$$

This last equation is true for every real number, which indicates that the given equation is an *identity*.   •

In general, an equation is an **identity** if it is true for all meaningful values of its variables. For instance, $3x + 4x = 7x$ and $x^2 - 3x + 2 = (x - 2)(x - 1)$ are identities. They are meaningful and true for all real numbers. The identity

$$\frac{2x^2 + 3x}{x} = 2x + 3$$

is meaningful and true for all real numbers except 0. An equation that is not an identity is called a **conditional equation**. A conditional equation is true for only *some* of the values for which the equation is meaningful. Thus, the equations solved in Examples 1 and 2 are conditional equations.

EXAMPLE 5   Decide whether the following equations are identities or conditional equations.
(a) $9p^2 - 25 = (3p + 5)(3p - 5)$
   Since the product of $3p + 5$ and $3p - 5$ is $9p^2 - 25$, the equation is meaningful and true for *every* value of p and is an identity.
(b) $\dfrac{(x - 3)(x + 2)}{x - 3} = x + 2$
   The equation is an identity since it is meaningful for all real numbers except 3 and is true for all such numbers.

(c) $5y - 4 = 11$

Choosing the value 3 as a replacement for y gives

$$5 \cdot \mathbf{3} - 4 = 11$$
$$11 = 11,$$

a true statement. On the other hand, $y = 4$ gives

$$5 \cdot \mathbf{4} - 4 = 11$$
$$16 = 11,$$

a false statement. Since the equation is meaningful for all real numbers and true for $y = 3$, but not true for $y = 4$, the equation is conditional. ●

Sometimes an equation has more than one letter. To solve for a specified variable, treat the other letters as constants. (As a general rule, letters from the beginning of the alphabet, such as a, b, and c, are used to represent constants, while letters such as x, y, and z represent variables.)

• • • • • • • • •

EXAMPLE 6   Solve the equation $3(2x - 5a) + 4b = 4x - 2$ for x.
Using the distributive property gives

$$6x - 15a + 4b = 4x - 2.$$

Treat x as the variable and the other letters as constants. Get all terms with x on one side and all terms without x on the other side.

$$6x - 4x = 15a - 4b - 2$$
$$2x = 15a - 4b - 2$$
$$x = \frac{15a - 4b - 2}{2} \quad ●$$

2.1 Exercises •

Decide whether each of the following equations is an identity or a conditional equation.

1. $x^2 + 5x = x(x + 5)$

2. $3y + 4 = 5(y - 2)$

3. $2(x - 7) = 5x + 3 - x$

4. $2x - 4 = 2(x - 2)$

5. $\dfrac{m + 3}{m} = 1 + \dfrac{3}{m}$

6. $\dfrac{p}{2 - p} = \dfrac{2}{p} - 1$

7. $4q^2 - 25 = (2q + 5)(2q - 5)$

8. $3(k + 2) - 5(k + 2) = -2k - 4$

Decide which of the following pairs of equations are equivalent.

9. $3x - 5 = 7$
$-6x + 10 = 14$

10. $-x = 2x + 3$
$-3x = 3$

11. $\dfrac{3x}{x - 1} = \dfrac{2}{x - 1}$
$3x = 2$

12. $\dfrac{x + 1}{12} = \dfrac{5}{12}$
$x + 1 = 5$

13. $\dfrac{x}{x - 2} = \dfrac{2}{x - 2}$
$x = 2$

14. $\dfrac{x + 3}{x + 1} = \dfrac{2}{x + 1}$
$x = -1$

15. $x = 4$
$x^2 = 16$

16. $z^2 = 9$
$z = 3$

Solve each of the following equations.

17. $4x - 1 = 15$

18. $-3y + 2 = 5$

19. $.2m - .5 = .1m + .7$

20. $.01p + 3.1 = 2.03p - 2.96$

21. $\dfrac{5}{6}k - 2k + \dfrac{1}{3} = \dfrac{2}{3}$

22. $\dfrac{3}{4} + \dfrac{1}{5}r - \dfrac{1}{2} = \dfrac{4}{5}r$

23. $3r + 2 - 5(r + 1) = 6r + 4$

24. $5(a + 3) + 4a - 5 = -(2a - 4)$

25. $2[m - (4 + 2m) + 3] = 2m + 2$

26. $4[2p - (3 - p) + 5] = -7p - 2$

27. $\dfrac{3x - 2}{7} = \dfrac{x + 2}{5}$

28. $\dfrac{2p + 5}{5} = \dfrac{p + 2}{3}$

29. $\dfrac{3k - 1}{4} = \dfrac{5k + 2}{8}$

30. $\dfrac{9x - 1}{6} = \dfrac{2x + 7}{3}$

31. $\dfrac{x}{3} - 7 = 6 - \dfrac{3x}{4}$

32. $\dfrac{y}{3} + 1 = \dfrac{2y}{5} - 4$

33. $\dfrac{1}{4p} + \dfrac{2}{p} = 3$

34. $\dfrac{2}{t} + 6 = \dfrac{5}{2t}$

35. $\dfrac{m}{2} - \dfrac{1}{m} = \dfrac{6m + 5}{12}$

36. $\dfrac{-3k}{2} + \dfrac{9k - 5}{6} = \dfrac{11k + 8}{k}$

37. $\dfrac{2r}{r - 1} = 5 + \dfrac{2}{r - 1}$

38. $\dfrac{3x}{x + 2} = \dfrac{1}{x + 2} - 4$

39. $\dfrac{5}{2a + 3} + \dfrac{1}{a - 6} = 0$

40. $\dfrac{2}{x + 1} = \dfrac{3}{2x - 5}$

41. $\dfrac{4}{x - 3} - \dfrac{8}{2x + 5} + \dfrac{3}{x - 3} = 0$

42. $\dfrac{5}{2p + 3} - \dfrac{3}{p - 2} = \dfrac{4}{2p + 3}$

43. $\dfrac{3}{2m + 4} = \dfrac{1}{m + 2} - 2$

44. $\dfrac{8}{3k - 9} - \dfrac{5}{k - 3} = 4$

45. $\dfrac{2p}{p - 2} = 3 + \dfrac{4}{p - 2}$

46. $\dfrac{5k}{k + 4} = 3 - \dfrac{20}{k + 4}$

47. $2(m + 1)(m - 1) = (2m + 3)(m - 2)$

48. $(2y - 1)(3y + 2) = 6(y + 2)^2$

49. $(3x - 4)^2 - 5 = 3(x + 5)(3x + 2)$

50. $(2x + 5)^2 = 3x^2 + (x + 3)^2$

Solve each equation in Exercises 51–60 for x.

51. $2(x - a) + b = 3x + a$

52. $5x - (2a + c) = a(x + 1)$

53. $ax + b = 3(x - a)$

54. $4a - ax = 3b + bx$

55. $\dfrac{x}{a - 1} = ax + 3$

56. $\dfrac{2a}{x - 1} = a - b$

57. $a^2x + 3x = 2a^2$

58. $ax + b^2 = bx - a^2$

59. $3x = (2x - 1)(m + 4)$

60. $-x = (5x + 3)(3k + 1)$

61. Find the error in the following.

$$x^2 + 2x - 15 = x^2 - 3x$$
$$(x + 5)(x - 3) = x(x - 3)$$
$$x + 5 = x$$
$$5 = 0$$

Find the value of k that will make each equation equivalent to x = 2.

62. $9x - 7 = k$

63. $-5x + 11x - 2 = k + 4$

64. $\dfrac{8}{k + x} = 4$

2.2 FORMULAS AND APPLICATIONS

Mathematics is an important problem-solving tool. Many times the solution of a problem depends on the use of a formula which expresses a relationship among several variables. For example, the formula

$$A = \frac{24f}{b(p + 1)} \qquad (*)$$

gives the approximate annual interest rate for a consumer loan paid off with monthly payments. Here f is the finance charge on the loan, p is the number of payments, and b is the original amount of the loan.

Suppose the number of payments, p, must be found when the other quantities are known. To do this, solve the equation for p by treating p as the variable and the other letters as constants. Begin by multiplying both sides of formula $(*)$ by $p + 1$.

$$(p + 1)A = (p + 1)\frac{24f}{b(p + 1)}$$

$$(p + 1)A = \frac{24f}{b}$$

Multiplying both sides by $1/A$ gives

$$\frac{1}{A}(p + 1)A = \frac{1}{A} \cdot \frac{24f}{b}$$

$$p + 1 = \frac{24f}{Ab}.$$

(Here we must assume $A \neq 0$. Why is this a very safe assumption?)

Finally, add -1 to both sides.

$$p = \frac{24f}{Ab} - 1$$

This process is called **solving for a specified variable.**

· · · · · · · · ·

EXAMPLE 1   Solve $J\left(\frac{x}{k} + a\right) = x$ for x.

To get all terms with x on one side of the equation and all terms without x on the other, first use the distributive property.

$$J\left(\frac{x}{k}\right) + Ja = x$$

Eliminate the denominator, k, by assuming $k \neq 0$ and multiplying both sides by k.

$$kJ\left(\frac{x}{k}\right) + kJa = kx$$

$$Jx + kJa = kx$$

Then add $-Jx$ to both sides to get the two terms with x together.

$$kJa = kx - Jx$$

$$kJa = x(k - J) \qquad \text{Factor the right side.}$$

Assuming $k \neq J$ permits multiplying both sides by $1/(k - J)$ to find

$$x = \frac{kJa}{k - J}. \quad \bullet$$

One of the main reasons for learning mathematics is to be able to use it in solving practical problems. However, for most students, learning how to apply mathematical skills to real situations is the most difficult task they face. In the rest of this section we give a few hints that may help you with applications.

A common difficulty with ''word problems'' is trying to do everything at once. It is usually best to attack the problem in stages.

Solving Word Problems

1. Read the problem carefully.

2. Decide on an unknown, usually the quantity you are asked to find, and name it with a variable that you *write down*. Many students, eager to get on with writing an equation, try to skip this step. But it is important. If you don't know what ''x'' represents, how can you write a meaningful equation or interpret a result?

3. Decide on variable expressions to represent any other unknowns in the problem. For example, if x represents the width of a rectangle and you know that the length is one more than twice the width, *write down* ''$1 + 2x =$ the length of the rectangle.''

4. If possible, draw a sketch or prepare a table showing the variables appearing in the problem and the relationships among them.

5. Use the information given in the problem to write a verbal equation. For example, if the problem states that the perimeter of a rectangle is 48 inches, *write down* ''$2 \cdot$ length $+ 2 \cdot$ width $= 48$.''

6. Use the results of steps 3 and 4 to convert the verbal equation into an algebraic equation in the variable from step 2.

7. Solve the equation from step 6.

8. Make sure you have answered the question in the problem.

9. Check your answer *in the words of the original problem.*

Notice how each of the steps listed is carried out in the examples on the following pages.

. . . .

EXAMPLE 2   If the length of a side of a square is increased by 3 centimeters, the perimeter of the new square is 40 centimeters more than twice the length of a side of the original square. Find the length of a side of the original square.

First, what should the variable represent? Since the length of a side of the original square is needed, let the variable represent that length. Write this down:

$$\text{Let } x = \text{length of side of original square.}$$

The length of the side of the new square can be expressed in terms of x:

$$x + 3 = \text{length of side of new square.}$$

Now draw a figure using the given information, as in Figure 1.

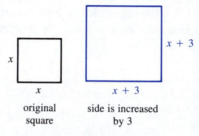

original   side is increased
square   by 3

Figure 1

Use the information given in the problem to write a verbal equation.

$$\begin{array}{l} \text{perimeter} \qquad\qquad \text{twice the length} \\ \text{of the new} = 40 + \text{of a side of the} \\ \text{square} \qquad\qquad\quad \text{original square} \end{array}$$

Transform the verbal equation into an algebraic equation.

$$4(x + 3) = 40 + \quad 2x$$

Solve the equation for the variable x.

$$4x + 12 = 40 + 2x$$
$$2x = 28$$
$$x = 14$$

Therefore, the length of a side of the original square is 14 centimeters.

Check the solution using the words of the problem.

Length of side of new square:	$14 + 3 = 17$ cm
Perimeter of new square:	$4(17) = 68$ cm
Twice length of a side of original square:	$2(14) = 28$ cm
40 + twice length of side of original square:	$40 + 28 = 68$ cm

Same ●

The next example is a constant velocity problem. The components *distance, rate,* and *time* are denoted by the letters *d*, *r*, and *t*, respectively. (The *rate* is also called the *speed* or *velocity*.) These variables are related by the equation

$$d = rt.$$

This equation is easily solved for *r* and *t*; $r = d/t$ and $t = d/r$.

. . . .

EXAMPLE 3   Chuck travels 80 km in the same time that Mary travels 180 km. Mary travels 50 km per hr faster than Chuck. Find the rate of each person.

$$\text{Let } x = \text{Chuck's rate,}$$
$$x + 50 = \text{Mary's rate.}$$

Summarize the facts in a table.

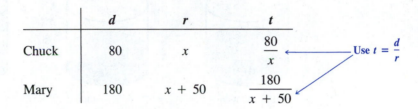

	d	*r*	*t*	
Chuck	80	x	$\dfrac{80}{x}$	Use $t = \dfrac{d}{r}$
Mary	180	$x + 50$	$\dfrac{180}{x + 50}$	

The unused information from the problem is "in the same time that."

$$
\begin{array}{c}
\text{Chuck's time} \\
\text{traveled}
\end{array}
=
\begin{array}{c}
\text{Mary's time} \\
\text{traveled}
\end{array}
$$

$$\frac{80}{x} = \frac{180}{x + 50}$$

$$x(x + 50) \cdot \frac{80}{x} = x(x + 50) \cdot \frac{180}{x + 50} \qquad \text{Multiply both sides by } x(x + 50).$$

$$80(x + 50) = 180x$$
$$80x + 4000 = 180x$$
$$4000 = 100x$$
$$40 = x \qquad \text{Chuck's rate}$$
$$50 + x = 50 + 40 = 90 \qquad \text{Mary's rate}$$

Therefore, Chuck's rate is 40 km per hr and Mary's rate is 90 km per hr.
  Check:

$$\text{Time traveled by Chuck:} \qquad 80/40 = 2$$

$$\text{Same}$$

$$\text{Time traveled by Mary:} \qquad 180/90 = 2 \qquad \bullet$$

· · · · · · · · ·

EXAMPLE 4 Elizabeth Thornton is a chemist. She needs a 20% solution of potassium permanganate. She has a 15% solution on hand, as well as a 30% solution. How many liters of the 15% solution should she add to 3 liters of the 30% solution to get the 20% solution?

Let x = number of liters of the 15% solution to be added.

Strength	Liters of solution	Liters of pure potassium permanganate
15%	x	$.15x$
30%	3	$.30(3)$
20%	$3 + x$	$.20(3 + x)$

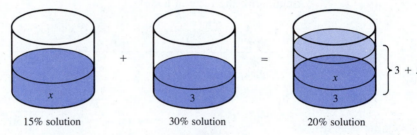

15% solution   30% solution   20% solution

Figure 2

$$
\begin{array}{ccc}
\text{liters of} & \text{liters of} & \text{liters of} \\
\text{pure pot. per.} + & \text{pure pot. per.} = & \text{pure pot. per.} \\
\text{in 15\% solution} & \text{in 30\% solution} & \text{in 20\% solution}
\end{array}
$$

$$.15x \quad + \quad .30(3) \quad = \quad .20(3 + x)$$
$$.15x + .90 = .60 + .20x$$
$$.30 = .05x$$
$$6 = x$$

Therefore, 6 liters of the 15% solution should be added.

Check:

Pure pot. per. contributed by 15% solution:	$.15(6) = .90$ liters
Pure pot. per. contributed by 30% solution:	$.30(3) = .90$ liters
Pure pot. per. in mixture:	$.90 + .90 = 1.8$ liters
Volume of mixture:	$6 + 3 = 9$ liters
Concentration of pot. per. in mixture:	$1.8/9 = .20$ or 20% ●

The next example involves work. The letters r, t, and A represent the rate that work is done, the time, and the amount of work accomplished, respectively. These variables are related by the equation

$$A = rt.$$

Amounts of work are often measured in terms of jobs accomplished. For instance, if one job is accomplished in t hours, then $A = 1$ and $r = 1/t$.

· · · · · · · · ·

EXAMPLE 5   One computer can do a job twice as fast as another. Working together, both computers can do the job in 8/3 hr. How long would it take the faster computer, working alone, to do the job?

Let t = numbers of hours for faster computer
working alone to complete the job,

$2t$ = numbers of hours for slower computer
working alone to complete the job.

	r	t	A
Faster computer	$\dfrac{1}{t}$	$\dfrac{8}{3}$	$\dfrac{1}{t} \cdot \dfrac{8}{3}$ ←— Use $A = rt$
Slower computer	$\dfrac{1}{2t}$	$\dfrac{8}{3}$	$\dfrac{1}{2t} \cdot \dfrac{8}{3}$

$$
\underset{\substack{\text{amount of work} \\ \text{done by faster} \\ \text{computer}}}{} + \underset{\substack{\text{amount of work} \\ \text{done by slower} \\ \text{computer}}}{} = \underset{\substack{\text{amount of work} \\ \text{done by both}}}{}
$$

$$
\frac{1}{t} \cdot \frac{8}{3} \quad + \quad \frac{1}{2t} \cdot \frac{8}{3} \quad = \quad 1
$$

$$
\frac{8}{3t} + \frac{8}{6t} = 1
$$

$$
\frac{8}{3} + \frac{8}{6} = t \qquad \text{Multiply both sides by } t.
$$

$$
4 = t
$$

The faster computer could do the entire job, working alone, in 4 hours.
Check:

Rate of faster computer: 1/4 job per hour

Rate of slower computer: 1/8 job per hour

Amount of job done by faster computer in 8/3 hours: $\quad \dfrac{1}{4} \cdot \dfrac{8}{3} = \dfrac{2}{3}$

Amount of job done by slower computer in 8/3 hours: $\quad \dfrac{1}{8} \cdot \dfrac{8}{3} = \dfrac{1}{3}$

Amount of job done working together: $\quad \dfrac{2}{3} + \dfrac{1}{3} = 1$ ●

When P dollars is invested at the annual interest rate r for t years, the amount of simple interest earned, I, is given by

$$
I = Prt.
$$

· · · · · · · · ·

EXAMPLE 6   John Miller receives a $14,000 bonus from his company. He invests part of the money in 6% tax-free bonds and the remainder at 15%. In one year he earns $1335 interest from the investment. Find the amount he has invested at each rate.

Let x = amount invested at 6%,

      $14,000 - x$ = amount invested at 15%.

| interest earned from 6% investment in one year | + | interest earned from 15% investment in one year | = | total interest earned in one year |

$$x(.06)(1) \qquad + \quad (14,000 - x)(.15)(1) = \qquad 1335$$

$$.06x + .15(14,000) - .15x = 1335$$

$$-.09x + 2100 = 1335$$

$$-.09x = -765$$

$$x = 8500 \qquad\qquad \text{Amount invested at 6%}$$

$$14,000 - x = 14,000 - 8500 = 5500 \quad \text{Amount invested at 15%}$$

Therefore, John Miller invested $8500 at 6% and $5500 at 15%.

Check:

Interest earned from 6% investment:	$(.06)8500 = \$510$
Interest earned from 15% investment:	$(.15)5500 = \$825$
Total interest earned:	$510 + 825 = \$1335$ ●

2.2 Exercises ·

Solve each of the following for the variable indicated. Assume all denominators are nonzero.

1. $PV = k$ for V

2. $F = whA$ for h

3. $V = lwh$ for l

4. $i = prt$ for p

5. $V = V_0 + gt$ for g

6. $s = s_0 + gt^2 + k$ for g

7. $s = \dfrac{1}{2} gt^2$ for g

8. $A = \dfrac{1}{2}(B + b)h$ for h

9. $A = \dfrac{1}{2}(B + b)h$ for B

10. $C = \dfrac{5}{9}(F - 32)$ for F

11. $S = 2\pi(r_1 + r_2)h$ for r_1

12. $A = P\left(1 + \dfrac{i}{m}\right)$ for m

13. $g = \dfrac{4\pi^2 l}{t^2}$ for l

14. $P = \dfrac{E^2 R}{r + R}$ for R

15. $S = 2\pi rh + 2\pi r^2$ for h

16. $u = f \cdot \dfrac{k(k + 1)}{n(n + 1)}$ for f

17. $A = \dfrac{24f}{b(p + 1)}$ for f

18. $A = \dfrac{24f}{b(p + 1)}$ for b

19. $\dfrac{1}{R} = \dfrac{1}{r_1} + \dfrac{1}{r_2}$ for R

20. $m = \dfrac{Ft}{v_1 - v_2}$ for v_2

Solve each of the following problems.

21. A triangle has a perimeter of 30 cm. Two sides of the triangle are both twice as long as the shortest side. Find the length of the shortest side.

22. The length of a rectangle is 3 cm less than twice the width. The perimeter is 54 cm. Find the width.

23. To fill a prescription a pharmacist must strengthen a 10% alcohol mixture to one that is 30% alcohol by adding pure (100%) alcohol to the mixture. How much pure alcohol should be added to 7 liters of the 10% mixture?

24. A student needs 10% hydrochloric acid for a chemistry experiment. How much 5% acid should be mixed with 60 ml of 20% acid to get a 10% solution?

25. An automobile radiator contains a ten-quart mixture of water and antifreeze that is 40% antifreeze. How much should the owner drain from his radiator and replace with pure antifreeze so that the radiator will contain 80% antifreeze?

26. Suppose the student (see Exercise 24) has only pure acid and 5% acid. How much pure acid should be added to the 5% acid to get 12 ml of 10% acid?

Exercises 27 and 28 depend on the idea of the octane rating *of gasoline, a measure of its antiknock qualities. In one measure of octane, a standard fuel is made with only two ingredients: heptane and isooctane. For this fuel, the octane rating is the percent of isooctane. An actual gasoline blend is then compared to a standard fuel. For example, a gasoline with an octane rating of 98 has the same antiknock properties as a standard fuel that is 98% isooctane.*

27. How many liters of 94-octane gasoline should be mixed with 200 liters of 99-octane gasoline to get a mixture that is 97-octane?

28. A service station has 92-octane and 98-octane gasoline. How many liters of the 92-octane gasoline should be added to the 98-octane gasoline to provide 12 liters of 96-octane gasoline needed for chemical research?

Solve each of the following problems.

29. On a vacation trip, Jose averaged 50 mph traveling from Amarillo to Flagstaff. Returning by a different route which covered the same number of miles, he averaged 55 mph. What is the distance between the two cities if his total traveling time was 32 hr?

30. Cindy left by plane to visit her mother in Hartford, 420 km away. Fifteen minutes later, her mother left to meet her at the airport. She drove the 20 km to the airport at 40 km per hr, arriving just as the plane taxied in. What was the speed of the plane?

31. Russ and Janet are running in the Apple Hill Fun Run. Russ runs at 7 mph, Janet at 5 mph. If they start at the same time, how long will it be before they are 1/2 mi apart?

32. If the run in Exercise 31 has a staggered start, and Janet starts first, with Russ starting 10 min later, how long will it be before he catches up with her?

33. Joann took 20 min to drive her boat upstream to waterski at her favorite spot. Coming back later in the day, at the same boat speed, took her 15 min. If the current in that part of the river is 5 km per hr, what was her boat speed?

34. Joe traveled against the wind in a small plane for 3 hr. The return trip with the wind took 2.8 hr. Find the speed of the wind if the speed of the plane in still air is 180 mph.

35. The distance between New York and London is 3469 miles. If an airplane has a cruising speed of 350 mph and a tail wind of 50 mph, how many miles out will it reach the point of no return? (A tail wind is a wind blowing in the same direction as the airplane. The point of no return is the point on the flight where it will take the same amount of time to fly on to the destination as to fly back to the starting point.)

36. Joe drove the 100 miles from Philadelphia to New York at an average speed of 45 mph and drove the return trip at an average speed of 36 mph.
   (a) What was his average speed for the round trip?
   (b) Why was the average speed closer to the lower speed?
   (c) Show that the average speed will be the same no matter what the length of the trip.

37. Mark can clean the house in 9 hr, while Wendy needs 6 hr. How long will it take them to clean the house if they work together?

38. Helen can paint a room in 5 hr. Jay can paint the same room in 4 hr. (He does a sloppier job.) How long will it take them to paint the room together?

39. Two chemical plants are polluting a river. If plant A produces a predetermined maximum amount of pollution in half the time as plant B, and together they produce the maximum pollution in 26 hr, how long will it take plant B alone?

40. A sewage treatment plant has two inlet pipes to its settling pond. One can fill the pond in 10 hr, the other in 12 hr. If the first pipe is open for 5 hr and then the second pipe is opened, how long will it take to fill the pond?

41. An inlet pipe can fill Dominic's pool in 5 hr, while an outlet pipe can empty it in 8 hr. In his haste to watch television, Dominic left both pipes open. How long did it take to fill the pool?

42. Suppose Dominic discovered his error (see Exercise 41) after an hour-long program. If he then closed the outlet pipe, how much longer would be needed to fill the pool?

43. A clock radio is on sale for $49. If the sale price is 15% less than the regular price, what was the regular price?

44. A shopkeeper prices his items 20% over their wholesale price. If a lamp is marked $74, what was its wholesale price?

45. Jim Marshall invests $20,000 received from an insurance settlement in two ways, some at 5% and some at 6%. Altogether, he makes $1080 per year interest. How much is invested at each rate?

46. Susan Macias received $52,000 profit from the sale of some land. She invested part at 7 1/2% interest and the rest at 5 1/2% interest. She earned a total of $3280 interest per year. How much did she invest at each rate?

47. Bill Cornett won $100,000 in a state lottery. He first paid income tax of 38% on the winnings. Of the rest, he invested some at 8 1/2% and some at 6%, making $5070 interest per year. How much is invested at each rate?

48. Mary Ellen Heise earned $48,000 from royalties on her cookbook. She paid a 38% income tax on these royalties. The balance was invested in two ways, at 7 1/2% and at 5 1/2%. The investments produced $1936.80 interest income per year. Find the amount invested at each rate.

49. Diane Gray bought two plots of land for a total of $120,000. On the first plot, she made a profit at 15%. On the second, she lost 10%. Her total profit was $5500. How much did she pay for each piece of land?

50. Suppose $10,000 is invested at 6%. How much additional money must be invested at 8% to produce a yield on the entire amount invested of 7.2%?

51. Kathryn Johnson earns take-home pay of $198 a week. If her deductions for retirement, union dues, medical plan, and so on amount to 26% of her wages, what is her weekly pay before deductions?

52. Barbara Burnett gives 10% of her net income to the church. This amounts to $80 a month. In addition, her paycheck deductions are 24% of her gross monthly income. What is her gross monthly income?

53. In planning his retirement, Byron Hopkins deposits some money at 6% with twice as much deposited at 5%. Find the amount deposited at each rate if the total annual interest income is $3200.

54. A church building fund has invested funds in two ways: part of the money at 5% and four times as much at 7%. Find the amount invested at each rate if the total annual income from interest is $4950.

55. A cashier has some $5 bills and some $10 bills. The total number of bills is n, and the total value of the money is v. Find the number of each kind of bill that the cashier has.

56. Let $0 < a < b < c < 100$. How many liters of $a\%$ solution should be mixed with m liters of $c\%$ solution to make a $b\%$ mixture?

57. Biologists can estimate the number of individuals of a species in an area. Suppose, for example, that 100 animals of the species are caught and marked. A period of time is permitted to elapse, and then b animals are caught. If c of these ($c \leq b$) are marked, show how biologists would estimate the total number of individuals in the area.

58. Suppose B dollars are invested, some at $m\%$ and the rest at $n\%$. If a total of I dollars in interest is earned per year, find the amount invested at each rate.

2.3 QUADRATIC EQUATIONS

An equation of the form $ax + b = 0$ is a first-degree equation because x has the exponent 1. In $ax^2 + bx + c = 0$ the highest exponent is 2. Hence it is a second-degree equation. If we consider a square having sides of length x, then $x \cdot x$ or x^2 is the area of the square. Consequently we often say "x squared" for x^2. The Latin word for square is *quadratum* so we refer to $ax^2 + bx + c$ as a *quadratic* and to second-degree equations as *quadratic equations*.

Quadratic Equation

> An equation that can be written in the form
> $$ax^2 + bx + c = 0,$$
> where a, b, and c are real numbers with $a \neq 0$, is a **quadratic equation.**

(Why is the restriction $a \neq 0$ necessary?) A quadratic equation written in the form $ax^2 + bx + c = 0$ is in **standard form.**

The simplest method of solving a quadratic equation, but one that is not always easily applied, is factoring. This method depends on the **zero-factor property** on the following page.

. . . .
Zero-Factor
Property
───────────

> If a and b are complex numbers, with $ab = 0$, then $a = 0$ or $b = 0$ or both. (Also, if $a = 0$ or $b = 0$, then $ab = 0$.)

In Example 1, the zero-factor property is used to solve a quadratic equation.

. . . .
─────────────────

EXAMPLE 1   Solve each equation.

(a) $6r^2 + 7r = 3$

First write the equation in standard form as

$$6r^2 + 7r - 3 = 0.$$

Now factor $6r^2 + 7r - 3$ to get

$$(3r - 1)(2r + 3) = 0.$$

By the zero-factor property, the product $(3r - 1)(2r + 3)$ can equal 0 only if

$$3r - 1 = 0 \quad \text{or} \quad 2r + 3 = 0.$$

Solve each of these linear equations separately to find that the solutions of the original equation are 1/3 and $-3/2$. Check these solutions by substituting back in the original equation.

(b) $x^2 = 5x$

To use the zero-factor property here, first get zero on one side.

$$x^2 = 5x$$
$$x^2 - 5x = 0 \qquad \text{Add } -5x \text{ to both sides.}$$
$$x(x - 5) = 0 \qquad \text{Factor.}$$
$$x = 0 \quad \text{or} \quad x - 5 = 0 \qquad \text{Set each factor equal to 0.}$$
$$x = 0 \quad \text{or} \qquad x = 5$$

Check that the solutions are 0 and 5 by substituting them in the original equation. ●

CAUTION   An error that we often make when solving equations like Example 1(b) is to begin by dividing both sides of the equation by x to obtain $x = 5$. Notice that if this is done, one of the two solutions is lost.

A quadratic equation of the form $x^2 = k$ can be solved by factoring using the following sequence of equivalent equations.

$$x^2 = k$$
$$x^2 - k = 0$$
$$(x - \sqrt{k})(x + \sqrt{k}) = 0$$
$$x - \sqrt{k} = 0 \quad \text{or} \quad x + \sqrt{k} = 0$$
$$x = \sqrt{k} \quad \text{or} \qquad x = -\sqrt{k}$$

This proves the following statement, sometimes called the **square root property,** which gives a direct way to solve equations of the form $x^2 = k$.

Square Root Property

> The solution set of $x^2 = k$ is
>
> $$\{\sqrt{k},\; -\sqrt{k}\}.$$

The solutions are often abbreviated as $\pm\sqrt{k}$. Both solutions are real if $k > 0$ and imaginary if $k < 0$. If $k = 0$, there is only one solution.

EXAMPLE 2   Solve each equation.

(a) $z^2 = 17$

The solution set is $\{\pm\sqrt{17}\}$.

(b) $m^2 = -25$

Since $\sqrt{-25} = 5i$, the solution set of $m^2 = -25$ is $\{\pm 5i\}$.

(c) $(y - 4)^2 = 12$

Use a generalization of the square root property, working as follows.

$$(y - 4)^2 = 12$$
$$y - 4 = \pm\sqrt{12}$$
$$y = 4 \pm \sqrt{12}$$
$$y = 4 \pm 2\sqrt{3} \quad \bullet$$

Completing the Square   As suggested by the equation in Example 2(c), a quadratic equation can be solved with the square root property by first writing the given equation in the form $(x + n)^2 = k$ for suitable numbers n and k. This has been known since Babylonian times (2000 B.C.). Euclid (*circa* 330–270 B.C.) described methods for solving quadratics by geometric figures. Hence we say "complete the square."

The next two examples show how to write a quadratic equation in the form $(x + n)^2 = k$ by *completing the square*. (We will use this process again in later chapters when drawing the graphs of equations.)

EXAMPLE 3   Solve $x^2 - 2x = 15$.

To rewrite this equation in the form $(x + n)^2 = k$, the left side must be rewritten as a perfect square trinomial. Expanding $(x + n)^2$ gives $x^2 + 2xn + n^2$. Get $x^2 - 2x$ in this form by first looking at the terms of first degree, $-2x$ and $2xn$. These terms are equal if

$$2xn = -2x$$

or

$$n = -1.$$

If $n = -1$, then $n^2 = (-1)^2 = 1$. Thus, $x^2 - 2x$ can be converted into a perfect square by adding 1, since

$$x^2 - 2x + 1 = (x - 1)^2.$$

If 1 is added to $x^2 - 2x$, the left side of the given equation, then 1 must also be added to the right side, with $x^2 - 2x = 15$ becoming

$$x^2 - 2x + 1 = 15 + 1,$$

or, after factoring on the left,

$$(x - 1)^2 = 16.$$

This equation, $(x - 1)^2 = 16$, is nothing more than the original equation, $x^2 - 2x = 15$, rewritten in an alternate form. In this new form, the equation can be solved by the square root property.

$$x - 1 = 4 \quad \text{or} \quad x - 1 = -4$$

There are two solutions:

$$x = 1 + 4 = 5 \quad \text{and} \quad x = 1 - 4 = -3. \quad \bullet$$

In Example 3 we rewrote the equation $x^2 - 2x = 15$ as $(x - 1)^2 = 16$ by completing the square, to get it in a form suitable for solution by the square root property. (A summary of the steps in completing the square is given after the next example.)

· · · · · · · · ·

EXAMPLE 4   Solve $9z^2 - 12z - 1 = 0$.

To rewrite this equation in the form $(z + n)^2 = k$, it is necessary that the coefficient of z^2 be 1. To get this coefficient, multiply both sides by 1/9.

$$z^2 - \frac{4}{3}z - \frac{1}{9} = 0$$

Now add 1/9 on both sides.

$$z^2 - \frac{4}{3}z = \frac{1}{9}$$

The left side must be rewritten as a perfect square, $(z + n)^2$. Since $(z + n)^2 = z^2 + 2zn + n^2$, the terms of first degree of $z^2 + 2zn + n^2$ and $z^2 - (4/3)z$ must satisfy

$$2zn = -\frac{4}{3}z$$

or

$$n = -\frac{2}{3}.$$

The value of n is always half the coefficient of the first degree term. If $n = -2/3$, then $n^2 = 4/9$, which should be added to both sides, giving

$$z^2 - \frac{4}{3}z + \frac{4}{9} = \frac{1}{9} + \frac{4}{9}.$$

Factoring on the left yields

$$\left(z - \frac{2}{3}\right)^2 = \frac{5}{9}.$$

Now use the square root property to find z.

$$z - \frac{2}{3} = \pm\sqrt{\frac{5}{9}}$$

$$z - \frac{2}{3} = \pm\frac{\sqrt{5}}{3}$$

$$z = \frac{2}{3} \pm \frac{\sqrt{5}}{3}$$

These two solutions can be written as

$$z = \frac{2 \pm \sqrt{5}}{3}. \quad \bullet$$

The process of *completing the square*, used in Examples 3 and 4, is summarized as follows.

Completing the Square

> To solve $ax^2 + bx + c = 0$, $a \neq 0$, by **completing the square:**
>
> **1.** If $a \neq 1$, multiply both sides by $1/a$. Then rewrite the equation so that the constant is alone on one side of the equals sign.
> **2.** Square half the coefficient of x, and add the square to both sides.
> **3.** Factor, and use the square root property.

Quadratic Formula The method of completing the square can be used to solve any quadratic equation. However, in the long run it is better to start with the general quadratic equation $ax^2 + bx + c = 0$, and, using the method of completing the square, solve this equation for x in terms of the constants a, b, and c. The result will be a general formula for solving any quadratic equation. To find this general formula, start with the fact that in a quadratic equation $a \neq 0$, and multiply both sides by $1/a$ to get

$$x^2 + \frac{b}{a}x + \frac{c}{a} = 0.$$

Add $-c/a$ to both sides.

$$x^2 + \frac{b}{a}x = -\frac{c}{a}$$

Now take half of b/a, and square the result.

$$\frac{1}{2} \cdot \frac{b}{a} = \frac{b}{2a} \quad \text{and} \quad \left(\frac{b}{2a}\right)^2 = \frac{b^2}{4a^2}$$

Add the square to both sides.

$$x^2 + \frac{b}{a}x + \frac{b^2}{4a^2} = \frac{b^2}{4a^2} - \frac{c}{a}$$

The expression on the left side of the equals sign can be written as the square of a binomial, while the expression on the right can be simplified. Doing all this yields

$$\left(x + \frac{b}{2a}\right)^2 = \frac{b^2 - 4ac}{4a^2}.$$

By the square root property, this last statement leads to

$$x + \frac{b}{2a} = \sqrt{\frac{b^2 - 4ac}{4a^2}} \quad \text{or} \quad x + \frac{b}{2a} = -\sqrt{\frac{b^2 - 4ac}{4a^2}}.$$

Since $4a^2 = (2a)^2$,

$$x + \frac{b}{2a} = \frac{\sqrt{b^2 - 4ac}}{|2a|} \quad \text{or} \quad x + \frac{b}{2a} = \frac{-\sqrt{b^2 - 4ac}}{|2a|}.$$

If $a > 0$, then $|2a| = 2a$, giving

$$x = \frac{-b + \sqrt{b^2 - 4ac}}{2a} \quad \text{or} \quad x = \frac{-b - \sqrt{b^2 - 4ac}}{2a}. \qquad (*)$$

If $a < 0$, then $|2a| = -2a$, giving the same two solutions as in $(*)$, except in reversed order. In either case, the solutions can be written as

$$x = \frac{-b + \sqrt{b^2 - 4ac}}{2a} \quad \text{or} \quad x = \frac{-b - \sqrt{b^2 - 4ac}}{2a}.$$

A more compact form of this result, called the **quadratic formula,** is given below.

Quadratic Formula

The solutions of the quadratic equation $ax^2 + bx + c = 0$, where $a \neq 0$, are

$$x = \frac{-b \pm \sqrt{b^2 - 4ac}}{2a}.$$

. . . .

EXAMPLE 5   Solve $x^2 - 4x + 1 = 0$.

Here $a = 1$, $b = -4$, and $c = 1$. Substitute these values into the quadratic formula.

$$x = \frac{-b \pm \sqrt{b^2 - 4ac}}{2a}$$

$$= \frac{-(-4) \pm \sqrt{(-4)^2 - 4(1)(1)}}{2(1)}$$

$$= \frac{4 \pm \sqrt{16 - 4}}{2} = \frac{4 \pm 2\sqrt{3}}{2}$$

$$= \frac{2(2 \pm \sqrt{3})}{2}$$

$$x = 2 \pm \sqrt{3}$$

The solutions are $x = 2 + \sqrt{3}$ and $x = 2 - \sqrt{3}$. ●

. . . .

EXAMPLE 6   Solve $2y^2 = y - 4$.

To find the values of a, b, and c, first rewrite the equation as $2y^2 - y + 4 = 0$. Then $a = 2$, $b = -1$, and $c = 4$.

$$y = \frac{-(-1) \pm \sqrt{(-1)^2 - 4(2)(4)}}{2(2)} \qquad \text{By quadratic formula}$$

$$= \frac{1 \pm \sqrt{1 - 32}}{4}$$

$$y = \frac{1 \pm \sqrt{-31}}{4} = \frac{1 \pm i\sqrt{31}}{4}$$

$$y = \frac{1}{4} \pm \frac{\sqrt{31}}{4}i \quad \text{in standard form.} \quad ●$$

The Discriminant   The quantity under the radical in the quadratic formula, $b^2 - 4ac$, is called the **discriminant.** The value of the discriminant determines whether the solutions of quadratic equations with real coefficients are real or nonreal and whether there are one or two solutions as follows.

. . . .

Discriminant (Real Coefficients)

Discriminant	Number of solutions	Kind of solutions
Positive	two	real
Zero	one	real
Negative	two	nonreal

. . . .

EXAMPLE 7 Find a value of k so that there is exactly one solution to the equation

$$16p^2 + kp + 25 = 0.$$

A quadratic equation with real coefficients will have exactly one solution if the discriminant is zero. Here, $a = 16$, $b = k$, and $c = 25$, giving the discriminant

$$b^2 - 4ac = k^2 - 4(16)(25) = k^2 - 1600.$$

The discriminant is 0 if $k^2 - 1600 = 0$

or if $k^2 = 1600,$

from which $k = \pm 40.$ ●

When the numbers a, b, and c are *integers*, the value of the discriminant can be used to determine whether the solution will be rational, irrational, or nonreal as follows.

. . . .

Discriminant (Integer Coefficients)

Discriminant	Number of solutions	Kind of solutions
Positive, perfect square	two	rational
Positive, but not a perfect square	two	irrational
Zero	one	rational
Negative	two	nonreal

CAUTION The restriction that a, b, and c be integers is important. For example, for the equation

$$x^2 - \sqrt{5}x - 1 = 0$$

the discriminant is $b^2 - 4ac = 5 + 4 = 9$, which would indicate two rational solutions. By the quadratic formula, however, the two solutions

$$x = \frac{\sqrt{5} \pm 3}{2}$$

are *irrational* numbers.

. . . .

EXAMPLE 8 Use the discriminant to determine whether the solutions of $5x^2 + 2x - 4 = 0$ are rational, irrational, or nonreal.

Since $a = 5$, $b = 2$, and $c = -4$, the discriminant is

$$b^2 - 4ac = (2)^2 - (4)(5)(-4) = 84.$$

The discriminant is positive, so there are two real number solutions. Since 84 is not a perfect square, the solutions will be irrational numbers. ●

Word problems often lead to quadratic equations, as in the next example.

· · · · · · · · ·

EXAMPLE 9   Michael wants to make an exposed gravel border of uniform width around a rectangular pool in his garden. The pool is 10 ft by 6 ft. He has enough material to cover 36 sq ft. How wide will the border be?

A sketch of the pool with border is shown in Figure 3.

Let x = the width of the border,

$6 + 2x$ = the width of the larger rectangle,

$10 + 2x$ = the length of the larger rectangle.

Since the area of the border is the difference of the areas of the larger rectangle and the pool,

$$\begin{matrix} \text{area of larger} \\ \text{rectangle} \end{matrix} - \begin{matrix} \text{area of} \\ \text{pool} \end{matrix} = \begin{matrix} \text{amount of} \\ \text{material.} \end{matrix}$$

$$(6 + 2x)(10 + 2x) - 10 \cdot 6 = 36$$

$$60 + 32x + 4x^2 - 60 = 36$$

$$4x^2 + 32x - 36 = 0$$

$$x^2 + 8x - 9 = 0 \qquad \text{Divide by 4.}$$

$$(x + 9)(x - 1) = 0$$

The solutions are -9 and 1. Since -9 cannot be the width of the border, the border must be 1 ft wide.

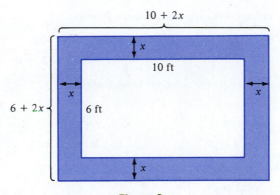

Figure 3

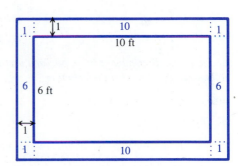

Figure 4

Check. Figure 4 shows the areas of the different parts of the border. The sum of the areas checks out to be 36, as required.   ●

2.3 Exercises

Solve the following equations by factoring.

1. $q^2 + 2q - 8 = 0$ **2.** $x^2 + 6x + 9 = 0$ **3.** $-q^2 + 3q + 4 = 0$ **4.** $x^2 - 7x = 0$

5. $10x^2 + 40x = 0$ **6.** $6r^2 + 7r = 3$ **7.** $-5k^2 = 5k - 30$ **8.** $8k^2 + 14k = -3$

Solve the following equations by the square root property.

9. $x^2 = 27$ **10.** $y^2 = 24$ **11.** $(m - 3)^2 = 5$ **12.** $(p + 2)^2 = 7$

13. $(3k - 1)^2 = 12$ **14.** $(4t + 1)^2 = 20$ **15.** $5x^2 - 20 = 0$ **16.** $9y^2 = 1$

Solve the following equations by completing the square.

17. $y^2 + 8y + 7 = 0$ **18.** $q^2 + 3q = \dfrac{7}{4}$ **19.** $3k^2 - 12k + 12 = 75$ **20.** $m^2 + 5m = 6$

21. $x^2 = 2x + 4$ **22.** $r^2 + 8r = -13$ **23.** $2p^2 = -(2p + 1)$ **24.** $9z^2 = 12z - 8$

Solve the following equations by the quadratic formula.

25. $y^2 = 3y - 1$ **26.** $-q^2 = 2q - 2$ **27.** $5p^2 + 1 = p$ **28.** $q^2 + 1 = q$

29. $9n^2 + 6n + 1 = 0$ **30.** $4x^2 + 4x + 1 = 0$ **31.** $m^2 - \sqrt{2}m - 1 = 0$

32. $z^2 - \sqrt{3}z - 2 = 0$ **33.** $\sqrt{2}p^2 - 3p + \sqrt{2} = 0$ **34.** $-\sqrt{6}k^2 - 2k + \sqrt{6} = 0$

Solve the following equations by any method.

35. $t^2 - t = 3$ **36.** $2s^2 - 2s = 3$ **37.** $n^2 + 4 = 3n$ **38.** $9p^2 = 25 + 30p$

39. $2m^2 = m - 1$ **40.** $3k^2 + 2 = k$ **41.** $\dfrac{2x - 3}{x + 1} = \dfrac{x + 1}{x + 2}$ **42.** $\dfrac{x}{x - 1} = \dfrac{3x}{x - 2}$

Solve each of the following equations by factoring first and then using the quadratic formula.

43. $x^3 - 1 = 0$ **44.** $x^3 + 64 = 0$ **45.** $x^3 - 64 = 0$

46. $x^3 + 1 = 0$ **47.** $8p^3 + 125 = 0$ **48.** $64r^3 - 343 = 0$

Evaluate the discriminant $b^2 - 4ac$ and use it to predict the type of solutions in each of the following. Do not solve the equations.

49. $x^2 + 8x + 16 = 0$ **50.** $x^2 - 5x + 4 = 0$ **51.** $3m^2 - 5m + 2 = 0$

52. $8y^2 = 14y - 3$ **53.** $2r^2 - 4r + 1 = 0$ **54.** $3z^2 = 4z - 5$

Find all values for k for which the following equations have exactly one solution.

55. $9x^2 + kx + 4 = 0$ **56.** $25m^2 - 10m + k = 0$ **57.** $y^2 + 11y + k = 0$

58. $z^2 - 5z + k = 0$ **59.** $kr^2 + (2k + 6)r + 16 = 0$ **60.** $ky^2 + 2(k + 4)y + 25 = 0$

Solve each of Exercises 61–66 for the indicated variable. Assume that all denominators are nonzero and that all variables represent positive real numbers.

61. $L = \dfrac{d^4 k}{h^2}$ for h **62.** $F = \dfrac{kMv^2}{r}$ for v **63.** $s = s_0 + gt^2 + k$ for t

64. $P = \dfrac{E^2 R}{(r + R)^2}$ for R **65.** $S = 2\pi rh + 2\pi r^2$ for r **66.** $pm^2 - 8qm + \dfrac{1}{r} = 0$ for m

67. Is it possible for the solution of a quadratic equation with real coefficients to consist of a single irrational number?

68. Is it possible for the solution of a quadratic equation with real coefficients to consist of one real and one nonreal root?

For each of the following numbers, find the values of a, b, and c for which the quadratic polynomial $ax^2 + bx + c$ *has those numbers as zeros.*

69. 4, 5 **70.** $-3, 2$ **71.** $1 + \sqrt{2}, 1 - \sqrt{2}$ **72.** $i, -i$

Solve the following problems.

73. A shopping center has a rectangular area of 40,000 square yards enclosed on three sides for a parking lot. The length is 200 yards more than twice the width. What are the dimensions of the lot?

74. Two integers have a sum of 10. The sum of the squares of the integers is 148. Find the integers.

75. An ecology center wants to set up an experimental garden. It has 300 meters of fencing to enclose a rectangular area of 5000 square meters. Find the dimensions of the rectangle.

76. Alfredo went into a frame-it-yourself shop. He wanted a frame 3 centimeters longer than wide. The frame he chose extends 1.5 centimeters beyond the picture on each side. Find the outside dimensions of the frame if the area of the unframed picture is 70 square centimeters.

77. Joan wants to buy a rug for a room that is 12 feet by 15 feet. She wants to leave a uniform strip of floor around the rug. She can afford 108 square feet of carpeting. What dimensions should the rug have?

78. Max can clean the garage in 9 hours less time than his brother Paul. Working together, they can do the job in 20 hours. How long would it take each one to do the job alone?

79. Dolores drives 10 mph faster than Steve. Both start at the same time for Atlanta from Chattanooga, a distance of about 100 miles. It takes Steve 1/3 of an hour longer than Dolores to make the trip. What is Steve's average speed?

80. Amy walks 1 mph faster than her friend Lisa. In a walk for charity, both walked the full distance of 24 miles. Lisa took 2 hours longer than Amy. What was Lisa's average speed?

Use the Pythagorean theorem, $c^2 = a^2 + b^2$, *to solve each of the following. Round to the nearest hundredth.*

81. One leg of a right triangle is 4 cm longer than the other leg. The hypotenuse (longest side) is 10 cm longer than the shorter leg. Find each of the three sides of the triangle.

82. A rectangle has a diagonal of 16 m. The width of the rectangle is 3.2 m less than the length. Find the length and width of the rectangle.

Let r_1 and r_2 be the solutions of the quadratic equation $ax^2 + bx + c = 0$. Show that the equations in Exercises 83 and 84 are true.

83. $r_1 + r_2 = -\dfrac{b}{a}$

84. $r_1 r_2 = \dfrac{c}{a}$

85. Suppose one solution of the equation $km^2 + 10m = 8$ is -4. Find the value of k, and the other solution.

For the equations in Exercises 86 and 87,   (a) solve for x in terms of y,   (b) solve for y in terms of x.

86. $4x^2 - 2xy + 3y^2 = 2$

87. $3y^2 + 4xy - 9x^2 = -1$

88. Find the error in the following solution.
$$(x - 1)(x - 3) = 5$$
$$x - 1 = 5 \quad \text{or} \quad x - 3 = 5$$
Solutions are $x = 6$ and $x = 8$.

2.4 EQUATIONS REDUCIBLE TO QUADRATICS

The equation $12m^4 - 11m^2 + 2 = 0$ is not a quadratic equation because of the m^4 term. However, it can be written as a quadratic equation by making the substitutions

$$x = m^2 \quad \text{and} \quad x^2 = m^4$$

which change the equation into

$$12x^2 - 11x + 2 = 0.$$

This quadratic equation can be solved for x, and then from $x = m^2$, the values of m, the solutions of the original equation, can be found.

An equation such as $12m^4 - 11m^2 + 2 = 0$ is said to be **quadratic in form** if it can be written as

$$au^2 + bu + c = 0,$$

where $a \neq 0$ and u is some algebraic expression.

. . . .

EXAMPLE 1   Solve $12m^4 - 11m^2 + 2 = 0$.

As mentioned above, this equation is quadratic in form. By making the substitution $x = m^2$, the equation becomes

$$12x^2 - 11x + 2 = 0,$$

which can be solved by factoring, as follows.

$$(3x - 2)(4x - 1) = 0$$
$$x = 2/3 \quad \text{or} \quad x = 1/4$$

To find m, use the fact that $x = m^2$ and replace x with 2/3 and then with 1/4.

$$m^2 = \frac{2}{3} \qquad \text{or} \qquad m^2 = \frac{1}{4}$$

$$m = \pm\sqrt{\frac{2}{3}} \qquad\qquad m = \pm\sqrt{\frac{1}{4}}$$

$$m = \pm\frac{\sqrt{2}}{\sqrt{3}}$$

$$m = \frac{\pm\sqrt{6}}{3} \qquad \text{or} \qquad m = \pm\frac{1}{2}$$

The given equation $12m^4 - 11m^2 + 2 = 0$ has the four solutions $-\sqrt{6}/3$, $\sqrt{6}/3$, $-1/2$, and $1/2$. As before, check these solutions by substituting back into the *original* equation. ●

· · · · · · · · ·

EXAMPLE 2   Solve $6p^{-2} + p^{-1} = 2$.
Let $u = p^{-1}$ and rearrange terms to get

$$6u^2 + u - 2 = 0.$$

Factor on the left, and then place each factor equal to 0.

$$(3u + 2)(2u - 1) = 0$$

$$3u + 2 = 0 \qquad \text{or} \quad 2u - 1 = 0$$

$$u = -\frac{2}{3} \qquad\qquad u = \frac{1}{2}$$

Since $u = p^{-1} = \dfrac{1}{p}$,

$$\frac{1}{p} = -\frac{2}{3} \quad \text{or} \qquad \frac{1}{p} = \frac{1}{2},$$

from which

$$p = -\frac{3}{2} \quad \text{or} \qquad p = 2. ●$$

· · · · · · · · ·

EXAMPLE 3   Solve $4p^4 - 16p^2 + 13 = 0$.
Let $u = p^2$ to get

$$4u^2 - 16u + 13 = 0.$$

To solve this equation, use the quadratic formula with $a = 4$, $b = -16$, and $c = 13$. Substitute these values into the quadratic formula.

$$u = \frac{-(-16) \pm \sqrt{(-16)^2 - 4(4)(13)}}{2(4)} = \frac{16 \pm \sqrt{48}}{8}$$

$$= \frac{16 \pm 4\sqrt{3}}{8} = \frac{4 \pm \sqrt{3}}{2}$$

Since $p^2 = u$, $p^2 = \dfrac{4 + \sqrt{3}}{2}$ or $p^2 = \dfrac{4 - \sqrt{3}}{2}$.

Finally, $p = \pm\sqrt{\dfrac{4 + \sqrt{3}}{2}}$ or $p = \pm\sqrt{\dfrac{4 - \sqrt{3}}{2}}$.

Rationalizing the denominators gives the four solutions

$$\frac{\pm\sqrt{8 + 2\sqrt{3}}}{2}, \quad \frac{\pm\sqrt{8 - 2\sqrt{3}}}{2}. \quad \bullet$$

To solve equations containing radicals or rational exponents, such as $x = \sqrt{15 - 2x}$, or $(x + 1)^{1/2} = x$, use the following result.

If P and Q are algebraic expressions, then every solution of the equation $P = Q$ is also a solution of the equation $(P)^n = (Q)^n$, for any positive integer n.

CAUTION Be very careful when using this result. It does not say that the equations $P = Q$ and $(P)^n = (Q)^n$ are equivalent; it says only that each solution of the original equation $P = Q$ is also a solution of the new equation $(P)^n = (Q)^n$. However, *the new equation may have more solutions than the original equation.*

For example, the solution set of the equation $x = -2$ is $\{-2\}$. Squaring both sides of the equation $x = -2$ gives the new equation $x^2 = 4$, which has solution set $\{-2, 2\}$. Since the solution sets are not equal, the equations are not equivalent. As this example shows, it is essential to check all proposed solutions back in the original equation.

EXAMPLE 4   Solve $x = \sqrt{15 - 2x}$.
The equation $x = \sqrt{15 - 2x}$ can be solved by squaring both sides as follows.

$$x^2 = (\sqrt{15 - 2x})^2$$
$$x^2 = 15 - 2x$$
$$x^2 + 2x - 15 = 0$$
$$(x + 5)(x - 3) = 0$$
$$x = -5 \quad \text{or} \quad x = 3$$

Now it is necessary to check the proposed solutions in the *original* equation,

$$x = \sqrt{15 - 2x}.$$

$$\text{If } x = -5, \text{ does } x = \sqrt{15 - 2x}? \qquad \text{If } x = 3, \text{ does } x = \sqrt{15 - 2x}?$$

$$-5 = \sqrt{15 + 10} \qquad\qquad\qquad 3 = \sqrt{15 - 6}$$

$$-5 = 5 \quad \text{False} \qquad\qquad\qquad 3 = 3 \quad \text{True}$$

As this check shows, only 3 is a solution of the given equation.   ●

· · · · · · · · ·

EXAMPLE 5   Solve $\sqrt{2x + 3} - \sqrt{x + 1} = 1$.

Separate the radicals by writing the equation as

$$\sqrt{2x + 3} = 1 + \sqrt{x + 1}.$$

Now square both sides. Be very careful when squaring on the right side of this equation. Recall that $(a + b)^2 = a^2 + 2ab + b^2$; replace a with 1 and b with $\sqrt{x + 1}$ to get the next equation, the result of squaring both sides of $\sqrt{2x + 3} = 1 + \sqrt{x + 1}$.

$$2x + 3 = 1 + 2\sqrt{x + 1} + x + 1$$

$$x + 1 = 2\sqrt{x + 1}$$

One side of the equation still contains a radical; to eliminate it, square both sides again.

$$x^2 + 2x + 1 = 4(x + 1)$$

$$x^2 - 2x - 3 = 0$$

$$(x - 3)(x + 1) = 0$$

$$x = 3 \quad \text{or} \quad x = -1$$

Check these proposed solutions in the original equation.

$$\text{If } x = 3, \text{ does } \sqrt{2x + 3} - \sqrt{x + 1} = 1?$$

$$\sqrt{9} - \sqrt{4} = 1$$

$$3 - 2 = 1 \quad \text{True}$$

$$\text{If } x = -1, \text{ does } \sqrt{2x + 3} - \sqrt{x + 1} = 1?$$

$$\sqrt{1} - \sqrt{0} = 1$$

$$1 - 0 = 1 \quad \text{True}$$

Both proposed solutions 3 and -1 are solutions of the original equation.   ●

· · · · · · · · ·

EXAMPLE 6   Solve $(5x^2 - 6)^{1/4} = x$.

Since the equation involves a fourth root, begin by raising both sides to the fourth power.

$$[(5x^2 - 6)^{1/4}]^4 = x^4$$

$$5x^2 - 6 = x^4$$

$$x^4 - 5x^2 + 6 = 0$$

Now substitute y for x^2.

$$y^2 - 5y + 6 = 0$$
$$(y - 3)(y - 2) = 0$$
$$y = 3 \quad \text{or} \quad y = 2$$

Since $y = x^2$,
$$x^2 = 3 \quad \text{or} \quad x^2 = 2$$
$$x = \pm\sqrt{3} \quad \text{or} \quad x = \pm\sqrt{2}.$$

Checking the four proposed solutions, $\sqrt{3}$, $-\sqrt{3}$, $\sqrt{2}$, and $-\sqrt{2}$, in the original equation shows that only $\sqrt{3}$ and $\sqrt{2}$ are solutions. ●

2.4 Exercises ···

Find all real solutions for each of the following equations.

1. $m^4 - 8m^2 + 15 = 0$

2. $3k^4 + 10k^2 - 25 = 0$

3. $2r^4 - 7r^2 + 5 = 0$

4. $4x^4 - 8x^2 + 3 = 0$

5. $(g - 2)^2 - 6(g - 2) + 8 = 0$

6. $-2(z - 4)^2 + 2(z - 4) + 3 = 0$

7. $6(k + 2)^4 - 11(k + 2)^2 + 4 = 0$

8. $8(m - 4)^4 - 10(m - 4)^2 + 3 = 0$

9. $7p^{-2} + 19p^{-1} = 6$

10. $5k^{-2} - 43k^{-1} = 18$

11. $(r - 1)^{2/3} + (r - 1)^{1/3} = 12$

12. $(y + 3)^{2/3} - 2(y + 3)^{1/3} - 3 = 0$

13. $5r = r^{-1}$

14. $a^{-1} = a^{-2}$

15. $\dfrac{17}{p^2 + 1} - \dfrac{22}{(p^2 + 1)^2} = 3$

16. $6 = \dfrac{7}{2y - 3} + \dfrac{3}{(2y - 3)^2}$

17. $r^3 - 13r^{3/2} + 40 = 0$

18. $a^3 - 8a^{3/2} + 7 = 0$

19. $1 + 3(r^2 - 1)^{-1} = 28(r^2 - 1)^{-2}$

20. $5(m^2 + 1)^{-2} = 4(m^2 + 1)^{-1} + 1$

21. $2(1 + 2\sqrt{x})^2 - (1 + 2\sqrt{x}) = 21$

22. $20(2 - \sqrt{m})^2 + 11(2 - \sqrt{m}) = 3$

23. $12\left(m - \dfrac{2}{m}\right)^2 - 13m + \dfrac{26}{m} = 14$

24. $6\left(2k - \dfrac{1}{k}\right)^2 - 22k + \dfrac{11}{k} = 35$

25. $\sqrt{2m + 1} = 2\sqrt{m}$

26. $3\sqrt{p} = \sqrt{8p + 16}$

27. $\sqrt{3z + 7} = 3z + 5$

28. $\sqrt{4r + 13} = 2r - 1$

29. $\sqrt{4k + 5} - 2 = 2k - 7$

30. $\sqrt{2t} + 4 = t$

31. $\sqrt[3]{4n + 3} = \sqrt[3]{2n - 1}$

32. $\sqrt[3]{2z} = \sqrt[3]{5z + 2}$

33. $\sqrt[3]{t^2 + 2t - 1} = \sqrt[3]{t^2 + 3}$

34. $\sqrt[3]{2x^2 - 5x + 4} = \sqrt[3]{2x^2}$

35. $(2r + 5)^{1/3} - (6r - 1)^{1/3} = 0$

36. $(3m + 7)^{1/3} - (4m + 2)^{1/3} = 0$

37. $\sqrt[4]{y^2 + 2y} - \sqrt[4]{3} = 0$

38. $\sqrt[4]{k^2 + 6k} - 2 = 0$

39. $(z^2 + 24z)^{1/4} - 3 = 0$

40. $(3t^2 + 52t)^{1/4} - 4 = 0$

41. $\sqrt{y} = \sqrt{y - 5} + 1$

42. $\sqrt{2m} = \sqrt{m + 7} - 1$

43. $\sqrt{r + 5} - 2 = \sqrt{r - 1}$

44. $\sqrt{m + 7} + 3 = \sqrt{m - 4}$

45. $\sqrt{y + 2} = \sqrt{2y + 5} - 1$

46. $\sqrt{4x + 1} = \sqrt{x - 1} + 2$

47. $\sqrt{3x + 4} - \sqrt{x + 1} = 1$

48. $\sqrt{2x + 6} - \sqrt{x + 2} = 1$

49. $\sqrt{5x - 1} + \sqrt{2 - x} = \sqrt{8x + 1}$

50. $\sqrt{x + 4} - \sqrt{x + 3} = \sqrt{3x + 10}$

51. $\sqrt{2\sqrt{7x + 2}} = \sqrt{3x + 2}$

52. $\sqrt{3\sqrt{2m + 3}} = \sqrt{5m - 6}$

53. $\sqrt{x + 2} = \sqrt{4 + 7\sqrt{x}}$

54. $3 - \sqrt{x} = \sqrt{2\sqrt{x} - 3}$

55. $(2r - 1)^{2/3} = r^{1/3}$

56. $(z - 3)^{2/5} = (4z)^{1/5}$

57. $p(2 + p)^{-1/2} + (2 + p)^{1/2} = 0$

58. $(2k - 9)^{-2/3} + 4(2k - 9)^{1/3} = 0$

2.5 INEQUALITIES

An equation says that two expressions are equal, while an **inequality** says that one expression is greater than, greater than or equal to, less than, or less than or equal to, another. As with equations, a value of the variable for which the inequality is true is a *solution* of the inequality; the set of all such solutions makes up the solution set of the inequality. Two inequalities with the same solution set are equivalent.

Inequalities are solved with the following properties of real numbers.

Properties of Inequalities

For real numbers a, b, and c,

(a) if $a < b$, then $a + c < b + c$

(b) if $a < b$, and if $c > 0$, then $ac < bc$

(c) if $a < b$, and if $c < 0$, then $ac > bc$.

Similar properties are valid if $<$ is replaced with $>$, $\leq$, or $\geq$.

CAUTION Pay careful attention to part (c): if both sides of an inequality are multiplied by a negative number, the direction of the inequality symbol must be reversed. For example, starting with the true statement $-3 < 5$, multiplying both sides by the positive number 2 gives

$$-3 \cdot 2 < 5 \cdot 2$$
$$-6 < 10,$$

still a true statement. On the other hand, starting with $-3 < 5$ and multiplying both sides by the *negative* number -2 gives a true result only if the direction of the inequality symbol is reversed.

$$-3(-2) > 5(-2)$$
$$6 > -10$$

To prove the properties of inequalities listed above, first recall from Chapter 1 that $a < b$ means that $b - a$ is positive. The proofs also depend on the fact that the sum or product of two positive numbers is positive.

To prove part (a), use the fact that $a < b$ to see that $b - a$ is positive. Rewrite $b - a$ as

$$b - a = (b + c) - (a + c).$$

Since $b - a$ is positive, $(b + c) - (a + c)$ must also be positive, so that

$$a + c < b + c.$$

To prove part (b), again the assumption $a < b$ makes $b - a$ positive. Since c is assumed positive, the product $(b - a)c$ is positive. By the distributive property,

$$(b - a)c = bc - ac,$$

and $bc - ac$ must be positive, giving

$$ac < bc.$$

For part (c), again we are given $a < b$, so that $b - a$ is positive. We are also told that $c < 0$. Since c is less than 0, then $0 - c$, or $-c$, must be positive. Thus,

$$(b - a)(-c) = -bc + ac = ac - bc$$

is positive. If $ac - bc$ is positive,

$$bc < ac.$$

Similar proofs could be given if $<$ is replaced with $>$, $\leq$, or $\geq$.

. . . .

EXAMPLE 1   Solve the inequality $-3x + 5 > -7$.
Use the properties of inequalities. First, add -5 to both sides.

$$-3x + 5 + (-5) > -7 + (-5)$$
$$-3x > -12$$

Now multiply both sides by $-1/3$. Since $-1/3 < 0$, reverse the direction of the inequality symbol.

$$-\frac{1}{3}(-3x) < -\frac{1}{3}(-12)$$

$$x < 4$$

The original inequality is satisfied by any real number less than 4. The solution set can be written $\{x \mid x < 4\}$. A graph of the solution set is shown in Figure 5, where the parenthesis is used to show that 4 itself does not belong to the solution set.   ●

Figure 5

The set $\{x | x < 4\}$, the solution set for the inequality in Example 1, is an example of an **interval.** A simplified notation, called **interval notation,** is used for writing intervals. With this notation, the interval in Example 1 can be written as just $(-\infty, 4)$. The symbol $-\infty$ is not a real number; the symbol is used as a convenience to show that the interval includes all real numbers less than 4. The interval $(-\infty, 4)$ is an example of an **open interval,** since the endpoint, 4, is not part of the interval. Examples of other sets written in interval notation are shown below. Square brackets are used to show that the given number *is* part of the graph. Whenever two real numbers a and b are used to write an interval, it is assumed that $a < b$.

Type of interval	Set	Interval notation	Graph
Open interval	$\{x | x > a\}$	$(a, +\infty)$	
	$\{x | a < x < b\}$	(a, b)	
	$\{x | x < b\}$	$(-\infty, b)$	
Half-open interval	$\{x | a < x \leq b\}$	$(a, b]$	
	$\{x | a \leq x < b\}$	$[a, b)$	
Closed interval	$\{x | a \leq x \leq b\}$	$[a, b]$	
	$\{x | x \geq a\}$	$[a, +\infty)$	
	$\{x | x \leq b\}$	$(-\infty, b]$	

EXAMPLE 2 Solve $4 - 3y \leq 7 + 2y$. Write the solution in interval notation and graph the solution on a number line.

Write the following series of equivalent inequalities.

$$4 - 3y \leq 7 + 2y$$
$$-4 - 2y + 4 - 3y \leq -4 - 2y + 7 + 2y$$
$$-5y \leq 3$$
$$(-1/5)(-5y) \geq (-1/5)(3)$$
$$y \geq -3/5$$

In set notation, the solution set is $\{y | y \geq -3/5\}$, while in interval notation the solution set is $[-3/5, +\infty)$. See Figure 6 for the graph of the solution set. ●

Figure 6

From now on, we shall write the solutions of all inequalities with interval notation.

The inequality $-2 < 5 + 3m < 20$ says that $5 + 3m$ is between -2 and 20. Solve this inequality using an extension of the properties of inequalities given above.

· · · · · · · · ·

EXAMPLE 3   Solve $-2 < 5 + 3m < 20$.
Write equivalent inequalities as follows.

$$-2 < 5 + 3m < 20$$
$$-7 < 3m < 15$$
$$-7/3 < m < 5$$

The solution, graphed in Figure 7, is the interval $(-7/3, 5)$.   ●

Figure 7

Quadratic Inequalities   Section 2.3 introduced quadratic equations. Now we will look at *quadratic inequalities*.

· ·

Quadratic Inequality

A **quadratic inequality** is an inequality that can be written in the form

$$ax^2 + bx + c < 0,$$

for real numbers $a \neq 0$, b, and c. (The symbol $<$ can be replaced with $>$, $\leq$, or $\geq$.)

Since quadratic equations usually have two solutions, while linear equations have just one, solving quadratic inequalities requires a little more work than solving linear inequalities.

· · · · · · · · ·

EXAMPLE 4   Solve the quadratic inequality $x^2 - x - 12 < 0$.
Begin by finding the values of x that satisfy $x^2 - x - 12 = 0$.

$$x^2 - x - 12 = 0$$
$$(x + 3)(x - 4) = 0$$
$$x = -3 \quad \text{and} \quad x = 4$$

These two points, -3 and 4, divide a number line into the three regions shown in Figure 8(a). If a point in region A, for example, leads to negative values for the polynomial $x^2 - x - 12$, then all points in region A will lead to negative values. The regions that make $x^2 - x - 12$ negative can be found by selecting a test point from each region and substituting it into the inequality.

In region A, choose -4: $(-4)^2 - (-4) - 12 = 8 > 0$

In region B, choose 0: $0^2 - 0 - 12 = -12 < 0$

In region C, choose 5: $5^2 - 5 - 12 = 8 > 0$

Only the points in region B, the interval $(-3, 4)$, make the expression $x^2 - x - 12$ negative. The graph of this solution is shown in Figure 8(b).   ●

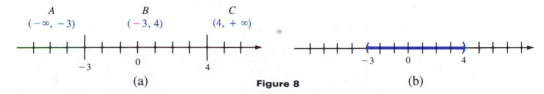

(a)

Figure 8

(b)

· · · · · · · · ·

EXAMPLE 5   Solve the inequality $2x^2 + 5x - 12 \geq 0$.

Find the values of x that satisfy $2x^2 + 5x - 12 = 0$.

$$2x^2 + 5x - 12 = 0$$
$$(2x - 3)(x + 4) = 0$$
$$x = 3/2 \quad \text{or} \quad x = -4$$

These two points divide the number line into the three regions shown in Figure 9. Choose -5, 0, and 2 as test points.

If $x = -5$, $2x^2 + 5x - 12 = 2(-5)^2 + 5(-5) - 12 = 13 > 0$.

If $x = 0$, $2x^2 + 5x - 12 = 2(0)^2 + 5(0) - 12 = -12 < 0$.

If $x = 2$, $2x^2 + 5x - 12 = 2(2)^2 + 5(2) - 12 = 6 > 0$.

The tests show that $2x^2 + 5x - 12 \geq 0$ in the interval $(-\infty, -4]$ and also in the interval $[3/2, +\infty)$. Since both of these intervals belong to the solution, the result can be written as the *union** of the two intervals, or as

$$(-\infty, -4] \cup [3/2, +\infty).$$

The graph of the solution is shown in Figure 9.   ●

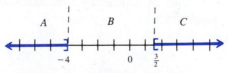

Figure 9

Rational Inequalities   The inequalities discussed in the remainder of this section involve quotients of algebraic expressions, and for this reason they are called **rational inequalities.** These inequalities can be solved in much the same way that quadratic inequalities are solved.

*The union of sets A and B, written $A \cup B$, is defined as $A \cup B = \{x|x$ is an element of A or x is an element of $B\}$.

. . . .

EXAMPLE 6   Solve the inequality $\dfrac{5}{x + 4} \geq 1$.

It is tempting to begin the solution by multiplying both sides of the inequality by $x + 4$, but to do this it would be necessary to consider whether $x + 4$ is positive or negative. Instead, subtract 1 from both sides of the inequality, getting

$$\frac{5}{x + 4} - 1 \geq 0.$$

Writing the left side as a single fraction gives

$$\frac{5 - (x + 4)}{x + 4} \geq 0 \quad \text{or} \quad \frac{1 - x}{x + 4} \geq 0.$$

Since the sign (positive or negative) depends on the signs of the numerator and denominator, the quotient (like the product) will change sign only when the denominator is 0 or when the numerator is 0. This happens when

$$1 - x = 0 \quad \text{or} \quad x + 4 = 0$$
$$x = 1 \qquad\qquad x = -4.$$

Now test a point in each of the intervals $(-\infty, -4)$, $(-4, 1)$, and $(1, +\infty)$. Values in the interval $(-4, 1)$ give a positive quotient and are part of the solution. With a quotient, the endpoints must be considered separately to make sure that no denominator is 0. With this inequality, -4 gives a 0 denominator while 1 satisfies the given inequality. In interval notation, the solution is $(-4, 1]$, as shown in Figure 10.   ●

Figure 10

⬛ **CAUTION**   As suggested by Example 6, we need to be very careful with endpoints of the intervals in the solution of rational inequalities.

. . . .

EXAMPLE 7   Solve $\dfrac{2x - 1}{3x + 4} < 5$.

Begin by subtracting 5 on both sides and combining the terms on the left into a single fraction.

$$\frac{2x - 1}{3x + 4} < 5$$

$$\frac{2x - 1}{3x + 4} - 5 < 0$$

$$\frac{2x - 1 - 5(3x + 4)}{3x + 4} < 0$$

$$\frac{-13x - 21}{3x + 4} < 0$$

Solve the equations

$$-13x - 21 = 0 \quad \text{and} \quad 3x + 4 = 0,$$

getting the solutions

$$x = -\frac{21}{13} \quad \text{and} \quad x = -\frac{4}{3}.$$

Use the values $-21/13$ and $-4/3$ to divide the number line into three intervals, and test a point in each region to see that values of x in the two intervals $(-\infty, -21/13)$ and $(-4/3, +\infty)$ make the quotient negative, as required. Neither endpoint satisfies the given inequality, so the solution set is written $(-\infty, -21/13) \cup (-4/3, +\infty)$. See Figure 11. ●

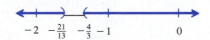

Figure 11

2.5 Exercises

Write each of the following in interval notation. Graph each interval.

1. $-1 < x < 4$

2. $x \geq -3$

3. $x < 0$

4. $2 > x \geq 1$

5. $-4 \geq x > -5$

6. $-9 > x$

Using the variable x, write each of the following intervals as an inequality.

7. $(-4, 3)$

8. $[2, 7)$

9. $(-\infty, -1]$

10. $(3, +\infty)$

11.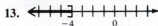

12.

13.

14.

Solve the following inequalities. Write the solutions in interval notation.

15. $2x + 1 \leq 9$

16. $3y - 2 \leq 10$

17. $-3p - 2 \leq 1$

18. $-5r + 3 \geq -2$

19. $6m - (2m + 3) \geq 4m + 5$

20. $2 - 4x + 5(x - 1) < -6(x - 2)$

21. $\dfrac{4x + 7}{-3} \leq 2x + 5$

22. $\dfrac{2z - 5}{-8} \leq 1 - z$

23. $2 \leq y + 1 \leq 5$

24. $-3 \leq 2t \leq 6$

25. $-10 > 3r + 2 > -16$

26. $4 > 6a + 5 > -1$

27. $-3 \leq \dfrac{x - 4}{-5} < 4$

28. $1 < \dfrac{4m - 5}{-2} < 9$

29. $2 \geq 5 - \dfrac{3}{4}m > -5$

30. $9 > 4 - \dfrac{1}{2}k \geq -1$

31. $y^2 - 10y + 25 < 25$

32. $m^2 + 6m + 9 < 9$

33. $x^2 - x \leq 6$

34. $r^2 + r < 12$

35. $2k^2 - 9k > -4$

36. $3n^2 < -10 - 13n$

37. $x^3 - 4x \leq 0$

38. $r^3 - 9r \geq 0$

Solve the following inequalities. Give the answers in interval notation.

39. $\dfrac{m-3}{m+5} \le 0$

40. $\dfrac{r+1}{r-4} > 0$

41. $\dfrac{k-1}{k+2} > 1$

42. $\dfrac{a-6}{a+2} < -1$

43. $\dfrac{3}{x-6} \le 2$

44. $\dfrac{6}{5-3x} \le 2$

45. $\dfrac{7}{k+2} \ge \dfrac{1}{k+2}$

46. $\dfrac{5}{p+1} > \dfrac{12}{p+1}$

47. $\dfrac{3}{2r-1} > \dfrac{-4}{r}$

48. $\dfrac{-5}{3h+2} \ge \dfrac{5}{h}$

49. $\dfrac{4}{y-2} \le \dfrac{3}{y-1}$

50. $\dfrac{4}{n+1} < \dfrac{2}{n+3}$

Solve the following rational inequalities using methods similar to those used above.

51. $\dfrac{2x-3}{x^2+1} \ge 0$

52. $\dfrac{9x-8}{4x^2+25} < 0$

53. $\dfrac{(3x-5)^2}{(2x-5)^3} > 0$

54. $\dfrac{(5x-3)^3}{(8x-25)^2} \le 0$

55. $\dfrac{(2x-3)(3x+8)}{(x-6)^3} \ge 0$

56. $\dfrac{(9x-11)(2x+7)}{(3x-8)^3} > 0$

Use the discriminant to find the values of k where the following equations have real solutions.

57. $x^2 - kx + 8 = 0$ **58.** $x^2 + kx - 5 = 0$ **59.** $x^2 + kx + 2k = 0$ **60.** $kx^2 + 4x + k = 0$

A sign graph for an algebraic expression in x is a number line labeled with plus and minus signs to indicate where the expression is positive and negative. Match each of the following algebraic expressions with its sign graph below.

61. $x - 2$

62. $-(x-2)$

63. $(x-2)^2$

64. $-(x-2)^2$

(a) $\xrightarrow{\hspace{1cm} + \quad | \quad - \hspace{1cm}}$
2

(b) $\xrightarrow{\hspace{1cm} + \quad | \quad + \hspace{1cm}}$
2

(c) $\xrightarrow{\hspace{1cm} - \quad | \quad - \hspace{1cm}}$
2

(d) $\xrightarrow{\hspace{1cm} - \quad | \quad + \hspace{1cm}}$
2

In Exercises 65–68, find a quadratic inequality having the given solution.

65. $(-\infty, 2) \cup (5, +\infty)$ **66.** $(2, 5)$ **67.** $[-4, 3]$ **68.** $(-\infty, -3] \cup [4, +\infty)$

In Exercises 69–72, find a rational inequality having the given solution.

69. $(-\infty, -3] \cup (0, +\infty)$ **70.** $[-1, 5)$ **71.** $(4, 9]$ **72.** $(-\infty, 4) \cup [9, +\infty)$

A product will break even or produce a profit only if the revenue from selling the product at least equals the cost of producing it. Find all x-intervals in Exercises 73 and 74 for which the product will at least break even.

73. The cost to produce x units of wire is $C = 50x + 5000$, while the revenue is $R = 60x$.

74. The cost to produce x units of squash is $C = 100x + 6000$, while the revenue is $R = 500x$.

75. The commodities market is very unstable; money can be made or lost quickly on investments in soybeans, wheat, and so on. Suppose that an investor kept track of her total profit, P, at time t, measured in months, after she began investing, and found that $P = 4t^2 - 29t + 30$. Find the time intervals where she has been ahead. (*Hint: $t > 0$ in this case.*)

76. Suppose the velocity of an object is given by $v = 2t^2 - 5t - 12$, where t is time in seconds. (Here, t can be positive or negative.) Find the time intervals where the velocity is negative.

77. Oliver's video club charges an annual fee of $30 and rents videos for $2 per day. Stan's video club has no annual fee, but charges $3 per day rental fee. Let x be the number of days of rentals during the year.
  (a) Express the cost of renting the videos from Oliver in terms of x.
  (b) Express the cost of renting the videos from Stan in terms of x.
  (c) Find all x-intervals for which renting from Stan is cheaper.

78. Two companies A and B offer you a sales position. Both jobs are essentially the same, but company A pays a straight 7% commission on sales and company B pays $100 per week plus 5% commission. Let x be the weekly sales.
  (a) Express a week's earnings from company A in terms of x.
  (b) Express a week's earnings from company B in terms of x.
  (c) Find all x-intervals for which company A pays a better salary.

79. The formula for converting from Celsius to Fahrenheit temperature is $F = 9C/5 + 32$. What temperature range in °F corresponds to 0° to 30°C?

80. A projectile is fired from ground level. After t seconds its height above the ground is $220t - 16t^2$ feet. For what time period is the projectile at least 624 feet above the ground?

Solve each inequality in Exercises 81 and 82.

81. $\dfrac{1}{2} < \dfrac{3}{q} < \dfrac{2}{3}$

82. $\dfrac{5}{8} \le \dfrac{2}{r} \le \dfrac{11}{12}$

83. If $a > b > 0$, show that $1/a < 1/b$.

84. If $a > b$, is it always true that $1/a < 1/b$?

85. Suppose $a > b > 0$. Show that $a^2 > b^2$.

86. Suppose $a > b > 0$. Show that $(\sqrt{a} - \sqrt{b})^2 > 0$ and that $b < \sqrt{ab} < \dfrac{a + b}{2}$.

87. Let $b > 0$. When is $b^2 > b$?

88. If $a < b$ and $c < d$, show that $a + c < b + d$.

2.6 ABSOLUTE VALUE EQUATIONS AND INEQUALITIES

In this section we discuss methods of solving equations and inequalities involving absolute value. Recall from Chapter 1 that the absolute value of a number a, written $|a|$, gives the distance from a to 0 on a number line. By this definition, the absolute value equation $|x| = 3$ can be solved by finding all real numbers at a distance of 3 units from 0. As shown in the graph of Figure 12, there are two numbers satisfying this condition, 3 and -3, making $\{3, -3\}$ the solution set of the equation $|x| = 3$. This idea leads to the following properties that are useful for solving absolute value problems.

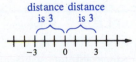

Figure 12

Properties of Absolute Value

If b is positive, then

$$|a| = b \quad \text{if and only if} \quad a = b \text{ or } a = -b.$$

For any values of a and b,

$$|a| = |b| \quad \text{if and only if} \quad a = b \text{ or } a = -b.$$

EXAMPLE 1   Solve $|p - 4| = 3$.

Let $a = p - 4$ and $b = 3$ in the first property above. (Note that $b = 3$ is positive as required.)

$$p - 4 = 3 \quad \text{or} \quad p - 4 = -3$$
$$p = 7 \qquad\qquad p = 1$$

Check that 7 and 1 are solutions of $|p - 4| = 3$.   ●

EXAMPLE 2   Solve $|4m - 3| = |m + 6|$.

Let $a = 4m - 3$ and $b = m + 6$ in the second property above.

$$4m - 3 = m + 6 \quad \text{or} \quad 4m - 3 = -(m + 6).$$

Solve each of these equations separately. Starting with $4m - 3 = m + 6$ gives

$$4m - 3 = m + 6$$
$$3m = 9$$
$$m = 3.$$

If $4m - 3 = -(m + 6)$, then

$$4m - 3 = -m - 6$$
$$5m = -3$$
$$m = -\frac{3}{5}.$$

The solutions of $|4m - 3| = |m + 6|$ are $-3/5$ and 3.   ●

Absolute Value Inequalities   The method used to solve absolute value equations can be generalized to solve inequalities with absolute value.

EXAMPLE 3   Solve each of the following.

(a) $|x| < 5$

Since absolute value gives the distance from a number to 0, the inequality $|x| < 5$ will be satisfied by all real numbers whose distance from 0 is less than

5. As shown in Figure 13, the solution includes all numbers from -5 to 5, or $-5 < x < 5$. In interval notation, the solution is written as the open interval $(-5, 5)$.

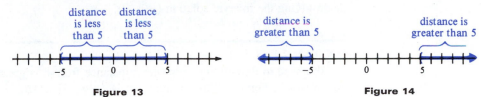

Figure 13   Figure 14

(b) $|x| > 5$

In a similar way, the solution of $|x| > 5$ is made up of all real numbers whose distance from 0 is greater than 5. This includes those numbers greater than 5 or those less than -5; that is,

$$x < -5 \quad \text{or} \quad x > 5.$$

In interval notation, the solution is written $(-\infty, -5) \cup (5, +\infty)$. A graph of the solution set is shown in Figure 14.   ●

The following properties of absolute value can be obtained from the definitions of absolute value and inequalities.

Additional Properties of Absolute Value

> If b is a positive number,
>
> **(a)** $|a| < b$ **if and only if** $-b < a < b;$
>
> **(b)** $|a| > b$ **if and only if** $a < -b$ **or** $a > b.$

NOTE   The $<$ symbol in part (a) leads to a single interval, $-b < a < b$, while the $>$ symbol in part (b) indicates two separate intervals.

EXAMPLE 4   Solve $|x - 2| < 5$.

To solve this inequality on the number line, find all real numbers whose distance from 2 is less than 5. As shown in Figure 15, the solution set is the interval $(-3, 7)$.

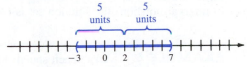

Figure 15

An algebraic solution of this inequality can be found using property (a) above. Let $a = x - 2$ and $b = 5$, so that $|x - 2| < 5$ if and only if

$$-5 < x - 2 < 5.$$

Adding 2 to each portion of this inequality produces

$$-3 < x < 7,$$

again giving the interval solution $(-3, 7)$. •

. . . .

EXAMPLE 5   Solve $|x - 8| \geq 1$.

We need to find all numbers whose distance from 8 is greater than or equal to 1. As shown in Figure 16 the solution set is $(-\infty, 7] \cup [9, +\infty)$. To find the solution using property (b) above, let $a = x - 8$ and $b = 1$ so that $|x - 8| \geq 1$ if and only if

$$x - 8 \leq -1 \quad \text{or} \quad x - 8 \geq 1.$$

Solve each inequality separately to get the same solution set, $(-\infty, 7] \cup [9, +\infty)$, as mentioned above. •

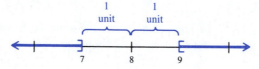

Figure 16

. . . .

EXAMPLE 6   Solve $|2 - 7m| - 1 > 4$.

In order to use the properties of absolute value given above, first add 1 to both sides; this gives

$$|2 - 7m| > 5.$$

Now use property (b) above. By this property, $|2 - 7m| > 5$ if and only if

$$2 - 7m < -5 \quad \text{or} \quad 2 - 7m > 5.$$

Solve each of these inequalities separately to get the solution set $(-\infty, -3/7) \cup (1, +\infty)$. •

. . . .

EXAMPLE 7   Solve $|2 - 5x| \geq -4$.

As stated in property (b) above, b must be a positive number, so property (b) does not apply here. However, since the absolute value of a number is always nonnegative, $|2 - 5x| \geq -4$ is always true. The solution set includes all real numbers. In interval notation, the solution set is $(-\infty, +\infty)$. •

. . . .

EXAMPLE 8   Write each statement using absolute value.

(a) k is not less than 5 units from 8.

Since the distance from k to 8, written $|k - 8|$ or $|8 - k|$, is not less than 5, the distance is greater than or equal to 5. Write this as

$$|k - 8| \geq 5.$$

(b) n is within .001 of 6.

This statement indicates that n may be .001 more than 6 or .001 less than 6. That is, the distance of n from 6 is no more than .001, written

$$|n - 6| \le .001. \quad \bullet$$

2.6 Exercises

Solve each of the following equations.

1. $|a - 2| = 1$ **2.** $|x - 3| = 2$ **3.** $|3m - 1| = 2$ **4.** $|4p + 2| = 5$

5. $|5 - 3x| = -3$ **6.** $|-3a + 7| = -3$ **7.** $\left|\dfrac{z - 4}{2}\right| = 5$ **8.** $\left|\dfrac{m + 2}{2}\right| = 7$

9. $\left|\dfrac{5}{r - 3}\right| = 10$ **10.** $\left|\dfrac{3}{2h - 1}\right| = 4$ **11.** $\left|\dfrac{6y + 1}{y - 1}\right| = 3$ **12.** $\left|\dfrac{3a - 4}{2a + 3}\right| = 1$

13. $|2k - 3| = |5k + 4|$ **14.** $|p + 1| = |3p - 1|$ **15.** $|4 - 3y| = |7 + 2y|$ **16.** $|2 + 5a| = |4 - 6a|$

Solve each of the following inequalities. Give the solution in interval notation.

17. $|x| \le 3$ **18.** $|y| \le 10$ **19.** $|m| > 1$ **20.** $|z| > 4$

21. $|a| < -2$ **22.** $|b| > -5$ **23.** $|x| - 3 \le 7$ **24.** $|r| + 3 \le 10$

25. $|2x + 5| < 3$ **26.** $\left|x - \dfrac{1}{2}\right| < 2$ **27.** $|3m - 2| > 4$ **28.** $|4x - 6| > 10$

29. $|3z + 1| \ge 7$ **30.** $|8b + 5| \ge 7$ **31.** $\left|5x + \dfrac{1}{2}\right| - 2 < 5$ **32.** $\left|x + \dfrac{2}{3}\right| + 1 < 4$

33. $\left|\dfrac{2x + 3}{x}\right| < 1$ **34.** $\left|\dfrac{7 - 4y}{y}\right| < 3$ **35.** $\left|\dfrac{6 + 2y}{y - 5}\right| > 2$ **36.** $\left|\dfrac{3 - 3p}{p + 4}\right| > 5$

37. $\left|\dfrac{2}{q - 2}\right| \le 4$ **38.** $\left|\dfrac{5}{t - 1}\right| \le 3$

Solve the equations in Exercises 39–48.

39. $|x + 1| = 2x$ **40.** $|m - 3| = 4m$ **41.** $|2k - 1| = k + 2$

42. $|3r + 5| = r + 3$ **43.** $|p| = |p|^2$ **44.** $5|m| = |m|^2$

45. $|1 - 6q|^2 - 4|1 - 6q| - 45 = 0$ **46.** $|6a - 5|^2 + 4|6a - 5| - 12 = 0$

47. $6|3r + 4|^2 + |3r + 4| = 2$ **48.** $5|3 - 4m|^2 - 6|3 - 4m| = 8$

49. Is $|a - b|^2$ always equal to $(b - a)^2$?

Solve each inequality in Exercises 50–53.

50. $|r^2 - 3| < 1$ **51.** $|m^2 - 7| \le 2$

52. $|z^2 - 8| \ge 7$ **53.** $|y^2 - 6| > 5$

54. Solve $|m + 2| \le |4m - 1|$ by first dividing each side by $|4m - 1|$.

55. Solve $|3x - 1| < 2|2x + 1|$.   **56.** Solve $|2k + 3| > 2|k - 1|$.

Write each statement in Exercises 57–64 using absolute value notation. For example, write
"k is at least 4 units from 1" as $|k - 1| \geq 4$.

57. x is within 4 units of 2.   **58.** m is no more than 8 units from 9.   **59.** z is no less than 2 units from 12.

60. p is at least 5 units from 9.   **61.** k is 6 units from 1.   **62.** r is 5 units from 3.

63. If x is within .0004 units of 2, then y is within .00001 units of 7.

64. y is within 10^{-6} units of 10 whenever x is within 2×10^{-4} units of 5.

65. If $|x - 2| < 3$, find the values of m and n such that $m < 3x + 5 < n$.

66. If $|x + 8| < 16$, find the values of p and q so that $p < 2x - 1 < q$.

67. If $|x - 1| < 10^{-6}$, show that $|7x - 7| < 10^{-5}$.

68. Suppose $|x - k| < 10^{-9}$ and show that $|9x - 9k| < 10^{-8}$.

2.7 VARIATION

In many applications of mathematics, it is necessary to write relationships between variables. For example, in chemistry the ideal gas law shows how temperature, pressure, and volume are related. In physics, various formulas in optics show how the focal length of a lens and the size of an image are related.

   If the quotient of two variables is constant, then one variable *varies directly* or is *directly proportional* to the other. This can be stated differently as follows.

Directly Proportional

> y **varies directly** as x, or y **is directly proportional** to x, means that a nonzero real number k (called the **constant of variation**) exists such that
>
> $$y = kx.$$

EXAMPLE 1   Suppose the value of y varies directly as the value of x, and that $y = 12$ when $x = 5$. Find y when $x = 21$.

   If y varies directly as x, then a real number k exists such that

$$y = kx.$$

To find the value of k, use the given information: $y = 12$ when $x = 5$. Replacing y with 12 and x with 5 gives

$$12 = k \cdot 5 \quad \text{or} \quad k = \frac{12}{5}.$$

In this example, then, the relationship between y and x is given by

$$y = \frac{12}{5}x.$$

Now find y when x is 21.

$$y = \frac{12}{5} \cdot 21 = \frac{252}{5}. \quad \bullet$$

Sometimes y varies as a power of x.

Varies as *n*th Power

> Let n be a positive real number. Then y **varies directly as the *n*th power** of x, or y is **directly proportional to the *n*th power** of x, if there exists a real number k such that
>
> $$y = kx^n.$$

The phrase "directly proportional" is sometimes abbreviated to just "proportional."

For example, the area of a square of side x is given by the formula $A = x^2$, so that the area *varies directly as the square* of the length of a side. Here $k = 1$.

The case where y increases as x decreases is an example of *inverse variation*.

Inverse Variation

> Let n be a positive real number. Then y **varies inversely as the *n*th power** of x means that there exists a real number k such that
>
> $$y = \frac{k}{x^n}.$$
>
> If $n = 1$, then $y = k/x$, and y **varies inversely** as x.

EXAMPLE 2   In a certain manufacturing process, the cost of producing a single item varies inversely as the square of the number of items produced. If 100 items are produced, each costs \$2. Find the cost per item if 400 items are produced.

We can let x represent the number of items produced and y the cost per item, and write

$$y = \frac{k}{x^2}$$

for some nonzero constant k. Since $y = 2$ when $x = 100$,

$$2 = \frac{k}{100^2} \quad \text{or} \quad k = 20{,}000.$$

Thus, the relationship between x and y is given by

$$y = \frac{20{,}000}{x^2}.$$

When 400 items are produced, the cost per item is given by

$$y = \frac{20{,}000}{400^2}$$

$$= .125, \text{ or } 12.5¢. \quad \bullet$$

One variable may depend on more than one other variable. Such variation is called *combined variation*. If a variable depends on the product of two or more other variables, we refer to that as *joint variation*.

<table>
<tr><td>**Joint**
Variation</td><td>y **varies jointly** as the nth power of x and the mth power of z if there exists a real number k such that

$$y = kx^n z^m.$$</td></tr>
</table>

The next example shows combined variation.

EXAMPLE 3   The weight a horizontal beam will support varies jointly as its width and the square of its height, and inversely as its length. A 16-foot-long beam, 4 in. wide and 12 in. high, can support 2000 lb. Find the load a beam that is 6 in. wide, 12 in. high, and 20 ft long will support.

Let W represent the weight supported by the beam and l, w, and h represent its length, width, and height respectively. Then W depends on the three variables, l, w, and h. Since W varies jointly as w and the square of h, and inversely as l,

$$W = \frac{kwh^2}{l}.$$

Find k by substituting the given values of W, w, h, and l into this equation. It is not necessary to express w, h, and l in the same units (feet or inches) first. Here, $W = 2000$, $w = 4$, $h = 12$, and $l = 16$.

$$2000 = \frac{k(4)(12)^2}{16}$$

$$32000 = 4(144)k$$

$$\frac{500}{9} = k$$

Therefore, the formula for W is

$$W = \frac{\frac{500}{9} wh^2}{l}$$

or

$$W = \frac{500wh^2}{9l}.$$

For $w = 6$, $h = 12$, and $l = 20$,

$$W = \frac{500(6)(144)}{9(20)}$$

$$W = 2400.$$

The beam will support a load of 2400 lb.   ●

The steps involved in solving a problem in variation can be summarized as follows.

Solving Variation Problems

1. Write, in an algebraic form, the general relationship among the variables. Use the constant k.
2. Substitute given values of the variables and find the value of k.
3. Substitute this value of k into the formula of Step 1, thus obtaining a specific formula.
4. Solve for the required unknown.

Notice how these steps were followed in Examples 1 through 3.

2.7 Exercises

Express each of the following as an equation.

1. a varies directly as b.

2. m is proportional to n.

3. x is inversely proportional to y.

4. p varies inversely as y.

5. r varies jointly as s and t.

6. R is proportional to m and p.

7. w is proportional to x^2 and inversely proportional to y.

8. c varies directly as d and inversely as f^2 and g.

Write each of the following formulas as an English phrase using the words varies *or* proportional.

9. $c = 2\pi r$, where c is the circumference of a circle of radius r

10. $d = s/5$, where d is the approximate distance (in miles) away of a storm and s is the number of seconds between seeing lightning and hearing thunder

11. $v = d/t$, where v is the average speed when traveling d miles in t hours

12. $d = 1/(4\pi nr^2)$, where d is the average distance a gas atom of radius r travels between collisions and n is the number of atoms per unit volume

13. $s = kx^3$, where s is the strength of a muscle of length x

14. $f = mv^2/r$, where f is the centripetal force of an object of mass m moving along a circle of radius r at velocity v

Solve each of the following.

15. If m varies directly as x and y, and $m = 10$ when $x = 4$ and $y = 7$, find m when $x = 11$ and $y = 8$.

16. Suppose m varies directly as z and p. If $m = 10$ when $z = 3$ and $p = 5$, find m when $z = 5$ and $p = 7$.

17. Let a be proportional to m and n^2, and inversely proportional to y^3. If $a = 9$ when $m = 4$, $n = 9$, and $y = 3$, find a if $m = 6$, $n = 2$, and $y = 5$.

18. If y varies directly as x, and inversely as m^2 and r^2, and $y = 5/3$ when $x = 1$, $m = 2$, and $r = 3$, find y if $x = 3$, $m = 1$, and $r = 8$.

19. Suppose m varies directly as p^2 and r^4. If p doubles and r triples, how does m change?

20. Let z vary directly as y^3 and inversely as x^2. If y doubles and x is halved, how does z change?

21. The distance a body falls from rest varies directly as the square of the time it falls (disregarding air resistance). If an object falls 1024 ft in 8 seconds, how far will it fall in 12 seconds?

22. Hooke's law for an elastic spring states that the distance a spring stretches varies directly as the force applied. If a force of 15 lb stretches a certain spring 8 in., how much will a force of 30 lb stretch the spring? See the figure.

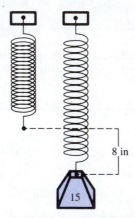

23. In electric current flow, it is found that the resistance (measured in units called ohms) offered by a fixed length of wire of a given material varies inversely as the square of the diameter of the wire. If a wire .01 inches in diameter has a resistance of .4 ohm, what is the resistance of a wire of the same length and material but .03 inches in diameter?

24. The illumination produced by a light source varies inversely as the square of the distance from the source. The illumination of a light source at 5 m is 70 candela. What is the illumination 12 m from the source?

25. The pressure exerted by a certain liquid at a given point is proportional to the depth of the point below the surface of the liquid. If the pressure 20 m below the surface is 70 km per sq cm, what pressure is exerted 40 m below the surface?

26. The distance that a person can see to the horizon from a point above the surface of the earth varies directly as the square root of the height. A person on a hill 121 m high can see for 15 km to the horizon. How far is the horizon from a hill 900 m high?

27. Simple interest varies jointly as principal and time. If $1000 left at interest for 2 yr earned $110, find the amount of interest earned by $5000 for 5 yr.

28. The volume of a right circular cylinder is jointly proportional to the square of the radius of the circular base and to the height. If the volume is 300 cu cm when the height is 10.62 cm and the radius is 3 cm, find the volume for a cylinder with a radius of 4 cm and a height of 15.92 cm.

29. The Downtown Construction Company is designing a building whose roof rests on round concrete pillars. The company's engineers know that the maximum load a cylindrical column of circular cross section can hold varies directly as the fourth power of the diameter and inversely as the square of the height. If a 9-meter column 1 m in diameter will support a load of 8 metric tons, how many metric tons will be supported by a column 12 m high and 2/3 m in diameter?

30. The force needed to keep a car from skidding on a curve varies inversely as the radius of the curve and jointly as the weight of the car and the square of the speed. It takes 3000 lb of force to keep a 2000-lb car from skidding on a curve of radius 500 ft at 30 mph. What force is needed to keep the same car from skidding on a curve of radius 800 ft at 60 mph?

31. The period of a pendulum varies directly as the square root of the length of the pendulum and inversely as the square root of the acceleration due to gravity. Find the period when the length is 121 cm and the acceleration due to gravity is 980 cm per sec^2, if the period is 6π seconds when the length is 289 cm and the acceleration due to gravity is 980 cm per sec^2.

32. The pressure on a point in a liquid is directly proportional to the distance from the surface to the point. In a certain liquid the pressure at a depth of 4 m is 60 kg per m^2. Find the pressure at a depth of 10 m.

33. The volume V of a gas varies directly as the temperature T and inversely as the pressure P. If V is 10 when T is 280 and P is 6, find V if T is 300 and P is 10.

34. Under certain conditions, the length of time that it takes for fruit to ripen during the growing season varies inversely as the average maximum temperature during the season. If it takes 25 days for fruit to ripen with an average maximum temperature of 80°, find the number of days it would take at 75°.

35. The horsepower needed to run a boat through water varies as the cube of the speed. If 80 horsepower are needed to go 15 km per hr in a certain boat, how many horsepower would be needed to go 30 km per hr?

36. According to Poiseuille's law, the resistance to flow of a blood vessel, R, is directly proportional to the length, l, and inversely proportional to the fourth power of the radius, r. If $R = 25$ when $l = 12$ and $r = .2$, find R as r increases to .3, while l is unchanged.

37. The Stefan-Boltzmann law says that the radiation of heat from an object is directly proportional to the fourth power of the Kelvin temperature of the object. For a certain object, $R = 213.73$ at room temperature (293° Kelvin). Find R if the temperature increases to 335° Kelvin.

38. Suppose a nuclear bomb is detonated at a certain site. The effects of the bomb will be felt over a distance from the point of detonation that is directly proportional to the cube root of the yield of the bomb. Suppose a 100-kiloton bomb has certain effects to a radius of 3 km from the point of detonation. Find the distance that the effects would be felt for a 1500-kiloton bomb.

39. When the brakes of a car are applied, the speed that the car was traveling is proportional to the square root of the distance that the car travels before coming to a stop. Suppose that a car moving at 60 km/hr will travel 18 m after the brakes are applied. Determine the formula giving speed as a function of the stopping distance.

40. Assume that a person's weight increases directly as the cube of his height. Find the weight of a 20-inch, 7-pound baby who grows up to be an adult 67 in. tall. (How reasonable does our assumption about weight and height seem?)

41. The cost of pizza varies directly as the area of the pizza. How does the cost vary with the radius of the pizza? If a pizza with a radius of 15 cm costs $7, find the cost of a pizza having a radius of 22.5 cm. (You might want to do some research at a nearby pizza establishment and see if this assumption is reasonable.)

2 | CHAPTER SUMMARY

Key Words

To understand the concepts presented in this chapter, you should know the meaning and use of the following words and expressions. For easy reference, the section in the chapter in which a word or expression was first used is given with each item.

2.1 solution
root
satisfy
equivalent equations
linear equation
empty set
identity
conditional equation
2.2 solving for a specified
variable

2.3 quadratic equation
zero-factor property
square root property
completing the square
quadratic formula
discriminant
2.4 quadratic in form
2.5 inequality
interval
interval notation
open interval

closed interval
quadratic inequality
rational inequality
union
2.6 absolute value inequality
properties of absolute
value
2.7 varies directly
directly proportional
constant of variation
varies inversely

Review Exercises

Solve each of the following equations.

1. $5y - 2(y + 4) = 3(2y + 1)$

2. $\dfrac{x - 3}{2} = \dfrac{2x + 1}{3}$

3. $\dfrac{p}{2} - \dfrac{3p}{4} = 8 + \dfrac{p}{3}$

4. $\dfrac{2r}{5} - \dfrac{r - 3}{10} = \dfrac{3r}{5}$

5. $\dfrac{2z}{5} - \dfrac{4z - 3}{10} = \dfrac{1 - z}{10}$

6. $\dfrac{p}{p + 2} - \dfrac{3}{4} = \dfrac{2}{p + 2}$

7. $(x - 3)(2x + 1) = 2(x + 2)(x - 4)$

8. $(3k + 1)^2 = 6k^2 + 3(k - 1)^2$

Solve for x.

9. $3(x + 2b) + a = 2x - 6$

10. $9x - 11(k + p) = x(a - 1)$

11. $\dfrac{x}{m - 2} = kx - 3$

12. $r^2x - 5x = 3r^2$

Solve each of the following for the indicated variable.

13. $2a + ay = 4y - 4a$ for y

14. $F = \dfrac{9}{5}C + 32$ for C

15. $A = P + Pi$ for P

16. $A = I\left(1 - \dfrac{j}{n}\right)$ for j

17. $\dfrac{1}{k} = \dfrac{1}{r_1} + \dfrac{1}{r_2}$ for r_1

18. $m = \dfrac{Ft}{\sqrt{I} - \sqrt{2}}$ for t

19. $V = \pi r^2 L$ for L

20. $P(r + R)^2 = E^2R$ for P

21. $\dfrac{xy^2 - 5xy + 4}{3x} = 2p$ for x

22. $\dfrac{zx^2 - 5x + z}{z + 1} = 9$ for z

Solve each of the following problems.

23. A stereo is on sale for 15% off. The sale price is $425. What was the original price?

24. To make a special mix for Valentine's Day, the owner of a candy store wants to combine chocolate hearts which sell for $5 per lb with candy kisses which sell for $3.50 per lb. How many pounds of each should be used to get 30 lb of a mix which can be sold for $4.50 per lb?

25. Two people are stuffing envelopes for a political campaign. Working together, they can stuff 5000 envelopes in 4 hr. If the first person worked alone, it would take 6 hr to stuff the envelopes. How long would it take the second person, working alone, to stuff the envelopes?

26. Maria can ride her bike to the university library in 20 min. The trip home, which is all uphill, takes her half an hour. If her rate is 8 mph slower on the return trip, how far does she live from the library?

Solve each equation.

27. $(b + 7)^2 = 5$

28. $(3y - 2)^2 = 8$

29. $2a^2 + a - 15 = 0$

30. $12x^2 = 8x - 1$

31. $2q^2 - 11q = 21$

32. $3x^2 + 2x = 16$

33. $2 - \dfrac{5}{p} = \dfrac{3}{p^2}$

34. $\dfrac{4}{m^2} = 2 + \dfrac{7}{m}$

Evaluate the discriminant for each of the following, and use it to predict the type of solutions for the equation.

35. $8y^2 = 2y - 6$ **36.** $6k^2 - 2k = 3$ **37.** $16r^2 + 3 = 26r$

38. $8p^2 + 10p = 7$ **39.** $25z^2 - 110z + 121 = 0$ **40.** $4y^2 - 8y + 17 = 0$

Solve each word problem.

41. Calvin wants to fence off a rectangular playground next to an apartment building. Since the building forms one boundary, he needs to fence only the other three sides. The area of the playground is to be 11,250 square meters. He has enough material to build 325 m of fence. Find the length and width of the playground.

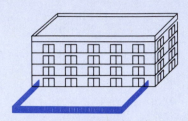

Exercise 41

42. Steve and Paula sell pies. It takes Paula one hour longer than Steve to bake a day's supply of pies. Working together, it takes them 6/5 hr to bake the pies. How long would it take Steve working alone?

Solve each equation.

43. $4a^4 + 3a^2 - 1 = 0$ **44.** $4(y - 2)^2 - 9(y - 2) + 2 = 0$ **45.** $(2z + 3)^{2/3} + (2z + 3)^{1/3} = 6$

46. $\sqrt{x} - 7 = 10$ **47.** $\sqrt{2p + 1} = 8$ **48.** $5\sqrt{m} = \sqrt{3m + 2}$

49. $\sqrt{4y - 2} = \sqrt{3y + 1}$ **50.** $\sqrt{2x + 3} = x + 2$ **51.** $\sqrt{p + 2} = 2 + p$

52. $\sqrt{k} = \sqrt{k + 3} - 1$ **53.** $\sqrt{x^2 + 3x} - 2 = 0$ **54.** $\sqrt[3]{2r} = \sqrt[3]{3r + 2}$

55. $\sqrt[3]{6y + 2} = \sqrt[3]{4y}$ **56.** $(x - 2)^{2/3} = x^{1/3}$ **57.** $\sqrt{3 + x} = \sqrt{3x + 7} - 2$

58. $\sqrt{4 + 3y} = \sqrt{y + 5} + 1$

Solve each inequality in Exercises 59–74. Write solutions in interval notation.

59. $-5z - 4 \geq 3(2z - 5)$ **60.** $-(4a + 5) < 3a - 2$ **61.** $3r - 4 + r > 2(r - 1)$

62. $7p - 2(p - 3) \leq 5(2 - p)$ **63.** $5 \leq 2x - 3 \leq 7$ **64.** $-8 < 3a - 5 < -1$

65. $-5 < \dfrac{2p - 1}{-3} \leq 2$ **66.** $3 < \dfrac{6z + 5}{-2} < 7$ **67.** $x^2 + 3x - 4 \leq 0$

68. $p^2 + 4p > 21$ **69.** $z^3 - 16z \leq 0$ **70.** $2r^3 - 3r^2 - 5r < 0$

71. $\dfrac{3a - 2}{a} > 4$ **72.** $\dfrac{5p + 2}{p} < -1$ **73.** $\dfrac{3}{r - 1} \leq \dfrac{5}{r + 3}$ **74.** $\dfrac{3}{x + 2} > \dfrac{2}{x - 4}$

75. If $a < b$, on what x-interval is $(x - a)(x - b)$ positive? Negative? Where is the product zero?

76. On what x-interval is $(x - a)^2$ positive? Negative? Where is it zero?

Work the following word problems.

77. Steve and Paula (from Exercise 42 above) have found that the profit from their pie shop is given by $P = -x^2 + 28x + 60$, where x is the number of units of pies sold daily. For what values of x is the profit positive?

78. A projectile is thrown upward. Its height in feet above the ground after t seconds is $320t - 16t^2$. **(a)** After how many seconds in the air will it hit the ground? **(b)** During what time interval is the projectile more than 576 ft above the ground?

Solve each equation.

79. $|2 - y| = 3$

80. $\left|\dfrac{r - 5}{3}\right| = 6$

81. $\left|\dfrac{7}{2 - 3a}\right| = 9$

82. $\left|\dfrac{8r - 1}{3r + 2}\right| = 7$

83. $|5r - 1| = |2r + 3|$

84. $|k + 7| = |k - 8|$

Solve each inequality in Exercises 85–94. Write solutions in interval notation.

85. $|m| \le 7$

86. $|z| > -1$

87. $|b| \le -1$

88. $|5m - 8| \le 2$

89. $|7k - 3| < 5$

90. $|2p - 1| > 2$

91. $|3r + 7| > 5$

92. $\left|\dfrac{1}{k + 3}\right| > 3$

93. $|3x + 2| < -4$

94. $|2x + 1| > -2$

95. Let $f = 2x\left[\dfrac{1}{2}(x^2 + 1)^{-1/2}(2x)\right] + 2(x^2 + 1)^{1/2}$. Find all intervals where **(a)** $f > 0$, and **(b)** $f < 0$.

96. Let $g = \dfrac{(x^2 + 5)^{1/2} - x[(1/2)(x^2 + 5)^{-1/2}(2x)]}{x^2 + 5}$. Find all intervals where **(a)** $g > 0$, and **(b)** $g < 0$.

Set up Exercises 97 and 98.

97. A book is to contain 36 sq in. of printed material per page, with margins of 1 in. along the sides, and $1\frac{1}{2}$ in. along the top and bottom. Let x represent the width of the printed area, and write an expression for the area of the entire page.

98. A hunter is at a point on a riverbank. He wants to get to his cabin, located 3 mi north and 8 mi west (see the figure). He can travel 5 mph on the river but only 2 mph on this very rocky ground. If he travels $8 - x$ mi along the river and then walks in a straight line to the cabin, find an expression for the total time that he travels.

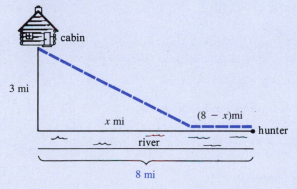

In Exercises 99 and 100, solve for x and express the solution in terms of intervals.

99. $\dfrac{1}{|x - 4|} < \dfrac{1}{|x + 7|}$

100. $\dfrac{1}{|x - 3|} - \dfrac{1}{|x + 4|} \ge 0$

101. Find the smallest value of M such that $|1/x| \le M$ for all x in the interval $[2, 7]$.

102. Find the smallest value of M such that $|1/(x + 7)| \le M$ for all x in the interval $(-4, 2)$.

Exercises 99–102 from *Calculus with Analytic Geometry* by Howard Anton, pp. 22, 23. Copyright © 1988 by Anton Textbooks, Inc. Reprinted by permission of John Wiley & Sons, Inc.

Write each of the statements below as an equation.

103. m varies directly as the square of z.

104. y varies inversely as r and directly as the cube of p.

105. Y varies jointly as M and the square of N and inversely as the cube of X.

106. A varies jointly as the third power of t and the fourth power of s, and inversely as p and the square of h.

Solve each problem below.

107. Suppose r varies directly as x and inversely as the square of y. If r is 10 when x is 5 and y is 3, find r when x is 12 and y is 4.

108. Suppose m varies jointly as n and the square of p, and inversely as q. If m is 20 when n is 5, p is 6, and q is 18, find m when n is 7, p is 11, and q is 2.

109. Suppose Z varies jointly as the square of J and the cube of M, and inversely as the fourth power of W. If Z is 125 when J is 3, M is 5 and W is 1, find Z if J is 2, M is 7, and W is 3.

110. The power a windmill obtains from the wind varies directly as the cube of the wind velocity. If a wind blowing at 10 km per hr produces 10,000 units of power, how much power is produced by a wind of 15 km per hr?

CUMULATIVE REVIEW EXERCISES
CHAPTERS 1 AND 2

[1.1] *Choose all words from the following list that describe the number given: (a) whole number,   (b) integer,   (c) rational number,   (d) irrational number,   (e) real number.*

1. $-11/3$ **2.** -5 **3.** $\sqrt{49}$ **4.** $\sqrt{3}$

Write the following numbers, in numerical order, from smallest to largest.

5. $|-8 + 2|, \quad -|3|, \quad -|-2|, \quad -|-2| + (-3), \quad -|-8| - |-6|$

6. $-2 - |-4|, \quad -3 + |2|, \quad -4, \quad -5 + |-3|$

Write each of the following without absolute value bars.

7. $|7 - \sqrt{5}|$ **8.** $|2 - \sqrt{11}|$ **9.** $|m - 3y|$, if $m/3 > y$ **10.** $|5 + y^2|$

[1.2] *Simplify each of the following. Write the results with only positive exponents. Assume all variables represent positive real numbers.*

11. $(-5)^{-3}$ **12.** $2^{-4} + 3^{-2}$ **13.** $(r^{-3})(r^{-2})(r^5)$

14. $\dfrac{p^3 z^2}{p^{-1} z^{-3}}$ **15.** $[(3^{-2})^2]^{-1}$ **16.** $\dfrac{(2x^3)^{-2}(2^2 x^5)^{-1}}{(2x^4)^{-3}}$

Rewrite each of the following, using only positive exponents. Assume that all variables represent positive real numbers, and that variables used as exponents represent rational numbers.

17. $32^{-6/5}$

18. $(625z^8)^{1/2}$

19. $(a - b)^{2/3} \cdot (a - b)^{-5/3}$ $(a > b)$

20. $(7k^{3/4}x^{1/8})(9k^{7/4}x^{3/8})$

21. $(3a^{-2/3}b^{5/3})(8a^2b^{-10/3})$

22. $z^{1+r} \cdot z^{3-2r}$

23. $\dfrac{r^{1/3} \cdot s^{5/3} \cdot t^{1/2}}{r^{-2/3} \cdot s^2 \cdot t^{-3/2}}$

24. $\dfrac{q^r \cdot q^{-5r}}{q^{-3r}}$

[1.3] *Find each of the following.*

25. $(-9m^2 + 11m - 2) + (4m^2 - 8m + 7)$

26. $(-7z^2 + 8z - 1) - (-4z^2 - 7z - 9)$

27. $(k - 7)(3k - 8)$

28. $(9w + 5)^2$

29. $(3k - 5)^3$

30. $(4k + 3)(2k^2 + 5k + 6)$

31. $(y + z + 2)(3y - 2z + 5)$

32. $(2r + 3s + 3)(3r - s + 2)$

33. $\dfrac{15x^4 + 30x^3 + 12x^2 - 9}{3x}$

34. $\dfrac{16a^6 + 24a^5 - 48a^4 + 12a}{8a^2}$

[1.4] *Factor as completely as possible.*

35. $3m^3 + 9m + 15m^5$

36. $r^2 + 15r + 54$

37. $6q^2 - q - 12$

38. $10b^2 - 19b - 15$

39. $8a^3 + 125$

40. $64p^6 - 27q^9$

41. $rs + rt - ps - pt$

42. $2m + 6 - am - 3a$

43. $(z - 4)^2 - (z + 4)^2$

44. $6(r + s)^2 + 13(r + s) - 5$

[1.5] *Perform each of the following operations.*

45. $\dfrac{2x - 2}{3} \cdot \dfrac{6x - 6}{(x - 1)^3}$

46. $\dfrac{3m - 15}{4m - 20} \cdot \dfrac{m^2 - 10m + 25}{12m - 60}$

47. $\dfrac{3z^2 + z - 2}{4z^2 - z - 5} \div \dfrac{3z^2 + 11z + 6}{4z^2 + 7z - 15}$

48. $\dfrac{1}{a + 1} - \dfrac{1}{a - 1}$

49. $\dfrac{m + 3}{m - 7} + \dfrac{m + 5}{2m - 14}$

50. $\dfrac{1}{x^2 + x - 12} - \dfrac{1}{x^2 - 7x + 12}$

[1.6] *Simplify each of the following. Assume that all variables represent positive real numbers.*

51. $\sqrt{1000}$

52. $-\sqrt[4]{32}$

53. $\sqrt{24 \cdot 3^2 \cdot 2^4}$

54. $\sqrt[3]{25 \cdot 3^4 \cdot 5^3}$

55. $\sqrt{50p^7q^8}$

56. $\sqrt[4]{1875h^5y^6q^9}$

57. $\dfrac{\sqrt[3]{a^3b^7c^7} \cdot \sqrt[3]{a^6b^8c^9}}{\sqrt[3]{a^7b^3c^5}}$

58. $\dfrac{15}{\sqrt{3}} - \dfrac{2}{\sqrt{27}} + \dfrac{4}{\sqrt{12}}$

59. $\dfrac{1}{2 - \sqrt{7}}$

60. $\dfrac{\sqrt{p} + \sqrt{p + 1}}{\sqrt{p} - \sqrt{p + 1}}$

[1.7] *Perform the following operations.*

61. $(-3 + 5i) - (-9 + 3i)$

62. $(-6 + 2i) + (-1 + 7i)$

63. $(1 + 3i)(2 - 5i)$

64. $(-2 + 5i)^2$

65. $i(2 + 3i)^2$

66. $(1 - 2i)^3$

67. $\dfrac{4 + 3i}{1 + i}$

68. $\dfrac{3 + 7i}{5 - 3i}$

69. $\dfrac{i}{3 + 2i}$

70. $\dfrac{5 - 2i}{12 + i} - \dfrac{2 - 9i}{4 + 3i}$

Simplify each expression in Exercises 71–74.

71. $\sqrt{-400}$ **72.** $-\sqrt{-39}$ **73.** $\sqrt{-3} \cdot \sqrt{-7}$ **74.** $\sqrt{-11} \cdot \sqrt{-6}$

75. Find i^{15}. **76.** Find i^{245}.

[2.1] *Solve each equation.*

77. $\dfrac{-7r}{2} + \dfrac{3r - 5}{4} = \dfrac{2r + 5}{4}$

78. $\dfrac{1}{z - 5} = 2 - \dfrac{3}{z - 5}$

79. $(3x - 4)^2 - 5 = 3(x + 5)(3x + 2)$

80. $\dfrac{x}{b - 1} = 5x + 3b$ for x.

81. $v = v_0 + gt$ for t

82. $s = s_0 + gt^2 + k$ for t^2

[2.2] *Solve each of the following.*

83. A triangle has a perimeter of 54 cm. Two of the sides of the triangle are equal in length, with the third side 6 cm shorter than either of the two equal sides. Find the lengths of the three sides of the triangle.

84. After a lottery win, John has $90,000 to invest. He puts part of the money in a certificate of deposit at 8%, and the rest into a real estate scheme paying 12%. The total annual income from the investments is $8800. How much does he have invested at each rate?

85. Suppose $40,000 is invested at 7%. How much additional money would have to be invested at 11% to make the yield on the entire amount equal to 9.4%?

86. A student needs 25% acid for an experiment. How many ml of 50% acid should be mixed with 40 ml of 15% acid to get the necessary 25% acid?

87. How many pounds of coffee worth $6 per lb should be mixed with 20 lb of coffee selling for $4.50 per lb to get a mixture that can be sold for $5 per lb?

88. Tom can run 6 mph, while Roy runs 4 mph. If they start running at the same time, how long will it take them to be 3/4 miles apart?

89. A boat can go 15 km upstream in the same time that it takes to go 27 km downstream. The speed of the current is 2 km per hr. Find the speed of the boat in still water.

90. An inlet pipe can fill a swimming pool in 1 day. An outlet can empty the pool in 36 hr. Suppose that both the inlet and the outlet were opened. How long would it take to fill the pool 5/8 full?

[2.3] *Solve each equation.*

91. $(5r - 3)^2 = 7$ **92.** $(7q + 2)^2 = 40$ **93.** $8k^2 + 14k + 3 = 0$

94. $2s^2 + 2s = 3$ **95.** $4z^2 - 4z - 5 = 0$ **96.** $12r^2 = 4r$

97. $x^2 - 2x + 2 = 0$ **98.** $9k^2 - 12k + 8 = 0$ **99.** $12y^2 - 4y + 3 = 0$

100. $25z^2 + 30z + 11 = 0$

Solve each of the following problems.

101. Find two consecutive odd integers whose product is -1.

102. The area of a field is 9600 m². One side is 40 m longer than the other side. Find the length and width of the field.

103. Person A can do a job in 5 hr. When A is working with person B, the job takes 3 hr. How long would it take B working alone to do the job?

104. One leg of a right triangle is 3 cm longer than three times the length of the shorter leg. The hypotenuse is 1 cm longer than the longer leg. Find the lengths of the sides of the triangle.

[2.4] *Solve each equation.*

105. $2z^4 - 7z^2 + 3 = 0$

106. $-(r + 1)^2 - 3(r + 1) + 3 = 0$

107. $(m - 1)^{2/3} + 3(m - 1)^{1/3} = 10$

108. $6z^{-2} + 7z^{-1} + 2 = 0$

109. $\sqrt{3s} - s = -6$

110. $\sqrt{r + 3} = \sqrt{2r - 1} - 1$

111. $(z^2 - 18z)^{1/4} = 0$

112. $p^{2/3} = 9p^{1/3}$

[2.5–2.6] *Solve each inequality. Write all solutions in interval notation.*

113. $12m - 17 \geq 8m + 7$

114. $-h \leq 6h + 30$

115. $-15 < -2y + 3 < -1$

116. $z^2 + 6z + 16 < 8$

117. $y^2 + 6y \geq 0$

118. $p^2 < -1$

119. $2t^3 - 2t^2 - 12t \leq 0$

120. $\dfrac{a - 6}{a + 2} < -1$

121. $\dfrac{3}{y + 6} \geq \dfrac{1}{y - 2}$

122. $|x - 1/2| < 2$

123. $|2x + 5| > 3$

124. $|3 - 5k| > 2$

Solve each of the following equations.

125. $|a - 2| = 1$

126. $\left|\dfrac{6y + 1}{y - 1}\right| = 3$

127. $|3z - 1| = |2z + 5|$

128. $|m + 2| = |m + 5|$

[2.7] *Write each of the statements below as an equation.*

129. b is inversely proportional to a.

130. r is proportional to s.

131. g varies jointly as the square of m and the cube of n.

132. h varies directly as the fifth power of t and inversely as the cube of v.

Solve each problem below.

133. Suppose u varies directly as x and y and inversely as the cube of z. If u is 15 when x is 6, y is 4, and z is 2, find u when x is 10, y is 20, and z is 5.

134. Let d be proportional to c^2 and m^3, and inversely proportional to h. If $d = 24$ when $c = 6$, $m = 1$, and $h = 3$, find d when $c = 4$, $m = 10$, and $h = 200$.

135. The kinetic energy of a moving object is proportional to the mass and the square of the velocity. If a 2000-kg car traveling at 30 meters per second has a kinetic energy of 900,000 units, how much kinetic energy is produced when the velocity is increased to 40 meters per second?

FUNCTIONS AND GRAPHS

The equations and inequalities in the previous chapter involved only *one* variable. However, it is very common for a practical application to use *two* variables, where the value of one variable depends on the value of the other. As an example, Figure 1(a) shows how the speed in miles per hour of a Porsche 928 depends on the time *t*

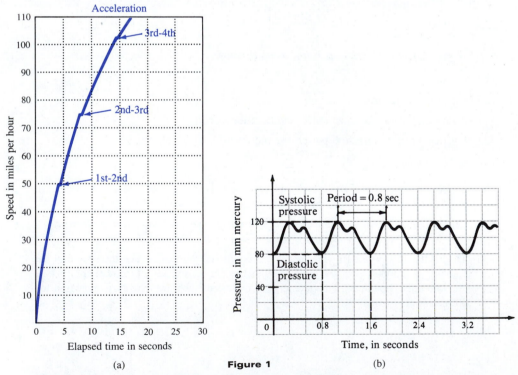

Figure 1(a): From *Road & Track,* April 1985. Reprinted with permission of *Road & Track* magazine.
Figure 1(b): From *Calculus for the Life Sciences* by Rodolfo De Sapio. W. H. Freeman and Company.
Copyright © 1978.

in seconds after the car has started from a dead stop. Figure 1(b) shows how a person's blood pressure depends on time. (Systolic and diastolic pressures are the upper and lower limits in the periodic changes in pressure that produce the pulse. The length of time between peaks is called the *period* of the pulse.)

Both Figures 1(a) and 1(b) show a *function,* a rule or procedure giving just one value of a variable from a given value of another variable. In each of these examples, a given value of time can be used to find just one value of the other variable, speed or blood pressure, respectively. The study of functions is a major theme of this text. Before getting to a more complete discussion of functions later in the chapter, we need to look at the topic of *graphing* with two variables, since functions are often studied by looking at their *graphs.*

3.1 RELATIONS AND GRAPHS

Chapter 1 showed a correspondence between real numbers and points on a number line, a correspondence set up by establishing a coordinate system for the line. This idea can be extended to two dimensions: in two dimensions the correspondence is between *pairs* of real numbers and points on a *plane.* One way to get this correspondence is by drawing perpendicular lines, one horizontal and one vertical. These lines intersect at a point O, called the **origin.** The horizontal line is the **x-axis,** and the vertical line is the **y-axis.**

Starting at the origin, the x-axis is made into a number line by placing positive numbers to the right and negative numbers to the left. The y-axis is made into a number line with positive numbers going up and negative numbers down. The plane into which the coordinate system is introduced is the **coordinate plane,** or **xy-plane.** The x-axis and y-axis divide the plane into four regions, or **quadrants,** labeled as shown in Figure 2. The points on the x-axis and y-axis themselves belong to no quadrant.

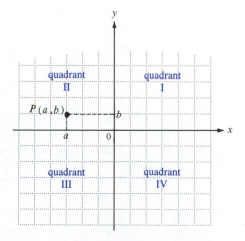

Figure 2

The *x*-axis and *y*-axis set up a **rectangular coordinate system,** or **Cartesian coordinate system,** named for one of its co-inventors, René Descartes; the other co-inventor was Pierre de Fermat. In his *Geometry (La géométrie)* of 1637, Descartes attempted to bring order to the science of geometry. He was constructing problems in the classic way by ruler and compass and wished to include curves that have an algebraic equation. But if he did so he would no longer be doing geometry. In the course of his investigations he developed the two-axis system and ordered pairs of numbers, which enabled him to have algebraic descriptions of geometrical curves. Other mathematicians translated his work and wrote commentaries, so that in less than a century the subject of analytic geometry with Cartesian coordinates was well developed. Descartes' (false) claim that one can tell everything about a curve from its equation helped motivate Isaac Newton (1642–1724) to develop results that eventually led to the invention of the calculus.

To find a pair of numbers corresponding to a given point *P* in Figure 2, draw a vertical line from *P* cutting the *x*-axis at *a*. Draw a horizontal line cutting the *y*-axis at *b*. Then point *P* has **coordinates** (*a*, *b*), where (*a*, *b*) is an **ordered pair** of numbers, two numbers written in parentheses in which the sequence of the numbers is important. For example, (3, 4) and (4, 3) in Figure 3 are not the same ordered pair since the sequence of the numbers is different. When an ordered pair represents values of *x* and *y,* the value for *x* is written first.

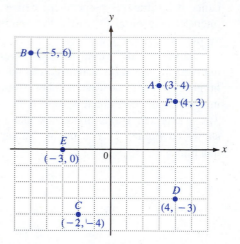

Figure 3

By this method, each point on the plane corresponds to just one ordered pair, and each ordered pair corresponds to exactly one point of the plane. The point is called the **graph** of the ordered pair, and the numbers in the ordered pair are the **coordinates** of the point.

For example, point *A* in Figure 3 corresponds to the ordered pair (3, 4). Also in Figure 3, point *B* corresponds to the ordered pair (−5, 6), *C* to (−2, −4), *D* to (4, −3), *E* to (−3, 0), and *F* to (4, 3).

The set of all points in the plane that correspond to a set of ordered pairs is the **graph** of the set of ordered pairs.

EXAMPLE 1 Graph the set of all points (x, y) satisfying the inequality $|x| \leq 2$.
As shown in the last chapter, $|x| \leq 2$ means

$$-2 \leq x \leq 2.$$

The graph of the set of all ordered pairs (x, y) whose x-coordinate satisfies the inequality $-2 \leq x \leq 2$ is bounded by vertical lines through $(2, 0)$ and $(-2, 0)$. The region satisfying $|x| \leq 2$ is shaded in Figure 4. ●

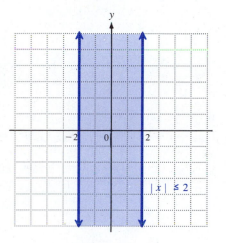

Figure 4

A set of ordered pairs is called a **relation.** The **domain** of the relation is the set of all first elements in the ordered pairs, and the **range** is the set of all second elements.

Although any set of ordered pairs is a relation, in mathematics we are most interested in those relations that are solution sets of equations. We may say that an equation *defines a relation,* or that it is the *equation of the relation.* For simplicity, we often refer to equations such as

$$y = 3x + 5 \quad \text{or} \quad x^2 + y^2 = 16$$

as relations, although technically the solution set of the equation is the relation.

EXAMPLE 2 For each relation defined below, give three ordered pairs that belong to the relation, and state the domain and range of the relation.

(a) $\{(2, 5), (7, -1), (10, 3), (-4, 0), (0, 5)\}$

Three ordered pairs from the relation are any three of the five ordered pairs in the set. The domain is the set of first elements,

$$\{2, 7, 10, -4, 0\},$$

and the range is the set of second elements,

$$\{5, -1, 3, 0\}.$$

(b) $y = 4x - 1$

To find an ordered pair of the relation, choose any number for x or y and substitute in the equation to get the corresponding value. For example, let $x = -2$. Then

$$y = 4(-2) - 1 = -9,$$

giving the ordered pair $(-2, -9)$. If $y = 3$, then

$$3 = 4x - 1$$
$$4 = 4x$$
$$1 = x,$$

and the ordered pair is $(1, 3)$. Verify that $(0, -1)$ also belongs to the relation. Since x and y can take any real-number values, both the domain and range are $(-\infty, +\infty)$.

(c) $x = \sqrt{y - 1}$

Verify that the ordered pairs $(1, 2)$, $(0, 1)$, and $(2, 5)$ belong to the relation. Since x equals the positive square root of $y - 1$, the domain is restricted to $[0, +\infty)$. Also, only nonnegative numbers have a square root, so the range is determined by the inequality

$$y - 1 \geq 0$$
$$y \geq 1,$$

giving $[1, +\infty)$ as the range.   ●

The set of all points in the plane corresponding to the ordered pairs of a relation is the **graph of the relation.** For now we can find graphs of relations only by identifying a reasonable number of ordered pairs, locating the corresponding points, and then guessing at the shape of the entire graph. Later we will develop better methods of identifying the graphs of many types of relations.

• • • • • • • • •

EXAMPLE 3   Graph $y = -4x + 3$.

Find several ordered pairs by selecting values for x (or y) and then finding the corresponding values of y (or x). For example, if $x = -3$, then $y = -4(-3) + 3 = 15$, producing the ordered pair $(-3, 15)$. Additional ordered pairs found in this way are given in the following table.

x	-3	-2	-1	0	1	2	3
y	15	11	7	3	-1	-5	-9
ordered pair	$(-3, 15)$	$(-2, 11)$	$(-1, 7)$	$(0, 3)$	$(1, -1)$	$(2, -5)$	$(3, -9)$

The ordered pairs from this table lead to the points that have been plotted in Figure 5(a). These points suggest that the entire graph is a straight line, as drawn in Figure 5(b). Notice that both the domain and the range are $(-\infty, +\infty)$.   ●

EXAMPLE 4   Graph $y = |x|$.

Start with a table.

x	-4	-3	-2	-1	0	1	2	3	4
y	4	3	2	1	0	1	2	3	4

Use this table to get the points in Figure 6. The graph drawn through these points is made up of portions of two straight lines. The domain is $(-\infty, +\infty)$ and the range is $[0, +\infty)$. ●

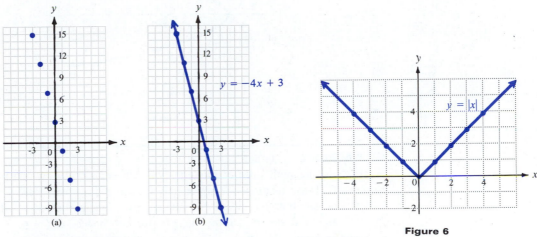

(a)

Figure 5

(b)

$y = -4x + 3$

$y = |x|$

Figure 6

CAUTION   There is a danger in the method used in Examples 3 and 4—we might choose a few values for x, find the corresponding values of y, begin to sketch a graph through these few points, and then make a completely wrong guess as to the shape of the graph. For example, choosing only -1, 0, and 1 as values of x in Example 4 above would produce only the three points $(-1, 1)$, $(0, 0)$, and $(1, 1)$. These three points would not have given enough information to determine the proper graph for $y = |x|$. However, this section involves only elementary graphs; when more complicated graphs are presented later, we will develop more accurate methods of working with them.

EXAMPLE 5   Graph $x = y^2$.

Since y is squared, it is probably easier to choose values of y and then to find the corresponding values of x. For example, choosing the value 2 for y gives $x = 2^2 = 4$. Choosing -2 for y gives $x = (-2)^2 = 4$, the same result. The following table shows the values of x corresponding to various values of y.

y	0	1	-1	2	-2	3	-3
x	0	1	1	4	4	9	9

The ordered pairs from this table were used to get the points plotted in Figure 7. (Don't forget that x always goes first in the ordered pair.) A smooth curve was then drawn through the resulting points. Here, y can take on any value, so the range is $(-\infty, +\infty)$. Since $x = y^2$, the domain is $[0, +\infty)$. ●

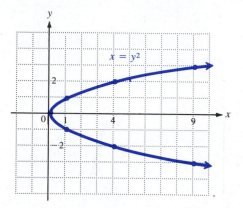

Figure 7

. . . .

EXAMPLE 6   Graph $y = \sqrt{x}$.

If we square both sides of this equation, we get $y^2 = x$, which was graphed in Example 5. These relations are not exactly the same, however. Here $y = \sqrt{x}$ implies $y \geq 0$ (the symbol $\sqrt{x}$ means the nonnegative square root of x). The domain is $[0, +\infty)$ as in Example 5, but the range is $[0, +\infty)$ instead of $(-\infty, +\infty)$. The graph is shown in Figure 8. ●

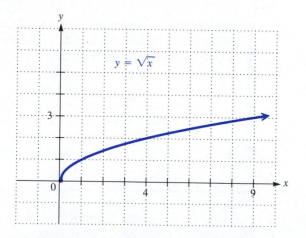

Figure 8

. . . .

EXAMPLE 7   Graph $y = \sqrt{x + 4}$.

The domain is determined by the fact that $x + 4 \geq 0$ or $x \geq -4$, giving $[-4, +\infty)$. Again, the range is $[0, +\infty)$. Selecting some values of x in the domain and calculating the corresponding *y*-values leads to the ordered pairs shown in the table.

x	−4	−3	0	5
y	0	1	2	3

Plotting these points and drawing a curve through them gives the graph in Figure 9. This graph, like the one in Figure 8, has an endpoint as indicated by the domain. ●

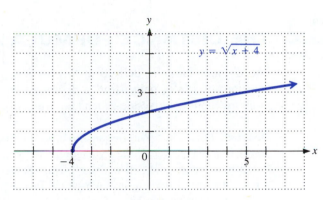

Figure 9

3.1 Exercises

If the point (a, b) is in the second quadrant, in what quadrant is each of the following points?

1. $(a, -b)$ **2.** $(-a, b)$ **3.** $(-a, -b)$ **4.** (b, a)

Graph the set of all points satisfying the following conditions for ordered pairs (x, y).

5. $x = 0$ **6.** $x > 0$ **7.** $y \le 0$ **8.** $y = 0$

9. $xy < 0$ **10.** $\dfrac{x}{y} > 0$ **11.** $|x| = 4,\ y \ge 2$ **12.** $|y| = 3,\ x \ge 4$

In Exercises 13–34, graph each relation and give its domain and range.

13. $y = 8x - 3$ **14.** $y = 2x + 7$ **15.** $y = 3x$ **16.** $y = -2x$

17. $4x = y^2$ **18.** $9x = y^2$ **19.** $16y^2 = -x$ **20.** $4y^2 = -x$

21. $y = |x| + 4$ **22.** $y = |x| - 3$ **23.** $x = |y|$ **24.** $x = |y| - 1$

25. $y = -|x + 1|$ **26.** $y = -|x - 2|$ **27.** $x = \sqrt{y}$ **28.** $x = \sqrt{y} + 1$

29. $x = \sqrt{y - 2}$ **30.** $x = \sqrt{y - 4}$ **31.** $y = \sqrt{2x + 4}$ **32.** $y = \sqrt{3x + 9}$

33. $y = -2\sqrt{x}$ **34.** $y = -\sqrt{x}$

35. A box has a square base with a side of measure *s* and height 2 in. less than the length of a side of the base. Write a relation giving the volume *V* in terms of *s*.

36. The length of a certain rectangular yard is 4 ft more than 3 times its width. Write a relation giving the length l in terms of the width, w.

37. The formula $A = (1/2)bh$ gives the area of a triangle with base b and height h. Write a relation expressing the area of a triangle in terms of its height if its height is 1/2 the length of its base.

38. The volume of a cone is found with the formula $V = (1/3)Bh$, where B is the area of the base of the cone and h is the height. Write a relation expressing the volume of a cone whose height is 1/2 the radius of the base.

3.2 THE DISTANCE FORMULA AND THE CIRCLE

The distance formula, derived in this section, is quite useful for determining the equations of sets of points from a description of the set. We will use this approach in this section to derive the general equation of a circle.

Distance Formula   We frequently need to find the distance between two given points in a plane. Recall the Pythagorean theorem: in a right triangle with legs a and b and hypotenuse c, $a^2 + b^2 = c^2$. This theorem may be used to obtain a formula for the distance between any two points in the plane. To get this distance formula, start with two points on a horizontal line, as in Figure 10(a). Use the symbol $P(x_1, y_1)$ to represent point P having coordinates (x_1, y_1) (read "x-sub-one, y-sub-one"). The distance between points $P(x_1, y_1)$ and $Q(x_2, y_1)$ can be found by subtracting the x-coordinates. (Absolute value is used to make sure that the distance is not negative—recall the work with distance in Chapter 1.) From this, the distance between points P and Q is $|x_1 - x_2|$. If $d(P, Q)$ represents the distance between P and Q, then

$$d(P, Q) = |x_1 - x_2|.$$

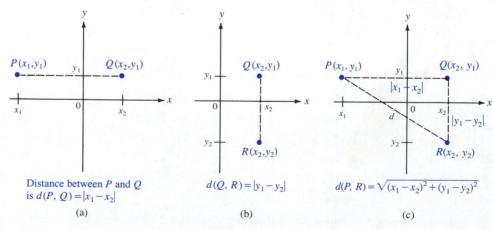

Distance between P and Q
is $d(P, Q) = |x_1 - x_2|$

(a)

$d(Q, R) = |y_1 - y_2|$

(b)

$d(P, R) = \sqrt{(x_1 - x_2)^2 + (y_1 - y_2)^2}$

(c)

Figure 10

Figure 10(b) shows points $Q(x_2, y_1)$ and $R(x_2, y_2)$ on a vertical line. To find the distance between Q and R, subtract the y-coordinates.

$$d(Q, R) = |y_1 - y_2|$$

Finally, Figure 10(c) shows two points, $P(x_1, y_1)$ and $R(x_2, y_2)$, which are *not* on a horizontal or vertical line. To find $d(P, R)$, construct the right triangle shown in the figure. One side of this triangle is horizontal and has length $|x_1 - x_2|$. The other side is vertical and has length $|y_1 - y_2|$. By the Pythagorean theorem,

$$[d(P, R)]^2 = |x_1 - x_2|^2 + |y_1 - y_2|^2.$$

Since $|x_1 - x_2|^2$ equals the expression $(x_1 - x_2)^2$ and $|y_1 - y_2|^2$ equals $(y_1 - y_2)^2$, the **distance formula** can be written as follows.

Distance Formula

Suppose $P(x_1, y_1)$ and $R(x_2, y_2)$ are two points in a coordinate plane. Then the distance between P and R, written $d(P, R)$, is

$$d(P, R) = \sqrt{(x_1 - x_2)^2 + (y_1 - y_2)^2}.$$

EXAMPLE 1   Find the distance between $P(-8\sqrt{3}, 4)$ and $Q(\sqrt{3}, -2)$.
According to the distance formula,

$$\begin{aligned}
d(P, Q) &= \sqrt{(-8\sqrt{3} - \sqrt{3})^2 + [4 - (-2)]^2} \\
&= \sqrt{(-9\sqrt{3})^2 + 6^2} \\
&= \sqrt{243 + 36} \\
&= \sqrt{279} \\
&= 3\sqrt{31}.
\end{aligned}$$

Using a calculator with a square root key gives 16.703 as an approximate value for $3\sqrt{31}$.   ●

EXAMPLE 2   Are the points $M(-2, 5)$, $N(12, 3)$, and $Q(10, -11)$ the vertices of a right triangle?
To decide whether or not the triangle determined by these three points is a right triangle, use the converse of the Pythagorean theorem: if the sides a, b, and c of a triangle satisfy $a^2 + b^2 = c^2$, then the triangle is a right triangle. A triangle with the three given points as vertices is shown in Figure 11 on the next page. This triangle is a right triangle if the square of the length of the hypotenuse equals the sum of the squares of the lengths of the legs. Use the distance formula to find the length of each side of the triangle.

$$d(M, N) = \sqrt{[12 - (-2)]^2 + (3 - 5)^2} = \sqrt{196 + 4} = \sqrt{200}$$

$$d(M, Q) = \sqrt{[10 - (-2)]^2 + (-11 - 5)^2} = \sqrt{144 + 256} = \sqrt{400} = 20$$

$$d(N, Q) = \sqrt{(10 - 12)^2 + (-11 - 3)^2} = \sqrt{4 + 196} = \sqrt{200}$$

Since

$$(\sqrt{200})^2 + (\sqrt{200})^2 = 200 + 200$$
$$= 400$$
$$= 20^2,$$
$$[d(M, Q)]^2 = [d(M, N)]^2 + [d(N, Q)]^2,$$

proving that the triangle is a right triangle with hypotenuse connecting M and Q. ●

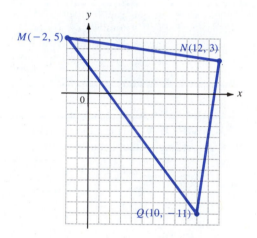

Figure 11

Circles A **circle** is the set of all points in a plane which lie a given distance from a given point. The given distance is the **radius** of the circle and the given point is the **center.** The equation of a circle can be found by using the distance formula.

For example, Figure 12 shows a circle of radius 3 with center at the origin. To find the equation of this circle, let (x, y) be any point on the circle. The distance between (x, y) and the center of the circle, $(0, 0)$, is given by

$$\sqrt{(x - 0)^2 + (y - 0)^2}.$$

Since this distance equals the radius, 3,

$$\sqrt{(x - 0)^2 + (y - 0)^2} = 3$$
$$\sqrt{x^2 + y^2} = 3$$
$$x^2 + y^2 = 9.$$

As suggested by the graph, the domain of this relation is $[-3, 3]$ and the range is $[-3, 3]$.

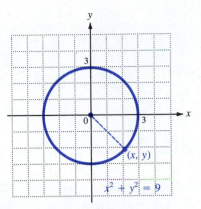

Figure 12

· · · · · · · · ·

EXAMPLE 3   Find an equation for the circle having radius 6 and center at $(-3, 4)$.

This circle is shown in Figure 13. Its equation can be found by using the distance formula. Start by letting (x, y) be any point on the circle. The distance from (x, y) to $(-3, 4)$ is given by

$$\sqrt{[x - (-3)]^2 + (y - 4)^2} = \sqrt{(x + 3)^2 + (y - 4)^2}.$$

This distance equals the radius, 6. Therefore,

$$\sqrt{(x + 3)^2 + (y - 4)^2} = 6$$

or

$$(x + 3)^2 + (y - 4)^2 = 36.$$

The graph in Figure 13 suggests that the domain is $[-9, 3]$ and the range is $[-2, 10]$.   ●

Generalizing the work of Example 3 to a circle of radius r and center (h, k) would give the following result.

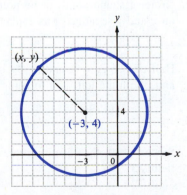

Figure 13

The circle with center (h, k) and radius r has equation

$$(x - h)^2 + (y - k)^2 = r^2,$$

the **center-radius form** of the equation of a circle. As a special case,

$$x^2 + y^2 = r^2$$

is the equation of a circle with radius r and center at the origin.

Starting with the center-radius form of the equation of a circle, $(x - h)^2 + (y - k)^2 = r^2$, and squaring $x - h$ and $y - k$ gives

$$x^2 - 2hx + h^2 + y^2 - 2ky + k^2 = r^2$$
$$x^2 + y^2 - 2hx - 2ky + (h^2 + k^2 - r^2) = 0.$$

Letting $-2h = D$, $-2k = E$, and $h^2 + k^2 - r^2 = F$, the equation becomes

$$x^2 + y^2 + Dx + Ey + F = 0, \qquad (*)$$

where D, E, and F are real numbers, the **general form** of the equation of a circle. Also, starting with an equation similar to $(*)$, the process of *completing the square* discussed in Section 2.3 can be used to get an equation of the form

$$(x - h)^2 + (y - k)^2 = m$$

for some number m. If $m > 0$, then $r^2 = m$, and the graph is that of a circle with radius $\sqrt{m}$. If $m = 0$, the graph is the single point (h, k), while the graph has no points if $m < 0$.

EXAMPLE 4   Decide if each equation has a circle as its graph.

(a) $x^2 - 6x + y^2 + 10y + 25 = 0$

Since this equation has the form of equation $(*)$ above, it either represents a circle, a single point, or no points at all. To decide which, complete the square on x and y separately, as explained in Section 2.3. Start with

$$(x^2 - 6x \quad) + (y^2 + 10y \quad) = -25.$$

Half of -6 is -3, and $(-3)^2 = 9$. Also, half of 10 is 5, and $5^2 = 25$. Add 9 and 25 on the left, and to compensate, add 9 and 25 on the right.

$$(x^2 - 6x + 9) + (y^2 + 10y + 25) = -25 + 9 + 25$$
$$(x - 3)^2 + (y + 5)^2 = 9$$

Since $9 > 0$, the equation represents a circle which has its center at $(3, -5)$ and radius 3.

(b) $x^2 + 10x + y^2 - 4y + 33 = 0$

Complete the square as above.

$$(x^2 + 10x + 25) + (y^2 - 4y + 4) = -33 + 25 + 4$$
$$(x + 5)^2 + (y - 2)^2 = -4$$

Since $-4 < 0$, there are no ordered pairs (x, y), with x and y real numbers, satisfying the equation. The graph of the given equation would contain no points.   ●

. . . .

EXAMPLE 5   Find the equation of a circle with center $(-2, 3)$ that has a diameter with one endpoint at $(1, 0)$.

The diameter of a circle goes through the center of the circle and has endpoints on the circle, so the point $(1, 0)$ is on the circle and satisfies its equation. Thus the radius is the distance between $(-2, 3)$ and $(1, 0)$.

$$r = \sqrt{(-3)^2 + 3^2} = \sqrt{18} = 3\sqrt{2}$$

Since the center is at $(-2, 3)$, the equation is

$$(x + 2)^2 + (y - 3)^2 = 18.   ●$$

Midpoint Formula   The distance formula is used to find the length of a line segment in a plane; the **midpoint formula** is used to find the coordinates of the midpoint of a line segment.

To develop this formula, let $A(x_1, y_1)$ and $B(x_2, y_2)$ be two different points in a plane (see Figure 14). Assume that A and B are not on a horizontal or vertical line. Let C be the intersection of the horizontal line through A and the vertical line through B. Let B' (read "B-prime") be the midpoint of segment AB. Draw a line through B' and parallel to segment BC. Let C' be the point where this line cuts segment AC. If the coordinates of B' are (x', y'), then C' has coordinates (x', y_1). Since B' is the midpoint of AB, point C' must be the midpoint of segment AC (why?), and

$$d(C, C') = d(C', A),$$

or

$$|x_2 - x'| = |x' - x_1|.$$

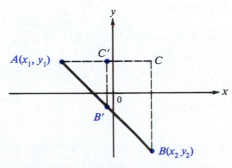

Figure 14

Because $d(C, C')$ and $d(C', A)$ must be positive the only solutions for this equation are found if

$$x_2 - x' = x' - x_1,$$

or

$$x_2 + x_1 = 2x'.$$

Finally

$$x' = \frac{x_1 + x_2}{2},$$

so that the x-coordinate of the midpoint is the average of the x-coordinates of the endpoints of the segment. In a similar manner, the y-coordinate of the midpoint is $(y_1 + y_2)/2$, proving the following result.

Midpoint Formula

The coordinates of the midpoint of the line segment with endpoints having coordinates (x_1, y_1) and (x_2, y_2) is

$$\left(\frac{x_1 + x_2}{2}, \frac{y_1 + y_2}{2} \right).$$

In words: the coordinates of the midpoint of a segment are found by finding the *average* of the x-coordinates and the *average* of the y-coordinates of the endpoints of the segment.

EXAMPLE 6   Find the coordinates of the midpoint M of the segment with endpoints having coordinates $(8, -4)$ and $(-9, 6)$.

Use the midpoint formula to find that the coordinates of M are

$$\left(\frac{8 + (-9)}{2}, \frac{-4 + 6}{2} \right) = \left(-\frac{1}{2}, 1 \right). \quad \bullet$$

EXAMPLE 7   A line segment has an endpoint with coordinates $(2, -8)$ and a midpoint with coordinates $(-1, -3)$. Find the coordinates of the other endpoint of the segment.

The x-coordinate of the midpoint is found from $(x_1 + x_2)/2$. Here, the x-coordinate of the midpoint is -1. To find x_2, use this formula, with $x_1 = 2$.

$$-1 = \frac{2 + x_2}{2}$$

$$-2 = 2 + x_2$$

$$-4 = x_2$$

In the same way, $y_2 = 2$; the endpoint has coordinates $(-4, 2)$. $\bullet$

The two formulas derived in this section can be used to prove various results from geometry.

EXAMPLE 8   Prove that the diagonals of a parallelogram bisect each other.

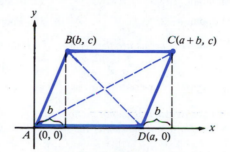

Figure 15

Figure 15 shows parallelogram *ABCD* having diagonals *AC* and *BD*. The figure has been placed on a coordinate system with *A* at the origin and side *AD* along the *x*-axis. We assign coordinates (*b, c*) to *B* and (*a*, 0) to *D*. Since *DC* is parallel to *AB* and is the same length as *AB*, use results from congruent triangles of geometry and write the coordinates of *C* as (*a* + *b, c*). Show that the diagonals bisect each other by showing that they have the same midpoint. By the midpoint formula,

$$\text{midpoint of } AC = \left(\frac{a + b + 0}{2}, \frac{c + 0}{2}\right) = \left(\frac{a + b}{2}, \frac{c}{2}\right)$$

$$\text{midpoint of } BC = \left(\frac{a + b}{2}, \frac{c + 0}{2}\right) = \left(\frac{a + b}{2}, \frac{c}{2}\right).$$

The midpoints are the same, so *AC* and *BD* bisect each other.   ●

NOTE   In proofs of this type, it is important to use the geometric properties of a figure to keep the number of variables used for the coordinates at a minimum. In Example 8, that is why we used the properties of congruent triangles to get the coordinates of point *C*.

3.2 Exercises

Find the distance d(P, Q) and the coordinates of the midpoint of segment PQ.

1. $P(5, 7)$, $Q(13, -1)$

2. $P(-2, 5)$, $Q(4, -3)$

3. $P(-8, -2)$, $Q(-3, -5)$

4. $P(-6, -10)$, $Q(6, 5)$

5. $P(\sqrt{2}, -\sqrt{5})$, $Q(3\sqrt{2}, 4\sqrt{5})$

6. $P(5\sqrt{7}, -\sqrt{3})$, $Q(-\sqrt{7}, 8\sqrt{3})$

Give the distance between the following points rounded to the nearest thousandth.

7. $(5, 7)$, $(2, 14)$

8. $(-4, 6)$, $(8, -5)$

9. $(3, -7)$, $(-5, 19)$

10. $(-9, -2)$, $(-1, -15)$

Find the coordinates of the other endpoint of the segments with endpoints and midpoints having coordinates as given.

11. Endpoint $(-3, 6)$, midpoint $(5, 8)$

12. Endpoint $(2, -8)$, midpoint $(3, -5)$

13. Endpoint $(6, -1)$, midpoint $(-2, 5)$

14. Endpoint $(-5, 3)$, midpoint $(-7, 6)$

Decide whether or not the following points are the vertices of a right triangle.

15. $(-2, 5)$, $(1, 5)$, $(1, 9)$

16. $(-9, -2)$, $(-1, -2)$, $(-9, 11)$

17. $(-4, 0)$, $(1, 3)$, $(-6, -2)$

18. $(-8, 2)$, $(5, -7)$, $(3, -9)$

Use the distance formula to decide whether or not the following points lie on a straight line. (Hint: The points lie on a straight line if the sum of the two smallest distances equals the largest distance.)

19. $(0, 7)$, $(3, -5)$, $(-2, 15)$

20. $(1, -4)$, $(2, 1)$, $(-1, -14)$

21. $(0, -9)$, $(3, 7)$, $(-2, -19)$

22. $(1, 3)$, $(5, -12)$, $(-1, 11)$

Find all values of x or y such that the distance between the given points is as indicated.

23. $(x, 7)$ and $(2, 3)$ is 5

24. $(5, y)$ and $(8, -1)$ is 5

25. $(3, y)$ and $(-2, 9)$ is 12

26. $(x, 11)$ and $(5, -4)$ is 17

27. (x, x) and $(2x, 0)$ is 4

28. (y, y) and $(0, 4y)$ is 6

Graph each of the following.

29. $x^2 + y^2 = 36$

30. $x^2 + y^2 = 81$

31. $(x - 2)^2 + y^2 = 36$

32. $x^2 + (y + 3)^2 = 49$

33. $(x - 4)^2 + (y + 3)^2 = 4$

34. $(x + 3)^2 + (y - 2)^2 = 16$

Find equations for each of the following circles.

35. Center $(1, 4)$, radius 3

36. Center $(-2, 5)$, radius 4

37. Center $(-8, 6)$, radius 5

38. Center $(3, -2)$, radius 2

39. Center $(-1, 2)$, passing through $(2, 6)$

40. Center $(2, -7)$, passing through $(-2, -4)$

41. Center $(-3, -2)$, tangent to the x-axis

42. Center $(5, -1)$, tangent to the y-axis

43. Endpoints of a diameter at $(1, 5)$ and $(7, 8)$

44. Endpoints of a diameter at $(3, -5)$ and $(-7, 2)$

For each of the following that are equations of circles, give the center and radius of the circle.

45. $x^2 + 6x + y^2 + 8y = -9$

46. $x^2 - 4x + y^2 + 12y + 4 = 0$

47. $x^2 + 8x + y^2 - 14y + 65 = 0$

48. $x^2 - 2x + y^2 = -1$

49. $2x^2 + 2y^2 = 4y + 96$

50. $3x^2 + 12x + 3y^2 = 63$

Use the appropriate formulas to find the circumference and area of each of the circles with equations as follows.

51. $(x - 2)^2 + (y + 4)^2 = 25$

52. $(x + 3)^2 + (y - 1)^2 = 9$

The unit circle *is a circle centered at the origin with a radius of 1. In Exercises 53 and 54, show that each point lies on the unit circle.*

53. $(-1/2, \sqrt{3}/2)$

54. $(-\sqrt{2}/2, -\sqrt{2}/2)$

55. Find all points (x, y) with $x = y$ that are 4 units from $(1, 3)$.

56. Find all points satisfying $x + y = 0$ that are 8 units from $(-2, 3)$.

57. Decide if each of the following points is *inside, on,* or *outside* the circle with center $(1, -4)$ and radius 6. **(a)** $(3, -2)$, **(b)** $(9, 1)$, **(c)** $(7, -4)$, **(d)** $(0, 9)$.

58. A circle is tangent to both axes, has its center in the third quadrant, and has a radius of $\sqrt{2}$. Find an equation for the circle.

59. Find the coordinates of a point whose distance from $(1, 0)$ is $\sqrt{10}$ and whose distance from $(5, 4)$ is $\sqrt{10}$.

60. Find the equation of the circle of smallest radius that contains the points $(1, 4)$ and $(-3, 2)$ within or on its boundary.

61. One circle has center at $(3, 4)$ and radius 5. A second circle has center at $(-1, 3)$ and radius 4. Do the circles cross?

62. Does the circle with radius 6 and center at $(0, 5)$ cross the circle with center at $(-5, -4)$ with radius 4?

63. One circle has center (a, b) and radius 4, while another has center (c, d) and radius s. State an inequality that is true if and only if the two circles intersect.

64. Find the coordinates of the points that divide the line segment joining $(2, 3)$ and $(10, 15)$ into four equal parts.

65. Find the coordinates of the points that divide the line segment joining $(4, 5)$ and $(10, 14)$ into three equal parts.

66. Show that the points $(-2, 2)$, $(13, 10)$, $(21, -5)$, and $(6, -13)$ are the vertices of a square.

67. Are the points $A(1, 1)$, $B(5, 2)$, $C(3, 4)$, $D(-1, 3)$ the vertices of a parallelogram (opposite sides equal in length)? Of a rhombus (all sides equal in length)?

Use the midpoint formula and distance formula, as necessary, to prove each of the following.

68. The midpoint of the hypotenuse of a right triangle is equally distant from all three vertices. (*Hint:* Place the right angle at the origin with the two legs on the axes.)

69. The diagonals of a rectangle are equal in length. (*Hint:* Place one corner of the rectangle at the origin with two sides on the axes.)

70. The line segment connecting the midpoints of two adjacent sides of any quadrilateral is the same length as the line segment connecting the midpoints of the other two sides.

3.3 FUNCTIONS

In the first section of this chapter, a relation was defined as a set of ordered pairs. In business, the price of an item often is directly related to the cost of producing the item. The relationship may be expressed as an equation. In such a situation it would be undesirable to have a particular cost produce more than one price. A special type of relation, which assigns exactly one range value to each value in the domain, is most suitable for applications and is so useful it is given a special name.

Definition of Function

A **function** is a relation that assigns to each element of a set X exactly one element of a set Y.

Set X is the domain and set Y is the range of the function.

EXAMPLE 1   Decide whether the following relations are functions.

(a) $\{(-2, 6), (12, 15), (6, -4), (6, 22), (17, 22)\}$

This relation is not a function because the domain element 6 is assigned to *two* range elements, -4 and 22. See Figure 16.

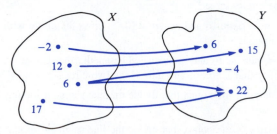

Figure 16

(b) $\{(5, 23), (14, 9), (15, -9), (26, -9)\}$

Since no x-value is assigned more than one y-value, this relation, shown in Figure 17, is a function. Note that although two x-values, 15 and 26, are assigned the same y-value, -9, this does not violate the definition of a function.

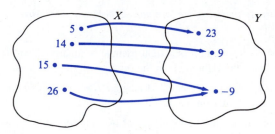

Figure 17

(c) The x^2 key on a calculator

Each real number input for x produces exactly one square, so this is an example of a function.

(d) The optical reader at the checkout counter in many stores that converts codes to prices

Since each code is assigned just one price, the set of ordered pairs (code, price) is a function.

(e) $\{(x, y) \mid x = |y|\}$

Most values of x are assigned two y-values by the equation $x = |y|$. For example, both (3, 3) and (3, -3) belong to this relation, so it is not a function.

(f) $\{(x, y) \mid y = x\}$

Typical ordered pairs are $(1, 1)$, $(-2, -2)$, $(3, 3)$, and so on. Each x-value is assigned just one y-value (which equals that x-value), so this relation is a function, sometimes called the **identity function.** •

We can think of a function as an input-output machine. If we input an element from the domain, the function (machine) outputs an element belonging to the range. See Figure 18. In fact, that is exactly how a calculator or computer works—as an input-output machine.

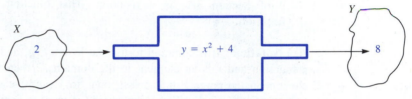

A function as an input-output machine

Figure 18

There is a quick way to tell from its graph if a relation is a function. Figure 19 shows two graphs. In the graph of part (a), for any x that might be chosen, exactly one value of y can be found, showing that this is the graph of a function. On the other hand, the graph in part (b) is not the graph of a function. For example, the vertical line through x_1 leads to two different values of y, namely y_1 and y_2. This example suggests the **vertical line test** for a function.

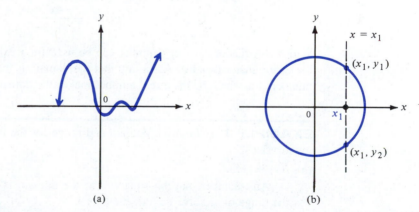

(a)   (b)

Figure 19

Vertical Line Test

If each vertical line intersects a graph in no more than one point, the graph is the graph of a function.

. . . .

EXAMPLE 2   Use graphs to decide whether the following relations are functions.

(a) $\{(x, y) | y = |x|\}$

This relation was graphed in Figure 6 in Section 3.1. By the vertical line test, the graph shows that $y = |x|$ defines a function, called the **absolute value function.**

(b) $\{(x, y) | y = \sqrt{x + 4}\}$

The vertical line test shows that this relation, whose graph is shown in Figure 9 of Section 3.1, is a function. This function is an example of a **root function.** ●

$f(x)$ Notation   It is common to use the letters f, g, and h to name functions. If f is a function and x is an element in the domain X, then $f(x)$ is used to represent the element in the range Y that corresponds to x, and we say $y = f(x)$. (The notation $f(x)$ is read "f of x".) For example, if f is used to name the function in Figure 17, then

$$f(5) = 23, \quad f(14) = 9, \quad f(15) = -9, \quad \text{and} \quad f(26) = -9.$$

For a given element x in set X, the corresponding element $f(x)$ in set Y is called the **value** or **image** of f at x. The set of all possible values of $f(x)$ makes up the range of the function. Throughout this book, if the domain for a function is not given, it will be assumed to be the largest possible set of real numbers for which $f(x)$ is a real number. For example, suppose function f is defined by

$$f(x) = \frac{-4x}{2x - 3}.$$

With this equation any real number can be used for x except $x = 3/2$, which makes the denominator equal 0. Assuming that the domain is the largest possible set of real numbers for which $f(x)$ is a real number makes the domain $(-\infty, 3/2) \cup (3/2, +\infty)$.

. . . .

EXAMPLE 3   Find the domain and range for the functions defined by the following rules.

(a) $f(x) = x^2$

Any number may be squared, so the domain is $(-\infty, +\infty)$. To determine the range of y, solve the equation for x.

$$f(x) = x^2$$
$$y = x^2$$
$$x = \pm\sqrt{y}$$

This shows that y can take any value in the interval $[0, +\infty)$ as x takes values in the interval $(-\infty, +\infty)$.

(b) $f(x) = \sqrt{x^2 + x - 6}$

The domain includes those values of x that make

$$x^2 + x - 6 \geq 0.$$

Factor. $$(x + 3)(x - 2) \geq 0$$

Solve this quadratic inequality to get $(-\infty, -3] \cup [2, +\infty)$ for the domain. The radical indicates the nonnegative square root, so the range is $[0, +\infty)$.   ●

For most of the functions in this book, the domain can be found with algebraic methods already discussed. The range, however, at this level of mathematics must often be found from the graph as shown in Figure 20.

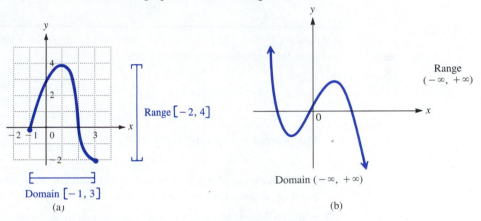

Figure 20

In Figure 20(a) the range is a closed interval $[-2, 4]$, but in Figure 20(b), the range is the set of real numbers $(-\infty, +\infty)$. The largest number in the range (if there is one) is called the **maximum** and the smallest number in the range (if there is one) is called the **minimum.** The function graphed in Figure 20(a) has a maximum of 4 and a minimum of -2. These values are important in many applications. One of the major applications of calculus is finding maximum and minimum function values.

 NOTE   Based on the definition of a function, functions f and g are **equal** if and only if f and g have exactly the same domains and $f(x) = g(x)$ for every value of x in the domain.

Suppose a function is defined by $f(x) = -3x + 2$. To emphasize that this statement is used to find values in the range of f, it is common to write

$$y = -3x + 2.$$

When a function is written in the form $y = f(x)$, x is called the **independent variable,** and y the **dependent variable.**

There is no reason to restrict the variables to x or y—different areas of study use different variables. For example, it is common to use t for time in physics, or p for price in economics. The function defined by $f(x) = -3x + 2$ above could just as well have been written $g(t) = -3t + 2$. Both functions have the real numbers as domain and assign the same value to each real number.

By definition, a function f is a rule that assigns to each element of one set exactly one element of a second set. For a particular value of x in the first set, the corresponding element in the second set is written $f(x)$. There is a distinction between f and $f(x)$: f is the function or rule, while $f(x)$ is the value obtained by applying the rule to an element x. However, it is very common to abbreviate

"the function defined by the rule $y = f(x)$"

as simply

"the function $y = f(x)$."

- - - - - - - - - -

EXAMPLE 4   Let $g(x) = 3\sqrt{x}$ and $f(x) = 1 + 4x$. Find each of the following and give the corresponding ordered pair.

(a) $g(16)$

To find $g(16)$, replace x in $g(x) = 3\sqrt{x}$ with 16.

$$g(16) = 3\sqrt{16} = 3 \cdot 4 = 12$$

The ordered pair is $(16, 12)$.

(b) $f(-3) = 1 + 4(-3) = -11$, and the ordered pair is $(-3, -11)$.

(c) $g(-4)$ does not exist; -4 is not in the domain of g since $\sqrt{-4}$ is not a real number. There is no ordered pair here.

(d) $g(m) = 3\sqrt{m}$, if m represents a nonnegative real number; the ordered pair is $(m, 3\sqrt{m})$.

(e) $g(f(3))$

First find $f(3)$, as follows.

$$f(3) = 1 + 4 \cdot 3 = 1 + 12 = 13$$

Now, $g(f(3)) = g(13) = 3\sqrt{13}$, and the corresponding ordered pair is $(3, 3\sqrt{13})$.

(f) $g(x + h)$

Substitute $x + h$ for x in $g(x) = 3\sqrt{x}$.

$$g(x + h) = 3\sqrt{x + h}.$$

The corresponding ordered pair is $(x + h, 3\sqrt{x + h})$.   ●

- - - - - - - - -

EXAMPLE 5   Let $f(x) = 2x^2 - 3x$. If h represents any nonzero number, then the quotient

$$\frac{f(x + h) - f(x)}{h}, \quad h \neq 0,$$

represents the slope of the line through $(x, f(x))$ and $(x + h, f(x + h))$. This expression is called a *difference quotient* and is used in calculus to determine the steepness of a curve at a point. Find and simplify the quotient.

To find $f(x + h)$, replace x in $f(x)$ with $x + h$.

$$f(x) = 2x^2 - 3x$$
$$f(x + h) = 2(x + h)^2 - 3(x + h) \qquad \text{Substitute } x + h \text{ for } x \text{ in } f(x).$$

Now, subtract $f(x)$, which equals $2x^2 - 3x$, to get $f(x + h) - f(x)$. Then

$$\frac{f(x + h) - f(x)}{h} = \frac{[2(x + h)^2 - 3(x + h)] - (2x^2 - 3x)}{h}$$

$$= \frac{2(x^2 + 2xh + h^2) - 3x - 3h - 2x^2 + 3x}{h}$$

$$= \frac{2x^2 + 4xh + 2h^2 - 3x - 3h - 2x^2 + 3x}{h}$$

$$= \frac{4xh + 2h^2 - 3h}{h}$$

$$= \frac{h(4x + 2h - 3)}{h}$$

$$= 4x + 2h - 3. \quad \bullet$$

CAUTION In Example 5, note that $f(x + h) \neq f(x) + f(h)$, since

$$f(x) + f(h) = 2x^2 - 3x + 2h^2 - 3h$$

and, as shown above,

$$f(x + h) = 2x^2 + 4xh + 2h^2 - 3x - 3h.$$

· · · · · · · · ·

EXAMPLE 6 A rectangle has the dimensions shown in Figure 21. Give the diagonal of the rectangle, d, as a function of the width, x.

By the Pythagorean theorem,

$$d^2 = x^2 + (x + 3)^2$$
$$= x^2 + x^2 + 6x + 9$$
$$= 2x^2 + 6x + 9$$
$$d = \sqrt{2x^2 + 6x + 9}.$$

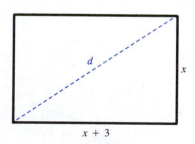

Figure 21

Only the positive square root is meaningful here, since d represents the diagonal of a rectangle. $\quad \bullet$

The concept of *function,* as with all of mathematics, required many years to be developed and clarified. Leibniz used the term function in 1694 to refer to any part of a curve, a point on it, the extent of its curvature, or its slope at a point. By the 18th century, mathematicians said a function was "an analytic expression composed in whatever way of a variable and numbers." In the second half of the 18th century a function was considered to be some sort of an equation, possibly with infinitely many terms.

The conception of a function as a correspondence, as in our text, required a new geometric, rather than algebraic, point of view. It was Lejeune Dirichlet (1805–59), a German mathematician, who wrote "$f(x)$ is a real function of a real variable x if, to every real number x, there corresponds a real number $f(x)$."

3.3 Exercises

Decide whether each of the following relations is a function.

1. $\{(4, 12), (7, -1), (5, -3), (8, 7), (2, 4)\}$ **2.** $\{(6, 5), (6, -1), (2, 4), (2, -4), (-5, 3)\}$

3. $\{(1, 1), (2, 2), (3, 3), (2, -2)\}$

4. $\{(-5, 3), (-4, 8), (-3, 2), (-2, 6), (-1, 12), (0, 0)\}$

5. The $\sqrt{x}$ key on a calculator   **6.** The $1/x$ key on a calculator

7. $\{(x, y)|x = y^2\}$ **8.** $\{(x, y)|y^2 = 3x + 1\}$ **9.** $\{(x, y)|2x - y = 1\}$

10. $\{(x, y)|y - x = 4\}$ **11.** $\{(x, y)|x^2 + y^2 = 4\}$ **12.** $\{(x, y)|x^2 - y^2 = 4\}$

For each of the following, find the indicated function values: (a) $f(-2),$ (b) $f(0),$
(c) $f(1),$ (d) $f(4).$

13.

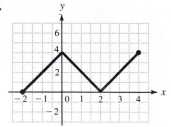

14.

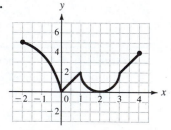

15.

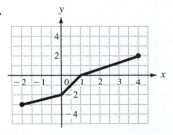

16.

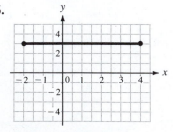

Let $f(x) = x^2 - 3,$ $g(x) = \sqrt{3x + 1},$ *and* $h(x) = \dfrac{x}{1 - x}.$ *Find each of the following.*

17. $f(0)$ **18.** $f(-1)$ **19.** $h(1)$ **20.** $h(-2)$

21. $g(5)$ **22.** $g(0)$ **23.** $f(-3) + 2$ **24.** $f(4) - 5$

25. $f(1) + g(1)$ **26.** $h(0) - f(0)$ **27.** $[g(1)]^2$ **28.** $[f(-1)]^2$

29. $f\left(\dfrac{1}{a}\right)$ **30.** $g(5a - 2)$ **31.** $\dfrac{1}{f(a)}$ **32.** $f(\sqrt{a})$

33. $\dfrac{f(5a)}{g(a)}$ **34.** $\sqrt{h(a)}$ **35.** $g\left(\dfrac{8}{3}\right)$ **36.** $h\left(\dfrac{1}{2}\right) \cdot f\left(\dfrac{1}{2}\right)$

Let $f(x) = \begin{cases} 2x^2 & \text{if } x < 0 \\ 3x & \text{if } 0 \le x \le 1 \\ |6 - x| & \text{if } x > 1. \end{cases}$ *Find each of the following.*

37. $f(-3)$ **38.** $f(1)$ **39.** $f(10)$ **40.** $f\left(\dfrac{1}{3}\right) + f(3)$

Give the domain and range of the functions defined or graphed in Exercises 41–56. Give only the domain in Exercises 45 and 46.

41. $g(x) = x^4$ **42.** $h(x) = (x - 2)^2$ **43.** $f(x) = \sqrt{8 + x}$ **44.** $f(x) = (3x + 2)^{1/2}$

45. $g(x) = \dfrac{2}{x^2 - 3x + 2}$ **46.** $h(x) = \dfrac{-4}{x^2 + 5x + 4}$ **47.** $r(x) = -\sqrt{x^2 - 4x - 5}$

48. $r(x) = -\sqrt{x^2 + 7x + 10}$ **49.** $f(x) = |x - 4|$ **50.** $k(x) = -|2x - 7|$

51.

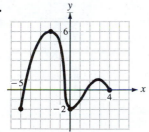

52.

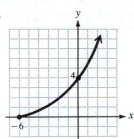

53.

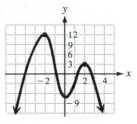

54.

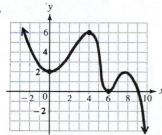

55.

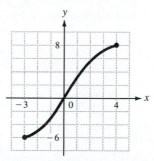

56.

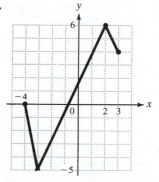

57. If $f(x) = x^2/x$ and $g(x) = x$, does $f(x) = g(x)$?

For the functions defined in Exercises 58–61, find   *(a)* $f(x + h)$,   *(b)* $f(x + h) - f(x)$, *and*   *(c)* $\dfrac{f(x + h) - f(x)}{h}$. *(Assume $h \ne 0$).*

58. $f(x) = x^2 - 4$ **59.** $f(x) = 8 - 3x^2$ **60.** $f(x) = 6x + 2$ **61.** $f(x) = 4x + 11$

62. A box is made from a piece of metal 12 by 16 inches by cutting squares of side x from each corner. (See the sketch.) Give the volume of the box as a function of x.

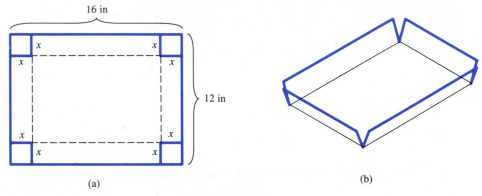

(a)

(b)

Exercise 62

63. A cone has a radius of 6 in. and a height of 9 in. (See the figure.) The cone is filled with water having a depth of h in. The radius of the surface of the water is r in. Use similar triangles to give r as a function of h.

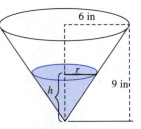

Exercise 63

When a function models a real-life situation, the domain is usually restricted by practical considerations. For instance, in the ''cut and fold a rectangle to obtain a box'' problem of Exercise 62, the value of x must be in the interval (0, 6). Determine a meaningful domain for the functions in Exercises 64–66.

64. The height of a ball thrown straight up into the air is $h(t) = -16t^2 + 50t + 5$ ft after t seconds.

65. According to the economic law of demand, the price that should be charged in order to sell x units of a certain commodity is $p(x) = 100 - .05x$ dollars.

66. If a worker earning ten dollars an hour works at least 40 hours a week, with time and a half for overtime, then her weekly salary for h hours of work is $s(h) = 400 + 15(h - 40)$ dollars.

67. An athletic field is shaped like a rectangle with semicircles of radius r attached to each end. See the figure. The perimeter of the field consists of a 440-yard track.
 (a) Show that each straight portion of the track must have length $220 - \pi r$ yd.
 (b) Express the area of the entire field as a function of r.
 (c) Determine a meaningful domain for the function in part (b).

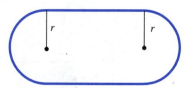

Exercise 67

68. Sixty feet of fencing is used to enclose a rectangular garden in which each side of the garden is at least 12 ft long and one side, of length x, is an extension of a 10-ft stone wall. See the figure.
 (a) Show that the length of each side perpendicular to the side with the stone wall is $25 - x$ ft.
 (b) Express the area of the garden as a function of x.
 (c) Determine a meaningful domain for the function in part (b).

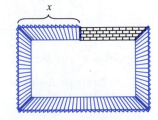

Exercise 68

69. A 40-cm wire is cut into two pieces. The left piece, of length x, is bent into a rectangle with a length that is three times its height. The other piece is bent into a circle. See the figure.
  (a) Show that the radius of the circle is $(40 - x)/(2\pi)$ cm.
  (b) Express the sum of the areas of the two geometric figures as a function of x.
  (c) Determine a meaningful domain for the function in part (b).

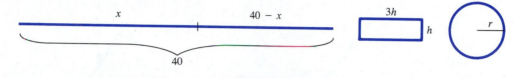

Exercise 69

70. Give the area of a circle as a function of its radius; also, give the circumference as a function of the radius.

71. Write the area of a circle as a function of the diameter of the circle; then write the circumference as a function of the diameter.

72. A rectangle is inscribed in a circle of radius r. Let x represent the length of one side of the rectangle. Give the area of the rectangle as a function of r.

73. The height of a cone is half the radius of the base. Give the volume of the cone as a function of the radius of the base.

3.4 LINEAR FUNCTIONS

In this section we begin the study of specific functions with *linear functions*. The name "linear" comes from the fact that the graph of every linear function is a straight line.

Definition of Linear Function

A function f is a **linear function** if
$$f(x) = ax + b,$$
for real numbers a and b, with $a \neq 0$.

A straight line is determined by two different points on the line. Points that are useful for sketching the graph of a linear function are determined by the *x-intercept* and the *y-intercept*. An **x-intercept** is an x-value (if any) of the point at which a graph crosses the x-axis, while a **y-intercept** is a y-value (if any) of the point at which a graph crosses the y-axis. The graph in Figure 22 has an x-intercept of x_1 and a y-intercept of y_1.

 CAUTION   If the graph goes through the origin, the x- and y-intercepts will give the same point. In this case an additional point must be found.

As suggested by the graph of a linear function $y = f(x)$ in Figure 22, x-intercepts can be found by setting $y = 0$ and solving for x, with y-intercepts found by setting $x = 0$ and solving for y. Example 1 shows how the intercepts are used to graph a line.

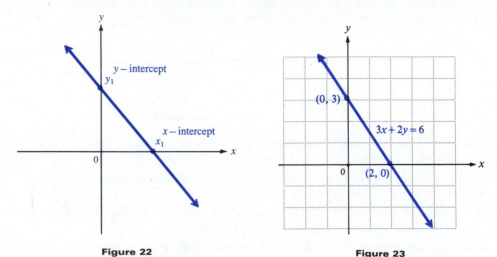

Figure 22   Figure 23

· · · · · · · · ·

EXAMPLE 1   Use the intercepts to graph $3x + 2y = 6$.

The equation can be written as a linear function by solving for y, to get $y = (-3/2)x + 3$, or $f(x) = (-3/2)x + 3$. Find the y-intercept by setting $x = 0$.

$$3 \cdot \mathbf{0} + 2y = 6$$
$$y = 3$$

For the x-intercept, set $y = 0$.

$$3x + 2 \cdot \mathbf{0} = 6$$
$$x = 2$$

Plotting $(0, 3)$ and $(2, 0)$ and drawing a straight line through the two points gives the graph in Figure 23.   ●

· · · · · · · · ·

EXAMPLE 2   Graph $y = -3$.

Think of $y = -3$ as $y = 0 \cdot x + -3$. This alternative form shows that for any value of x that might be chosen, y is $0 \cdot x + -3 = 0 + -3 = -3$. Since y always equals -3, the value of y can never be 0. Thus, the graph has no x-intercept. The only way a straight line can have no x-intercept is by being parallel to the x-axis, as shown in Figure 24.   ●

A linear function defined by $y = b$, such as $y = -3$ in Example 2, is called a **constant function.**

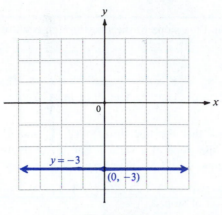

Figure 24

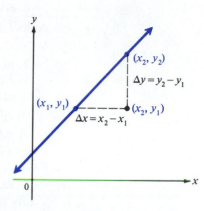

Figure 25

Slope   An important characteristic of a straight line is its *slope,* a numerical measure of the steepness of the line that tells how much y changes for 1 unit of change in x. To find this measure, start with the nonvertical line through the two distinct points (x_1, y_1) and (x_2, y_2), as shown in Figure 25. (Since the points are distinct, $x_1 \neq x_2$). The difference $x_2 - x_1$ is called the **change in x** and denoted by Δx (read "delta x"), where Δ is the Greek letter *delta*. In the same way, the change in y is $\Delta y = y_2 - y_1$.

The slope of a nonvertical line, usually symbolized with the letter m, is defined as the quotient of the change in y and the change in x.

Definition of Slope

The **slope m** of the line through (x_1, y_1) and (x_2, y_2) is

$$m = \frac{\Delta y}{\Delta x} = \frac{y_2 - y_1}{x_2 - x_1}, \quad \Delta x \neq 0.$$

Since either point can be designated as (x_1, y_1), the slope formula also can be written

$$\frac{y_1 - y_2}{x_1 - x_2}, \quad \Delta x \neq 0.$$

The slope of a line can be found only if the line is nonvertical. This guarantees that $x_2 \neq x_1$ so that the denominator $x_2 - x_1 \neq 0$. It is not possible to define the slope of a vertical line.

The slope of a vertical line is undefined.

. . . .

EXAMPLE 3   Find the slope of the line through each of the following pairs of points.

(a) $(-4, 8)$, $(2, -3)$

Let $x_1 = -4$, $y_1 = 8$, $x_2 = 2$, and $y_2 = -3$. Then $\Delta y = -3 - 8 = -11$ and $\Delta x = 2 - (-4) = 6$. The slope $m = \Delta y/\Delta x = -11/6$.

(b) $(2, 7)$, $(2, -4)$

A sketch shows that the line through $(2, 7)$ and $(2, -4)$ is vertical. As mentioned above, the slope of a vertical line is not defined. (An attempt to use the definition of slope would produce a zero denominator.)

(c) $(5, -3)$ and $(-2, -3)$

By the definition of slope,

$$m = \frac{-3 - (-3)}{-2 - 5} = \frac{0}{-7} = 0. \quad \bullet$$

The line through the two points in Example 3(c) is a horizontal line; this example suggests the following generalization.

. . . .

The slope of a horizontal line is 0.

Figure 26 shows lines of various slopes. As the figure shows, a line with a positive slope is increasing (as x increases), while a line with a negative slope is decreasing (as x increases).

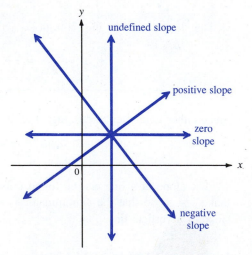

Figure 26

Figure 27 shows four points $A(x_1, y_1)$, $B(x_2, y_2)$, $C(x_3, y_3)$, and $D(x_4, y_4)$, all on the same straight line. Right triangles ABE and CDF have been completed. The lines

AE and *CF* are parallel, as are the lines *EB* and *FD*, making triangles *ABE* and *CDF* similar. Since similar triangles have corresponding sides proportional,

$$\frac{y_2 - y_1}{x_2 - x_1} = \frac{y_4 - y_3}{x_4 - x_3}.$$

The quotient on the left is the one used in the definition of slope, and it equals the quotient on the right for any choice of four distinct points on the line, showing that the slope of a line is the same regardless of the pair of distinct points chosen.

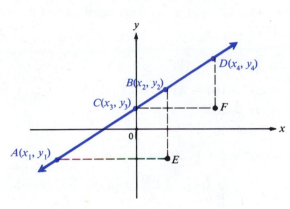

Figure 27

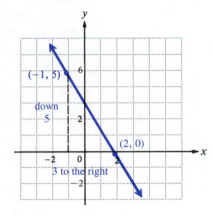

Figure 28

CAUTION When you are calculating slope, be careful to subtract the *y*-co-ordinates and the *x*-coordinates in the same order.

Give correct slope:

$$\frac{y_2 - y_1}{x_2 - x_1} \quad \text{and} \quad \frac{y_1 - y_2}{x_1 - x_2}$$

Give incorrect slope:

$$\frac{y_2 - y_1}{x_1 - x_2} \quad \text{and} \quad \frac{y_1 - y_2}{x_2 - x_1}$$

EXAMPLE 4 Graph the line passing through $(-1, 5)$ and having slope $-5/3$.
First locate the point $(-1, 5)$, as shown in Figure 28. The slope of this line is $\Delta y/\Delta x = -5/3$, so a change of 3 units horizontally produces a change of -5 units vertically. Starting at $(-1, 5)$ and going 5 units down and 3 units to the right gives a second point, $(2, 0)$, which can then be used to complete the graph. ●

Straight line graphs are often good choices for **supply-demand curves.** In most supply-demand curves, the supply of an item goes up as the price goes up, while the demand goes down as the price goes up.

EXAMPLE 5 Suppose the price *p* and demand *x* for hand calculators are related by

$$p = 30 - \frac{2}{3}x, \qquad \text{Demand equation}$$

where x is in millions and p is in dollars. That is, to sell x million hand calculators, the price per calculator must be $30 - (2/3)x$ dollars. Also, suppose the price p and the supply x for hand calculators are related by

$$p = \frac{5}{6}x. \qquad \text{Supply equation}$$

That is, to stimulate the production of x million calculators, the price per calculator must be $(5/6)x$ dollars. Find the equilibrium price, where supply and demand are equal.

Since both expressions in x equal p, set them equal to find the value of x that gives the equilibrium price.

$$\frac{5}{6}x = 30 - \frac{2}{3}x$$

$$5x = 180 - 4x \qquad \text{Multiply by 6.}$$

$$9x = 180$$

$$x = 20$$

From the supply equation, when $x = 20$, the price p is

$$p = \frac{5}{6}(20) = \frac{100}{6} \approx 16.67,$$

so the equilibrium price is $16.67. Figure 29 shows the graphs of the two lines and the point where supply and demand are equal.   ●

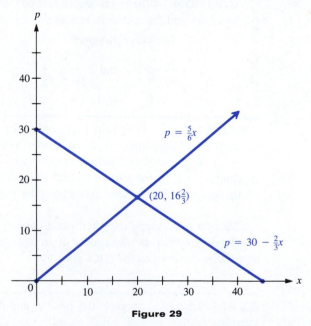

Figure 29

Some applications require a function whose graph has parts of two or more lines. The next examples show such functions, called **functions defined piecewise.**

· · · · · · · · · ·

EXAMPLE 6   Graph the functions defined as follows.

(a) $f(x) = \begin{cases} x + 1 & \text{if } x > 2 \\ -2x + 5 & \text{if } x \leq 2 \end{cases}$

For $x > 2$, graph $f(x) = x + 1$. For $x \leq 2$, graph $f(x) = -2x + 5$, as shown in Figure 30. A solid circle at $(2, 1)$ shows that this point is part of the graph, while the open circle at $(2, 3)$ shows that that point is not part of the graph.

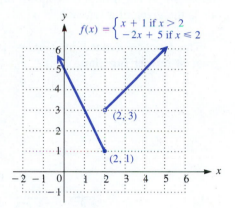

Figure 30

(b) $f(x) = \begin{cases} 2x + 3 & \text{if } x \leq 1 \\ -x + 6 & \text{if } x > 1 \end{cases}$

Graph $f(x) = 2x + 3$ for $x \leq 1$. For $x > 1$, graph $f(x) = -x + 6$. The graph consists of the two pieces shown in Figure 31. The two lines meet at the point $(1, 5)$, which is an endpoint of the first line.   ●

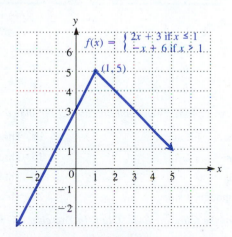

Figure 31

. . . .

EXAMPLE 7   The symbol $[x]$ is used to represent the greatest integer less than or equal to x. For example, $[8.4] = 8$, $[-5] = -5$, $[\pi] = 3$, $[-6.9] = -7$, and so on. In general,

$$\text{for } 0 \le x < 1, f(x) = 0,$$
$$\text{for } 1 \le x < 2, f(x) = 1,$$
$$\text{for } 2 \le x < 3, f(x) = 2,$$

and so on. A graph of $f(x)$ is shown in Figure 32. The vertical line test shows that this is the graph of a function.   ●

The function defined by $f(x) = [x]$, discussed in Example 7, is called the **greatest-integer function.**

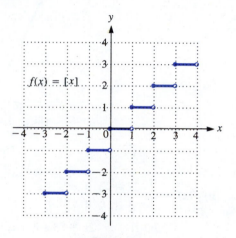

Figure 32

3.4 Exercises ·

Graph each of the following.

1. $y = 3x + 1$ **2.** $2x + y = 4$ **3.** $2x + 5y = 10$ **4.** $3x + 4y = 12$

5. $x = 2$ **6.** $y = -3$ **7.** $y = -5$ **8.** $x = 6$

9. $y = 3x$ **10.** $x = -2y$ **11.** $x = -5y$ **12.** $y = x$

Graph the line passing through the given point and having the indicated slope.

13. Through $(-1, 3)$, $m = 3/2$ **14.** Through $(-2, 8)$, $m = -1$

15. Through $(3, -4)$, $m = -1/3$ **16.** Through $(-2, -3)$, $m = -3/4$

17. Through $(-1, 4)$, $m = 0$ **18.** Through $(2, -5)$, $m = 0$

19. Through $(3, 2/3)$, undefined slope   **20.** Through $(9/4, 2)$, undefined slope

Find the slope of each of the following lines. (Hint: In Exercises 25–32, find two points on the line.)

21. Through $(-2, 3)$ and $(-1, 2)$ **22.** Through $(8, 4)$ and $(-1, -3)$ **23.** Through $(-4, -3)$ and $(5, 0)$

24. Through (2, 3) and (2, 7)   **25.** $y = 2x$ **26.** $y = -2x + 1$

27. $3x + 4y = 6$ **28.** $2x + y = 8$ **29.** $y = 4$

30. $x = -6$ **31.** The x-axis   **32.** The y-axis

For each of the following slopes, identify the line below having that slope.

33. 1/2   **34.** −2   **35.** 0   **36.** −1/2   **37.** 2   **38.** Undefined

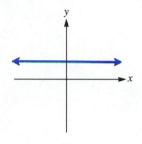

(a)

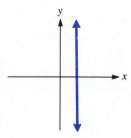

(b)

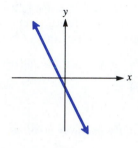

(c)

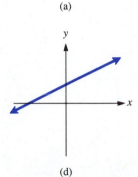

(d)

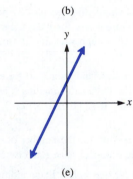

(e)

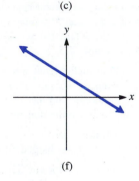

(f)

Use slopes to decide whether or not the following points lie on a straight line. (Hint: In Exercise 39, find the slopes of the lines through M and N and through M and P. If these slopes are the same, then the points lie on a straight line.)

39. $M(1, -2), N(3, -18), P(-2, 22)$ **40.** $A(0, -2), B(3, 7), C(-4, -14)$

Graph the functions defined in Exercises 41–44.

41. $f(x) = \begin{cases} x - 1 & \text{if } x \le 3 \\ 2 & \text{if } x > 3 \end{cases}$ **42.** $h(x) = \begin{cases} -2 & \text{if } x \ge 1 \\ 2 & \text{if } x < 1 \end{cases}$

43. $p(x) = \begin{cases} |x| & \text{if } x > -2 \\ x & \text{if } x \le -2 \end{cases}$ **44.** $r(x) = \begin{cases} |x| - 1 & \text{if } x > -1 \\ x - 1 & \text{if } x \le -1 \end{cases}$

45. When a diabetic takes long-acting insulin, the insulin reaches its peak effect on the blood sugar level in about three hours. This effect remains fairly constant for five hours, then declines, and is very low until the next injection. In a typical patient, the level of insulin might be given by the following function.

$$i(t) = \begin{cases} 40t + 100 & \text{if } 0 \le t \le 3 \\ 220 & \text{if } 3 < t \le 8 \\ -80t + 860 & \text{if } 8 < t \le 10 \\ 60 & \text{if } 10 < t \le 24 \end{cases}$$

Here $i(t)$ is the blood sugar level, in appropriate units, at time t measured in hours from the time of the injection. Suppose a patient takes insulin at 6 A.M. Find the blood sugar level at each of the following times: **(a)** 7 A.M., **(b)** 9 A.M., **(c)** 10 A.M., **(d)** noon, **(e)** 2 P.M., **(f)** 5 P.M., **(g)** midnight. **(h)** Graph $y = i(t)$.

46. To rent a midsized car from Avis costs $27 per day or fraction of a day. If you pick up the car in Lansing, and drop it in West Lafayette, there is a fixed $25 dropoff charge. Let $C(x)$ represent the cost of renting the car for x days, taking it from Lansing to West Lafayette. Find each of the following:
(a) $C(3/4)$, **(b)** $C(9/10)$, **(c)** $C(1)$, **(d)** $C(1\ 5/8)$, **(e)** $C\ (2.4)$.
(f) Graph $y = C(x)$. **(g)** Is C a function? **(h)** Is C a linear function?

Graph the functions defined as follows.

47. $f(x) = [-x]$

48. $f(x) = [2x]$

49. $g(x) = [2x - 1]$

50. $h(x) = [3x + 1]$

51. $k(x) = [3x]$

52. $r(x) = [3x] + 1$

Solve each of the following problems.

53. Suppose a chain-saw rental firm charges a fixed $4 sharpening fee plus $7 per day or fraction of a day. Let $S(x)$ represent the cost of renting a saw for x days. Find the following: **(a)** $S(1)$, **(b)** $S(1.25)$, **(c)** $S(3.5)$. **(d)** Graph $y = S(x)$. **(e)** Give the domain and range of S.

54. It costs $.25 to mail a letter weighing one ounce or less, and then $.20 for each additional ounce or fraction of an ounce. Let $L(x)$ be the cost of mailing a letter weighing x ounces. Find the following: **(a)** $L(.75)$, **(b)** $L(1.6)$, **(c)** $L(4)$. **(d)** Graph $y = L(x)$. **(e)** Give the domain and range of L.

55. Use the greatest integer function and write an expression for the number of ounces for which postage will be charged on a letter weighing x ounces (See Exercise 54).

56. A car rental cost $37 for one day, which includes 50 free miles. Each additional 25 miles or portion costs $10. Graph the ordered pairs (miles, cost) for a one-day rental. Use the greatest integer function to write an expression for the cost of a one-day rental for x miles.

57. Suppose that the demand and price for a certain model of electric can opener are related by $p = 16 - (5/4)x$, where p is price and x is demand, in appropriate units. Find the price when the demand is at the following levels: **(a)** 0 units, **(b)** 4 units, **(c)** 8 units. Find the demand for the electric can opener at the following prices: **(d)** $6, **(e)** $11, **(f)** $16. **(g)** Graph $p = 16 - (5/4)x$. Suppose that the price and supply of the item above are related by $p = (3/4)x$, where x represents the supply and p the price. Find the supply at the following prices: **(h)** $0, **(i)** $10, **(j)** $20.
(k) Graph $p = (3/4)x$ on the same axes used for part (g). **(l)** Find the equilibrium supply (the supply at the point where the supply and demand curves cross). **(m)** Find the equilibrium price (the price at the equilibrium supply).

58. Let the supply and demand equations for strawberry-flavored licorice be as follows.

$$\text{Supply: } p = (3/2)x \quad \text{and} \quad \text{Demand: } p = 81 - (3/4)x$$

(a) Graph these on the same axes. **(b)** Find the equilibrium demand. (See Exercise 57.) **(c)** Find the equilibrium price.

In Exercises 59–62, decide which of the following numbers gives the slope of the linear function f. (a) 3, (b) 1, (c) 1/3, (d) −1/3, (e) −1, (f) −3. (Hint: Use the slope formula and consider the sign of the differences in the numerator and denominator as well as their comparative sizes.)

59.
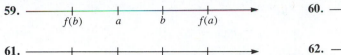

60.
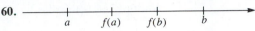

61.

62.
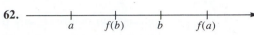

3.5 EQUATIONS OF A LINE

In the previous section we assumed that the graph of a linear function is a straight line. This section develops various forms for the equation of a line.

The first equation given is for a **vertical line,** the only line that is not the graph of a function. The vertical line through the point $(k, 0)$ goes through all points of the form (k, y); this fact is used to determine the equation of a vertical line.

**Equation of a
Vertical Line**

> For any value of y, an equation of the vertical line through (k, y) is
>
> $$x = k.$$

For example, the vertical line through $(-6, 0)$ in Figure 33 has equation $x = -6$.

Point-Slope Form Suppose now that a line has slope m and goes through the fixed point (x_1, y_1), as in Figure 34. Let (x, y) be any other point on this line. By the definition of slope, the slope of this line is

$$\frac{y - y_1}{x - x_1}.$$

Figure 33

Figure 34

Since the slope of the line is m,

$$\frac{y - y_1}{x - x_1} = m.$$

Multiplying both sides by $x - x_1$ gives

$$y - y_1 = m(x - x_1). \qquad (*)$$

Since (x, y) represents any point on the line except (x_1, y_1), and since (x_1, y_1) also satisfies equation $(*)$, all points on the line satisfy equation $(*)$. Thus, equation $(*)$ is the equation of the given line.

Point-Slope Form

The line with slope m passing through the point (x_1, y_1) has equation

$$y - y_1 = m(x - x_1),$$

the **point-slope form** of the equation of a line.

EXAMPLE 1   Write an equation of the line through $(-4, 1)$ with slope -3.
    Here $x_1 = -4$, $y_1 = 1$, and $m = -3$. Use the point-slope form of the equation of a line.

$$y - \mathbf{1} = -\mathbf{3}[x - (-\mathbf{4})]$$
$$y - 1 = -3(x + 4)$$
$$y - 1 = -3x - 12$$
$$y = -3x - 11$$

The final equation is written in the form of a linear function. ●

EXAMPLE 2   Find an equation of the line through $(-3, 2)$ and $(2, -4)$.
    First find the slope, using the definition.

$$m = \frac{-4 - 2}{2 - (-3)} = -\frac{6}{5}$$

Either $(-3, 2)$ or $(2, -4)$ can be used for (x_1, y_1). Choosing $x_1 = -3$ and $y_1 = 2$, the point-slope form gives

$$y - 2 = -\frac{6}{5}[x - (-3)]$$

$$5(y - 2) = -6(x + 3)$$
$$5y - 10 = -6x - 18$$
$$5y = -6x - 8$$
$$y = -\frac{6}{5}x - \frac{8}{5}.$$

Verify that the equation is the same if $(2, -4)$ is used instead of $(-3, 2)$ in the point-slope form.   ●

Slope-Intercept Form   As a special case of the point-slope form of the equation of a line, suppose that a line passes through the point $(0, b)$, so the line has y-intercept b. If the line has slope m, then using the point-slope form with $x_1 = 0$ and $y_1 = b$ gives

$$y - y_1 = m(x - x_1)$$
$$y - b = m(x - 0)$$
$$y = mx + b$$

as an equation of the line. This result is called the *slope-intercept form* of the equation of a line since it shows the slope and the y-intercept.

Slope-Intercept Form

The line with slope m and y-intercept b has equation

$$y = mx + b,$$

the **slope-intercept form** of the equation of a line.

The slope-intercept form was used in the previous section to define a linear function, $f(x) = ax + b$.

EXAMPLE 3   Find the slope and y-intercept of $3x - y = 2$. Graph the line.

First write $3x - y = 2$ in the slope-intercept form, $y = mx + b$. Do this by solving for y, getting $y = 3x - 2$. From this form of the equation, the slope is $m = 3$ and the y-intercept is $b = -2$. Draw the graph by first locating the y-intercept, as in Figure 35. Then, as in the previous section, use the slope of 3, or 3/1, to get a second point on the graph, by going 1 unit to the right and 3 units up.   ●

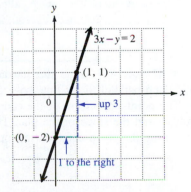

Figure 35

As mentioned in the previous section, a horizontal line has a slope of 0. Letting $m = 0$ in the slope intercept form of the equation of a line gives $y = 0x + b$, or $y = b$, as an equation of a **horizontal line.**

Equation of a Horizontal Line

> For any value of x, an equation of the horizontal line through (x, b) is
>
> $$y = b.$$

Since the slope-intercept form $y = mx + b$ can be written as $mx - 1y + b = 0$ and the vertical-line equation $x = k$ can be written as $1x + 0y - k = 0$, we see that every line has an equation of the form $ax + by + c = 0$, where a and b are not both 0. Conversely, assuming $b \neq 0$, solving $ax + by + c = 0$ for y gives $y = (-a/b)x - c/b$, the equation of a line with slope $-a/b$ and y-intercept $-c/b$. If $b = 0$, solving for x gives $x = -c/a$, a vertical line. In any case, the equation $ax + by + c = 0$ has a line for its graph. This equation is called the **standard form** of the equation of a line.

Standard Form

> If a and b are not both 0, then the equation $ax + by + c = 0$ has a line for its graph. Also, any line has an equation of the form $ax + by + c = 0$.

Parallel and Perpendicular Lines One application of slope involves deciding whether or not two lines are parallel. Since two parallel lines are equally "steep," they should have the same slope. Also, two lines with the same "steepness" are parallel. Slope is undefined for vertical lines, but we know from geometry that any two vertical lines are parallel.

Parallel Lines

> Two nonvertical lines are parallel if and only if they have the same slope.

EXAMPLE 4   Find the standard-form equation of the line through the point $(3, 5)$ and parallel to the line $2x + 5y = 4$.

The point-slope form of the equation of a line requires a point that the line goes through and the slope of the line. The line here goes through $(3, 5)$. Find the slope by writing the given equation in slope-intercept form (that is, solve for y).

$$2x + 5y = 4$$

$$y = -\frac{2}{5}x + \frac{4}{5}$$

The slope is $-2/5$. Since the lines are parallel, the slope of the line whose equation is needed must also be $-2/5$. Substitute $m = -2/5$, $x_1 = 3$, and $y_1 = 5$ into the point-slope form.

$$y - y_1 = m(x - x_1)$$

$$y - 5 = -\frac{2}{5}(x - 3)$$

$$5(y - 5) = -2(x - 3)$$

$$5y - 25 = -2x + 6$$

$$2x + 5y - 31 = 0 \quad \bullet$$

The slopes of perpendicular lines are also related. The product of the slopes of two perpendicular lines is -1. This means that the slopes of two perpendicular lines are negative reciprocals of one another. We give an outline of a proof here.

Figure 36 shows the perpendicular lines, L_1 and L_2. Suppose L_1 has equation $y = m_1x + b_1$ and L_2 has equation $y = m_2x + b_2$. By the slope formula,

$$m_1 = \frac{MQ}{PQ} = \frac{MQ}{1} = MQ.$$

Similarly, $m_2 = -QN$. (As shown in Figure 36, $m_1 > 0$, $m_2 < 0$.) From similar triangles MPQ and PNQ,

$$\frac{m_1}{1} = \frac{1}{-m_2}$$

$$-m_1m_2 = 1$$

$$m_1m_2 = -1,$$

as required.

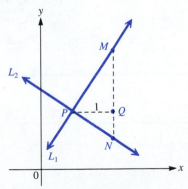

Figure 36

Perpendicular Lines

Two nonvertical lines are perpendicular if and only if the product of their slopes is -1.

EXAMPLE 5   Find the standard-form equation of the line L through $(3, 0)$ and perpendicular to $5x - y = 4$.

Use the point-slope form to get the equation of line L. To find the slope of line L, first find the slope of $5x - y = 4$ by solving for y.

$$5x - y = 4$$
$$y = 5x - 4$$

This line has a slope of 5. Since the lines are perpendicular, if line L has slope m, then $5m = -1$ and $m = -1/5$. Now use the point-slope form to find the required equation

$$y - 0 = -\frac{1}{5}(x - 3)$$
$$5y = -x + 3$$
$$x + 5y - 3 = 0 \quad \bullet$$

The various forms of the equation of a line can be summarized as follows.

Equations of a Line

General equation	Description of line
$ax + by + c = 0$	**Standard form** x-intercept $-c/a$, y-intercept $-c/b$, slope $-a/b$ (if $a \neq 0$ and $b \neq 0$)
$x = k$	**Vertical line** x-intercept k, no y-intercept, line has undefined slope
$y = k$	**Horizontal line** No x-intercept, y-intercept k, line has slope 0
$y = mx + b$	**Slope-intercept form** Slope is m, y-intercept is b
$y - y_1 = m(x - x_1)$	**Point-slope form** Slope is m, line passes through (x_1, y_1)

3.5 Exercises

Write an equation in standard form for each of the following lines.

1. Through $(1, 3)$, $m = -2$

2. Through $(2, 4)$, $m = -1$

3. Through $(-5, 4)$, $m = -3/2$

4. Through $(-4, 3)$, $m = 3/4$

5. $f(2) = 0$, $m = -3/4$

6. $f(0) = -1$, $m = 3/5$

7. $f(-3) = 2$, $m = 0$

8. $f(6) = 1$, $m = 0$

9. $f(-8) = 1$, undefined slope

10. $f(0) = 4$, undefined slope

11. Through $(4, 2)$ and $(1, 3)$

12. Through $(8, -1)$ and $(4, 3)$

13. Through $(-1, 3)$ and $(3, 4)$

14. Through $(6, 0)$ and $(3, 2)$

15. x-intercept 3, y-intercept -2

16. x-intercept -2, y-intercept 4

17. x-intercept -5, no y-intercept

18. y-intercept 3, no x-intercept

19. Horizontal, $f(8) = 7$

20. Vertical, $f(-6) = 5$

21. Through $(-1, 4)$, parallel to $x + 3y = 5$

22. Through $(3, -2)$, parallel to $2x - y = 5$

23. Through $(2, -5)$, parallel to $y - 4 = 2x$

24. Through $(0, 5)$, parallel to $3x - 7y = 8$

25. $f(3) = -4$, perpendicular to $x + y = 4$

26. $f(-2) = 6$, perpendicular to $2x - 3y = 5$

27. $f(1) = 6$, perpendicular to $3x + 5y = 1$

28. $f(-2) = 0$, perpendicular to $8x - 3y = 7$

29. x-intercept -2, parallel to $y = 2x$

30. y-intercept 3, perpendicular to $x = 4$

Consider the linear equation $3x - ky = 5$. For the conditions in Exercises 31–33, find the value of k.

31. The line passes through the point $(3, 2)$.

32. The line has slope 1/2.

33. The line has y-intercept 1.

34. Two lines intersect at the point $(-2, 3)$ and are each perpendicular to one of the coordinate axes. What are the equations of the lines?

For each linear function in Exercises 35–40, identify its graph in (a) – (f) below.

35. $f(x) = 1 - x$

36. $f(x) = \dfrac{1}{2}x$

37. $f(x) = 2$

38. $f(x) = -x - 1$

39. $f(x) = 2x - 1$

40. $f(x) = 1 - 2x$

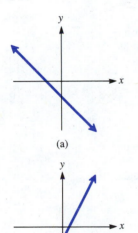

(a)

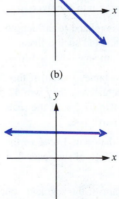

(b)

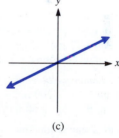

(c)

(d)

(e)

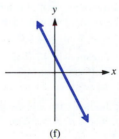

(f)

41. A linear function has $f(2) = 3$ and $f(3) = 4$. Find $f(x)$ and $f(7)$.

42. If $f(5) = -2$, $f(1) = 0$, and f is a linear function, find $f(x)$ and $f(-4)$.

43. Find k so that the line through $(4, -1)$ and $(k, 2)$ is as follows.
(a) Parallel to $3y + 2x = 6$
(b) Perpendicular to $2y - 5x = 1$

44. Find r so that the line through $(2, 6)$ and $(-4, r)$ is as follows.
(a) Parallel to $2x - 3y = 4$
(b) Perpendicular to $x + 2y = 1$

45. Use slopes to show that the quadrilateral with vertices at $(1, 3)$, $(-5/2, 2)$, $(-7/2, 4)$, and $(2, 1)$ is a parallelogram (has opposite sides parallel).

46. Use slopes to show that the square with vertices at $(-2, 5)$, $(4, 5)$, $(4, -1)$, and $(-2, -1)$ has diagonals that are perpendicular.

Find and simplify the slopes of the lines through the points $(x, f(x))$ and $(x + h, f(x + h))$ for the functions f defined by the following rules.

47. $f(x) = 2x - 1$ **48.** $f(x) = 4 - x$ **49.** $f(x) = 1 - x^2$ **50.** $f(x) = x^2 + 2$

Many real situations can be described approximately by a straight-line graph. One way to find the equation of such a straight line is to use two typical data points from the graph and the point-slope form of the equation of a line. In each of the following problems, assume that the data can be approximated fairly closely by a straight line. Use the given information to find the equation of the line. Find the slope of each line.

51. A company finds that it can make a total of 20 solar heaters for $13,900, while 10 heaters cost $7500. Let y be the total cost to produce x solar heaters.

52. The sales of a small company were $27,000 in its second year of operation and $63,000 in its fifth year. Let y represent sales in year x.

53. When a certain industrial pollutant is introduced into a river, the reproduction of catfish declines. In a given period of time, three tons of the pollutant results in a fish population of 37,000. Also, 12 tons of pollutant produce a fish population of 28,000. Let y be the fish population when x tons of pollutant are introduced into the river.

54. In the snake *Lampropelbis polyzona*, total length y is related to tail length x in the domain 30 mm $\leq x \leq$ 200 mm by a linear function. Find such a linear function, if a snake 455 mm long has a 60-mm tail, and a 1050-mm snake has a 140-mm tail.

55. According to research done by the political scientist James March, if the Democrats win 45% of the two-party vote for the House of Representatives, they win 42.5% of the seats. If the Democrats win 55% of the vote, they win 67.5% of the seats. Let y be the percent of seats won, and x the percent of the two-party vote.

56. If the Republicans win 45% of the two-party vote, they win 32.5% of the seats (see Exercise 55). If they win 60% of the vote, they get 70% of the seats. Let y represent the percent of the seats, and x the percent of the vote.

57. Show that the line $y = x$ is the perpendicular bisector of the segment connecting (a, b) and (b, a), where $a \neq b$. (*Hint:* Use the midpoint formula and the slope formula.)

58. Let $y = m_1x + b_1$ and $y = m_2x + b_2$ be the equations of two nonvertical lines.   (a) Set $m_1x + b_1$ equal to $m_2x + b_2$ and solve for x to find the x-coordinate of the point of intersection of the lines.   (b) If $m_1 = m_2$, the lines must be parallel. Why?

3.6 QUADRATIC FUNCTIONS

In earlier sections of this chapter, we have discussed several first-degree functions, where the highest power of the variable is 1. This section looks at functions of degree 2, called *quadratic functions*.

Definition of Quadratic Function

A function f is a **quadratic function** if

$$f(x) = ax^2 + bx + c,$$

where a, b, and c are real numbers with $a \neq 0$.

The simplest quadratic function is given by $f(x) = x^2$ with $a = 1$, $b = 0$, and $c = 0$. To find some points on the graph of this function, choose some values for x and find the corresponding values for $f(x)$, as in the chart with Figure 37. Then plot these points, and draw a smooth curve through them. (The reason for drawing a smooth curve depends on ideas from calculus). This graph is called a **parabola.** Every quadratic function has a graph that is a parabola.

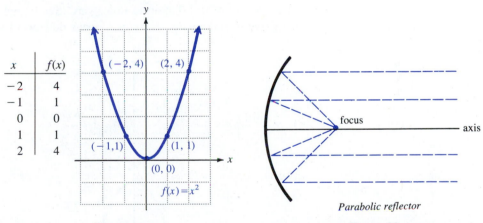

x	$f(x)$
-2	4
-1	1
0	0
1	1
2	4

Figure 37

Parabolic reflector

Figure 38

Parabolas are symmetric about a line (the y-axis in Figure 37). Intuitively, this means that if the graph were folded along the line of symmetry, the two sides would coincide. We discuss symmetry in more detail later in this chapter. The line of symmetry for a parabola is called the **axis** of the parabola. The point where the axis intersects the parabola is the **vertex** of the parabola.

Parabolas have many practical applications. For example, the reflectors of solar ovens and flashlights are made by revolving a parabola about its axis. The **focus** of a parabola is a point on its axis that determines the curvature. See Figure 38. (The focus is discussed in more detail in Chapter 10). When the parabolic reflector of a solar oven is aimed at the sun, the light rays bounce off the reflector and collect at the focus, creating an intense temperature at that point. On the other hand, when a lightbulb is placed at the focus of a parabolic reflector, light rays reflect out parallel to the axis.

Starting with $f(x) = x^2$, there are several possible ways to get a more general expression for a quadratic function:

$f(x) = ax^2$	Multiply by a positive or negative coefficient.
$f(x) = x^2 + k$	Add a positive or negative constant.
$f(x) = (x - h)^2$	Replace x with $x - h$, where h is a constant.
$f(x) = a(x - h)^2 + k$	Do all of the above.

The graph of each of these quadratic functions is still a parabola, but modified from that of $f(x) = x^2$. The next few examples show how these changes modify the graphs of the functions. The first example shows the result of changing $f(x) = x^2$ to $f(x) = ax^2$, for a nonzero constant a.

· · · · · · · · ·

EXAMPLE 1   Graph the functions defined as follows.

(a) $g(x) = -x^2$

For a given value of x, the corresponding value of $g(x)$ will be the negative of what it was for $f(x) = x^2$. Because of this, the graph of $g(x) = -x^2$ is the same shape as that of $f(x)$, but opens downward. See Figure 39. This is generally true; the graph of $f(x) = ax^2 + bx + c$ opens downward whenever $a < 0$.

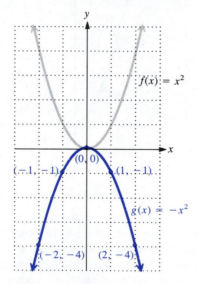

Figure 39

(b) $g(x) = \dfrac{1}{2}x^2$

Choose a value of x, and then find $g(x)$. The coefficient of 1/2 will cause the resulting value of $g(x)$ to be smaller than for $f(x) = x^2$, making the parabola "broader" than $f(x) = x^2$. See Figure 40. In both parabolas of this example, the axis is the vertical line $x = 0$ and the vertex is the origin $(0, 0)$.   ●

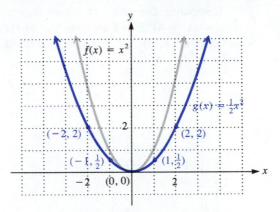

Figure 40

The next few examples show the results of horizontal and vertical shifts, called **translations,** of the graph of $f(x) = x^2$.

● ● ● ● ● ● ● ● ●

EXAMPLE 2   Graph $g(x) = x^2 - 4$.

By comparing the tables of values for $g(x) = x^2 - 4$ and $f(x) = x^2$ shown with Figure 41, we can see that for corresponding x-values, the y-values of g are each 4 less than those for f. Thus, the graph of $f(x) = x^2 - 4$ is the same as that of $f(x) = x^2$, but translated 4 units down. See Figure 41. The vertex of this parabola (here the lowest point) is at $(0, -4)$. The axis of the parabola is the vertical line $x = 0$.   ●

$g(x) = x^2 - 4$		$f(x) = x^2$	
x	y	x	y
-2	0	-2	4
-1	-3	-1	1
0	-4	0	0
1	-3	1	1
2	0	2	4

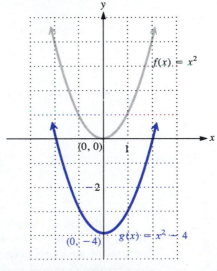

Figure 41

. . . .

EXAMPLE 3   Graph $g(x) = (x - 4)^2$.

Comparing the tables of values shown with Figure 42 shows that the graph of $g(x) = (x - 4)^2$ is the same as that of $f(x) = x^2$, but translated 4 units right. The vertex is at (4, 0). As shown in Figure 42, the axis of this parabola is the vertical line $x = 4$. ●

$g(x) = (x - 4)^2$ $f(x) = x^2$

x	y
2	4
3	1
4	0
5	1
6	4

x	y
-2	4
-1	1
0	0
1	1
2	4

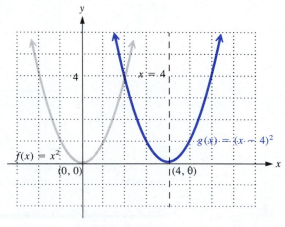

Figure 42

. . . .

EXAMPLE 4   Graph $f(x) = -(x + 3)^2 + 1$.

The vertex of this parabola is translated 3 units to the left and 1 unit up. The parabola opens downward. The vertex, the point $(-3, 1)$, is the *highest* point on the graph. The axis is the line $x = -3$. See Figure 43. ●

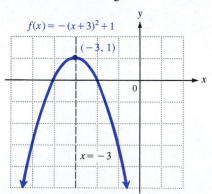

Figure 43

The process of *completing the square* (discussed in Chapter 2) can be used to change $f(x) = ax^2 + bx + c$ to the form $f(x) = a(x + h)^2 + k$, so that the vertex and axis of the graph may be identified from the equation. Follow the steps given in the next two examples to do this.

EXAMPLE 5   Graph $f(x) = x^2 - 6x + 7$.

To graph this parabola, $x^2 - 6x + 7$ must be rewritten in the form $(x - h)^2 + k$. Start as follows.

$$f(x) = (x^2 - 6x \quad) + 7$$

As shown earlier, a number must be added inside the parentheses to get a perfect square trinomial. To find this number, take half the coefficient of x and then square the result. Half of -6 is -3, and $(-3)^2$ is 9. Now add and subtract 9 inside the parentheses. (This is the same as adding 0.)

$$f(x) = (x^2 - 6x + 9 - 9) + 7$$

Group the terms and factor as follows.

$$f(x) = (x^2 - 6x + 9) - 9 + 7$$
$$f(x) = (x - 3)^2 - 2$$

This result shows that the vertex of the parabola is $(3, -2)$ and the axis is the line $x = 3$. The graph is shown in Figure 44.   ●

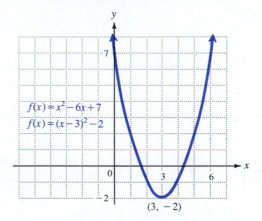

Figure 44

NOTE   In Example 5, we added and subtracted 9 *on the same side* of the equation to complete the square. This differs from adding the same number to *each side of the equation,* as when we completed the square in Chapter 2. Here, since we want just y on one side of the equation, we had to change slightly that step in the process of completing the square.

EXAMPLE 6   Graph $f(x) = -3x^2 - 2x + 1$.

To complete the square, first factor out -3.

$$f(x) = -3\left(x^2 + \frac{2}{3}x \quad\right) + 1$$

(This is necessary to make the coefficient of x^2 equal 1.) Half the coefficient of x is

1/3, and $(1/3)^2 = 1/9$. Add and subtract 1/9 inside the parentheses as follows.

$$f(x) = -3\left(x^2 + \frac{2}{3}x + \frac{1}{9} - \frac{1}{9}\right) + 1$$

Use the distributive property and simplify.

$$f(x) = -3\left(x^2 + \frac{2}{3}x + \frac{1}{9}\right) - 3\left(-\frac{1}{9}\right) + 1$$

$$f(x) = -3\left(x^2 + \frac{2}{3}x + \frac{1}{9}\right) + \frac{1}{3} + 1$$

$$f(x) = -3\left(x^2 + \frac{2}{3}x + \frac{1}{9}\right) + \frac{4}{3}$$

Factor. $$f(x) = -3\left(x + \frac{1}{3}\right)^2 + \frac{4}{3}$$

Now the equation of the parabola is written in the form $f(x) = a(x - h)^2 + k$. In this form, the equation shows that the axis of the parabola is the vertical line

$$x + \frac{1}{3} = 0 \quad \text{or} \quad x = -\frac{1}{3}$$

and that the vertex is $(-1/3, 4/3)$. Additional points can be found by substituting x-values near the vertex into the original equation. For example, $(-2, -7)$ and $(1/2, -3/4)$ are on the graph, shown in Figure 45. The intercepts are often good additional points to find. Here, the y-intercept is

$$y = -3(0)^2 - 2(0) + 1 = 1,$$

giving the point $(0, 1)$. The x-intercepts are found by solving the equation

$$0 = -3x^2 - 2x + 1.$$

From the quadratic formula,

$$x = \frac{2 \pm \sqrt{4 + 12}}{-6},$$

giving -1 and $1/3$ as the x-intercepts.   ●

We can now generalize the work above to get a formula for the vertex of a parabola. Starting with the general quadratic function, $f(x) = ax^2 + bx + c$, and completing the square will change the function to the form $f(x) = a(x - h)^2 + k$. Begin by factoring a from the first two terms.

$$f(x) = ax^2 + bx + c$$

$$= a\left(x^2 + \frac{b}{a}x \quad\right) + c$$

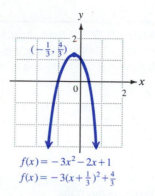

$$f(x) = -3x^2 - 2x + 1$$
$$f(x) = -3(x + \tfrac{1}{3})^2 + \tfrac{4}{3}$$

Figure 45

Now add $\left(\dfrac{1}{2} \cdot \dfrac{b}{a}\right)^2 = \dfrac{b^2}{4a^2}$ in the parentheses and subtract $a\left(\dfrac{b^2}{4a^2}\right)$ from c.

$$f(x) = a\left(x^2 + \frac{b}{a}x + \frac{b^2}{4a^2}\right) + c - a\left(\frac{b^2}{4a^2}\right)$$

$$= a\left(x + \frac{b}{2a}\right)^2 + c - \frac{b^2}{4a}$$

Comparing the last result with $f(x) = a(x - h)^2 + k$ shows that

$$h = -\frac{b}{2a} \quad \text{and} \quad k = c - \frac{b^2}{4a}.$$

Letting $x = h$ in $f(x) = a(x - h)^2 + k$ gives

$$f(h) = a(h - h)^2 + k = k,$$

so $k = f(h)$, or $k = f(-b/2a)$.

The following statement summarizes this discussion.

Graph of a Quadratic Function

The quadratic function defined by $f(x) = ax^2 + bx + c$ can be written in the form

$$y = f(x) = a(x - h)^2 + k, \quad a \neq 0,$$

where
$$h = -\frac{b}{2a} \quad \text{and} \quad k = f(h).$$

The graph of f

1. is a parabola with vertex (h, k), and the vertical line $x = h$ as axis,
2. opens upward if $a > 0$ and downward if $a < 0$,
3. is broader than $y = x^2$ if $|a| < 1$ and is narrower than $y = x^2$ if $|a| > 1$.

The vertex and axis of a parabola can be found from its equation either by completing the square or by memorizing the formula $h = -b/2a$ and letting $k = f(h)$.

· · · · · · · · ·

EXAMPLE 7   Find the axis and the vertex of the parabola having equation $f(x) = 2x^2 + 4x + 5$ using the formula given above.

Here $a = 2$, $b = 4$, and $c = 5$. The axis of the parabola is the vertical line

$$x = h = -\frac{b}{2a}$$

$$= -\frac{4}{2(2)}$$

$$x = -1.$$

The vertex is the point

$$(-1, f(-1)) = (-1, 3). \quad \bullet$$

The fact that the vertex of a parabola of the form $f(x) = ax^2 + bx + c$ is the highest or lowest point on the graph can be used in applications to find a maximum or a minimum value.

· · · · · · · · ·

EXAMPLE 8   Ms. Whitney owns and operates Aunt Emma's Pie Shop. She has hired a consultant to analyze her business operations. The consultant tells her that her profit $P(x)$ is given by

$$P(x) = 120x - x^2,$$

where x is the number of units of pies that she makes. How many units of pies should be made to maximize profit? What is the maximum possible profit?

The profit function can be rewritten as $P(x) = -x^2 + 120x + 0$, a quadratic function with $a = -1$, $b = 120$, and $c = 0$. Complete the square to find that the vertex of the parabola is (60, 3600). Since $a < 0$, the parabola opens downward and the vertex is the highest point on the graph, producing a *maximum* rather than a minimum. Figure 46 shows that portion of the profit function in quadrant I. (Why is quadrant I the only one of interest here?) The maximum profit of $3600 is made when 60 units of pies are made. In this case, profit increases as more and more pies are made up to 60 units and then decreases as more and more pies are made past this point. $\bullet$

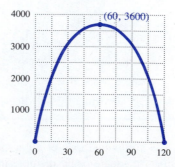

Figure 46

3.6 Exercises ··

1. Graph the functions defined as follows on the same coordinate system.

   (a) $f(x) = 2x^2$ (b) $f(x) = 3x^2$ (c) $f(x) = \frac{1}{2}x^2$ (d) $f(x) = \frac{1}{3}x^2$

   (e) How does the coefficient affect the shape of the graph?

2. Graph the functions defined as follows on the same coordinate system.
   (a) $f(x) = x^2 + 2$ (b) $f(x) = x^2 - 1$ (c) $f(x) = x^2 + 1$ (d) $f(x) = x^2 - 2$
   (e) How do these graphs differ from the graph of $f(x) = x^2$?

3. Graph the functions defined as follows on the same coordinate system.
   (a) $f(x) = (x - 2)^2$ (b) $f(x) = (x + 1)^2$ (c) $f(x) = (x + 3)^2$ (d) $f(x) = (x - 4)^2$
   (e) How do these graphs differ from the graph of $f(x) = x^2$?

4. Give the range of each of the functions defined as follows.

   (a) $f(x) = (x - 1)^2 + 5$ (b) $f(x) = -(x + 3)^2 - 1$

   (c) $g(x) = -2(x + 4)^2 - 3$ (d) $h(x) = -\frac{1}{2}(x - 5)^2 + 2$

Graph the functions defined as follows. Give the vertex, axis, x-intercepts, and y-intercepts of each graph.

5. $f(x) = (x - 2)^2$
6. $f(x) = (x + 4)^2$
7. $g(x) = (x + 3)^2 - 4$

8. $h(x) = (x - 5)^2 - 4$
9. $k(x) = -2(x + 3)^2 + 2$
10. $F(x) = -3(x - 2)^2 + 1$

11. $H(x) = -\frac{1}{2}(x + 1)^2 - 3$
12. $G(x) = \frac{2}{3}(x - 2)^2 - 1$
13. $f(x) = x^2 - 2x + 3$

14. $f(x) = x^2 + 6x + 5$
15. $g(x) = -x^2 - 4x + 2$
16. $k(x) = -x^2 + 6x - 6$

17. $f(x) = 2x^2 - 4x + 5$
18. $g(x) = -3x^2 + 24x - 46$

The figure shows the graph of a quadratic function, f.

19. What is the maximum value of $f(x)$?

20. How many solutions are there to the equation $f(x) = 1$?

21. How many solutions are there to the equation $f(x) = 5$?

22. For what value of x will $f(x)$ be as large as possible?

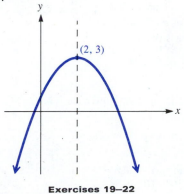

Exercises 19–22

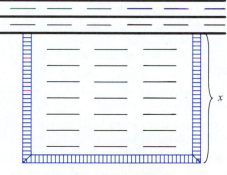

Exercise 23

23. Glenview College wants to construct a rectangular parking lot on land bordered on one side by a highway. It has 320 ft of fencing which it will use to fence off the other three sides. What should be the dimensions of the lot if the enclosed area is to be a maximum? See the figure. (*Hint:* Let x represent the width of the lot and let $320 - 2x$ represent the length. Graph the area parabola, $A = x(320 - 2x)$, and investigate the vertex.)

24. The revenue of a charter bus company depends on the number of unsold seats. If the revenue $R(x)$, is given by $R(x) = 5000 + 50x - x^2$, where x is the number of unsold seats, find the maximum revenue and the number of unsold seats which produce maximum revenue.

25. The number of mosquitoes, $M(x)$, in millions, in a certain area of Kentucky depends on the June rainfall, x, in inches, approximately as $M(x) = 10x - x^2$. Find the rainfall that will produce the maximum number of mosquitoes.

26. If an object is thrown upward with an initial velocity of 32 feet per second, then its height after t seconds is given by $h = 32t - 16t^2$. Find the maximum height attained by the object. Find the number of seconds it takes the object to hit the ground.

27. Find two numbers whose sum is 20 and whose product is a maximum. (*Hint:* Let x and $20 - x$ be the two numbers, and write an equation for the product.)

28. A charter flight charges a fare of $200 per person plus $4 per person for each unsold seat on the plane. If the plane holds 100 passengers, and if x represents the number of unsold seats, find the following.
   (a) An expression for the total revenue received for the flight (*Hint:* Multiply the number of people flying, $100 - x$, by the price per ticket.)
   (b) The graph for the expression in part (a)
   (c) The number of unsold seats that will produce the maximum revenue
   (d) The maximum revenue

29. The demand for a certain type of cosmetic is given by $p = 500 - x$, where p is the price per unit when x units are demanded.
   (a) Find the revenue, $R(x)$, obtained when x units are demanded. (*Hint:* Revenue = number of units demanded $\times$ price per unit.)
   (b) Graph the revenue function defined by $y = R(x)$.
   (c) From the graph of the revenue function, estimate the price that will produce maximum revenue.
   (d) What is the maximum revenue?

30. Between the months of June and October, the percent of maximum possible chlorophyll production in a leaf is approximated by $C(x)$, where $C(x) = 10x + 50$. Here x is time in months with $x = 1$ representing June. From October through December, $C(x)$ is approximated by $C(x) = -20(x - 5)^2 + 100$. Find the percent of maximum possible chlorophyll production in each of the following months: **(a)** June, **(b)** July, **(c)** September, **(d)** October, **(e)** November, **(f)** December. **(g)** Sketch a graph of $y = C(x)$, from June through December. In what month is chlorophyll production a maximum?

31. An arch is shaped like a parabola. It is 30 m wide at the base and 15 m high. How wide is the arch 10 m from the ground?

32. A culvert is shaped like a parabola, 18 cm across the top and 12 cm deep. How wide is the culvert 8 cm from the top?

33. Suppose that a quadratic function with $a > 0$ is written in the form $f(x) = a(x - h)^2 + k$. Match each of the items (a), (b), and (c) with one of the items (A), (B), or (C).
   (a) k is positive   **(A)** $f(x)$ intersects the x-axis at only one point
   (b) k is negative   **(B)** $f(x)$ does not intersect the x-axis
   (c) k is zero   **(C)** $f(x)$ intersects the x-axis twice

The figures below show several possible graphs of $f(x) = ax^2 + bx + c$. *For the restrictions on a, b, and c in Exercises 34–37, select the possible corresponding graph.*

34. $a < 0, b^2 - 4ac = 0$

35. $a > 0, b^2 - 4ac < 0$

36. $a > 0, b^2 - 4ac > 0$

37. $a > 0, b^2 - 4ac = 0$

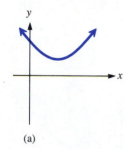

(a)

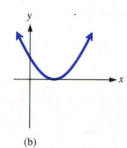

(b)

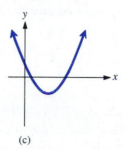

(c)

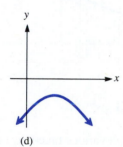

(d)

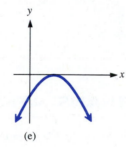

(e)

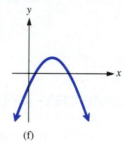

(f)

38. Find a value of c so that $y = x^2 - 10x + c$ has exactly one x-intercept.

39. Find b so that $y = x^2 + bx + 9$ has exactly one x-intercept.

40. Find the quadratic function having x-intercepts 2 and 5, and y-intercept 5.

In Exercises 41–42, let a parabola have equation $y = ax^2 + bx + c$.

41. From the discriminant introduced in Section 2.3, when does the graph of the parabola not cross the x-axis?

42. When does the graph cross the x-axis in two distinct points?

43. Let $f(x) = a(x - h)^2 + k$, and show that $f(h + x) = f(h - x)$. Why does this show that the parabola is symmetric to its axis?

44. A parabola can be defined as the set of all points in the plane equally distant from a given point (called the *focus*) and a given line (called the *directrix*) not containing the point. See the figure.
  (a) Suppose (x, y) is any point on the parabola. Suppose the line mentioned in the definition is $y = -p$. Find the distance between (x, y) and the line $y = -p$. (The distance from a point to a line is the length of the perpendicular from the point to the line.)
  (b) If $y = -p$ is the line mentioned in the definition, why is it reasonable to use $(0, p)$ as the given point? (See the figure.) Find the distance from (x, y) to $(0, p)$.
  (c) Find an equation for the parabola in the figure.

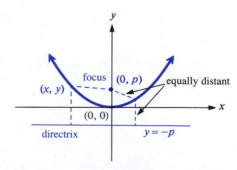

Exercise 44

Use the geometric definition of parabola given in Exercise 44 to find the equation of each parabola having the given point as focus.

45. $(0, 3)$, directrix $y = -3$ **46.** $(0, -5)$, directrix $y = 5$ **47.** $(-2, 0)$, directrix $x = 2$

48. $(8, 0)$, directrix $x = -8$ **49.** $(3, 6)$, vertex $(3, 4)$ **50.** $(-5, 2)$, vertex $(-5, 5)$

51. Use the results of Exercise 44 to find the equation of a parabola with vertex at (h, k) and axis $x = h$.

52. Find the largest possible value of y if $y = -(x - 2)^2 + 9$. Then find the following.
   (a) The largest possible value of $\sqrt{-(x - 2)^2 + 9}$
   (b) The smallest possible value of $1/[-(x - 2)^2 + 9]$

53. Find the smallest possible value of y if $y = 3 + (x + 5)^2$. Then find the following.
   (a) The smallest possible value of $\sqrt{3 + (x + 5)^2}$
   (b) The largest possible value of $1/[3 + (x + 5)^2]$

54. The figure shows a portion of the graph of $f(x) = x^2 + 3x + 1$ and a rectangle with its base on the x-axis and a vertex on the graph. What is the area of the rectangle?

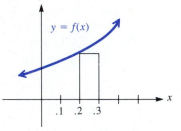

3.7 SYMMETRY AND TRANSLATIONS

Two of the main objectives of this course are to learn to graph various functions and to recognize and be able to write an equation for $f(x)$ from the graph of f. Several graphing techniques are presented in this section that generalize the methods used to graph quadratic functions in Section 3.6. Later chapters will introduce techniques for graphing additional specific functions and for writing the equations of functions whose graphs are given.

Symmetry   The graph shown in Figure 47(a) is cut in half by the y-axis with each half the mirror image of the other half. A graph with this property is said to be *symmetric with respect to the y-axis*. As this graph suggests, a graph is symmetric with respect to the y-axis if the point $(-x, y)$ is on the graph whenever (x, y) is on the graph.

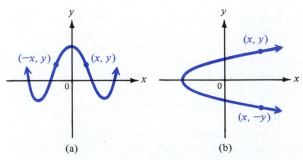

Figure 47

If the graph in Figure 47(b) were folded in half along the x-axis, the portion at the top would exactly match the portion at the bottom. Such a graph is *symmetric with respect to the x-axis:* the point $(x, -y)$ is on the graph whenever the point (x, y) is on the graph.

The following test tells when a graph is symmetric with respect to the x-axis or y-axis.

Symmetry with Respect to an Axis

The graph of an equation is **symmetric with respect to the y-axis** if the replacement of x with $-x$ results in an equivalent equation.

The graph of an equation is **symmetric with respect to the x-axis** if the replacement of y with $-y$ results in an equivalent equation.

EXAMPLE 1   Test for symmetry with respect to the x-axis or the y-axis.

(a) $y = x^2 + 4$

Replace x with $-x$.

$$y = x^2 + 4 \text{ becomes } y = (-x)^2 + 4 = x^2 + 4$$

The result is the same as the original equation, so the graph, shown in Figure 48, is symmetric with respect to the y-axis. Check that the graph is *not* symmetric with respect to the x-axis.

(b) $x = y^2 - 3$

Replace y with $-y$ to get $x = (-y)^2 - 3 = y^2 - 3$, the same as the original equation. The graph is symmetric with respect to the x-axis as shown in Figure 49. Is the graph symmetric with respect to the y-axis?

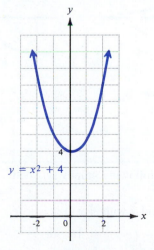

Figure 48

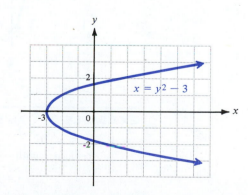

Figure 49

(c) $x^2 + y^2 = 16$

Here,

$$(-x)^2 + y^2 = 16 \quad \text{and} \quad x^2 + (-y)^2 = 16$$

both become

$$x^2 + y^2 = 16.$$

Thus the graph, a circle of radius 4 centered at the origin, is symmetric with respect to both axes.

(d) $2x + y = 4$

Replace x with $-x$ and then y with $-y$; in neither case does an equivalent equation result. This graph is symmetric with respect to neither the x-axis nor the y-axis. •

Another kind of symmetry is found when a graph can be rotated 180° about the origin, with the result coinciding exactly with the original graph. Symmetry of this type is called *symmetry with respect to the origin*. It turns out that rotating a graph 180° is equivalent to saying that the point $(-x, -y)$ is on the graph whenever (x, y) is on the graph. Figure 50 shows two graphs that are symmetric with respect to the origin.

Symmetry with Respect to the Origin

> The graph of an equation is **symmetric with respect to the origin** if the replacement of both x with $-x$ and y with $-y$ results in an equivalent equation.

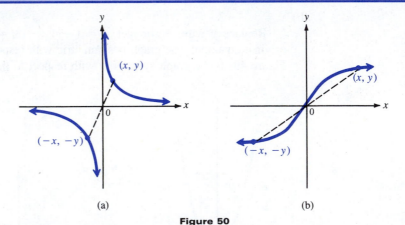

(a) (b)

Figure 50

EXAMPLE 2 Are the following graphs symmetric with respect to the origin?

(a) $x^2 + y^2 = 16$

Replace x with $-x$ and y with $-y$ to get

$$(-x)^2 + (-y)^2 = 16 \quad \text{or} \quad x^2 + y^2 = 16,$$

an equivalent equation. This graph, shown in Figure 51, is symmetric with respect to the origin.

(b) $y = x^3$

Replace x with $-x$ and y with $-y$ to get

$$-y = (-x)^3, \quad \text{or} \quad -y = -x^3, \quad \text{or} \quad y = x^3,$$

an equivalent equation. The graph, symmetric with respect to the origin, is shown in Figure 52.   ●

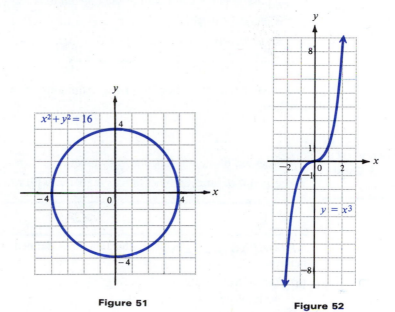

Figure 51   **Figure 52**

A graph symmetric with respect to both the x- and y-axes is automatically symmetric with respect to the origin. (See Exercise 48.) However, a graph symmetric to the origin need not be symmetric to either axis. (See Figure 52.) Of the three types of symmetry—with respect to the x-axis, the y-axis, and the origin—a graph possessing any two must have the third type also. (See Exercise 49.)

The various tests for symmetry are summarized below.

Tests for Symmetry

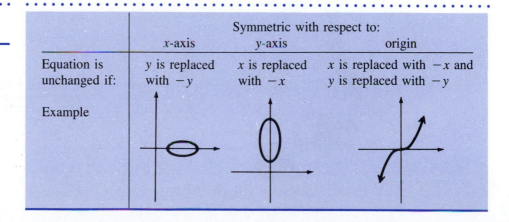

	Symmetric with respect to:		
	x-axis	y-axis	origin
Equation is unchanged if:	y is replaced with $-y$	x is replaced with $-x$	x is replaced with $-x$ and y is replaced with $-y$
Example			

Odd and Even Functions A function whose graph is symmetric with respect to the y-axis is an even function, while a function having a graph symmetric with respect to the origin is an odd function.

Definition of Odd and Even Functions

Suppose x and $-x$ are both in the domain of a function f. Then

f is an **odd function** if $f(-x) = -f(x)$ for every x in the domain of f;

f is an **even function** if $f(-x) = f(x)$ for every x in the domain of f.

The next statements follow from the definitions of symmetry and of odd and even functions.

Symmetry of Odd and Even Functions

The graph of an odd function is symmetric with respect to the origin;

the graph of an even function is symmetric with respect to the y-axis.

EXAMPLE 3 Decide if the functions defined as follows are odd, even, or neither.

(a) $f(x) = 8x^4 - 3x^2$

Replace x with $-x$.

$$f(-x) = 8(-x)^4 - 3(-x)^2$$
$$= 8x^4 - 3x^2$$
$$= f(x)$$

Since $f(x) = f(-x)$ for each x in the domain of the function, it is symmetric with respect to the y-axis and f is an even function.

(b) $f(x) = 6x^3 - 9x$

Here

$$f(-x) = 6(-x)^3 - 9(-x)$$
$$= -6x^3 + 9x$$
$$= -f(x).$$

This function is symmetric with respect to the origin and is odd.

(c) $f(x) = \dfrac{1}{x - 3}$

Replacing x with $-x$ produces

$$f(-x) = \frac{1}{-x - 3}.$$

This equals neither

$$f(x) = \frac{1}{x - 3}$$

nor $$-f(x) = -\frac{1}{x - 3} = \frac{1}{-x + 3}.$$

This function is neither odd nor even.   ●

The exponents on the terms in parts (a) and (b) of Example 3 show the origin of the names *odd* and *even*. However, other functions will be introduced later that are odd or even but do not have terms with exponents. Actually, most functions are neither odd nor even.

Translations   In Section 3.6, we saw that the graph of the function defined by $g(x) = a(x + h)^2 + k$ was the same shape as the graph of $f(x) = ax^2$, but was translated vertically and horizontally if $h \neq 0$ and $k \neq 0$. More generally, the graph of a function g, defined by $g(x) = f(x) + c$, where c is a real number, can be found from the graph of the function f as follows. For every point (x, y) on the graph of f, there will be a corresponding point $(x, y + c)$ on the graph of g. The new graph will be the same as the graph of f, but moved, or *translated*, c units upward if c is positive, or $|c|$ units downward if c is negative. The resulting graph is called a **vertical translation.** Figure 53 shows a graph of a function f and two different vertical translations of it.

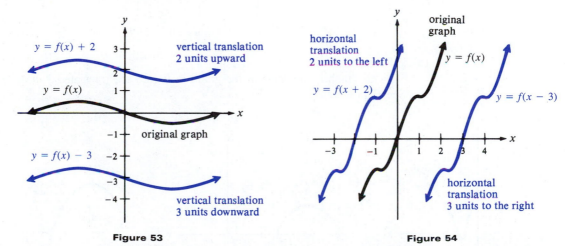

Figure 53   Figure 54

If a function g is defined by $y = f(x - c)$, for each ordered pair (x, y) of f, there will be a corresponding ordered pair $(x - c, y)$ on the graph of g. This has the effect of translating the graph of f horizontally; c units to the right if c is positive and $|c|$ units to the left if c is negative. Figure 54 shows the graph of $y = f(x)$ along with the graphs of $y = f(x - 3)$ and $y = f(x + 2)$; each of these graphs is obtained from that of $y = f(x)$ by a **horizontal translation.**

· ·

Translations of the Graph of a Function

Let f be a function, and let c be a positive number.

To graph	Shift the graph of $y = f(x)$ by c units
$y = f(x) + c$	upward
$y = f(x) - c$	downward
$y = f(x + c)$	left
$y = f(x - c)$	right

· · · · · · · · · · ·

EXAMPLE 4   A graph of a function defined by $y = f(x)$ is shown in Figure 55. Use this graph to find each of the following graphs.

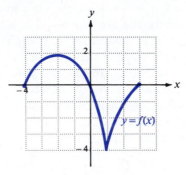

Figure 55

(a) $g(x) = f(x) + 3$

This graph is the same as the graph in Figure 55, translated 3 units upward. See Figure 56(a).

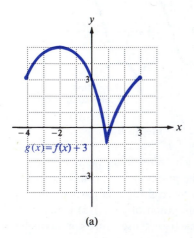

(a)

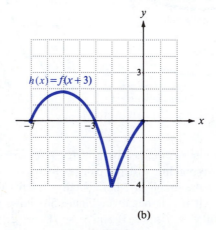

(b)

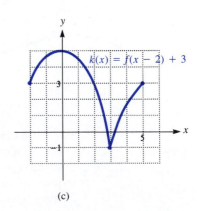

(c)

Figure 56

(b) $h(x) = f(x + 3)$

To get the graph of $y = f(x + 3)$, translate the graph of $y = f(x)$ three units to the left. See Figure 56(b).

(c) $k(x) = f(x - 2) + 3$

This graph will look like the graph of $f(x)$ translated 2 units to the right and 3 units up, as shown in Figure 56(c). ●

In Section 3.6 we saw that the graph of $f(x) = a(x + h)^2 + k$ was broader than that of $f(x) = (x + h)^2 + k$ if $|a| < 1$ and narrower if $|a| > 1$. Also, if $a < 0$, the two parabolas open in opposite directions. This idea can be generalized.

Let f be a function and a a nonzero number. The graph of $y = a \cdot f(x)$ is broader than the graph of $y = f(x)$ if $|a| < 1$, and the graph of $y = a \cdot f(x)$ is narrower than the graph of $y = f(x)$ if $|a| > 1$. Also, if $a < 0$, the graph of $y = a \cdot f(x)$ is the reflection of the graph of $y = |a| \cdot f(x)$ about the x-axis.

EXAMPLE 5   Graph $h(x) = |2x - 4|$.

Factor out 2 as follows.

$$\begin{aligned} h(x) &= |2x - 4| \\ &= |2(x - 2)| \\ &= |2| \cdot |x - 2| \\ &= 2|x - 2| \end{aligned}$$

The graph of h will be narrower than the graph of $f(x) = |x - 2|$, because $2 > 1$. See Figure 57. ●

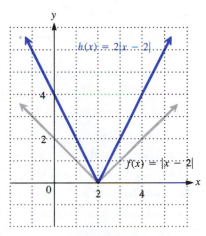

Figure 57

3.7 Exercises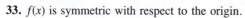

Plot the following points, and then plot the points that are symmetric to the given point with respect to the (a) x-axis, (b) y-axis, (c) origin.

1. $(5, -3)$ **2.** $(-6, 1)$ **3.** $(-4, -2)$ **4.** $(-8, 0)$

Decide whether each of the following equations have graphs that are symmetric with respect to the x-axis, the y-axis, or the origin.

5. $3x^2 - y^2 = 8$ **6.** $5y^2 - x^2 = 6$ **7.** $x^2y = 5$ **8.** $y = \dfrac{-1}{x^2 + 9}$

9. $y = 3x + 5$ **10.** $x = y^2$ **11.** $|y| = x$ **12.** $xy = 1$

Use symmetry, point plotting, and translations to help graph each of the following.

13. $|y| = 2(x - 1)$ **14.** $|x| = |y - 1|$ **15.** $|y| = |x + 2|$ **16.** $xy = 4$

17. $\dfrac{x}{y} = -1$ **18.** $\dfrac{x}{y} = 2$ **19.** $y = x^3$ **20.** $y = -x^3$

21. $y + 5 = (x - 2)^3$ **22.** $y - 2 = (x + 1)^3$ **23.** $x^2 = y^2$ **24.** $y = \dfrac{1}{1 + x^2}$

A graph of $y = x^2$ is shown in the figure. Explain how each of the graphs in Exercises 25–32 could be obtained from the graph shown. Sketch each graph.

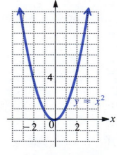

25. $y = x^2 + 2$ **26.** $y = x^2 - 7$ **27.** $y = -(x - 4)^2$

28. $y = -(x + 5)^2$ **29.** $y = (x - 2)^2 - 3$ **30.** $y = (x + 1)^2 + 5$

31. $y = -(x + 4)^2 + 2$ **32.** $y = -(x - 3)^2 - 1$

Exercises 25–32

Suppose that $f(2) = 3$. For each assumption in Exercises 33–35, find another value of the function.

33. $f(x)$ is symmetric with respect to the origin.

34. $f(x)$ is symmetric with respect to the y-axis.

35. $f(x)$ is symmetric with respect to the line $x = 6$.

36. Find the function whose graph can be obtained by translating the graph of $f(x) = 5x - 4$ up 2 units and left 3 units.

37. Find the function whose graph can be obtained by translating the graph of $f(x) = 4 - 5x$ down 2 units and right 3 units.

Explain why each of the following statements is true.

38. The function $f(x) = 0$ is both even and odd.

39. A nonzero function cannot be both even and odd.

40. If (a, b) is on the graph of an even function, then so is $(-a, b)$.

41. If (a, b) is on the graph of an odd function, then so is $(-a, -b)$.

42. Complete the left half of the graph of $f(x)$ in the figure for each of the following conditions:
(a) $f(-x) = f(x)$,   (b) $f(-x) = -f(x)$.

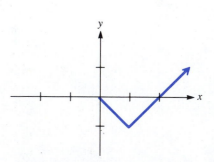

Exercise 42

43. Complete the right half of the graph of $f(x)$ in the figure for each of the following conditions: **(a)** $f(x)$ is odd, **(b)** $f(x)$ is even.

In Exercises 44 and 45, let F be some algebraic expression involving x as the only variable.

44. Suppose the equation $y = F$ is changed to $y = -F$. What is the relationship between the graphs of the two equations?

45. Suppose the equation $y = F$ is changed to $y = c \cdot F$, for some constant c. What is the effect on the graph of $y = F$? Discuss the effect depending on whether $c > 0$ or $c < 0$, and $|c| > 1$ or $|c| < 1$.

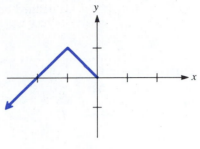

Exercise 43

Sketch examples of graphs which are as follows.

46. Symmetric to the x-axis but not the y-axis

47. Symmetric to the origin but to neither the x-axis nor the y-axis

Prove each statement below.

48. A graph symmetric with respect to both the x-axis and the y-axis is also symmetric to the origin.

49. A graph possessing two of the three types of symmetry, with respect to the x-axis, y-axis, and origin, must possess the third type of symmetry also.

Answer true *or* false *in Exercises 50–53.*

50. The zero function (where $f(x) = 0$) is symmetric with respect to the x-axis, the y-axis, and the origin.

51. A nonzero function cannot be symmetric with respect to the x-axis.

52. The graph of an even function is symmetric with respect to the y-axis.

53. The graph of an odd function is symmetric with respect to the x-axis.

3.8 ALGEBRA OF FUNCTIONS

When a company accountant sits down to estimate the firm's overhead, the first step might be to find functions representing the cost of materials, labor charges, equipment maintenance, and so on. The sum of these various functions could then be used to find the total overhead for the company. Methods of combining functions are discussed in this section.

Given two functions f and g, their *sum*, written $f + g$, is defined by

$$(f + g)(x) = f(x) + g(x),$$

for all x such that both $f(x)$ and $g(x)$ exist. Similar definitions can be given for the difference, $f - g$, product, fg, and quotient, f/g, of functions; however, the quotient,

$$\left(\frac{f}{g}\right)(x) = \frac{f(x)}{g(x)},$$

is defined only for values of x where both $f(x)$ and $g(x)$ exist, and $g(x) \neq 0$. The various operations on functions are defined below.

Definition of Operations on Functions

Given two functions f and g, then for all values of x for which both $f(x)$ and $g(x)$ exist, the functions $f + g$, $f - g$, fg, and f/g are defined as follows.

Sum	$(f + g)(x) = f(x) + g(x)$
Difference	$(f - g)(x) = f(x) - g(x)$
Product	$(fg)(x) = f(x) \cdot g(x)$
Quotient	$\left(\dfrac{f}{g}\right)(x) = \dfrac{f(x)}{g(x)}, \quad g(x) \neq 0$

NOTE The condition $g(x) \neq 0$ in the definition of the quotient means that the domain of $\left(\dfrac{f}{g}\right)(x)$ consists of all values of x for which $g(x)$ is not zero. The condition does not mean that $g(x)$ is a function that is never zero.

EXAMPLE 1 Let $f(x) = x^2 + 1$, and $g(x) = 3x + 5$. Find each of the following.

(a) $(f + g)(1)$

Since $f(1) = 2$ and $g(1) = 8$, use the definition above to get

$$(f + g)(1) = f(1) + g(1)$$
$$= 2 + 8$$
$$= 10.$$

(b) $(f - g)(-3) = f(-3) - g(-3)$
$$= 10 - (-4) = 14$$

(c) $(fg)(5) = f(5) \cdot g(5) = 26 \cdot 20 = 520$

(d) $\left(\dfrac{f}{g}\right)(0) = \dfrac{f(0)}{g(0)} = \dfrac{1}{5}$ ●

EXAMPLE 2 Let $f(x) = 8x - 9$ and $g(x) = \sqrt{2x - 1}$.

(a) $(f + g)(x) = f(x) + g(x) = 8x - 9 + \sqrt{2x - 1}$

(b) $(f - g)(x) = f(x) - g(x) = 8x - 9 - \sqrt{2x - 1}$

(c) $(fg)(x) = f(x) \cdot g(x) = (8x - 9)\sqrt{2x - 1}$

(d) $\left(\dfrac{f}{g}\right)(x) = \dfrac{f(x)}{g(x)} = \dfrac{8x - 9}{\sqrt{2x - 1}}$ ●

In Example 2, the domain of f is the set of all real numbers, while the domain of g, where $g(x) = \sqrt{2x - 1}$, includes just those real numbers that make $2x - 1 \geq 0$; the domain of g is the interval $[1/2, +\infty)$. The domains of $f + g$, $f - g$, and fg are thus $[1/2, +\infty)$. With f/g, the denominator cannot be zero, so the value $1/2$ is excluded from the domain. The domain of f/g is $(1/2, +\infty)$.

The domains of $f + g$, $f - g$, fg, and f/g are summarized below. (Recall that the intersection of two sets is the set of all elements belonging to *both sets*.)

Domains

> For functions f and g, the domains of $f + g$, $f - g$, and fg include all real numbers in the intersection of the domains of f and g, while the domain of f/g includes those real numbers in the intersection of the domains of f and g for which $g(x) \neq 0$.

Composition of Functions   The sketch in Figure 58 shows a function f that assigns to each element x of set X some element y of set Y. Suppose also that a function g takes each element of set Y and assigns a value z of set Z. Using both f and g, then, an element x in X is assigned an element z in Z. The result of this process is a new function h, which takes an element x in X and assigns an element z in Z.

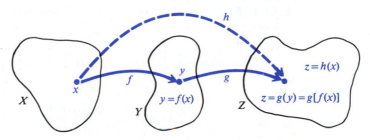

Figure 58

This function h is called the **composition** of functions g and f, written $g \circ f$, (read "g of f"), and defined as follows.

Definition of Composition of Functions

> Let f and g be functions. The **composite function,** or **composition,** of g and f, written $g \circ f$, is defined by
>
> $$(g \circ f)(x) = g(f(x)),$$
>
> for all x in the domain of f such that $f(x)$ is in the domain of g.

The notation $g(f(x))$ is read "g of f of x".

· · · · · · · · ·

EXAMPLE 3   Given $f(x) = 2x - 1$ and $g(x) = \dfrac{4}{x - 1}$, find each of the following.

(a) $f(g(2))$

First find $g(2)$. Since

$$g(x) = \frac{4}{x - 1},$$

$$g(2) = \frac{4}{2 - 1} = \frac{4}{1} = 4.$$

Now find $f(g(2)) = f(4)$

$$f(x) = 2x - 1$$

$$f(g(2)) = f(4) = 2(4) - 1 = 7$$

(b) $g(f(-3))$

$$f(-3) = 2(-3) - 1 = -7$$

$$g(f(-3)) = g(-7) = \frac{4}{-7 - 1} = \frac{4}{-8} = -\frac{1}{2} \quad \bullet$$

· · · · · · · · ·

EXAMPLE 4   Let $f(x) = 4x + 1$ and $g(x) = 2x^2 + 5x$. Find each of the following.

(a) $(g \circ f)(x)$

By definition, $(g \circ f)(x) = g(f(x))$. Use the given functions as follows.

$$
\begin{aligned}
(g \circ f)(x) &= g(f(x)) \\
&= g(4x + 1) \\
&= 2(4x + 1)^2 + 5(4x + 1) \\
&= 2(16x^2 + 8x + 1) + 20x + 5 \\
&= 32x^2 + 16x + 2 + 20x + 5 \\
&= 32x^2 + 36x + 7
\end{aligned}
$$

(b) $(f \circ g)(x)$

Use the definition above with f and g interchanged, so that $(f \circ g)(x)$ becomes $f(g(x))$.

$$
\begin{aligned}
(f \circ g)(x) &= f(g(x)) \\
&= f(2x^2 + 5x) \\
&= 4(2x^2 + 5x) + 1 \\
&= 8x^2 + 20x + 1 \quad \bullet
\end{aligned}
$$

As this example shows, it is not always true that $f \circ g = g \circ f$. In fact, $f \circ g$ is very rarely equal to $g \circ f$. In Example 4, the domain of both composite functions is the set of all real numbers.

▌**CAUTION** In general, the composite function $f \circ g$ is not the same as the product $f \cdot g$. For example, with f and g defined as in Example 4,
$$f \circ g = 8x^2 + 20x + 1$$
▌ but $$f \cdot g = (4x + 1)(2x^2 + 5x) = 8x^3 + 22x^2 + 5x.$$

- - - - - - - - -

EXAMPLE 5 Let $f(x) = 1/x$ and $g(x) = \sqrt{3 - x}$. Find $f \circ g$ and $g \circ f$. Give the domain of each.

First find $f \circ g$.
$$(f \circ g)(x) = f(g(x)) = f(\sqrt{3 - x})$$
$$= \frac{1}{\sqrt{3 - x}}$$

The denominator $\sqrt{3 - x}$ is a nonzero real number only when $3 - x > 0$, or $x < 3$, so that the domain of $f \circ g$ is the interval $(-\infty, 3)$.

Use the same functions to find $g \circ f$, as follows.
$$(g \circ f)(x) = g(f(x)) = g\left(\frac{1}{x}\right)$$
$$= \sqrt{3 - \frac{1}{x}} = \sqrt{\frac{3x - 1}{x}}$$

The domain of $g \circ f$ is the set of all real numbers x such that $x \neq 0$ and $3 - f(x) \geq 0$. Since $3 - f(x) = 3 - (1/x) = (3x - 1)/x$, this is the same as $(3x - 1)/x \geq 0$ and $x \neq 0$. By the methods in Section 2.5, the set $(-\infty, 0) \cup [1/3, +\infty]$ is the domain of $g \circ f$. ●

In Example 5 we found $(f \circ g)(x)$ given $f(x)$ and $g(x)$. Sometimes it is necessary to work backwards and find $f(x)$ and $g(x)$ given $(f \circ g)(x)$. This skill is useful in calculus.

- - - - - - - - -

EXAMPLE 6 Find functions f and g such that
$$(f \circ g)(x) = (x^2 - 5)^3 - 4(x^2 - 5) + 3.$$

One pair of functions that will work is
$$f(x) = x^3 - 4x + 3 \quad \text{and} \quad g(x) = x^2 - 5.$$
Then $$(f \circ g)(x) = f(g(x))$$
$$= f(x^2 - 5)$$
$$= (x^2 - 5)^3 - 4(x^2 - 5) + 3.$$

Other pairs of functions f and g might also work. For instance,
$$f(x) = (x - 5)^3 - 4(x - 5) + 3$$
and $$g(x) = x^2. ●$$

3.8 Exercises

For each of the pairs of functions defined as follows, find $f + g$, $f - g$, fg, and f/g. Give the domain of each.

1. $f(x) = 4x - 1$, $g(x) = 6x + 3$

2. $f(x) = 9 - 2x$, $g(x) = -5x + 2$

3. $f(x) = 3x^2 - 2x$, $g(x) = x^2 - 2x + 1$

4. $f(x) = 6x^2 - 11x$, $g(x) = x^2 - 4x - 5$

5. $f(x) = \sqrt{2x + 5}$, $g(x) = \sqrt{4x - 9}$

6. $f(x) = \sqrt{11x - 3}$, $g(x) = \sqrt{2x - 15}$

Let $f(x) = 4x^2 - 2x$ and let $g(x) = 8x + 1$. Find each of the following.

7. $(f + g)(3)$

8. $(f + g)(-5)$

9. $(fg)(4)$

10. $(fg)(-3)$

11. $\left(\dfrac{f}{g}\right)(-1)$

12. $\left(\dfrac{f}{g}\right)(4)$

13. $(f + g)(m)$

14. $(f - g)(2k)$

15. $(f \circ g)(2)$

16. $(f \circ g)(-5)$

17. $(g \circ f)(2)$

18. $(g \circ f)(-5)$

19. $(f \circ g)(k)$

20. $(g \circ f)(5z)$

Find $f \circ g$ and $g \circ f$ for each of the pairs of functions defined as follows.

21. $f(x) = 8x + 12$, $g(x) = 3x - 1$

22. $f(x) = -6x + 9$, $g(x) = 5x + 7$

23. $f(x) = 5x + 3$, $g(x) = -x^2 + 4x + 3$

24. $f(x) = 4x^2 + 2x + 8$, $g(x) = x + 5$

25. $f(x) = \dfrac{1}{x}$, $g(x) = x^2$

26. $f(x) = \dfrac{2}{x^4}$, $g(x) = 2 - x$

27. $f(x) = \sqrt{x + 2}$, $g(x) = 8x^2 - 6$

28. $f(x) = 9x^2 - 11x$, $g(x) = 2\sqrt{x + 2}$

29. $f(x) = \dfrac{1}{x - 5}$, $g(x) = \dfrac{2}{x}$

30. $f(x) = \dfrac{8}{x - 6}$, $g(x) = \dfrac{4}{3x}$

The graphs of functions f and g are shown. Use these graphs to find the values in Exercises 31–38.

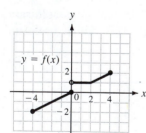

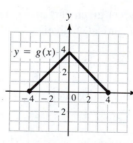

31. $f(1) + g(1)$

32. $f(4) - g(3)$

33. $f(-2) \cdot g(4)$

34. $\dfrac{f(4)}{g(2)}$

35. $(f \circ g)(2)$

36. $(g \circ f)(2)$

37. $(g \circ f)(-4)$

38. $(f \circ g)(-2)$

39. The graphs of functions f and g are shown. Draw the graph of $f \circ g$.

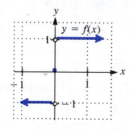

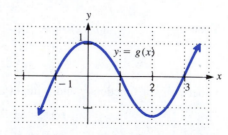

For each of the pairs of functions defined as follows, show that $(f \circ g)(x) = x$ and
$(g \circ f)(x) = x$.

40. $f(x) = 3x, \quad g(x) = \dfrac{1}{3}x$

41. $f(x) = \dfrac{3}{4}x, \quad g(x) = \dfrac{4}{3}x$

42. $f(x) = 8x - 11, \quad g(x) = \dfrac{x + 11}{8}$

43. $f(x) = \dfrac{x - 3}{4}, \quad g(x) = 4x + 3$

44. $f(x) = x^3 + 6, \quad g(x) = \sqrt[3]{x - 6}$

45. $f(x) = \sqrt[5]{x - 9}, \quad g(x) = x^5 + 9$

In Exercises 46–51, a function h is defined. Find functions f and g such that $h(x) =$
$(f \circ g)(x)$. *Many such pairs of functions exist.*

46. $h(x) = (6x - 1)^2$

47. $h(x) = (11x^2 + 12x)^2$

48. $h(x) = \sqrt{x^2 - 1}$

49. $h(x) = \dfrac{1}{x^2 + 2}$

50. $h(x) = \dfrac{(x - 2)^2 + 1}{5 - (x - 2)^2}$

51. $h(x) = (x + 2)^3 - 3(x + 2)^2$

52. Suppose the population P of a certain species of fish depends on the number x (in hundreds) of a smaller kind of fish which serves as its food supply, so that $P(x) = 2x^2 + 1$. Suppose, also, that the number x (in hundreds) of the smaller species of fish depends upon the amount a (in appropriate units) of its food supply, a kind of plankton. Suppose $x = f(a) = 3a + 2$. Find $(P \circ f)(a)$, the relationship between the population P of the large fish and the amount a of plankton available.

53. Suppose that for a certain brand of vacuum cleaner the number that will be sold is given by $D(p) = (-p^2/100) + 500$, where p is the price in dollars. If the price, in terms of the cost, c, is expressed as $p(c) = 2c - 10$, find the demand in terms of the cost.

54. An oil well off the Gulf Coast is leaking, with the leak spreading oil over the surface of the gulf as a circle. At any time t, in minutes, after the beginning of the leak, the radius of the circular oil slick on the water surface is $r(t) = 4t$ ft. Let $A(r) = \pi r^2$ represent the area of a circle of radius r. Find and interpret $(A \circ r)(t)$.

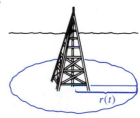

Exercise 54

55. When a thermal inversion layer is over a city (such as happens often in Los Angeles), pollutants cannot rise vertically but are trapped below the layer and must disperse horizontally. Assume that a factory smokestack begins emitting a pollutant at 8 A.M. Assume that the pollutant disperses horizontally, forming a circle. If t represents the time, in hours, since the factory began emitting pollutants ($t = 0$ represents 8 A.M.), assume that the radius of the circle of pollution is $r(t) = 2t$ mi. Let $A(r) = \pi r^2$ represent the area of a circle of radius r. Find and interpret $(A \circ r)(t)$.

56. Suppose $g(x) = x - 5$.
  (a) For any function f, the graph of $f \circ g$ is a translation of the graph of f. Describe the translation.
  (b) For any function f, the graph of $g \circ f$ is a translation of the graph of f. Describe the translation.

57. Suppose $g(x) = -x$. How are the graphs of $f \circ g$ and $g \circ f$ related to the graph of f? (*Hint:* See Exercise 56.)

58. Let $f(x) = x/(x - 1)$ for $x \neq 1$. Show that $f \circ f = x$.

59. If $f(x) = 1 + x$ and $g(x) = 1/(1 + x)$, what is the domain of $f(x) \cdot g(x)$?

Recall that even and odd functions were defined in the previous section.

60. Let f be any function. Show that the function $g(x) = (1/2)[f(x) + f(-x)]$ is even.

61. Let f be any function. Show that the function $h(x) = (1/2)[f(x) - f(-x)]$ is odd.

62. Use the results of the previous two exercises to show that any function may be expressed as the sum of an odd function and an even function.

63. Show that the composite of two odd functions is odd.

64. Show that the composite of an even function and an odd function is odd.

3.9 INVERSE FUNCTIONS

The operations of addition and subtraction "undo" one another. That is, adding 5 to a number and then subtracting 5 from the sum gives back the original number. Certain pairs of functions also "undo" one another. To have this property the functions must both be *one-to-one functions*.

One-to-one Functions Given the function $y = x^2$, it is possible for two different values of x to lead to the same value of y. For example, the value $y = 4$ is obtained from either of two values of x: both $2^2 = 4$ and $(-2)^2 = 4$. See Figure 59(a). On the other hand, for the function $y = 6x$, a given value of y can be found from exactly one value of x. For this function, if $y = 30$, then $x = 5$; there is no other value of x that will produce a value of 30 for y. See Figure 59(b).

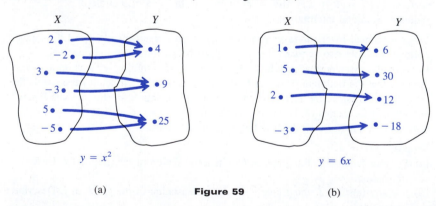

(a)   **Figure 59**   (b)

This second function, $y = 6x$, is an example of a one-to-one function. A function f is **one-to-one** if $f(a) = f(b)$ implies that $a = b$. In a one-to-one function, each x-value corresponds to exactly one y-value and each y-value corresponds to exactly one x-value. In Figure 60(a), each y-value is assigned to exactly one x-value, so the function is one-to-one. The function in Figure 60(b) has one y-value assigned to *two* x-values, which indicates that it is not a one-to-one function.

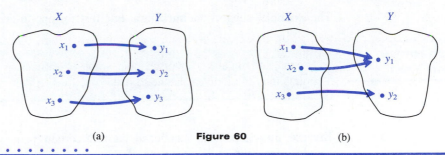

(a)   **Figure 60**   (b)

EXAMPLE 1   Decide whether the functions defined as follows are one-to-one.

(a) $f(x) = -4x + 12$

     For this function, any value of y, say $y = b$, corresponds to exactly one x-value, since if $y = b$,

$$b = -4x + 12$$
$$4x = 12 - b$$
$$x = \frac{12 - b}{4}.$$

Thus, the function is one-to-one.

(b) $g(x) = \sqrt{25 - x^2}$

     Suppose $y = k$. Then $g(x) = k$, and

$$k = \sqrt{25 - x^2}$$
$$k^2 = 25 - x^2$$
$$x^2 = 25 - k^2$$
$$x = \sqrt{25 - k^2} \quad \text{or} \quad -\sqrt{25 - k^2}.$$

Here $y = k$ corresponds to more than one x-value, so g is not one-to-one.   ●

     There is a useful graphical test for deciding whether or not a function is one-to-one. Figure 61(a) shows the graph of a function defined by $y = f(x)$ cut by a horizontal line. As the graph suggests, $f(x_1) = f(x_2) = f(x_3)$, even though x_1, x_2, and x_3 are all different points. Since one value of y corresponds to more than one value of x, the function is not one-to-one. On the other hand, drawing horizontal lines on the graph of Figure 61(b) shows that a given value of y can be obtained from only one value of x, so this function is one-to-one.

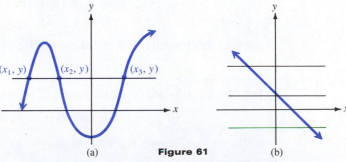

(a)   **Figure 61**   (b)

The examples suggest the **horizontal line test** for one-to-one functions.

**Horizontal
Line Test**

> If each horizontal line intersects the graph of a function in no more than one point, then the function is one-to-one.

Inverse Functions As mentioned earlier, certain pairs of one-to-one functions "undo" one another. For example, if

$$f(x) = 8x + 5 \quad \text{and} \quad g(x) = \frac{x - 5}{8}$$

$$f(\mathbf{10}) = 8 \cdot 10 + 5 = \mathbf{85} \quad \text{and} \quad g(\mathbf{85}) = \frac{85 - 5}{8} = \mathbf{10}.$$

Starting with 10, we "applied" function f and then "applied" function g to the result, which gave back the number 10. See Figure 62.

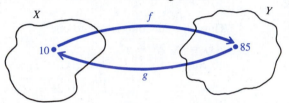

Figure 62

Similarly, for these same functions, check that

$$f(3) = 29 \quad \text{and} \quad g(29) = 3,$$
$$f(-5) = -35 \quad \text{and} \quad g(-35) = -5,$$
$$g(2) = -\frac{3}{8} \quad \text{and} \quad f\left(-\frac{3}{8}\right) = 2.$$

In particular, for these functions,

$$f(g(13)) = f(1) = 13 \quad \text{and} \quad g(f(13)) = g(109) = 13.$$

In fact, for *any* value of x,

$$f(g(x)) = x \quad \text{and} \quad g(f(x)) = x,$$
or $\qquad (f \circ g)(x) = x \quad \text{and} \quad (g \circ f)(x) = x.$

Because of this property, g is called the *inverse* of f.

**Definition of
Inverse
Function**

> Let f be a one-to-one function. Then g is the **inverse function** of f if
>
> $$(f \circ g)(x) = x \quad \text{for every } x \text{ in the domain of } g,$$
>
> and $\qquad (g \circ f)(x) = x \quad \text{for every } x \text{ in the domain of } f.$

A special notation is often used for inverse functions: if g is the inverse of a function f, then g is written as f^{-1} (read "f-inverse.") In the example above, $f(x) = 8x + 5$, and $g(x) = f^{-1}(x) = (x - 5)/8$.

CAUTION Do not confuse the -1 in f^{-1} with a negative exponent. The symbol f^{-1} does not represent $1/f$; it is used to name the function that is the inverse of function f.

The definition of inverse function can be used to show that the domain of f equals the range of f^{-1}, and the range of f equals the domain of f^{-1}. See Figure 63.

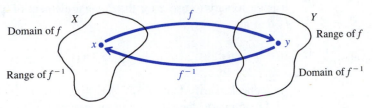

Figure 63

• • • • • • • • •

EXAMPLE 2   Let functions f and g be defined by $f(x) = x^3 - 1$ and $g(x) = \sqrt[3]{x + 1}$, respectively. Is g the inverse function of f?

First check that f is one-to-one. Since it is, now find $(f \circ g)(x)$ and $(g \circ f)(x)$.

$$(f \circ g)(x) = f[g(x)] = (\sqrt[3]{x + 1})^3 - 1$$
$$= x + 1 - 1$$
$$= x$$

$$(g \circ f)(x) = g[f(x)] = \sqrt[3]{(x^3 - 1) + 1}$$
$$= \sqrt[3]{x^3}$$
$$= x$$

Since both $(f \circ g)(x) = x$ and $(g \circ f)(x) = x$, function g is indeed the inverse of function f, so that f^{-1} is given by

$$f^{-1}(x) = \sqrt[3]{x + 1}. \quad \bullet$$

CAUTION Keep in mind that a function f can have an inverse function f^{-1} if and only if f is one-to-one.

Given a one-to-one function f and an x in the domain of f, the corresponding value in the range of f is found by means of the equation $y = f(x)$. With the inverse function f^{-1}, the value of y can be used to produce x, since $x = f^{-1}(y)$. Therefore, the equation for f^{-1} can be found by solving $y = f(x)$ for x.

For example, let $f(x) = 7x - 2$. Then $y = 7x - 2$. The function f is one-to-one, so that f^{-1} exists. Solve $y = 7x - 2$ for x, as follows.

$$y = 7x - 2$$
$$7x = y + 2$$
$$x = \frac{y + 2}{7}$$

Since $x = f^{-1}(y)$,

$$f^{-1}(y) = \frac{y + 2}{7}.$$

It is customary to use x for the domain element of a function, so replace y with x in f^{-1} to get

$$f^{-1}(x) = \frac{x + 2}{7}.$$

As a check, verify that $(f \circ f^{-1})(x) = x$ and $(f^{-1} \circ f)(x) = x$.

In summary, the equation of an inverse function can be found with the following steps.

Finding an Equation for f^{-1}

1. Check that the function f defined by $y = f(x)$ is a one-to-one function.
2. Solve for x. Let $x = f^{-1}(y)$.
3. Exchange x and y to get $y = f^{-1}(x)$.
4. Check that $(f \circ f^{-1})(x) = x$ and $(f^{-1} \circ f)(x) = x$.

EXAMPLE 3   For each of the functions defined as follows, find any inverse functions.

(a) $f(x) = \dfrac{4x + 6}{5}$

This function is one-to-one and so has an inverse. Let $y = f(x)$, and solve for x.

$$y = \frac{4x + 6}{5}$$
$$5y = 4x + 6$$
$$5y - 6 = 4x$$
$$\frac{5y - 6}{4} = x$$

Finally, exchange x and y, and let $y = f^{-1}(x)$, to get

$$\frac{5x - 6}{4} = y,$$

or $$f^{-1}(x) = \frac{5x - 6}{4}.$$

Check this result as follows.

$$(f \circ f^{-1})(x) = f\left(\frac{5x - 6}{4}\right)$$

$$= \frac{4\left(\dfrac{5x - 6}{4}\right) + 6}{5} = x$$

$$(f^{-1} \circ f)(x) = f^{-1}\left(\frac{4x + 6}{5}\right)$$

$$= \frac{5\left(\dfrac{4x + 6}{5}\right) - 6}{4} = x$$

Therefore, $f^{-1}(x)$ is the correct inverse of $f(x)$.

(b) $f(x) = x^3 - 1$

Two different values of $f(x)$ correspond to two different values of x, so the function is one-to-one and has an inverse. To find the inverse, first solve $y = x^3 - 1$ for x, as follows.

$$y = x^3 - 1$$
$$y + 1 = x^3$$
$$\sqrt[3]{y + 1} = x$$

Exchange x and y.

$$\sqrt[3]{x + 1} = y,$$

or

$$f^{-1}(x) = \sqrt[3]{x + 1}$$

We demonstrated in Example 2 that $f^{-1}(x) = \sqrt[3]{x + 1}$ is the correct inverse of $f(x) = x^3 - 1$.

(c) $f(x) = x^2$

The two different x-values 4 and -4 give the same value of y, namely 16, showing that the function is not one-to-one and has no inverse function. If the domain of f is restricted to $[0, +\infty)$, then f would have the inverse function

$$f^{-1}(x) = \sqrt{x}$$

with range $[0, +\infty)$. On the other hand, if the domain of f is restricted to $(-\infty, 0]$, the inverse would be

$$f^{-1}(x) = -\sqrt{x}$$

with range $(-\infty, 0]$. See Figure 64. ●

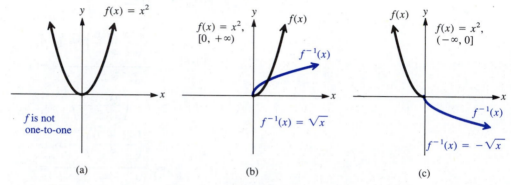

(a) (b) (c)

Figure 64

Suppose f and f^{-1} are inverse functions of each other. Suppose that $f(a) = b$, for real numbers a and b. Then by the definition of inverse, $f^{-1}(b) = a$. This shows that if a point (a, b) is on the graph of f, then (b, a) will belong to the graph of f^{-1}. As shown in Figure 65, the points (a, b) and (b, a) are *symmetric with respect to the line* $y = x$. Thus, the graph of f^{-1} can be obtained from the graph of f by reflecting the graph of f about the line $y = x$, that is, by interchanging x and y in the ordered pairs.

For example, Figure 66 shows the graph of $f(x) = x^3 - 1$ as a solid line and the graph of $f^{-1}(x) = \sqrt[3]{x + 1}$ as a dashed line. These graphs are symmetric with respect to the line $y = x$.

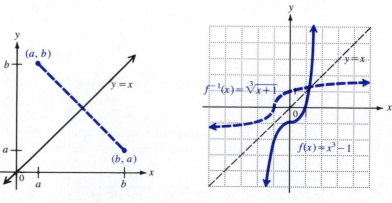

Figure 65

Figure 66

. . . .

EXAMPLE 4   Let $f(x) = \sqrt{x + 5}$ with domain $[-5, +\infty)$. Find $f^{-1}(x)$.

As shown by the graph in Figure 67, the function f is one-to-one and has an inverse function. To find this inverse function, start with $y = \sqrt{x + 5}$ and solve for x.

$$y = \sqrt{x + 5}$$
$$y^2 = x + 5$$
$$y^2 - 5 = x$$

Exchanging x and y gives

$$x^2 - 5 = y.$$

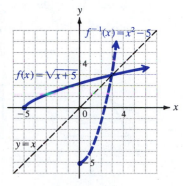

Figure 67

We cannot give just $x^2 - 5$ as $f^{-1}(x)$. In the definition of f above, the domain was given as $[-5, +\infty)$. The range of f is $[0, +\infty)$. Since the range of f equals the domain of f^{-1}, the definition of the function f^{-1} must be given as

$$f^{-1}(x) = x^2 - 5, \quad \text{domain } [0, +\infty).$$

As a check, the range of f^{-1}, $[-5, +\infty)$ equals the domain of f. Verify that $(f \circ f^{-1})(x) = x$ and $(f^{-1} \circ f)(x) = x.$ ●

3.9 Exercises . . . .

Which of the functions graphed or defined as follows are one-to-one?

1.

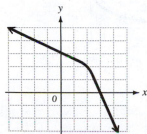

2.

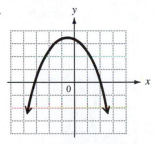

3.

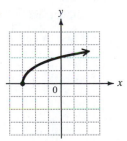

4.

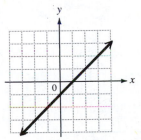

5.

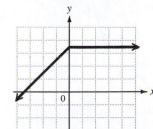

6.

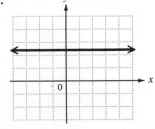

7. $y = 4x - 5$ **8.** $y = -x$ **9.** $y = -x^2$ **10.** $y = (x - 2)^2$

11. $y = \sqrt{36 - x^2}$ **12.** $y = -\sqrt{100 - x^2}$ **13.** $y = |25 - x^2|$ **14.** $y = -|16 - x^2|$

15. $y = x^3 - 1$ **16.** $y = -\sqrt[3]{x + 5}$ **17.** $y = \dfrac{1}{x + 2}$ **18.** $y = \dfrac{-4}{x - 8}$

19. $y = 9$ **20.** $y = -4$

Which of the pairs of functions graphed or defined as follows are inverses of each other?

21.

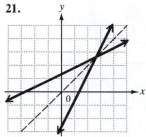

22.

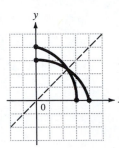

23.

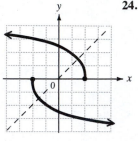

24.

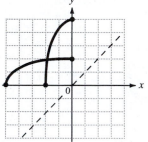

25. $f(x) = 2x + 4, \quad g(x) = \dfrac{1}{2}x - 2$

26. $f(x) = 5x - 5, \quad g(x) = \dfrac{1}{5}x + 1$

27. $f(x) = \dfrac{1}{x + 1}, \quad g(x) = \dfrac{x - 9}{12}$

28. $f(x) = \dfrac{1}{x + 1}, \quad g(x) = \dfrac{1 - x}{x}$

29. $f(x) = \dfrac{2}{x + 6}, \quad g(x) = \dfrac{6x + 2}{x}$

30. $f(x) = \dfrac{1}{x}, \quad g(x) = \dfrac{1}{x}$

31. $f(x) = x^2 + 3$, domain $[0, +\infty)$, and $g(x) = \sqrt{x - 3}$, domain $[3, +\infty)$

32. $f(x) = \sqrt{x + 8}$, domain $[-8, +\infty)$, and $g(x) = x^2 - 8$, domain $[0, +\infty)$

33. $f(x) = -|x + 5|$, domain $[-5, +\infty)$, and $g(x) = |x - 5|$, domain $[5, +\infty)$

34. $f(x) = |x - 1|$, domain $[-1, +\infty)$, and $g(x) = |x + 1|$, domain $[1, +\infty)$

Graph the inverse of each one-to-one function.

35.

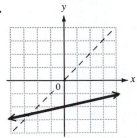

36.

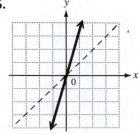

37.

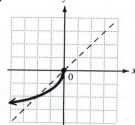

38.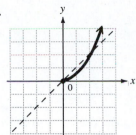

39.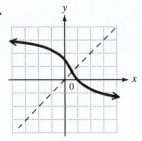

40.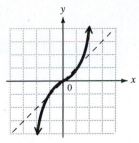

For each function defined as follows that is one-to-one, write an equation for the inverse function in the form $y = f^{-1}(x)$.

41. $y = 4x - 5$

42. $y = 3x - 4$

43. $y = -\frac{2}{5}x$

44. $y = \frac{1}{3}x$

45. $y = -x^3 - 2$

46. $y = x^3 + 1$

47. $y = -x^2 + 2$

48. $y = x^2$

49. $y = \frac{4}{x}$

50. $y = \frac{1}{x}$

51. $y = (x - 3)^2$, domain $(-\infty, 3]$

52. $y = \sqrt{2 - x}$, domain $(-\infty, 2]$

53. $f(x) = \sqrt{6 + x}$, domain $[-6, +\infty)$

54. $f(x) = (x - 2)^2 + 1$, domain $[2, +\infty)$

The graph of a function f is shown in the figure. Use the graph to find each of the following values.

55. $f^{-1}(4)$

56. $f^{-1}(2)$

57. $f^{-1}(0)$

58. $f^{-1}(-2)$

59. $f^{-1}(-3)$

60. $f^{-1}(-4)$

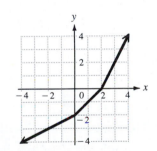

Exercises 55–60

Let $f(x) = x^2 + 5x$ for $x \geq -5/2$. Find the value of the expression in Exercises 61 and 62, rounding to the nearest hundredth.

61. $f^{-1}(7)$

62. $f^{-1}(-3)$

63. Suppose that $f(x)$ is the number of cars that can be built for x dollars. What does $f^{-1}(1000)$ represent?

64. Suppose that $f(r)$ is the volume (in cubic inches) of a sphere of radius r inches. What does $f^{-1}(5)$ represent?

65. Suppose that the point $(1, 4)$ is on the graph of an odd function that is its own inverse. Find three other points on the graph of the function.

66. If a line has slope a, what is the slope of its reflection in the line $y = x$?

67. Find $f^{-1}(f(2))$, where $f(2) = 3$.

Let f be a function having an inverse. Prove the statements in Exercises 68 and 69.

68. Every x-intercept of f is a y-intercept of f^{-1}.

69. Every y-intercept of f is an x-intercept of f^{-1}.

70. Let f be an odd one-to-one function. What can you say about f^{-1}?

71. Let f be an even function. What can you say about f^{-1}?

72. Give an example of a function f such that $f = f^{-1}$.

3 | CHAPTER SUMMARY

Key Words

To understand the concepts presented in this chapter, you should know the meaning and use of the following words and expressions. For easy reference, the section in the chapter in which a word or expression was first used is given with each item.

3.1 rectangular coordinate
    system
  Cartesian coordinate
    system
  coordinates
  ordered pairs
  graph
  relation
  domain
  range
3.2 distance formula
  circle
  radius
  midpoint formula
3.3 function

identity function
vertical line test
absolute value function
root function
$f(x)$ notation
maximum
minimum
independent variable
dependent variable
3.4 linear function
intercepts
$\Delta x, \Delta y$
slope
functions defined piecewise
greatest integer function

3.5 point-slope form
  slope-intercept form
  standard form
3.6 quadratic function
  parabola
  axis
  vertex
3.7 symmetry
  odd function
  even function
  translation
3.8 composition of functions
3.9 one-to-one function
  horizontal line test
  inverse functions

Review Exercises

In Exercises 1 and 2, find d(P, Q) and the midpoint of segment PQ.

1. $P(3, -1)$ and $Q(-4, 5)$ 2. $P(-8, 2)$ and $Q(3, -7)$

3. Find the other endpoint of a line segment having one end at $(-5, 7)$ and having its midpoint at $(1, -3)$.

4. Are the points $(5, 7)$, $(3, 9)$, $(6, 8)$ the vertices of a right triangle?

5. Find all possible values of k so that $(-1, 2)$, $(-10, 5)$, and $(-4, k)$ are the vertices of a right triangle.

6. Find all possible values of x so that the distance between $(x, -9)$ and $(3, -5)$ is 6.

7. Find all points (x, y) with $x = 6$ so that (x, y) is 4 units from $(1, 3)$.

8. Find all points (x, y) with $x + y = 0$ so that (x, y) is 6 units from $(-2, 3)$.

Prove each of the following.

9. The medians to the two equal sides of an isosceles triangle are equal in length.

10. The line segment connecting midpoints of two sides of a triangle is half as long as the third side.

Find equations for each of the circles below.

11. Center $(-2, 3)$, radius 5

12. Center $(\sqrt{5}, -\sqrt{7})$, radius $\sqrt{3}$

13. Center $(-8, 1)$, passing through $(0, 16)$

14. Center $(3, -6)$, tangent to the x-axis

Find the center and radius of each of the following that are circles.

15. $x^2 - 4x + y^2 + 6y + 12 = 0$

16. $x^2 - 6x + y^2 - 10y + 30 = 0$

17. $x^2 + 7x + y^2 + 3y + 1 = 0$

18. $x^2 + 11x + y^2 - 5y + 46 = 0$

Tell whether each of the following is true or false.

19. The number a is in the domain of a relation if and only if the vertical line $x = a$ intersects the graph of the relation.

20. The number b is in the range of a relation if and only if the horizontal line $y = b$ intersects the graph of the relation.

Decide whether the following curves are graphs of functions. Give the domain and range of each relation.

21.

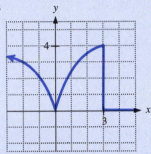

22.

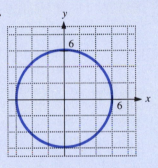

23.

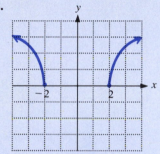

24.

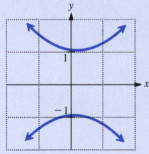

25.

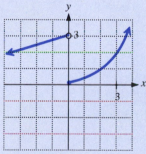

26.

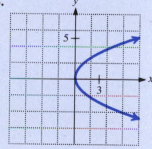

Identify any equations that define y as a function of x.

27. $x + y = 4$

28. $x = \dfrac{1}{2}y^2$

29. $y = 3 - x^2$

30. $y = \dfrac{-8}{x}$

31. $y = \sqrt{x - 7}$

32. $x = |y - 1|$

Give the domain of the functions defined as follows.

33. $y = -4 + |x|$

34. $y = 3x^2 - 1$

35. $y = (x - 4)^2$

36. $y = \dfrac{8 + x}{8 - x}$

37. $y = -\sqrt{\dfrac{5}{x^2 + 9}}$

38. $y = \sqrt{49 - x^2}$

Find the slope for each of the following lines.

39. Through $(8, 7)$ and $(1/2, -2)$

40. Through $(2, -2)$ and $(3, -4)$

41. $9x - 4y = 2$

42. $11x + 2y = 3$

43. $x - 2 = 0$

44. $y + 6 = 0$

Graph each of the following lines.

45. $3x + 7y = 14$

46. $2x - 5y = 5$

47. $3y = x$

48. $y = 3$

49. Through $(2, -4)$, $m = 3/4$

50. Through $(-3, -2)$, $m = -1$

For each line in Exercises 51–60, write an equation in standard form.

51. Through $(-2, 4)$ and $(1, 3)$

52. Through $(-2/3, -1)$ and $(0, 4)$

53. Through $(3, -5)$ with slope -2

54. Through $(-4, 4)$ with slope $3/2$

55. x-intercept -3, y-intercept 5

56. No x-intercept, y-intercept $3/4$

57. Through $(2, -1)$, parallel to $3x - y = 1$

58. Through $(3, -5)$ parallel to $y = 4$

59. Through $(0, 5)$, perpendicular to $8x + 5y = 3$

60. Through $(-7, 4)$, perpendicular to $y = 8$

61. The college fieldhouse used for graduation has 5000 seats available for family and friends of the graduates. Use the greatest integer function to write an expression for the number of tickets that can be allocated to each of x graduates.

Graph each of the functions defined as follows. Give the vertex and axis in Exercises 64–71.

62. $f(x) = \begin{cases} 3x + 1 & \text{if } x < 2 \\ -x + 4 & \text{if } x \geq 2 \end{cases}$

63. $f(x) = \begin{cases} -4x + 2 & \text{if } x \leq 1 \\ 3x - 5 & \text{if } x > 1 \end{cases}$

64. $f(x) = x^2 - 4$

65. $f(x) = 6 - x^2$

66. $f(x) = 3(x + 1)^2 - 5$

67. $f(x) = -\dfrac{1}{4}(x - 2)^2 + 3$

68. $f(x) = x^2 - 4x + 2$

69. $f(x) = -x^2 - 7x - 10$

70. $f(x) = -3x^2 - 12x - 1$

71. $f(x) = 4x^2 - 4x + 3$

In Exercises 72–75, consider the graph of $f(x) = a(x - h)^2 + k$, with a positive.

72. What is the y-coordinate of the lowest point of the graph?

73. What is the x-coordinate of the lowest point of the graph?

74. What is the y-intercept of the graph?

75. Under what condition, involving the letters a, h, or k, will the graph have one or more x-intercepts? If so, express the x-intercept(s) in terms of a, h, and k.

76. If a is positive, what is the smallest value of $ax^2 + bx + c$?

Use parabolas to work each of the following problems.

77. Find two numbers whose sum is 11 and whose product is a maximum.

78. Find two numbers having a sum of 40 such that the sum of the square of one and twice the square of the other is minimum.

79. Find the rectangular region of maximum area that can be enclosed with 180 meters of fencing.

80. Find the rectangular region of maximum area that can be enclosed with 180 meters of fencing if no fencing is needed along one side of the region.

Decide whether the equations in Exercises 81–88 have graphs that are symmetric to the x-axis, the y-axis, or the origin.

81. $3y^2 - 5x^2 = 15$ **82.** $x + y^2 = 8$ **83.** $y^3 = x + 1$ **84.** $x^2 = y^3$

85. $|y| = -x$ **86.** $|x + 2| = |y - 3|$ **87.** $|x| = |y|$ **88.** $xy = 8$

89. Graph $y = |x|$, and use the result in Exercises 90–93.

Using your graph from Exercise 89, explain how each of the following graphs could be obtained. Sketch each graph.

90. $y = |x| - 3$ **91.** $y = -|x|$ **92.** $y = -|x| - 2$ **93.** $y = 2|x - 3| - 4$

Let $f(x) = 2x + 3$. Find the function corresponding to each of the following reflections of the graph of $f(x)$.

94. About the x-axis **95.** About the y-axis **96.** About the origin **97.** About the line $y = x$

The graph of a function f is shown in the figure. Sketch the graph of each function defined as follows.

98. $y = f(x) + 3$ **99.** $y = f(x) - 4$

100. $y = f(x - 2)$ **101.** $y = f(x + 4)$

102. $y = f(x + 3) - 2$ **103.** $y = f(x - 1) + 4$

104. $y = |f(x)|$

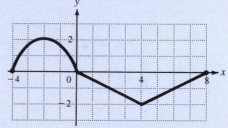

Exercises 98–104

Decide whether the functions defined as follows are odd, even, or neither.

105. $g(x) = \dfrac{x}{|x|}$ $(x \neq 0)$ **106.** $h(x) = 9x^6 + 5|x|$

107. $f(x) = x$ **108.** $f(x) = -x^4 + 9x^2 - 3x^6$

Let $f(x) = 3x^2 - 4$ and $g(x) = x^2 - 3x - 4$. Find each of the following.

109. $(f + g)(x)$ **110.** $(fg)(x)$ **111.** $(f - g)(4)$ **112.** $(f + g)(-4)$

113. $(f + g)(2k)$ **114.** $(fg)(1 + r)$ **115.** $(f/g)(3)$ **116.** $(f/g)(-1)$

117. The domain of $(fg)(x)$ **118.** The domain of $(f/g)(x)$

Let $f(x) = \sqrt{x - 2}$ and $g(x) = x^2$. Find each of the following.

119. $(f \circ g)(x)$ **120.** $(g \circ f)(x)$ **121.** $(f \circ g)(-6)$

122. $(f \circ g)(2)$ **123.** $(g \circ f)(3)$ **124.** $(g \circ f)(24)$

Which of the functions graphed or defined as follows are one-to-one?

125.

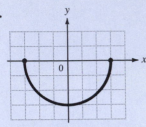

126.

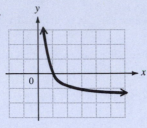

127.

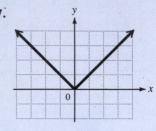

128. $y = \dfrac{8x - 9}{5}$

129. $y = -x^2 + 11$

130. $y = \sqrt{100 - x^2}$

For each of the functions defined in Exercises 131–34, write an equation for the inverse function in the form $y = f^{-1}(x)$ and then graph f and f^{-1}.

131. $f(x) = 12x + 3$

132. $f(x) = x^3 - 3$

133. $f(x) = x^2 - 6$

134. $f(x) = \sqrt{25 - x^2}$, domain [0, 5]

135. Suppose that $f(t)$ is the world population t years after 1990. What does $f^{-1}(6{,}000{,}000{,}000)$ represent?

136. Find the point of intersection of $y = f(x)$ and $y = f^{-1}(x)$ where $f(x) = 3x - 5$.

137. Cylindrical cans make the most efficient use of materials when their height is the same as the diameter of their top.
   (a) Express the volume V of such a can as a function of the diameter d of its top.
   (b) Express the surface area S of such a can as a function of the diameter d of its top.
       (*Hint:* The curved side is made from a rectangle whose length is the circumference of the top of the can.)

138. A baseball diamond is a square 90 ft long on each side. Casey runs a constant 30 ft per sec whether he hits a ground ball or a home run. Today in his first time at bat, he hit a home run. Write an expression that measures his line-of-sight distance from second base as a function of the time t, in seconds, after he left home plate.*

139. Alice, on vacation in Canada, found that she got a 15% premium on her U.S. money. When she returned, she discovered that there was a 15% discount on converting her Canadian money back into U.S. currency. Describe each conversion function. Show that one is not the inverse of the other; that is, show that after converting both ways, Alice lost money.*

*Exercises 138–39: From *Calculus*, 4th Edition by Stanley I. Grossman. Copyright © 1988 by Harcourt Brace Jovanovich, Inc. Reprinted with permission of Harcourt Brace Jovanovich, Inc. and the author.

POLYNOMIAL AND RATIONAL FUNCTIONS

Polynomial functions are the simplest type of function, since a polynomial involves only the operations of addition, subtraction, and multiplication. In calculus it is shown how more complicated functions can be closely approximated by polynomial functions.

Definition of Polynomial Function

A **polynomial function of degree** n, where n is a nonnegative integer, is a function defined by an expression of the form

$$f(x) = a_n x^n + a_{n-1} x^{n-1} + \ldots + a_1 x + a_0,$$

where $a_n, a_{n-1}, \ldots, a_1$, and a_0 are complex numbers, with $a_n \neq 0$.*

Constant functions (where $f(x) = a_0$), linear functions (with $f(x) = a_1 x + a_0$), and quadratic functions (with $f(x) = a_2 x^2 + a_1 x + a_0$) are polynomial functions of degree 0, 1, and 2, respectively, that we have discussed earlier. The discussion in this chapter generalizes our earlier work to higher degree polynomial functions. For example,

$$f(x) = 3x^3 + 2x + 5 \quad \text{and} \quad f(x) = x^4 - 2x^3 + 4x - 1$$

are polynomial functions of degree 3 and 4. The number a_n is the **leading coefficient** of $f(x)$. If all the coefficients of a polynomial expression are 0, the polynomial is 0 and is called the **zero polynomial.** The zero polynomial has no degree. However, a polynomial $f(x) = a_0$ for a nonzero complex number a_0 has degree 0.

In this chapter we discuss the graphs of polynomial functions of degree higher than 2 and methods of finding, or at least approximating, the values of x that satisfy $f(x) = 0$, called the **zeros** of $f(x)$. The chapter ends with a section on *rational functions,* which are defined as quotients of polynomials.

*Remember, *complex numbers* are imaginary numbers like $5i$ or $-i\sqrt{3}$, real numbers, or numbers such as $4 - 2i$ or $6 + i\sqrt{7}$.

4.1 SYNTHETIC DIVISION

The quotient of two polynomials was found in Chapter 1 with a division algorithm similar to that used to divide with whole numbers.

Division Algorithm

> Let $f(x)$ and $g(x)$ be polynomials with $g(x)$ of lower degree than $f(x)$ and $g(x)$ of degree one or more. There exist unique polynomials $q(x)$ and $r(x)$ such that
>
> $$f(x) = g(x) \cdot q(x) + r(x),$$
>
> where either $r(x) = 0$ or the degree of $r(x)$ is less than the degree of $g(x)$.

Recall that $q(x)$ is the quotient polynomial and $r(x)$ is the remainder polynomial or the remainder.

A shortcut method of performing long division with certain polynomials, called *synthetic division,* will be useful in applying the theorems presented in the next few sections. The method is used only when a polynomial is divided by a first-degree binomial of the form $x - k$, where the coefficient of x is 1. To illustrate, notice the example worked on the left below. On the right the division process is simplified by omitting all variables and writing only coefficients, with 0 used to represent the coefficient of any missing terms. Since the coefficient of x in the divisor is always 1 in these divisions, it too can be omitted. These omissions simplify the problem as shown on the right below.

$$
\begin{array}{r}
3x^2 + 10x + 40 \\
x - 4 \overline{)3x^3 - 2x^2 \qquad\qquad - 150} \\
\underline{3x^3 - 12x^2} \\
10x^2 \\
\underline{10x^2 - 40x} \\
40x - 150 \\
\underline{40x - 160} \\
10
\end{array}
\qquad
\begin{array}{r}
3 \quad 10 \quad 40 \\
-4 \overline{)3 - 2 + 0 - 150} \\
\underline{3 - 12} \\
10 \\
\underline{10 - 40} \\
40 - 150 \\
\underline{40 - 160} \\
10
\end{array}
$$

The numbers in color that are repetitions of the numbers directly above them can also be omitted.

$$
\begin{array}{r}
3 \quad 10 \quad 40 \\
-4 \overline{)3 - 2 + 0 - 150} \\
\underline{- 12} \\
10 \\
\underline{- 40} \\
40 - 150 \\
\underline{- 160} \\
10
\end{array}
$$

The entire problem can now be condensed vertically, and the top row of numbers omitted since it duplicates the bottom row if the 3 is brought down.

$$
\begin{array}{r|rrr}
-4 & 3 & -2 & 0 & -150 \\
& & -12 & -40 & -160 \\
\hline
& 3 & 10 & 40 & 10
\end{array}
$$

The rest of the bottom row is obtained by subtracting -12, -40, and -160 from the corresponding terms above.

With synthetic division it is useful to change the sign of the divisor, so the -4 at the left is changed to 4, which also changes the sign of the numbers in the second row. To compensate for this change, subtraction is changed to addition. Doing this gives the following result:

$$
\begin{array}{r|rrr}
4 & 3 & -2 & 0 & -150 \\
& & 12 & 40 & 160 \\
\hline
& 3 & 10 & 40 & 10
\end{array}
$$

In summary, to use synthetic division to divide a polynomial by a binomial of the form $x - k$, begin by writing the coefficients of the polynomial in decreasing powers of the variable, using 0 as the coefficient of any missing powers. The number k is written to the left in the same row. In the example above, $x - k$ is $x - 4$, so k is 4. Next bring down the leading coefficient of the polynomial, 3 in the example above, as the first number in the last row. Multiply the 3 by 4 to get the first number in the second row, 12. Add 12 to -2; this gives 10, the second number in the third row. Multiply 10 by 4 to get 40, the next number in the second row. Add 40 to 0 to get the third number in the third row, and so on. This process of multiplying each result in the third row by k and adding the product to the number in the next column is repeated until there is a number in the last row for each coefficient in the first row.

CAUTION   To avoid incorrect results, it is essential to use a 0 for any missing terms, including a missing constant, when you set up the division.

· · · · · · · · ·

EXAMPLE 1   Use synthetic division to divide $5x^3 - 6x^2 - 28x - 2$ by $x + 2$.

Begin by writing $-2 \underline{|} \, 5 \quad -6 \quad -28 \quad -2.$

The 2 is changed to -2 since k is found by writing $x + 2$ as $x - (-2)$. Next, bring down the 5.

$$
\begin{array}{r|rrr}
-2 & 5 & -6 & -28 & -2 \\
& & & & \\
\hline
& 5 & & &
\end{array}
$$

Now, multiply -2 by 5 to get -10, and add it to the -6 in the first row. The result is -16.

$$
\begin{array}{r|rrr}
-2 & 5 & -6 & -28 & -2 \\
& & -10 & & \\
\hline
& 5 & -16 & &
\end{array}
$$

Next, $(-2)(-16) = 32$. Add this to the -28 in the first row.

$$
\begin{array}{r|rrrr}
-2 & 5 & -6 & -28 & -2 \\
& & -10 & 32 & \\
\hline
& 5 & -16 & 4 &
\end{array}
$$

Finally, $(-2)(4) = -8$, which is added to the -2 to get -10.

$$
\begin{array}{r|rrrr}
-2 & 5 & -6 & -28 & -2 \\
& & -10 & 32 & -8 \\
\hline
& 5 & -16 & 4 & -10
\end{array}
$$

The coefficients of the quotient polynomial and the remainder are read directly from the bottom row. Since the degree of the quotient will always be one less than the degree of the polynomial to be divided,

$$\frac{5x^3 - 6x^2 - 28x - 2}{x + 2} = 5x^2 - 16x + 4 + \frac{-10}{x + 2}. \quad \bullet$$

The result of the division in Example 1 can be written as

$$5x^3 - 6x^2 - 28x - 2 = (x + 2)(5x^2 - 16x + 4) + (-10)$$

by multiplying both sides by the denominator $x + 2$. The following theorem is a generalization of the division process illustrated above.

. . . .

For any polynomial $f(x)$ and any complex number k, there exists a unique polynomial $q(x)$ and number r such that

$$f(x) = (x - k)q(x) + r.$$

For example, in the synthetic division above,

$$5x^3 - 6x^2 - 28x - 2 = (x + 2) \; (5x^2 - 16x + 4) + (-10).$$

$$\underbrace{}_{f(x)} \quad = \quad \underbrace{(x - k)}_{} \cdot \quad \underbrace{q(x)}_{} \quad + \quad \underbrace{r}_{}$$

This theorem is a special case of the division algorithm given earlier. Here $g(x)$ is the first-degree polynomial $x - k$.

Evaluating $f(k)$ By the division algorithm, $f(x) = (x - k)q(x) + r$. This equality is true for all complex values of x, so it is true for $x = k$. Replacing x with k gives

$$f(k) = (k - k)q(k) + r$$
$$f(k) = r.$$

This proves the following **remainder theorem,** which gives a new method of evaluating polynomial functions.

Remainder Theorem

If the polynomial $f(x)$ is divided by $x - k$, the remainder is $f(k)$.

As an illustration of this theorem, we have seen that when the polynomial $f(x) = 5x^3 - 6x^2 - 28x - 2$ is divided by $x + 2$ or $x - (-2)$, the remainder is -10. Now substituting -2 for x in $f(x)$ gives

$$f(-2) = (-2 + 2)[5(-2)^2 - 16(-2) + 4] + (-10)$$
$$= 0[5(-2)^2 - 16(-2) + 4] + (-10)$$
$$= -10.$$

As shown here, the simpler way to find the value of a polynomial is often by using synthetic division. By the remainder theorem, instead of replacing x by -2 to find $f(-2)$, divide $f(x)$ by $x + 2$ using synthetic division as in Example 1. Then $f(-2)$ is the remainder, -10.

$$
\begin{array}{r|rrrr}
-2 & 5 & -6 & -28 & -2 \\
   &   & -10 & 32 & -8 \\
\hline
   & 5 & -16 & 4 & -10 \quad \leftarrow f(-2)
\end{array}
$$

EXAMPLE 2   Let $f(x) = -x^4 + 3x^2 - 4x - 5$. Find $f(-3)$.

Use the remainder theorem and synthetic division.

$$
\begin{array}{r|rrrrr}
-3 & -1 & 0 & 3 & -4 & -5 \\
   &   & 3 & -9 & 18 & -42 \\
\hline
   & -1 & 3 & -6 & 14 & -47
\end{array}
$$

The remainder when $f(x)$ is divided by $x - (-3) = x + 3$ is -47, so $f(-3) = -47$. ●

The remainder theorem gives a quick way to decide if a number k is a zero of a polynomial $f(x)$. Use synthetic division to find $f(k)$; if the remainder is zero, then $f(k) = 0$ and k is a zero of $f(x)$. A zero of $f(x)$ is also called a **root** or **solution** of the equation $f(x) = 0$.

EXAMPLE 3   Decide whether or not the given number is a zero of the given polynomial.

(a) 2; $f(x) = x^3 - 4x^2 + 9x - 10$

Use synthetic division.

$$
\begin{array}{r|rrrr}
2 & 1 & -4 & 9 & -10 \\
  &   & 2 & -4 & 10 \\
\hline
  & 1 & -2 & 5 & 0
\end{array}
$$

Since the remainder is 0, $f(2) = 0$, and 2 is a zero of the polynomial $f(x) = x^3 - 4x^2 + 9x - 10$.

(b) -4; $f(x) = x^4 + x^2 - 3x + 1$

Remember to use a coefficient of 0 for the missing x^3 term in the synthetic division.

$$
\begin{array}{r|rrrrr}
-4 & 1 & 0 & 1 & -3 & 1 \\
   &   & -4 & 16 & -68 & 284 \\
\hline
   & 1 & -4 & 17 & -71 & 285
\end{array}
$$

The remainder is not 0, so -4 is not a zero of $f(x) = x^4 + x^2 - 3x + 1$. In fact, $f(-4) = 285$.

(c) $1 + 2i$; $f(x) = x^4 - 2x^3 + 4x^2 + 2x - 5$

Use synthetic division and operations with complex numbers.

$$
\begin{array}{r|rrrrr}
1 + 2i & 1 & -2 & 4 & 2 & -5 \\
       &   & 1 + 2i & -5 & -1 - 2i & 5 \\
\hline
       & 1 & -1 + 2i & -1 & 1 - 2i & 0
\end{array}
$$

Since the remainder is zero, $1 + 2i$ is a zero of the given polynomial. ●

4.1 Exercises

Use synthetic division to perform each of the following divisions.

1. $\dfrac{x^3 + 2x^2 - 17x - 10}{x + 5}$

2. $\dfrac{a^4 + 4a^3 + 2a^2 + 9a + 4}{a + 4}$

3. $\dfrac{m^4 - 3m^3 - 4m^2 + 12m}{m - 2}$

4. $\dfrac{p^4 - 3p^3 - 5p^2 + 2p - 16}{p + 2}$

5. $\dfrac{3x^3 - 11x^2 - 20x + 3}{x - 5}$

6. $\dfrac{4p^3 + 8p^2 - 16p - 9}{p + 3}$

7. $\dfrac{4m^3 - 3m - 2}{m + 1}$

8. $\dfrac{3q^3 - 4q + 2}{q - 1}$

9. $\dfrac{x^5 + 3x^4 + 2x^3 + 2x^2 + 3x + 1}{x + 2}$

10. $\dfrac{m^6 - 3m^4 + 2m^3 - 6m^2 - 5m + 3}{m + 2}$

11. $\dfrac{\frac{1}{3}x^3 - \frac{2}{9}x^2 + \frac{1}{27}x + 1}{x - \frac{1}{3}}$

12. $\dfrac{x^3 + x^2 + \frac{1}{2}x + \frac{1}{8}}{x + \frac{1}{2}}$

13. $\dfrac{y^3 - 1}{y - 1}$

14. $\dfrac{r^5 - 1}{r - 1}$

Express each polynomial in the form $f(x) = (x - k)q(x) + r$ for the given value of k.

15. $f(x) = 2x^3 + x^2 + x - 8$; $k = 1$

16. $f(x) = 2x^3 + 3x^2 + 4x - 10$; $k = -1$

17. $f(x) = -x^3 + 2x^2 + 4$; $k = -2$

18. $f(x) = -4x^3 + 2x^2 - 3x - 10$; $k = 2$

19. $f(x) = 4x^4 - 3x^3 - 20x^2 - x$; $k = 3$

20. $f(x) = 2x^4 + x^3 - 15x^2 + 3x$; $k = -3$

For each of the following polynomials, use the remainder theorem and synthetic division to find f(k).

21. $k = 5;$ $f(x) = -x^2 + 2x + 7$

22. $k = -3;$ $f(x) = 3x^2 + 8x + 5$

23. $k = 3;$ $f(x) = x^2 - 4x + 5$

24. $k = -2;$ $f(x) = x^2 + 5x + 6$

25. $k = -1;$ $f(x) = x^3 - 4x^2 + 2x + 1$

26. $k = 2;$ $f(x) = 2x^3 - 3x^2 - 5x + 4$

27. $k = 2 + i;$ $f(x) = x^2 - 5x + 1$

28. $k = 3 - 2i;$ $f(x) = x^2 - x + 3$

Use synthetic division to decide whether or not the given number is a zero of the given polynomial.

29. 4; $f(r) = 2r^3 - 6r^2 - 9r + 4$

30. $-6;$ $f(y) = 2y^3 + 9y^2 - 16y + 12$

31. $-5;$ $f(x) = x^3 + 7x^2 + 10x$

32. 3; $f(y) = 2y^3 - 3y^2 - 5y$

33. 2/5; $f(z) = 5z^4 + 2z^3 - z + 3$

34. 1/2; $f(w) = 2w^4 - 3w^2 + 4$

35. $2 + i;$ $f(k) = k^2 + 3k + 4$

36. $1 - 2i;$ $f(z) = z^2 - 3z + 5$

37. $i;$ $f(x) = x^3 + 2ix^2 + 2x + i$

38. $-i;$ $f(p) = p^3 - ip^2 + 3p + 5i$

39. Find the remainder when the polynomial $x^{99} - 2x^{52} + x^2$ is divided by $x + 1$.

40. Find the remainder when the polynomial $x^{100} + 3x^{25} - 3$ is divided by $x - i$.

41. Find a value of k so that when the polynomial $x^2 - 6x + 15$ is divided by $x - k$ the remainder is 6.

42. Find a value of k so that $(x^2 + 4x + 7) \div (x - k)$ has a remainder of 3.

4.2   COMPLEX ZEROS OF POLYNOMIAL FUNCTIONS

By the remainder theorem, if $f(k) = 0$, then the remainder when $f(x)$ is divided by $x - k$ is zero. This means that $x - k$ is a factor of $f(x)$. Conversely, if $x - k$ is a factor of $f(x)$, then $f(k)$ must equal 0. This is summarized in the following **factor theorem.**

Factor Theorem

> The polynomial $x - k$ is a factor of the polynomial $f(x)$ if and only if $f(k) = 0$.

EXAMPLE 1   Is $x - 1$ a factor of $f(x) = 2x^4 + 3x^2 - 5x + 7$?

By the factor theorem, $x - 1$ will be a factor of $f(x)$ only if $f(1) = 0$. Use synthetic division and the remainder theorem to decide.

$$
\begin{array}{r|rrrrr}
1 & 2 & 0 & 3 & -5 & 7 \\
  &   & 2 & 2 & 5 & 0 \\
\hline
  & 2 & 2 & 5 & 0 & 7
\end{array}
$$

Since the remainder is 7, $f(1) = 7$ and not 0, so $x - 1$ is not a factor of $f(x)$.   ●

. . . .

EXAMPLE 2   Is $x - i$ a factor of

$$f(x) = 3x^3 + (-4 - 3i)x^2 + (5 + 4i)x - 5i?$$

The only way $x - i$ can be a factor of $f(x)$ is for $f(i)$ to be 0. To see if this is the case, use synthetic division.

$$
\begin{array}{r|rrrr}
i & 3 & -4 - 3i & 5 + 4i & -5i \\
  &   & 3i & -4i & 5i \\
\hline
  & 3 & -4 & 5 & 0
\end{array}
$$

The remainder is 0, so $f(i) = 0$, and $x - i$ is a factor of $f(x)$.   ●

The factor theorem can be used to factor a polynomial of higher degree into linear factors. Linear factors are factors of the form $ax - b$ for complex numbers a and b.

. . . .

EXAMPLE 3   Factor $f(x)$ into linear factors given that k is a zero of $f(x)$.

(a) $f(x) = 6x^3 + 19x^2 + 2x - 3;\quad k = -3$

Since $k = -3$ is a zero of $f(x)$, $x - (-3) = x + 3$ is a factor. Use synthetic division to divide $f(x)$ by $x + 3$.

$$
\begin{array}{r|rrrr}
-3 & 6 & 19 & 2 & -3 \\
   &   & -18 & -3 & 3 \\
\hline
   & 6 & 1 & -1 & 0
\end{array}
$$

The quotient is $6x^2 + x - 1$, so

$$f(x) = (x + 3)(6x^2 + x - 1).$$

Factor $6x^2 + x - 1$ as $(2x + 1)(3x - 1)$ to get

$$f(x) = (x + 3)(2x + 1)(3x - 1),$$

where all factors are linear.

(b) $f(x) = 3x^3 + (-1 + 3i)x^2 + (-12 + 5i)x + 4 - 2i;\quad k = 2 - i$

One factor is $x - (2 - i)$ or $x - 2 + i$. Divide $f(x)$ by $x - (2 - i)$.

$$
\begin{array}{r|rrrr}
2 - i & 3 & -1 + 3i & -12 + 5i & 4 - 2i \\
      &   & 6 - 3i & 10 - 5i & -4 + 2i \\
\hline
      & 3 & 5 & -2 & 0
\end{array}
$$

By the division algorithm,

$$f(x) = (x - 2 + i)(3x^2 + 5x - 2).$$

Factor $3x^2 + 5x - 2$ as $(3x - 1)(x + 2)$; then a linear factored form of $f(x)$ is

$$f(x) = (x - 2 + i)(3x - 1)(x + 2).\quad ●$$

The next theorem says that every polynomial of degree 1 or more has a zero, which means that every such polynomial can be factored. The theorem was first proved by Carl Friedrich Gauss (1777–1855) as part of his doctoral dissertation completed in 1799 when he was 22. This theorem, which Gauss named the "fundamental theorem of algebra," had challenged the world's finest mathematicians for at least 200 years. Peter Rothe had stated it in 1608, followed by Albert Girard in 1629 and René Descartes in 1637. Jean LeRond D'Alembert (1717–83) thought he had a proof in 1746; consequently it is known in France today as D'Alembert's theorem. Two of the world's greatest mathematicians, Euler and Lagrange, had attempted unsuccessfully to solve it in 1749 and 1772, respectively. Their errors were noted in Gauss's dissertation. Gauss's proof uses advanced mathematical concepts outside of the field of algebra. To this day, no purely algebraic proof has been discovered.

Fundamental Theorem of Algebra

Every polynomial of degree 1 or more has at least one complex zero.

From the fundamental theorem, if $f(x)$ is of degree 1 or more then there is some number k_1 such that $f(k_1) = 0$. By the factor theorem, then

$$f(x) = (x - k_1) \cdot q_1(x)$$

for some polynomial $q_1(x)$. If $q_1(x)$ is of degree 1 or more, the fundamental theorem and the factor theorem can be used to factor $q_1(x)$ in the same way. There is some number k_2 such that $q_1(k_2) = 0$, so that

$$q_1(x) = (x - k_2) q_2(x)$$

and $$f(x) = (x - k_1)(x - k_2)q_2(x).$$

Assuming that $f(x)$ has degree n and repeating this process n times gives

$$f(x) = a(x - k_1)(x - k_2) \ldots (x - k_n),$$

where a is the leading coefficient of $f(x)$. Each of these factors leads to a zero of $f(x)$, so $f(x)$ has the n zeros $k_1, k_2, k_3, \ldots, k_n$. This result suggests the next theorem.

Theorem

A polynomial of degree n has at most n distinct zeros.

The theorem says that there exist *at most* n distinct zeros. For example, the polynomial $f(x) = x^3 + 3x^2 + 3x + 1 = (x + 1)^3$ is of degree 3 but has only one zero, -1. Actually, the zero -1 occurs three times, since there are three factors of $x + 1$; this zero is called a **zero of multiplicity 3.**

· · · · · · · · · ·

EXAMPLE 4   Find a polynomial $f(x)$ of degree 3 that satisfies the following conditions.

(a) Zeros of -1, 2, and 4; $f(1) = 3$

These three zeros give $x - (-1) = x + 1$, $x - 2$, and $x - 4$ as factors of $f(x)$. Since $f(x)$ is to be of degree 3, these are the only possible factors by the theorem just above. Therefore, $f(x)$ has the form

$$f(x) = a(x + 1)(x - 2)(x - 4)$$

for some real number a. To find a, use the fact that $f(1) = 3$.

$$f(1) = a(1 + 1)(1 - 2)(1 - 4) = 3$$
$$a(2)(-1)(-3) = 3$$
$$6a = 3$$
$$a = \frac{1}{2}$$

Thus, $$f(x) = \frac{1}{2}(x + 1)(x - 2)(x - 4),$$

or, by multiplication,

$$f(x) = \frac{1}{2}x^3 - \frac{5}{2}x^2 + x + 4.$$

(b) -2 is a zero of multiplicity 3; $f(-1) = 4$

The polynomial $f(x)$ has the form

$$f(x) = a(x + 2)(x + 2)(x + 2)$$
$$= a(x + 2)^3.$$

Since $f(-1) = 4$,

$$f(-1) = a(-1 + 2)^3 = 4,$$
$$a(1)^3 = 4$$

or $$a = 4,$$

and $f(x) = 4(x + 2)^3 = 4x^3 + 24x^2 + 48x + 32.$ ●

NOTE   In Example 4(a), we cannot clear the denominators in $f(x)$ by multiplying both sides by 2, because the result would equal $2 \cdot f(x)$, not $f(x)$ itself.

The remainder theorem can be used to show that both $2 + i$ and $2 - i$ are zeros of $f(x) = x^3 - x^2 - 7x + 15$. In general, if $a + bi$ is a zero of a polynomial function with *real* coefficients, then so is $a - bi$. To prove this requires the following properties of complex conjugates. Let $z = a + bi$, and write $\bar{z}$ for the conjugate of z, so that $\bar{z} = a - bi$. For example, if $z = -5 + 2i$, then $\bar{z} = -5 - 2i$. The proof of the following equalities is left for the exercises (see Exercises 49–52).

Properties of Conjugates

For any complex numbers c and d,

$$\overline{c + d} = \overline{c} + \overline{d}$$
$$\overline{c \cdot d} = \overline{c} \cdot \overline{d}$$
$$\overline{c^n} = (\overline{c})^n.$$

If the complex number z is a zero of $f(x)$, then the conjugate of z, $\overline{z}$, is also a zero of $f(x)$. This is shown by starting with the polynomial

$$f(x) = a_n x^n + a_{n-1} x^{n-1} + \ldots + a_1 x + a_0,$$

where all coefficients are real numbers. If $z = a + bi$ is a zero of $f(x)$, then

$$f(z) = a_n z^n + a_{n-1} z^{n-1} + \ldots + a_1 z + a_0 = 0.$$

Taking the conjugate of both sides of this last equation gives

$$\overline{a_n z^n + a_{n-1} z^{n-1} + \ldots + a_1 z + a_0} = \overline{0}.$$

Using generalizations of the properties $\overline{c + d} = \overline{c} + \overline{d}$ and $\overline{c \cdot d} = \overline{c} \cdot \overline{d}$ gives

$$\overline{a_n z^n} + \overline{a_{n-1} z^{n-1}} + \ldots + \overline{a_1 z} + \overline{a_0} = \overline{0}$$

or

$$\overline{a_n}\, \overline{z^n} + \overline{a_{n-1}}\, \overline{z^{n-1}} + \ldots + \overline{a_1}\, \overline{z} + \overline{a_0} = \overline{0}.$$

Now use the third property from above and the fact that for any real number a, $\overline{a} = a$, to get

$$a_n (\overline{z})^n + a_{n-1} (\overline{z})^{n-1} + \ldots + a_1 (\overline{z}) + a_0 = 0.$$

Hence $\overline{z}$ is also a zero of $f(x)$, which completes the proof of the **conjugate zeros theorem.**

Conjugate Zeros Theorem

If $f(x)$ is a polynomial *having only real coefficients* and if $a + bi$ is a zero of $f(x)$, where a and b are real numbers, then $a - bi$ is also a zero of $f(x)$.

 CAUTION   The requirement that the polynomial have only real coefficients is *essential*. For example, $f(x) = x - (1 + i)$ has $1 + i$ as a zero, but the conjugate $1 - i$ is not a zero.

EXAMPLE 5   Find a polynomial of lowest degree having only real coefficients and zeros 3 and $2 + i$.

The complex number $2 - i$ also must be a zero, so the polynomial has at least three zeros, 3, $2 + i$, and $2 - i$. For the polynomial to be of lowest degree these must be the only zeros. By the factor theorem there must be three factors, $x - 3$, $x - (2 + i)$, and $x - (2 - i)$. A polynomial of lowest degree is

$$f(x) = (x - 3)[x - (2 + i)][x - (2 - i)]$$
$$= (x - 3)(x - 2 - i)(x - 2 + i)$$
$$= x^3 - 7x^2 + 17x - 15.$$

Other polynomials, such as $2(x^3 - 7x^2 + 17x - 15)$ or $\sqrt{5}(x^3 - 7x^2 + 17x - 15)$, for example, also satisfy the given conditions on zeros. The information on zeros given in the problem is not enough to give a specific value for the leading coefficient. ●

The theorem on conjugate zeros is important in helping predict the number of real zeros of polynomials with real coefficients. A polynomial with real coefficients of odd degree n, where $n \geq 1$, must have at least one real zero (since zeros of the form $a + bi$, where $b \neq 0$, occur in conjugate pairs.) On the other hand, a polynomial with real coefficients of even degree n may have no real zeros.

· · · · · · · · ·

EXAMPLE 6   Find all zeros of $f(x) = x^4 - 7x^3 + 18x^2 - 22x + 12$, given that $1 - i$ is a zero.

Since the polynomial has only real coefficients and since $1 - i$ is a zero, by the conjugate zeros theorem $1 + i$ is also a zero. To find the remaining zeros, first divide the original polynomial by $x - (1 - i)$.

$$\underline{1 - i\,|1 \qquad -7 \qquad 18 \qquad -22 \qquad 12}$$
$$\underline{\qquad\qquad 1 - i \quad -7 + 5i \quad 16 - 6i \quad -12}$$
$$\qquad 1 \quad -6 - i \quad 11 + 5i \quad -6 - 6i \qquad 0$$

By the factor theorem, since $x = 1 - i$ is a zero of $f(x)$, $x - (1 - i)$ is a factor and $f(x)$ can be written as

$$f(x) = [x - (1 - i)][x^3 + (-6 - i)x^2 + (11 + 5i)x + (-6 - 6i)].$$

We know that $x = 1 + i$ is also a zero of $f(x)$ so that

$$f(x) = [x - (1 - i)][x - (1 + i)]q(x).$$

Thus, $x^3 + (-6 - i)x^2 + (11 + 5i)x + (-6 - 6i) = (x - (1 + i))q(x)$.

Use synthetic division to find $q(x)$.

$$\underline{1 + i\,|1 \qquad -6 - i \qquad 11 + 5i \qquad -6 - 6i}$$
$$\underline{\qquad\qquad 1 + i \qquad -5 - 5i \qquad 6 + 6i}$$
$$\qquad 1 \qquad -5 \qquad\quad 6 \qquad\qquad 0$$

Since $q(x) = x^2 - 5x + 6$, $f(x)$ can be written as

$$f(x) = [x - (1 - i)][x - (1 + i)](x^2 - 5x + 6).$$

Now find the zeros of the quadratic polynomial $x^2 - 5x + 6$. Factoring the polynomial shows that the zeros are 2 and 3, so the four zeros of $f(x)$ are $1 - i$, $1 + i$, 2, and 3. ●

4.2 Exercises

Use the factor theorem to decide whether or not the second polynomial is a factor of the first.

1. $4x^2 + 2x + 42$; $x - 3$

2. $-3x^2 - 4x + 2$; $x + 2$

3. $x^3 + 2x^2 - 3$; $x - 1$

4. $2x^3 + x + 2$; $x + 1$

5. $2x^4 + 5x^3 - 2x^2 + 5x + 3$; $x + 3$

6. $5x^4 + 16x^3 - 15x^2 + 8x + 16$; $x + 4$

For each of the following, find a polynomial of degree 3 with only real coefficients which satisfies the given conditions.

7. Zeros of -3, -1, and 4; $f(2) = 5$

8. Zeros of 1, -1, and 0; $f(2) = -3$

9. Zeros of -2, 1, and 0; $f(-1) = -1$

10. Zeros of 2, 5, and -3; $f(1) = -4$

11. Zeros of 3, i, and $-i$; $f(2) = 50$

12. Zeros of -2, i, and $-i$; $f(-3) = 30$

For each of the following polynomials, find all zeros and their multiplicities.

13. $f(x) = 3(x - 2)(x + 3)(x - 1 + i)$

14. $f(x) = 5(x + 1 - \sqrt{2})(2x + 5)x^2$

15. $f(x) = 4x^7 + x^5$

16. $f(x) = (x + 1)^2(x - 2)(x + 1)^3$

17. $f(x) = (x^2 + x - 2)^5(x - 1 + \sqrt{3})^2$

18. $f(x) = (5x - 2)^3(x^2 + 9)^2$

For each of the following, find a polynomial of lowest degree with only real coefficients having the given zeros.

19. $5 + i$ and $5 - i$

20. $3 - 2i$ and $3 + 2i$

21. $1 + \sqrt{2}$, $1 - \sqrt{2}$, and 1

22. $1 - \sqrt{3}$, $1 + \sqrt{3}$, and -2

23. $2 + i$, $2 - i$, 3, and -1

24. $3 + 2i$, $3 - 2i$, -1, and 2

25. 2 and $3 + i$

26. -1 and $4 - 2i$

27. $1 + 2i$, 2 (multiplicity 2)

28. $2 + i$, -3 (multiplicity 2)

For each of the following polynomials, one zero is given. Find all others. $(x-k)$ $(x-)$, $\boxed{1}$ $1 \ -5 \ 17 \ -15$

29. $f(x) = x^3 - x^2 - 4x - 6$; 3

30. $f(x) = x^3 - 5x^2 + 17x - 13$; $\boxed{1}$

31. $f(x) = 2x^3 - 2x^2 - x - 6$; 2

32. $f(x) = 2x^3 - 5x^2 + 6x - 2$; $1 + i$ $1 -4 \ 13$ 0

33. $f(x) = x^4 + 5x^2 + 4$; $-i$

34. $f(x) = x^4 + 10x^3 + 27x^2 + 10x + 26$; i

$x^2 - 4x + 13 = 0$

Factor $f(x)$ into linear factors given that k is a zero of $f(x)$.

35. $f(x) = 2x^3 - 3x^2 - 17x + 30$; $k = 2$

36. $f(x) = 2x^3 - 3x^2 - 5x + 6$; $k = 1$

37. $f(x) = 6x^3 + 25x^2 + 3x - 4$; $k = -4$

38. $f(x) = 8x^3 + 50x^2 + 47x - 15$; $k = -5$

39. $f(x) = x^3 + (7 - 3i)x^2 + (12 - 21i)x - 36i$; $k = 3i$

40. $f(x) = 2x^3 + (3 + 2i)x^2 + (1 + 3i)x + i$; $k = -i$

41. Show that -2 is a zero of multiplicity 2 of $f(x) = x^4 + 2x^3 - 7x^2 - 20x - 12$ and find all other complex zeros. Then write $f(x)$ in factored form.

42. Show that -1 is a zero of multiplicity 3 of $f(x) = x^5 + 9x^4 + 33x^3 + 55x^2 + 42x + 12$, and find all other complex zeros. Then write $f(x)$ in factored form.

43. What are the possible numbers of real zeros (counting multiplicities) for a polynomial with real coefficients of degree five?

44. Tell why a polynomial of degree four with real coefficients has either 0, 2, or 4 real zeros, counting multiplicities.

45. The displacement at time t of a particle moving along a straight line is given by $s(t) = t^3 - 2t^2 - 5t + 6$, where t is in seconds and s is measured in centimeters. The displacement is 0 after 1 second has elapsed. At what other times is the displacement 0?

46. For headphone radios, the cost function (in thousands of dollars) is given by $C(x) = 2x^3 - 9x^2 + 17x - 4$, and the revenue function (in thousands of dollars) is given by $R(x) = 5x$, where x is the number of items (in hundred thousands) produced. Cost equals revenue if 200,000 items are produced ($x = 2$). Find all other break-even points.

47. Explain why it is not possible for a polynomial of degree 3 with real coefficients to have zeros of 1, 2, and $1 + i$.

48. Show that the zeros of $f(x) = x^3 + ix^2 - (7 - i)x + (6 - 6i)$ are $1 - i$, 2, and -3. Does the conjugate zeros theorem apply? Why?

If c and d are complex numbers, prove each of the following statements. (Hint: Let $c = a + bi$ and $d = m + ni$ and form all the conjugates, the sums, and the products.)

49. $\overline{c + d} = \overline{c} + \overline{d}$

50. $\overline{cd} = \overline{c} \cdot \overline{d}$.

51. $\overline{a} = a$ for any real number a.

52. $\overline{c^n} = (\overline{c})^n$

4.3 RATIONAL ZEROS OF POLYNOMIAL FUNCTIONS

By the fundamental theorem of algebra, every polynomial of degree 1 or more has a zero. However, the fundamental theorem merely says that a zero exists. It gives no help at all in identifying zeros. Other theorems can be used to find any rational zeros of polynomials with rational coefficients or to find decimal approximations of any irrational zeros.

The **rational zeros theorem** gives a useful method for finding a set of possible zeros of a polynomial with integer coefficients.

Rational Zeros Theorem

Let

$$f(x) = a_n x^n + a_{n-1} x^{n-1} + \ldots + a_1 x + a_0, \quad a_n \neq 0,$$

be a polynomial with only integer coefficients. If p/q is a rational number written in lowest terms and if p/q is a zero of $f(x)$, then p is a factor of the constant term a_0 and q is a factor of the leading coefficient a_n.

This theorem can be proved as follows. Given that $x = p/q$ (with p/q in lowest terms) is a zero of $f(x)$, then $qx - p$ is a factor of $f(x)$. If the other factor is $b_{n-1}x^{n-1} + \ldots + b_0$, then

$$(qx - p)(b_{n-1}x^{n-1} + \ldots + b_0) = f(x)$$
$$(qx - p)(b_{n-1}x^{n-1} + \ldots + b_0) = a_n x^n + \ldots + a_0.$$

Therefore,

$$qb_{n-1} = a_n \quad \text{and} \quad -pb_0 = a_0.$$

From $qb_{n-1} = a_n$, q is a factor of a_n and from $-pb_0 = a_0$, p is a factor of a_0.

⋯⋯⋯⋯

EXAMPLE 1   Find all rational zeros of $f(x) = 2x^3 - x^2 - 2x + 1$.

Since $f(x)$ has only integer coefficients, the rational zeros theorem applies. For this polynomial, $a_0 = 1$ and $a_n = a_3 = 2$. By the rational zeros theorem, p must be a factor of $a_0 = 1$ so that p is either 1 or -1. Also q must be a factor of $a_3 = 2$, so q is ± 1 or ± 2. Any rational zeros are in the form p/q, and so must be either ± 1 or $\pm 1/2$. Use synthetic division to check out these possibilities.

$$
\begin{array}{r|rrrr}
1 & 2 & -1 & -2 & 1 \\
  &   & 2 & 1 & -1 \\
\hline
  & 2 & 1 & -1 & 0
\end{array}
$$

One zero is 1. To find any other zeros, set the quotient polynomial, $2x^2 + x - 1$, equal to zero.

$$2x^2 + x - 1 = 0$$
$$(2x - 1)(x + 1) = 0$$
$$x = \frac{1}{2} \quad \text{or} \quad x = -1$$

The rational zeros of $f(x)$ are -1, $1/2$, and 1.   ●

▌ **NOTE**   When the quotient is a second-degree polynomial, either factoring or ▌ the quadratic formula can be used to get any remaining zeros.

⋯⋯⋯⋯

EXAMPLE 2   Find all rational zeros of $f(x) = 2x^4 - 11x^3 + 14x^2 - 11x + 12$.

All coefficients are integers, so the rational zeros theorem can be used. If p/q is to be a rational zero of $p(x)$, by the rational zeros theorem p must be a factor of $a_0 = 12$ and q must be a factor of $a_4 = 2$. The possible values of p are ± 1, ± 2, ± 3, ± 4, ± 6, or ± 12, while q must be ± 1 or ± 2. The possible rational zeros are found by forming all possible quotients of the form p/q; then any rational zero of $f(x)$ will come from the list

$$\pm 1, \quad \pm 1/2, \quad \pm 2, \quad \pm 3, \quad \pm 3/2, \quad \pm 4, \quad \pm 6, \quad \pm 12.$$

Though it is not certain whether any of these numbers are zeros, if $f(x)$ has any rational zeros, they will be in the list above. These proposed zeros can be checked by synthetic division. A search of the possibilities leads to the discovery that 4 is a rational zero of $f(x)$.

$$\begin{array}{r|rrrrr} 4 & 2 & -11 & 14 & -11 & 12 \\ & & 8 & -12 & 8 & -12 \\ \hline & 2 & -3 & 2 & -3 & 0 \end{array}$$

As a fringe benefit of this calculation, zeros of the simpler polynomial $q(x) = 2x^3 - 3x^2 + 2x - 3$ can now be sought. Any rational zero of $q(x)$ will have a numerator of ± 3 or ± 1, with a denominator of ± 1 or ± 2. Thus, any rational zeros of $q(x)$ will come from the list

$$\pm 3, \quad \pm 3/2, \quad \pm 1, \quad \pm 1/2.$$

Again use synthetic division and trial and error to find that 3/2 is a zero.

$$\begin{array}{r|rrrr} \frac{3}{2} & 2 & -3 & 2 & -3 \\ & & 3 & 0 & 3 \\ \hline & 2 & 0 & 2 & 0 \end{array}$$

The quotient is $2x^2 + 2$, which, by the quadratic formula, has i and $-i$ as zeros. They are, however, complex zeros. The rational zeros of

$$f(x) = 2x^4 - 11x^3 + 14x^2 - 11x + 12$$

· · · · · · · · · ·

EXAMPLE 3   Find all rational zeros of $f(x) = 6x^4 + 7x^3 - 12x^2 - 3x + 2$, and factor the polynomial.

For a rational number p/q to be a zero of $f(x)$, p must be a factor of $a_0 = 2$ and q must be a factor of $a_4 = 6$. Thus, p can be ± 1 or ± 2 and q can be ± 1, ± 2, ± 3, or ± 6. The rational zeros, p/q, must be from the following list.

$$\pm 1, \quad \pm 2, \quad \pm 1/2, \quad \pm 1/3, \quad \pm 1/6, \quad \pm 2/3$$

Check 1 first because it is easy.

$$\begin{array}{r|rrrrr} 1 & 6 & 7 & -12 & -3 & 2 \\ & & 6 & 13 & 1 & -2 \\ \hline & 6 & 13 & 1 & -2 & 0 \end{array}$$

The 0 remainder shows that 1 is a zero. Now use the quotient polynomial $6x^3 + 13x^2 + x - 2$ and synthetic division to find that -2 is also a zero.

$$\begin{array}{r|rrrr} -2 & 6 & 13 & 1 & -2 \\ & & -12 & -2 & 2 \\ \hline & 6 & 1 & -1 & 0 \end{array}$$

The new quotient polynomial is $6x^2 + x - 1$. Use the quadratic formula or factor to solve the equation $6x^2 + x - 1 = 0$. The remaining two zeros are 1/3 and $-1/2$.

Factor the polynomial $f(x)$ in the following way. Since the four zeros of $f(x) = 6x^4 + 7x^3 - 12x^2 - 3x + 2$ are 1, -2, 1/3, and $-1/2$, the factors are $x - 1$, $x + 2$, $x - 1/3$, and $x + 1/2$, and

$$f(x) = a(x - 1)(x + 2)\left(x - \frac{1}{3}\right)\left(x + \frac{1}{2}\right).$$

Since the leading coefficient of $f(x)$ is 6, let $a = 6$. Then

$$f(x) = 6(x - 1)(x + 2)\left(x - \frac{1}{3}\right)\left(x + \frac{1}{2}\right)$$

$$= (x - 1)(x + 2)(3)\left(x - \frac{1}{3}\right)(2)\left(x + \frac{1}{2}\right)$$

$$= (x - 1)(x + 2)(3x - 1)(2x + 1). \quad \bullet$$

To find any rational zeros of a polynomial with fractional coefficients, first multiply the polynomial by a number that will clear it of all fractional coefficients. Then use the rational zeros theorem, which requires only integer coefficients.

· · · · · · · · · ·

EXAMPLE 4   Find all rational zeros of

$$f(x) = x^4 - \frac{1}{6}x^3 + \frac{2}{3}x^2 - \frac{1}{6}x - \frac{1}{3}.$$

To find the values of x that make $f(x) = 0$, or

$$x^4 - \frac{1}{6}x^3 + \frac{2}{3}x^2 - \frac{1}{6}x - \frac{1}{3} = 0,$$

first multiply both sides by 6 to eliminate all fractions. This gives

$$6x^4 - x^3 + 4x^2 - x - 2 = 0.$$

This polynomial will have the same zeros as $f(x)$. The possible rational zeros are of the form p/q where p is ± 1 or ± 2, and q is ± 1, ± 2, ± 3, or ± 6. Then p/q may be

$$\pm 1, \quad \pm 2, \quad \pm \frac{1}{2}, \quad \pm \frac{1}{3}, \quad \pm \frac{1}{6}, \quad \text{or} \quad \pm \frac{2}{3}.$$

Use synthetic division to find that $-1/2$ and $2/3$ are zeros.

$$
\begin{array}{r|rrrrr}
-\dfrac{1}{2} & 6 & -1 & 4 & -1 & -2 \\
 & & -3 & 2 & -3 & 2 \\
\hline
 & 6 & -4 & 6 & -4 & 0
\end{array}
\qquad
\begin{array}{r|rrrr}
\dfrac{2}{3} & 6 & -4 & 6 & -4 \\
 & & 4 & 0 & 4 \\
\hline
 & 6 & 0 & 6 & 0
\end{array}
$$

The final quotient is $q(x) = 6x^2 + 6 = 6(x^2 + 1)$. The zeros of this polynomial are i and $-i$. Since these zeros are not rational numbers, there are just two rational zeros: $-1/2$ and $2/3$. •

CAUTION Remember, the rational zeros theorem can be used only if the coefficients of $f(x)$ are integers. Polynomial functions with rational coefficients can be rewritten with integer coefficients in order to use the theorem, but the theorem cannot be used with polynomial functions with irrational or imaginary coefficients.

4.3 Exercises

Give all possible rational zeros for the following polynomials.

1. $f(x) = 6x^3 + 17x^2 - 31x - 1$

2. $f(x) = 15x^3 + 61x^2 + 2x - 1$

3. $f(x) = 2x^3 + 7x^2 + 12x - 8$

4. $f(x) = 2x^3 + 20x^2 + 68x - 40$

5. $f(x) = x^4 + 4x^3 + 3x^2 - 10x + 50$

6. $f(x) = x^4 - 2x^3 + x^2 + 18$

Find all rational zeros of the following polynomials.

7. $f(x) = x^3 - 2x^2 - 13x - 10$

8. $f(x) = x^3 + 5x^2 + 2x - 8$

9. $f(x) = x^3 + 6x^2 - x - 30$

10. $f(x) = x^3 - x^2 - 10x - 8$

11. $f(x) = x^3 + 9x^2 - 14x - 24$

12. $f(x) = x^3 + 3x^2 - 4x - 12$

13. $f(x) = x^4 + 9x^3 + 21x^2 - x - 30$

14. $f(x) = x^4 + 4x^3 - 7x^2 - 34x - 24$

Find the rational zeros of the following polynomials; then write each polynomial in factored form with each factor having only integer coefficients.

15. $f(x) = 6x^3 + 17x^2 - 31x - 12$

16. $f(x) = 15x^3 + 61x^2 + 2x - 8$

17. $f(x) = 2x^3 + 7x^2 + 12x - 8$

18. $f(x) = 2x^3 + 20x^2 + 68x - 40$

19. $f(x) = x^4 + 4x^3 + 3x^2 - 10x + 50$

20. $f(x) = x^4 - 2x^3 + x^2 + 18$

21. $f(x) = x^4 + 2x^3 - 13x^2 - 38x - 24$

22. $f(x) = 6x^4 + x^3 - 7x^2 - x + 1$

23. $f(x) = x^5 + 3x^4 - 5x^3 - 11x^2 + 12$

24. $f(x) = 4x^5 + 4x^4 - 37x^3 - 37x^2 + 9x + 9$

Find all rational zeros of the following polynomials.

25. $f(x) = x^3 - \dfrac{4}{3}x^2 - \dfrac{13}{3}x - 2$

26. $f(x) = x^3 + x^2 - \dfrac{16}{9}x + \dfrac{4}{9}$

27. $f(x) = x^4 + \dfrac{1}{4}x^3 + \dfrac{11}{4}x^2 + x - 5$

28. $f(x) = \dfrac{10}{7}x^4 - x^3 - 7x^2 + 5x - \dfrac{5}{7}$

29. $f(x) = \dfrac{1}{3}x^5 + x^4 - \dfrac{5}{3}x^3 - \dfrac{11}{3}x^2 + 4$

30. $f(x) = x^5 + x^4 - \dfrac{37}{4}x^2 + \dfrac{9}{4}x + \dfrac{9}{4}$

Find all integer zeros of the following equations.

31. $6x^3 - 19x^2 + 2x + 3 = 0$

32. $12x^3 - 5x^2 - 9x + 1 = 0$

33. A rectangle of area 42 sq in. has its base on the x-axis and its upper corners on the parabola $y = 16 - x^2$. Find the dimensions of the rectangle. (*Hint:* What are the coordinates of P?)

34. The width of an open rectangular box is twice its height and its length is 10 more than its height. Find the dimensions of the box if its volume is 234 cu in.

35. Show that $f(x) = x^2 - 2$ has no rational zeros, so $\sqrt{2}$ must be irrational.

36. Show that $f(x) = x^2 - 5$ has no rational zeros, so $\sqrt{5}$ must be irrational.

37. Show that $f(x) = x^4 + 5x^2 + 4$ has no rational zeros.

38. Show that $f(x) = x^5 - 3x^3 + 5$ has no rational zeros.

39. Show that any integer zeros of a polynomial must be factors of the constant term a_0.

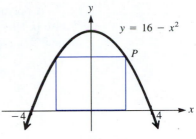

Exercise 33

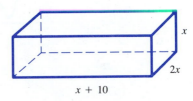

Exercise 34

4.4 GRAPHS OF POLYNOMIAL FUNCTIONS

We have already discussed the graphs of polynomial functions of degree 0 to 2. In this section we show how to graph polynomial functions of degree 3 or more. The domains will be restricted to real numbers, since we will be graphing on the real-number plane. Without calculus, it is necessary to plot many points to determine the graph of a polynomial function. The theorems in this chapter will be helpful in deciding which points to plot.

The domain of every polynomial function is the set of all real numbers. The range of a polynomial function of odd degree is also the set of all real numbers. Some typical graphs of polynomial functions of odd degree are shown in Figure 1. These graphs suggest that for every polynomial function f of odd degree there is at least one real value of x that makes $f(x) = 0$. The zeros are the x-intercepts of the graph.

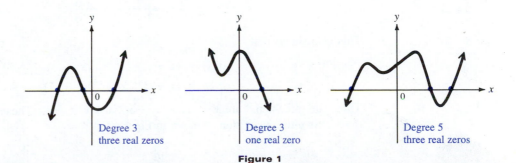

| Degree 3 | Degree 3 | Degree 5 |
| three real zeros | one real zero | three real zeros |

Figure 1

A polynomial function of even degree will have a range of the form $(-\infty, k]$ or else $[k, +\infty)$ for some real number k. Figure 2 shows two typical graphs of polynomial functions of even degree.

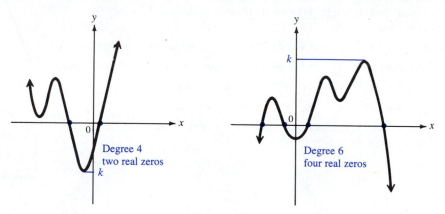

Degree 4
two real zeros

Degree 6
four real zeros

Figure 2

The graphs in Figures 1 and 2 show that polynomial functions often have *turning points* where the function changes from increasing to decreasing or from decreasing to increasing. A polynomial function of degree n has at most $n - 1$ turning points with at least one turning point between each pair of successive zeros. The graphs shown above illustrate this.

If a polynomial function is defined by an expression that can be factored, its graph can be approximated without plotting very many points; this method is shown in the next examples.

EXAMPLE 1   Graph $f(x) = 2x^3 + 5x^2 - x - 6$.

If $f(x)$ has rational zeros, it can be factored. Any rational zeros will be of the form p/q where p is ± 1, ± 2, ± 3, or ± 6, and q is ± 1 or ± 2. Possible zeros are ± 1, ± 2, ± 3, ± 6, $\pm 1/2$, or $\pm 3/2$. Use synthetic division to see that 1 is a zero.

$$
\begin{array}{r|rrr}
1 & 2 & 5 & -1 & -6 \\
  &   & 2 & 7 & 6 \\
\hline
  & 2 & 7 & 6 & 0
\end{array}
$$

This also shows that

$$
\begin{aligned}
f(x) &= (x - 1)(2x^2 + 7x + 6) \\
     &= (x - 1)(2x + 3)(x + 2).
\end{aligned}
$$

Thus, the three zeros are $x = 1$, $x = -3/2$ and $x = -2$. These three zeros divide the x-axis into four regions, shown in Figure 3.

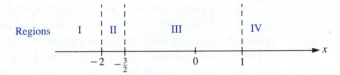

Figure 3

In any of these regions, the values of $f(x)$ are either always positive or always negative. To find the sign of $f(x)$ in each region, select an x-value in each region and substitute it into the equation for $f(x)$ to determine if the values of the function are positive or negative in that region. A typical selection of test points and the results of the tests are shown below.

Region	Test point	Value of $f(x)$	Sign of $f(x)$
I $(-\infty, -2)$	-3	-12	negative
II $(-2, -3/2)$	$-7/4$	$11/32$	positive
III $(-3/2, 1)$	0	-6	negative
IV $(1, +\infty)$	2	28	positive

Plot the three zeros and the test points and connect them with a smooth curve to get the graph. When the values of $f(x)$ are negative, the graph is below the x-axis, and when $f(x)$ takes on positive values, the graph is above the x-axis, as shown in Figure 4. The sketch could be improved by plotting additional points in each region.   ●

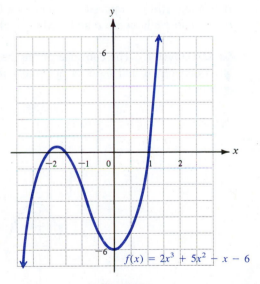

$$f(x) = 2x^3 + 5x^2 - x - 6$$

Figure 4

If a polynomial function has no rational zeros, it cannot be factored, so another method must be used to select useful points to plot. Much of the work in locating irrational real zeros uses the following result, which is related to the fact that graphs of polynomial functions are unbroken curves, with no gaps or sudden jumps. The proof requires advanced methods, so it is not given here. Figure 5 illustrates the theorem.

Intermediate Value Theorem for Polynomials

If $f(x)$ is a polynomial with only real coefficients, and if for real numbers a and b, the values $f(a)$ and $f(b)$ are opposite in sign, then there exists at least one real zero between a and b.

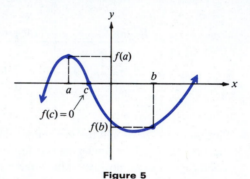

Figure 5

This theorem helps to identify intervals where zeros of polynomials are located. If $f(a)$ and $f(b)$ are opposite in sign, then 0 is between $f(a)$ and $f(b)$, and there must be a number c between a and b where $f(c) = 0$.

CAUTION The converse of the theorem is false. The fact that $f(c) = 0$, for c between a and b, does not imply that 0 is between $f(a)$ and $f(b)$. See Figure 6.

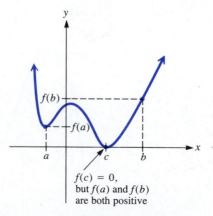

$f(c) = 0$,
but $f(a)$ and $f(b)$
are both positive

Figure 6

EXAMPLE 2   Show that $f(x) = x^3 - 2x^2 - x + 1$ has a real zero between 2 and 3.

Use synthetic division to find $f(2)$ and $f(3)$.

$$\begin{array}{r|rrrr} 2 & 1 & -2 & -1 & 1 \\ & & 2 & 0 & -2 \\ \hline & 1 & 0 & -1 & -1 \end{array} \qquad \begin{array}{r|rrrr} 3 & 1 & -2 & -1 & 1 \\ & & 3 & 3 & 6 \\ \hline & 1 & 1 & 2 & 7 \end{array}$$

Since $f(2)$ is negative but $f(3)$ is positive, there must be a real zero between 2 and 3.   ●

The intermediate value theorem for polynomials is helpful in limiting the search for real zeros to smaller and smaller intervals. In Example 2 the theorem was used to verify that there is a real zero between 2 and 3. The theorem could then be used repeatedly to locate the zero more accurately.

As suggested by the graphs of Figure 7, if $y = f(x)$ is a polynomial of degree $n \geq 1$ and $|x|$ gets larger and larger, then so will $|y|$. The next theorem, the *boundedness theorem,* shows how the bottom row of a synthetic division can be used to express this idea by placing upper and lower bounds on the possible real zeros of a polynomial.

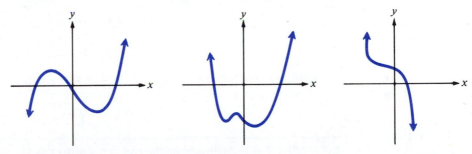

Figure 7

Boundedness Theorem

Let $f(x)$ be a polynomial of degree $n \geq 1$ with real coefficients and with a positive leading coefficient. If $f(x)$ is divided synthetically by $x - c$ and

(a) if $c > 0$ and all numbers in the bottom row of the synthetic division are nonnegative, then $f(x)$ has no zero greater than c;

(b) if $c < 0$ and the numbers in the bottom row of the synthetic division alternate in sign (with 0 considered positive or negative, as needed), then $f(x)$ has no zero less than c.

An outline of the proof of part (a) is given here. The proof for part (b) is similar. By the division algorithm, if $f(x)$ is divided by $x - c$, then

$$f(x) = (x - c)q(x) + r,$$

where all coefficients of $q(x)$ are nonnegative, $r \geq 0$, and $c > 0$. If $x > c$, then $x - c > 0$. Since $q(x) > 0$, and $r \geq 0$,

$$f(x) = (x - c)q(x) + r > 0.$$

This means that $f(x)$ will never be 0 for $x > c$.

· · · · · · · · ·

EXAMPLE 3   Show that the real zeros of $f(x) = 2x^4 - 5x^3 + 3x + 1$ satisfy the following conditions.

(a) No real zero is greater than 3.

Since $f(x)$ has real coefficients and the leading coefficient, 2, is positive, the boundedness theorem can be used. Divide $f(x)$ synthetically by $x - 3$.

$$
\begin{array}{r|rrrrr}
3 & 2 & -5 & 0 & 3 & 1 \\
  &   & 6 & 3 & 9 & 36 \\
\hline
  & 2 & 1 & 3 & 12 & 37
\end{array}
$$

Since $3 > 0$ and all numbers in the last row of the synthetic division are nonnegative, $f(x)$ has no real zero greater than 3.

(b) No real zero is less than -1.

Divide $f(x)$ by $x + 1$.

$$
\begin{array}{r|rrrrr}
-1 & 2 & -5 & 0 & 3 & 1 \\
   &   & -2 & 7 & -7 & 4 \\
\hline
   & 2 & -7 & 7 & -4 & 5
\end{array}
$$

Here $-1 < 0$ and the numbers in the last row alternate in sign, so $f(x)$ has no zero less than -1.   ●

· · · · · · · · ·

EXAMPLE 4   Approximate the real zeros of $f(x) = x^4 - 6x^3 + 8x^2 + 2x - 1$.

First, check for rational zeros. The only possible rational zeros are ± 1.

$$
\begin{array}{r|rrrrr}
1 & 1 & -6 & 8 & 2 & -1 \\
  &   & 1 & -5 & 3 & 5 \\
\hline
  & 1 & -5 & 3 & 5 & 4
\end{array}
\qquad
\begin{array}{r|rrrrr}
-1 & 1 & -6 & 8 & 2 & -1 \\
   &   & -1 & 7 & -15 & 13 \\
\hline
   & 1 & -7 & 15 & -13 & 12
\end{array}
$$

Neither 1 nor -1 is a zero, so the polynomial has no rational zeros. Now use the two theorems of this section to search in some consistent way for the location of irrational real zeros.

The leading coefficient of $f(x)$ is positive and the numbers in the last row of the second synthetic division above alternate in sign. Since $-1 < 0$, by the boundedness theorem -1 is less than any zero of $f(x)$. (It has been shown that -1 is not a zero of $f(x)$.) Also, $f(-1) = 12 > 0$. By substitution, or synthetic division, $f(0) = -1 < 0$. Thus $f(0)$ and $f(-1)$ have opposite sign, and by the intermediate value theorem, there is at least one real zero between -1 and 0.

Try $c = -.5$ in the synthetic division. Divide $f(x)$ by $x + .5$.

$$
\begin{array}{r|rrrr}
-.5 & 1 & -6 & 8 & 2 & -1 \\
 & & -.5 & 3.25 & -5.625 & 1.8125 \\
\hline
 & 1 & -6.5 & 11.25 & -3.625 & .8125
\end{array}
$$

Since $f(-.5) > 0$ and $f(0) < 0$, there is a real zero between $-.5$ and 0. Try $c = -.4$.

$$
\begin{array}{r|rrrr}
-.4 & 1 & -6 & 8 & 2 & -1 \\
 & & -.4 & 2.56 & -4.224 & .8896 \\
\hline
 & 1 & -6.4 & 10.56 & -2.224 & -.1104
\end{array}
$$

Since $f(-.5)$ is positive, but $f(-.4)$ is negative, there is a zero between $-.5$ and $-.4$. The value of $f(-.4)$ is closer to zero than $f(-.5)$, so it is probably safe to say that, to one decimal place of accuracy, $-.4$ is an approximation to a real zero of $f(x)$. A more accurate result can be found, if desired, by continuing this process.

To locate the remaining real zeros of $f(x)$, continue in the same way. Use synthetic division to find $f(1)$, $f(2)$, $f(3)$, and so on, until there is a change in sign. It is helpful to use the shortened form of synthetic division shown below. Only the last row of the synthetic division is shown for each division. The first row of the chart is used for each division and the work in the second row of the division is done mentally.

x					$f(x)$	
	1	−6	8	2	−1	
−1	1	−7	15	−13	12	←—Zero between −1 and 0
0	1	−6	8	2	−1	←—Zero between 0 and 1
1	1	−5	3	5	4	
2	1	−4	0	2	3	←—Zero between 2 and 3
3	1	−3	−1	−1	−4	←—Zero between 3 and 4
4	1	−2	0	2	7	

Since the polynomial is of degree 4, there are no more than 4 zeros. Expand the table to approximate the real zeros to the nearest tenth. For example, for the zero between 0 and 1, the work might go as follows. Start halfway between 0 and 1 with $x = .5$. Since $f(.5) > 0$ and $f(0) < 0$, try $x = .4$ next.

x					$f(x)$	
	1	−6	8	2	−1	
.5	1	−5.5	5.25	4.63	1.31	
.4	1	−5.6	5.76	4.30	.72	
.3	1	−5.7	6.29	3.89	.17	←—Zero between .3 and .2
.2	1	−5.8	6.84	3.37	−.33	

The value $f(.3) = .17$ is closer to 0 than $f(.2) = -.33$, so to the nearest tenth, the zero is .3. Use synthetic division to verify that the remaining two zeros are approximately 2.4 and 3.7. ●

In *Geometry,* published as an appendix to *A Discourse on the Method of Rightly Conducting the Reason and Seeking Truth in the Sciences* (1637), René Descartes stated for the first time a general rule, which had been useful earlier in specific cases. The rule gives a method to determine limits to the number of positive and the number of negative roots of a polynomial. This labor-shortening rule became known as *Descartes' rule of signs* because of his exposition. His *Geometry* had a profound effect on mathematicians. It is the oldest mathematical text that we can read without having difficulty with the notation.

**Descartes'
Rule of Signs**

Let $f(x)$ be a polynomial with real coefficients and terms in descending order.

(a) The number of positive real zeros of $f(x)$ either is equal to the number of variations in sign occurring in the coefficients of $f(x)$, or else is less than the number of variations by a positive even integer.

(b) The number of negative real zeros of $f(x)$ either equals the number of variations in sign of $f(-x)$, or else is less than the number of variations by a positive even integer.

In the theorem, the number of variations in sign of the coefficients of $f(x)$ or $f(-x)$ refers to changes from positive to negative in successive terms of the polynomial. Missing terms (those with 0 coefficients) are counted as no change in sign and can be ignored.

For the purposes of this theorem, zeros of multiplicity k count as k zeros. For example,

$$f(x) = (x - 1)^4$$
$$= +\; x^4 - 4x^3 + 6x^2 - 4x + 1$$

has 4 changes of sign. By Descartes' rule of signs, $f(x)$ has 4, 2, or 0 positive real zeros. In this case, there are 4; each of the 4 positive real zeros is 1. Also,

$$f(-x) = x^4 + 4x^3 + 6x^2 + 4x + 1$$

which has 0 changes in sign, indicating there are 0 negative real zeros, as expected.

The polynomial in Example 4,

$$f(x) = x^4 - 6x^3 + 8x^2 + 2x - 1,$$

has 3 variations in sign:

$$+ x^4 - 6x^3 + 8x^2 + 2x - 1$$

By Descartes' rule of signs, $f(x)$ has either 3 or $3 - 2 = 1$ positive real zeros. Example 4 showed that $f(x)$ has 3 positive real zeros. Since $f(x)$ is of degree 4 and has 3 positive real zeros, it must have 1 negative real zero, which corresponds to the result in Example 4. This could be verified with part (b) of Descartes' rule of signs. Since

$$f(-x) = (-x)^4 - 6(-x)^3 + 8(-x)^2 + 2(-x) - 1$$
$$= x^4 + 6x^3 + 8x^2 - 2x - 1$$

has only one variation in sign, $f(x)$ has only one negative real zero.

· · · · · · · · ·

EXAMPLE 5   Apply Descartes' rule of signs to

$$f(x) = x^5 + 5x^4 + 3x^2 + 2x + 1.$$

The polynomial $f(x)$ has no variations in sign and so has no positive real zeros. Here

$$f(-x) = -x^5 + 5x^4 + 3x^2 - 2x + 1,$$

with three variations in sign, so $f(x)$ has either 3 or 1 negative real zeros. The other zeros are complex numbers.   ●

· · · · · · · · ·

EXAMPLE 6   Graph the function defined by $f(x) = 8x^3 - 12x^2 + 2x + 1$.

By Descartes' rule of signs, $f(x)$ has 2 or 0 positive real zeros and 1 negative real zero, so the graph of $f(x)$ will have 2 or 0 positive x-intercepts and 1 negative x-intercept. To find some ordered pairs belonging to the graph, use synthetic division to evaluate, say $f(3)$, in hopes of finding a number greater than all zeros of $f(x)$.

$$\begin{array}{r|rrrr} 3 & 8 & -12 & 2 & 1 \\ & & 24 & 36 & 114 \\ \hline & 8 & 12 & 38 & 115 \end{array}$$

Since $f(3) = 115$, the point $(3, 115)$ belongs to the graph. Also, since the bottom row is all positive, there are no zeros greater than 3. Now find $f(-1)$.

$$\begin{array}{r|rrrr} -1 & 8 & -12 & 2 & 1 \\ & & -8 & 20 & -22 \\ \hline & 8 & -20 & 22 & -21 \end{array}$$

From this result, the point $(-1, -21)$ belongs to the graph. Since the signs in the last row alternate, -1 is less than any zero of $f(x)$. This shows that the x-intercepts are between -1 and 3. Using the shortened form of synthetic division shown in the chart below, find several points between -1 and 3, to help in sketching the graph.

x				$f(x)$	Ordered pair	
	8	-12	2	1		
3	8	12	38	115	(3, 115)	
2	8	4	10	21	(2, 21)	←**Zero**
1	8	-4	-2	-1	(1, -1)	←**Zero**
0	8	-12	2	1	(0, 1)	←**Zero**
-1	8	-20	22	-21	($-1, -21$)	

All three possible real zeros have been located. Since the numbers in the row for $x = 2$ are all positive and $2 > 0$, 2 is greater than or equal to any zero of $f(x)$ and the top row of the chart is not needed.

By the intermediate value theorem, there is a zero between 0 and 1 and between -1 and 0, as well as between 1 and 2. To get the graph, plot the points from the chart and then draw a continuous curve through them, as in Figure 8.   ●

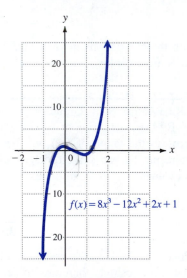

$$f(x) = 8x^3 - 12x^2 + 2x + 1$$

Figure 8

· · · · · · · · ·

EXAMPLE 7   Graph the function defined by $f(x) = 3x^4 - 14x^3 + 24x - 3$.

To draw this graph, shown in Figure 9, first use Descartes' rule of signs to see that there are 3 or 1 positive real zeros and 1 negative real zero. Thus, there are 3 or 1 positive x-intercepts and 1 negative x-intercept. To find points to plot, use synthetic division to make a table like the one shown below. Start with $x = 0$ and work up through the positive integers until a row with all positive numbers is found. Then work down through the negative integers until a row with alternating signs is found.

x					$f(x)$	Ordered pair	
	3	−14	0	24	−3		
5	3	1	5	49	242	(5, 242)	←**Zero**
4	3	−2	−8	−8	−35	(4, −35)	
3	3	−5	−15	−21	−66	(3, −66)	
2	3	−8	−16	−8	−19	(2, −19)	←**Zero**
1	3	−11	−11	13	10	(1, 10)	←**Zero**
0	3	−14	0	54	−3	(0, −3)	
−1	3	−17	17	7	−10	(−1, −10)	←**Zero**
−2	3	−20	40	−56	109	(−2, 109)	

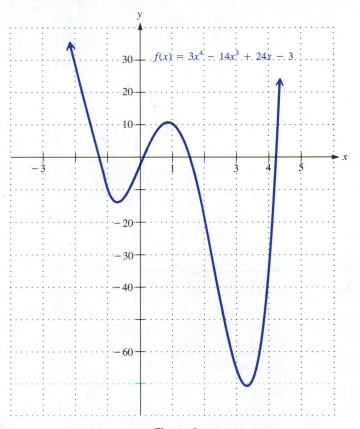

$$f(x) = 3x^4 - 14x^3 + 24x - 3$$

Figure 9

Since the row in the chart for $x = 5$ contains all positive numbers, the polynomial has no zero greater than 5. Also, since the row for $x = -2$ has numbers which alternate in sign, there is no zero less than -2. By the changes in sign of $f(x)$, the polynomial has zeros between -1 and -2, 0 and 1, 1 and 2, and 4 and 5. Three are positive and one is negative, as predicted. Each of these zeros is an x-intercept on the graph. Plotting the points found above and drawing a continuous curve through them gives the graph shown in Figure 9. ●

CAUTION As the graph in Figure 10 suggests, it is possible to have several zeros between a pair of consecutive integers or two or more zeros without a sign change. Predicting the number of positive and negative zeros ahead of time will help to avoid missing any zeros in such cases.

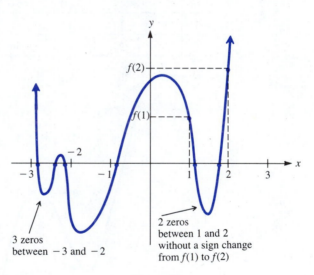

Figure 10

4.4 Exercises

After factoring the polynomial and locating its zeros, sketch the graph of each function.

1. $f(x) = x^4 - 3x^2 + 2$ **2.** $f(x) = 3x^3 + 8x^2 + 3x - 2$ **3.** $f(x) = 2x^3 - 9x^2 - 2x + 24$

4. $f(x) = x^4 - 2x^2 - 3$ **5.** $f(x) = 2x^4 + 5x^3 - 5x - 2$ **6.** $f(x) = 3x^4 + 7x^3 - 9x^2 - 7x + 6$

Use the intermediate value theorem for polynomials to show that the following polynomials have a real zero between the numbers given.

7. $f(x) = 3x^2 - 2x - 6$;   1 and 2   **8.** $f(x) = x^3 + x^2 - 5x - 5$;   2 and 3

9. $f(x) = 2x^3 - 8x^2 + x + 16$;   2 and 2.5   **10.** $f(x) = 3x^3 + 7x^2 - 4$;   1/2 and 1

11. $f(x) = 2x^4 - 4x^2 + 3x - 6$;   2 and 1.5   **12.** $f(x) = x^4 - 4x^3 - x + 1$;   1 and .3

Show that the real zeros of each of the following polynomials satisfy the given conditions.

13. $f(x) = x^4 - x^3 + 3x^2 - 8x + 8$; no real zero greater than 2

14. $f(x) = 2x^5 - x^4 + 2x^3 - 2x^2 + 4x - 4$; no real zero greater than 1

15. $f(x) = x^4 + x^3 - x^2 + 3$; no real zero less than -2

16. $f(x) = x^5 + 2x^3 - 2x^2 + 5x + 5$; no real zero less than -1

Apply Descartes' rule of signs to determine the possible number of positive and negative real zeros of the following polynomials.

17. $f(x) = 5x^3 + 2x^2 - x + 1$ **18.** $f(x) = -x^3 + 6x^2 - 3$

19. $f(x) = -2x^4 + 4x^3 - 5x^2 + x - 4$ **20.** $f(x) = x^4 + 2x^3 + x^2 + 1$

For each of the following polynomials, approximate each zero as a decimal to the nearest tenth.

21. $f(x) = x^3 + 3x^2 - 2x - 6$ **22.** $f(x) = x^3 + x^2 - 5x - 5$

23. $f(x) = x^3 + 6x - 13$ **24.** $f(x) = 4x^3 - 3x^2 + 4x - 5$

25. $f(x) = -2x^4 - x^2 + x - 5$ **26.** $f(x) = -x^4 + 2x^3 + 3x^2 + 6$

The following polynomials have zeros in the given intervals. Approximate these zeros to the nearest hundredth.

27. $f(x) = x^4 + x^3 - 6x^2 - 20x - 16$; [3.2, 3.3] and [$-1.4$, -1.1]

28. $f(x) = x^4 - 2x^3 - 2x^2 - 18x + 5$; [.2, .4] and [3.7, 3.8]

Graph each of the functions defined as follows.

29. $f(x) = x^3 - 7x - 6$ **30.** $f(x) = x^3 + x^2 - 4x - 4$

31. $f(x) = x^4 - 5x^2 + 6$ **32.** $f(x) = x^3 - 3x^2 - x + 3$

33. $f(x) = 6x^3 + 11x^2 - x - 6$ **34.** $f(x) = x^4 - 2x^2 - 8$

35. $f(x) = -x^3 + 6x^2 - x - 14$ **36.** $f(x) = 6x^4 - x^3 - 23x^2 - 4x + 12$

Approximate the solution of each inequality in Exercises 37–40 to the nearest tenth.

37. $2x^3 - 3x^2 - 12x + 1 \le 0$ **38.** $3x^3 - 3x^2 + 1 \ge 0$

39. $x^4 - 18x^2 + 5 \ge 0$ **40.** $x^4 - 8x^2 - 1 \le 0$

41. The polynomial function A defined by $A(x) = -.015x^3 + 1.06x$ gives the approximate alcohol concentration (in tenths of a percent) in an average person's bloodstream x hr after drinking about 8 oz of 100-proof whiskey. The function is approximately valid for x in the interval [0, 8]. **(a)** Graph $y = A(x)$. **(b)** Using the graph you drew for part (a), estimate the time of maximum alcohol concentration. **(c)** In one state, a person is legally drunk if the blood alcohol concentration exceeds .15%. Use the graph of part (a) to estimate the period in which this average person is legally drunk.

42. A technique for measuring cardiac output depends on the concentration of a dye in the bloodstream after a known amount is injected into a vein near the heart. For a normal heart, the concentration of the dye in the bloodstream at time x (in sec) is given by $g(x) = -.006x^4 + .14x^3 - .05x^2 + 2x$. Graph $g(x)$.

43. The pressure of the oil in a reservoir tends to drop with time. By taking sample pressure readings for a particular oil reservoir, petroleum engineers have found that the change in pressure is given by $P(t) = t^3 - 25t^2 + 200t$, where t is time in years from the date of the first reading. **(a)** Graph $P(t)$. **(b)** Use the graph from part (a) to decide for what time period the change in pressure (drop) is increasing; decreasing.

44. During the early part of the twentieth century, the deer population of the Kaibab Plateau in Arizona experienced a rapid increase, because hunters had reduced the number of natural predators and because the deer were protected from hunters. The increase in population depleted the food resources and eventually caused the population to decline. For the period from 1905 to 1930, the deer population was approximated by $D(x) = -.125x^5 + 3.125x^4 + 4000$, where x is time in years from 1905. **(a)** Use a calculator to find enough points and graph $D(x)$. **(b)** From the graph, over what period of time (from 1905 to 1930) was the population increasing? relatively stable? decreasing?

45. Give an example of a polynomial function that is never negative and has -3 and 2 as zeros.

46. Give an example of a polynomial function that has -3 and 2 as zeros and is positive only between -3 and 2.

Determine the domain and range of the functions defined in Exercises 47 and 48.

47. $f(x) = \sqrt{x^3 - 3x^2 - 4x}$

48. $f(x) = \sqrt{x^3 - x}$

49. Explain why a seventh-degree polynomial with some terms missing, that is, with one or more zero coefficients, cannot have seven positive zeros.

50. Show that a polynomial of the form $x^4 \ldots + 1$ must have an even number of positive zeros and that a polynomial of the form $x^4 \ldots - 1$ must have an odd number of positive zeros.

4.5 RATIONAL FUNCTIONS

A rational expression is a fraction that is the quotient of two polynomials. A function defined by a rational expression is called a *rational function*.

**Definition of
Rational
Function**

If $p(x)$ and $q(x)$ are polynomials with $q(x) \neq 0$, then

$$f(x) = \frac{p(x)}{q(x)}$$

is a **rational function.**

Since any values of x such that $q(x) = 0$ are excluded from the domain, a rational function usually has a graph that has one or more breaks in it.

The simplest rational function with a variable denominator is defined by

$$f(x) = \frac{1}{x}.$$

The domain of this function is the set of all real numbers except 0. The number 0 cannot be used as a value of x, but it is helpful to find the values of $f(x)$ for several values of x close to 0. The following table shows what happens to $f(x)$ as x gets closer and closer to 0 from either side.

x approaches 0

x	-1	$-.1$	$-.01$	$-.001$	.001	.01	.1	1
$f(x)$	-1	-10	-100	-1000	1000	100	10	1

$|f(x)|$ **gets larger and larger**

The table suggests that $|f(x)|$ gets larger and larger as x gets closer and closer to 0, written in symbols as

$$|f(x)| \to \infty \text{ as } x \to 0.$$

(The symbol $x \to 0$ means that x approaches as close to desired to 0, without ever being equal to 0.) Since x cannot equal 0, the graph of $f(x) = 1/x$ will never intersect the vertical line $x = 0$. The line $x = 0$, the y-axis, is called a *vertical asymptote* for the graph. The graph gets closer and closer to the y-axis as x gets closer and closer to 0.

On the other hand, as $|x|$ gets larger and larger, the values of $f(x) = 1/x$ get closer and closer to 0. (See the table.)

x	$-10,000$	-1000	-100	-10	10	100	1000	10,000
$f(x)$	$-.0001$	$-.001$	$-.01$	$-.1$	.1	.01	.001	.0001

Letting $|x|$ get larger and larger without bound (written $|x| \to \infty$) causes the graph of $y = 1/x$ to move closer and closer to the horizontal line $y = 0$, the x-axis. That is,

$$\lim_{|x| \to \infty} f(x) = 0,$$

read as "the limit of $f(x)$ as $|x|$ becomes infinitely large is zero." The line $y = 0$ is called a *horizontal asymptote*.

To graph $f(x)$, first replace x with $-x$, getting $f(-x) = 1/(-x) = -1/x = -f(x)$, showing that $f(x) = 1/x$ is an odd function, symmetric with respect to the origin. Choosing some positive values of x and finding the corresponding values of $f(x)$ gives the first-quadrant part of the graph shown in Figure 11 on the following page. The other part of the graph (in the third quadrant) can be found by symmetry.

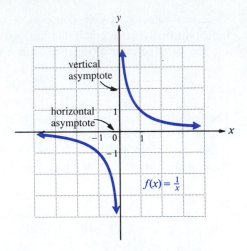

$f(x) = \frac{1}{x}$

Figure 11

• • • • • • • • •

EXAMPLE 1   Graph $f(x) = -\dfrac{2}{x}$.

Rewrite $f(x)$ as

$$f(x) = -2 \cdot \frac{1}{x}.$$

Compared to $f(x) = 1/x$, the graph will be reflected about the x-axis (because of the negative sign) and each point will be twice as far from the x-axis. See the graph in Figure 12. The y-axis is the vertical asymptote and the horizontal asymptote is $y = 0$, or the x-axis.   ●

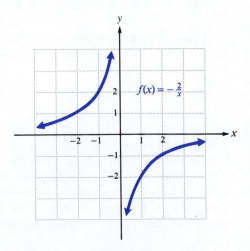

$f(x) = -\frac{2}{x}$

Figure 12

EXAMPLE 2   Graph $f(x) = \dfrac{2}{1+x}$.

The domain of this function is the set of all real numbers except -1. Since $x \neq -1$, the graph cannot cross the line $x = -1$, which is thus a vertical asymptote. The horizontal asymptote is $y = 0$. As shown in Figure 13, the graph is similar to that of $f(x) = 1/x$, translated 1 unit to the left. The y-intercept is 2.   ●

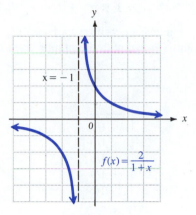

Figure 13

The examples above suggest the following definitions of vertical and horizontal asymptotes.

Definition of Asymptotes

For the rational function defined by $y = f(x)$, and for real numbers a and b,

if $\lim\limits_{x \to a} |f(x)| = \infty$, then the line $x = a$ is a **vertical asymptote;**

if $\lim\limits_{|x| \to \infty} f(x) = b$, then the line $y = b$ is a **horizontal asymptote.**

When the rational function $f(x) = p(x)/q(x)$ is written in lowest terms, vertical asymptotes occur at values of x that make the denominator 0. Therefore, to find all vertical asymptotes, find all solutions to $q(x) = 0$.

EXAMPLE 3   Graph $f(x) = \dfrac{1}{(x-1)(x+4)}$.

Find the vertical asymptotes by setting the denominator equal to 0 and solving for x.

$$(x-1)(x+4) = 0$$
$$x = 1 \quad \text{or} \quad x = -4$$

As $|x|$ gets larger and larger, $|(x - 1)(x + 4)|$ also gets larger and larger, bringing y closer and closer to 0. This means that the x-axis is a horizontal asymptote. The vertical asymptotes divide the x-axis into three regions. Find the sign of $f(x)$ in each region by using a test point as was done with polynomials earlier in this chapter.

Region	Test Point	Value of $f(x)$	Sign of $f(x)$
$x < -4$	-5	1/6	+
$-4 < x < 1$	-1	$-1/6$	$-$
$1 < x$	2	1/6	+

Finding the y-intercept by letting $x = 0$ and then plotting it and a few additional points (shown with the graph) gives the result shown in Figure 14. ●

x	y
-5	1/6
-2	$-1/6$
-1	$-1/6$
0	$-1/4$
2	1/6

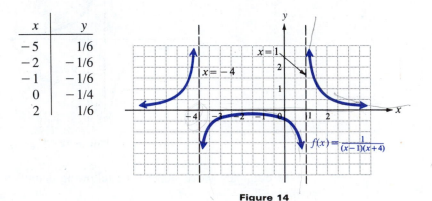

Figure 14

• • • • • • • • • •

EXAMPLE 4   Graph $f(x) = \dfrac{x + 1}{(2x - 1)(x + 3)}$.

The graph has as vertical asymptotes the lines $x = 1/2$ and $x = -3$. To find any horizontal asymptote, multiply the factors in the denominator.

$$f(x) = \frac{x + 1}{(2x - 1)(x + 3)} = \frac{x + 1}{2x^2 + 5x - 3}$$

Now divide each term in the numerator and denominator by x^2, since 2 is the highest exponent on x.

$$f(x) = \frac{\dfrac{x}{x^2} + \dfrac{1}{x^2}}{\dfrac{2x^2}{x^2} + \dfrac{5x}{x^2} - \dfrac{3}{x^2}} = \frac{\dfrac{1}{x} + \dfrac{1}{x^2}}{2 + \dfrac{5}{x} - \dfrac{3}{x^2}}$$

As $|x|$ gets larger and larger, the quotients $1/x$, $1/x^2$, $5/x$, and $3/x^2$ all approach 0.

For example,

$$\text{if } x = 10, 100, 1000, \ldots,$$

$$\text{then } \frac{1}{x} = \frac{1}{10}, \frac{1}{100}, \frac{1}{1000}, \ldots,$$

$$\text{and } \frac{1}{x^2} = \frac{1}{100}, \frac{1}{10000}, \frac{1}{1000000}, \ldots$$

Thus, as $|x| \to \infty$, the value of $f(x)$ approaches

$$\frac{0 + 0}{2 + 0 - 0} = \frac{0}{2} = 0.$$

The line $y = 0$ is therefore a horizontal asymptote.

Replacing x with 0 gives $-1/3$ as the y-intercept. For the x-intercept, the value of y must be 0. The only way that y can equal 0 is if the numerator, $x + 1$, is 0. This happens when $x = -1$, making -1 the x-intercept.

Since this graph has an x-intercept, consider the sign of $f(x)$ in each region determined by a vertical asymptote or an x-intercept. Here there are four regions to be considered as shown in the chart below.

Region	Test Point	Value of $f(x)$	Sign of $f(x)$
$(-\infty, -3)$	-4	$-1/3$	$-$
$(-3, -1)$	-2	$1/5$	$+$
$(-1, 1/2)$	0	$-1/3$	$-$
$(1/2, +\infty)$	2	$1/5$	$+$

Using the asymptotes and intercepts and plotting a few points (shown in the table of values) gives the graph in Figure 15. ●

x	y
-4	$-1/3$
-2	$1/5$
-1	0
0	$-1/3$
1	$1/2$
2	$1/5$

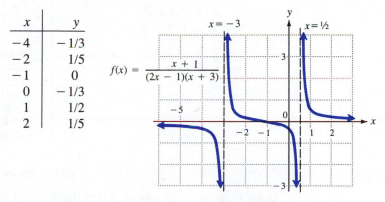

$$f(x) = \frac{x + 1}{(2x - 1)(x + 3)}$$

Figure 15

EXAMPLE 5 Graph $f(x) = \dfrac{2x - 5}{x - 3}$.

The vertical asymptote is the line $x = 3$. Work as in the previous example to find any horizontal asymptote. Since the highest exponent on x is 1, divide each term in the numerator and denominator by x.

$$f(x) = \frac{2x - 5}{x - 3} = \frac{\dfrac{2x}{x} - \dfrac{5}{x}}{\dfrac{x}{x} - \dfrac{3}{x}} = \frac{2 - \dfrac{5}{x}}{1 - \dfrac{3}{x}}$$

As $|x|$ gets larger and larger, both $3/x$ and $5/x$ approach 0.

$$\frac{2 - 0}{1 - 0} = \frac{2}{1} = 2,$$

showing that the line $y = 2$ is a horizontal asymptote.

Letting $x = 0$ gives the y-intercept $5/3$. The x-intercept is found when the numerator, $2x - 5$, is 0, so the x-intercept is $5/2$. The following chart shows the sign of $f(x)$ in the regions determined by the x-intercept and the vertical asymptote.

Region	Test Point	Value of $f(x)$	Sign of $f(x)$
$(-\infty, 5/2)$	0	5/3	+
$(5/2, 3)$	11/4	-2	$-$
$(3, +\infty)$	4	3	+

Using this information and plotting the ordered pairs shown with the graph produced the graph of Figure 16.   ●

x	y
-2	9/5
0	5/3
2	1
5/2	0
11/4	-2
4	3

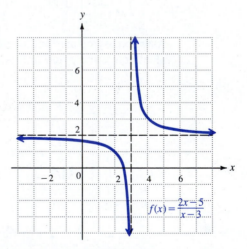

$$f(x) = \frac{2x - 5}{x - 3}$$

Figure 16

There is an alternate way to sketch the graph in Example 5. Use synthetic division to divide $2x - 5$ by $x - 3$, getting

$$\frac{2x - 5}{x - 3} = 2 + \frac{1}{x - 3}.$$

This result shows that the graph of $f(x) = (2x - 5)/(x - 3)$ is the same as that of $1/(x - 3)$, translated 2 units upward. Since the graph of $1/(x - 3)$ is the same as that of $1/x$ translated 3 units to the right, the final graph is that of $1/x$ translated 3 units to the right and 2 units upward.

. . . .

EXAMPLE 6   Graph $f(x) = \dfrac{3(x + 1)(x - 2)}{(x + 4)^2}$.

The only vertical asymptote is the line $x = -4$. Find any horizontal asymptotes by multiplying the factors in the numerator and denominator.

$$f(x) = \frac{3x^2 - 3x - 6}{x^2 + 8x + 16}$$

Now divide the numerator and denominator by x^2.

$$f(x) = \frac{\dfrac{3x^2}{x^2} - \dfrac{3x}{x^2} - \dfrac{6}{x^2}}{\dfrac{x^2}{x^2} + \dfrac{8x}{x^2} + \dfrac{16}{x^2}} = \frac{3 - \dfrac{3}{x} - \dfrac{6}{x^2}}{1 + \dfrac{8}{x} + \dfrac{16}{x^2}}$$

Letting $|x|$ get larger and larger shows that $y = 3/1$, or $y = 3$, is a horizontal asymptote. The y-intercept is $-3/8$, and the x-intercepts are -1 and 2. The sign of $f(x)$ in each region is shown below.

Region	Test Point	Value of $f(x)$	Sign of $f(x)$
$(-\infty, -4)$	-8	13.125	$+$
$(-4, -1)$	-2	3	$+$
$(-1, 2)$	0	$-3/8$	$-$
$(2, +\infty)$	4	.46875	$+$

See the final graph in Figure 17. ●

x	y
-12	7.2
-8	13.1
-2	3
-1	0
0	$-.4$
2	0
4	.5
12	1.5

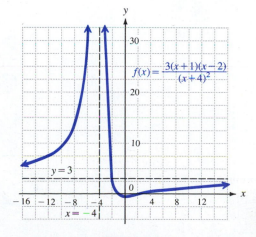

Figure 17

In Figure 17, the graph of the function actually crosses the horizontal asymptote. While it is possible for the graph of a rational function to cross a horizontal asymptote, such a graph can never cross a vertical asymptote. The reason that a graph can cross a horizontal asymptote comes from the definition—only values of x where $|x|$ is relatively large are used in finding horizontal asymptotes. The value of x where the graph crosses the horizontal asymptote of Figure 17 is found by letting $f(x) = 3$, since the horizontal asymptote is the line $y = 3$. Solving $f(x) = 3$ gives $x = -2$, so the graph crosses the horizontal asymptote at $(-2, 3)$.

The next example shows a rational function defined by an expression having a numerator with higher degree than the denominator. Polynomial division must be used for these functions.

· · · · · · · · ·

EXAMPLE 7   Graph $f(x) = \dfrac{x^2 + 1}{x - 2}$.

The vertical asymptote is the line $x = 2$. Dividing the numerator and denominator by x^2 as in the examples above would show that the graph has no horizontal asymptote. Since the numerator has higher degree than the denominator, divide to rewrite the equation for the function in another form.

$$
\underline{2|}\begin{array}{ccc} 1 & 0 & 1 \\ & 2 & 4 \\ \hline 1 & 2 & 5 \end{array}
$$

By this result,

$$
f(x) = \frac{x^2 + 1}{x - 2} = x + 2 + \frac{5}{x - 2}.
$$

For very large values of $|x|$, $5/(x - 2)$ is close to 0, and the graph approaches the line $y = x + 2$. This line is an **oblique asymptote** (neither vertical nor horizontal) for the graph. The y-intercept is $-1/2$. There is no x-intercept. Using the asymptotes, the y-intercept, and additional points as needed leads to the graph in Figure 18.   ●

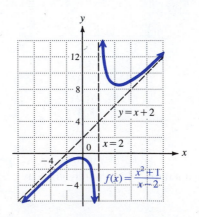

Figure 18

In general, if the degree of the numerator is exactly one more than the degree of the denominator, the rational function will have an oblique asymptote. The equation of this asymptote is found by dividing the numerator by the denominator and dropping the remainder, since the remainder becomes close to 0 for large values of $|x|$.

To summarize: when graphing a rational function, use the following procedure. (Assume that the rational function is defined by an expression written in lowest terms.)

Graphing Rational Functions

1. Find any vertical asymptotes by setting the denominator equal to 0 and solving for x. If a is a zero of the denominator, then the line $x = a$ will be a vertical asymptote. Draw $x = a$ as a dashed line.

2. Determine if there is a horizontal or oblique asymptote. If so, draw it as a dashed line. There are three possibilities:

   (a) If the numerator has lower degree than the denominator, there is a horizontal asymptote, $y = 0$.

   (b) If the numerator and denominator have the same degree, and the function is defined by an expression of the form

   $$f(x) = \frac{a_n x^n + \cdots + a_0}{b_n x^n + \cdots + b_0}, \quad b_n \neq 0,$$

   dividing numerator and denominator by x^n produces the horizontal asymptote

   $$y = \frac{a_n}{b_n}.$$

   (c) If the numerator is exactly one degree more than the denominator, there will be an oblique asymptote. To find it, divide the numerator by the denominator and drop any remainder. The rest of the quotient gives the equation of the asymptote.

3. Find and plot the x-intercepts.

4. Find the regions determined by the vertical asymptotes and x-intercepts. Use a test point in each region to find the sign of $f(x)$ in that region.

5. Plot the y-intercept (if any) and the points from step 4.

6. Complete the sketch.

It is fairly common in calculus to draw the graph of a rational function whose numerator and denominator have a common factor. (This is done with functions whose graphs are not "continuous" because of "holes" in the graphs.) Work as in the next example.

EXAMPLE 8   Graph $f(x) = \dfrac{(x-1)(x^2-3)}{x-1}$.

First,

$$\frac{(x-1)(x^2-3)}{x-1} = x^2 - 3, \quad \text{if } x \neq 1.$$

This means that the graph of f is the same as that of the parabola $y = x^2 - 3$, except at $x = 1$. At $x = 1$, there is no value of $f(x)$, so that the graph has no point for $x = 1$. See Figure 19. (Note that $x = 1$ is not a vertical asymptote since the magnitude of $f(x)$ does not get larger and larger as x approaches 1. Instead, $f(x)$ gets closer and closer to -2. ●

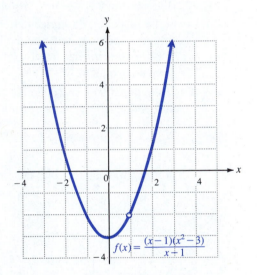

$$f(x) = \frac{(x-1)(x^2-3)}{x-1}$$

Figure 19

4.5 Exercises

For Exercises 1–12, find any vertical, horizontal, or oblique asymptotes.

1. $f(x) = \dfrac{2}{x-5}$ **2.** $f(x) = \dfrac{-1}{x+2}$ **3.** $f(x) = \dfrac{-8}{3x-7}$ **4.** $f(x) = \dfrac{5}{4x-9}$

5. $f(x) = \dfrac{3x-5}{2x+9}$ **6.** $f(x) = \dfrac{4x+3}{3x-7}$ **7.** $f(x) = \dfrac{2}{x^2-4x+3}$ **8.** $f(x) = \dfrac{-5}{x^2-3x-10}$

9. $f(x) = \dfrac{x^2-1}{x+3}$ **10.** $f(x) = \dfrac{x^2+4}{x-1}$ **11.** $f(x) = \dfrac{(x-3)(x+1)}{(x+2)(2x-5)}$ **12.** $f(x) = \dfrac{3(x+2)(x-4)}{(5x-1)(x-5)}$

13. Sketch the following graphs and compare them with the graph of $f(x) = 1/x^2$.

    (a) $f(x) = \dfrac{1}{(x-3)^2}$ **(b)** $f(x) = \dfrac{-2}{x^2}$ **(c)** $f(x) = \dfrac{-2}{(x-3)^2}$

Graph each function defined in Exercises 14–37.

14. $f(x) = \dfrac{4}{5 + 3x}$

15. $f(x) = \dfrac{3}{(x + 4)^2}$

16. $f(x) = \dfrac{3x}{(x + 1)(x - 2)}$

17. $f(x) = \dfrac{2x + 1}{(x + 2)(x + 4)}$

18. $f(x) = \dfrac{5x}{x^2 - 1}$

19. $f(x) = \dfrac{-x}{x^2 - 4}$

20. $f(x) = \dfrac{3x}{x - 1}$

21. $f(x) = \dfrac{x - 5}{x + 3}$

22. $f(x) = \dfrac{3x}{x^2 - 16}$

23. $f(x) = \dfrac{x}{x^2 - 9}$

24. $f(x) = \dfrac{x^2 - 5}{x + 2}$

25. $f(x) = \dfrac{x^2 - 3x + 2}{x - 3}$

26. $f(x) = \dfrac{x^2 + 1}{x + 3}$

27. $f(x) = \dfrac{2x^2 + 3}{x - 4}$

28. $f(x) = \dfrac{2x(x - 2)}{x^2 + x - 2}$

29. $f(x) = \dfrac{(x - 1)^2}{x^2 - 2x - 3}$

30. $f(x) = \dfrac{(x + 4)(x - 1)}{x^2 + 1}$

31. $f(x) = \dfrac{(x - 5)(x - 2)}{x^2 + 9}$

32. $f(x) = \dfrac{1}{x^2 + 1}$

33. $f(x) = \dfrac{-9}{x^2 + 9}$

34. $f(x) = \dfrac{(x - 1)(x + 2)}{x + 2}$

35. $f(x) = \dfrac{(2x - 3)(x - 4)}{x - 4}$

36. $f(x) = \dfrac{(x + 2)(x^2 + 1)}{x + 2}$

37. $f(x) = \dfrac{(x + 3)(x^2 - 4)}{x + 3}$

38. The figures below show the four ways that a rational function can approach the vertical line $x = 2$ as an asymptote. Identify the graph of each of the following rational functions.

(a) $f(x) = \dfrac{1}{(x - 2)^2}$

(b) $f(x) = \dfrac{1}{x - 2}$

(c) $f(x) = \dfrac{-1}{x - 2}$

(d) $f(x) = \dfrac{-1}{(x - 2)^2}$

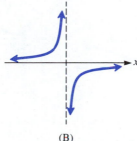

(A)

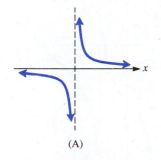

(B)

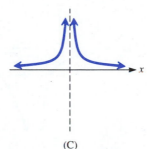

(C)

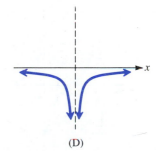

(D)

39. Antique-car fans often enter their cars in a *concours d'elegance* in which a maximum of 100 points can be awarded to a particular car. Points are awarded for the general attractiveness of the car. The function defined by $C(x) = \dfrac{10x}{49(101 - x)}$ expresses the cost, in thousands of dollars, of restoring a car so that it will win x points. **(a)** Graph the function. **(b)** Would a more reasonable domain for C be $(0, 100)$ or $[0, 100)$? Why?

40. In situations involving environmental pollution, a cost-benefit model expresses cost as a function of the percentage of pollutant removed from the environment. Suppose a cost-benefit model is expressed as $y = 6.7x/(100 - x)$, where y is the cost in thousands of dollars of removing x percent of a certain pollutant. **(a)** Graph the function. **(b)** Is it possible, according to this function, to remove all of the pollutant?

*An idealized version of the **Laffer curve,** named for economist Arthur Laffer, is shown here. According to this curve, increasing a tax rate, say from x_1 percent to x_2 percent on the graph, can actually lead to a decrease in government revenue. All economists agree on the endpoints, 0 revenue at tax rates of both 0% and 100%, but there is much disagreement on the location of the rate x_1 that produces maximum revenue.*

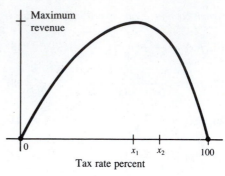

41. Suppose an economist studying the Laffer curve produces the rational function $y = \dfrac{60x - 6000}{x - 120}$, where y is government revenue in millions of dollars from a tax rate of x percent, with the function valid for $50 \leq x \leq 100$. Find the revenue from the following tax rates: **(a)** 50%, **(b)** 60%, **(c)** 80%, **(d)** 100%. **(e)** Graph the function.

Exercises 41, 42

42. Suppose our economist studies a different tax, this time producing $y = \dfrac{80x - 8000}{x - 110}$, with y giving government revenue in tens of millions of dollars for a tax rate of x percent, with the function valid for $55 \leq x \leq 100$. Find the revenue for the following tax rates: **(a)** 55%, **(b)** 60%, **(c)** 70%, **(d)** 90%, **(e)** 100%. **(f)** Graph the function.

43. Graph $f(x) = \dfrac{x^2(x - 2)}{x^3 + 5x^2 + 3x - 9}$.

44. Graph $f(x) = \dfrac{(x - 3)(x + 1)^2}{x^3 - 2x^2 + x}$.

For each of the following graphs, find an equation for a possible corresponding rational function.

45.

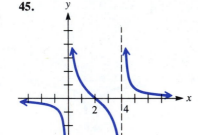

46.

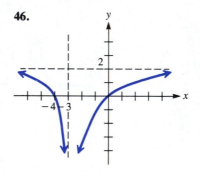

47.

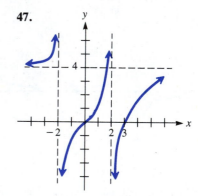

4 CHAPTER SUMMARY

Key Words

polynomial function
leading coefficient
zero polynomial
zero of a function
4.1 division algorithm
dividend
divisor
quotient polynomial

remainder polynomial
synthetic division
4.2 factor theorem
fundamental theorem
multiplicity
conjugate zeros theorem
4.3 rational zeros theorem

4.4 intermediate value
theorem
boundedness theorem
Descartes' rule of signs
4.5 rational function
vertical asymptote
horizontal asymptote
oblique asymptote

Review Exercises

Use synthetic division to find q(x) and r for each of the following.

1. $\dfrac{2x^3 + 3x^2 - 4x + 1}{x - 1}$

2. $\dfrac{4x^3 + 3x^2 + 3x + 5}{x - 5}$

3. $\dfrac{2x^3 - 4x + 6}{x - 2}$

4. $\dfrac{3x^4 - x^2 + x - 1}{x + 1}$

Use synthetic division to find f(2) for each of the following.

5. $f(x) = x^3 + 3x^2 - 5x + 1$

6. $f(x) = 2x^3 - 4x^2 + 3x - 10$

7. $f(x) = 5x^4 - 12x^2 + 2x - 8$

8. $f(x) = x^5 - 3x^2 + 2x - 4$

In Exercises 9–12, find a polynomial of lowest degree having the following zeros.

9. $-1, 4, 7$

10. $8, 2, 3$

11. $-\sqrt{7}, \sqrt{7}, 2, -1$

12. $1 + \sqrt{5}, 1 - \sqrt{5}, -4, 1$

13. Is -1 a zero of $f(x) = 2x^4 + x^3 - 4x^2 + 3x + 1$?

14. Is -2 a zero of $f(x) = 2x^4 + x^3 - 4x^2 + 3x + 1$?

15. Is $x + 1$ a factor of $f(x) = x^3 + 2x^2 + 3x - 1$?

16. Is $x + 1$ a factor of $f(x) = 2x^3 - x^2 + x + 4$?

17. Find a polynomial of degree 3 with -2, 1, and 4 as zeros, and $f(2) = 16$.

18. Find a polynomial of degree 4 with 1, -1, and $3i$ as zeros, and $f(2) = 39$.

19. Find a lowest-degree polynomial with real coefficients having zeros 2, -2, and $-i$.

20. Find a lowest-degree polynomial with real coefficients having zeros of 2, -3, and $5i$.

21. Give an example of a fourth-degree polynomial having exactly two distinct real zeros and sketch its graph.

22. Give an example of a cubic polynomial having exactly two distinct real zeros and sketch its graph.

23. Find all zeros of $f(x) = x^4 - 3x^3 - 8x^2 + 22x - 24$, given that $1 - i$ is a zero.

24. Find all zeros of $f(x) = x^4 + x^3 - x^2 + x - 2$, given that 1 is a zero.

In Exercises 25–27, find all rational zeros.

25. $f(x) = 2x^3 - 9x^2 - 6x + 5$ **26.** $f(x) = 3x^3 - 10x^2 - 27x + 10$ **27.** $f(x) = x^3 - \dfrac{17}{6}x^2 - \dfrac{13}{3}x - \dfrac{4}{3}$

28. Use a polynomial to show that $\sqrt{11}$ is irrational.

29. Show that $f(x) = x^3 - 9x^2 + 2x - 5$ has no rational zeros.

30. Find all values of r for which the polynomial $f(x) = x^3 + 2x^2 + rx + 3$ has a rational zero.

31. Find a value of r such that when the polynomial $x^3 - 3x^2 + rx - 4$ is divided by $x - 2$, the remainder is 5.

Show that the polynomials in Exercises 32–35 have real zeros satisfying the given conditions.

32. $f(x) = 3x^3 - 8x^2 + x + 2$, zero in $[-1, 0]$ and $[2, 3]$

33. $f(x) = 4x^3 - 37x^2 + 50x + 60$, zero in $[2, 3]$ and $[7, 8]$

34. $f(x) = 2x^4 - x^3 - 21x^2 + 51x - 36$, no real zero greater than 4

35. $f(x) = 6x^4 + 13x^3 - 11x^2 - 3x + 5$, no real zero greater than 1 or less than -3

36. The function $f(x) = 1/x$ is negative at $x = -1$ and positive at $x = 1$, but has no zero between -1 and 1. Explain why this does not contradict the intermediate value theorem.

Approximate the real zeros of each of the following as a decimal to the nearest tenth. Then graph the function defined by the given expression.

37. $f(x) = 2x^3 - 11x^2 - 2x + 2$ **38.** $f(x) = x^4 - 4x^3 - 5x^2 + 14x - 15$

Graph each function defined in Exercises 39 and 40.

39. $f(x) = 3x^4 + 4x^2 + 7x$ **40.** $f(x) = 3x^3 - 2x^2 - 7x - 2$

41. (a) Find the number of positive and negative zeros of $f(x) = x^3 + 3x^2 - 4x - 2$.
(b) Show that $f(x)$ has a zero between -4 and -3. Approximate this zero to the nearest tenth.   (c) Graph $y = f(x)$.

42. Without solving, explain why the equation $x^4 + 2x^3 + 7x - 3 = 0$ has two real roots and two imaginary roots.

Graph each function defined in Exercises 43–48.

43. $f(x) = \dfrac{2}{3x - 1}$ **44.** $f(x) = \dfrac{4x - 2}{3x + 1}$ **45.** $f(x) = \dfrac{6x}{(x - 1)(x + 2)}$

46. $f(x) = \dfrac{2x}{x^2 - 1}$ **47.** $f(x) = \dfrac{x^2 + 4}{x + 2}$ **48.** $f(x) = \dfrac{x^2 - 1}{x}$

49. (a) Sketch the graph of a function that has the line $x = 3$ as a vertical asymptote, the line $y = 2$ as a horizontal asymptote, and the x-intercepts 1 and 4.
(b) Find an equation for a possible corresponding rational function.

50. (a) Sketch the graph of a function that has the lines $x = -2$ and $x = 2$ as vertical asymptotes, has the x-axis as a horizontal asymptote, has 0 as an x-intercept, and is never negative.
(b) Find an equation for a possible corresponding rational function.

5

EXPONENTIAL AND LOGARITHMIC FUNCTIONS

In this chapter we will study two kinds of functions that are quite different from those studied before. The functions discussed earlier are examples of *algebraic functions*. The functions in this chapter are not algebraic, but are *transcendental functions,* functions that "transcend" or go beyond the basic operations of addition, subtraction, multiplication, division, and taking roots. Many applications of mathematics, particularly to growth and decay of populations, involve the closely interrelated exponential and logarithmic functions introduced in this chapter. As shown later, these two types of functions are inverses of each other.

5.1 EXPONENTIAL FUNCTIONS

The following problem illustrates the behavior of exponential functions. A person contracts to work for a period of thirty days and is given the option of receiving a lump sum of $1000 or receiving $.01 on the first day of work, $.02 on the second, $.04 on the third, $.08 on the fourth, and so on, with each day's pay double that of the previous day. While the first choice may appear more lucrative (after all, the *total* amount earned by the second method after seven days is only $1.27), the second method actually will earn the worker over five million dollars on the thirtieth day alone, and over ten million dollars for the month!

A function that defines the amount of payment on the nth day in the example above is

$$f(n) = .01(2^{n-1}).$$

The base 2 appears as a result of the doubling that occurs. A function in which the variable appears in the exponent is an *exponential function*. Exponential functions tend to increase or decrease rapidly. The function in the example above is an increasing function. An example of a decreasing function is one that gives the amount of radioactive substance present as time passes, since the amount gradually diminishes.

As shown in Chapter 1, if $a > 0$, the symbol a^m can be defined for any rational value of m. In this section, the definition of a^m is extended to include all *real*, and not just rational, values of the exponent m. For example, what is meant by $2^{\sqrt{3}}$? The exponent, $\sqrt{3}$, can be approximated more and more closely by the numbers 1.7, 1.73, 1.732, and so on, making it reasonable that $2^{\sqrt{3}}$ should be approximated more and more closely by the numbers $2^{1.7}$, $2^{1.73}$, $2^{1.732}$, and so on. (Recall, for example, that $2^{1.7} = 2^{17/10}$, which means $\sqrt[10]{2^{17}}$.) In fact, this is exactly how the number $2^{\sqrt{3}}$ is defined in a more advanced course. To show that this assumption is reasonable, Figure 1 gives the graphs of the function $f(x) = 2^x$ with three different domains.

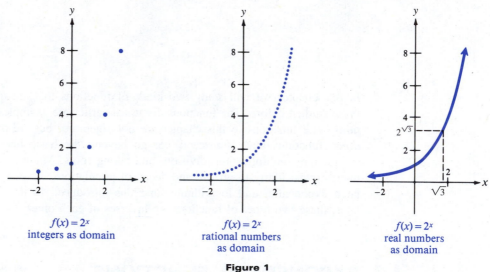

$f(x) = 2^x$
integers as domain

$f(x) = 2^x$
rational numbers
as domain

$f(x) = 2^x$
real numbers
as domain

Figure 1

We shall assume that the meaning given to real exponents at the beginning of this section is such that all previous rules and theorems for exponents are valid for real number exponents as well as rational ones. In addition to the rules for exponents presented earlier, the following additional properties will prove useful. First, any given real value of x leads to exactly one value of 2^x. For example,

$$2^2 = 4, \quad 2^{-3} = 1/8, \quad \text{and} \quad 2^{1/2} = \sqrt{2} \approx 1.4142.$$

Furthermore, if $\qquad 3^x = 3^4, \quad$ then $\quad x = 4.$
And if $\qquad p^2 = 3^2, \quad$ then $\quad p = 3.$

Also $\qquad 4^2 < 4^3 \quad$ but $\quad \left(\dfrac{1}{2}\right)^2 > \left(\dfrac{1}{2}\right)^3,$

so that when $a > 1$, increasing the exponent on a leads to a *larger* number, but if $0 < a < 1$, increasing the exponent on a leads to a *smaller* number.

These properties are generalized in the next theorem. No proof of these properties is given, since the proof requires more advanced mathematics than that of this course.

For any real number $a > 0$, $a \neq 1$, and any real number x, the following are true.

1. a^x is a unique real number.
2. $a^b = a^c$ if and only if $b = c$.
3. If $a > 1$ and $m < n$, then $a^m < a^n$.
4. If $0 < a < 1$ and $m < n$, then $a^m > a^n$.

Part 1 of the theorem requires $a > 0$ so that a^x is always defined. For example, $(-6)^x$ is not a real number if $x = 1/2$. If $a > 0$, then a^x will always be positive, since a is positive. For part 2 to hold, a must not equal 1 since $1^4 = 1^5$, even though $4 \neq 5$. Figure 2(a) in Example 4 below illustrates part 3: the base, 2, of the exponential 2^x is greater than 1, and as the x-values increase, so do the values of 2^x. Part 4 of the theorem is shown in Figure 2(b) where the base, 1/2, is between 0 and 1, and as the x-values increase, $(1/2)^x$ decreases.

The theorem on exponentials can be used to solve equations with variable exponents, as shown in the next example.

EXAMPLE 1 Solve $\left(\dfrac{1}{3}\right)^x = 81$.

First, write 1/3 as 3^{-1}, so that $(1/3)^x = (3^{-1})^x = 3^{-x}$. Since $81 = 3^4$,

$$\left(\frac{1}{3}\right)^x = 81$$

becomes

$$3^{-x} = 3^4.$$

By part 2 of the theorem on exponentials,

$$-x = 4$$
$$x = -4.$$

(Section 5.4 shows a method for solving equations of this type when both sides cannot be written as powers of the same number.) ●

EXAMPLE 2 Find b if $b^{4/3} = 81$.

By the definition of rational exponent,

$$b^{4/3} = (b^{1/3})^4 = 81.$$

Then

$$b^{1/3} = \pm\sqrt[4]{81}$$
$$b = (\pm\sqrt[4]{81})^3$$
$$b = (\pm 3)^3$$
$$b = 27 \quad \text{or} \quad b = -27.$$

Check both proposed solutions in the original equation. Since

$$27^{4/3} = (27^{1/3})^4 = 3^4 = 81$$

and

$$(-27)^{4/3} = [(-27)^{1/3}]^4 = (-3)^4 = 81,$$

both solutions are correct. •

CAUTION Remember that raising both sides of an equation to the same power may produce false "solutions." Thus, it was *necessary* to check both answers in Example 2.

A one-to-one exponential function f can now be defined with domain the set of all real numbers (and not just the rationals).

Definition of Exponential Function

The function f defined by

$$f(x) = a^x, \quad a > 0 \text{ and } a \neq 1,$$

is the **exponential function with base a**.

If $a = 1$, the function is the constant function given by $f(x) = 1$.

EXAMPLE 3   If $f(x) = 2^x$, find each of the following.

(a) $f(-1)$

Replace x with -1.

$$f(-1) = 2^{-1} = \frac{1}{2}$$

(b) $f(3) = 2^3 = 8$

(c) $f(5/2) = 2^{5/2} = (2^5)^{1/2} = 32^{1/2} = \sqrt{32} = 4\sqrt{2}$ •

EXAMPLE 4   Graph the exponential functions defined as follows.

(a) $f(x) = 2^x$

The base of this exponential function is 2. The y-intercept is

$$y = 2^x = 2^0 = 1.$$

Some ordered pairs that satisfy the equation are $(-2, 1/4)$, $(-1, 1/2)$, $(0, 1)$, $(1, 2)$, $(2, 4)$, and $(3, 8)$. Since $2^x > 0$ for all x and $2^x \to 0$ as $x \to -\infty$, the x-axis is a horizontal asymptote. Plotting these points and then drawing a smooth curve through them gives the graph in Figure 2(a). As the graph suggests, the domain of the function is $(-\infty, +\infty)$ and the range is $(0, +\infty)$. The function is increasing on its entire domain, and is one-to-one by the horizontal line test.

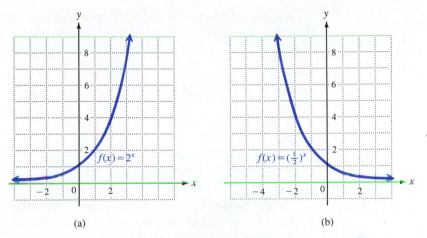

Figure 2

(b) $f(x) = (1/2)^x$

Again the y-intercept is at $(0, 1)$ and the x-axis is a horizontal asymptote. Plot a few ordered pairs and draw a smooth curve through them. For example, $(-3, 8)$, $(-2, 4)$, $(-1, 2)$, $(0, 1)$, and $(1, 1/2)$ are on the graph shown in Figure 2(b). Like the function in part (a), this function also has domain $(-\infty, +\infty)$ and range $(0, +\infty)$ and is one-to-one. The graph is decreasing on the entire domain. ●

Starting with $f(x) = 2^x$ and replacing x with $-x$, gives $f(-x) = 2^{-x} = (2^{-1})^x = (1/2)^x$. For this reason, the graphs of $f(x) = 2^x$ and $f(x) = (1/2)^x$ are reflections of each other about the y-axis. This is suggested by the graphs in Figures 2(a) and (b).

The graph of $f(x) = 2^x$ is typical of graphs of $f(x) = a^x$ where $a > 1$. For larger values of a, the graphs rise more steeply, but the general shape is similar to the graph in Figure 2(a). When $0 < a < 1$ the graph decreases like the graph of $f(x) = (1/2)^x$ in Figure 2(b). In Figure 3, the graphs of several typical exponential functions illustrate these facts.

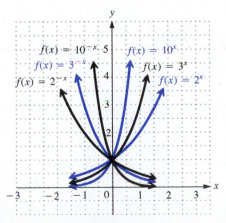

Figure 3

We can use composition of functions to produce more general exponential functions. Suppose $h(u) = ka^u$, where k is a constant and $u = g(x)$. Let $f(x) = h(g(x))$. Then

$$f(x) = h(g(x)) = ka^{g(x)}.$$

For example, if $a = 5$, $g(x) = 2x + 3$, and $k = 4$, then

$$f(x) = 4 \cdot 5^{2x + 3}.$$

· · · · · · · · ·

EXAMPLE 5   Graph each of the following.

(a) $f(x) = -2^x$

The negative sign causes the graph to be reflected about the x-axis, just as $f(x) = -x^2$ was as compared to $f(x) = x^2$. The domain is the real numbers and the range is $(-\infty, 0)$. See Figure 4(a).

(b) $f(x) = 2^{x+3}$

The graph will be the graph of $f(x) = 2^x$ translated 3 units to the left as shown in Figure 4(b).

(c) $f(x) = 2^x + 3$

This graph is that of $f(x) = 2^x$ translated 3 units upward. See Figure 4(c). ●

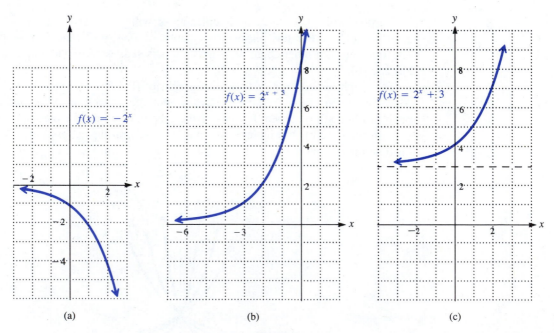

(a) (b) (c)

Figure 4

EXAMPLE 6   Graph $f(x) = 2^{-x^2}$.

Since

$$f(x) = 2^{-x^2} = 1/(2^{x^2}),$$

$0 < y \leq 1$ for all values of x. The y-intercept is at $(0, 1)$. The x-axis is a horizontal asymptote. Replacing x with $-x$ shows that the graph is symmetric with respect to the y-axis. Plotting the y-intercept and other points, such as $(1/4, .96)$, $(1/2, .84)$, $(1, 1/2)$, and $(2, 1/16)$, and drawing a smooth curve through them gives the graph of Figure 5. It is necessary to plot several points close to $(0, 1)$ to determine the correct shape of the graph there. This type of "bell-shaped" curve is important in statistics.   ●

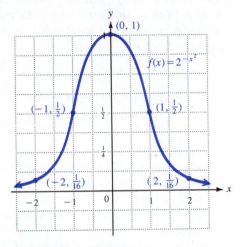

Figure 5

The formula for compound interest (interest paid on both principal and interest) is an important application of exponential functions.

**Compound
Interest**

If P dollars is deposited in an account paying an annual rate of interest r compounded (paid) n times per year, the account will contain

$$A = P\left(1 + \frac{r}{n}\right)^{nt}$$

dollars after t years.

. . . .

EXAMPLE 7   Suppose $1000 is deposited in an account paying 8% per year compounded quarterly (four times a year). Find the total amount in the account after 10 years if no withdrawals are made. Find the amount of interest earned.

Use the compound interest formula from above with $P = 1000$, $r = .08$, $n = 4$, and $t = 10$.

$$A = P\left(1 + \frac{r}{n}\right)^{nt}$$

$$A = 1000\left(1 + \frac{.08}{4}\right)^{4(10)}$$

$$= 1000(1 + .02)^{40} = 1000(1.02)^{40}$$

The number $(1.02)^{40}$ can be found in financial tables or by using a calculator with a y^x key. With a calculator, enter 1.02, press the key labeled y^x (or x^y), and then enter 40 and press the "=" key. To five decimal places, $(1.02)^{40} = 2.20804$. The amount on deposit after 10 years is

$$A = 1000(1.02)^{40} = 1000(2.20804) = 2208.04,$$

or $2208.04. The amount of interest earned is

$$\$2208.04 - \$1000.00 = \$1208.04. \quad \bullet$$

The Number e   Perhaps the single most useful base for an exponential function is the irrational number e. Base e exponential functions provide a good model for many natural, as well as economic, phenomena. The letter e was chosen to represent this number in honor of the Swiss mathematician Leonard Euler (pronounced "oiler") (1707–83). Applications of the exponential function with base e are given later in this chapter.

To see one way the number e is used, begin with the formula for compound interest. Suppose that a lucky investment produces annual interest of 100%, so that $r = 1.00$, or $r = 1$. Suppose also that only $1 can be deposited at this rate, and for only one year. Then $P = 1$ and $t = 1$. Substituting into the formula for compound interest gives

$$P\left(1 + \frac{r}{n}\right)^{nt} = 1\left(1 + \frac{1}{n}\right)^{n(1)}$$

$$= \left(1 + \frac{1}{n}\right)^{n}.$$

If interest is compounded annually, making $n = 1$, the total amount on deposit is

$$\left(1 + \frac{1}{n}\right)^{n} = \left(1 + \frac{1}{1}\right)^{1}$$

$$= 2^1 = 2,$$

so an investment of $1 becomes $2 in one year. As interest is compounded more and more often, the value of this expression will increase.

A calculator with a y^x key was used to get the following results. These results have been rounded when necessary to five decimal places.

Frequency of Compounding	n	$\left(1 + \dfrac{1}{n}\right)^n$
annually	1	2
semiannually	2	2.25
quarterly	4	2.44141
monthly	12	2.61304
daily	365	2.71457
hourly	8760	2.71813
every minute	525,600	2.71828

The table suggests that as n increases, the value of $(1 + 1/n)^n$ gets closer and closer to some fixed number. It turns out that this is indeed the case. This fixed number is called e.

Value of e

To nine decimal places,

$$e \approx 2.718281828.$$

Scientific and business calculators give values of e^x. To obtain a specific value, key in the number x and then press the key labeled e^x. (If your calculator does not have an e^x key, press the INV key and then the ln x key. The reason that this method works will be apparent in Section 2 of this chapter.) Also, Table 1 in this book gives various values of e^x.

In Figure 6, the functions defined by $f(x) = 2^x$, $f(x) = e^x$, and $f(x) = 3^x$ are graphed for comparison.

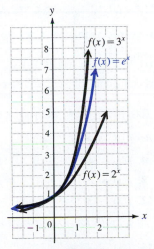

Figure 6

It can be shown that in many situations involving growth or decay of a population, the amount or number present at time t can be closely approximated by an exponential function with base e. The next example illustrates exponential growth.

. . . .

EXAMPLE 8 Suppose the population of a midwestern city is approximated by

$$P(t) = 10,000e^{.04t},$$

where t represents time measured in years. Find the population of the city at the following times: **(a)** $t = 0$, **(b)** $t = 5$.

(a) The population at time $t = 0$ is

$$
\begin{aligned}
P(0) &= 10,000e^{(.04)0} \\
&= 10,000e^0 \\
&= 10,000(1) \\
&= 10,000,
\end{aligned}
$$

written $P_0 = 10,000$.

(b) The population of the city at year $t = 5$ is

$$
\begin{aligned}
P(5) &= 10,000e^{(.04)5} \\
&= 10,000e^{.2}.
\end{aligned}
$$

The number $e^{.2}$ can be found in Table 1 or by using a suitable calculator. By either of these methods, $e^{.2} = 1.22140$ (to five decimal places), so that

$$P(5) = 10,000(1.22140) = 12,214.$$

In five years the population of the city will be about 12,200. ●

5.1 Exercises ∙∙

1. Graph each of the functions defined as follows. Compare the graphs to that of $f(x) = 2^x$.
   (a) $f(x) = 2^x + 1$ (b) $f(x) = 2^x - 4$ (c) $f(x) = 2^{x+1}$ (d) $f(x) = 2^{x-4}$

2. Graph each of the functions defined as follows. Compare the graphs to that of $f(x) = 3^{-x}$.
   (a) $f(x) = 3^{-x} - 2$ (b) $f(x) = 3^{-x} + 4$ (c) $f(x) = 3^{-x-2}$ (d) $f(x) = 3^{-x+4}$

Graph each of the functions defined as follows.

3. $f(x) = 3^x$ 4. $f(x) = 4^x$ 5. $f(x) = (3/2)^x$ 6. $f(x) = 3^{-x}$

7. $f(x) = e^{-x}$ 8. $f(x) = e^x$ 9. $f(x) = 2^{x-1}$ 10. $f(x) = e^{1-x}$

11. $f(x) = e^{|x|}$ 12. $f(x) = 2^{-|x|}$

Use a calculator to help graph each of the functions defined as follows.

13. $f(x) = x \cdot 2^x$ 14. $f(x) = x \cdot e^{-x}$

Solve each equation in Exercises 15–32.

15. $4^x = 2$

16. $125^r = 5$

17. $\left(\dfrac{1}{2}\right)^k = 4$

18. $\left(\dfrac{2}{3}\right)^x = \dfrac{9}{4}$

19. $2^{3-y} = 8$

20. $5^{2p+1} = 25$

21. $\dfrac{1}{27} = b^{-3}$

22. $\dfrac{1}{81} = k^{-4}$

23. $4 = r^{2/3}$

24. $z^{5/2} = 32$

25. $27^{4z} = 9^{z+1}$

26. $32^t = 16^{1-t}$

27. $\left(\dfrac{1}{2}\right)^{-x} = \left(\dfrac{1}{4}\right)^{x+1}$

28. $\left(\dfrac{2}{3}\right)^{k-1} = \left(\dfrac{81}{16}\right)^{k+1}$

29. $4^{|x|} = 64$

30. $3^{-|x|} = \dfrac{1}{27}$

31. $e^{-5x} = (e^2)^x$

32. $e^{3(1+x)} = e^{-8x}$

33. Suppose \$10,000 is left at interest for 3 years at 12%. Find the final amount on deposit if the interest is compounded **(a)** annually, **(b)** quarterly, **(c)** daily (365 days).

34. Find the final amount on deposit if \$5800 is left at interest for 6 years at 13% and interest is compounded **(a)** annually, **(b)** semiannually, **(c)** daily (365 days).

35. Suppose the population of a city is given by $P(t)$, where $P(t) = 1,000,000e^{.02t}$, and t represents time measured in years from some initial year. Find each of the following: **(a)** $P(0)$, **(b)** $P(2)$, **(c)** $P(4)$, **(d)** $P(10)$. **(e)** Graph $y = P(t)$.

36. Suppose the quantity in grams of a radioactive substance present at time t is $Q(t) = 500e^{-.05t}$. Let t be time measured in days from some initial day. Find the quantity present at each of the following times: **(a)** $t = 0$, **(b)** $t = 4$, **(c)** $t = 8$, **(d)** $t = 20$. **(e)** Graph $y = Q(t)$.

Give an equation of the form $f(x) = a^x$ to define the exponential function whose graph contains the given point.

37. $(3, 8)$

38. $(-2, 9)$

39. $(-.5, .2)$

40. $(2/3, 4)$

Use properties of exponents to write each of the following in the form $f(t) = ka^t$ where k is a constant. (Hint: Recall $4^{x+y} = 4^x \cdot 4^y$.)

41. $f(t) = 2^{4t+3}$

42. $f(t) = 2^{7-t}$

43. $f(t) = \left(\dfrac{1}{3}\right)^{2t-1}$

44. $f(t) = 9^{.5t+2}$

Simplify each expression. (Expressions like this arise in calculus from a technique called the quotient rule, used to determine the shape of a curve.)

45. $\dfrac{(e^{-x} - e^x)(-e^{-x} + e^x) - (e^{-x} + e^x)(-e^{-x} - e^x)}{(e^{-x} - e^x)^2}$

46. $\dfrac{(e^x + e^{2x})(2e^{2x} + 3e^{3x}) - (e^{2x} + e^{3x})(e^x + 2e^{2x})}{(e^x + e^{2x})^2}$

Suppose f is an exponential function of the form $f(x) = a^x$ and $f(3) = 2$. Determine the function values in Exercises 47–49.

47. $f(-3)$

48. $f(0)$

49. $f(9)$

50. What two points on the graph of $f(x) = a^x$ can be found without any computation?

The pressure of the atmosphere, p(h) in pounds per square inch, is given by $p(h) = p_0 e^{-kh}$, where h is the height above sea level and p_0 and k are constants. The pressure at sea level is 15 pounds per square inch and the pressure is 9 pounds per square inch at a height of 12,000 feet. Use this information in Exercises 51 and 52.

51. Find the pressure at an altitude of 6000 feet.

52. What would be the pressure encountered by a spaceship at an altitude of 150,000 feet?

53. For $a > 1$, how does the graph of $y = a^x$ change as a increases? What if $0 < a < 1$?

54. When defining an exponential function, why did we require $a > 0$?

Any points where the graphs of functions f and g cross give solutions of the equation $f(x) = g(x)$. Use this idea to estimate the number of solutions of the following equations.

55. $x = 2^x$ **56.** $2^{-x} = -x$ **57.** $3^{-x} = 1 - 2x$ **58.** $3x + 2 = 4^x$

Let $f(x) = a^x$ define an exponential function of base a.

59. Is f odd, even, or neither?

60. Prove that $f(m + n) = f(m) \cdot f(n)$ for any real numbers m and n.

In calculus, it is shown that

$$e^x = 1 + x + \frac{x^2}{2 \cdot 1} + \frac{x^3}{3 \cdot 2 \cdot 1} + \frac{x^4}{4 \cdot 3 \cdot 2 \cdot 1} + \frac{x^5}{5 \cdot 4 \cdot 3 \cdot 2 \cdot 1} + \cdots$$

61. Use the terms shown here and replace x with 1 to approximate $e^1 = e$ to three decimal places. Then check your results in Table 1 or with a calculator.

62. Use the terms shown here and replace x with $-.05$ to approximate $e^{-.05}$ to four decimal places. Check your results in Table 1 or with a calculator.

5.2 LOGARITHMIC FUNCTIONS

Exponential functions defined by $f(x) = a^x$ for all positive values of a, where $a \neq 1$, were discussed in the previous section. As mentioned there, exponential functions are one-to-one, and so have inverse functions. In this section we discuss the inverses of exponential functions. The equation of the inverse comes from exchanging x and y. Doing this with $y = a^x$ gives

$$x = a^y$$

as the inverse of the exponential function defined by $f(x) = a^x$. To solve $x = a^y$ for y, we use the following statement. For all real numbers y, and all positive numbers a and x, where $a \neq 1$,

$$y = \log_a x \quad \textbf{if and only if} \quad x = a^y.$$

Log is an abbreviation for *logarithm*. Read $\log_a x$ as "the logarithm of x to the base a."

This key statement should be memorized. It is important to remember the location of the base and exponent in each part.

$$\overset{\text{exponent}}{\downarrow} \qquad \qquad \qquad \overset{\text{exponent}}{\downarrow}$$

$$\text{logarithmic form: } y = \log_a x \qquad \text{exponential form: } a^y = x$$

$$\underset{\uparrow}{\text{base}} \qquad \qquad \qquad \underset{\uparrow}{\text{base}}$$

The discussion above suggests the following definition of logarithm.

Definition of Logarithm

A **logarithm** is an exponent: $\log_a x$ is the exponent on the base a that will yield the number x.

EXAMPLE 1   The chart below shows several pairs of equivalent statements. The same statement is written in both exponential and logarithmic forms.

Exponential form	Logarithmic form
$2^3 = 8$	$\log_2 8 = 3$
$(1/2)^{-4} = 16$	$\log_{1/2} 16 = -4$
$10^5 = 100{,}000$	$\log_{10} 100{,}000 = 5$
$3^{-4} = 1/81$	$\log_3 (1/81) = -4$
$5^1 = 5$	$\log_5 5 = 1$
$(3/4)^0 = 1$	$\log_{3/4} 1 = 0$ ●

Equations with logarithms often can be solved by rewriting them in exponential form as shown in the next example.

EXAMPLE 2   Solve each of the following equations.

(a) $\log_x \dfrac{8}{27} = 3$

First, write the expression in exponential form; then solve.

$$x^3 = \frac{8}{27}$$

$$x^3 = \left(\frac{2}{3}\right)^3$$

$$x = \frac{2}{3}$$

The solution is 2/3.

(b) $\log_4 x = 5/2$

In exponential form, the given statement becomes

$$4^{5/2} = x$$
$$(4^{1/2})^5 = x$$
$$2^5 = x$$
$$32 = x.$$

The solution is 32. ●

The logarithmic function with base a is defined as follows.

Definition of Logarithmic Function

If $a > 0$, $a \neq 1$, and $x > 0$, then the function f defined by

$$f(x) = \log_a x$$

is the **logarithmic function with base a.**

Exponential and logarithmic functions are inverses of each other. Since the domain of an exponential function is the set of all real numbers, the range of a logarithmic function will also be the set of all real numbers. In the same way, both the range of an exponential function and the domain of a logarithmic function are the set of all positive real numbers, so logarithms can be found for positive numbers only.

EXAMPLE 3   Graph the logarithmic functions defined as follows.

(a) $f(x) = \log_2 x$

One way to graph a logarithmic function is to begin with its inverse function. Here, the inverse has equation $y = 2^x$. The graph of the equation $y = 2^x$ is shown with a dashed curve in Figure 7(a). To get the graph of $y = \log_2 x$, reverse the ordered pairs used to graph $y = 2^x$. For example, the pairs $(-1, 1/2)$, $(0, 1)$, $(1, 2)$, and $(2, 4)$ for $y = 2^x$, become $(1/2, -1)$, $(1, 0)$, $(2, 1)$, and $(4, 2)$ for $x = 2^y$ or $y = \log_2 x$. The graph of the equation $y = \log_2 x$ is shown as a solid curve. As the graph suggests, the function defined by $f(x) = \log_2 x$ is increasing for all its domain, is one-to-one, and has the y-axis **as a vertical asymptote.**

(b) $f(x) = \log_{1/2} x$

The graph of the equation of the inverse, $y = (1/2)^x$, is shown with a dashed **curve in Figure** 7(b) for comparison. The graph of $y = \log_{1/2} x$, shown as a solid **curve, is** found by graphing the equivalent equation $x = (1/2)^y$. It is easiest to **choose** y-values and calculate x-values. Some ordered pairs that satisfy $x = (1/2)^y$ **are** $(4, -2)$, $(2, -1)$, $(1, 0)$, and $(1/2, 1)$. The function defined by $f(x) = \log_{1/2} x$ is decreasing for all its domain, is one-to-one, and has the y-axis **for a vertical asymptote.** ●

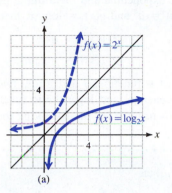

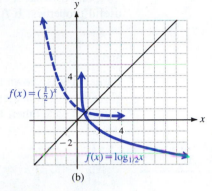

(a)   (b)

Figure 7

· · · · · · · · ·

EXAMPLE 4   Graph $f(x) = \log_2 (x - 1)$.

The graph of $f(x) = \log_2 (x - 1)$ will be the graph of $f(x) = \log_2 x$, translated one unit to the right. The vertical asymptote is $x = 1$. The domain of the function defined by $f(x) = \log_2 (x - 1)$ is $(1, +\infty)$, since logarithms can be found only for positive numbers. To find some ordered pairs to plot, use the equivalent equation

$$x - 1 = 2^y \quad \text{or} \quad x = 2^y + 1,$$

choosing values for y and calculating the corresponding x-values. See Figure 8.   ●

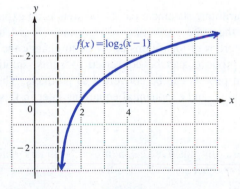

Figure 8

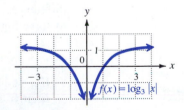

Figure 9

· · · · · · · · ·

EXAMPLE 5   Graph $f(x) = \log_3 |x|$.

Use the definition of logarithm to write the equation $y = \log_3 |x|$ as $|x| = 3^y$. Then choose values for y and find the corresponding x-values. Some ordered pairs for this equation are shown below; the graph is given in Figure 9.   ●

x	-3	-1	$-1/3$	$1/3$	1	3
y	1	0	-1	-1	0	1

Compositions of the exponential and logarithmic functions can be used to get two useful properties. If $f(x) = a^x$ and $g(x) = \log_a x$, then

$$f[g(x)] = a^{\log_a x}$$

and
$$g[f(x)] = \log_a a^x.$$

Earlier it was shown that if functions f and g are inverses of each other, $f[g(x)] = g[f(x)] = x$. Since exponential and logarithmic functions of the same base are inverses of each other, this result gives the following theorem.

Theorem

If $a > 0$ and $a \neq 1$, then

$$\log_a a^x = x.$$

Also, if $x > 0$, then

$$a^{\log_a x} = x.$$

EXAMPLE 6   Simplify each of the following.

(a) $\log_5 5^3 = 3$

(b) $7^{\log_7 10} = 10$

(c) $\log_e e^{x^2} = x^2$ ●

Since a logarithmic statement can be written as an exponential statement, it is not surprising that there are properties of logarithms based on the properties of exponents. The properties of logarithms allow us to change the form of logarithmic statements so that products can be converted to sums, quotients can be converted to differences, and powers can be converted to products. These properties can be used to solve logarithmic and exponential equations later in this chapter.

Properties of Logarithms

If x and y are any positive real numbers, r is any real number, and a is any positive real number, $a \neq 1$, then the following are true.

1. $\log_a xy = \log_a x + \log_a y$ **2.** $\log_a \dfrac{x}{y} = \log_a x - \log_a y$

3. $\log_a x^r = r \cdot \log_a x$ **4.** $\log_a a = 1$

5. $\log_a 1 = 0$

To prove property 1 of logarithms, let $m = \log_a x$ and $n = \log_a y$. Then, changing to exponential form,

$$a^m = x \quad \text{and} \quad a^n = y.$$

Multiplication gives

$$a^m \cdot a^n = xy.$$

By a property of exponents.

$$a^{m+n} = xy.$$

Now write this statement in exponential form.

$$\log_a xy = m + n$$

Since $m = \log_a x$ and $n = \log_a y$,

$$\log_a xy = \log_a x + \log_a y.$$

To prove property 2, use m and n as defined above. Then

$$\frac{a^m}{a^n} = \frac{x}{y}.$$

Since

$$\frac{a^m}{a^n} = a^{m-n},$$

then

$$a^{m-n} = \frac{x}{y}.$$

In logarithmic form, this statement can be written

$$\log_a \frac{x}{y} = m - n,$$

or

$$\log_a \frac{x}{y} = \log_a x - \log_a y.$$

For property 3,

$$(a^m)^r = x^r \quad \text{or} \quad a^{mr} = x^r.$$

Again using logarithmic form,

$$\log_a x^r = mr,$$

or

$$\log_a x^r = r \cdot \log_a x.$$

Finally, 4 and 5 follow directly from the exponential form since $a^1 = a$ and $a^0 = 1$.

The properties of logarithms are useful for rewriting expressions with logarithms in different forms, as shown in the next examples.

· · · · · · · · ·

EXAMPLE 7   Assuming all variables represent positive real numbers, use the properties of logarithms to write each of the following in a different form.

(a) $\log_6 7 \cdot 9 = \log_6 7 + \log_6 9$

(b) $\log_9 \dfrac{15}{7} = \log_9 15 - \log_9 7$

(c) $\log_5 \sqrt{8} = \log_5 8^{1/2} = \dfrac{1}{2} \log_5 8$

(d) $\log_a \dfrac{mnq}{p^2} = \log_a mnq - \log_a p^2$

$\qquad = \log_a m + \log_a n + \log_a q - 2 \log_a p$

(e) $\log_a \sqrt[3]{m^2} = \log_a m^{2/3} = \dfrac{2}{3} \log_a m$

(f) $\log_b \sqrt[n]{\dfrac{x^3 y^5}{z^m}} = \dfrac{1}{n} \log_b \dfrac{x^3 y^5}{z^m}$

$\qquad = \dfrac{1}{n} (\log_b x^3 + \log_b y^5 - \log_b z^m)$

$\qquad = \dfrac{3}{n} \log_b x + \dfrac{5}{n} \log_b y - \dfrac{m}{n} \log_b z$ ●

· · · · · · · · ·

EXAMPLE 8 Use the properties of logarithms to write each of the following as a single logarithm with a coefficient of 1. Assume all variables represent positive real numbers.

(a) $\log_3 (x + 2) + \log_3 x - \log_3 2 = \log_3 \dfrac{(x + 2)x}{2}$

(b) $2 \log_a m - 3 \log_a n = \log_a m^2 - \log_a n^3 = \log_a \dfrac{m^2}{n^3}$

(c) $\dfrac{1}{2} \log_b m + \dfrac{3}{2} \log_b 2n - \log_b m^2 n$

$\qquad = \log_b m^{1/2} + \log_b (2n)^{3/2} - \log_b m^2 n$

$\qquad = \log_b \dfrac{m^{1/2}(2n)^{3/2}}{m^2 n} = \log_b \dfrac{2^{3/2} n^{1/2}}{m^{3/2}}$ ●

· · · · · · · · ·

EXAMPLE 9 Assume $\log_{10} 2 = .3010$. Find the base 10 logarithms of 4 and 5.

By the properties of logarithms,

$$\log_{10} 4 = \log_{10} 2^2 = 2 \log_{10} 2 = 2(.3010) = .6020$$

$$\log_{10} 5 = \log_{10} \dfrac{10}{2} = \log_{10} 10 - \log_{10} 2 = 1 - .3010 = .6990. \quad ●$$

The properties of logarithms were developed over a twenty-five-year period by the Scot John Napier of Merchiston (1550–1617). In 1614 Napier published *Mirifici* *logarithmorum canonis descriptio* (A Description of the Wonderful Law of Logarithms), a work that is second only to Newton's *Principia* in the history of British mathematics. Napier's work, which astonished the scientific world, contains the first description of logarithms, the first table of logarithms, and the first use of the word "logarithm" (meaning "number of the ratio"). Napier had been studying sequences of numbers obtained by repeated multiplication (exponentiation) and an ancient trigonometric formula that reduces multiplication to addition.* Professor Henry Briggs of the University of London was so impressed by Napier's achievement that he traveled to Edinburgh to meet him. Together they determined that a log table would be more useful if the base was 10 rather than Napier's base of $(1 - 1/10^7)^{10^7}$, which is approximately $1/e$.

5.2 Exercises

For each of the following statements, write an equivalent statement in logarithmic form.

1. $10^4 = 10,000$ **2.** $8^2 = 64$ **3.** $(1/2)^{-4} = 16$ **4.** $(2/3)^{-3} = 27/8$

For each of the following statements, write an equivalent statement in exponential form.

5. $\log_6 36 = 2$ **6.** $\log_5 5 = 1$ **7.** $\log_{10} .0001 = -4$ **8.** $\log_3 \sqrt[3]{9} = 2/3$

Find the value of each of the following. Assume all variables represent positive real numbers.

9. $\log_5 25$ **10.** $\log_3 81$ **11.** $\log_8 8$ **12.** $\log_7 1$

13. $\log_4 \dfrac{\sqrt[3]{4}}{2}$ **14.** $\log_9 \dfrac{\sqrt[4]{27}}{3}$ **15.** $\log_6 36^4$ **16.** $\log_5 125^2$

17. $\log_e e^4$ **18.** $\log_e \dfrac{1}{e}$ **19.** $\log_e \sqrt{e}$ **20.** $\log_e e^9$

21. $e^{\log_e 2}$ **22.** $e^{\log_e 5}$ **23.** $2^{\log_2 9}$ **24.** $8^{\log_8 11}$

25. $3^{\log_2 4}$ **26.** $2^{\log_5 1/5}$

Solve each of the following equations.

27. $\log_x 25 = -2$ **28.** $\log_x \dfrac{1}{16} = -2$ **29.** $\log_9 27 = m$ **30.** $\log_8 4 = z$

31. $\log_y 8 = \dfrac{3}{4}$ **32.** $\log_r 7 = 1/2$

Write each of the following as a sum, difference, or product of logarithms. Simplify the result if possible. Assume all variables represent positive real numbers.

33. $\log_2 \dfrac{6x}{y}$ **34.** $\log_3 \dfrac{4p}{q}$ **35.** $\log_5 \dfrac{5\sqrt{7}}{3}$ **36.** $\log_2 \dfrac{2\sqrt{3}}{5}$

37. $\log_4 (2x + 5y)$ **38.** $\log_6 (7m + 3q)$ **39.** $\log_k \dfrac{pq^2}{m}$ **40.** $\log_z \dfrac{x^5 y^3}{3}$

*The trigonometric formula that he used is the fourth equation in Section 7.5.

Write each of the expressions in Exercises 41–48 as a single logarithm with a coefficient of 1. Assume that all variables represent positive real numbers.

41. $\log_a x + \log_a y - \log_a m$

42. $(\log_b k - \log_b m) - \log_b a$

43. $2 \log_m a - 3 \log_m b^2$

44. $\frac{1}{2} \log_y p^3 q^4 - \frac{2}{3} \log_y p^4 q^3$

45. $-\frac{3}{4} \log_x a^6 b^8 + \frac{2}{3} \log_x a^9 b^3$

46. $\log_a (pq^2) + 2 \log_a (p/q)$

47. $\log_b (x + 2) + \log_b 7x - \log_b 8$

48. $\log_h (4m + 1) + \log_h 2m - \log_h 3$

49. Graph $f(x) = \log_3 x$ and $f(x) = 3^x$ on the same axes.

50. Graph $f(x) = \log_4 x$ and $f(x) = 4^x$ on the same axes.

51. Graph each of the following equations. Compare the graphs to that of $y = \log_2 x$.
   (a) $y = (\log_2 x) + 3$ **(b)** $y = \log_2 (x + 3)$ **(c)** $y = |\log_2 (x + 3)|$

52. Graph each of the following equations. Compare the graphs to that of $y = \log_{1/2} x$.
   (a) $y = (\log_{1/2} x) - 2$ **(b)** $y = \log_{1/2} (x - 2)$ **(c)** $y = |\log_{1/2} (x - 2)|$

Graph each of the functions defined as follows. Give any x- and y-intercepts for each graph.

53. $f(x) = \log_5 x$

54. $f(x) = \log_{10} x$

55. $f(x) = \log_{1/2} (1 - x)$

56. $f(x) = \log_{1/3} (3 - x)$

57. $f(x) = \log_3 (x + 3) - 2$

58. $f(x) = \log_2 (x - 1) + 2$

Given $\log_{10} 2 = .3010$ and $\log_{10} 3 = .4771$, evaluate the logarithms in Exercises 59–62 without using calculators or tables.

59. $\log_{10} 6$

60. $\log_{10} 12$

61. $\log_{10} 20$

62. $\log_{10} 36$

63. Which is larger, $\log_7 2$ or $\log_6 2$?

64. Which is larger, $\log_{1/3} 2$ or $\log_{1/4} 2$?

Suppose f is a logarithmic function and $f(3) = 2$. Determine the function values in Exercises 65–67.

65. $f(1/3)$

66. $f(1)$

67. $f(9)$

68. Does the fact that $(5, 4)$ is on the graph of the logarithmic function of base a mean that $5 = \log_a 4$ or that $4 = \log_a 5$?

69. The population of an animal species that is introduced into a certain area may grow rapidly at first but then grow more slowly as time goes on. A logarithmic function can provide an excellent description of such growth. Suppose that the population of foxes in an area t months after the foxes were introduced there is $F(t) = 500 \log_{10} (2t + 3)$. Use a calculator with a log key to find the population of foxes at the following times:
(a) when they are first released into the area (that is, when $t = 0$),   **(b)** after 3 months,   **(c)** after 15 months.   **(d)** Graph $y = F(t)$.

70. The loudness of sounds is measured in a unit called a *decibel*. To measure with this unit, we first assign an intensity of I_0 to a very faint sound, called the *threshold sound*. If a particular sound has intensity I, then the decibel rating of this louder sound is

$$d = 10 \log_{10} \frac{I}{I_0}.$$

Find the decibel ratings of the following sounds, having intensities as given. (You will need a calculator with a log key.) Round answers to the nearest whole number.
(a) Whisper, $115I_0$
(b) Busy street, $9,500,000I_0$
(c) Heavy truck, 20 m away, $1,200,000,000I_0$
(d) Rock music, $895,000,000,000I_0$
(e) Jetliner at takeoff, $109,000,000,000,000I_0$

71. If the intensity of a sound is doubled, by how much is the decibel rating increased? (See Exercise 70.)

72. The *Richter scale rating* of an earthquake of intensity I is given by $\log_{10}(I/I_0)$, where I_0 is the intensity of an earthquake of a certain (small) size. Find the Richter scale ratings of earthquakes having the following intensities: (a) $1000I_0$, (b) $1,000,000I_0$, (c) $100,000,000I_0$.

73. The San Francisco earthquake of 1906 had a Richter scale rating of 8.6. Use a calculator with a y^x key to express the intensity of this earthquake as a multiple of I_0 (see Exercise 72).

74. How much more powerful is an earthquake with a Richter scale rating of 8.6 than one with a rating of 8.2?

5.3 EVALUATING LOGARITHMS

Common Logarithms As we saw in the previous section, logarithms were developed between 1590 and 1614 by John Napier, as an aid for numerical calculations. Since our number system uses base 10, logarithms to base 10 were the most convenient for calculations. Base 10 logarithms are called **common logarithms.** The common logarithm of the number x, $\log_{10} x$, is often abbreviated as log x. This convention started when these logarithms were used extensively for calculation. (However, some advanced books use log x as an abbreviation for $\log_e x$.)

A calculator or a table of common logarithms can be used to find decimal approximations for base 10 logarithms. A table of logarithms, Table 2, and an appendix on its use are provided in the back of this text. With a calculator, simply enter the number and press the log key. Some calculators may also require the use of the second-function key.

· · · · · · · · ·

EXAMPLE 1   Use a calculator to find the following logarithms.
(a) $\log_{10} 42.7$

Enter 42.7 and press the log key. (With some calculators, the $\log_{10}$ key is a second-function key.) You should get

$$\log_{10} 42.7 \approx 1.63043$$

to five decimal places.
(b) $\log_{10} .06294 \approx -1.20107$ ●

Sometimes the logarithm of a number is known and it is necessary to find the number.

· · · · · · · · ·

EXAMPLE 2   Find x if $\log_{10} x = 2.2741$.
Write $\log_{10} x = 2.2741$ in exponential form.

$$x = 10^{2.2741}$$

The y^x key can be used to find $10^{2.2741}$.

$$10^{2.2741} \approx 187.97 \quad \bullet$$

Some calculators have a key marked 10^x for solving problems as in Example 2, while other calculators require using the INV key with the log key. This works because 10^x is the inverse of $\log_{10} x$.

Natural Logarithms   In most practical applications of logarithms today, the number $e \approx 2.71828$ is used as base. Logarithms to base e are called **natural logarithms,** since they occur in many natural-world applications, such as those involving growth and decay. The abbreviation ln x ("el en" x) is used for the natural logarithm of x, so that $\log_e x = \ln x$. The graph of $f(x) = e^x$ was given in Figure 6. Since the functions defined by $f(x) = e^x$ and $f(x) = \ln x$ are inverses, the graph of $f(x) = \ln x$, shown in Figure 10, can be found by reflecting the graph of $f(x) = e^x$ about the line $y = x$.

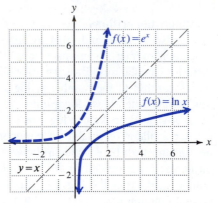

Figure 10

The following useful results for the natural logarithm function are direct applications of the properties of logarithms given earlier.

· · · · · · · · · · · · ·   ·

**Properties of
Natural
Logarithms**

1. $e^{\ln x} = x$ if $x > 0$	**2.** $\ln e^x = x$
3. $\ln 1 = 0$	**4.** $\ln e = 1$

· · · · · · · · ·

EXAMPLE 3   Evaluate each expression using a property of natural logarithms.
(a) $e^{\ln 5} = 5$ by property 1.

(b) $e^{\ln \sqrt{3}} = \sqrt{3}$

(c) $\ln e^2 = 2$ by property 2.

(d) $\ln e^{-1.5} = -1.5$

(e) $8 \ln 1 - 16 \ln e + 2 \ln e^4$

Since $\ln 1 = 0$, $\ln e = 1$, and $\ln e^4 = 4$,

$$8 \ln 1 - 16 \ln e + 2 \ln e^4 = 8(0) - 16(1) + 2(4) = -8. \quad \bullet$$

Values of natural logarithms can be found with a calculator which has an ln key or with a table of natural logarithms. Natural logarithms are given in Table 1. Reading directly from this table,

$$\ln 55 = 4.0073,$$
$$\ln 1.9 = .6419,$$
and
$$\ln .4 = -.9163.$$

· · · · · · · · ·

EXAMPLE 4   Use a calculator or Table 1 to find the following logarithms.

(a) ln 85

With a calculator, enter 85, press the ln key, and read the result, 4.4427. Table 1 does not give ln 85. However, the value of ln 85 can be found using the properties of logarithms.

$$\ln 85 = \ln \mathbf{(8.5 \times 10)}$$
$$= \ln \mathbf{8.5} + \ln \mathbf{10}$$
$$\approx 2.1401 + 2.3026$$
$$= 4.4427$$

A result found in this way is sometimes slightly different from the answer found using a calculator, due to rounding error.

(b) ln 36

A calculator gives $\ln 36 = 3.5835$. To use the table, first use properties of logarithms, since 36 is not listed in Table 1.

$$\ln 36 = \ln \mathbf{6}^2 = \mathbf{2} \ln \mathbf{6}$$
$$\approx 2(1.7918) = 3.5836$$

Alternatively, ln 36 can be found as follows.

$$\ln 36 = \ln 9 \cdot 4 = \ln 9 + \ln 4 = 2.1972 + 1.3863 = 3.5835 \quad \bullet$$

Logarithms to Other Bases   A calculator or a table will give the values of either natural logarithms (base e) or common logarithms (base 10). Sometimes, however, it is convenient to use logarithms to other bases. For example, base 2 logarithms are important in computer science. The **change-of-base theorem** can be used to convert logarithms from one base to another.

Change-of-Base Theorem

If x is any positive number and if a and b are positive real numbers, $a \neq 1$, $b \neq 1$, then

$$\log_a x = \frac{\log_b x}{\log_b a}.$$

To prove this result, use the definition of logarithm to write $y = \log_a x$ as $x = a^y$ or $x = a^{\log_a x}$ (for positive x and positive a, $a \neq 1$). Now take base b logarithms of both sides of this last equation.

$$\log_b x = \log_b a^{\log_a x}$$

or

$$\log_b x = (\log_a x)(\log_b a), \qquad \log_b a^r = r \log_b a$$

from which

$$\log_a x = \frac{\log_b x}{\log_b a}.$$

NOTE As an aid in remembering the change-of-base theorem, notice that x is above a on both sides of the equation.

EXAMPLE 5 Use the change-of-base theorem to approximate each of the following. Round to the nearest hundredth.

(a) $\log_5 27$

With natural logarithms as the base b logarithms, $x = 27$, $a = 5$, and $b = e$. Substituting into the change-of-base theorem gives

$$\log_5 27 = \frac{\ln 27}{\ln 5} \approx \frac{3.2958}{1.6094} \approx 2.05.$$

To check, use a calculator with a y^x key to verify that $5^{2.05} \approx 27$.

(b) $\log_2 .1$

Use natural logarithms, with $x = .1$ and $a = 2$.

$$\log_2 .1 = \frac{\ln .1}{\ln 2} \approx \frac{-2.3026}{.6931} \approx -3.32 \quad \bullet$$

EXAMPLE 6 One measure of the diversity of the species in an ecological community is given by

$$H = -[P_1 \log_2 P_1 + P_2 \log_2 P_2 + \ldots + P_n \log_2 P_n],$$

where $P_1, P_2, \ldots, P_n$ are the proportions of a sample belonging to each of n species found in the sample. For example, in a community with two species, where there are 90 of one species and 10 of the other, $P_1 = 90/100 = .9$ and $P_2 = 10/100 = .1$, so

$$H = -[.9 \log_2 .9 + .1 \log_2 .1].$$

The value of $\log_2 .1$ was found in Example 5(b) above. Now find $\log_2 .9$.

$$\log_2 .9 = \frac{\ln .9}{\ln 2} \approx \frac{-.1054}{.6931} \approx -.152$$

Therefore, $\quad H \approx -[(.9)(-.152) + (.1)(-3.32)] \approx .47.$ ●

· · · · · · · · ·

EXAMPLE 7   The cost in dollars, $C(x)$, of manufacturing x picture frames, where x is measured in thousands, is

$$C(x) = 5000 + 2000 \log_{10} (x + 1).$$

Find the cost of manufacturing 19,000 frames.
     To find the cost of producing 19,000 frames, let $x = 19$. This gives

$$\begin{aligned}
C(19) &= 5000 + 2000 \log_{10} (19 + 1) \\
&= 5000 + 2000 \log_{10} 20 \\
&\approx 5000 + 2000(1.3010) \\
&= 7602.
\end{aligned}$$

Thus, 19,000 frames cost a total of about \$7600 to produce.   ●

· · · · · · · · ·

EXAMPLE 8   In chemistry, the pH of a solution is defined as

$$pH = -\log_{10} [H^+],$$

where $[H^+]$ is the hydrogen ion concentration in moles per liter. The number pH is a measure of the acidity or alkalinity of solutions. Pure water has a pH of 7.0, with values greater than that indicating alkalinity and values less than 7.0 indicating acidity. Find the following.

(a) The pH of milk with $[H^+] = 4 \times 10^{-7}$

$$\begin{aligned}
pH &= -\log_{10} [H^+] \\
&= -\log_{10} (4 \times 10^{-7}) \\
&= -(\log_{10} 4 + \log_{10} 10^{-7}) \\
&\approx -(.6021 - 7) \\
&= -.6021 + 7 \\
&\approx 6.4
\end{aligned}$$

It is customary to round pH values to the nearest tenth.

(b) The hydrogen ion concentration of shampoo with pH $= 5.5$

$$\begin{aligned}
pH &= -\log_{10} [H^+] \\
5.5 &= -\log_{10} [H^+] \\
-5.5 &= \log_{10} [H^+]
\end{aligned}$$

Write this expression as

$$[H^+] = 10^{-5.5}$$
$$\approx 3.2 \times 10^{-6}$$

rounded to two significant digits. ●

5.3 Exercises ···

Find each of the following logarithms to four decimal places.

1. $\log_{10} .006532$ **2.** $\log_{10} 52.79$ **3.** $\log_{10} 104.3$ **4.** $\log_{10} 26,570$

5. $\log_{10} .9638$ **6.** $\log_{10} 4000$ **7.** $\log_{10} 271,000$ **8.** $\log_{10} .00517$

Use the properties of logarithms to evaluate the following.

9. $\ln e$ **10.** $\ln 1$ **11.** $\ln e^4$ **12.** $\ln e^2$

13. $\ln e^{-3}$ **14.** $\ln e^{-.017}$ **15.** $e^{\ln 1.2}$ **16.** $e^{\ln .003}$

17. $e^{2 \ln 3}$ **18.** $\ln (1/e)$ **19.** $\ln (1/e^{.5})$ **20.** $e^{(\ln 2)3}$

21. $e^{\ln 2 + \ln 3}$ **22.** $\dfrac{\ln 25}{\ln \sqrt{5}}$ **23.** $(e^5)^{\ln 2}$ **24.** $\ln (6 + 2) - \ln 2$

Find each of the following logarithms to four decimal places. Use a calculator or Table 1.

25. $\ln 4$ **26.** $\ln 6$ **27.** $\ln 17$ **28.** $\ln 29$

29. $\ln 350$ **30.** $\ln 900$ **31.** $\ln 81,000$ **32.** $\ln 121,000$

Find each of the following logarithms to the nearest hundredth.

33. $\log_5 10$ **34.** $\log_9 12$ **35.** $\log_{1/2} 3$ **36.** $\log_{12} 62$

37. $\log_{100} 83$ **38.** $\log_{200} 175$ **39.** $\log_{2.9} 7.5$ **40.** $\log_{5.8} 12.7$

Graph each of the functions defined in Exercises 41–44. Give any x- or y-intercepts.

41. $f(x) = \log_{10} x$ **42.** $f(x) = |\ln x|$ **43.** $f(x) = \ln (x - 1) + 1$ **44.** $f(x) = \ln (x + 2) - 2$

45. If (b, c) is on the graph of the natural logarithmic function, does $\ln b = c$ or $\ln c = b$?

46. The function $f(x) = \ln |x|$ plays a prominent role in calculus. Give its domain, range, and symmetries.

Use the properties of logarithms to relate the graphs of each of the following functions to the graph of $f(x) = \ln x$.

47. $f(x) = \ln e^2 x$ **48.** $f(x) = \ln (x/e)$

Work the following problems. Refer to Example 6 for Exercises 49 and 50.

49. Suppose a sample of a small community shows two species with 50 individuals each. Find the index of diversity H.

50. A virgin forest in northwestern Pennsylvania has 4 species of large trees with the following proportions of each: hemlock, .521; beech, .324; birch, .081; maple, .074. Find the index of diversity H.

51. The number of years, n, since two independently evolving languages split off from a common ancestral language is approximated by $n \approx -7600 \log r$, where r is the proportion of words from the ancestral language common to both languages. Find n if **(a)** $r = .9$, **(b)** $r = .3$. **(c)** How many years have elapsed since the split if half of the words of the ancestral language are common to both languages?

Find the pH of each of the following substances, using the given hydronium ion concentration.

52. Grapefruit, 6.3×10^{-4}

53. Crackers, 3.9×10^{-9}

54. Limes, 1.6×10^{-2}

55. Sodium hydroxide (lye), 3.2×10^{-14}

Find $[H^+]$ for each of the following substances, using the given pH.

56. Soda pop, 2.7

57. Wine, 3.4

58. Beer, 4.8

59. Drinking water, 6.5

60. Suppose the number of rabbits in a colony is $y = y_0 e^{.4t}$, where t represents time in months and y_0 is the rabbit population when $t = 0$.
   (a) If $y_0 = 100$, find the number of rabbits present at time $t = 4$.
   (b) Find y_0 if there are 30 rabbits after 4 months.

61. A Midwestern city finds its residents moving to the suburbs. Its population is declining according to the relationship $P = P_0 e^{-.04t}$, where t is time measured in years and P_0 is the population at time $t = 0$. Assume $P_0 = 1,000,000$.
   (a) Find the population at time $t = 1$.
   (b) If the decline continues at the same rate, what will the population be after 10 years?

62. In the central Sierra Nevada Mountains of California, the percent of moisture that falls as snow rather than rain is approximated reasonably well by $p = 86.3 \ln h - 680$, where p is the percent of snow at an altitude h (in feet). (Assume $h \geq 3000$.) Find the percent of moisture that falls as snow at the following altitudes:
   (a) 3000 ft, **(b)** 4000 ft, **(c)** 7000 ft. **(d)** Graph p.

In Exercises 63 and 64, assume that a and b represent positive numbers other than 1, and x is any real number.

63. Show that $1/\log_a b = \log_b a$.

64. Show that $a^x = e^{x \ln a}$.

5.4 EXPONENTIAL AND LOGARITHMIC EQUATIONS

In Section 1 of this chapter we solved exponential equations such as $(1/3)^x = 81$ by writing each side of the equation as a power of 3. That method cannot be used to solve an equation such as $7^x = 12$, since 12 cannot easily be written as a power of 7. However, the equation $7^x = 12$ can be solved by taking the logarithm of each side.

Property of Logarithms

If $x > 0$, $y > 0$, $a > 0$, and $a \neq 1$, then

$$x = y \quad \text{if and only if} \quad \log_a x = \log_a y.$$

The reason that $\log_a x = \log_b x$ implies $x = y$ is because a logarithmic function is one-to-one.

EXAMPLE 1   Solve the equation $7^x = 12$.

While the result above is valid for any appropriate base a, the best practical base to use is either base e or base 10, since these are the logarithms most easily approximated with a table or a calculator. Taking base e (natural) logarithms of both sides gives

$$\ln 7^x = \ln 12$$
$$x \cdot \ln 7 = \ln 12$$
$$x = \frac{\ln 12}{\ln 7}.$$

To get a decimal approximation for x, use Table 1 or a calculator.

$$x = \frac{\ln 12}{\ln 7} \approx \frac{2.4849}{1.9459}$$

Using a calculator to divide 2.4849 by 1.9459 gives

$$x \approx 1.2770.$$

A calculator with a y^x key can be used to check this answer. Evaluate $7^{1.2770}$; the result should be approximately 12.   ●

▞ **CAUTION**   A common mistake is to confuse $(\ln 12)/(\ln 7)$ with $\ln(12/7)$. However, as shown in Example 1,

$$\frac{\ln 12}{\ln 7} \neq \ln \frac{12}{7} = \ln 12 - \ln 7.$$

EXAMPLE 2   Solve $3^{2x-1} = 4^{x+2}$.

Taking natural logarithms on both sides gives

$$\ln 3^{2x-1} = \ln 4^{x+2}$$

Now use a property of logarithms.

$$(2x - 1) \ln 3 = (x + 2) \ln 4$$
$$2x \ln 3 - \ln 3 = x \ln 4 + 2 \ln 4 \quad \text{Distributive property}$$
$$2x \ln 3 - x \ln 4 = 2 \ln 4 + \ln 3$$

Factor out x on the left to get

$$x(2 \ln 3 - \ln 4) = 2 \ln 4 + \ln 3$$

or

$$x = \frac{2 \ln 4 + \ln 3}{2 \ln 3 - \ln 4}.$$

By properties of logarithms,

$$x = \frac{\ln 16 + \ln 3}{\ln 9 - \ln 4}$$

or, finally,

$$x = \frac{\ln 48}{\ln \dfrac{9}{4}}.$$

This quotient could be approximated by a decimal if desired.

$$x = \frac{\ln 48}{\ln 2.25} \approx \frac{3.8712}{.8109} \approx 4.774$$

To the nearest thousandth, the solution is 4.774. To find $\ln 2.25$ with Table 1, write $\ln 2.25$ as $\ln 1.5^2 = 2 \ln 1.5$. ●

· · · · · · · · ·

EXAMPLE 3   Solve $e^{x^2} = 200$.

Take natural logarithms on both sides; then use properties of logarithms.

$$e^{x^2} = 200$$
$$\ln e^{x^2} = \ln 200$$

Since $\ln e^{x^2} = x^2$,

$$x^2 = \ln 200$$
$$x = \pm \sqrt{\ln 200}$$
$$x \approx \pm 2.302$$

to the nearest thousandth. ●

· · · · · · · · ·

EXAMPLE 4   Solve $3 = 5(1 - e^x)$.

First solve for e^x.

$$3 = 5(1 - e^x)$$
$$\frac{3}{5} = 1 - e^x$$
$$e^x = 1 - \frac{3}{5} = \frac{2}{5}$$

Now take the natural logarithm on each side.

$$\ln e^x = \ln \frac{2}{5}$$

Since $\ln e^x = x$, $\qquad x = \ln \frac{2}{5} = \ln .4 \approx -.916$

to the nearest thousandth. •

The properties of logarithms given earlier are useful in solving logarithmic equations, as shown in the next examples.

. . . .

EXAMPLE 5   Solve $\log_a (x + 6) - \log_a (x + 2) = \log_a x$.
Using a property of logarithms, rewrite the equation as

$$\log_a \frac{x + 6}{x + 2} = \log_a x.$$

Then, since logarithmic functions are one-to-one,

$$\frac{x + 6}{x + 2} = x$$
$$x + 6 = x(x + 2)$$
$$x + 6 = x^2 + 2x$$
$$x^2 + x - 6 = 0.$$

Factoring gives $\qquad\qquad (x - 2)(x + 3) = 0$

so that $\qquad\qquad\qquad x = 2 \quad \text{or} \quad x = -3.$

$\text{Log}_a x$ cannot be evaluated for $x = -3$, since the domain of $\log_a x$ includes only positive numbers. Verify by substitution that $x = 2$ is the only solution. •

. . . .

EXAMPLE 6   Solve $\log_{10} (3x + 2) + \log_{10} (x - 1) = 1$.
By a property of logarithms,

$$\log_{10} (3x + 2)(x - 1) = 1.$$

Write this expression in exponential form.

$$(3x + 2)(x - 1) = 10^1$$
$$3x^2 - x - 2 = 10$$
$$3x^2 - x - 12 = 0$$

Now use the quadratic formula to arrive at

$$x = \frac{1 \pm \sqrt{1 + 144}}{6}.$$

If $x = (1 - \sqrt{145})/6$, then $x - 1 < 0$ and $\log (x - 1)$ does not exist. For this reason, $(1 - \sqrt{145})/6$ must be discarded as a solution. A calculator can help to show that $(1 + \sqrt{145})/6$ is the only solution.   ●

. . . .

EXAMPLE 7   Solve $\log_3 (3m^2)^{1/4} - 1 = 2$

Solve for the logarithm first.

$$\log_3 (3m^2)^{1/4} - 1 = 2$$
$$\log_3 (3m^2)^{1/4} = 3$$

Now write the expression in exponential form.

$$(3m^2)^{1/4} = 3^3$$
$$3^{1/4} m^{1/2} = 3^3$$
$$m^{1/2} = 3^{11/4}$$

Square both sides to get

$$m = (3^{11/4})^2 = 3^{11/2}$$

or

$$m \approx 420.888.   ●$$

. . . .

EXAMPLE 8   Suppose

$$P(t) = 10,000e^{.4t}$$

gives the population of a city at time t (in years). In how many years will the population double?

If the population is doubled, it will be 20,000. To find the time t when that occurs, let $P(t) = 20,000$ and substitute this value into the given function. Then divide by 10,000 on both sides of the equation.

$$P(t) = 10,000e^{.4t}$$
$$20,000 = 10,000e^{.4t}$$
$$2 = e^{.4t}$$

Now take natural logarithms on both sides.

$$\ln 2 = \ln e^{.4t}$$

Since $\ln e^{.4t} = .4t$,

$$\ln 2 = .4t$$

with

$$t = \frac{\ln 2}{.4}.$$

Use a calculator or Table 1 to find ln 2 and then use a calculator to find the quotient. This gives $t = 1.733$ to the nearest thousandth. The population of the city will double in about 1 3/4 years.   ●

. . . .

EXAMPLE 9   Solve the equation $y = \dfrac{1 - e^x}{1 - e^{-x}}$ for x in terms of y.

Begin by multiplying both sides by $1 - e^{-x}$.

$$y(1 - e^{-x}) = 1 - e^x$$

$$y - ye^{-x} = 1 - e^x$$

$$e^x + y - 1 - ye^{-x} = 0 \qquad \text{Get 0 alone on one side.}$$

$$e^{2x} + (y - 1)e^x - y = 0 \qquad \text{Multiply both sides by } e^x.$$

Rewrite this equation as

$$(e^x)^2 + (y - 1)e^x - y = 0,$$

a quadratic equation in e^x. Solve this equation by using the quadratic formula with $a = 1$, $b = y - 1$, and $c = -y$.

$$e^x = \frac{-(y - 1) \pm \sqrt{(y - 1)^2 - 4(1)(-y)}}{2(1)}$$

$$= \frac{-(y - 1) \pm \sqrt{y^2 - 2y + 1 + 4y}}{2}$$

$$= \frac{-(y - 1) \pm \sqrt{y^2 + 2y + 1}}{2}$$

$$= \frac{-(y - 1) \pm \sqrt{(y + 1)^2}}{2}$$

$$e^x = \frac{-y + 1 \pm (y + 1)}{2}$$

First, use the $+$ sign.

$$e^x = \frac{-y + 1 + y + 1}{2} = \frac{2}{2} = 1$$

If $e^x = 1$, then $x = 0$ and $e^{-x} = 1$ also. This leads to a zero denominator in the original equation, so the $+$ sign leads to no solution.

Try the $-$ sign.

$$e^x = \frac{-y + 1 - y - 1}{2} = \frac{-2y}{2} = -y$$

If $e^x = -y$, take natural logarithms on both sides to get

$$\ln e^x = \ln(-y)$$

$$x = \ln(-y).$$

This result will be satisfied by all values of y less than 0.   ●

CAUTION   In solving equations like the one in Example 9 that involve a^x (for any value of a), always solve first for a^x. Then use logarithms to get x.

A summary of methods of solving exponential or logarithmic equations follows.

Solving Exponential or Logarithmic Equations

In summary, to solve an exponential or logarithmic equation, first use the properties of algebra to change the given equation into one of the following forms, where a and b are real numbers with appropriate restrictions.

1. $a^{f(x)} = b$

   To solve, take logarithms on both sides.

2. $\log_a f(x) = b$

   Solve by changing to the exponential form $a^b = f(x)$.

3. $\log_a f(x) = \log_a g(x)$

   From the given equation, obtain the equation $f(x) = g(x)$, then solve algebraically.

4. In a more complicated equation, such as the one in Example 9, it is necessary to first solve for $e^{f(x)}$ or $\log_a f(x)$ and then solve the resulting equation using one of the methods given above.

5.4 Exercises

Solve the following equations. Give answers as decimals rounded to the nearest thousandth.

1. $3^x = 6$
2. $4^x = 12$
3. $6^{1-2k} = 8$
4. $2^{k-3} = 11$
5. $4^{3m-1} = 12^{m+2}$
6. $3^{2m-5} = 13^{m-1}$
7. $e^{k-1} = 4$
8. $e^{2-y} = 12$
9. $2e^{5a+2} = 8$
10. $10e^{3z-7} = 5$
11. $2^x = -3$
12. $(1/4)^p = -4$
13. $100(1 + .02)^{3+n} = 150$
14. $500(1 + .05)^{p/4} = 200$
15. $2^{x^2-1} = 12$
16. $3^{2-x^2} = 4$
17. $2(e^x + 1) = 10$
18. $5(e^{2x} - 2) = 15$
19. $\log (t - 1) = 1$
20. $\log q^2 = 1$
21. $\log (x - 3) = 1 - \log x$
22. $\log (z - 6) = 2 - \log (z + 15)$
23. $\ln (y + 2) = \ln (y - 7) + \ln 4$
24. $\ln p - \ln (p + 1) = \ln 5$
25. $\ln (5 + 4y) - \ln (3 + y) = \ln 3$
26. $\log_4 (z + 3) + \log_4 (z - 3) = 1$
27. $\log_3 (a - 3) = 1 + \log_3 (a + 1)$
28. $\log w + \log (3w - 13) = 1$
29. $\log_2 \sqrt{2y^2 - 1} = 1/2$
30. $5^{\log_5 (x+1)} = 9$
31. $\log_e e^{(x^3-2)} = 6$
32. $\log_x (5x - 6) = 2$
33. $7^{x \log_7 2} = 8$
34. $\log_2 (\log_2 x) = 1$
35. $\log z = \sqrt{\log z}$
36. $\log x^2 = (\log x)^2$

37. The amount of a radioactive specimen present at time t (measured in seconds) is $A(t) = 5000(10)^{-.02t}$, where $A(t)$ is measured in grams. Find the half-life of the specimen, that is, the time it will take for exactly half the specimen to remain.

A large cloud of radioactive debris from a nuclear explosion has floated over the Pacific Northwest, contaminating much of the hay supply. Consequently, farmers in the area are concerned that the cows who eat this hay will give contaminated milk. The percent of the initial amount of radioactive iodine still present in the hay after t days is approximated by $P(t) = 100 e^{-.1t}$, *where t is time measured in days.*

38. Some scientists feel that the hay is safe after the level of radioactive iodine has declined to 10% of the original amount. Find the number of days before the hay could be used.

39. Other scientists believe that the hay is not safe until the level of radioactive iodine has declined to 1% of the original amount. Find the number of days this would take.

Solve the following equations for x. (Hint: In Exercises 42–45 multiply by e^x.)

40. $2^{2x} - 3 \cdot 2^x + 2 = 0$

41. $5^{2x} + 2 \cdot 5^x - 3 = 0$

42. $e^x - 5 + 6e^{-x} = 0$

43. $e^x - 6 + 5e^{-x} = 0$

44. $y = \dfrac{e^x + e^{-x}}{2}$

45. $y = \dfrac{e^x - e^{-x}}{2}$

Solve each of the following equations for the indicated variables. Use logarithms to the appropriate bases.

46. $P = P_0 e^{kt/1000}$, for t

47. $I = \dfrac{E}{R}(1 - e^{-Rt/2})$, for t

48. $T = T_0 + (T_1 - T_0) 10^{-kt}$, for t

49. $A = \dfrac{Pr}{1 - (1 + r)^{-t}}$, for t

50. $(\log_{10} x) - y = \log_{10}(3x - 1)$, for x

51. $\log_{10}(x - y) = \log_{10}(3x - 1)$, for x

Find the formula, domain, and range of $f^{-1}(x)$.

52. $f(x) = \log_5(x - 2)$

53. $f(x) = \log_{10}(2x + 6)$

54. $f(x) = 5e^{2x+6}$

55. $f(x) = 3 \cdot 10^{7-x}$

Solve the inequalities in Exercises 56–61 for x. The inequalities in Exercises 60 and 61 are studied in calculus to determine where certain functions are increasing.

56. $\log_2 x < 4$

57. $\log_3 x > -1$

58. $\log_x 16 < 4$

59. $\log_x .1 < -1$

60. $xe^x + e^x > 0$

61. $\dfrac{e^x - xe^x}{e^{2x}} > 0$

62. Recall (from Exercise 70 of Section 5.2) the formula for the decibel rating of a sound, $d = 10 \log(I/I_0)$. Solve this formula for I.

63. Solve the formula for compound interest, $A = P\left(1 + \dfrac{r}{n}\right)^{nt}$, for t.

64. Use natural logarithms to solve $A = Pe^{rt}$ for t.

65. The growth of bacteria in food products makes it necessary to time-date some products (such as milk) so that they will be sold and consumed before the bacteria count is too high. Suppose for a certain product that the number of bacteria present is given by $f(t) = 500e^{.1t}$, under certain storage conditions, where t is time in days after packing of the product and the value of $f(t)$ is in millions. Find the number of bacteria present at each of the following times: **(a)** 2 days, **(b)** 1 week, **(c)** 2 weeks. **(d)** Suppose the product cannot be safely eaten after the bacteria count reaches 3,000,000,000. How long will this take? **(e)** If $t = 0$ corresponds to January 1, what date should be placed on the product?

5.5 EXPONENTIAL GROWTH AND DECAY

In many situations that occur in biology, economics, and the social sciences, a quantity changes at a rate proportional to the amount present. In such cases the amount present at time t is a function of t called the **exponential growth function.**

Exponential Growth Function

Let y_0 be the amount or number present at time $t = 0$. Then, under certain conditions, the amount present at any time t is given by

$$y = y_0 e^{kt},$$

where k is a constant.

In this section we see how to determine the constant k of the function from given data. When $k > 0$, the function describes growth; when $k < 0$, the function describes decay. Radioactive decay is an important application; it has been shown that radioactive substances decay exponentially—that is, in the function

$$y = y_0 e^{kt},$$

k is a negative number.

EXAMPLE 1   If 600 grams of a radioactive substance are present initially and 3 years later only 300 grams remain, how much of the substance will be present after 6 years?

To express the situation as an exponential equation,

$$y = y_0 e^{kt},$$

we first find y_0 and then find k. From the statement of the problem, $y = 600$ when $t = 0$ (that is, initially), so

$$600 = y_0 e^{k(0)}$$
$$600 = y_0,$$

giving the exponential decay equation

$$y = 600 \, e^{kt}.$$

Since there are 300 grams after 3 years, use the fact that $y = 300$ when $t = 3$ to find k.

$$300 = 600 \, e^{3k}$$
$$\frac{1}{2} = e^{3k}$$

Take natural logarithms on both sides, then solve for k.

$$\ln \frac{1}{2} = \ln e^{3k}$$

$$\ln .5 = 3k \qquad \text{Since } \ln e^x = x$$

$$\frac{\ln .5}{3} = k$$

$$k \approx -.231,$$

giving $\qquad y = 600e^{-.231t}$

as the exponential decay equation. To find the amount present after 6 years let $t = 6$.

$$y = 600 \, e^{-.231(6)} \approx 600 \, e^{-1.386} \approx 150$$

After 6 years, about 150 grams of the substance remain.   ●

As mentioned in the exercises for the previous section, the *half-life* of a radioactive substance is the time it takes for half of a given amount of the substance to decay.

· · · · · · · ·

EXAMPLE 2   The amount in grams of a certain radioactive substance at time t is given by

$$y = y_0 \, e^{-.1t},$$

where t is in days. Find the half-life of the substance.

Find the time t when y will equal $y_0/2$. That is, solve the equation

$$\frac{y_0}{2} = y_0 \, e^{-.1t}$$

for t. Divide both sides by y_0.

$$\frac{1}{2} = e^{-.1t}$$

Now take natural logarithms on both sides and solve for t.

$$\ln \frac{1}{2} = \ln e^{-.1t}$$

$$\ln \frac{1}{2} = -.1t \qquad \text{Using } \ln e^x = x$$

$$t = \frac{-\ln 1/2}{.1}$$

From Table 1 or a calculator, $\ln 1/2 = \ln .5 = -.6931$, so that

$$t \approx 6.9 \text{ days.}   ●$$

· · · · · · · · ·

EXAMPLE 3   Carbon 14, also known as radiocarbon, is a radioactive form of carbon that is found in all living plants and animals. After a plant or animal dies, the radiocarbon disintegrates with a half-life of 5600 years. Scientists can determine the age of the remains by comparing the amount of radiocarbon with the amounts present in living plants and animals. This technique is called *carbon dating*. The amount of radiocarbon present after t years is given by

$$y = y_0 e^{-(\ln 2)(1/5600)t},$$

where y_0 is the amount present in living plants and animals.

(a) Verify the formula for $t = 5600$.

Substitute 5600 for t. Then

$$y = y_0 e^{-(\ln 2)(1/5600)5600} = y_0 e^{-\ln 2} = y_0(e^{\ln 2})^{-1} = y_0 2^{-1} = \frac{1}{2}y_0.$$

This result is correct. Since 5600 years is the half-life of carbon 14, half of the initial amount should remain after 5600 years.

(b) A round table hanging in Winchester Castle (in England) was alleged to belong to King Arthur, who lived in the 5th century. A chemical analysis recently showed that the table had 91% of the amount of radiocarbon present in living wood. How old is the table?

The amount of radiocarbon present in the round table after t years is $.91y_0$. Therefore, in the equation

$$y = y_0 e^{-(\ln 2)(1/5600)t}$$

replace y with $.91y_0$ and solve for t.

$$.91y_0 = y_0 e^{-(\ln 2)(1/5600)t}$$
$$.91 = e^{-(\ln 2)(1/5600)t}$$
$$\ln .91 = \ln e^{-(\ln 2)(1/5600)t}$$
$$\ln .91 = -(\ln 2)(1/5600)t$$
$$t = \frac{(5600)\ln .91}{-\ln 2} \approx 760$$

The table is about 760 years old and therefore could not have belonged to King Arthur.   ●

The compound interest formula

$$A = P\left(1 + \frac{r}{n}\right)^{nt}$$

was discussed in Section 1 of this chapter. The table presented there shows that increasing the frequency of compounding makes smaller and smaller differences in the amount of interest earned. In fact, it can be shown that even if interest is compounded at intervals of time as small as one chooses (such as each hour, each minute, or each second), the total amount of interest earned will be only slightly more than for daily compounding. This is true even for a process called **continuous compounding,** which can be described loosely as compounding every instant. As suggested in Section 1, the value of the expression $(1 + 1/n)^n$ approaches e as n gets larger. Because of this, the formula for continuous compounding involves the number e.

Continuous Compounding

> If P dollars is deposited at a rate of interest r compounded continuously for t years, the final amount on deposit is
>
> $$A = Pe^{rt}$$
>
> dollars.

EXAMPLE 4   Suppose $5000 is deposited in an account paying 8% compounded continuously for five years. Find the total amount on deposit at the end of five years.

Let $P = 5000$, $t = 5$, and $r = .08$. Then

$$A = \mathbf{5000e^{.08(5)}} = 5000e^{.4}.$$

From Table 1 or a calculator, $e^{.4} \approx 1.49182$, and

$$A = 5000(1.49182) = 7459.10,$$

or $7459.10. Check that daily compounding would have produced a compound amount about 30¢ less.   ●

EXAMPLE 5   How long will it take for the money in an account that is compounded continuously at 8% interest to double?

Use the formula for continuous compounding, $A = Pe^{rt}$, to find the time t that makes $A = 2P$. Substitute $2P$ for A and $.08$ for r; then solve for t.

$$A = Pe^{rt}$$
$$2P = Pe^{.08t}$$
$$2 = e^{.08t}$$

Taking natural logarithms on both sides gives

$$\ln 2 = \ln e^{.08t}.$$

Use the property $\ln e^x = x$ to get $\ln e^{.08t} = .08t$, and

$$\ln 2 = .08t$$
$$\frac{\ln 2}{.08} = t$$
$$8.664 = t.$$

It will take about 8 2/3 years for the amount to double.   ●

5.5 Exercises ·

1. A population of lice is growing exponentially. After 2 months the population has increased from 100 to 125. In how many months will the population reach 500 lice?

2. A population of bacteria in a culture is increasing exponentially. The original culture of 25,000 bacteria contains 40,000 bacteria after 10 hours. How long will it be until there are 60,000 bacteria in the culture?

3. A radioactive substance is decaying exponentially. The substance is reduced from 800 grams to 400 grams after 4 days. How much remains after 6 days?

4. When a bactericide is introduced into a culture of bacteria, the number of bacteria decreases exponentially. After 9 hours there are only 20,000 bacteria. In how many hours will the original population of 50,000 bacteria be reduced to half?

5. The amount of a certain chemical that will dissolve in a solution increases exponentially as the temperature is increased. At 0°C 1 gram dissolved and at a temperature of 10°C 11 grams dissolved. At what temperature will 15 grams dissolve?

6. The amount of a certain radioactive specimen present at time t (in days) decreases exponentially. If 5000 grams decreased to 4000 grams in 5 days, find the half-life of the specimen.

Exercises 7–9 refer to the carbon dating process of Example 3 in the text.

7. Suppose an Egyptian mummy is discovered in which the amount of carbon 14 is only half of the original amount. About how long ago did the Egyptian die?

8. If an object contains 1/4 of the amount of carbon 14 as it originally did, how old is the object? How old if the amount is 1/8?

9. The Lascaux caves of France contain prehistoric paintings of animals. Charcoal found in these caves contains only about 15% of the amount of carbon 14 in living trees. Estimate the age of the paintings.

Nuclear energy derived from radioactive isotopes can be used to supply power to space vehicles. Suppose that the output of the radioactive power supply for a certain satellite is $P(t) = 30 \, e^{-t/250}$ watts, where t is the time in days.*

10. How much power will be available at the end of one year? (Ignore leap year.)

11. How long will it take for the power to drop to half its original strength?

12. The equipment aboard the satellite requires 10 watts of power to operate properly. What is the operational life of the satellite?

Find each of the amounts in Exercises 13–18, assuming continuous compounding.

13. $2,000 at 8% for 1 year   14. $2,000 at 8% for 5 years   15. $12,700 at 10% for 3 years

16. $175.25 at 11% for 8 years   17. $5,800 at 13% for 6 years   18. $10,000 at 12% for 3 years

19. Assuming an inflation rate of 5% compounded continuously, how long will it take for prices to double?

20. In Exercise 19, how long will it take if the inflation rate is 9%?

*Bernice Kastner, Ph.D., SPACEMATHEMATICS. National Aeronautics and Space Administration (NASA), 1985.

Solve $A = Pe^{rt}$ for the given variable.

21. r **22.** t

Newton's law of cooling says that the rate at which a body cools is proportional to the difference in temperature between the body and the environment into which it is introduced. The temperature $f(t)$ of the body at time t in appropriate units after being introduced into an environment having constant temperature T_0 is $f(t) = T_0 + Ce^{-kt}$, where C and k are constants. Use this result in Exercises 23–26.

23. A piece of metal is heated to 300°C and then placed in a cooling liquid at 50°C. After 4 minutes, the metal has cooled to 175°C. Find its temperature after 12 minutes.

24. Boiling water, at 100°C, is placed in a freezer at 0°C. The temperature of the water is 50°C after 24 minutes. Find the temperature of the water after 96 minutes.

25. A volcano discharges lava at 800°C. The surrounding air has a temperature of 20°C. The lava cools to 410°C in five hours. Find its temperature after 15 hours.

26. Min refuses to drink coffee cooler than 95°F. She makes coffee with a temperature of 170°F in a room with a temperature of 70°F. The coffee cools to 120°F in 10 minutes. What is the longest time she can let the coffee sit before she drinks the coffee?

Many environmental situations place effective limits on the growth of the number of an organism in an area. Many such limited growth situations are described by the logistic function, *defined by*

$$G(t) = \frac{mG_0}{G_0 + (m - G_0)e^{-kmt}}$$

where G_0 is the initial number present, m is the maximum possible size of the population, and k is a positive constant. Assume $G_0 = 1000$, $m = 2500$, $k = .0004$, and t is time in decades (10-year periods).

27. Find $G(1)$.   **28.** Find $G(2)$.

29. When will the population double?   **30.** When will the population reach 5000?

5 | CHAPTER SUMMARY

Key Words

5.1 exponential function
compound interest
5.2 logarithm
logarithmic function

5.3 common logarithms
natural logarithms
5.5 exponential growth function
continuous compounding

Review Exercises

Graph each of the functions defined as follows.

1. $f(x) = 2^x$

2. $f(x) = 2^{-x}$

3. $f(x) = (1/2)^{x+1}$

4. $f(x) = 4^x + 4^{-x}$

5. $f(x) = (x - 1)e^{-x}$

6. $f(x) = \log_2 (x - 1)$

7. $f(x) = \log_3 2x$

8. $f(x) = \ln x^{-2}$

Solve each of the following equations.

9. $8^p = 32$

10. $9^{2y-1} = 27^y$

11. $\dfrac{8}{27} = b^{-3}$

12. $\dfrac{1}{2} = \left(\dfrac{b}{4}\right)^{1/4}$

The amount of a certain radioactive material, in grams, present after t days, is given by $A(t) = 800e^{-.04t}$. Find A(t) for the following values of t.

13. $t = 0$

14. $t = 5$

How much would $1200 amount to at 10% compounded continuously for the following number of years?

15. 4 years

16. 10 years

17. Historically, the consumption of electricity has increased at a continuous rate of 6% per year. If it continued to increase at this rate, find the number of years before exactly twice as much electricity would be needed.

18. Suppose a conservation campaign together with higher rates caused demand for electricity to increase at only 2% per year. (See Exercise 17.) Find the number of years before twice as much electricity would be needed as is needed today.

Write each of the following expressions in logarithmic form.

19. $2^5 = 32$

20. $100^{1/2} = 10$

21. $(1/16)^{1/4} = 1/2$

22. $(3/4)^{-1} = 4/3$

23. $10^{.4771} = 3$

24. $e^{2.4849} = 12$

25. $e^1 = 1.1052$

26. $2^{2.322} = 5$

In Exercises 27–30, write each logarithm in exponential form.

27. $\log_{10} .001 = -3$

28. $\log_2 \sqrt{32} = 5/2$

29. $\log 3.45 = .537819$

30. $\ln 45 = 3.806662$

31. What is the base of the logarithmic function whose graph contains the point $(81, 4)$?

32. What is the base of the exponential function whose graph contains the point $(-3, 1/8)$?

Use properties of logarithms to write each of the following as a sum, difference, or product of logarithms. Assume all variables represent positive real numbers.

33. $\log_3 \dfrac{mn}{p}$

34. $\log_2 \dfrac{\sqrt{5}}{3}$

35. $\ln x^2 y^4 \sqrt[5]{m^3 p}$

36. $\log_7 (7k + 5r^2)$

Find each of the following logarithms to four decimal places.

37. $\log_{10} 8.47$

38. $\log_{10} .00421$

39. $\log_{10} 1050$

40. $\log_{10} 69,800$

41. $\ln e^{-5.3}$

42. $\ln e^{.04}$

43. $\ln 89$

44. $\ln .000050$

45. $\ln 8$

46. $\log_{3.4} 15.8$

47. $\log_{1/2} 9.45$

48. $\log_3 769$

Solve each of the following equations. Round to the nearest thousandth.

49. $5^r = 11$

50. $10^{2r-3} = 17$

51. $e^{p+1} = 10$

52. $(1/2)^{3k+1} = 3$

53. $6^{2-m} = 2^{3m+1}$

54. $4(1 + e^x) = 8$

55. $\log_{64} y = 1/3$

56. $\log_2 (y + 3) = 5$

57. $\ln 6x - \ln (x + 1) = \ln 4$

58. $\log_{16} \sqrt{x + 1} = 1/4$

59. $\log (3p - 1) = 1 - \log p$

60. $\log_2 (2b - 1)^2 = 4$

Solve for the indicated variable.

61. $y = \dfrac{5^x - 5^{-x}}{2}$, for x

62. $y = \dfrac{1}{2(5^x - 5^{-x})}$, for x

63. $2 \log_a (x - 2) = 1 + \log_a (x + 1)$, for x

64. $r = r_0 e^{nt}$, for t

65. $N = a + b \ln \dfrac{c}{d}$, for c

66. $P = \dfrac{k}{1 + e^{-rt}}$, for t

The height, in meters, of the members of a certain tribe is approximated by $h(t) = .5 + \log t$, where t is the tribe member's age in years, and $1 \leq t \leq 20$. Find the height of a tribe member of the ages in Exercises 67–70.

67. 2 years

68. 5 years

69. 10 years

70. 20 years

71. A person learning certain skills involving repetition tends to learn quickly at first. Then learning tapers off and approaches some upper limit. Suppose the number of symbols per minute a keypunch operator can produce is given by $p(t) = 250 - 120(2.8)^{-.5t}$, where t is the number of months the operator has been in training. Find each of the following: **(a)** $p(2)$, **(b)** $p(4)$, **(c)** $p(10)$. **(d)** Graph $p(t)$.

72. The concentration of pollutants, in grams per liter, in the east fork of the Big Weasel River is approximated by $P(x) = .04e^{-4x}$, where x is the number of miles downstream from a paper mill that the measurement is taken. Find each of the following: **(a)** $P(.5)$, **(b)** $P(1)$, **(c)** the concentration of pollutants 2 miles downstream.

73. A population is growing according to the growth law $y = 2e^{.02t}$. Match each of the questions (a), (b), and (c) with one of the solutions (A), (B), or (C).

(a) How long will it take for the population to triple?   **(A)** Solve $2e^{.02t} = 3$ for t.

(b) When will the population reach 3 million?   **(B)** Solve $2e^{.02t} = 3 \cdot 2$ for t.

(c) How large will the population be in 3 years?   **(C)** Evaluate $2e^{.02(3)}$.

74. The population of the world is expected to double in the next 44 years. Without solving for the growth constant k, determine how much the population will increase in half that time.

75. If the world population continues to grow at the current rate, by what factor will the population grow in the next 220 years? See Exercise 74. (*Hint:* $220 = 5 \cdot 44$.)

76. Give the property that justifies each step of the following derivation. Let a be any number.

(a) (a, e^a) is on the graph of $f(x) = e^x$.

(b) (e^a, a) is on the graph of $g(x) = \ln x$.

(c) $\ln e^a = a$

77. Find the domain and range of the function defined by $f(x) = \log_2 (x^2 - x - 2)$.

If R dollars is deposited at the end of each year in an account paying a rate of interest r per year compounded annually, then after t years the account will contain a total of

$$R\left[\frac{(1 + r)^t - 1}{r}\right]$$

dollars. In Exercises 78–80, find the final amount on deposit. Use logarithms or a calculator. (Such a sequence of payments is called an annuity.*)*

78. $800, 12%, 10 years   **79.** $1500, 14%, 7 years   **80.** $375, 10%, 12 years

81. Manuel deposits $10,000 at the end of each year for 12 years in an account paying 12% compounded annually. He then puts this total amount on deposit in another account paying 10% compounded semiannually for another 9 years. Find the total amount on deposit after the entire 21-year period.

82. Scott Hardy deposits $12,000 at the end of each year for 8 years in an account paying 14% compounded annually. He then leaves the money alone with no further deposits for an additional 6 years. Find the total amount on deposit after the entire 14-year period.

83. A radioactive substance is decaying exponentially. A sample of the substance is reduced from 500 grams to 400 grams after 4 days.
   (a) How much is left after 10 days?
   (b) How long will it take for the substance to decay to 100 grams?

84. The population of a boomtown is increasing exponentially. There were 10,000 people in town when the boom began. Two years later the population had reached 12,000. Assume this growth rate continues.
   (a) What will be the population after 5 years?
   (b) How long will it take for the population to double?

85. Correct the mistakes in the following.

$$\log_5 125 - \log_5 25 = \frac{\log_5 125}{\log_5 25} = \log_5 \left(\frac{125}{25}\right) = \log_5 5 = 1$$

6 TRIGONOMETRIC FUNCTIONS

Many different types of functions have been discussed throughout this book, including linear, quadratic, polynomial, exponential, and logarithmic. This chapter introduces the trigonometric functions, which differ in a fundamental way from those previously studied: the trigonometric functions describe a *periodic* or *repetitive* relationship.

An example of a periodic relationship is shown by this electrocardiogram, a graph of the human heartbeat. The EKG shows electrical impulses from the heart. Each small square represents .04 seconds, and each large square represents .2 seconds. How often does this (abnormal) heart beat?

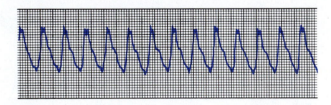

Trigonometric functions describe many natural phenomena, making them important in the study of electronics, optics, heat, X-rays, acoustics, and seismology, for example. Trigonometric functions occur again and again in calculus and are key to the study of navigation and surveying.

6.1 THE SINE AND COSINE FUNCTIONS

A circle with center at the origin and radius one unit is called a **unit circle.** Consider starting at the point $(1, 0)$ on the unit circle and letting (x, y) be any other point on the circle. Measuring in a counterclockwise direction along the arc of the circle from

(1, 0) to (x, y) gives a positive real number s as the length of this arc. There is also a negative number that gives the length of the arc from (1, 0) to (x, y), measured in a clockwise direction. An arc with $s > 0$ is shown in Figure 1.

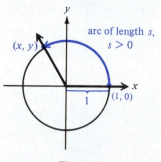

Figure 1

The circumference of a circle of radius r is given by $C = 2\pi r$. Since $r = 1$, the circumference of a unit circle is 2π, so that s may take values from 0 to 2π or from -2π to 0. The possible values of s can be extended to include all real numbers by allowing arcs which wrap around the circle more than once. For example, $s = 3\pi$ would correspond to the same point on the circle, $(-1, 0)$, as $s = \pi$ or $s = -\pi$. Figure 2 shows how s can take any real number as a value by wrapping a real number line around the unit circle.

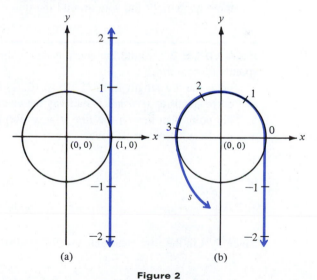

(a)   (b)

Figure 2

· · · · · · · · · ·

EXAMPLE 1   Find the coordinates of the points on the unit circle corresponding to the following arc lengths.

(a) $\pi/2$

Since $\pi/2$ is one-fourth of the circumference of the unit circle, an arc of length $\pi/2$ starting at (1, 0) ends at (0, 1). See Figure 3 on the next page.

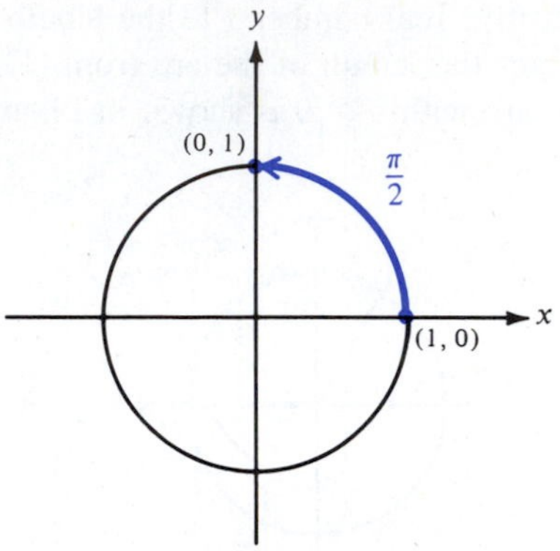

Figure 3

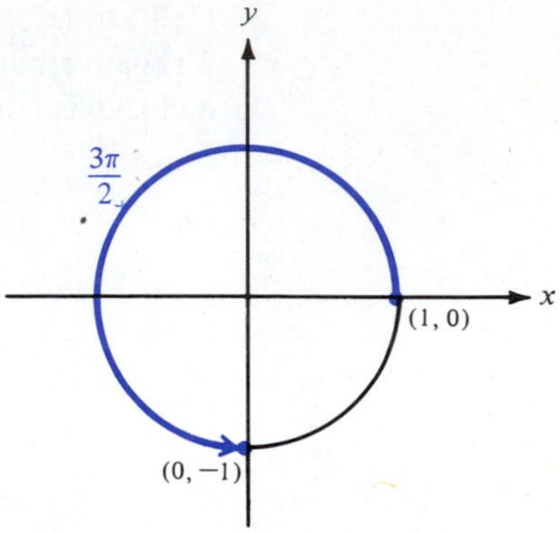

Figure 4

(b) $\dfrac{3\pi}{2}$

As Figure 4 shows, the arc of length $3\pi/2$ corresponds to $(0, -1)$. This result can be reasoned in two ways. Thinking of the length of the arc as $3(\pi/2)$, the point is located by going one-quarter of the way around the unit circle three times. Alternately, thinking of the length of the arc as $(3/4)(2\pi)$, the point is three-quarters of the way around the unit circle. ●

· · · · · · · · ·

EXAMPLE 2   Find the coordinates of the point P on the unit circle that corresponds to $s = \pi/4$.

Since the arc length from $(1, 0)$ to $(0, 1)$ is $\pi/2$, the point P on the unit circle that corresponds to $s = \pi/4$ is halfway between $(1, 0)$ and $(0, 1)$, as shown in Figure 5. This point also lies on the line $y = x$. Replacing y with x in the equation of the unit circle, $x^2 + y^2 = 1$, gives

$$x^2 + x^2 = 1,$$

or
$$2x^2 = 1$$

$$x^2 = \frac{1}{2}.$$

Since P is in the first quadrant, use the positive square root. This gives

$$x = \frac{1}{\sqrt{2}} = \frac{\sqrt{2}}{2}.$$

Also, since $y = x$,
$$y = \frac{\sqrt{2}}{2}.$$

The point P on the unit circle that corresponds to $s = \pi/4$ is $(\sqrt{2}/2, \sqrt{2}/2)$. ●

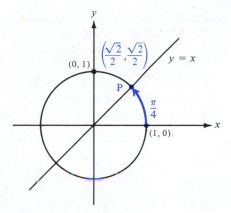

Figure 5

· · · · · · · · ·

EXAMPLE 3   Find the coordinates of the point Q on the unit circle that corresponds to each of the following.

(a) $s = 3\pi/4$

Figure 6 shows the point Q and the point P from Example 2. Since $3\pi/4 = \pi - \pi/4$, by symmetry Q has the same y-coordinate as P, but an x-coordinate with opposite sign. Thus, the coordinates of Q are $(-\sqrt{2}/2, \sqrt{2}/2)$.

(b) $s = -\pi/4$

As Figure 7 shows, this time the coordinates of Q are $(\sqrt{2}/2, -\sqrt{2}/2)$.

(c) $s = 5\pi/4$

Since $5\pi/4 = \pi + \pi/4$, use symmetry again to find that the coordinates are $(-\sqrt{2}/2, -\sqrt{2}/2)$. See Figure 8.   ●

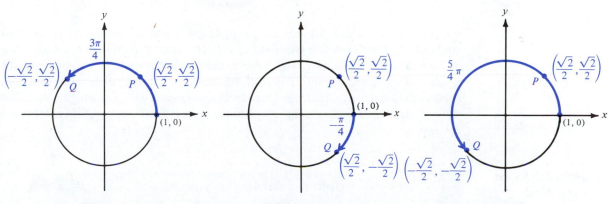

Figure 6   **Figure 7**   **Figure 8**

· · · · · · · · ·

EXAMPLE 4 Find the coordinates of the point on the unit circle that corresponds to $s = \pi/3$. Refer to Figure 9(a).

Since $\pi/3$ is $1/3$ of π, angle POR is one-third of $180°$, or $60°$. Then $\angle OPR$ is $30°$ because the angles in the triangle must have a sum of $180°$. Since OP is a radius of the unit circle, its length is 1. The length of OR is $1/2$ because triangle OPR is half of an equilateral triangle. See Figure 9(b). Since

$$x^2 + y^2 = 1,$$

$$\left(\frac{1}{2}\right)^2 + y^2 = 1$$

$$y^2 = \frac{3}{4}$$

$$y = \frac{\sqrt{3}}{2}.$$

Thus, the coordinates of P are $(1/2, \sqrt{3}/2)$. ●

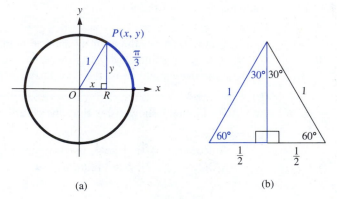

(a)   (b)

Figure 9

The correspondence between an arc of length s and a point (x, y) on the unit circle can be used to define two functions. Each value of s leads to a unique value of x and a unique value of y. The function that associates each arc length s with the y-value of the corresponding point on the unit circle is called the **sine function,** abbreviated $\sin s = y$. Also, the function that associates each arc length with the x-value of the corresponding point is called the **cosine function,** which is abbreviated $\cos s = x$.

· · · · · · · · · · · ·   ·

Definition of Sine and Cosine Functions

If s is any real number, and the point (x, y) on the unit circle corresponds to an arc of length s, then

$$\sin s = y \quad \text{and} \quad \cos s = x.$$

By these definitions, the coordinates of the point associated with an arc of length s can be written as $(\cos s, \sin s)$.

Both the sine and cosine functions have the set of real numbers as domain. The values of $\cos s$ (or x) range from 1 when $s = 0$ to -1 when $s = \pi$. Cos s is 0 when s is $\pi/2$ or $3\pi/2$. Also, sin s ranges from -1 when $s = 3\pi/2$ to 1 when $s = \pi/2$, with $\sin s = 0$ when s is 0 or π.

Because the circumference of the unit circle is 2π, adding 2π to any value of s leaves the corresponding values of $\sin s$ and $\cos s$ unchanged.

Period of Sine and Cosine

For every real number s,

$$\sin s = \sin(s + 2\pi)$$

$$\text{and} \quad \cos s = \cos(s + 2\pi).$$

Sine and cosine, which describe a cyclic relationship, are examples of **periodic functions.** For sine and cosine, the number 2π is called the **period** of the function, because it is the smallest positive number that satisfies the statements in the box. While it is also true, for example, that $\sin s = \sin(s + 4\pi)$ and $\sin s = \sin(s + 6\pi)$, the period is 2π since 2π can be shown to be the *smallest* positive number that can be used.

EXAMPLE 5   Find each of the following function values.

(a) $\sin 0$

If $s = 0$, the corresponding point is $(1, 0)$. Since $\sin s$ is the y-coordinate of the point, $\sin 0 = 0$.

(b) $\cos \pi/2$

The point corresponding to $s = \pi/2$ is $(0, 1)$. Since, by definition, $\cos s$ is the x-coordinate of the point, $\cos \pi/2 = 0$.

(c) $\cos \pi/4$

Example 2 showed that $(\sqrt{2}/2, \sqrt{2}/2)$ corresponds to $\pi/4$. From this, the value of $\cos \pi/4 = \sqrt{2}/2$. (Also, $\sin \pi/4 = \sqrt{2}/2$.) ●

While sine and cosine functions will not be graphed in detail until later, it is beneficial to sketch the basic graphs now. This is done with the definitions of sine and cosine in terms of the unit circle. A unit circle with various arc lengths s is shown in Figure 10 on the next page. The vertical dashed lines give the corresponding values of y, which is $\sin s$. Projecting horizontally gives points on the graph of $y = \sin s$. Figure 10 shows only *one period* of the graph; the complete graph would extend indefinitely to the right and to the left. (The process used to get the graph in Figure 10 could have come from measuring successive positions of bicycle pedals, just one application of sine.) A similar process can be used to obtain the graph of

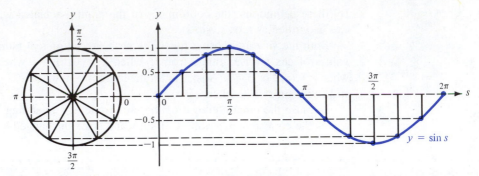

Figure 10

$y = \cos s$, shown in Figure 11. Again, the portion shown is only one period of the graph.

The following chart summarizes the intervals where sine and cosine are increasing and decreasing. The results in the chart can be found by inspecting the graphs of Figure 10 and Figure 11.

Interval	$[0, \pi/2]$	$[\pi/2, \pi]$	$[\pi, 3\pi/2]$	$[3\pi/2, 2\pi]$
Quadrant	**I**	**II**	**III**	**IV**
$\sin s$	increases from 0 to 1	decreases from 1 to 0	decreases from 0 to -1	increases from -1 to 0
$\cos s$	decreases from 1 to 0	decreases from 0 to -1	increases from -1 to 0	increases from 0 to 1

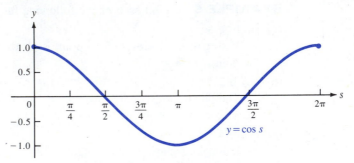

Figure 11

The word *sine* originates from the term for a half-chord. During the second century A.D., the astronomer Ptolemy was working in Alexandria at the great library where many scholars gathered. There he wrote *Almagest (The Greatest),* which included "a table of chords in a circle" in which he calculated the lengths of chords of arcs of different sizes. This list, which is essentially a table of sines, proved to be a powerful computational tool for astronomers.

The word *chord* is *jiva* in Sanskrit. The Hindu astronomers decided half-chord, or *anda-jiva,* would be more convenient. *Jiva* was transliterated into Arabic as *j-y-b* because vowels are not shown in Semitic languages. When the Latin translator Gherardo of Cremona (*circa* 1150) saw *j-y-b* he read it as *jayb* and translated it as *sinus*. The English translation of *jayb* is *bay* or *cove,* and the Latin translation of *cove* is *sinus*. The connection between a half-chord and the sine is shown in Figure 12.

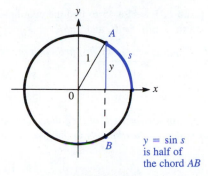

$y = \sin s$
is half of
the chord AB

Figure 12

6.1 Exercises

For each of the following arc lengths, find the coordinates of the corresponding point on the unit circle.

1. $13\pi/4$ **2.** $15\pi/4$ **3.** π **4.** -3π

5. 4π **6.** 17π **7.** -2π **8.** $-\dfrac{\pi}{2}$

9. $\dfrac{5\pi}{2}$ **10.** $\dfrac{11\pi}{2}$ **11.** $\dfrac{-11\pi}{4}$ **12.** $\dfrac{-7\pi}{4}$

13. 2.25π **14.** 1.5π **15.** $.75\pi$ **16.** -2.25π

In each of the following exercises, the point that corresponds to arc length s is given. Use symmetry to find the coordinates of the point that corresponds to   (a) $-s$,   (b) $s + 2\pi$,   (c) $s + \pi$,   (d) $\pi - s$.

17. $\left(\dfrac{2}{3}, \dfrac{\sqrt{5}}{3}\right)$ **18.** $\left(\dfrac{5}{13}, \dfrac{12}{13}\right)$ **19.** $\left(\dfrac{4}{5}, \dfrac{3}{5}\right)$ **20.** $\left(\dfrac{3}{4}, \dfrac{\sqrt{7}}{4}\right)$

21. $\left(-\dfrac{1}{2}, \dfrac{\sqrt{3}}{2}\right)$ **22.** $\left(\dfrac{\sqrt{3}}{2}, -\dfrac{1}{2}\right)$ **23.** $\left(-\dfrac{2}{5}, \dfrac{-\sqrt{21}}{5}\right)$ **24.** $\left(\dfrac{-3}{5}, \dfrac{-4}{5}\right)$

For each of the following arc lengths, find sin s and cos s.

25. 0 **26.** $\dfrac{\pi}{2}$ **27.** $\dfrac{\pi}{4}$ **28.** $\dfrac{3\pi}{4}$

29. $\dfrac{3\pi}{2}$ **30.** 987π **31.** -1423π **32.** -2π

33. $-\dfrac{\pi}{4}$ **34.** $\dfrac{-\pi}{2}$ **35.** $\dfrac{-3\pi}{4}$ **36.** $\dfrac{-7\pi}{4}$

Identify the quadrant in which arcs having the lengths in Exercises 37–44 would terminate.

37. $s = 6$ **38.** $s = 3/2$ **39.** $s = 10$ **40.** $s = 18$

41. $s = 36$ **42.** $s = 92$ **43.** $s = -896.1$ **44.** $s = -1046.001$

45. Use a 30°–60° right triangle as in Example 4 to find the coordinates of the point on the unit circle that corresponds to $\pi/6$.

46. Use symmetry and the result of Exercise 45 to find the coordinates of the points on the unit circle that correspond to $-\pi/6$, $5\pi/6$, and $7\pi/6$.

Using the results of Example 4 and Exercises 45–46, find each of the following.

47. $\sin \pi/3$ **48.** $\cos \pi/3$ **49.** $\cos 2\pi/3$ **50.** $\sin 2\pi/3$

51. $\sin 4\pi/3$ **52.** $\cos -\pi/3$ **53.** $\sin 5\pi/6$ **54.** $\cos 5\pi/6$

Find the value of x or y as appropriate for each of the following points on the unit circle. Then, assuming that the real number s corresponds to the given point, find sin s and cos s.

55. $\left(\dfrac{3}{5}, y\right), y > 0$ **56.** $\left(x, \dfrac{7}{25}\right), x > 0$

57. $\left(x, \dfrac{5}{8}\right), x < 0$ **58.** $\left(\dfrac{2}{3}, y\right), y < 0$

59. $\left(-\dfrac{1}{\sqrt{13}}, y\right), y > 0$ **60.** $\left(x, -\dfrac{2}{\sqrt{15}}\right), x < 0$

61. $\left(\dfrac{3}{\sqrt{11}}, y\right), y < 0$ **62.** $\left(x, \dfrac{3}{\sqrt{19}}\right), x > 0$

63. $\left(x, \dfrac{b}{\sqrt{a^2 + b^2}}\right), x < 0, a < 0$ **64.** $\left(\dfrac{2p}{\sqrt{4p^2 + t^2}}, y\right), y > 0, t < 0$

65. Determine which of the following numbers is closest to $\cos 2$: $.4$, $.6$, 0, $-.4$, or $-.6$.

66. Determine which of the following numbers is closest to $\sin 2$: $.9$, $.1$, 0, $-.9$, or $-.1$.

67. Explain why the sine function is odd.

68. Explain why the cosine function is even.

Determine the sign of each number in Exercises 69–72.

69. $\sin 4$ **70.** $\cos 3$ **71.** $\cos 6$ **72.** $\sin 5$

73. For what values of s does $\sin s = \cos s$?

6.2 FURTHER TRIGONOMETRIC FUNCTIONS

In the previous section, the sine and cosine functions were defined as $\sin s = y$ and $\cos s = x$, where (x, y) is the point on the unit circle corresponding to an arc of length s. Four additional functions can now be derived from these two basic functions: the **tangent**, **cotangent**, **cosecant**, and **secant** functions, abbreviated as **tan**, **cot**, **csc**, and **sec**, respectively. These six functions are called the **trigonometric functions** or the **circular functions**. (The definitions of sine and cosine are included below for reference.)

Let s be any real number and let the point (x, y) on the unit circle correspond to s. Then

$$\sin s = y \qquad\qquad \csc s = \frac{1}{y} \quad (y \neq 0)$$

$$\cos s = x \qquad\qquad \sec s = \frac{1}{x} \quad (x \neq 0)$$

$$\tan s = \frac{y}{x} \quad (x \neq 0) \qquad\qquad \cot s = \frac{x}{y} \quad (y \neq 0).$$

EXAMPLE 1   Find $\tan \pi/4$, $\cot \pi/4$, $\csc \pi/4$, and $\sec \pi/4$.

Example 2 of the last section showed that the point $(\sqrt{2}/2, \sqrt{2}/2)$ corresponds to an arc of length $s = \pi/4$, so that $x = \sqrt{2}/2$ and $y = \sqrt{2}/2$. By their definitions,

$$\tan \frac{\pi}{4} = \frac{y}{x} = \frac{\sqrt{2}/2}{\sqrt{2}/2} = 1$$

$$\cot \frac{\pi}{4} = \frac{x}{y} = \frac{\sqrt{2}/2}{\sqrt{2}/2} = 1$$

$$\csc \frac{\pi}{4} = \frac{1}{y} = \frac{1}{\sqrt{2}/2} = \frac{2}{\sqrt{2}} = \sqrt{2}$$

$$\sec \frac{\pi}{4} = \frac{1}{x} = \frac{1}{\sqrt{2}/2} = \frac{2}{\sqrt{2}} = \sqrt{2}. \quad \bullet$$

Like the sine and cosine functions, the four new trigonometric functions are periodic. The cosecant and secant functions have the same period as sine and cosine, 2π. The tangent and cotangent functions, however, have a period of π, as shown in the next chapter.

EXAMPLE 2   Find the values of the trigonometric functions for an arc of length $\pi/2$.

The point which corresponds to an arc of length $\pi/2$ is $(0, 1)$ (see Figure 13 on the next page), so that $x = 0$ and $y = 1$. Use the definitions of the various functions.

$$\sin \frac{\pi}{2} = y = 1 \qquad\qquad \cot \frac{\pi}{2} = \frac{x}{y} = \frac{0}{1} = 0$$

$$\cos \frac{\pi}{2} = x = 0 \qquad\qquad \csc \frac{\pi}{2} = \frac{1}{y} = \frac{1}{1} = 1$$

$$\tan \frac{\pi}{2} = \frac{y}{x} = \frac{1}{0} \text{ (undefined)} \qquad\qquad \sec \frac{\pi}{2} = \frac{1}{x} = \frac{1}{0} \text{ (undefined)} \quad \bullet$$

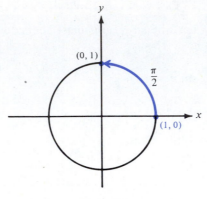

Figure 13

Several important properties of the trigonometric functions can be obtained from the definitions of the functions. First, recall that the *reciprocal* of a nonzero number a is $1/a$. (There is no reciprocal for 0.) The numbers $1/x$ and $x/1$ are reciprocals. From the definitions of trigonometric functions, $1/x$ is sec s and $x/1$ (or x) is cos s, so that cos s and sec s are reciprocals. In the same way, sine and cosecant are reciprocals as are tangent and cotangent. In summary, we have the following relationships, called the **reciprocal identities.**

Reciprocal Identities

$$\sin s = \frac{1}{\csc s} \quad \text{and} \quad \csc s = \frac{1}{\sin s}$$

$$\cos s = \frac{1}{\sec s} \quad \text{and} \quad \sec s = \frac{1}{\cos s}$$

$$\tan s = \frac{1}{\cot s} \quad \text{and} \quad \cot s = \frac{1}{\tan s}$$

These identities hold whenever the denominators are not zero.

EXAMPLE 3   Suppose tan $s = 3/4$. Find cot s.
Since cot $s = 1/\tan s$,

$$\cot s = \frac{1}{3/4} = \frac{4}{3}. \quad \bullet$$

The definitions can be used to determine the signs of the values of the trigonometric functions in each of the four quadrants. For example, if s terminates in quadrant I, then both x and y are positive, and all the trigonometric functions have positive values. In quadrant II, x is negative, and y is positive, so cos $s = x$ is negative, sin $s = y$ is positive, tan $s = y/x$ is negative, and so on. The signs of the values of the trigonometric functions in the various quadrants are summarized as follows.

Signs of Values of Trigonometric Functions in Quadrants

II	I
Sine and cosecant have positive values.	All functions have positive values.
III	IV
Tangent and cotangent have positive values.	Cosine and secant have positive values.

EXAMPLE 4 Suppose s terminates in quadrant II and $\sin s = 2/3$. Find the values of the other trigonometric functions.

Since $\sin s = y$, then $y = 2/3$. To find x, use the fact that on the unit circle $x^2 + y^2 = 1$.

$$x^2 + \left(\frac{2}{3}\right)^2 = 1$$

$$x^2 + \frac{4}{9} = 1$$

$$x^2 = \frac{5}{9}$$

$$x = \frac{\pm\sqrt{5}}{3}$$

Since s terminates in quadrant II, x must be negative, so $x = -\sqrt{5}/3$.

The values of the remaining functions can now be found from their definitions.

$$\cos s = -\frac{\sqrt{5}}{3}$$

$$\tan s = \frac{2/3}{-\sqrt{5}/3} = \frac{2}{-\sqrt{5}} = \frac{-2\sqrt{5}}{5}$$

$$\cot s = \frac{-\sqrt{5}/3}{2/3} = \frac{-\sqrt{5}}{2}$$

$$\sec s = \frac{1}{-\sqrt{5}/3} = \frac{-3}{\sqrt{5}} = \frac{-3\sqrt{5}}{5}$$

$$\csc s = \frac{1}{2/3} = \frac{3}{2} \quad \bullet$$

Several other relationships among the trigonometric functions can be derived from the definitions of these functions. The equation of the unit circle is $x^2 + y^2 = 1$; since $\cos s = x$ and $\sin s = y$,

$$(\cos s)^2 + (\sin s)^2 = 1.$$

It is customary to write $(\sin s)^2$ as $\sin^2 s$, giving $\sin^2 s + \cos^2 s = 1$. Starting with $x^2 + y^2 = 1$, divide both sides by x^2, and then by y^2, to get two additional identities: $1 + \tan^2 s = \sec^2 s$, and $\cot^2 s + 1 = \csc^2 s$. Finally, two more identities, for $\tan s$ and $\cot s$ are derived from the equalities $\cos s = x$ and $\sin s = y$. These last few identities (whose proofs are included as Exercises 71–74 below), make up the **fundamental identities.**

Fundamental Identities

$$\sin^2 s + \cos^2 s = 1 \qquad\qquad \tan s = \frac{\sin s}{\cos s} \quad (\cos s \neq 0)$$

$$1 + \tan^2 s = \sec^2 s$$

$$\cot^2 s + 1 = \csc^2 s \qquad\qquad \cot s = \frac{\cos s}{\sin s} \quad (\sin s \neq 0)$$

These relationships can be used to find all values of the trigonometric functions for a particular value of s, given the value of one function and the quadrant in which the arc of length s terminates.

EXAMPLE 5 Suppose that $\tan t = 1/4$, and t terminates in quadrant III. Find the other five trigonometric function values for t.

Since $\cot t = 1/\tan t$,

$$\cot t = \frac{1}{1/4} = 4.$$

Use the identity $1 + \tan^2 t = \sec^2 t$ and substitute $1/4$ for $\tan t$.

$$1 + \left(\frac{1}{4}\right)^2 = \sec^2 t$$

$$1 + \frac{1}{16} = \sec^2 t$$

$$\frac{17}{16} = \sec^2 t$$

The arc of length t terminates in quadrant III, where $\sec t$ has negative values, so taking square roots of both sides of $17/16 = \sec^2 t$ gives

$$-\frac{\sqrt{17}}{4} = \sec t.$$

Find $\cos t$ from the identity $\cos t = 1/\sec t$.

$$\cos t = -\frac{4\sqrt{17}}{17}$$

Finally, use $\tan t = \sin t/\cos t$ or $\sin t = (\tan t)(\cos t)$ to get

$$\sin t = \left(\frac{1}{4}\right)\left(-\frac{4\sqrt{17}}{17}\right) = -\frac{\sqrt{17}}{17}.$$

The reciprocal of $\sin t$ gives $\csc t$.

$$\csc t = 1/\sin t = -\sqrt{17} \quad \bullet$$

6.2 Exercises

For each of the following, find tan s, cot s, sec s, and csc s. (Do not use tables or a calculator.)

1. $\sin s = 1/2$, $\cos s = \sqrt{3}/2$ **2.** $\sin s = 3/4$, $\cos s = \sqrt{7}/4$ **3.** $\sin s = 4/5$, $\cos s = -3/5$

4. $\sin s = -1/2$, $\cos s = -\sqrt{3}/2$ **5.** $\sin s = -\sqrt{3}/2$, $\cos s = 1/2$ **6.** $\sin s = 12/13$, $\cos s = 5/13$

For each of the following, find the values of the six trigonometric functions. For Exercises 13–18, use the results of Example 4 and Exercises 45 and 46 of the previous section. (Do not use tables or calculator.)

7. π **8.** $3\pi/2$ **9.** $3\pi/4$ **10.** $5\pi/4$

11. $-\pi/4$ **12.** $-\pi/2$ **13.** $\pi/6$ **14.** $\pi/3$

15. $2\pi/3$ **16.** $5\pi/6$ **17.** $7\pi/6$ **18.** $4\pi/3$

Complete the following table of signs of the values of the trigonometric functions. (Do not use tables or calculator.)

	Quadrant	sin	cos	tan	cot	sec	csc
19.	I	+	+			+	+
20.	II	+		−	−		
21.	III						
22.	IV			−			−

Find the value of x or y, as appropriate, for each of the following points on the unit circle. Then, assuming that the real number s corresponds to the given point, find the six trigonometric function values for s.

23. $\left(x, \dfrac{1}{4}\right)$, $x > 0$ **24.** $\left(\dfrac{3}{8}, y\right)$, $y < 0$ **25.** $\left(-\dfrac{3}{7}, y\right)$, $y > 0$

26. $\left(x, -\dfrac{4}{9}\right)$, $x > 0$ **27.** $\left(x, -\dfrac{\sqrt{3}}{2}\right)$, $x > 0$ **28.** $\left(x, \dfrac{\sqrt{2}}{2}\right)$, $x < 0$

29. $\left(-\dfrac{1}{\sqrt{7}}, y\right)$, $y > 0$ **30.** $\left(x, -\dfrac{3}{\sqrt{11}}\right)$, $x < 0$ **31.** $\left(x, \dfrac{a}{\sqrt{a^2 + b^2}}\right)$, $x > 0$, $b > 0$

32. $\left(\dfrac{t}{\sqrt{4p^2 + t^2}}, y\right)$, $y < 0$, $p > 0$ **33.** (x, q), $x < 0$, $0 < q < 1$ **34.** (p, y), $y > 0$, $-1 < p < 0$

Decide whether each of the following statements is possible *or* impossible. *(Use the range for sine and cosine given in the previous section, and the reciprocal identities.)*

35. $\sin t = 2$

36. $\cos s = -1.001$

37. $\csc t = 2$

38. $\sec s = -1.001$

39. $\cos s + 1 = .6$

40. $\tan t - 1 = 4.2$

41. $\sin t = 1/2$ and $\csc t = 2$

42. $\cos s = 3/4$ and $\sec s = 4/3$

43. $\tan s = 2$ and $\cot s = -2$

44. $\sec t = 1/5$ and $\cos t = 2$

Decide what quadrant(s) s must terminate in to satisfy the following conditions for $0 \le s < 2\pi$.

45. $\sin s > 0$, $\cos s < 0$

46. $\cos s > 0$, $\tan s > 0$

47. $\sec s < 0$, $\csc s < 0$

48. $\tan s > 0$, $\cot s > 0$

49. $\cos s < 0$

50. $\tan s > 0$

51. $\csc s < 0$

52. $\sin s > 0$

For each of the following find the values of the other trigonometric functions.

53. $\cos s = \dfrac{-4}{9}$, s terminates in quadrant II

54. $\csc s = 2$, s terminates in quadrant II

55. $\tan s = \dfrac{3}{2}$, $\csc s = \dfrac{\sqrt{13}}{3}$

56. $\sec s = -2$, $\cot s = \dfrac{\sqrt{3}}{3}$

57. $\sin t = \dfrac{\sqrt{5}}{5}$, $\cos t < 0$

58. $\tan t = -\dfrac{3}{5}$, $\sec t > 0$

59. $\sec s = -\sqrt{7}/2$, $\tan s = \sqrt{3}/2$

60. $\csc t = -\sqrt{17}/3$, with $\cot t = 2\sqrt{2}/3$

61. $\sin s = a$, s is in quadrant I

62. $\tan t = m$, t is in quadrant III

Find a value of s between 0 *and* π *for which cos s has the same value as each of the following.*

63. $\cos \dfrac{6\pi}{5}$

64. $\cos \dfrac{8\pi}{5}$

65. $\cos -\dfrac{\pi}{4}$

66. $\cos -\dfrac{3\pi}{4}$

Determine the sign of each of the following numbers.

67. $\tan 1$

68. $\cot 2$

69. $\sec 3$

70. $\csc 4$

Use the definitions of the trigonometric functions to prove the following statements.

71. $1 + \tan^2 s = \sec^2 s$

72. $\cot^2 s + 1 = \csc^2 s$

73. $\tan s = \dfrac{\sin s}{\cos s}$, $\cos s \neq 0$

74. $\cot s = \dfrac{\cos s}{\sin s}$, $\sin s \neq 0$

75. The figure shows a quarter circle of radius 1 with FB tangent to the circle at B and EC tangent at C. Give each of the following lengths in terms of s:
(a) OA, (b) AG, (c) OD, (d) BF, (e) arc BG, (f) arc CG.
(Note: Part (d) shows how the tangent function got its name.)

76. In the figure, triangle OCD is similar to triangle GAO, so
$$\frac{OG}{OD} = \frac{AG}{CO}.$$

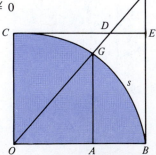

Since $OG = 1$, $CO = 1$, and $AG = \sin s$,

$$\frac{1}{OD} = \frac{\sin s}{1} \quad \text{and} \quad OD = \frac{1}{\sin s} = \csc s.$$

Give the following lengths in terms of s: **(a)** OF, **(b)** CD.

Using the fact that the cosine function is even and the sine function is odd, determine whether each of the following functions is even or odd.

77. $\csc s$ **78.** $\sec s$ **79.** $\tan s$ **80.** $\cot s$

6.3 ANGLES AND THE UNIT CIRCLE

A very basic idea in trigonometry is that of *angle*. To determine an angle, start with the top drawing of Figure 14, showing a line through the two points A and B. This line is named **line AB.** The portion of the line between A and B, including points A and B themselves, is called **line segment AB.** The portion of line AB that starts at A and continues through B, and on past B, is called **ray AB.** Point A is the endpoint of the ray.

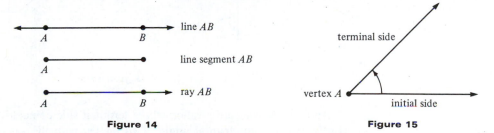

Figure 14

Figure 15

An **angle** is formed by rotating a ray around its endpoint. The initial position of the ray is the **initial side** of the angle, while the location of the ray at the end of its rotation is the **terminal side** of the angle. The endpoint of the ray is the **vertex** of the angle. Figure 15 shows the initial and terminal sides of an angle with vertex A.

If the rotation of an angle is counterclockwise, the angle is **positive.** If the rotation is clockwise, the angle is **negative.** Figure 16 shows angles of both types.

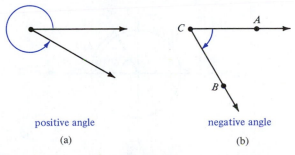

positive angle

(a)

negative angle

(b)

Figure 16

An angle can be named by using the name of its vertex. For example, the angle in Figure 16(b) can be called angle *C*. Also, an angle can be named by using three letters. For example, the angle in Figure 16(b) could be named angle *ACB*. (Put the vertex in the middle.)

An angle is in **standard position** if its vertex is at the origin of a coordinate system and its initial side is along the positive *x*-axis. The two angles in Figure 17 are in standard position. An angle in standard position is said to lie in the quadrant where its terminal side lies.

. . . .

EXAMPLE 1   Find the quadrants for the angles in Figure 17.

The angle in Figure 17(a) is a quadrant I angle. The angle in Figure 17(b) is a quadrant II angle.   ●

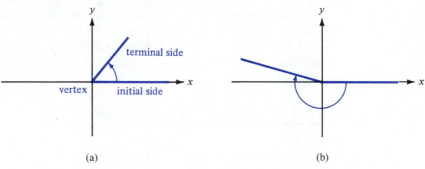

(a)   (b)

Figure 17

An angle in standard position whose terminal side coincides with the *x*-axis or *y*-axis is called a **quadrantal angle.** Two angles with the same initial side and the same terminal side, but different amounts of rotation, are called **coterminal angles.** Figure 18 shows two examples of coterminal angles.

Two systems are commonly used to measure angles, degree and radian measure. Around 100 B.C. Babylonian astronomers divided the circle into 360 equal parts that

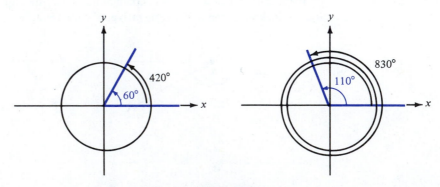

Figure 18

we call degrees. Today we use degree measure in most applications of trigonometry. To use degree measure, assign 360 degrees to the rotation of a ray through a complete circle. As Figure 19 shows, the terminal side corresponds with its initial side when it makes a complete rotation. **One degree, 1°,** represents 1/360 of a rotation. One sixtieth of a degree is called a *minute,* and one sixtieth of a minute is a *second.* The measure 12° 42′ 38″ represents 12 degrees, 42 minutes, 38 seconds.

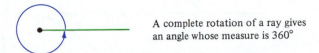

A complete rotation of a ray gives an angle whose measure is 360°

Figure 19

An angle having a measure between 0° and 90° is called an **acute angle.** An angle whose measure is exactly 90° is a **right angle.** An angle measuring more than 90° but less than 180° is an **obtuse angle,** and an angle of exactly 180° is a **straight angle.**

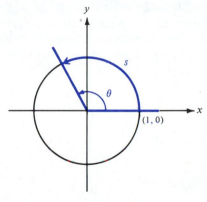

Figure 20

Degree measure is based on an arbitrary assignment of 360° to one complete rotation of a ray. In advanced work in trigonometry, and in calculus, angles are measured in a more natural system of measurement: radians. Radian measure simplifies many formulas. To see how to measure an angle in radians, start with Figure 20, which shows an angle θ (θ is the Greek letter *theta*) in standard position on a unit circle. As shown in the figure, angle θ determines an arc of positive length s on the unit circle. This real number s can be used as a measure of θ.

$$\theta = s \text{ radians}$$

This can be stated simply as $\theta = s$.

Several angles and their radian measures are shown in Figure 21.

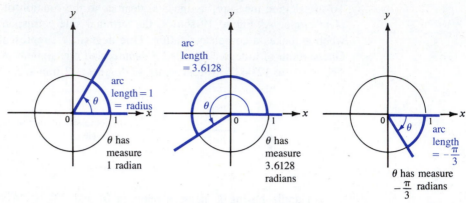

Figure 21

While the radian measure of an angle was defined in terms of a unit circle, radian measures can be found for angles from a circle of any positive radius r. To see how, start with the **central angle** (an angle whose vertex is the center of a circle) θ in Figure 22. Let θ cut an arc of length s' on the unit circle and an arc of length s on the circle of radius r, where $r > 0$. By the definition of radian measure above, $\theta = s'$. From geometry, the arc lengths s' and s have the same ratios as the radii of their respective circles, or

$$\frac{s}{s'} = \frac{r}{1},$$

from which

$$s' = \frac{s}{r}.$$

Since $\theta = s'$,

$$\theta = \frac{s}{r}.$$

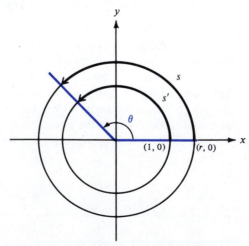

Figure 22

Radian Measure

Suppose a circle has a radius $r > 0$. Let θ be a central angle of the circle. If θ cuts off an arc of length s on the circle, then the radian measure of θ is

$$\theta = \frac{s}{r}.$$

The radian measure of an angle is the ratio of the arc length cut by the angle to the radius of the circle. In this ratio, the units of measure "divide out," leaving only a number. For this reason, radian measure is just a real number—there are no units associated with a radian measure.

EXAMPLE 2   Find the radian measure of a central angle that cuts off an arc of length 8 inches on a circle with a radius of 5 inches.

Start with $\theta = s/r$ and replace s with 8 and r with 5.

$$\theta = \frac{8}{5} = 1.6 \text{ radians} \quad \bullet$$

An angle of measure 360° would correspond to an arc that went entirely around the unit circle. This makes the radian measure of a 360° angle 2π, or

$$360° = 2\pi \text{ radians},$$

giving a basis for comparing degree measure and radian measure. Dividing both sides of this last result by 2 gives the following correspondence.

Degree-Radian Correspondence

$$180° = \pi \text{ radians}$$

Since π radians $= 180°$, divide both sides by π to find that

$$1 \text{ radian} = \frac{180°}{\pi},$$

or, approximately, 1 radian $= 57.3°$.

On the other hand, dividing the equation $180° = \pi$ radians on both sides by 180 gives

$$1° = \frac{\pi}{180} \text{ radians},$$

or, approximately, $1° = .0174533$ radians.

· · · · · · · · ·

EXAMPLE 3   Convert each of the following degree measures to radians.

(a) 45°

Since 1° = $\pi/180$ radians,

$$45° = 45\left(\frac{\pi}{180}\right) \text{ radians } = \frac{45\pi}{180} \text{ radians } = \frac{\pi}{4} \text{ radians.}$$

The word *radian* is often omitted, so the result could be written as just 45° = $\pi/4$.

(b) 240°

$$240° = 240\left(\frac{\pi}{180}\right) = \frac{4\pi}{3}. \quad \bullet$$

· · · · · · · · ·

EXAMPLE 4   Convert each of the following radian measures to degrees.

(a) $\dfrac{9\pi}{4}$

Since 1 radian = $180°/\pi$,

$$\frac{9\pi}{4} \text{ radians } = \frac{9\pi}{4}\left(\frac{180°}{\pi}\right) = 405°.$$

(b) $\dfrac{11\pi}{3}$ radians $= \dfrac{11\pi}{3}\left(\dfrac{180°}{\pi}\right) = 660°$

(c) 3 radians $= 3\left(\dfrac{180°}{\pi}\right) = \dfrac{540°}{\pi} \approx 171.9°$ $\bullet$

Many calculators will convert back and forth from radian measure to degree measure. Since most calculators work with decimal degrees rather than degrees, minutes, and seconds, we will use decimal degrees in this text.

A calculator could be used to show that 30 radians is about the same as 1719°. Figure 23 shows angles of measure 30 radians and 30 degrees; the figure shows that these angles are not at all equal in size.

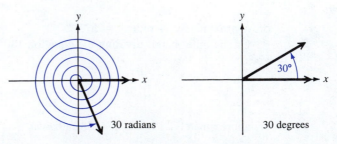

Figure 23

The relationship $\theta = s/r$, found above, gives a way to find an arc length on a circle when the central angle and radius are known. This is useful in applications as the next example shows. (In this example, $\theta = s/r$ is rewritten as $s = r\theta$.)

· · · · · · · · · ·

EXAMPLE 5   Reno, Nevada, is approximately due north of Los Angeles. The latitude of Reno is 40° N, while that of Los Angeles is 34° N. (The N means that the location is north of the equator.) If the radius of the earth is 6400 km, find the north-south distance between the two cities.

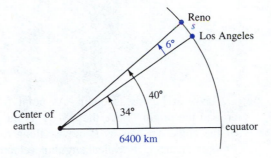

Figure 24

Latitude gives the measure of a central angle with vertex at the earth's center whose initial side goes through the earth's equator and whose terminal side goes through the location in question. As shown in Figure 24, the central angle between Reno and Los Angeles is 6°. The distance between the two cities can thus be found by the formula $s = r\theta$, after 6° is first converted to radians.

$$6° = 6\left(\frac{\pi}{180}\right) = \frac{\pi}{30} \text{ radians}$$

The distance between the two cities is

$$s = r\theta$$

$$s = 6400\left(\frac{\pi}{30}\right) \text{ km}$$

$$\approx 670 \text{ km.} \quad \bullet$$

Linear and Angular Velocity   Radian measure is very useful for discussing linear and angular velocity of a point. To see how, suppose that point P moves at a constant speed along a circle of radius r and center O. See Figure 25 on the next page. The measure of how fast the position of P is changing is called **linear velocity.** If v represents linear velocity, then

$$v = \frac{s}{t},$$

where s is the length of the arc cut by point P at time t. (This formula is just a restatement of the familiar result $d = rt$.)

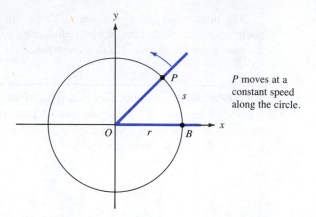

P moves at a constant speed along the circle.

Figure 25

Look at Figure 25 again. As point *P* moves along the circle, ray *OP* rotates around the origin. Since the ray *OP* is the terminal side of angle *POB*, the measure of the angle changes as *P* moves along the circle. The measure of how fast angle *POB* is changing is called **angular velocity.** Angular velocity, written ω (ω is the Greek letter *omega*), can be given as

$$\omega = \frac{\theta}{t}, \quad \theta \text{ in radians,}$$

where θ is the measure of angle *POB* at time *t*. The angle measure θ must be in radians, with ω expressed as radians per unit of time. Angular velocity is used in physics and engineering, among other applications.

As shown above, the length *s* of the arc cut on a circle of radius *r* by a central angle of measure θ radians is $s = r\theta$. Using this formula, the formula for linear velocity, $v = s/t$, becomes

$$v = \frac{r\theta}{t} \quad \text{or} \quad v = r\omega.$$

This last formula relates linear and angular velocity.

It was mentioned above that a radian is a "pure number," with no units associated with it. This is why the product of the length *r*, measured in units such as centimeters, and ω, measured in units such as radians per second, is velocity, *v*, measured in units such as centimeters per second.

· · · · · · · · ·

EXAMPLE 6   A belt runs a drive wheel of radius 5 inches at 70 revolutions per minute.

(a) Find the angular velocity of the drive wheel in radians per minute; in radians per second.

One revolution goes through 2π radians. The wheel makes 70 revolutions per minute, so the angular velocity is

$$\omega = 2\pi(70) = 140\pi \text{ radians per minute.}$$

Since there are 60 seconds in a minute, the angular velocity in radians per second is

$$\omega = \frac{140\pi}{60} = \frac{7}{3}\pi.$$

(b) Find the linear velocity of the belt in inches per second.

The linear velocity of the belt is the same as the linear velocity of a point on the circumference of the drive wheel.

$$v = r\omega = 5 \cdot \frac{7}{3}\pi = \frac{35}{3}\pi \text{ inches per second.} \quad \bullet$$

The formulas for velocity are summarized below.

Angular and Linear Velocity

Angular velocity: $\omega = \dfrac{\theta}{t}$ (ω in radians per unit time)

Linear velocity: $v = \dfrac{s}{t}$

$v = \dfrac{r\theta}{t}$ (θ in radians)

$v = r\omega$

EXAMPLE 7   Suppose that point P is on a circle with a radius of 10 centimeters, and ray OP is rotating with angular velocity of $\pi/18$ radians per second.

(a) Find the angle generated by OP in 6 seconds.

The angular velocity is $\pi/18$ radians per second. Since $\omega = \theta/t$, then in 6 seconds

$$\frac{\pi}{18} = \frac{\theta}{6},$$

or $\theta = 6(\pi/18) = \pi/3$ radians.

(b) Find the distance traveled by P along the circle in 6 seconds.

In 6 seconds P generates an angle of $\pi/3$ radians. Since $s = r\theta$,

$$s = 10\left(\frac{\pi}{3}\right) = \frac{10\pi}{3} \text{ centimeters.}$$

(c) Find the linear velocity of P.

Since $v = r\omega$, then in 6 seconds

$$v = 10 \cdot \frac{\pi}{18} = \frac{5\pi}{9} \text{ centimeters per second.} \quad \bullet$$

6.3 Exercises

Find the angles of smallest positive measure coterminal with the following angles.

1. $-40°$ **2.** $-98°$ **3.** $-125°$ **4.** $-203°$

5. $450°$ **6.** $489°$ **7.** $539°$ **8.** $699°$

Convert each of the following degree measures to radians. Leave answers as multiples of π.

9. $60°$ **10.** $30°$ **11.** $90°$ **12.** $120°$

13. $135°$ **14.** $270°$ **15.** $300°$ **16.** $390°$

17. $405°$ **18.** $20°$ **19.** $140°$ **20.** $320°$

Convert each of the following radian measures to degrees.

21. $\pi/3$ **22.** $8\pi/3$ **23.** $7\pi/4$ **24.** $2\pi/3$

25. $11\pi/6$ **26.** $15\pi/4$ **27.** $-\pi/6$ **28.** $-\pi/4$

29. 5π **30.** 7π **31.** $7\pi/20$ **32.** $17\pi/20$

Convert each of the following degree measures to radians. Round to four decimal places.

33. $139.1667°$ **34.** $174.8333°$ **35.** $64.4833°$ **36.** $85.0667°$

37. $-29.7100°$ **38.** $-157.1858°$ **39.** $-209.7708°$ **40.** $-387.0858°$

In Exercises 41–46 convert each radian measure to degrees. Write answers with four decimal places.

41. 2 **42.** 5 **43.** 1.74

44. 3.06 **45.** $.0912$ **46.** $.3417$

47. Find the angle in radians formed by the hands of a clock at 1:30.

48. Suppose that the minute hand of a clock is 2 in. long. Determine the distance traveled by the tip of the minute hand in 10.5 hr.

49. Where does the line segment from the origin to the point $(-3, 4)$ intersect the unit circle?

50. Where does the line segment from the origin to the point $(12, -3)$ intersect the unit circle?

Give the degree and radian measure for each of the following angles.

51. 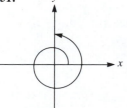   **52.** 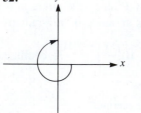   **53.** 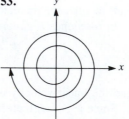   **54.**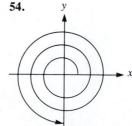

Find the measure of the central angle in radians for each of the following.

55. $r = 8$ in., $s = 12$ in.

56. $r = 18$ mm, $s = 6$ mm

57. $r = 16.4$ m, $s = 20.1$ m

58. $r = 5.80$ cm, $s = 12.3$ cm

59. $r = 1.93470$ cm, $s = 5.98421$ cm

60. $r = 294.893$ m, $s = 122.097$ m

Find the distance in miles between the following pairs of cities whose latitudes are given. Assume the cities are on a north-south line and that the radius of the earth is 4.0×10^3 mi. Give answers to two significant digits.

61. Grand Portage, Minnesota, 44° N, and New Orleans, Louisiana, 30° N

62. Farmersville, California, 36° N, and Penticton, British Columbia, 49° N

63. New York City, 41° N, and Lima, Peru, 12° S

64. Halifax, Nova Scotia, 45° N, and Buenos Aires, Argentina, 34° S

Find the arc length cut by each of the following angles.

65. $r = 8.00$ in., $\theta = \pi$ radians

66. $r = 72.0$ ft, $\theta = \pi/8$ radians

67. $r = 12.3$ cm, $\theta = 2\pi/3$ radians

68. $r = .892$ cm, $\theta = 11\pi/10$ radians

69. $r = 4.82$ m, $\theta = 60°$

70. $r = 71.9$ cm, $\theta = 135°$

Work the following exercises.

71. **(a)** How many inches will the weight on the left rise if the pulley is rotated through an angle of 71.8°?   **(b)** Through what angle must the pulley be rotated to raise the weight 6 in.?

72. Find the radius of the pulley on the right if a rotation of 51.6° raises the weight 11.4 cm.

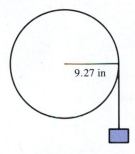

9.27 in

Exercise 71

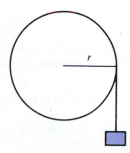

r

Exercise 72

Find the value of the indicated variable in each of the following.

73. θ, if $\omega = \pi/4$ radian per min, $t = 5$ min

74. θ, if $\omega = 2\pi/3$ radians per sec, $t = 3$ sec

75. ω, if $\theta = 2\pi/5$ radians, $t = 10$ sec

76. ω, if $\theta = 3\pi/4$ radians, $t = 8$ sec

77. t, if $\theta = 3\pi/8$ radians, $\omega = \pi/24$ radians per min

78. t, if $\theta = 2\pi/9$ radians, $\omega = 5\pi/27$ radians per min

79. v, if $r = 8$ cm, $\omega = 9\pi/5$ radians per sec

80. v, if $r = 12$ m, $\omega = 2\pi/3$ radians per sec

The formula $\omega = \theta/t$ can be rewritten as $\theta = \omega t$. Using ωt for θ changes $s = r\theta$ to $s = r\omega t$. Use this formula to find the values of the missing variables in each of the following.

81. $r = 6$ cm, $\omega = \pi/3$ radians per sec, $t = 9$ sec

82. $r = 9$ yd, $\omega = 2\pi/5$ radians per sec, $t = 12$ sec

83. $s = 6\pi$ cm, $r = 2$ cm, $\omega = \pi/4$ radians per sec

84. $s = 12\pi/5$ m, $r = 3/2$ m, $\omega = 2\pi/5$ radians per sec

85. $s = 3\pi/4$ km, $r = 2$ km, $t = 4$ sec

86. $s = 8\pi/9$ m, $r = 4/3$ m, $t = 12$ sec

87. A model airplane flies in a horizontal circle of radius 30 ft attached to a horizontal string. The airplane makes one revolution every 6 sec.
(a) Find the angular velocity of the airplane.
(b) How far does the airplane travel in 3 min?
(c) Find the linear velocity of the airplane.

88. The tires of a bicycle have a radius of 13 in. and are turning at the rate of 200 revolutions per minute. How fast is the bicycle traveling in miles per hour? (Note: There are 5280 ft in a mile.)

89. A phonograph record turns at the rate of 33 1/3 revolutions per minute. How fast is the needle moving when the record begins playing, at which time the needle is about 6 in. from the center of the record? When the needle is more than 6 in. from the center, is the needle moving faster or slower?

90. Find ω for the hour hand of a clock.

91. Find ω for the second hand of a clock.

92. Find the distance traveled in 9 sec. by a point on the edge of a circle of radius 6 cm, which is rotating through $\pi/3$ radians per sec.

93. A point on the edge of a circle travels $8\pi/9$ m in 12 sec. The radius of the circle is 4/3 m. Find ω.

94. Find the number of seconds it would take for a point on the edge of a circle to move $12\pi/5$ m, if the radius of the circle is 1.5 m and the circle is rotating through $2\pi/5$ radians per sec.

95. The earth revolves on its axis once every 24 hr. Assuming that the earth's radius is 6400 km, find the following.
(a) Angular velocity of the earth in radians per day and radians per hr
(b) Linear velocity at the North Pole or South Pole
(c) Linear velocity at Quito, Equador, a city on the equator
(d) Linear velocity at Salem, Oregon (halfway from the equator to the North Pole)

96. Eratosthenes (*circa* 230 B.C.) made a famous measurement of the earth.* He observed at Syene (the modern Aswan), at noon and at the summer solstice, that a vertical stick had no shadow, while at Alexandria (on the same meridian as Syene) the sun's rays were inclined 1/50 of a complete circle to the vertical. See the figure. He then calculated the circumference of the earth from the known distance of 5000 stades between Alexandria

---

*From *A Survey of Geometry*, Vol. I by Howard Eves. Copyright © 1963 by Allyn & Bacon, Inc.

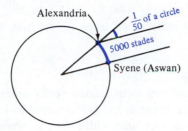

and Syene. Obtain Eratosthenes' result of 250,000 stades for the circumference of the earth. There is reason to suppose that a stade is about equal to 516.7 feet. Assuming this, calculate from the above result the polar diameter of the earth in miles. (The actual polar diameter of the earth, to the nearest mile, is 7900 miles.)

6.4 TRIGONOMETRIC FUNCTIONS OF ANGLES

The trigonometric functions defined so far have sets of real numbers for their domains. These domains can be extended to include angle measures by using the radian measure of the angle. For example, Figure 26 shows an angle with a measure θ in degrees or s in radians, and by the various definitions given so far, $\sin \theta = \sin s$, $\cos \theta = \cos s$, and so on. Thus, the trigonometric functions lead to the same function values whether the domain represents arc lengths or angle measures.

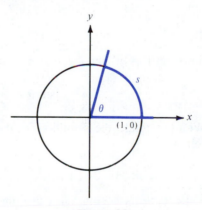

Figure 26

· · · · · · · · · ·

EXAMPLE 1   Find the following trigonometric function values.

(a) $\sin 90°$

As shown in the previous section, $90° = \pi/2$ radians. Then

$$\sin 90° = \sin \frac{\pi}{2} = 1.$$

(b) tan 45°

Since 45° $= \pi/4$ radians, and using results from Section 6.2,

$$\tan 45° = \tan \frac{\pi}{4} = 1. \quad \bullet$$

The trigonometric functions were defined earlier in terms of the coordinates of a point P on the unit circle. These definitions can now be extended so that P need not be on the unit circle. To do this, start with Figure 27 showing an angle θ in standard position, a point P' on the unit circle, and a point P that lies r units from the origin. Triangles OPQ and $OP'Q'$ are both right triangles and OP has length r. From the figure,

$$\frac{y'}{1} = \frac{y}{r} \quad \text{and} \quad \frac{x'}{1} = \frac{x}{r}.$$

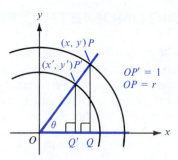

Figure 27

Since $\sin \theta = y'$ and $\cos \theta = x'$, this last result gives

$$\sin \theta = \frac{y}{r} \quad \text{and} \quad \cos \theta = \frac{x}{r}.$$

With this result, the definitions of the trigonometric functions can be generalized as follows.

Definition of Trigonometric Functions of an Angle

Let (x, y) be a point other than the origin on the terminal side of an angle θ in standard position. Let r be the distance from the origin to (x, y). Then the **trigonometric functions** of θ are defined as follows. (Assume no denominators are 0.)

$$\sin \theta = \frac{y}{r} \qquad \cos \theta = \frac{x}{r} \qquad \tan \theta = \frac{y}{x}$$

$$\csc \theta = \frac{r}{y} \qquad \sec \theta = \frac{r}{x} \qquad \cot \theta = \frac{x}{y}$$

With these new definitions of the trigonometric functions, the values of the trigonometric functions for first quadrant angles can be thought of as ratios of the sides of a right triangle. (A right triangle has a 90° angle.)

. . . .

EXAMPLE 2   The terminal side of an angle α goes through the point (8, 15). Find the values of the trigonometric functions of α.

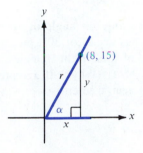

Figure 28

Figure 28 shows angle α and the triangle formed by dropping a perpendicular from the point (8, 15) to the x-axis. Since (8, 15) is on the terminal side of the angle, $x = 8$ and $y = 15$. Find r with the Pythagorean theorem.

$$r^2 = x^2 + y^2 \quad \text{or} \quad r = \sqrt{x^2 + y^2}$$

(Recall that $\sqrt{a}$ represents the *nonnegative* square root of a.) With $x = 8$ and $y = 15$,

$$r = \sqrt{8^2 + 15^2}$$
$$= \sqrt{64 + 225}$$
$$r = \sqrt{289} = 17.$$

The values of the trigonometric functions of angle α are now found by the definitions given above.

$$\sin \alpha = \frac{y}{r} = \frac{15}{17} \qquad \cos \alpha = \frac{x}{r} = \frac{8}{17} \qquad \tan \alpha = \frac{y}{x} = \frac{15}{8}$$

$$\csc \alpha = \frac{r}{y} = \frac{17}{15} \qquad \sec \alpha = \frac{r}{x} = \frac{17}{8} \qquad \cot \alpha = \frac{x}{y} = \frac{8}{15} \quad \bullet$$

The definitions of the trigonometric functions of angles, together with some results from geometry, can be used to find the values of the trigonometric functions for 30°, 45°, and 60°. The values of the trigonometric functions of 30° and 60° are found from a 30°–60° right triangle. Such a triangle can be obtained from an **equilateral triangle,** a triangle with all sides equal in length. Each angle of an equilateral triangle has a measure of 60°. See Figure 29(a) on the next page.

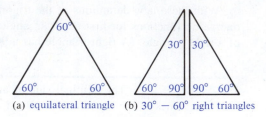

(a) equilateral triangle   (b) $30° - 60°$ right triangles

Figure 29

Bisecting one angle of an equilateral triangle gives two right triangles, each of which has angles of 30°, 60°, and 90°, as shown in Figure 29(b). If the hypotenuse of one of these right triangles has a length of 2, then the shortest side will have a length of 1. (Why?) If x represents the length of the medium side, then, by the Pythagorean theorem,

$$2^2 = 1^2 + x^2$$
$$4 = 1 + x^2$$
$$3 = x^2$$
$$\sqrt{3} = x.$$

The length of the medium side is $\sqrt{3}$.

30°–60° Right Triangles

In a 30°–60° right triangle, the hypotenuse is always twice as long as the shortest side (the side opposite the 30° angle), and the medium side is $\sqrt{3}$ times as long as the shortest side.

EXAMPLE 3   Find the trigonometric function values for 30°.

Start with a 30° angle in standard position, as shown in Figure 30. Choose a point P on the terminal side of the angle so that $r = 2$. By the work above, P will have coordinates $(\sqrt{3}, 1)$, with $x = \sqrt{3}$, $y = 1$, and $r = 2$. By the definitions of the trigonometric functions,

$$\sin 30° = \frac{1}{2} \qquad \tan 30° = \frac{\sqrt{3}}{3} \qquad \sec 30° = \frac{2\sqrt{3}}{3}$$

$$\cos 30° = \frac{\sqrt{3}}{2} \qquad \cot 30° = \sqrt{3} \qquad \csc 30° = 2. \quad \bullet$$

If you have a calculator that finds trigonometric function values at the touch of a key, you may wonder why we spend so much time in finding values for special angles. We do this because a calculator gives only *approximate* values in most cases, while we sometimes need *exact* values. For example, a calculator might give the tangent of 30° as

$$\tan 30° \approx .5773502692$$

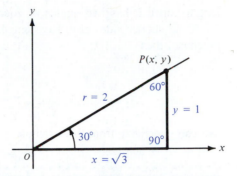

Figure 30

($\approx$ means "is approximately equal to"); however, we found the *exact* value:

$$\tan 30° = \frac{\sqrt{3}}{3}.$$

Since an exact value is frequently more useful than an approximation, you should be able to give exact values of all the trigonometric functions for the special angles.

EXAMPLE 4   Find the values of the trigonometric functions for 210°.

Draw an angle of 210° in standard position, as shown in Figure 31. Choose point *P* on the terminal side of the angle so that $r = 2$. From the 30°–60° right triangle in Figure 31, the coordinates of point *P* are $(-\sqrt{3}, -1)$. This makes $x = -\sqrt{3}$, $y = -1$, and $r = 2$, with

$$\sin 210° = -\frac{1}{2} \qquad \tan 210° = \frac{\sqrt{3}}{3} \qquad \sec 210° = -\frac{2\sqrt{3}}{3}$$

$$\cos 210° = -\frac{\sqrt{3}}{2} \qquad \cot 210° = \sqrt{3} \qquad \csc 210° = -2. \quad \bullet$$

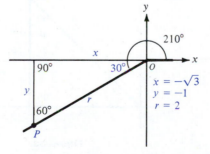

Figure 31

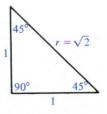

45° − 45° right triangle

Figure 32

In the first section of this chapter, we found the trigonometric function values for $\pi/4$, $3\pi/4$, $-\pi/4$, and so on, by using a unit circle. Since $\pi/4$ corresponds to an angle of 45°, we also can find these function values by using a 45°–45° right triangle, such as the one that is shown in Figure 32. This triangle has two sides of equal length (since it has two angles of equal measure) and so is an **isosceles** triangle. If the shorter sides each have length 1 and if r represents the length of the hypotenuse, then

$$1^2 + 1^2 = r^2$$
$$2 = r^2$$
$$\sqrt{2} = r.$$

We can generalize from this example.

45°–45° Right Triangles

In a 45°–45° right triangle, the hypotenuse is $\sqrt{2}$ times as long as either of the shorter sides.

EXAMPLE 5   Find the values of the trigonometric functions for 45°.

Place a 45° angle in standard position, as in Figure 33. Choose point P on the terminal side of the angle so that $r = \sqrt{2}$. Then the coordinates of P become $(1, 1)$. Using the definitions of the trigonometric functions with $x = 1$, $y = 1$, and $r = \sqrt{2}$ gives

$$\sin 45° = \frac{\sqrt{2}}{2} \qquad \tan 45° = 1 \qquad \sec 45° = \sqrt{2}$$

$$\cos 45° = \frac{\sqrt{2}}{2} \qquad \cot 45° = 1 \qquad \csc 45° = \sqrt{2}.$$

The denominators were rationalized for the values of $\sin 45°$ and $\cos 45°$.   ●

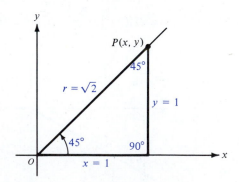

Figure 33

A summary of the trigonometric function values for certain special angles is given inside the back cover of this book.

Figure 34 shows an acute angle θ in standard position. A right triangle has been drawn. By the work above, $\sin \theta = y/r$. It is convenient to call y the length of the *side opposite* angle θ, with r the length of the *hypotenuse*. Also, x is the length of the *side adjacent* to θ. We can use these terms to restate the trigonometric functions of an acute angle.

Trigonometric Functions of an Acute Angle

If θ is an acute angle of a right triangle, then

$$\sin \theta = \frac{\text{side opposite}}{\text{hypotenuse}} \qquad \csc \theta = \frac{\text{hypotenuse}}{\text{side opposite}}$$

$$\cos \theta = \frac{\text{side adjacent}}{\text{hypotenuse}} \qquad \sec \theta = \frac{\text{hypotenuse}}{\text{side adjacent}}$$

$$\tan \theta = \frac{\text{side opposite}}{\text{side adjacent}} \qquad \cot \theta = \frac{\text{side adjacent}}{\text{side opposite}}.$$

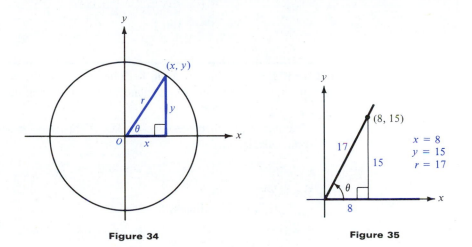

Figure 34

Figure 35

EXAMPLE 6 Find the values of $\sin \theta$, $\cos \theta$, and $\tan \theta$ for angle θ of Figure 35.

As shown in the figure, the length of the hypotenuse is 17, the length of the side opposite θ is 15, and the length of the side adjacent to θ is 8, with

$$\sin \theta = \frac{15}{17}, \quad \cos \theta = \frac{8}{17}, \quad \text{and} \quad \tan \theta = \frac{15}{8}. \quad \bullet$$

Using a Calculator Up to now, exact values of the trigonometric functions have been found only for certain special values of θ. The methods used to find these special values were based primarily on geometry. For angles whose function values cannot easily be found by geometry, approximate values can be found with a calculator or by using special tables, such as Table 3. Although we shall assume that a calculator will be used, a brief discussion of the use of Table 3 and the use of *reference angles* is given in the Appendix. (The appendix also shows how to convert decimal degrees to degrees and minutes.)

CAUTION   A common mistake with calculator use is to give an answer to an accuracy greater than is warranted by the original data. If a measurement is given to 3 significant digits, then any calculations made with it cannot have greater accuracy than that and should be rounded accordingly. The following chart shows some numbers and the number of *significant digits* in each number.

Number	Number of significant digits
29.6	3
1.39	3
.000096	2
.03	1
100.2	4

EXAMPLE 7   Find each of the following. Use the appropriate number of significant digits. Make sure the calculator is set for degree measure.

(a) sin 49.2°

Enter 49.2°, then press the *sin* key. (This may be a second function key on some calculators.)

$$\sin 49.2° \approx .757$$

(b) cos 148.5°

Enter 148.5° and press the *cos* key.

$$\cos 148.5° \approx -.8526$$

(c) sec 97.97694°

Calculators do not have *sec* keys. However,

$$\sec \theta = \frac{1}{\cos \theta}$$

for all angles θ when $\cos \theta$ is not 0°. Use this identity to find sec 97.97694° by pressing the *cos* key and then the $1/x$ key to get the reciprocal.

$$\sec 97.97694° = \frac{1}{\cos 97.97694°}$$

$$\approx -7.205933 \quad \bullet$$

EXAMPLE 8 Find each function value. Use the appropriate number of significant digits. Set the calculator in radian mode.

(a) $\tan 2.14159 \approx -1.55742$

(b) cot .5739

Find $\cot \theta$ from the identity $\cot \theta = 1/\tan \theta$. Enter .5739, press the key for tangent, then the $1/x$ key to get

$$\cot .5739 = 1.547. \quad \bullet$$

6.4 Exercises

Find the values of the six trigonometric functions for the following angles. Do not use tables or a calculator.

1. 120°
2. 135°
3. 150°
4. 225°
5. $4\pi/3$
6. $5\pi/3$
7. 330°
8. 390°
9. 420°
10. 495°
11. 510°
12. 570°
13. π
14. $3\pi/2$
15. $-\pi/2$
16. $-\pi$

Complete the following table. Do not use tables or a calculator.

	θ in degrees	θ in radians	$\sin\theta$	$\cos\theta$	$\tan\theta$	$\cot\theta$	$\sec\theta$	$\csc\theta$
17.	30°	$\pi/6$	1/2	$\sqrt{3}/2$			$2\sqrt{3}/3$	2
18.	45°	$\pi/4$			1	1		
19.	60°	$\pi/3$		1/2	$\sqrt{3}$		2	
20.	120°	$2\pi/3$	$\sqrt{3}/2$		$-\sqrt{3}$			$2\sqrt{3}/3$
21.	135°	$3\pi/4$	$\sqrt{2}/2$	$-\sqrt{2}/2$			$-\sqrt{2}$	$\sqrt{2}$
22.	150°	$5\pi/6$		$-\sqrt{3}/2$	$-\sqrt{3}/3$			2
23.	210°	$7\pi/6$	$-1/2$		$\sqrt{3}/3$	$\sqrt{3}$		-2
24.	240°	$4\pi/3$	$-\sqrt{3}/2$	$-1/2$			-2	$-2\sqrt{3}/3$

Find the values of the six trigonometric functions for θ where the point given below is on the terminal side of angle θ in standard position.

25. $(-3, 4)$
26. $(-4, -3)$
27. $(24, 7)$
28. $(-7, 24)$
29. $(-9, -12)$
30. $(5, -12)$
31. $(2\sqrt{2}, -2\sqrt{2})$
32. $(-2\sqrt{2}, 2\sqrt{2})$
33. $(\sqrt{5}, -2)$
34. $(-\sqrt{7}, \sqrt{2})$
35. $(-\sqrt{13}, \sqrt{3})$
36. $(-\sqrt{11}, -\sqrt{5})$

37. The terminal side lies on the line $y = 5x$, with $x < 0$.

38. The terminal side lies on the line $y + 3x = 0$, with $x > 0$.

Find the values of the six trigonometric functions of θ in the following right triangles.

39.

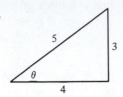

40.

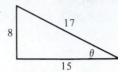

Evaluate each of the following. (Do not use tables or calculator.)

41. $\sin^2 120° + \cos^2 120°$

42. $\cot^2 135° - \sin 30° + 4 \tan 45°$

43. $\sin^2 225° - \cos^2 270° + \tan 60°$

44. $\cot^2 90° - \sec^2 180° + \csc^2 135°$

45. $\cos^2 60° + \sec^2 150° - \csc^2 210°$

46. $\cot^2 135° + \tan^4 60° - \sin^4 180°$

Answer true *or* false *for each of the following.*

47. $\sin 30° + \sin 60° = \sin(30° + 60°)$

48. $\sin(30° + 60°) = \sin 30° \cdot \cos 60° + \sin 60° \cdot \cos 30°$

49. $\cos 60° = 2 \cos^2 30° - 1$

50. $\cos 60° = 2 \cos 30°$

51. $\sin 120° = \sin 150° - \sin 30°$

52. $\sin 210° = \sin 180° + \sin 30°$

53. $\sin 120° = \sin 180° \cdot \cos 60° - \sin 60° \cdot \cos 180°$

54. $\cos 300° = \cos 240° \cdot \cos 60° - \sin 240° \cdot \sin 60°$

55. $\cos 150° = \cos 120° \cdot \cos 30° - \sin 120° \cdot \sin 30°$

56. $\sin 120° = 2 \sin 60° \cdot \cos 60°$

Find all values of the angle θ, when 0° ≤ θ < 360°, for which the following are true.

57. $\sin \theta = \dfrac{1}{2}$

58. $\cos \theta = \dfrac{\sqrt{3}}{2}$

59. $\tan \theta = \sqrt{3}$

60. $\sec \theta = \sqrt{2}$

61. $\cos \theta = -\dfrac{1}{2}$

62. $\cot \theta = -\dfrac{\sqrt{3}}{3}$

63. $\cot \theta$ is undefined

64. $\csc \theta$ is undefined

Use a calculator to find the following values.

65. $\sin 39.33°$

66. $\cos 58.67°$

67. $\sin (-38.8°)$

68. $\csc (-168.5°)$

69. $\cos (-124.83°)$

70. $\sec (274.49°)$

71. $\sec 1.9024$

72. $\cot 3.1998$

73. $\sin 7.5833$

74. $\tan 6.4754$

75. $\cos (-4.0230)$

76. $\cot (-3.8429)$

Use a calculator with sine and tangent keys and find each of the following. (Be sure to set the machine for degree measure.) Then explain why these answers are not really "correct" if the exact value has been requested.

77. $\sin 45°$

78. $\tan 60°$

Without using a calculator or table, decide which of the following is greater.

79. $\sin 50°$ or $\tan 50°$

80. $\sin 43°$ or $\sin 44°$

81. $\tan 1$ or $\tan 2$

82. $\cos .5$ or $\cot .5$

83. $\cos 5°$ or $\cos^2 5°$

84. $\sin 33°$ or $\sec 33°$

Does there exist an angle θ with the following properties?

85. $\cos \theta = -.6$ and $\sin \theta = .8$

86. $\cos \theta = \dfrac{2}{3}$ and $\sin \theta = \dfrac{3}{4}$

Suppose that $90° < \theta < 180°$. Find the sign of each of the following.

87. $\sin 2\theta$

88. $\tan \dfrac{\theta}{2}$

89. $\cot (\theta + 180°)$

90. $\cos (-\theta)$

Find the exact value of each labeled part in each of the following figures.

91.

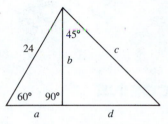

92.

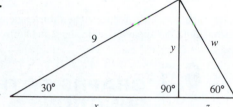

93.

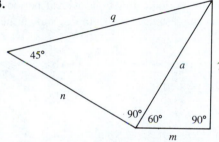

94.

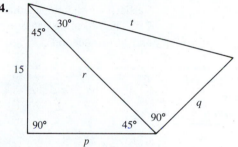

When a light ray travels from one medium, such as air, to another medium, such as water or glass, the speed of the light changes, and the direction that the ray is traveling changes. (This is why a fish under water is in a different position than it appears to be.) These changes are given by Snell's law, $c_1/c_2 = \sin \theta_1/\sin \theta_2$, where c_1 is the speed in the first medium, c_2 is the speed in the second medium, and θ_1 and θ_2 are the angles shown in the figure.

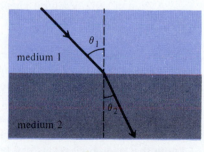

If this medium is less dense, light travels at a faster speed, c_1.

If this medium is more dense, light travels at a slower speed, c_2.

In the following exercises, assume that $c_1 = 3 \times 10^8$ m per sec. Find the speed of light in the second medium.

95. $\theta_1 = 46°, \quad \theta_2 = 31°$

96. $\theta_1 = 39°, \quad \theta_2 = 28°$

Find θ_2 for the following values of θ_1 and c_2. Round to the nearest degree.

97. $\theta_1 = 40°, \quad c_2 = 1.5 \times 10^8$ m per sec

98. $\theta_1 = 62°, \quad c_2 = 2.6 \times 10^8$ m per sec

The figure shows a fish's view of the world above the surface of the water.* Suppose that a light ray comes from the horizon, enters the water, and strikes the fish's eye.

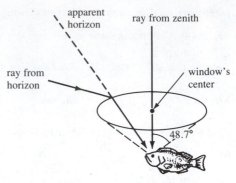

99. Let us assume that this ray gives a value of 90° for angle θ_1 in the formula for Snell's law. (In a practical situation this angle would probably be a little less than 90°.) The speed of light in water is about 2.254×10^8 m per sec. Find angle θ_2.

    (Your result should have been about 48.7°. This means that a fish sees the world above the water as a cone, making an angle of 48.7° with the vertical.)

100. Suppose that an object is located at a true angle of 29.6° above the horizon. Find the apparent angle above the horizon to a fish.

6.5 GRAPHS OF THE SINE AND COSINE FUNCTIONS

Earlier, we mentioned that sine and cosine are periodic functions with a period of 2π. The repeating nature of these functions is suggested by their graphs, shown there and in this section.

Definition of Periodic Function

A **periodic function** is a function with the property that

$$f(x) = f(x + p),$$

for every real number x in the domain of f and for some positive real number p. The smallest positive value of p is called the **period.**

Although certain natural phenomena were known to be periodic since prehistoric times, it was not until 1737 that the mathematical study of trigonometric functions was begun by Euler. In his *Introductio in analysin infinitorum* (1748), he showed by the use of identities that all values of the circular functions can be determined from the values of the functions for angles from 0° to 30°.

The sine function, defined by $y = \sin x$, has all real numbers for its domain, with range $[-1, 1]$.† Since the sine function has period 2π, we can sketch its graph by first concentrating on the values of x between 0 and 2π. Then we can repeat this portion of the graph for other values of x. Also, since the sine function is odd, we know that additional points can be found by its symmetry with respect to the origin.

As shown earlier, for x-values from 0 to $\pi/2$, $\sin x$ increases from 0 to 1, and for x-values from $\pi/2$ to π, $\sin x$ decreases from 1 back to 0. Also, $\sin x$ is negative for $(\pi, 2\pi)$. Therefore, for x-values from π to $3\pi/2$, $\sin x$ decreases from 0 to -1, and for x-values from $3\pi/2$ to 2π, $\sin x$ increases from -1 back to 0. These facts

*From "The Amateur Scientist" by Jearl Walker in *Scientific American*, March, 1984. Copyright © 1984 by Scientific American, Inc. All rights reserved. Reprinted by permission.

†In this section, x is used as the variable (as in $\sin x$), instead of s or θ to allow graphs to be drawn on the familiar xy-coordinate system.

are summarized in the table of values shown below. (The value $\sqrt{2}/2$ has been rounded to the nearest tenth.)

x	0	$\pi/4$	$\pi/2$	$3\pi/4$	π	$5\pi/4$	$3\pi/2$	$7\pi/4$	2π
$\sin x$	0	.7	1	.7	0	$-.7$	-1	$-.7$	0

By plotting the points from the table of values and connecting them with a smooth curve, we get the solid portion of the graph of Figure 36. Since the sine function is periodic and has all real numbers as domain, the graph continues in both directions indefinitely, as indicated by the dashed lines.

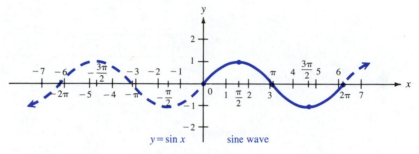

$y = \sin x$ sine wave

Figure 36

The same scale is used on both axes so as not to distort the shape of the graph. Since the period of $y = \sin x$ is 2π, it is convenient to use subdivisions of 2π on the x-axis. The more familiar x-values, 1, 2, 3, 4, and so on, are still present on the x-axis in Figure 36 but are usually not shown to avoid cluttering the graph. This graph is sometimes called a **sine wave** or **sinusoid.** You should memorize the shape of this basic graph and be able to sketch it quickly. The key points of the graph (shown in Figure 36) are $(0, 0)$, $(\pi/2, 1)$, $(\pi, 0)$, $(3\pi/2, -1)$, and $(2\pi, 0)$. Sketch the graph by plotting these five points and connecting them with the characteristic sine wave.

The graph of $y = \cos x$ can be found in the same way as the graph of $y = \sin x$. The domain of cosine is the set of all real numbers and the range of cosine is $[-1, 1]$. Cosine is an even function, so the graph will be symmetric about the y-axis. Here the key points are $(0, 1)$, $(\pi/2, 0)$, $(\pi, -1)$, $(3\pi/2, 0)$, and $(2\pi, 1)$. The graph of $y = \cos x$ has the same shape as the graph of $y = \sin x$. In fact, it is the sine wave, shifted $\pi/2$ units to the left. (See Figure 37.)

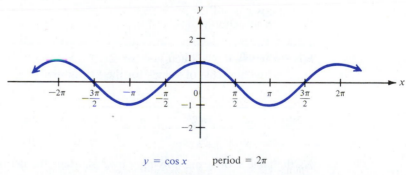

$y = \cos x$ period $= 2\pi$

Figure 37

The next few examples show graphs that are "stretched" or "compressed" either vertically or horizontally, or both, when compared with the graphs of $y = \sin x$ or $y = \cos x$.

· · · · · · · · · ·

EXAMPLE 1   Graph $y = 2 \sin x$.

For a given value of x, the value of y is twice as large as it would be for $y = \sin x$, as shown in the table of values, changing the range to $[-2, 2]$. See Figure 38, which also shows a graph of $y = \sin x$ for comparison.

x	0	$\dfrac{\pi}{2}$	π	$\dfrac{3\pi}{2}$	2π
$\sin x$	0	1	0	-1	0
$2 \sin x$	0	2	0	-2	0

●

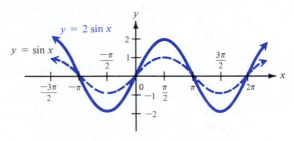

Figure 38

The method in Example 1 could be used to get the following results.

· · · · · · · · · · · · ·

Amplitude of Sine and Cosine

For $a \neq 0$, the graph of $y = a \sin x$ or $y = a \cos x$ will have the same shape as $y = \sin x$ or $y = \cos x$, respectively, except the range will be $-|a| \le y \le |a|$. The number $|a|$ is called the **amplitude.**

No matter what the value of the amplitude, the period of both $y = a \sin x$ and $y = a \cos x$ is still 2π.

· · · · · · · · · ·

EXAMPLE 2   Graph $y = \sin 2x$.

Start with a table of values for the key points.

x	0	$\pi/4$	$\pi/2$	$3\pi/4$	π	$5\pi/4$	$3\pi/2$	$7\pi/4$
$2x$	0	$\pi/2$	π	$3\pi/2$	2π	$5\pi/2$	3π	$7\pi/2$
$\sin 2x$	0	1	0	-1	0	1	0	-1

As the table shows, multiplying x by 2 shortens the period by half. The amplitude is not changed. Figure 39 shows the graph of $y = \sin 2x$. The graph of $y = \sin x$ is included again for comparison. ●

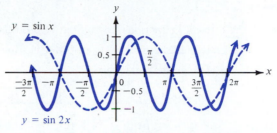

Figure 39

Example 2 suggests the following generalization. (See Exercises 61 and 62.)

Period of Sine and Cosine

> The graph of $y = \sin bx$ will look like that of $y = \sin x$, but with period $|2\pi/b|$. Also, the graph of $y = \cos bx$ looks like that of $y = \cos x$, but with period $|2\pi/b|$.

The graphs of $\sin bx$ and $\cos bx$ are compressed versions of the graphs of $\sin x$ and $\cos x$ when $|b| > 1$, and are stretched versions when $|b| < 1$.

EXAMPLE 3 Graph $y = \cos \dfrac{2}{3}x$.

The period here is $|2\pi/(2/3)| = 3\pi$. Thus, the key points will start at 0 and end at 3π. The other key points occur at $(1/4)(3\pi) = 3\pi/4$, $(1/2)(3\pi) = 3\pi/2$, and $(3/4)(3\pi) = 9\pi/4$. Since this is a cosine graph, a peak occurs at $x = 0$ and a low point in the middle of the period at $3\pi/2$. Since the amplitude is 1, the key points are $(0, 1)$, $(3\pi/4, 0)$, $(3\pi/2, -1)$, $(9\pi/4, 0)$, and $(3\pi, 1)$. Plot these points and draw the typical cosine graph through them as shown in Figure 40. ●

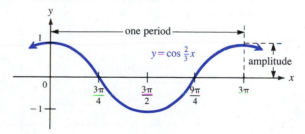

Figure 40

When graphing $y = a \sin bx$ or $y = a \cos bx$, we assume $b > 0$. If an equation has $b < 0$, the identities of the next chapter can be used to change the equation to one where $b > 0$.

Graphing Sine and Cosine

The following steps can be used to graph $y = a \sin bx$ or $y = a \cos bx$ where $b > 0$.

1. Find the period, $2\pi/b$. Start at 0 on the x-axis and lay off a distance $2\pi/b$.
2. Divide the interval from 0 to $2\pi/b$ into four equal parts.
3. Locate the x-intercepts.

Equation	x-intercepts
$y = a \sin bx$	$0, \ \dfrac{\pi}{b}, \ \dfrac{2\pi}{b}$ (beginning, middle, and end of interval)
$y = a \cos bx$	$\dfrac{\pi}{2b}, \ \dfrac{3\pi}{2b}$ (one-quarter and three-quarter points of interval)

4. Locate the points where the graph reaches maximum and minimum values.

Equation	Graph has a maximum when x is
$y = a \sin bx$	$\dfrac{\pi}{2b}$ (for $a > 0$) $\quad$ or $\quad \dfrac{3\pi}{2b}$ (for $a < 0$)
$y = a \cos bx$	0 and $\dfrac{2\pi}{b}$ (for $a > 0$) $\quad$ or $\quad \dfrac{\pi}{b}$ (for $a < 0$)

5. Use the tables or a calculator to find as many additional points as needed. Then sketch the graph.
6. Draw additional periods of the graph, to the right and to the left, as needed.

EXAMPLE 4 Graph $y = -2 \sin 3x$.

The period is $2\pi/3$. The amplitude is $|-2| = 2$. Sketch the graph using the steps shown in Figure 41. Notice how the minus sign affects the location of the maximum and minimum points. ●

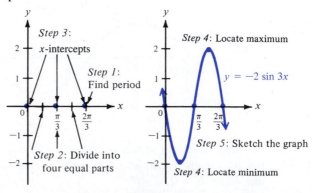

Figure 41

EXAMPLE 5   Graph $y = 3 \cos \dfrac{1}{2} x$.

The period is $2\pi/(1/2) = 4\pi$. The amplitude is 3. Perform the following steps as shown in Figure 42.

Step 1   Locate 4π on the x-axis.

Step 2   Divide the interval from 0 to 4π into 4 parts, locating key points at 0, π, 2π, 3π, and 4π.

Step 3   Since this is a cosine graph, the x-intercepts are π and 3π.

Step 4   The graph has its maximum value, 3, at 0 and 4π. The graph has its minimum value, -3, at 2π.

Step 5   Sketch the typical cosine curve through the points found in the steps above. ●

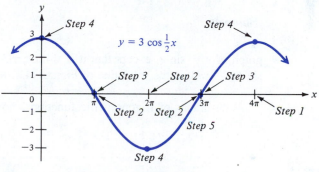

Figure 42

Translations   Recall the discussion of translations in Section 3.7. The ideas discussed there can now be applied to trigonometric functions. The graph of $y = c + a \sin bx$ or $y = c + a \cos bx$ is shifted vertically when compared with $y = a \sin bx$ or $y = a \cos bx$, respectively. The next example shows how to draw a graph involving these vertical translations.

EXAMPLE 6   Graph $y = 3 - 2 \cos 3x$.

The values of y will be 3 greater than the corresponding values of y obtained from $y = -2 \cos 3x$. This means that the graph of $y = 3 - 2 \cos 3x$ is the same as the graph of $y = -2 \cos 3x$, except with a vertical translation of 3 units upward. See Figure 43 on the next page. ●

The next few examples show sine and cosine graphs that include *horizontal* translations.

EXAMPLE 7   Graph $y = \sin \left(x - \dfrac{\pi}{3} \right)$.

Based on our work with translations in Chapter 3, this graph is the same as the

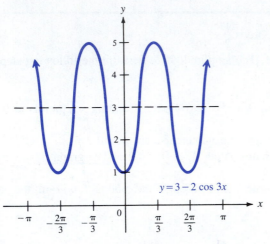

Figure 43

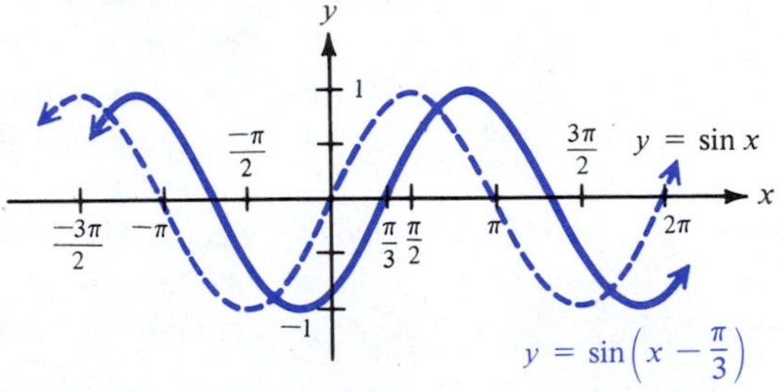

Figure 44

graph of $y = \sin x$, except that it is shifted $\pi/3$ units to the right. The graph of $y = \sin (x - \pi/3)$ is the solid line of Figure 44.   ●

With graphs of trigonometric functions, horizontal translations are called *phase shifts*.

Phase Shift

> The graph of $y = \sin (x - d)$ has the shape of the basic sine graph $y = \sin x$, but with a translation of $|d|$ units—to the left if $d < 0$ and to the right if $d > 0$. The number d is the **phase shift** of the graph.
>
> Also, $y = \cos (x - d)$ has the shape of $y = \cos x$ but is translated $|d|$ units—to the left if $d < 0$ and to the right if $d > 0$. Again, d is the phase shift.

The next example shows a graph of the form $y = a \cos b(x - d)$. Such graphs have an amplitude of $|a|$, and both a phase shift (if $d \neq 0$) and a period different from 2π (if $b \neq 1$).

EXAMPLE 8   Graph $y = -2 \cos (3x + \pi)$.

The amplitude is $|-2| = 2$, and the period is $2\pi/3$. To find the phase shift, write $-2 \cos (3x + \pi)$ in the form $a \cos b(x - d)$.

$$-2 \cos (3x + \pi) = -2 \cos 3\left(x + \frac{\pi}{3} \right) = -2 \cos 3\left[x - \left(-\frac{\pi}{3} \right) \right]$$

Since $d = -\pi/3$, the phase shift is $|-\pi/3| = \pi/3$ to the left. To draw the graph, the effect of the phase shift must be considered first. Start at $-\pi/3$ (which replaces 0 as the starting point). Since the period is $2\pi/3$, one period will end at $-\pi/3 + $

$2\pi/3 = \pi/3$. Now locate the other key points. The midpoint will be halfway between $-\pi/3$ and $\pi/3$, at 0. The one-quarter point will be at $-\pi/6$ and the three-quarter point will be at $\pi/6$. Because of the negative sign, *minimum* values of -2 occur at $-\pi/3$ and $\pi/3$, and the maximum value of 2 occurs at 0. See the graph in Figure 45.   ●

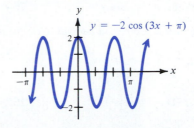

Figure 45

Sine and Cosine Graphs

The following summary is useful for work with sine and cosine graphs. Assume $b > 0$.

Function defined by	$y = c + a \sin b(x - d)$ or $y = c + a \cos b(x - d)$
Amplitude	$\lvert a \rvert$
Period	$\dfrac{2\pi}{b}$
Vertical translation	up c units if $c > 0$ down $\lvert c \rvert$ units if $c < 0$
Phase shift (horizontal translation)	d units to the right if $d > 0$ $\lvert d \rvert$ units to the left if $d < 0$

EXAMPLE 9   One example of a phase shift occurs in electrical work. A simple alternating current circuit is shown in Figure 46. The relationship between voltage V and current I in the circuit is also shown in the figure.

As this graph shows, current and voltage are *out of phase* by 90°. In this example, current *leads* the voltage by 90°, or voltage *lags* by 90°.   ●

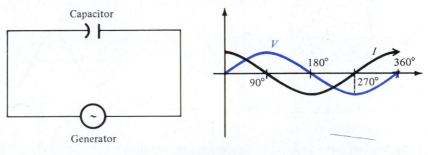

Figure 46

6.5 Exercises

Graph the following over the interval $[-2\pi, 2\pi]$. *Identify the amplitude.*

1. $y = 2 \cos x$
2. $y = 3 \sin x$
3. $y = \dfrac{2}{3} \sin x$
4. $y = \dfrac{3}{4} \cos x$

5. $y = -\cos x$
6. $y = -\sin x$
7. $y = -2 \sin x$
8. $y = -3 \cos x$

Graph each of the following over a two-period interval. Give the period, the amplitude, and any vertical translations.

9. $y = \sin \dfrac{1}{2}x$
10. $y = \sin \dfrac{2}{3}x$
11. $y = \cos \dfrac{1}{3}x$
12. $y = \cos \dfrac{3}{4}x$

13. $y = \sin 3x$
14. $y = \cos 3x$
15. $y = -\sin 4x$
16. $y = -\cos 6x$

17. $y = 2 \sin \dfrac{1}{4}x$
18. $y = 3 \sin 2x$
19. $y = -2 \cos 3x$
20. $y = -5 \cos 2x$

21. $y = \dfrac{1}{2} \sin 3x$
22. $y = \dfrac{2}{3} \cos \dfrac{1}{2}x$
23. $y = 2 - \cos x$
24. $y = 1 + \sin x$

25. $y = -3 + 2 \sin x$
26. $y = 2 - 3 \cos x$
27. $y = \cos \pi x$
28. $y = -\sin \pi x$

For the following, find the amplitude, the period, any vertical translation, and any phase shift. Draw the graph over a one-period interval.

29. $y = \cos\left(x - \dfrac{\pi}{2}\right)$
30. $y = \sin\left(x + \dfrac{\pi}{4}\right)$
31. $y = 3 \sin\left(x + \dfrac{3\pi}{2}\right)$

32. $y = 2 \cos\left(x - \dfrac{\pi}{3}\right)$
33. $y = \dfrac{3}{2} \sin 2\left(x - \dfrac{\pi}{4}\right)$
34. $y = -\dfrac{1}{2} \cos 4\left(x + \dfrac{\pi}{2}\right)$

35. $y = 3 \cos(4x + \pi)$
36. $y = -4 \sin(2x - \pi)$
37. $y = \dfrac{1}{2} \cos\left(\dfrac{1}{2}x - \dfrac{\pi}{4}\right)$

38. $y = -\dfrac{1}{4} \sin\left(\dfrac{3}{4}x + \dfrac{\pi}{8}\right)$
39. $y = -3 + 2 \sin\left(x - \dfrac{\pi}{2}\right)$
40. $y = 4 - 3 \cos(x + \pi)$

For each of the graphs in Exercises 41–44, give the equation of a sine function having that graph.

41.
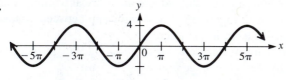

42.
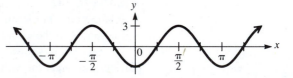

43.
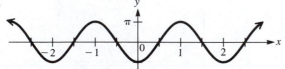

44.
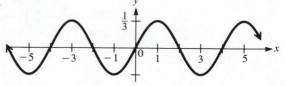

45. What is the maximum value of $y = 5 - 4 \sin 3x$?

46. What is the minimum value of $y = 2 + 3 \cos 4x$?

47. Is the function defined by $f(x) = \sin x$ even, odd, or neither?

48. Is the function defined by $f(x) = \cos x$ even, odd, or neither?

49. Graph $y = \sin\left(x - \dfrac{\pi}{2}\right)$. What trigonometric function has the same graph?

50. Graph $y = \cos\left(x - \dfrac{\pi}{2}\right)$. What trigonometric function has the same graph?

Pure sounds produce single sine waves on an oscilloscope. Find the amplitude and period of each sine wave in the following photographs. On the vertical scale, each square represents .5, and on the horizontal scale each square represents $\pi/6$.

51.
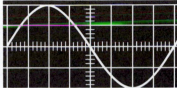

52.
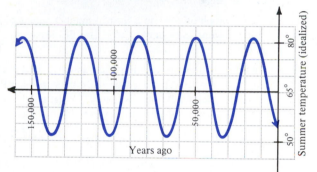

Scientists believe that the average annual temperature in a given location is periodic. The overall temperature at a given place during a given season fluctuates as time goes on, from colder to warmer, and back to colder. The graph shows an idealized description of the summer temperature for several thousand years of a location at the latitude of Anchorage.

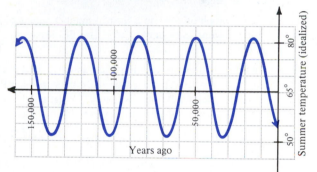

53. Find the period of this graph.

54. What is the trend of the temperature now?

Many of the activities of living organisms are periodic. For example, the graph at right shows the time that flying squirrels begin their evening activity.

55. Find the amplitude of this graph.

56. Find the period.

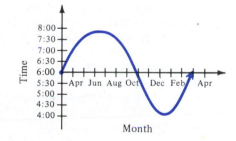

The graph shown here gives the variation in blood pressure for a typical person. Systolic and diastolic pressures are the upper and lower limits of the periodic changes in pressure which produce the pulse. The length of time between peaks is called the period of the pulse.

57. Find the amplitude of the graph.

58. Find the pulse rate for this person—the number of pulse beats in one minute.

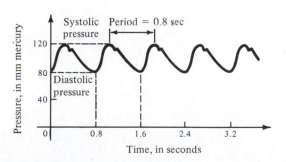

59. The voltage E in an electrical circuit is given by $E = 5 \cos 120\pi t$ where t is time measured in seconds.
(a) Find the amplitude and period.
(b) How many cycles are completed in one second? (The number of cycles completed in one second is the *frequency*.)
(c) Find E when $t = 0$; .03; .06; .09; .12.
(d) Graph E on $[0, 1/30]$.

60. For another electrical circuit, the voltage E is given by $E = 3.8 \cos 40\pi t$ where t is time measured in seconds.
(a) Find the amplitude and the period.
(b) Find the frequency. See Exercise 59(b).
(c) Find E when $t = .02, .04, .08, .12, .14$.
(d) Graph one period of E.

61. To find the period of $y = \sin bx$, where $b > 0$, first observe that as bx varies from 0 to 2π, we get one period of the graph of $y = \sin bx$. Show that x must therefore vary from 0 to $2\pi/b$, so that the period of $y = \sin bx$ is $2\pi/b$.

62. In Exercise 61, show that the period of $y = \sin bx$ is $|2\pi/b|$, no matter whether b is positive or negative.

63. Sketch the graph of $y = \sin x$ for real number values of x from 0 to .2 in increments of .02. On the same axes, draw $y = x$. Use this sketch to argue that for small values of x, $\sin x \approx x$.

64. The quotient $(\sin x)/x$ is very important in calculus. Find the value of this quotient, starting with $x = .1$. Then let x decrease by steps of .01 until $x = .01$ is reached. (Why can't we let x decrease all the way to 0?) What number does the quotient seem to be approaching as x gets closer and closer to 0?

Graph two periods of each function in Exercises 65–68.

65. $y = \sin^2 x$ **66.** $y = \sin^2 2x$ **67.** $y = \cos^2 x$ **68.** $y = 4 \cos^2 x$

69. If a bumper of a car is given a firm downward push and then released, the shock absorbers of the car cause it to bounce back and then quickly return to stable position, producing an example of *damped oscillatory motion*. Such motion often can be represented by an expression of the form $y = a \cdot e^{-kt} \cdot \sin b(t + c)$, where a, k, b, and c are constants, with $k > 0$, t represents time, and e is the base of natural logarithms discussed in Chapter 5. Graphs of $y = e^{-t}$ and $y = -e^{-t}$ for $t > 0$ are shown in the figure. Use these graphs to obtain the graph of $y = e^{-t} \cdot \sin t$.

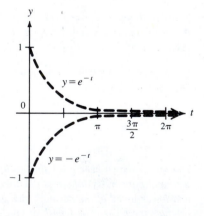

70. Motion that gets out of control can be represented by an expression of the form $y = a \cdot e^{kt} \cdot \sin b(t + c)$, with variables as given in Exercise 69. Use the graphs of $y = e^t$ and $y = -e^t$ in the figure to graph $y = e^t \cdot \sin t$.

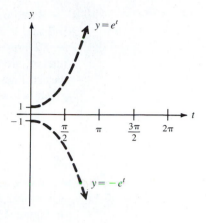

Write each of the following as the composite of two functions where the inner function is linear.

71. $y = 5 \cos (3x - \pi)$

72. $y = -\sin 2\left(x - \dfrac{\pi}{3}\right)$

73. $y = \sin^2(2x + 3)$

74. $y = \cos^2(4 - x)$

6.6 GRAPHS OF THE OTHER TRIGONOMETRIC FUNCTIONS

This section continues the discussion of graphs of the trigonometric functions, beginning with the graph of $y = \tan x$. Since $\tan x = \sin x/\cos x$, the sine and cosine functions provide useful information about the tangent function. The values of $\sin x$ and $\cos x$ are both positive in quadrant I and both negative in quadrant III, so $\tan x$ is positive in those quadrants. In quadrants II and IV, $\sin x$ and $\cos x$ have opposite signs and so $\tan x$ is negative there. As shown in the next chapter, the period of the tangent function is π. Hence,

$$\tan x = \tan (x + \pi)$$

for all x. Therefore, $y = \tan x$ need be graphed only within an interval of π units. A convenient interval for this purpose is $(-\pi/2, \pi/2)$ since $\tan x$ exists for all values in the interval. There is no point on the graph of $y = \tan x$ at $x = -\pi/2$ or $x = \pi/2$.

Since $\tan (-x) = -\tan x$, the tangent function is an odd function, with symmetry about the origin. In the interval $(0, \pi/2)$, the values of $\tan x$ are positive. As x goes from 0 to $\pi/2$, a calculator or Table 3 shows that the values of $\tan x$ get larger and larger without bound. As x goes from $-\pi/2$ up to 0, the values of $\tan x$ are negative and approach 0. These results are summarized in the following chart.

Tangent Function

As x increases from	$\tan x$
0 to $\pi/2$	increases from 0 without bound
$-\pi/2$ to 0	increases to 0

Based on these results, the graph of $y = \tan x$ approaches the vertical lines $x = \pi/2$ and $x = -\pi/2$ but never touches them, so that the lines $x = \pi/2$ and $x = -\pi/2$ are vertical asymptotes. In fact, the lines $x = \pi/2 + k\pi$, where k is an integer, are all vertical asymptotes. These asymptotes are indicated on the graph with dashed lines. (See Figure 47.) A table of values for $\tan x$, for x in $(-\pi/2, \pi/2)$, is given below.

x	$-\dfrac{\pi}{3}$	$-\dfrac{\pi}{4}$	$-\dfrac{\pi}{6}$	0	$\dfrac{\pi}{6}$	$\dfrac{\pi}{4}$	$\dfrac{\pi}{3}$
$\tan x$	-1.7	-1	$-.6$	0	$.6$	1	1.7

When we plot the points from the table and let the graph approach the asymptotes at $x = \pi/2$ and $x = -\pi/2$, we get the portion of the graph shown by the solid curve in Figure 47. More of the graph can be sketched by repeating the same curve, as shown in the figure. This basic graph, like the graphs for sine and cosine, should be memorized. Convenient key points for the tangent graph are $(-\pi/4, -1)$, $(0, 0)$, $(\pi/4, 1)$, with vertical asymptotes at $-\pi/2$ and $\pi/2$. (The idea of *amplitude*, discussed earlier, applies only to sine and cosine and is not used here.)

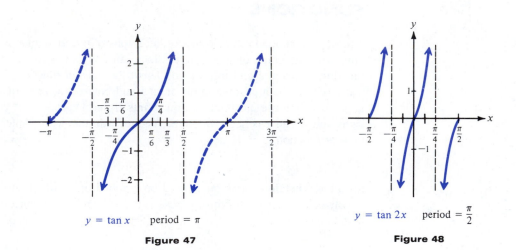

$y = \tan x$ period $= \pi$

Figure 47

$y = \tan 2x$ period $= \dfrac{\pi}{2}$

Figure 48

EXAMPLE 1   Graph $y = \tan 2x$.

Multiplying x by 2 changes the period to $\pi/2$, so one period goes from $-\pi/4$ to $\pi/4$. Key points are found by first dividing the interval $[-\pi/4, \pi/4]$ into quarters with tick marks at $-\pi/8$, 0, and $\pi/8$. The key points of the function are $(-\pi/8, -1)$, $(0, 0)$, and $(\pi/8, 1)$ with the vertical asymptotes at $-\pi/4$ and $\pi/4$. Sketching the typical tangent graph through these points gives the graph shown in Figure 48 where two periods are graphed.   ●

Example 1 suggests the following generalization.

Period of Tangent

> If $b > 0$, the graph of $y = \tan bx$ has period π/b.

The fact that $\cot x = 1/(\tan x)$ can be used to find the graph of $y = \cot x$. The period of cotangent, like tangent, is π. The domain of cotangent excludes $0 + k\pi$, where k is any integer, making the vertical lines $x = k\pi$ asymptotes. Any values of x that lead to asymptotes for $\tan x$ will make $\cot x = 0$, so that $\cot(-\pi/2) = 0$, $\cot \pi/2 = 0$, $\cot 3\pi/2 = 0$, and so on. The values of $\tan x$ increase as x goes from $-\pi/2$ to $\pi/2$. Since $\cot x = 1/(\tan x)$, the values of $\cot x$ decrease as x goes from $-\pi/2$ to 0 and from 0 to $\pi/2$. Like the graph of $y = \tan x$, the graph of $y = \cot x$ is symmetric with respect to the origin. By using these facts and plotting points as necessary, we get the graph of one period of $y = \cot x$ as shown with a solid curve in Figure 49. (An additional period is shown as a dashed curve.)

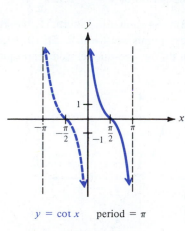

$y = \cot x$ period $= \pi$

Figure 49

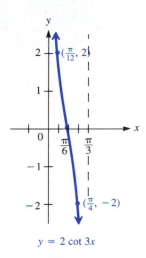

$y = 2 \cot 3x$

Figure 50

EXAMPLE 2   Graph one period of $y = 2 \cot 3x$.

Because the coefficient of x is 3, the period is $\pi/3$, and one period can be drawn between 0 and $\pi/3$. Sketch the vertical asymptotes $x = 0$ and $x = \pi/3$. The key points will be located at $x = (1/4)(\pi/3) = \pi/12$, $x = (1/2)(\pi/3) = \pi/6$, and $x = (3/4)(\pi/3) = \pi/4$. Because of the coefficient 2, the first and third key points have y-values of ± 2 instead of ± 1. Thus, the key points are $(\pi/12, 2)$, $(\pi/6, 0)$, and $(\pi/4, -2)$. Complete the graph by sketching the cotangent-shaped graph through these points and approaching the asymptotes. See Figure 50.   ●

The steps in graphing one period of $y = a \tan bx$, where $b > 0$, are summarized on the next page. (The steps for graphing $y = a \cot bx$ are similar.)

Graphing Tangent

1. Find the period, π/b. (Assume $b > 0$.)

2. Start at 0 on the x-axis and lay off two intervals, each with length half the period. One interval goes to the left and the other goes to the right of 0.

3. Draw the asymptotes as vertical dashed lines at the endpoints of the interval of Step 2.

4. Locate a point at $(0, 0)$.

5. Locate the one-quarter and three-quarter points whose x-coordinates are halfway between 0 and the ends of the intervals in Step 2. Sketch the typical tangent graph through the key points.

6. Draw additional periods, both to the right and to the left, as needed.

EXAMPLE 3   Graph one period of $y = -3 \tan \dfrac{1}{2} x$.

The period is $\pi/(1/2) = 2\pi$. As shown in Figure 51, the asymptotes are the lines $x = -\pi$ and $x = \pi$. The key points are $(-\pi/2, 3)$, $(0, 0)$, and $(\pi/2, -3)$. Note the effect of the negative sign on the one-quarter and three-quarter y-values.   ●

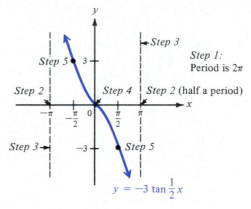

Figure 51

The main features of the graph of $y = \csc x$ are related to those of the sine graph, since $\csc x = 1/\sin x$. When $\sin x = 0$, $\csc x$ is undefined. This occurs at $x = n\pi$, so the graph of $y = \csc x$ is restricted to values of $x \neq n\pi$, where n is an integer. This means that the lines $x = n\pi$ are asymptotes. Since $\csc x = 1/(\sin x)$, the period is 2π, the same as for sine. When $\sin x = 1$, the value of $\csc x$ is also 1, and as suggested by the table of values shown on the next page, when $0 < \sin x < 1$, then $\csc x > 1$. Also, if $-1 < \sin x < 0$, then $\csc x < -1$ and $\csc x = -1$ when $\sin x = -1$. By using this information and plotting a few points, we find that the graph takes the shape of the solid curve in Figure 52. To show how the two graphs are related, the graph of $y = \sin x$ is also shown as a dashed curve.

x	$-\dfrac{\pi}{2}$	$-\dfrac{\pi}{3}$	$-\dfrac{\pi}{4}$	$-\dfrac{\pi}{6}$	0	$\dfrac{\pi}{6}$	$\dfrac{\pi}{4}$	$\dfrac{\pi}{3}$	$\dfrac{\pi}{2}$
$\sin x$	-1	$-\dfrac{\sqrt{3}}{2}$	$-\dfrac{\sqrt{2}}{2}$	$-\dfrac{1}{2}$	0	$\dfrac{1}{2}$	$\dfrac{\sqrt{2}}{2}$	$\dfrac{\sqrt{3}}{2}$	1
$\csc x$	-1	$-\dfrac{2\sqrt{3}}{3}$	$-\sqrt{2}$	-2	$-$	2	$\sqrt{2}$	$\dfrac{2\sqrt{3}}{3}$	1

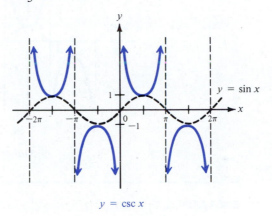

Figure 52

· · · · · · · · · ·

EXAMPLE 4   Graph two periods of $y = \dfrac{3}{2}\csc\left(x - \dfrac{\pi}{2}\right)$.

   Compared with $y = \csc x$, this graph is translated $\pi/2$ units to the right. Locate the key points for $y = (3/2)\sin(x - \pi/2)$: $(\pi/2, 0)$, $(\pi, 3/2)$, $(3\pi/2, 0)$, $(2\pi, -3/2)$, and $(5\pi/2, 0)$. Sketch the asymptotes of the cosecant graph where the y-values of the sine graph are 0, that is, at $x = \pi/2$, $x = 3\pi/2$, $x = 5\pi/2$, and, in general, at $x = n\pi/2$ for any odd integer n. The other key points of the sine graph are also on the cosecant graph. Complete the graph as shown in Figure 53. There are no values of y between 3/2 and $-3/2$; note how this relates to the increased amplitude of $y = (3/2) \sin x$ as compared with $y = \sin x$.   ●

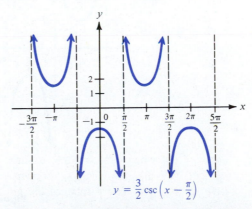

$$y = \frac{3}{2}\csc\left(x - \frac{\pi}{2}\right)$$

Figure 53

The graph of $y = \sec x$ in Figure 54 is related to the cosine graph in the same way that the graph of $y = \csc x$ is related to the sine graph.

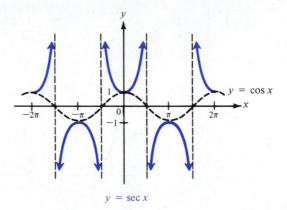

$y = \sec x$

Figure 54

· · · · · · · · ·

EXAMPLE 5   Graph $y = \sec 2\left(x + \dfrac{\pi}{4}\right)$.

The graph is translated $\pi/4$ units to the left compared to the graph of $y = \sec 2x$. The period is $2\pi/2 = \pi$. The key points for the graph of $y = \cos 2(x + \pi/4)$ are $(-\pi/4, 1)$, $(0, 0)$, $(\pi/4, -1)$, $(\pi/2, 0)$, $(3\pi/4, 1)$. The asymptotes for the secant graph will be $x = 0$, $x = \pi/2$, and, in general, $x = n\pi/2$ for any integer n. The points $(-\pi/4, 1)$, $(\pi/4, -1)$ and $(3\pi/4, 1)$ are also on the secant graph. The completed graph is shown in Figure 55.   ●

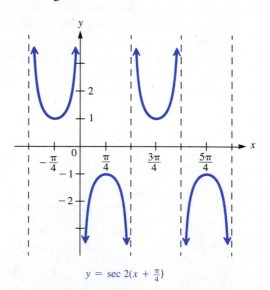

$y = \sec 2(x + \frac{\pi}{4})$

Figure 55

Additional Graphical Techniques (Optional)   Trigonometric functions which are the sum of two or more functions can be graphed with a method called **addition of ordinates.** An **ordinate** is the y-value of an ordered pair. For example, in the ordered pair $(\pi, -1)$, the number -1 is the ordinate. This graphing method is best described by examples.

· · · · · · · · ·

EXAMPLE 6   Graph $y = x + \sin x$.

Begin by graphing $y = x$ and $y = \sin x$ on the same coordinate axes. Figure 56 shows the two graphs. Then select some x-values, and for these values add the two corresponding ordinates to get the ordinate of the sum, $x + \sin x$. For example, when $x = 0$, both ordinates are 0, so that $P_1 = (0, 0)$ is a point on the graph of $y = x + \sin x$. When $x = \pi/2$, the ordinates are $\pi/2$ and 1. Their sum is $\pi/2 + 1$, which is approximately 2.6, and $P_2 = (\pi/2, \pi/2 + 1)$, or approximately $(1.6, 2.6)$, is on the graph. At $x = 3\pi/2$, the sum $x + \sin x$ is $3\pi/2 + (-1)$, or approximately 3.7, with $P_3 = (3\pi/2, 3\pi/2 - 1)$, or approximately $(4.7, 3.7)$, on the graph. As many points as necessary can be located in this way. The graph is then completed by drawing a smooth curve through the points. The graph of $y = x + \sin x$ is shown in color in Figure 56.   ●

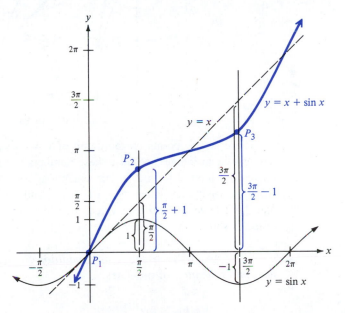

Figure 56

As shown on the graph in Figure 56, to get the graph of $y = x + \sin x$ the ordinates are actually treated as line segments. For example, the ordinate of P_2 is found by adding the lengths of the two line segments that represent the ordinates of x and $\sin x$ at $\pi/2$. The same is true for the ordinate of P_3 as well as for each of the other ordinates.

········

EXAMPLE 7   Graph $y = \cos x - \tan x$.

Think of $y = \cos x - \tan x$ as $y = \cos x + (-\tan x)$. Start by graphing $y = \cos x$ and $y = -\tan x$ on the same axes, as in Figure 57. At $x = 0$, the ordinates are 1 and 0, so the ordinate of $y = \cos x + (-\tan x)$ is $1 + 0 = 1$ and the point $(0, 1)$ is on the graph. At any point where the graphs of $\cos x$ and $-\tan x$ intersect, the ordinate is doubled. See, for example, P_1 and P_2 on the graph in Figure 57.

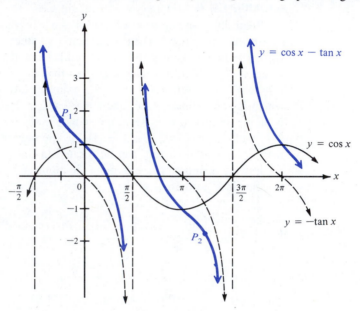

Figure 57

There are no points on the graph of $y = \cos x - \tan x$ when $x = \pi/2 + n\pi$, where n is an integer, because these numbers are not in the domain of tangent. This makes the lines $x = \pi/2 + n\pi$ asymptotes, so that as x approaches $\pi/2$ and $3\pi/2$ from the right, the values of y get larger and larger. Also, when x approaches $\pi/2$ and $3\pi/2$ from the left, the values of y get smaller and smaller. A portion of this graph is shown in color in Figure 57.   ●

While Examples 6 and 7 used *addition* of ordinates to find graphs, Example 8 uses *multiplication* of ordinates.

········

EXAMPLE 8   Graph $y = \dfrac{1}{x} \sin x$ for x in $[-\pi, \pi]$.

For all values of x, $-1 \le \sin x \le 1$. For $x > 0$, we also have $1/x > 0$, with

$$-1 \cdot \frac{1}{x} \le \frac{1}{x} \sin x \le \frac{1}{x} \cdot 1,$$

or
$$-\frac{1}{x} \le \frac{1}{x} \sin x \le \frac{1}{x}.$$

This inequality shows that the graph of $y = (1/x) \sin x$ is always bounded by the graphs of $y = 1/x$ and $y = -1/x$ (for positive x). The graph of $y = (1/x) \sin x$ will touch the graph of $y = 1/x$ or $y = -1/x$ whenever $|\sin x| = 1$, which happens for $x = \pi/2 + n\pi$, for n an integer. The x-intercepts occur when $\sin x = 0$ or when $x = 0 + n\pi$ for n an integer. Use a calculator to get additional points on the graph as needed. See also Exercise 64 of the previous section, which suggests that the product $(1/x)\sin x$ approaches 1 as x approaches 0. The open circle on the final graph, shown in Figure 58, shows that the value of the function is undefined when $x = 0$. ●

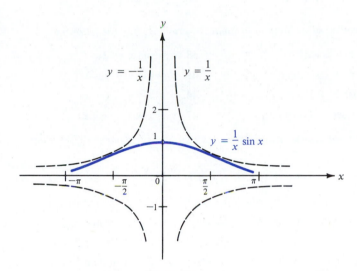

Figure 58

6.6 Exercises

Graph each of the following over the interval $[-2\pi, 2\pi]$.

1. $y = 2 \tan x$ **2.** $y = -\tan x$ **3.** $y = -\cot x$ **4.** $y = \dfrac{1}{2} \cot x$

5. $y = 1 + \tan x$ **6.** $y = -2 + \tan x$ **7.** $y = -1 + 2 \tan x$ **8.** $y = 3 + \dfrac{1}{2} \tan x$

9. $y = 1 - \cot x$ **10.** $y = -2 - \cot x$ **11.** $y = -2 \csc x$ **12.** $y = -\dfrac{1}{2} \csc x$

13. $y = -\sec x$ **14.** $y = -3 \sec x$ **15.** $y = -2 - \csc x$ **16.** $y = 1 - \sec x$

Graph each of the following over a two-period interval. Identify the period.

17. $y = \tan 3x$ **18.** $y = 2 \tan \dfrac{1}{4} x$ **19.** $y = \cot 3x$ **20.** $y = -\cot \dfrac{1}{2} x$

21. $y = \csc 4x$ **22.** $y = \csc \dfrac{1}{4} x$ **23.** $y = \sec \dfrac{1}{2} x$ **24.** $y = -\sec 4x$

Graph each of the following over a one-period interval. Identify the period.

25. $y = \tan\left(x - \dfrac{\pi}{4}\right)$

26. $y = \cot\left(x + \dfrac{3\pi}{4}\right)$

27. $y = \sec\left(x + \dfrac{\pi}{4}\right)$

28. $y = \csc\left(x + \dfrac{\pi}{3}\right)$

29. $y = 3\csc\left(x + \dfrac{3\pi}{2}\right)$

30. $y = \csc\left(\dfrac{1}{2}x - \dfrac{\pi}{4}\right)$

31. $y = 2 + 3\sec(2x - \pi)$

32. $y = \tan\left(\dfrac{1}{2}x + \dfrac{\pi}{3}\right)$

33. $y = 1 - \dfrac{1}{2}\csc\left(x - \dfrac{3\pi}{4}\right)$

34. $y = 1 - 2\cot\left(x + \dfrac{\pi}{2}\right)$

35. $y = \dfrac{2}{3}\tan\left(\dfrac{3}{4}x - \pi\right) - 2$

36. $y = 2 + \dfrac{1}{4}\sec\left(\dfrac{1}{2}x + \pi\right)$

Use the method of addition of ordinates to graph each of the following.

37. $y = x + \cos x$

38. $y = \sin x - 2x$

39. $y = 3x - \cos 2x$

40. $y = x + 2 - \sin x$

41. $y = \sin x + \sin 2x$

42. $y = \cos x - \cos(1/2)x$

43. $y = \sin x + \tan x$

44. $y = \sin x + \csc x$

45. $y = 2\cos x - \sec x$

46. $y = 2\sec x + \sin x$

47. $y = \cos x + \cot x$

48. $y = \sin x - 2\cos x$

For each of the following conditions, give all of the trigonometric functions that satisfy the condition.

49. $f(x - \pi) = f(x)$

50. $f(-x) = -f(x)$

51. $f(-x) = f(x)$

52. $f(x + \pi) = -f(x)$

53. Find the value of b for which the graph of $y = 5\tan bx$ has period $\pi/5$.

54. Find the value of b for which the graph of $y = -2\sec b(x - 3)$ has period 6π.

55. Which of the trigonometric functions can attain the value 1/2?

56. If c is any number, how many solutions does the equation $c = \tan x$ have in the interval $(-2\pi, 2\pi]$?

57. A rotating beacon is located at point A next to a long wall. (See the figure.) The beacon is 4 m from the wall. The distance d is given by $d = 4\tan 2\pi t$, where t is time measured in seconds since the beacon started rotating. (When $t = 0$, the beacon is aimed at point R. When the beacon is aimed to the right of R, the value of d is positive; d is negative if the beacon is aimed to the left of R.) Find d for the following times: **(a)** $t = 0$, **(b)** $t = .4$, **(c)** $t = .8$, **(d)** $t = 1.2$. **(e)** Why is .25 a meaningless value for t? **(f)** What is a meaningful domain for t?

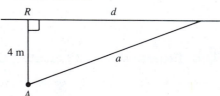

Exercises 57, 58

58. In the figure, the distance a is given by $a = 4|\sec 2\pi t|$. Find a for the following times: **(a)** $t = 0$, **(b)** $t = .86$, **(c)** $t = 1.24$. **(d)** Why are the absolute value bars needed here, but not in the equation giving d?

59. Let a person h_1 ft tall stand d ft from a statue h_2 ft tall, where $h_2 > h_1$. Let θ be the angle of elevation from eye level to the top of the statue. See the figure.
(a) Show that $d = (h_2 - h_1)\cot\theta$.
(b) Let $h_2 = 55$ and $h_1 = 5$. Graph d for $0° < \theta < 90°$.

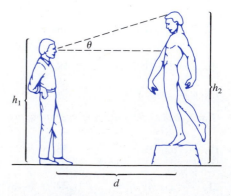

Exercise 59

60. The quotient $y = \sin x/\cos x$ does not exist if $x = -\pi/2$ or if $x = \pi/2$. Start at $x = -1.4$ and evaluate the quotient with x increasing by .2 until $x = 1.4$ is reached. Plot the values obtained. What graph is suggested?

Graph each of the following.

61. $y = x \sin x$ **62.** $y = x \cos x$ **63.** $y = 2^{-x} \sin x$ **64.** $y = x^2 \sin x$

6.7 INVERSE TRIGONOMETRIC FUNCTIONS

The definition of the inverse of a one-to-one function was given earlier. In this section we define the inverses of trigonometric functions. In looking for the inverse of $f(x) = \sin x$, recall that a function must be one-to-one before it can have an inverse function. The graph of sine in Figure 59 shows that $f(x) = \sin x$ is not a one-to-one function since different values of x can lead to the *same* value of y. However, by suitably restricting the domain of the sine function, a one-to-one function can be defined. It is customary to restrict the domain of $f(x) = \sin x$ to the interval $[-\pi/2, \pi/2]$ and define a new function with this domain as $f(x) = \text{Sin } x$.

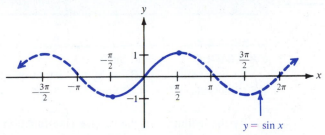

Figure 59

The graph of $y = \text{Sin } x$ is shown in Figure 60(a). Reflecting the graph of $y = \text{Sin } x$ about the 45° line $y = x$ gives the graph of the inverse function, shown in Figure 60(b).

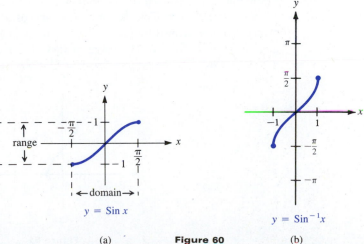

$y = \text{Sin } x$

(a)   **Figure 60**   (b)

$y = \text{Sin}^{-1}x$

The equation of the inverse of $y = \text{Sin } x$ is found by first exchanging x and y to get $x = \text{Sin } y$. This equation then is solved for y by writing $\mathbf{y = Sin^{-1}x}$, read "the inverse sine of x."

> ◢ **CAUTION** Although $\text{Sin}^n x = (\text{Sin } x)^n$ when $n \neq -1$, $\text{Sin}^{-1}x$ has the special meaning given above. The reciprocal $1/\text{Sin } x$ may be written as $(\text{Sin } x)^{-1}$.

As Figure 60(b) suggests, the domain of $y = \text{Sin}^{-1}x$ is $[-1, 1]$, while the domain of $y = \text{Sin } x$ is the range of $y = \text{Sin}^{-1}x$, $[-\pi/2, \pi/2]$. Other domains of $y = \text{Sin } x$ could be chosen so long as the result is a function; however, this domain is universally agreed upon since it centers on the origin.

An alternative notation for $\text{Sin}^{-1}x$ is **arcsin x.** The name *arcsine* comes from the fact that if $y = \text{arcsin } x$, then y is the arc length on the unit circle that corresponds to the number x.

<table>
<tr><td>

Definition of Inverse Sine Function

</td><td>

$y = \mathbf{Sin^{-1}}x$ or $y = \mathbf{arcsin}\ x$ if and only if $x = \mathbf{Sin}\ y,$

where $-1 \leq x \leq 1$ and $-\pi/2 \leq y \leq \pi/2$.

</td></tr>
</table>

EXAMPLE 1   Find $\text{Sin}^{-1} 1/2$.

Let $y = \text{Sin}^{-1} 1/2$. Then by the definition of the inverse sine function, $\text{Sin } y = 1/2$. Since $\text{Sin } \pi/6 = 1/2$, $\text{Sin}^{-1} 1/2 = \pi/6$.   ●

By the definition of the inverse sine function,

$$\text{Sin } (\text{Sin}^{-1} x) = x \quad \text{and} \quad \text{Sin}^{-1} (\text{Sin } y) = y$$

only for values of x satisfying $-1 \leq x \leq 1$ and y satisfying $-\pi/2 \leq y \leq \pi/2$. For example, since $7\pi/6$ is not in the interval $[-\pi/2, \pi/2]$, $\text{Sin}^{-1} (\sin 7\pi/6) \neq 7\pi/6$. (Show that $\text{Sin}^{-1} (\sin 7\pi/6) = -\pi/6$.)

We define $\mathbf{f(x) = Cos^{-1}}x$ (or $f(x) = \mathbf{arccos}\ x$) as we did $\text{Sin}^{-1}x$, with the one-to-one function defined by $f(x) = \text{Cos } x$, for x in $[0, \pi]$. This domain becomes the range of $f(x) = \text{Cos}^{-1}x$. The range of $f(x) = \text{Cos } x$, $[-1, 1]$, becomes the domain of $f(x) = \text{Cos}^{-1}x$. The graph of $f(x) = \text{Cos}^{-1}(x)$ is shown in Figure 61.

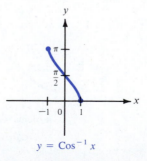

$y = \text{Cos}^{-1} x$

Figure 61

For each of the other trigonometric functions, an inverse function can be defined by a suitable restriction on the domain, just as with sine and cosine. The inverse trigonometric functions, their domains, and their ranges follow.

Definitions of Inverse Trigonometric Functions

Function	Domain	Range
$y = \operatorname{Sin}^{-1} x$	$[-1, 1]$	$[-\pi/2, \pi/2]$
$y = \operatorname{Cos}^{-1} x$	$[-1, 1]$	$[0, \pi]$
$y = \operatorname{Tan}^{-1} x$	$(-\infty, +\infty)$	$(-\pi/2, \pi/2)$
$y = \operatorname{Cot}^{-1} x$	$(-\infty, +\infty)$	$(0, \pi)$
$y = \operatorname{Sec}^{-1} x$	$(-\infty, -1] \cup [1, +\infty)$	$[0, \pi], \quad y \neq \pi/2^*$
$y = \operatorname{Csc}^{-1} x$	$(-\infty, -1] \cup [1, +\infty)$	$[-\pi/2, \pi/2], \quad y \neq 0^*$

The graphs of $f(x) = \operatorname{Tan}^{-1}x$ and $f(x) = \operatorname{Sec}^{-1}x$ shown in Figures 62 and 63 were found by reflecting the graphs of $f(x) = \operatorname{Tan} x$ and $f(x) = \operatorname{Sec} x$, with restricted domains, about the line $y = x$. The graphs of $f(x) = \operatorname{Cot}^{-1} x$ and $f(x) = \operatorname{Csc}^{-1}x$ are left for Exercises 75 and 78.

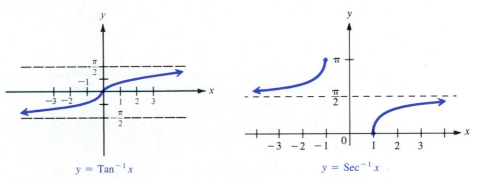

$y = \operatorname{Tan}^{-1}x$

Figure 62

$y = \operatorname{Sec}^{-1}x$

Figure 63

EXAMPLE 2   Find $\operatorname{Cos}^{-1}(-\sqrt{2}/2)$.

Because of the restriction $0 \leq \operatorname{Cos}^{-1} x \leq \pi$, the angle represented by $\operatorname{Cos}^{-1} x$ can terminate only in quadrants I and II. Since $-\sqrt{2}/2$ is negative, the angle is restricted to quadrant II. Let $y = \operatorname{Cos}^{-1}(-\sqrt{2}/2)$. Then $\operatorname{Cos} y = -\sqrt{2}/2$. In quadrant II, $\operatorname{Cos} 3\pi/4 = -\sqrt{2}/2$, so

$$\operatorname{Cos}^{-1}\left(-\frac{\sqrt{2}}{2}\right) = \frac{3\pi}{4}. \quad \bullet$$

*Sec^{-1} and Csc^{-1} are sometimes defined with a different range.

· · · · · · · · ·

EXAMPLE 3   Find sin $(\text{Tan}^{-1}(-3/2))$ without using tables or a calculator.

Let $y = \text{Tan}^{-1}(-3/2)$, so that Tan $y = -3/2$. Since the inverse tangent function has values only in quadrants I and IV ($-\pi/2 < \text{Tan}^{-1} x < \pi/2$), and since $x = -3/2$, work in quadrant IV. Sketch y in quadrant IV and label a triangle as shown in Figure 64. The hypotenuse has a length of $\sqrt{13}$, so that sin $y = -3/\sqrt{13}$, or

$$\sin\left(\text{Tan}^{-1}\left(-\frac{3}{2}\right)\right) = \frac{-3}{\sqrt{13}} = \frac{-3\sqrt{13}}{13}. \bullet$$

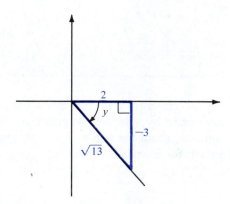

Figure 64

· · · · · · · · ·

EXAMPLE 4   Solve the equation $y = 2 \text{Sin}^{-1}(x + 4)$ for x.

First multiply by 1/2 to get

$$\frac{y}{2} = \text{Sin}^{-1}(x + 4).$$

Use the definition of the inverse sine function to get

$$\text{Sin}\frac{y}{2} = x + 4$$

or

$$x = \left(\text{Sin}\frac{y}{2}\right) - 4. \bullet$$

· · · · · · · ·

EXAMPLE 5   Write sec $(\text{Tan}^{-1} x)$ as an algebraic expression.

Let $\text{Tan}^{-1} x = u$. Then Tan $u = x$. Since $-\pi/2 < \text{Tan}^{-1} x < \pi/2$, u is in quadrant I or quadrant IV. As shown in Figure 65, in either case the value of secant is positive, so

$$\sec (\text{Tan}^{-1} x) = \sec u$$
$$= \sqrt{x^2 + 1}. \bullet$$

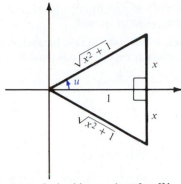

u can be in either quadrant I or IV

Figure 65

As mentioned earlier, an alternative notation for inverse trigonometric functions is used in some books.

<table>
<tr><td rowspan="4">**Alternative Notations for Inverse Trigonometric Functions**</td><td>$\text{Sin}^{-1}x$ is written **arcsin** x.</td><td>$\text{Csc}^{-1}x$ is written **arccsc** x.</td></tr>
<tr><td>$\text{Cos}^{-1}x$ is written **arccos** x.</td><td>$\text{Sec}^{-1}x$ is written **arcsec** x.</td></tr>
<tr><td>$\text{Tan}^{-1}x$ is written **arctan** x.</td><td>$\text{Cot}^{-1}x$ is written **arccot** x.</td></tr>
</table>

In the exercise sets, these notations will be used interchangeably.

6.7 Exercises

For the following, give the value of y in radians without using tables or a calculator.

1. $y = \text{Sin}^{-1}\left(-\dfrac{\sqrt{3}}{2}\right)$

2. $y = \text{Cos}^{-1}\dfrac{\sqrt{3}}{2}$

3. $y = \text{Tan}^{-1} 1$

4. $y = \text{Tan}^{-1}(-1)$

5. $y = \text{Sin}^{-1}(-1)$

6. $y = \text{Cos}^{-1}(-1)$

7. $y = \text{Cos}^{-1} 1/2$

8. $y = \text{Sin}^{-1}\left(-\dfrac{\sqrt{2}}{2}\right)$

9. $y = \arccos\left(-\dfrac{\sqrt{2}}{2}\right)$

10. $y = \arctan\dfrac{\sqrt{3}}{3}$

11. $y = \arctan(-\sqrt{3})$

12. $y = \arccos(-1/2)$

For the following, give the value in degrees. (Round to the nearest hundredth.)

13. $y = \text{Sin}^{-1}(-.1334)$

14. $y = \text{Cos}^{-1}(-.1334)$

15. $y = \text{Cos}^{-1}(-.3987)$

16. $y = \text{Sin}^{-1}.7790$

17. $y = \text{Cos}^{-1}.9272$

18. $y = \text{Tan}^{-1} 1.767$

19. $y = \text{Tan}^{-1} 1.111$

20. $y = \text{Sin}^{-1}.8192$

21. $y = \arctan(-.9217)$

22. $y = \arctan(-.2867)$

Give the values of the expressions in Exercises 23–26 in radians to four significant digits.

23. $\text{Sin}^{-1}.7214$

24. $\text{Cos}^{-1}.3004$

25. $\text{Tan}^{-1}(-4.114)$

26. $\text{Sin}^{-1}(-.9946)$

27. Enter 1.003 in your calculator and push the keys for inverse sine. The machine will tell you that something is wrong. What is wrong?

28. Enter 1.003 in your calculator and push the keys for inverse tangent. This time, unlike in Exercise 27, you get an answer easily. What is the difference?

Give the value of each of the following without using tables or a calculator.

29. $\tan\left(\arccos \dfrac{3}{4}\right)$

30. $\sin\left(\arccos \dfrac{1}{4}\right)$

31. $\cos\left(\text{Tan}^{-1}\left(-2\right)\right)$

32. $\sec\left(\text{Sin}^{-1}\left(-\dfrac{1}{5}\right)\right)$

33. $\cot\left(\arcsin\left(-\dfrac{2}{3}\right)\right)$

34. $\cos\left(\arctan \dfrac{8}{3}\right)$

35. $\cos\left(\arccos \dfrac{1}{2}\right)$

36. $\sin\left(\arcsin \dfrac{\sqrt{3}}{2}\right)$

37. $\tan(\text{Tan}^{-1}\left(-1\right))$

38. $\cot\left(\text{Cot}^{-1}\left(-\sqrt{3}\right)\right)$

39. $\sec\left(\text{Sec}^{-1} 2\right)$

40. $\csc\left(\text{Csc}^{-1} \sqrt{2}\right)$

41. $\arccos\left(\cos \dfrac{\pi}{4}\right)$

42. $\arctan\left(\tan\left(-\dfrac{\pi}{4}\right)\right)$

43. $\text{Sin}^{-1}\left(\sin(2\pi + 1)\right)$

44. $\text{Cos}^{-1}\left(\cos\left(-\dfrac{\pi}{5}\right)\right)$

45. $\text{Sin}^{-1}\left(\sin \pi\right)$

46. $\text{Tan}^{-1}\left(\tan 2\pi/3\right)$

47. $\text{Cos}^{-1}\left(\cos 5\pi/4\right)$

48. $\text{Sin}^{-1}\left(\sin 3\pi/4\right)$

Use the inverse key of your calculator together with the trigonometric function keys to find each of the following to six decimal places.

49. $\cos(\arctan .3)$

50. $\sin(\arccos .75)$

51. $\tan(\arcsin .1225)$

52. $\cot(\arccos .5823)$

Solve each of the following equations for x.

53. $y = 4 \text{ Sin}^{-1} x$

54. $3y = \text{Cos}^{-1} x$

55. $2y = \text{Tan}^{-1} 2x$

56. $y = 3 \text{ Sin}^{-1} (x/2)$

57. $y = \text{Sin}^{-1} (x + 2)$

58. $y = \text{Tan}^{-1} (2x - 1)$

Write each expression in Exercises 59–70 as an algebraic expression.

59. $\sin(\arccos u)$

60. $\tan(\arccos u)$

61. $\sec(\text{Cot}^{-1} u)$

62. $\csc(\text{Sec}^{-1} u)$

63. $\cot(\arcsin u)$

64. $\cos(\arcsin u)$

65. $\sin\left(\text{Sec}^{-1} \dfrac{u}{2}\right)$

66. $\cos\left(\text{Tan}^{-1} \dfrac{3}{u}\right)$

67. $\tan\left(\arcsin \dfrac{u}{\sqrt{u^2 + 2}}\right)$

68. $\cos\left(\arccos \dfrac{u}{\sqrt{u^2 + 5}}\right)$

69. $\sec\left(\text{arccot} \dfrac{\sqrt{4 - u^2}}{u}\right)$

70. $\csc\left(\arctan \dfrac{\sqrt{9 - u^2}}{u}\right)$

71. Determine the values of x for which the following are true.   **(a)** $\text{Cos}^{-1}(\cos x) = x$ **(b)** $\cos(\text{Cos}^{-1} x) = x$

72. Determine the values of x for which the following are true.   **(a)** $\text{Sin}^{-1}(\sin x) = x$ **(b)** $\sin(\text{Sin}^{-1} x) = x$

Give the domain of each of the following functions.

73. $\text{Cos}^{-1}(x^2 - 3)$

74. $\text{Sin}^{-1} \sqrt{x}$

Graph each equation in Exercises 75–80, and give the domain and range.

75. $y = \text{Cot}^{-1} x$

76. $y = \text{Tan}^{-1} 2x$

77. $y = \text{arcsec } 2x$

78. $y = \text{arccsc } x$

79. $y = 2 \text{ Cos}^{-1} x$

80. $y = \text{Sin}^{-1} \dfrac{x}{2}$

81. Enter 1.74283 in your calculator (set for radians), and press the sine key. Then press the keys for inverse sine. You get 1.398763 instead of 1.74283. What happened?

Tell whether the statements in Exercises 82–89 are true or false.

82. $2 \operatorname{Cos}^{-1} x = \operatorname{Cos}^{-1} 2x$

83. $\operatorname{Cot}^{-1} x = 1/\operatorname{Tan}^{-1} x$

84. $\operatorname{Tan}^{-1} x = \operatorname{Sin}^{-1} x / \operatorname{Cos}^{-1} x$

85. $\operatorname{Sin}^{-1}(-x) = -\operatorname{Sin}^{-1} x$

86. $y = \operatorname{Sin}^{-1} x$ defines an even function.

87. $y = \operatorname{Cos}^{-1} x$ defines an even function.

88. $y = x \cdot \operatorname{Sin}^{-1} x$ defines an odd function.

89. $y = x^2 \cdot \operatorname{Cos}^{-1} x$ defines an even function.

90. Suppose that an airplane flying faster than sound goes directly over you. Assume that the plane is flying level. At the instant that you feel the sonic boom from the plane, the angle of elevation to the plane is given by $\alpha = 2 \arcsin(1/m)$, where m is the Mach number of the plane's speed. Find α for each of the following values of m to the nearest degree: **(a)** $m = 1.2$, **(b)** $m = 1.5$, **(c)** $m = 2$, **(d)** $m = 2.5$.

91. A painting 1 m high and 3 m from the floor will cut off an angle θ to an observer, where

$$\theta = \operatorname{Tan}^{-1}\left(\frac{x}{x^2 + 2}\right).$$

Assume that the observer is x m from the wall where the painting is displayed and that the eyes of the observer are 2 m above the ground. See the figure. Rounding to the nearest degree, find the value of θ for the following values of x: **(a)** 1, **(b)** 2, **(c)** 3.

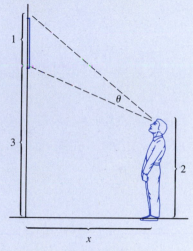

Exercise 91

6 | CHAPTER SUMMARY

Key Words

6.1 unit circle
sine
cosine
periodic function

6.2 tangent
cotangent
cosecant
secant

reciprocal identities

6.3 angle
vertex
standard position
quadrantal angle
coterminal angles
degree measure
radian measure

linear velocity
angular velocity

6.5 period
amplitude
phase shift

6.6 addition of ordinates

6.7 inverse trigonometric
functions

Review Exercises

For each of the following arc lengths, find the coordinates of the corresponding point on the unit circle.

1. $-\pi/4$

2. $2\pi/3$

3. 3π

4. $-\pi$

For each of the following values of s, find sin s, cos s, and tan s.

5. $5\pi/4$ **6.** $5\pi/3$ **7.** $-\pi/3$ **8.** -4π

The point $(-2/3, \sqrt{5}/3)$ is the endpoint of arc s on the unit circle. Find the coordinates of the endpoints of the following arcs.

9. $-s$ **10.** $\pi + s$ **11.** $\pi - s$ **12.** $2\pi - s$

Solve the following equations for $0 \le x \le 2\pi$. (Hint: Use the wrapping definition of cosine and sine discussed in Section 1 of this chapter.)

13. $\cos x = \cos (x + 1)$ **14.** $\sin x = \sin(x + 1)$

Give the quadrant in which θ terminates.

15. $\sin \theta > 0, \cos \theta < 0$ **16.** $\cos \theta > 0, \cot \theta < 0$ **17.** $\tan \theta > 0, \sin \theta < 0$ **18.** $\sec \theta < 0, \tan \theta < 0$

For each of the following, find the values of the other trigonometric functions.

19. $\sin s = 2/3$, s terminates in quadrant I   **20.** $\cos s = 4/5$, s terminates in quadrant IV

21. $\sec s = \sqrt{5}$, $\cot s = -1/2$ **22.** $\tan s = -4/3$, $\sec s = -5/3$

Work each of the following exercises.

23. A pulley is rotating 320 times per minute. Through how many degrees does a point on the edge of the pulley move in 2/3 sec?

24. The propeller of a speedboat rotates 650 times per minute. Through how many degrees will a point on the edge of the propeller rotate in 2.4 sec?

Convert radian measures to degrees, and degree measures to radians.

25. $3\pi/4$ **26.** $4\pi/5$ **27.** $31\pi/5$ **28.** $-11\pi/18$

29. $270°$ **30.** $480°$ **31.** $1020°$ **32.** $2000°$

Find the exact value of each of the following.

33. $\sin \pi/3$ **34.** $\cos 2\pi/3$ **35.** $\tan 4\pi/3$

36. $\cot 390°$ **37.** $\sec 900°$ **38.** $\cot (-1020°)$

Find the sine, cosine, and tangent of each of the following angles.

39.

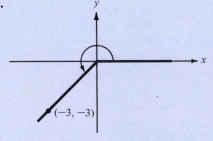

40.

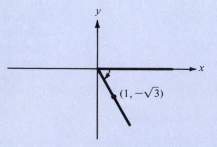

Find the value of each of the following.

41. $\tan 235°$ **42.** $\sec (-87°)$ **43.** $\sin 247.1°$

44. sec 28.7°   **45.** cot 5.842   **46.** cos (−3.1998)

Find a value of θ, where 0° ≤ θ ≤ 90°, for each of the following.

47. sin θ = .8258   **48.** cot θ = 1.124   **49.** cos θ = .9754   **50.** sec θ = 1.263

Find s in each of the following. Assume that 0 ≤ s ≤ π/2.

51. cos s = .9250   **52.** tan s = 4.011   **53.** sin s = .4924   **54.** csc s = 1.236

Find the lengths of the missing sides in each of the following right triangles.

55.

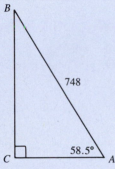

56.

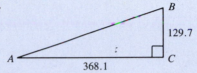

57. $A = 39.72°, b = 38.97$ m   **58.** $B = 47.53°, b = 298.6$ m

Solve each of the following problems.

59. Find t if $\theta = 5\pi/12$ radians and $\omega = 8\pi/9$ radians per sec.

60. Find θ if $t = 12$ sec. and $\omega = 9$ radians per sec.

61. Find ω if $t = 8$ sec. and $\theta = 2\pi/5$ radians.

62. Find ω if $s = 12\pi/25$ ft, $r = 3/5$ ft, and $t = 15$ sec.

63. Find s if $r = 11.46$ cm, $\omega = 4.283$ radians per sec., and $t = 5.813$ sec.

64. Find the linear velocity of a point on the edge of a flywheel of radius 7 m if the flywheel is rotating 90 times per sec.

For each of the following give the amplitude, period, vertical translation, and phase shift as applicable.

65. $y = -\dfrac{1}{2}\cos 3x$ **66.** $y = 2\sin 5x$ **67.** $y = 1 + 2\sin\dfrac{1}{4}x$ **68.** $y = 3 - \dfrac{1}{4}\cos\dfrac{2}{3}x$

69. $y = 3\cos\left(x + \dfrac{\pi}{2}\right)$ **70.** $y = -\sin\left(x - \dfrac{3\pi}{4}\right)$ **71.** $y = \dfrac{1}{2}\csc\left(2x - \dfrac{\pi}{4}\right)$ **72.** $y = 2\sec(\pi x - 2\pi)$

Graph the equations in Exercises 73–92 over a one-period interval.

73. $y = 3\sin x$ **74.** $y = \dfrac{1}{2}\sec x$ **75.** $y = -\tan x$ **76.** $y = -2\cos x$

77. $y = 2 + \cot x$ **78.** $y = -1 + \csc x$ **79.** $y = \sin 2x$ **80.** $y = \tan 3x$

81. $y = 3\cos 2x$ **82.** $y = \dfrac{1}{2}\cot 3x$ **83.** $y = \cos\left(x - \dfrac{\pi}{4}\right)$

84. $y = \tan\left(x - \dfrac{\pi}{2}\right)$ **85.** $y = \sec\left(2x + \dfrac{\pi}{3}\right)$ **86.** $y = \sin\left(3x + \dfrac{\pi}{2}\right)$

87. $y = 1 + 2\cos 3x$ **88.** $y = -1 - 3\sin 2x$ **89.** $y = 2\sin \pi x$

90. $y = -\dfrac{1}{2}\cos(\pi x - \pi)$ **91.** $y = 1 - 2\sec\left(x - \dfrac{\pi}{4}\right)$ **92.** $y = -\csc(2x - \pi) + 1$

93. Which is larger, $3°$ or 3 radians?

94. Suppose that f is a sine function with period 10 and $f(5) = 2$. Explain why $f(25) = 2$.

95. Suppose that f is a sine function with period π and $f(6\pi/5) = 1$. Explain why $f(-4\pi/5) = 1$.

Graph the equations in Exercises 96–99 using the method of addition of ordinates.

96. $y = \tan x - x$ **97.** $y = \dfrac{1}{2}x + \cos x$ **98.** $y = \sin x + \cos x$ **99.** $y = \tan x + \cot x$

100. The figure shows the population of lynx and hares in Canada for the years 1847–1903. The hares are food for the lynx. An increase in hare population causes an increase in lynx population some time later. The increasing lynx population then causes a decline in hare population.
   (a) Estimate the length of one period.
   (b) Estimate maximum and minimum hare population.

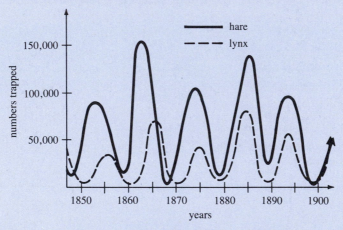

Exercise 100

101. The amount of pollution in the air fluctuates with the seasons. It is lower after heavy spring rains and higher after periods of little rain. In addition to this seasonal fluctuation, the long-term trend is upward. An idealized graph of this situation is shown in the figure. Trigonometric functions can be used to describe the fluctuating part of the pollution levels. Powers of the number e (the base of natural logarithms is e; to six decimal places, $e = 2.718282$) can be used to show the long-term growth. In fact, the pollution level in a certain area might be given by

$$P(t) = 7(1 - \cos 2\pi t)(t + 10) + 100e^{.2t},$$

where t is time in years, with $t = 0$ representing January 1 of the base year. Thus, July 1 of the same year would be represented by $t = .5$, and October 1 of the

following year would be represented by $t = 1.75$. Find the pollution
levels on the following dates:
(a) January 1, base year, (b) July 1, base year,
(c) January 1, following year, (d) July 1, following year.

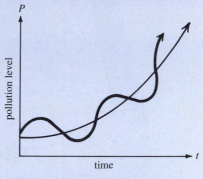

pollution level

time

Exercise 101

Find each of the following. Give answers in radians.

102. $\sin^{-1} \sqrt{2}/2$

103. $\text{Cos}^{-1} (-1/2)$

104. $\text{Tan}^{-1} (-\sqrt{3})$

105. arctan 1.780

106. arcsin $(-.6604)$

107. $\text{Cos}^{-1} .8039$

Find each of the following without using tables or a calculator.

108. $\sin\left(\text{Sin}^{-1} \dfrac{1}{2}\right)$

109. $\tan\left(\text{Tan}^{-1} \dfrac{2}{3}\right)$

110. $\cos (\arccos (-1))$

111. $\sin\left(\arcsin\left(-\dfrac{\sqrt{3}}{2}\right)\right)$

112. $\arccos\left(\cos \dfrac{3\pi}{4}\right)$

113. $\text{arcsec} (\sec \pi)$

Write each of the following as an expression in u.

114. $\sin (\text{Tan}^{-1} u)$

115. $\cos\left(\arctan \dfrac{u}{\sqrt{1 - u^2}}\right)$

116. $\tan\left(\text{arcsec} \dfrac{\sqrt{u^2 + 1}}{u}\right)$

In Exercises 117–19, graph each function and give the domain and range.

117. $y = \text{Sin}^{-1} x$

118. $y = \text{Cos}^{-1} x$

119. $y = \text{arccot } x$

120. In the study of alternating current in electricity, instantaneous voltage is given by $e = E_{max} \sin 2\pi ft$, where f is the number of cycles per second, E_{max} is the maximum voltage, and t is time in seconds. (a) Solve the equation for t. (b) Find the smallest positive value of t in radians if $E_{max} = 12$, $e = 5$, and $f = 100$.

121. Many computer languages such as BASIC and FORTRAN have only the arctangent function available. To use the other inverse trigonometric functions, it is necessary to express them in terms of arctangent. This can be done as follows.
(a) Let $u = \arcsin x$. Solve the equation for x in terms of u.
(b) Use the result of part (a) to label the three sides of the triangle of the figure in terms of x.
(c) Use the triangle from part (b) to write an equation for $\tan u$ in terms of x.
(d) Solve the equation from (c) for u.
(e) Use your equation from part (d) to calculate arcsin $(1/2)$. Compare the answer with the actual value of arcsin $(1/2)$.

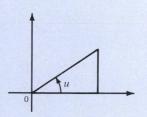

Exercise 121

7

TRIGONOMETRIC IDENTITIES AND EQUATIONS

A conditional equation, such as $2x + 1 = 9$ or $m^2 - 2m = 3$, is true for only certain values in the domain of its variable. For example, $2x + 1 = 9$ is true only for $x = 4$, and $m^2 - 2m = 3$ is true only for $m = 3$ and $m = -1$. On the other hand, an **identity** is an equation that is true for *every* value in the domain of its variable. Examples of identities include

$$5(x + 3) = 5x + 15 \quad \text{and} \quad (a + b)^2 = a^2 + 2ab + b^2.$$

This chapter discusses conditional equations and identities involving trigonometric functions. The domain of the variable is assumed to be all values for which both sides of the equation are meaningful.

7.1 FUNDAMENTAL IDENTITIES

This section introduces trigonometric identities that are important in many applications of trigonometry, but especially in the study of calculus. To begin, the fundamental identities introduced earlier are restated here. For all values of s for which denominators are not 0:

Reciprocal identities
$$\sin s = \frac{1}{\csc s} \qquad \cos s = \frac{1}{\sec s} \qquad \tan s = \frac{1}{\cot s}$$

$$\csc s = \frac{1}{\sin s} \qquad \sec s = \frac{1}{\cos s} \qquad \cot s = \frac{1}{\tan s}$$

Quotient identities
$$\tan s = \frac{\sin s}{\cos s} \qquad \cot s = \frac{\cos s}{\sin s}$$

Pythagorean identities
$$\sin^2 s + \cos^2 s = 1 \qquad \tan^2 s + 1 = \sec^2 s$$
$$1 + \cot^2 s = \csc^2 s.$$

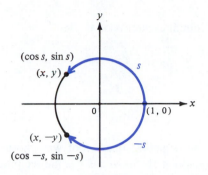

The unit circle in Figure 1 shows an arc of length s starting at $(1, 0)$, with endpoint (x, y). The same figure also shows an arc of length $-s$, again starting at $(1, 0)$. This arc has endpoint $(x, -y)$. Since $\sin s = y$,

$$\sin (-s) = -y = -\sin s,$$

$$\cos (-s) = x = \cos s,$$

and
$$\tan (-s) = \frac{-y}{x} = -\tan s.$$

The reciprocal identities can be used to find $\csc (-s)$, $\sec (-s)$, and $\cot (-s)$.

Negative Angle Identities

$\sin (-s) = -\sin s$	$\cos (-s) = \cos s$	$\tan (-s) = -\tan s$
$\csc (-s) = -\csc s$	$\sec (-s) = \sec s$	$\cot (-s) = -\cot s$

These identities show that the sine and tangent functions and their reciprocals are odd, while the cosine function and its reciprocal are even.

Trigonometric identities are useful in several ways. One use of trigonometric identities is to find values of the other trigonometric functions from the value of a given trigonometric function. For example, given a value of $\tan s$, the value of $\cot s$ can be found by using the identity $\cot s = 1/\tan s$. In fact, given $\tan s$ and the quadrant in which s terminates, the values of all the other trigonometric functions can be found by using identities, as shown below.

EXAMPLE 1   If $\tan s = -5/3$ and s is in quadrant II, find the values of the other trigonometric functions.

From the identity $\cot s = 1/\tan s$, $\cot s = -3/5$. Next, find $\sec s$ using the identity $\tan^2 s + 1 = \sec^2 s$.

$$\left(-\frac{5}{3}\right)^2 + 1 = \sec^2 s$$

$$\frac{25}{9} + 1 = \sec^2 s$$

Combining terms on the left,
$$\frac{34}{9} = \sec^2 s$$

$$-\sqrt{\frac{34}{9}} = \sec s$$

$$-\frac{\sqrt{34}}{3} = \sec s.$$

The negative square root was chosen because the values of $\sec s$ are negative in quadrant II. Since $\cos s$ is the reciprocal of $\sec s$,

$$\cos s = \frac{-3}{\sqrt{34}} = \frac{-3\sqrt{34}}{34},$$

after rationalizing the denominator. Now find $\sin s$ by using the identity $\sin^2 s + \cos^2 s = 1$.

$$\sin^2 s + \left(\frac{-3\sqrt{34}}{34}\right)^2 = 1$$

$$\sin^2 s = 1 - \frac{9}{34} = \frac{25}{34}$$

$$\sin s = \frac{5}{\sqrt{34}} \quad \text{or} \quad \sin s = \frac{5\sqrt{34}}{34}$$

The positive square root was used because the values of $\sin s$ are positive in quadrant II. Finally, $\csc s = \sqrt{34}/5$ since $\csc s$ is the reciprocal of $\sin s$. ●

· · · · · · · · ·

EXAMPLE 2   Express $\cos x$ in terms of $\tan x$.
   Since $\cos x$ and $\tan x$ are each related to $\sec x$ by an identity, start with $\tan^2 x + 1 = \sec^2 x$. Then

$$\frac{1}{\tan^2 x + 1} = \frac{1}{\sec^2 x} \quad \text{or} \quad \frac{1}{\tan^2 x + 1} = \cos^2 x.$$

Take the square root of both sides.

$$\pm\sqrt{\frac{1}{\tan^2 x + 1}} = \cos x$$

$$\cos x = \frac{\pm 1}{\sqrt{\tan^2 x + 1}}$$

Rationalize the denominator to get

$$\cos x = \frac{\pm\sqrt{\tan^2 x + 1}}{\tan^2 x + 1}.$$

Choose the $+$ or the $-$ sign, depending on the quadrant of x. ●

Another use of identities is to simplify trigonometric expressions by substituting one half of an identity for the other half. For example, the fact that $\sin^2 \theta + \cos^2 \theta$ equals 1 is used in the following example.

EXAMPLE 3   Use the fundamental identities to write $\tan(-\theta) + \cot(-\theta)$ in terms of $\sin \theta$ and $\cos \theta$. Then simplify.

From the fundamental identities,

$$\tan(-\theta) + \cot(-\theta) = \frac{\sin(-\theta)}{\cos(-\theta)} + \frac{\cos(-\theta)}{\sin(-\theta)}.$$

Use the negative angle identities on the right, then simplify the expression by adding the two fractions on the right side, using the common denominator $\cos \theta \sin \theta$.

$$\tan(-\theta) + \cot(-\theta) = \frac{-\sin \theta}{\cos \theta} + \frac{\cos \theta}{-\sin \theta}$$

$$= \frac{-\sin \theta}{\cos \theta} - \frac{\cos \theta}{\sin \theta}$$

$$= \frac{-\sin^2 \theta - \cos^2 \theta}{\cos \theta \sin \theta}$$

$$= \frac{-(\sin^2 \theta + \cos^2 \theta)}{\cos \theta \sin \theta}$$

Now substitute 1 for $\sin^2 \theta + \cos^2 \theta$, to get

$$\tan(-\theta) + \cot(-\theta) = \frac{-1}{\cos \theta \sin \theta}. \quad \bullet$$

EXAMPLE 4   Remove the radical in the expression $\sqrt{9 + x^2}$ by replacing x with $3 \tan \theta$, where $0 < \theta < \pi/2$. This kind of substitution is useful in calculus.

Letting $x = 3 \tan \theta$ gives

$$\sqrt{9 + x^2} = \sqrt{9 + (\mathbf{3 \tan \theta})^2} = \sqrt{9 + 9 \tan^2 \theta}$$

$$= \sqrt{9(1 + \tan^2 \theta)} = 3\sqrt{1 + \tan^2 \theta}$$

$$= 3\sqrt{\sec^2 \theta}.$$

On the interval $0 < \theta < \pi/2$, the value of $\sec \theta$ is positive, giving

$$\sqrt{9 + x^2} = 3 \sec \theta. \quad \bullet$$

In Example 4, since $x = 3 \tan \theta$,

$$\tan \theta = \frac{x}{3}.$$

In a right triangle, tan θ is the ratio of the length of the side opposite the angle to the side adjacent to the angle. This definition was used to label the right triangle in Figure 2, and the Pythagorean theorem was used to find the length of the hypotenuse. From the right triangle in Figure 2,

$$\sin \theta = \frac{x}{\sqrt{9 + x^2}}, \quad \cos \theta = \frac{3}{\sqrt{9 + x^2}},$$

and so on.

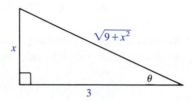

Figure 2

7.1 Exercises

Find sin s in Exercises 1–4.

1. $\cos s = 3/4$, s in quadrant I

2. $\cot s = -1/3$, s in quadrant IV

3. $\cos (-s) = \sqrt{5}/5$, $\tan s < 0$

4. $\tan s = -\sqrt{7}/2$, $\sec s > 0$

5. Find $\tan \theta$ if $\cos \theta = -2/5$, and $\sin \theta < 0$.

6. Find $\csc \alpha$ if $\tan \alpha = 6$, and $\cos \alpha > 0$.

Use the fundamental identities to find the values of the remaining five trigonometric functions of θ.

7. $\sin \theta = \dfrac{2}{3}$, θ in quadrant II

8. $\cos \theta = \dfrac{1}{5}$, θ in quadrant I

9. $\tan (-\theta) = \dfrac{1}{4}$, θ in quadrant IV

10. $\tan (-\theta) = -\dfrac{2}{3}$, θ in quadrant III

11. $\sec \theta = -3$, θ in quadrant II

12. $\csc \theta = -\dfrac{5}{2}$, θ in quadrant III

13. $\cot \theta = \dfrac{4}{3}$, $\sin \theta > 0$

14. $\sin \theta = -\dfrac{4}{5}$, $\cos \theta < 0$

15. $\sec \theta = \dfrac{4}{3}$, $\sin \theta < 0$

16. $\cos \theta = -\dfrac{1}{4}$, $\sin \theta > 0$

For each trigonometric expression in List I, choose the expression from List II that completes a fundamental identity.

List I	List II
17. $\dfrac{\cos x}{\sin x}$	**(a)** $\sin^2 x + \cos^2 x$
18. $\tan x$	**(b)** $\cot x$
19. $\cos(-x)$	**(c)** $\sec^2 x$
20. $\tan^2 x + 1$	**(d)** $\dfrac{\sin x}{\cos x}$
21. 1	**(e)** $\cos x$

For each expression in List I, choose the expression from List II that completes an identity. You will have to rewrite one or both expressions, using a fundamental identity, to recognize the matches.

List I	List II
22. $-\tan x \cos x$	**(a)** $\dfrac{\sin^2 x}{\cos^2 x}$
23. $\sec^2 x - 1$	**(b)** $\dfrac{1}{\sec^2 x}$
24. $\dfrac{\sec x}{\csc x}$	**(c)** $\sin(-x)$
25. $1 + \sin^2 x$	**(d)** $\csc^2 x - \cot^2 x + \sin^2 x$
26. $\cos^2 x$	**(e)** $\tan x$

In each of the following, use the fundamental identities to get an equivalent expression involving only sines and cosines and then simplify it.

27. $\csc^2 \beta - \cot^2 \beta$ **28.** $\dfrac{\tan(-\theta)}{\sec \theta}$ **29.** $\tan(-\alpha)\cos(-\alpha)$ **30.** $\cot^2 x(1 + \tan^2 x)$

31. $\tan^2 \theta - \dfrac{\sec^2 \theta}{\csc^2 \theta}$ **32.** $\dfrac{\tan x \csc x}{\sec x}$ **33.** $\sec \theta + \tan \theta$ **34.** $\dfrac{\sec \alpha}{\tan \alpha + \cot \alpha}$

35. $\sec^2 t - \tan^2 t$ **36.** $\csc^2 \gamma + \sec^2 \gamma$ **37.** $\cot^2 \beta - \csc^2 \beta$ **38.** $1 + \cot^2 \alpha$

39. $\dfrac{1 + \tan^2 \theta}{\cot^2 \theta}$ **40.** $\dfrac{1 - \sin^2 t}{\csc^2 t}$

41. $\cot^2 \beta \sin^2 \beta + \tan^2 \beta \cos^2 \beta$ **42.** $\sec^2 x + \cos^2 x$

43. $\dfrac{\cot^2 \alpha + \csc^2 \alpha}{\cos^2 \alpha}$ **44.** $1 - \tan^4 \theta$ **45.** $1 - \cot^4 s$ **46.** $\tan^4 \gamma - \cot^4 \gamma$

Complete this chart so that each expression in the column at the left is written in terms of the expressions given across the top.

	sin θ	cos θ	tan θ	cot θ	sec θ	csc θ
47. sin θ	sin θ	$\pm\sqrt{1-\cos^2\theta}$	$\dfrac{\pm\tan\theta}{\sqrt{1+\tan^2\theta}}$			$\dfrac{1}{\csc\theta}$
48. cos θ		cos θ	$\dfrac{\pm\sqrt{\tan^2\theta+1}}{\tan^2\theta+1}$		$\dfrac{1}{\sec\theta}$	
49. tan θ			tan θ	$\dfrac{1}{\cot\theta}$		
50. cot θ			$\dfrac{1}{\tan\theta}$	cot θ	$\dfrac{\pm\sqrt{\sec^2\theta-1}}{\sec^2\theta-1}$	
51. sec θ		$\dfrac{1}{\cos\theta}$			sec θ	
52. csc θ	$\dfrac{1}{\sin\theta}$					csc θ

53. Suppose that $\cos\theta = x/(x+1)$. Find $\sin\theta$.

54. Find $\tan\alpha$ if $\sec\alpha = (p+4)/p$.

Show that each of the following is not an identity by replacing the variables with numbers that show the result to be false.

55. $(\sin s + \cos s)^2 = 1$

56. $(\tan s + 1)^2 = \sec^2 s$

57. $2\sin s = \sin 2s$

58. $\sin x = \sqrt{1-\cos^2 x}$

59. $\sin^3 x + \cos^3 x = 1$

60. $\sin x + \sin y = \sin(x+y)$

Use the indicated substitution to remove the radical in the given expression in Exercises 61–66. Assume $0 < \theta < \pi/2$. Then find the indicated values.

61. $\sqrt{16+9x^2}$, let $x = \dfrac{4}{3}\tan\theta$; find $\sin\theta$ and $\cos\theta$

62. $\sqrt{x^2-25}$, let $x = 5\sec\theta$; find $\sin\theta$ and $\tan\theta$

63. $\sqrt{(1-x^2)^3}$, let $x = \cos\theta$; find $\sin\theta$ and $\tan\theta$

64. $\dfrac{\sqrt{x^2-9}}{x}$, let $x = 3\sec\theta$; find $\sin\theta$ and $\tan\theta$

65. $x^2\sqrt{1+16x^2}$, let $x = \dfrac{1}{4}\tan\theta$; find $\sin\theta$ and $\cos\theta$

66. $x^2\sqrt{9+x^2}$, let $x = 3\tan\theta$; find $\sin\theta$ and $\cos\theta$

67. Let $\cos x = 1/5$. Find all possible values for $\dfrac{\sec x - \tan x}{\sin x}$.

68. Let $\csc x = -3$. Find all possible values for $\dfrac{\sin x + \cos x}{\sec x}$.

Prove the following for first-quadrant values of s.

69. $\log\sin s = -\log\csc s$

70. $\log\tan s = \log\sin s - \log\cos s$

7.2 VERIFYING TRIGONOMETRIC IDENTITIES

One of the skills required for more advanced work in mathematics (and especially in calculus) is the ability to use the trigonometric identities to write trigonometric expressions in alternate forms. This skill is developed by using the fundamental identities to verify that a trigonometric equation is an identity (for those values of the variable for which it is defined). Here are some hints that may help you get started.

Techniques for Verifying Identities

1. Memorize the basic identities. Whenever you see either half of a basic identity, the other half should come to mind. Also, be aware of equivalent forms of the fundamental identities. For example, $\sin^2 \theta = 1 - \cos^2 \theta$ is an alternative form of $\sin^2 \theta + \cos^2 \theta = 1$.

2. Try to rewrite the more complicated side of the equation so that it is identical to the simpler side.

3. It is often helpful to express all other trigonometric functions in the equation in terms of sine and cosine and then simplify the result.

4. You should usually perform any factoring or indicated algebraic operations. For example, the expression $\sin^2 x + 2 \sin x + 1$ can be factored as $(\sin x + 1)^2$. The sum or difference of two trigonometric expressions, such as

$$\frac{1}{\sin \theta} + \frac{1}{\cos \theta}$$

can be found in the same way as for any other rational expression. In this example,

$$\frac{1}{\sin \theta} + \frac{1}{\cos \theta} = \frac{\cos \theta}{\sin \theta \cos \theta} + \frac{\sin \theta}{\sin \theta \cos \theta} = \frac{\cos \theta + \sin \theta}{\sin \theta \cos \theta}.$$

5. Keep in mind the side you are not changing as you select substitutions. It represents your goal. For example, to verify the identity

$$\tan^2 x + 1 = \frac{1}{\cos^2 x},$$

try to think of an identity that relates $\tan x$ to $\cos x$. Here, since $\sec x = 1/\cos x$ and $\sec^2 x = \tan^2 x + 1$, the secant function is the best link between the two sides of the equation.

6. If an expression contains $1 + \sin x$, multiplying both numerator and denominator by $1 - \sin x$ would give $1 - \sin^2 x$, which would be replaced with $\cos^2 x$.

These hints are used in the examples in this section.

CAUTION Verifying identities is not the same as solving equations. Techniques used in solving equations, such as adding the same terms to both sides, or multiplying both sides by the same term, are not valid for work with identities since we are starting with a statement (to be verified) that may not be true.

. . . .

EXAMPLE 1   Verify that $\cot s + 1 = \csc s(\cos s + \sin s)$ is an identity.

Use the fundamental identities to rewrite one side of the equation so that it is identical to the other side. Since the right side is more complicated, it is a good idea to start with it. Begin by changing $\csc s$ on the right to $1/\sin s$. Then use trigonometric identities until the right side matches the left.

$$\csc s(\cos s + \sin s) = \frac{1}{\sin s}(\cos s + \sin s)$$

$$= \frac{\cos s}{\sin s} + \frac{\sin s}{\sin s} = \cot s + 1$$

The original statement is an identity because the right side equals the left side, for all values of s for which $\sin s$ is not zero. ●

. . . .

EXAMPLE 2   Verify that $\tan^2 \alpha(1 + \cot^2 \alpha) = \dfrac{1}{1 - \sin^2 \alpha}$ is an identity.

Working with the left side gives the following.

$$\tan^2 \alpha(1 + \cot^2 \alpha) = \tan^2 \alpha + \tan^2 \alpha \cot^2 \alpha$$

$$= \tan^2 \alpha + \tan^2 \alpha \cdot \frac{1}{\tan^2 \alpha}$$

$$= \tan^2 \alpha + 1 = \sec^2 \alpha$$

$$= \frac{1}{\cos^2 \alpha} = \frac{1}{1 - \sin^2 \alpha} \quad ●$$

. . . .

EXAMPLE 3   Show that $\dfrac{\tan t - \cot t}{\sin t \cos t} = \sec^2 t - \csc^2 t.$

Work with the left side.

$$\frac{\tan t - \cot t}{\sin t \cos t} = \frac{\tan t}{\sin t \cos t} - \frac{\cot t}{\sin t \cos t}$$

$$= \tan t \cdot \frac{1}{\sin t \cos t} - \cot t \cdot \frac{1}{\sin t \cos t}$$

$$= \frac{\sin t}{\cos t} \cdot \frac{1}{\sin t \cdot \cos t} - \frac{\cos t}{\sin t} \cdot \frac{1}{\sin t \cos t}$$

$$= \frac{1}{\cos^2 t} - \frac{1}{\sin^2 t} = \sec^2 t - \csc^2 t \quad ●$$

. . . .

EXAMPLE 4   Show that $\dfrac{\sin \theta}{1 + \cos \theta} = \dfrac{1 - \cos \theta}{\sin \theta}$.

Use suggestion 6 for verifying identities and multiply numerator and denominator on the left by $1 - \cos \theta$.

$$\frac{\sin \theta}{1 + \cos \theta} = \frac{\sin \theta}{1 + \cos \theta} \cdot \frac{1 - \cos \theta}{1 - \cos \theta}$$

$$= \frac{\sin \theta \, (1 - \cos \theta)}{1 - \cos^2 \theta}$$

Now use the identity $1 - \cos^2 \theta = \sin^2 \theta$ to get

$$\frac{\sin \theta}{1 + \cos \theta} = \frac{\sin \theta \, (1 - \cos \theta)}{\sin^2 \theta}$$

$$= \frac{1 - \cos \theta}{\sin \theta}. \quad \bullet$$

. . . .

EXAMPLE 5   Verify the identity

$$\sin \theta \cos \theta \left(\frac{\tan \theta - \cot \theta}{\sin \theta \cos \theta} \right) = \sin \theta \cos \theta \, (\sec^2 \theta - \csc^2 \theta).$$

Simplify the left side first.

$$\sin \theta \cos \theta \left(\frac{\tan \theta - \cot \theta}{\sin \theta \cos \theta} \right) = \tan \theta - \cot \theta$$

Now work on the right side.

$$\sin \theta \cos \theta \, (\sec^2 \theta - \csc^2 \theta)$$

$$= \sin \theta \cos \theta \sec^2 \theta - \sin \theta \cos \theta \csc^2 \theta$$

$$= \sin \theta \cos \theta \, \frac{1}{\cos^2 \theta} - \sin \theta \cos \theta \, \frac{1}{\sin^2 \theta}$$

$$= \frac{\sin \theta}{\cos \theta} - \frac{\cos \theta}{\sin \theta}$$

$$= \tan \theta - \cot \theta$$

Since both sides equal $\tan \theta - \cot \theta$, they are equal to each other and

$$\sin \theta \cos \theta \left(\frac{\tan \theta - \cot \theta}{\sin \theta \cos \theta} \right) = \sin \theta \cos \theta \, (\sec^2 \theta - \csc^2 \theta). \quad \bullet$$

7.2 Exercises

For each of the following, perform the indicated operations and simplify the result.

1. $\tan \theta + \dfrac{1}{\tan \theta}$

2. $\dfrac{\cos x}{\sin x} + \dfrac{\sin x}{\cos x}$

3. $\cot s(\tan s + \sin s)$

4. $\sec \beta \,(\cos \beta + \sin \beta)$

5. $\dfrac{1}{\csc^2 \theta} + \dfrac{1}{\sec^2 \theta}$

6. $\dfrac{1}{\sin \alpha - 1} - \dfrac{1}{\sin \alpha + 1}$

7. $\dfrac{\cos x}{\sec x} + \dfrac{\sin x}{\csc x}$

8. $\dfrac{\cos \gamma}{\sin \gamma} + \dfrac{\sin \gamma}{1 + \cos \gamma}$

9. $(1 + \sin t)^2 + \cos^2 t$

10. $(1 + \tan s)^2 - 2 \tan s$

Factor each of the following trigonometric expressions.

11. $\sin^2 \gamma - 1$

12. $9 \sec^2 \theta - 25$

13. $(\sin x + 1)^2 - (\sin x - 1)^2$

14. $(\tan x + \cot x)^2 - (\tan x - \cot x)^2$

15. $2 \sin^2 x + 3 \sin x + 1$

16. $4 \tan^2 \beta + \tan \beta - 3$

17. $4 \sec^2 x + 3 \sec x - 1$

18. $2 \csc^2 x + 7 \csc x - 30$

19. $\cos^4 x + 2 \cos^2 x + 1$

20. $\cot^4 x + 3 \cot^2 x + 2$

Use the fundamental identities to simplify each of the given expressions.

21. $\tan \theta \cos \theta$

22. $\cot \alpha \sin \alpha$

23. $\sec r \cos r$

24. $\cot t \tan t$

25. $\dfrac{\sin \beta \tan \beta}{\cos \beta}$

26. $\dfrac{\csc \theta \sec \theta}{\cot \theta}$

27. $\sec^2 x - 1$

28. $\csc^2 t - 1$

29. $\dfrac{\sin^2 x}{\cos^2 x} + \sin x \csc x$

30. $\dfrac{1}{\tan^2 \alpha} + \cot \alpha \tan \alpha$

Verify each of the following trigonometric identities.

31. $1 - \sec \alpha \cos \alpha = \tan \alpha \cot \alpha - 1$

32. $\csc^4 \theta = \cot^4 \theta + 2 \cot^2 \theta + 1$

33. $\dfrac{\sin^2 \theta}{\cos^2 \theta} = \sec^2 \theta - 1$

34. $\cot \beta \sin \beta = \cos \beta$

35. $\sin^2 \alpha + \tan^2 \alpha + \cos^2 \alpha = \sec^2 \alpha$

36. $\sin^2 s - 1 = -\cos^2 s$

37. $\dfrac{\sin^2 \gamma}{\cos \gamma} = \sec \gamma - \cos \gamma$

38. $(1 + \tan^2 x) \cos^2 x = 1$

39. $\cot s + \tan s = \sec s \csc s$

40. $\dfrac{\cos \alpha}{\sec \alpha} + \dfrac{\sin \alpha}{\csc \alpha} = \sec^2 \alpha - \tan^2 \alpha$

41. $\sin^4 \theta - \cos^4 \theta = 2 \sin^2 \theta - 1$

42. $\dfrac{1 + \sin x}{\cos x} = \dfrac{\cos x}{1 - \sin x}$

43. $(1 - \cos^2 \alpha)(1 + \cos^2 \alpha) = 2 \sin^2 \alpha - \sin^4 \alpha$

44. $\dfrac{(\sec \theta - \tan \theta)^2 + 1}{\sec \theta \csc \theta - \tan \theta \csc \theta} = 2 \tan \theta$

45. $\dfrac{\cos \theta + 1}{\tan^2 \theta} = \dfrac{\cos \theta}{\sec \theta - 1}$

46. $\dfrac{1}{\sec \alpha - \tan \alpha} = \sec \alpha + \tan \alpha$

47. $\dfrac{1}{1 - \sin \theta} + \dfrac{1}{1 + \sin \theta} = 2 \sec^2 \theta$

48. $\dfrac{1 - \cos x}{1 + \cos x} = (\cot x - \csc x)^2$

49. $\dfrac{\tan s}{1 + \cos s} + \dfrac{\sin s}{1 - \cos s} = \cot s + \sec s \csc s$

50. $\dfrac{1}{\tan \alpha - \sec \alpha} + \dfrac{1}{\tan \alpha + \sec \alpha} = -2 \tan \alpha$

51. $\dfrac{\cot \alpha + 1}{\cot \alpha - 1} = \dfrac{1 + \tan \alpha}{1 - \tan \alpha}$

52. $\dfrac{\csc \theta + \cot \theta}{\tan \theta + \sin \theta} = \cot \theta \csc \theta$

53. $\sin^2 \alpha \sec^2 \alpha + \sin^2 \alpha \csc^2 \alpha = \sec^2 \alpha$

54. $\sec^4 x - \sec^2 x = \tan^4 x + \tan^2 x$

55. $\sin \theta + \cos \theta = \dfrac{\sin \theta}{1 - \dfrac{\cos \theta}{\sin \theta}} + \dfrac{\cos \theta}{1 - \dfrac{\sin \theta}{\cos \theta}}$

56. $\dfrac{\sin \theta}{1 - \cos \theta} - \dfrac{\sin \theta \cos \theta}{1 + \cos \theta} = \csc \theta + \csc \theta \cos^2 \theta$

57. $\dfrac{\cot^2 t - 1}{1 + \cot^2 t} = 1 - 2 \sin^2 t$

58. $\dfrac{\tan^2 t - 1}{\sec^2 t} = \dfrac{\tan t - \cot t}{\tan t + \cot t}$

59. $(\sin s + \cos s)^2 \cdot \csc s = 2 \cos s + \dfrac{1}{\sin s}$

60. $\dfrac{\sin^3 t - \cos^3 t}{\sin t - \cos t} = 1 + \sin t \cos t$

61. $\dfrac{1 + \cos x}{1 - \cos x} - \dfrac{1 - \cos x}{1 + \cos x} = 4 \cot x \csc x$

62. $\dfrac{\sin^4 \alpha - \cos^4 \alpha}{\sin^2 \alpha - \cos^2 \alpha} = 1$

63. $\dfrac{\cot^2 x + \sec^2 x + 1}{\cot^2 x} = \sec^4 x$

64. $\dfrac{\sec^4 s - \tan^4 s}{1 + 2 \tan^2 s} = \sec^2 s - \tan^2 s$

Given a complicated equation involving trigonometric functions, it is a good idea to decide whether it is likely to be an identity before trying to prove that it is. Substitute $s = 1$ and $s = 2$ into each of the following. (Be sure to use radians.) If you get the same results on both sides of the equation, it may *be an identity. Then prove that it is.*

65. $\dfrac{2 + 5 \cos s}{\sin s} = 2 \csc s + 5 \cot s$

66. $1 + \cot^2 s = \dfrac{\sec^2 s}{\sec^2 s - 1}$

67. $\dfrac{\tan s - \cot s}{\tan s + \cot s} = 2 \sin^2 s$

68. $\dfrac{1}{1 + \sin s} + \dfrac{1}{1 - \sin s} = \sec^2 s$

69. $\dfrac{1 - \tan^2 s}{1 + \tan^2 s} = \cos^2 s - \sin s$

70. $\dfrac{\sin^3 s - \cos^3 s}{\sin s - \cos s} = \sin^2 s + 2 \sin s \cos s + \cos^2 s$

Show that the statements in Exercises 71–74 are not *identities for real numbers s and t.*

71. $\sin (\csc s) = 1$

72. $\sqrt{\cos^2 s} = \cos s$

73. $\csc t = \sqrt{1 + \cot^2 t}$

74. $\sin t = \sqrt{1 - \cos^2 t}$

75. Let $\tan \theta = t$ and show that $\sin \theta \cos \theta = \dfrac{t}{t^2 + 1}$.

76. When does $\sin x = \sqrt{1 - \cos^2 x}$?

7.3 SUM AND DIFFERENCE IDENTITIES

The trigonometric identities discussed so far have involved just one angle. In this section and the next, we introduce trigonometric identities that involve *two* angles.

Cosine of Sum or Difference   Several examples presented throughout this book have shown that $\cos (A - B)$ does not equal $\cos A - \cos B$. For example, if $A = \pi/2$ and $B = 0$,

$$\cos (A - B) = \cos (\pi/2 - 0) = \cos \pi/2 = 0,$$

while $\cos A - \cos B = \cos \pi/2 - \cos 0 = 0 - 1 = -1.$

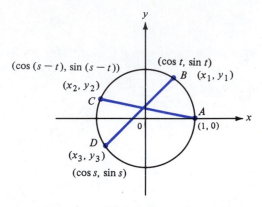

Figure 3

The actual formula for cos $(A - B)$ is derived in this section. Start by letting $A(1, 0)$, $B(x_1, y_1)$, $C(x_2, y_2)$, and $D(x_3, y_3)$ be points on the unit circle such that the length of arc AD is s, the length of arc AB is t, and the length of arc AC is $s - t$. See Figure 3. Here we assume that $0 < t < s < 2\pi$ although the results are valid for any values of s and t. Arc BD also has length $s - t$, so that arcs AC and BD are equal in length. Since these arcs are equal in length, line segments AC and BD must also have equal length. By the distance formula,

$$\sqrt{(x_2 - 1)^2 + (y_2 - 0)^2} = \sqrt{(x_3 - x_1)^2 + (y_3 - y_1)^2}.$$

By squaring both sides and simplifying, we get

$$x_2^2 - 2x_2 + 1 + y_2^2 = x_3^2 - 2x_3x_1 + x_1^2 + y_3^2 - 2y_3y_1 + y_1^2. \qquad (*)$$

Since points B, C, and D are on the unit circle,

$$x_1^2 + y_1^2 = 1, \quad x_2^2 + y_2^2 = 1, \quad \text{and} \quad x_3^2 + y_3^2 = 1.$$

Substituting these results into equation $(*)$ gives

$$2 - 2x_2 = 2 - 2x_3x_1 - 2y_3y_1$$

or

$$x_2 = x_3x_1 + y_3y_1.$$

Since $x_2 = \cos(s - t)$, $x_3 = \cos s$, $x_1 = \cos t$, $y_3 = \sin s$, and $y_1 = \sin t$, we have shown the following.

**Cosine of
Difference of
Two Angles**

$$\cos(s - t) = \cos s \cos t + \sin s \sin t$$

EXAMPLE 1   Find the value of cos 15°.

To find cos 15°, write 15° as the difference of two angles which have known function values. Since the trigonometric function values of both 45° and 30° are known, write 15° as 45° − 30°. Then use the identity for the cosine of the difference of two angles.

$$\cos 15° = \cos (45° - 30°) = \cos 45° \cos 30° + \sin 45° \sin 30°$$

$$= \frac{\sqrt{2}}{2} \cdot \frac{\sqrt{3}}{2} + \frac{\sqrt{2}}{2} \cdot \frac{1}{2} = \frac{\sqrt{6} + \sqrt{2}}{4} \quad \bullet$$

We can get a formula for cos (s + t) by using the result for cos (s − t) and writing s + t as s − (−t).

$$\cos (s + t) = \cos [s - (-t)]$$

$$= \cos s \cos (-t) + \sin s \sin (-t)$$

From earlier work, cos (−t) = cos t, and sin (−t) = −sin t. Making these substitutions gives the following.

Cosine of Sum of Two Angles

$$\cos (s + t) = \cos s \cos t - \sin s \sin t$$

EXAMPLE 2 Find $\cos \dfrac{5}{12} \pi$.

$$\cos \frac{5}{12} \pi = \cos \left(\frac{\pi}{6} + \frac{\pi}{4} \right) = \cos \frac{\pi}{6} \cos \frac{\pi}{4} - \sin \frac{\pi}{6} \sin \frac{\pi}{4}$$

$$= \frac{\sqrt{3}}{2} \cdot \frac{\sqrt{2}}{2} - \frac{1}{2} \cdot \frac{\sqrt{2}}{2} = \frac{\sqrt{6} - \sqrt{2}}{4} \quad \bullet$$

EXAMPLE 3 Suppose sin x = 1/2, cos y = −12/13, and x and y are both in quadrant II. Find cos (x + y).

By the identity above, cos (x + y) = cos x cos y − sin x sin y. The values of sin x and cos y are given, so that cos (x + y) can be found if cos x and sin y are known. Cos x can be found by using the fact that $\sin^2 x + \cos^2 x = 1$. Substitute 1/2 for sin x.

$$\sin^2 x + \cos^2 x = 1$$

$$\left(\frac{1}{2} \right)^2 + \cos^2 x = 1$$

$$\frac{1}{4} + \cos^2 x = 1$$

$$\cos^2 x = \frac{3}{4}$$

Since x is in quadrant II, cos x is negative, so cos $x = -\sqrt{3}/2$. Find sin y similarly.

$$\sin^2 y + \cos^2 y = 1$$

$$\sin^2 y + \left(-\frac{12}{13}\right)^2 = 1$$

$$\sin^2 y + \frac{144}{169} = 1$$

$$\sin^2 y = \frac{25}{169}$$

Since y is in quadrant II, sin $y = 5/13$. Now find cos $(x + y)$.

$$\cos (x + y) = \cos x \cos y - \sin x \sin y$$

$$= -\frac{\sqrt{3}}{2} \cdot \left(-\frac{12}{13}\right) - \frac{1}{2} \cdot \frac{5}{13}$$

$$= \frac{12\sqrt{3}}{26} - \frac{5}{26}$$

$$\cos (x + y) = \frac{12\sqrt{3} - 5}{26} \qquad \bullet$$

Cofunction Identities   The identities for the cosine of the sum and difference of two angles can be used to derive other identities. For example, substituting $\pi/2$ for s in the identity for cos $(s - t)$ gives

$$\cos (\pi/2 - t) = \cos \pi/2 \cdot \cos t + \sin \pi/2 \cdot \sin t$$

$$= 0 \cdot \cos t + 1 \cdot \sin t$$

$$\cos (\pi/2 - t) = \sin t.$$

This result is true for any value of t since the identity for cos $(s - t)$ is true for any values of s and t. The identity cos $(\pi/2 - t) = \sin t$ is a **cofunction identity.** The cofunction identities are listed below.

**Cofunction
Identities**

$\cos (\pi/2 - t) = \sin t$	$\sin (\pi/2 - t) = \cos t$
$\tan (\pi/2 - t) = \cot t$	$\cot (\pi/2 - t) = \tan t$
$\sec (\pi/2 - t) = \csc t$	$\csc (\pi/2 - t) = \sec t$

The derivation of some of the cofunction identities is included in Exercises 58 and 59. Because of these identities, sine and cosine are called **cofunctions,** as are tangent and cotangent, and secant and cosecant. A right triangle can be used as an aid in remembering the cofunction identities although it applies only to the special case with $t < \pi/2$. In Figure 4,

$$\sin t = \frac{b}{c} \text{ and } \cos\left(\frac{\pi}{2} - t\right) = \frac{b}{c}.$$

Also,

$$\tan t = \frac{b}{a} \text{ and } \cot\left(\frac{\pi}{2} - t\right) = \frac{b}{a},$$

and so on. The source of the "co" in cofunction is the fact that angles t and $\pi/2 - t$ are complementary angles. The definition of complementary angles is extended now to *any* pair of angles whose sum is $\pi/2$.

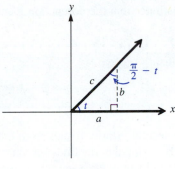

Figure 4

EXAMPLE 4   Find numbers s and θ that satisfy the following.

(a) $\cot s = \tan \pi/12$

Since tangent and cotangent are cofunctions,

$$\cot s = \tan (\pi/2 - s) = \tan \pi/12.$$

One solution to this equation is found by letting $\pi/2 - s = \pi/12$. Then $s = 5\pi/12$.

(b) $\sin \theta = \cos (-30°)$

In a similar way,

$$\sin \theta = \cos (90° - \theta) = \cos (-30°).$$
$$90° - \theta = -30°$$
$$\theta = 120°.$$

Hence, $120°$ is one solution of the equation.   ●

EXAMPLE 5   Write $\cos (\pi - t)$ as a function of t.

Use the identity for $\cos (s - t)$. Replace s with π.

$$\cos (\pi - t) = \cos \pi \cdot \cos t + \sin \pi \cdot \sin t$$
$$= (-1) \cdot \cos t + (0) \cdot \sin t = -\cos t \quad ●$$

Sine and Tangent of Sum and Difference Formulas for $\sin (s + t)$ and $\sin (s - t)$ can be derived from the results developed above. From a cofunction identity,

$$\sin (s + t) = \cos \left[\frac{\pi}{2} - (s + t) \right]$$

$$= \cos \left[\left(\frac{\pi}{2} - s \right) - t \right].$$

Using the identity for $\cos (s - t)$ gives

$$\sin (s + t) = \cos \left(\frac{\pi}{2} - s \right) \cdot \cos t + \sin \left(\frac{\pi}{2} - s \right) \cdot \sin t.$$

Substitute from the cofunction identities again to get the following.

Sine of Sum or Difference of Two Angles

$$\sin (s + t) = \sin s \cos t + \cos s \sin t$$

$$\sin (s - t) = \sin s \cos t - \cos s \sin t$$

The second identity comes from the first by writing $\sin (s - t)$ as $\sin [s + (-t)]$.

By using the identities for $\sin (s + t)$, $\cos (s + t)$, $\sin (s - t)$, and $\cos (s - t)$, and the identity $\tan x = \sin x / \cos x$, we have the following new identities.

Tangent of Sum or Difference of Two Angles

$$\tan (s + t) = \frac{\tan s + \tan t}{1 - \tan s \tan t}$$

$$\tan (s - t) = \frac{\tan s - \tan t}{1 + \tan s \tan t}$$

We show the proof for the first of these two identities. The proof of the other is very similar. Start with

$$\tan (s + t) = \frac{\sin (s + t)}{\cos (s + t)}$$

$$= \frac{\sin s \cos t + \cos s \sin t}{\cos s \cos t - \sin s \sin t}.$$

To express this result in terms of the tangent function, multiply both numerator and denominator by $1/(\cos s \cos t)$.

$$\tan (s + t) = \frac{\dfrac{\sin s \cos t + \cos s \sin t}{1}}{\dfrac{\cos s \cos t - \sin s \sin t}{1}} \cdot \frac{\dfrac{1}{\cos s \cos t}}{\dfrac{1}{\cos s \cos t}}$$

$$\tan (s + t) = \dfrac{\dfrac{\sin s \cos t}{\cos s \cos t} + \dfrac{\cos s \sin t}{\cos s \cos t}}{\dfrac{\cos s \cos t}{\cos s \cos t} - \dfrac{\sin s \sin t}{\cos s \cos t}}$$

Using the identity $\tan x = \sin x/\cos x$ gives

$$\tan (s + t) = \frac{\tan s + \tan t}{1 - \tan s \tan t}.$$

EXAMPLE 6   Use identities to find the exact value of each of the following.

(a) $\sin 75°$

$$
\begin{aligned}
\sin 75° &= \sin (45° + 30°) \\
&= \sin 45° \cos 30° + \cos 45° \sin 30° \\
&= \frac{\sqrt{2}}{2} \cdot \frac{\sqrt{3}}{2} + \frac{\sqrt{2}}{2} \cdot \frac{1}{2} = \frac{\sqrt{6} + \sqrt{2}}{4}
\end{aligned}
$$

(b) $\tan \dfrac{7}{12} \pi$

$$\tan \frac{7}{12} \pi = \tan \left(\frac{\pi}{3} + \frac{\pi}{4} \right)$$

$$= \frac{\tan \dfrac{\pi}{3} + \tan \dfrac{\pi}{4}}{1 - \tan \dfrac{\pi}{3} \tan \dfrac{\pi}{4}} = \frac{\sqrt{3} + 1}{1 - \sqrt{3} \cdot 1}$$

To simplify this result, rationalize the denominator by multiplying numerator and denominator by $1 + \sqrt{3}$.

$$
\begin{aligned}
\tan \frac{7}{12} \pi &= \frac{\sqrt{3} + 1}{1 - \sqrt{3}} \cdot \frac{1 + \sqrt{3}}{1 + \sqrt{3}} \\
&= \frac{3 + 2\sqrt{3} + 1}{1 - 3}
\end{aligned}
$$

$$\tan \frac{7}{12} \pi = -2 - \sqrt{3} \quad \bullet$$

EXAMPLE 7   If $\sin s = 4/5$ and $\cos t = -5/13$, where s is in quadrant II and t is in quadrant III, find each of the following.

(a) $\sin (s + t)$

Use the identity for the sine of the sum of two angles,

$$\sin (s + t) = \sin s \cos t + \cos s \sin t.$$

In addition to the given values of sin s and cos t, this identity requires values of cos s and sin t. The first of these can be found by using $\sin^2 s + \cos^2 s = 1$.

$$\sin^2 s + \cos^2 s = 1$$

$$\frac{16}{25} + \cos^2 s = 1$$

$$\cos^2 s = \frac{9}{25}$$

$$\cos s = -\frac{3}{5} \qquad s \text{ is in quadrant II}$$

In the same way, find sin $t = -12/13$. Now use the formula for sin $(s + t)$.

$$\sin (s + t) = \frac{4}{5}\left(-\frac{5}{13}\right) + \left(-\frac{3}{5}\right)\left(-\frac{12}{13}\right)$$

$$= -\frac{20}{65} + \frac{36}{65} = \frac{16}{65}.$$

(b) tan $(s + t)$

Use the results of part (a) to get tan $s = -4/3$ and tan $t = 12/5$. Then use the identity for tan $(s + t)$.

$$\tan (s + t) = \frac{\tan s + \tan t}{1 - \tan s \tan t}$$

$$\tan (s + t) = \frac{-\dfrac{4}{3} + \dfrac{12}{5}}{1 - \left(-\dfrac{4}{3}\right)\left(\dfrac{12}{5}\right)} = \frac{\dfrac{16}{15}}{1 + \dfrac{48}{15}} = \frac{\dfrac{16}{15}}{\dfrac{63}{15}} = \frac{16}{63}. \qquad \bullet$$

· · · · · · · ·

EXAMPLE 8   Write each of the following as a function of θ.

(a) sin $(30° + \theta)$

By the identity for sin $(s + t)$,

$$\sin (30° + \theta) = \sin 30° \cos \theta + \cos 30° \sin \theta$$

$$\sin (30° + \theta) = \frac{1}{2} \cos \theta + \frac{\sqrt{3}}{2} \sin \theta.$$

(b) $\tan (45° - \theta) = \dfrac{\tan 45° - \tan \theta}{1 + \tan 45° \tan \theta}$

$$\tan (45° - \theta) = \frac{1 - \tan \theta}{1 + \tan \theta} \qquad \bullet$$

7.3 Exercises ···

Write each of the following in terms of the cofunction of a complementary angle.

1. tan 87° **2.** sin 15° **3.** cos $\pi/12$ **4.** sin $2\pi/5$

5. csc $(-14.4°)$ **6.** sin 142.23° **7.** sin $5\pi/8$ **8.** cot $9\pi/10$

9. sec 12 **10.** tan 1.43 **11.** cot 176.9814° **12.** sin 98.0142°

Use the cofunction identities to find a value of θ that makes each of the following true.

13. $\tan \theta = \cot (45° + 2\theta)$ **14.** $\sin \theta = \cos (2\theta - 10°)$ **15.** $\sec \theta = \csc (\theta/2 + 20°)$

16. $\cos \theta = \sin (3\theta + 10°)$ **17.** $\sin (3\theta - 15°) = \cos (\theta + 25°)$ **18.** $\cot (\theta - 10°) = \tan (2\theta + 20°)$

Find the value of each of the following without using a calculator or table.

19. cos 285° **20.** cos $(-15°)$ **21.** sin $(-105°)$

22. sin 75° **23.** tan $11\pi/12$ **24.** tan $(-5\pi/12)$

25. $\cos 40° \cos 50° - \sin 40° \sin 50°$ **26.** $\sin 80° \cos 35° - \cos 80° \sin 35°$

27. $\sin (-10°) \cos 35° - \cos (-10°) \sin 35°$ **28.** $\cos \pi/5 \cos 3\pi/10 - \sin \pi/5 \sin 3\pi/10$

29. $\dfrac{\tan 80° - \tan (-55°)}{1 + \tan 80° \tan (-55°)}$ **30.** $\dfrac{\tan 5\pi/12 + \tan \pi/4}{1 - \tan 5\pi/12 \tan \pi/4}$

31. $\cos 2\pi/5 \cos \pi/10 - \sin 2\pi/5 \sin \pi/10$ **32.** $\sin 7\pi/9 \cos 2\pi/9 + \cos 7\pi/9 \sin 2\pi/9$

Write each of the following as a function of θ or x.

33. $\cos (\pi/6 + \theta)$ **34.** $\cos (\pi/4 - \theta)$ **35.** $\sin (\pi/3 + \theta)$

36. $\sin (\theta - \pi/6)$ **37.** $\tan (3\pi/2 - x)$ **38.** $\tan (x + \pi/4)$

For each of the following, find $\cos (s + t)$, $\sin (s - t)$, and $\tan (s + t)$.

39. $\cos s = -1/5$ and $\sin t = 3/5$, s and t in quadrant II

40. $\sin s = 2/3$ and $\sin t = -1/3$, s in quadrant II and t in quadrant IV

41. $\sin s = 3/5$ and $\sin t = -12/13$, s in quadrant I and t in quadrant III

42. $\cos s = -8/17$ and $\cos t = -3/5$, s and t in quadrant III

43. $\cos s = -15/17$ and $\sin t = 4/5$, s in quadrant II and t in quadrant I

44. $\sin s = -8/17$ and $\cos t = -8/17$, s and t in quadrant III

45. $\sin s = \sqrt{5}/7$ and $\sin t = \sqrt{6}/8$, s and t in quadrant I

46. $\cos s = \sqrt{2}/4$ and $\sin t = -\sqrt{5}/6$, s and t in quadrant IV

Verify each of the identities in Exercises 47–56.

47. $\cos (\pi/2 + x) = -\sin x$ **48.** $\sin (\pi/2 + x) = \cos x$

49. $\cos 2x = \cos^2 x - \sin^2 x$ **50.** $\sin 2x = 2 \sin x \cos x$
    (*Hint:* $\cos 2x = \cos (x + x)$.)

51. $\tan (x - y) - \tan (y - x) = \dfrac{2(\tan x - \tan y)}{1 + \tan x \tan y}$ **52.** $\dfrac{\cos (\alpha - \beta)}{\cos \alpha \sin \beta} = \tan \alpha + \cot \beta$

53. $\dfrac{\tan (x + y) - \tan y}{1 + \tan (x + y) \tan y} = \tan x$

54. $\dfrac{\sin (x - y)}{\sin (x + y)} = \dfrac{\tan x - \tan y}{\tan x + \tan y}$

55. $\cos (\alpha + \beta) \cos (\alpha - \beta) = 1 - \sin^2 \alpha - \sin^2 \beta$

56. $\cos 4x \cos 7x - \sin 4x \sin 7x = \cos 11x$

57. What happens when you try to evaluate the following?

$$\frac{\tan 65.902° + \tan 24.098°}{1 - \tan 65.902° \tan 24.098°}$$

Why does this happen?

58. Use the identity $\cos (\pi/2 - t) = \sin t$; replace t with $\pi/2 - t$, and derive the identity $\cos t = \sin (\pi/2 - t)$.

59. Derive each identity. **(a)** $\tan t = \cot (\pi/2 - t)$ **(b)** $\csc t = \sec (\pi/2 - t)$

60. Let $f(x) = \cos x$. Prove that $\dfrac{f(x + h) - f(x)}{h} = \cos x \left(\dfrac{\cos h - 1}{h} \right) - \sin x \left(\dfrac{\sin h}{h} \right)$.

61. Let $f(x) = \sin x$. Show that $\dfrac{f(x + h) - f(x)}{h} = \sin x \left(\dfrac{\cos h - 1}{h} \right) + \cos x \left(\dfrac{\sin h}{h} \right)$.

62. The slope of a line is defined as the ratio of the vertical change to the horizontal change. As shown in the first sketch, the tangent of the *angle of inclination* θ is given by the ratio of the side opposite to the side adjacent. This ratio is the same as that used in finding the slope, m, so that $m = \tan \theta$.

   In the second sketch, let the two lines have angles of inclination α and β, and slopes m_1 and m_2, respectively. Let θ be the smallest positive angle between the lines. Show that

$$\tan \theta = \frac{m_2 - m_1}{1 + m_1 m_2}.$$

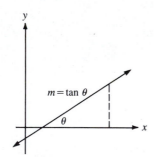
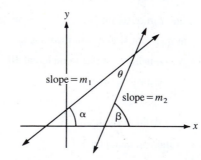

Exercises 62–64

Use the result from Exercise 62 to find the angle between the following pairs of lines. Round to the nearest tenth of a degree.

63. $x + y = 9$, $2x + y = -1$

64. $5x - 2y + 4 = 0$, $3x + 5y = 6$

For Exercises 65–68, use the identities in this section to evaluate each of the following.

65. $\cos (\mathrm{Sin}^{-1} 8/17 + \mathrm{Tan}^{-1} 3/4)$

66. $\cos (\mathrm{Tan}^{-1} 5/12 - \mathrm{Cos}^{-1} 4/5)$

67. $\sin (\mathrm{Cos}^{-1} 5/13 + \mathrm{Tan}^{-1} 3/4)$

68. $\sin (\mathrm{Sin}^{-1} 2/3 + \mathrm{Tan}^{-1} 1/4)$

69. If the straight line distance of an object from a starting point is given by $s(t) = 3 \cos(t + \pi/4)$, where t represents time, the motion is called *simple harmonic motion*. Use an identity to write $s(t)$ in terms of $\cos t$ and $\sin t$.

70. The equation $s(t) = 2 \sin(t + \pi/3)$ is another example of simple harmonic motion (see Exercise 69). Use an identity to rewrite the equation in terms of $\cos t$ and $\sin t$.

7.4 MULTIPLE-ANGLE IDENTITIES

Some special cases of the identities for the sum of two angles are used often enough to be expressed as separate identities. These are the identities that result from the sum identities when $s = t$, so that $s + t = 2s$. These identities are called **double-angle identities.** For example, in the identity for $\cos(s + t)$, let $t = s$ to derive an expression for $\cos 2s$.

$$\cos 2s = \cos(s + s)$$
$$= \cos s \cos s - \sin s \sin s$$
$$= \cos^2 s - \sin^2 s$$

Substitution from either $\cos^2 s = 1 - \sin^2 s$ or $\sin^2 s = 1 - \cos^2 s$ leads to two alternate forms for this identity, as shown below.

Cosine of Double Angle

$$\cos 2s = \cos^2 s - \sin^2 s$$
$$\cos 2s = 1 - 2 \sin^2 s$$
$$\cos 2s = 2 \cos^2 s - 1$$

Now find an identity for $\sin 2s$ by starting with the one for $\sin(s + t)$.

$$\sin 2s = \sin(s + s)$$
$$= \sin s \cos s + \cos s \sin s$$

Sine of Double Angle

$$\sin 2s = 2 \sin s \cos s$$

Finally, find $\tan 2s$ from the identity for $\tan(s + t)$.

$$\tan 2s = \tan(s + s)$$
$$= \frac{\tan s + \tan s}{1 - \tan s \tan s}$$

Tangent of Double Angle

$$\tan 2s = \frac{2 \tan s}{1 - \tan^2 s}$$

CAUTION   Be careful to avoid such common errors as equating cos $2s$ and $2 \cos s$. The identities given above show that

$$\cos 2s \neq 2 \cos s,$$

$$\sin 2s \neq 2 \sin s,$$

and $\tan 2s \neq 2 \tan s.$

Also, be careful when you divide these expressions by 2. For example,

$$\frac{\sin 2t}{2} \neq \sin t.$$

Always use the *identities* to rewrite a trigonometric function of a multiple of x, such as cos $2x$, *before* carrying out any other operations.

EXAMPLE 1   Simplify each of the following.

(a) sin 15° cos 15°

The product of the sine and cosine of the same angle suggests the identity for sin $2s$: sin $2s = 2 \sin s \cos s$. With this identity,

$$\sin 15° \cos 15° = \left(\frac{1}{2}\right)(2) \sin 15° \cos 15°$$

$$= \frac{1}{2}(2 \sin 15° \cos 15°)$$

$$= \frac{1}{2}(\sin 2 \cdot 15°)$$

$$= \frac{1}{2} \sin 30°$$

$$= \frac{1}{2} \cdot \frac{1}{2} = \frac{1}{4}.$$

(b) $\cos^2 7x - \sin^2 7x$

This expression suggests an identity for cos $2s$: cos $2s = \cos^2 s - \sin^2 s$. Substituting $7x$ for s gives

$$\cos^2 7x - \sin^2 7x = \cos 2(7x)$$

$$= \cos 14x. \quad \bullet$$

The double-angle identities for $2s$ can be used to find values of the trigonometric functions of s as shown in the following example.

EXAMPLE 2   Find the values of the six trigonometric functions of θ if $\cos 2\theta$ = 4/5 and θ terminates in quadrant II.

Use one of the double-angle identities for cosine.

$$\cos 2\theta = 1 - 2 \sin^2 \theta$$

$$\frac{4}{5} = 1 - 2 \sin^2 \theta$$

$$-\frac{1}{5} = -2 \sin^2 \theta$$

$$\frac{1}{10} = \sin^2 \theta$$

$$\sin \theta = \frac{\sqrt{10}}{10}$$

The positive square root was chosen since θ terminates in quadrant II. The values of $\cos \theta$ and $\tan \theta$ can now be found using the fundamental identities.

$$\sin^2 \theta + \cos^2 \theta = 1$$

$$\frac{1}{10} + \cos^2 \theta = 1$$

$$\cos^2 \theta = \frac{9}{10}$$

$$\cos \theta = \frac{-3}{\sqrt{10}} \qquad \text{Since } \theta \text{ is in quadrant II}$$

$$\cos \theta = \frac{-3\sqrt{10}}{10}$$

Verify that $\tan \theta = \sin \theta/\cos \theta = -1/3$. Use reciprocals to find that $\csc \theta = \sqrt{10}$, $\sec \theta = -\sqrt{10}/3$, and $\cot \theta = -3$.   ●

EXAMPLE 3   Given $\cos \theta = 3/5$, where $3\pi/2 < \theta < 2\pi$, find $\cos 2\theta$, $\sin 2\theta$, and $\tan 2\theta$.

From $\cos \theta = 3/5$, the identity $\sin^2 \theta + \cos^2 \theta = 1$ leads to $\sin \theta = \pm 4/5$. Since θ terminates in quadrant IV, $\sin \theta = -4/5$. Then, using the double-angle identities,

$$\sin 2\theta = 2 \sin \theta \cos \theta = 2 \left(-\frac{4}{5} \right) \left(\frac{3}{5} \right) = -\frac{24}{25}.$$

Also, $$\cos 2\theta = \cos^2 \theta - \sin^2 \theta = \frac{9}{25} - \frac{16}{25} = -\frac{7}{25}$$

$$\tan 2\theta = \frac{\sin 2\theta}{\cos 2\theta} = \frac{-24/25}{-7/25} = \frac{24}{7}.$$

As an alternative way of finding $\tan 2\theta$, start with $\sin \theta = -4/5$ and $\cos \theta = 3/5$, to get $\tan \theta = -4/3$.

$$\tan 2\theta = \frac{2 \tan \theta}{1 - \tan^2 \theta} = \frac{2\left(-\dfrac{4}{3}\right)}{1 - \dfrac{16}{9}} = \frac{-\dfrac{8}{3}}{-\dfrac{7}{9}} = \frac{24}{7} \quad \bullet$$

The double-angle identities, together with the sum and difference identities, allow trigonometric functions of multiple values of s to be rewritten in terms of s.

EXAMPLE 4   Write $\sin 3s$ in terms of $\sin s$.

$$
\begin{aligned}
\sin 3s &= \sin (2s + s) & &3s = 2s + s \\
&= \sin 2s \cos s + \cos 2s \sin s & &\text{Sum identity for sine} \\
&= (2 \sin s \cos s) \cos s + (\cos^2 s - \sin^2 s) \sin s & &\text{Double-angle identities} \\
&= 2 \sin s \cos^2 s + \cos^2 s \sin s - \sin^3 s & &\text{Multiply.} \\
&= 2 \sin s(1 - \sin^2 s) + (1 - \sin^2 s) \sin s - \sin^3 s & &\cos^2 s = 1 - \sin^2 s \\
&= 2 \sin s - 2 \sin^3 s + \sin s - \sin^3 s - \sin^3 s & &\text{Distributive property} \\
\sin 3s &= 3 \sin s - 4 \sin^3 s & &\text{Combine terms.} \quad \bullet
\end{aligned}
$$

From the alternate forms of the double-angle identity for cosine, three additional identities can be derived by substitution and algebraic rearrangement. These **half-angle identities,** listed below, are used in the study of calculus.

Half-angle Identities

$$\cos \frac{s}{2} = \pm \sqrt{\frac{1 + \cos s}{2}} \qquad \sin \frac{s}{2} = \pm \sqrt{\frac{1 - \cos s}{2}}$$

$$\tan \frac{s}{2} = \pm \sqrt{\frac{1 - \cos s}{1 + \cos s}} \qquad \tan \frac{s}{2} = \frac{\sin s}{1 + \cos s}$$

$$\text{and} \quad \tan \frac{s}{2} = \frac{1 - \cos s}{\sin s}$$

In these identities, the plus or minus sign is selected according to the quadrant in which $s/2$ terminates. For example, if s represents an angle of $324°$, then $s/2 = 162°$, which lies in quadrant II. In quadrant II, $\cos s/2$ and $\tan s/2$ are negative, and $\sin s/2$ is positive.

To derive the identity for $\sin s/2$, start with the identity

$$\cos 2x = 1 - 2 \sin^2 x.$$

Now solve for sin x: $\quad 2 \sin^2 x = 1 - \cos 2x$

$$\sin x = \pm \sqrt{\frac{1 - \cos 2x}{2}}.$$

Let $2x = s$, so that $x = s/2$, and substitute into this last expression.

$$\sin \frac{s}{2} = \pm \sqrt{\frac{1 - \cos s}{2}}$$

The identity for cos $s/2$ from the box above is derived in a similar way, by starting with the double-angle identity $\cos 2x = 2 \cos^2 x - 1$. One identity for tan $s/2$ comes from the half-angle identities for sine and cosine.

$$\tan \frac{s}{2} = \frac{\pm \sqrt{\dfrac{1 - \cos s}{2}}}{\pm \sqrt{\dfrac{1 + \cos s}{2}}} = \pm \sqrt{\frac{1 - \cos s}{1 + \cos s}}$$

The other two identities for tan $s/2$ given above are proven as in Exercise 86 by using the results of Example 4 of Section 2 of this chapter.

· · · · · · · · ·

EXAMPLE 5   Find cos 112.5°.

Since $112.5° = 225°/2$, use the identity for cos $s/2$ with $s = 225°$. Since 112.5° is in quadrant II, where cosine is negative, the minus sign must appear before the radical.

$$\cos 112.5° = \cos \frac{225°}{2} = -\sqrt{\frac{1 + \cos 225°}{2}}$$

$$= -\sqrt{\frac{1 - \dfrac{\sqrt{2}}{2}}{2}} = -\sqrt{\frac{2 - \sqrt{2}}{4}} = -\frac{\sqrt{2 - \sqrt{2}}}{2} \quad \bullet$$

· · · · · · · ·

EXAMPLE 6   Find tan 22.5°.

Use the identity $\tan \dfrac{s}{2} = \dfrac{\sin s}{1 + \cos s}$, with $s = 45°$.

$$\tan 22.5° = \tan \frac{45°}{2} = \frac{\sin 45°}{1 + \cos 45°}$$

$$= \frac{\dfrac{\sqrt{2}}{2}}{1 + \dfrac{\sqrt{2}}{2}} = \frac{\sqrt{2}}{2 + \sqrt{2}} = \sqrt{2} - 1$$

(Here the denominator was rationalized.) See also Exercise 85. $\bullet$

· · · · · · · · ·

EXAMPLE 7   Simplify each of the following.

(a) $\pm \sqrt{\dfrac{1 + \cos 12x}{2}}$

Start with the identity for $\cos s/2$,

$$\cos s/2 = \pm \sqrt{\dfrac{1 + \cos s}{2}},$$

and replace s with $12x$ to get

$$\pm \sqrt{\dfrac{1 + \cos \mathbf{12x}}{2}} = \cos \dfrac{\mathbf{12x}}{2} = \cos 6x.$$

(b) $\dfrac{1 - \cos 5\alpha}{\sin 5\alpha}$

Use the third identity for $\tan s/2$ given above to get

$$\dfrac{1 - \cos 5\alpha}{\sin 5\alpha} = \tan \dfrac{5\alpha}{2}. \quad \bullet$$

· · · · · · · · ·

EXAMPLE 8   Given $\cos s = 2/3$, with $3\pi/2 < s < 2\pi$, find $\cos s/2$, $\sin s/2$, and $\tan s/2$.

Since $3\pi/2 < s < 2\pi$, $3\pi/4 < s/2 < \pi$. Thus, $s/2$ terminates in quadrant II and $\cos s/2$ and $\tan s/2$ are negative, while $\sin s/2$ is positive. From the half-angle identities,

$$\sin \dfrac{s}{2} = \sqrt{\dfrac{1 - \dfrac{2}{3}}{2}} = \sqrt{\dfrac{1}{6}} = \dfrac{\sqrt{6}}{6}$$

$$\cos \dfrac{s}{2} = -\sqrt{\dfrac{1 + \dfrac{2}{3}}{2}} = -\sqrt{\dfrac{5}{6}} = -\dfrac{\sqrt{30}}{6}$$

$$\tan \dfrac{s}{2} = \dfrac{\dfrac{\sqrt{6}}{6}}{-\dfrac{\sqrt{30}}{6}} = -\dfrac{\sqrt{5}}{5}. \quad \bullet$$

7.4 Exercises ·

Write each of the following as a single trigonometric function value.

1. $\sqrt{\dfrac{1 - \cos 40°}{2}}$

2. $\sqrt{\dfrac{1 + \cos 76°}{2}}$

3. $\sqrt{\dfrac{1 - \cos 147°}{1 + \cos 147°}}$

4. $\sqrt{\dfrac{1 + \cos 165°}{1 - \cos 165°}}$

5. $\dfrac{1 - \cos 59.74°}{\sin 59.74°}$　　　**6.** $\dfrac{\sin 158.2°}{1 + \cos 158.2°}$　　　**7.** $\pm\sqrt{\dfrac{1 + \cos 18x}{2}}$　　　**8.** $\pm\sqrt{\dfrac{1 + \cos 20\alpha}{2}}$

9. $\pm\sqrt{\dfrac{1 - \cos 8\theta}{1 + \cos 8\theta}}$　　**10.** $\pm\sqrt{\dfrac{1 - \cos 5A}{1 + \cos 5A}}$　　**11.** $\pm\sqrt{\dfrac{1 + \cos x/4}{2}}$　　**12.** $\pm\sqrt{\dfrac{1 - \cos 3\theta/5}{2}}$

Write each of the following as a single trigonometric function value or as a single number.

13. $1 - 2 \sin^2 15°$　　　**14.** $\dfrac{2 \tan 15°}{1 - \tan^2 15°}$　　　**15.** $2 \sin \pi/3 \cos \pi/3$　　　**16.** $\dfrac{2 \tan \pi/3}{1 - \tan^2 \pi/3}$

17. $\sin \pi/8 \cos \pi/8$　　**18.** $\cos^2 \pi/8 - 1/2$　　**19.** $\dfrac{\tan 51°}{1 - \tan^2 51°}$　　**20.** $\dfrac{1}{4} - \dfrac{1}{2} \sin^2 47.1°$

21. $\dfrac{1}{8} \sin 29.5° \cos 29.5°$　　**22.** $\sin^2 2\pi/5 - \cos^2 2\pi/5$　　**23.** $\cos^2 2\alpha - \sin^2 2\alpha$　　**24.** $\dfrac{\tan 2y/5}{1 - \tan^2 2y/5}$

Determine whether the positive or negative square root should be selected.

25. $\sin 195° = \pm\sqrt{\dfrac{1 - \cos 390°}{2}}$　　　　　**26.** $\cos 58° = \pm\sqrt{\dfrac{1 + \cos 116°}{2}}$

27. $\tan 225° = \pm\sqrt{\dfrac{1 - \cos 450°}{1 + \cos 450°}}$　　　**28.** $\sin (-10°) = \pm\sqrt{\dfrac{1 - \cos (-20°)}{2}}$

Use the identities of this section to find the values of sine, cosine, and tangent for each of the following.

29. $\theta = 22.5°$ 　　　　　　　　　　　**30.** $\theta = 15°$

31. $\theta = 195°$ 　　　　　　　　　　　**32.** $x = -\pi/8$

33. $x = 5\pi/2$ 　　　　　　　　　　　　**34.** $x = 3\pi/2$

Find values of k and t to make the following statements true.

35. $8 \sin 3\theta \cos 3\theta = k \sin t\theta$ 　　　　　**36.** $2 \cos^2 3\theta - 2 \sin^2 3\theta = k \cos t\theta$

37. $3 - 3 \cos 4\theta = k \sin^2 t\theta$ 　　　　　　**38.** $\dfrac{\tan 2\theta}{1 - \tan^2 2\theta} = k \tan t\theta$

Find each of the following.

39. $\cos x$, if $\cos 2x = -5/12$ and $\pi/2 < x < \pi$ 　　　**40.** $\sin x$, if $\cos 2x = 2/3$ and $\pi < x < 3\pi/2$

41. $\cos \alpha/2$, if $\cos \alpha = -1/4$ and $\pi < \alpha < 3\pi/2$ 　　**42.** $\sin \beta/2$, if $\cos \beta = 3/4$ and $3\pi/2 < \beta < 2\pi$

Use the identities of this section to find values of the six trigonometric functions for each of the following.

43. x, given $\cos 2x = -5/12$ and $\pi/2 < x < \pi$ 　　　**44.** t, given $\cos 2t = 2/3$ and $\pi/2 < t < \pi$

45. 2θ, given $\sin \theta = 2/5$ and $\cos \theta < 0$ 　　　　**46.** 2β, given $\cos \beta = -12/13$ and $\sin \beta > 0$

47. $2x$, given $\tan x = 2$ and $\cos x > 0$ 　　　　　　**48.** $2x$, given $\tan x = 5/3$ and $\sin x < 0$

49. $\alpha/2$, given $\cos \alpha = 1/3$ and $\sin \alpha < 0$ 　　　　**50.** $\theta/2$, given $\cos \theta = -2/3$ and $\sin \theta > 0$

Verify that each of the following equations is an identity.

51. $(\sin \gamma + \cos \gamma)^2 = \sin 2\gamma + 1$

52. $\cos 2s = \cos^4 s - \sin^4 s$

53. $\cot s + \tan s = 2 \csc 2s$

54. $\sin 4\alpha = 4 \sin \alpha \cos \alpha \cos 2\alpha$

55. $\dfrac{1 + \cos 2x}{\sin 2x} = \cot x$

56. $\sec^2 \dfrac{x}{2} = \dfrac{2}{1 + \cos x}$

57. $\cot^2 \dfrac{x}{2} = \dfrac{(1 + \cos x)^2}{\sin^2 x}$

58. $\sin^2 \dfrac{x}{2} = \dfrac{\tan x - \sin x}{2 \tan x}$

59. $\dfrac{\sin 2x}{2 \sin x} = \cos^2 \dfrac{x}{2} - \sin^2 \dfrac{x}{2}$

60. $\dfrac{2}{1 + \cos x} - \tan^2 \dfrac{x}{2} = 1$

61. $\tan 8k - \tan 8k \cdot \tan^2 4k = 2 \tan 4k$

62. $\sin 2\gamma = \dfrac{2 \tan \gamma}{1 + \tan^2 \gamma}$

63. $\cos 2y = \dfrac{2 - \sec^2 y}{\sec^2 y}$

64. $\dfrac{2 \cos 2\alpha}{\sin 2\alpha} = \cot \alpha - \tan \alpha$

65. $\sin 2\alpha \cos 2\alpha = \sin 2\alpha - 4 \sin^3 \alpha \cos \alpha$

66. $\dfrac{\tan \dfrac{x}{2} + \cot \dfrac{x}{2}}{\cot \dfrac{x}{2} - \tan \dfrac{x}{2}} = \sec x$

67. $1 - \tan^2 \dfrac{\theta}{2} = \dfrac{2 \cos \theta}{1 + \cos \theta}$

68. $\cos x = \dfrac{1 - \tan^2 \dfrac{x}{2}}{1 + \tan^2 \dfrac{x}{2}}$

Express the following as trigonometric functions of x.

69. $\tan^2 2x$

70. $\cos^2 2x$

71. $\cos 3x$

72. $\sin 4x$

An airplane flying faster than sound sends out sound waves that form a cone, as shown in the figure. The cone intersects the ground to form a hyperbola. As this hyperbola passes over a particular point on the ground, a sonic boom is heard at that point.

If α is the angle at the vertex of the cone, then $\sin (\alpha/2) = 1/m$ where m is the Mach number of the plane. (We assume $m > 1$.) The Mach number is the ratio of the speed of the plane and the speed of sound. For example, a speed of Mach 1.4 means that the plane is flying 1.4 times the speed of sound. Find α or m, as necessary, for each of the following.

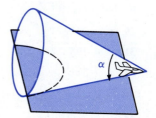

Exercises 73–78

73. $m = 3/2$

74. $m = 5/4$

75. $m = 2$

76. $m = 5/2$

77. $\alpha = 30°$

78. $\alpha = 60°$

Let $\sin s = -0.481143$, with $7\pi/4 < s < 2\pi$. Find each value in Exercises 79–84.

79. $\sin 2s$

80. $\sin \dfrac{1}{2}s$

81. $\cos 2s$

82. $\cos \dfrac{1}{2}s$

83. $\tan 2s$

84. $\tan \dfrac{1}{2}s$

85. In Example 6 the identity $\tan \dfrac{s}{2} = \dfrac{\sin s}{1 + \cos s}$ was used to find that $\tan 22.5° = \sqrt{2} - 1$.

(a) Find $\tan 22.5°$ with the identity $\tan s/2 = \pm\sqrt{(1 - \cos s)/(1 + \cos s)}$.

(b) Show that both answers are the same.

86. Go through the following steps to prove that $\tan \dfrac{A}{2} = \dfrac{\sin A}{1 + \cos A}$.

(a) Start with $\tan A/2 = \pm\sqrt{(1 - \cos A)/(1 + \cos A)}$, and multiply numerator and denominator by $\sqrt{1 + \cos A}$ to show that

$$\tan \frac{A}{2} = \pm \left| \frac{\sin A}{1 + \cos A} \right|.$$

(b) Show that $1 + \cos A \geq 0$, giving

$$\tan \frac{A}{2} = \frac{\pm |\sin A|}{1 + \cos A}.$$

(c) By considering the quadrant in which A lies, show that $\tan A/2$ and $\sin A$ have the same sign, with

$$\tan \frac{A}{2} = \frac{\sin A}{1 + \cos A}.$$

Simplify each of the following.

87. $\sin^2 2x + \cos^2 2x$ **88.** $\sin^2 \dfrac{x}{2} + \cos^2 \dfrac{x}{2}$ **89.** $1 + \tan^2 4x$ **90.** $\cot^2 \dfrac{x}{3} + 1$

7.5 SUM AND PRODUCT IDENTITIES; REDUCTION IDENTITY

One group of identities in this section can be used to rewrite a product of two functions as a sum or difference. Another group can be used to rewrite a sum or difference of two functions as a product. Some of these identities can also be used to rewrite an expression involving both sine and cosine functions as one with only one of these functions. In the next section on conditional equations, the need for this kind of change will become clear. These identities are also useful in graphing and in calculus.

The identities of this section all result from the sum and difference identities for sine and cosine. Adding the identities for $\sin (s + t)$ and $\sin (s - t)$ gives

$$\begin{aligned}
\sin (s + t) &= \sin s \cos t + \cos s \sin t \\
\underline{\sin (s - t)} &= \underline{\sin s \cos t - \cos s \sin t} \\
\sin (s + t) + \sin (s - t) &= 2 \sin s \cos t.
\end{aligned}$$

This last statement can be expressed as follows.

$$\sin s \cos t = \frac{1}{2} [\sin (s + t) + \sin (s - t)]$$

Subtracting sin $(s - t)$ from sin $(s + t)$ gives the next result.

$$\cos s \sin t = \frac{1}{2} [\sin (s + t) - \sin (s - t)]$$

Using the identities for cos $(s + t)$ and cos $(s - t)$ similarly gives the following.

$$\cos s \cos t = \frac{1}{2} [\cos (s + t) + \cos (s - t)]$$

$$\sin s \sin t = \frac{1}{2} [\cos (s - t) - \cos (s + t)]$$

EXAMPLE 1   Rewrite cos 2θ sin θ as the sum or difference of two functions.
By the identity for cos s sin t,

$$\cos 2\theta \sin \theta = \frac{1}{2} (\sin 3\theta - \sin \theta)$$

$$\cos 2\theta \sin \theta = \frac{1}{2} \sin 3\theta - \frac{1}{2} \sin \theta. \quad \bullet$$

EXAMPLE 2   Evaluate cos 15° cos 45°.
Use the identity for cos s cos t.

$$\cos 15° \cos 45° = \frac{1}{2} [\cos (15° + 45°) + \cos (15° - 45°)]$$

$$= \frac{1}{2} [\cos 60° + \cos (-30°)] = \frac{1}{2} (\cos 60° + \cos 30°)$$

$$= \frac{1}{2} \left(\frac{1}{2} + \frac{\sqrt{3}}{2} \right) = \frac{1 + \sqrt{3}}{4} \quad \bullet$$

From these new identities further identities can be found that are used in calculus to rewrite a sum of trigonometric functions as a product. To begin, let $s + t = x$, and let $s - t = y$. Then adding x and y gives

$$x + y = (s + t) + (s - t) = 2s$$

or
$$s = \frac{x + y}{2}.$$

Subtracting y from x produces

$$x - y = (s + t) - (s - t) = 2t$$

from which

$$t = \frac{x - y}{2}.$$

With these results, the identity

$$\sin s \cos t = \frac{1}{2} [\sin (s + t) + \sin (s - t)]$$

becomes

$$\sin \left(\frac{x + y}{2}\right) \cos \left(\frac{x - y}{2}\right) = \frac{1}{2} (\sin x + \sin y).$$

By reversing the equality and multiplying both sides by 2, we get the following.

$$\sin x + \sin y = 2 \sin \left(\frac{x + y}{2}\right) \cos \left(\frac{x - y}{2}\right)$$

The following three identities can be obtained in a similar way.

$$\sin x - \sin y = 2 \cos \left(\frac{x + y}{2}\right) \sin \left(\frac{x - y}{2}\right)$$

$$\cos x + \cos y = 2 \cos \left(\frac{x + y}{2}\right) \cos \left(\frac{x - y}{2}\right)$$

$$\cos x - \cos y = -2 \sin \left(\frac{x + y}{2}\right) \sin \left(\frac{x - y}{2}\right)$$

EXAMPLE 3   Write $\sin 2\gamma - \sin 4\gamma$ as a product of two functions.
Use the identity for $\sin x - \sin y$.

$$\sin 2\gamma - \sin 4\gamma = 2 \cos \left(\frac{2\gamma + 4\gamma}{2}\right) \sin \left(\frac{2\gamma - 4\gamma}{2}\right)$$

$$= 2 \cos \frac{6\gamma}{2} \sin \frac{-2\gamma}{2}$$

$$= 2 \cos 3\gamma \sin (-\gamma)$$

$$\sin 2\gamma - \sin 4\gamma = -2 \cos 3\gamma \sin \gamma \quad \bullet$$

· · · · · · · · · ·

EXAMPLE 4   Verify that $\dfrac{\sin 3s + \sin s}{\cos s + \cos 3s} = \tan 2s$ is an identity.

Work on the left side as follows.

$$\frac{\sin 3s + \sin s}{\cos s + \cos 3s} = \frac{2 \sin\left(\dfrac{3s + s}{2}\right) \cos\left(\dfrac{3s - s}{2}\right)}{2 \cos\left(\dfrac{s + 3s}{2}\right) \cos\left(\dfrac{s - 3s}{2}\right)}$$

$$= \frac{\sin 2s \cos s}{\cos 2s \cos(-s)} = \frac{\sin 2s}{\cos 2s} = \tan 2s \quad \bullet$$

Reduction Identity   Equations of the form $y = a \sin x + b \cos x$ occur so often in mathematics that it is useful to know how to rewrite them in simpler form. In particular, these equations must often be graphed, which is easier to do using the procedure developed here. Figure 5 shows a circle of radius r with the point (a, b) on the circle and on the terminal side of angle α. The circle has equation $x^2 + y^2 = r^2$, so that

$$a^2 + b^2 = r^2, \quad \text{or} \quad r = \sqrt{a^2 + b^2}.$$

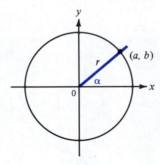

Figure 5

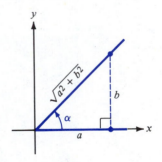

Figure 6

From Figure 6,

$$\sin \alpha = \frac{b}{r} = \frac{b}{\sqrt{a^2 + b^2}} \quad \text{and} \quad \cos \alpha = \frac{a}{r} = \frac{a}{\sqrt{a^2 + b^2}}.$$

Now rewrite $a \sin x + b \cos x$ as follows.

$$a \sin x + b \cos x = \frac{a}{1} \cdot \frac{\sqrt{a^2 + b^2}}{\sqrt{a^2 + b^2}} \sin x + \frac{b}{1} \cdot \frac{\sqrt{a^2 + b^2}}{\sqrt{a^2 + b^2}} \cos x$$

$$= \sqrt{a^2 + b^2} \left(\frac{a}{\sqrt{a^2 + b^2}} \sin x + \frac{b}{\sqrt{a^2 + b^2}} \cos x \right)$$

$$= \sqrt{a^2 + b^2} \left(\sin x \frac{a}{\sqrt{a^2 + b^2}} + \cos x \frac{b}{\sqrt{a^2 + b^2}} \right)$$

Substitute $\sin \alpha$ and $\cos \alpha$ from above.

$$= \sqrt{a^2 + b^2} \ (\sin x \cos \alpha + \cos x \sin \alpha)$$

The quantity in parentheses equals $\sin (x + \alpha)$.

$$a \sin x + b \cos x = \sqrt{a^2 + b^2} \sin (x + \alpha)$$

This result, the **reduction identity,** is summarized as follows.

Reduction Identity

$$a \sin x + b \cos x = \sqrt{a^2 + b^2} \sin (x + \alpha),$$

where $\quad \sin \alpha = \dfrac{b}{\sqrt{a^2 + b^2}} \quad$ and $\quad \cos \alpha = \dfrac{a}{\sqrt{a^2 + b^2}}.$

EXAMPLE 5 Rewrite $\dfrac{1}{2} \sin \theta + \dfrac{\sqrt{3}}{2} \cos \theta$ using the reduction identity.

From the identity above, with $a = \dfrac{1}{2}$ and $b = \dfrac{\sqrt{3}}{2}$,

$$a \sin \theta + b \cos \theta = \sqrt{a^2 + b^2} \sin (\theta + \alpha)$$

becomes $\qquad \dfrac{1}{2} \sin \theta + \dfrac{\sqrt{3}}{2} \cos \theta = 1 \cdot \sin (\theta + \alpha),$

where angle α satisfies the conditions

$$\sin \alpha = \frac{b}{\sqrt{a^2 + b^2}} = \frac{\sqrt{3}}{2} \quad \text{and} \quad \cos \alpha = \frac{a}{\sqrt{a^2 + b^2}} = \frac{1}{2}.$$

The smallest possible positive value of α that satisfies both of these conditions is $\alpha = 60°$. Thus

$$\frac{1}{2} \sin \theta + \frac{\sqrt{3}}{2} \cos \theta = \sin (\theta + 60°). \quad \bullet$$

The reduction identity of this section is useful for graphs that involve sums of sines and cosines. It can be used instead of the method of addition of ordinates discussed in the previous chapter.

EXAMPLE 6 Graph $y = \sin x + \cos x$.

First rewrite $\sin x + \cos x$ using the reduction identity. Since $a = b = 1$, $\sqrt{a^2 + b^2} = \sqrt{2}$, and

$$\sin x + \cos x = \sqrt{2} \sin (x + \alpha).$$

To find α, let

$$\sin \alpha = \frac{b}{\sqrt{a^2 + b^2}} = \frac{1}{\sqrt{2}} \quad \text{and} \quad \cos \alpha = \frac{a}{\sqrt{a^2 + b^2}} = \frac{1}{\sqrt{2}}.$$

The smallest positive angle satisfying these conditions is $\pi/4$, so that

$$y = \sin x + \cos x = \sqrt{2} \sin \left(x + \frac{\pi}{4} \right).$$

The last expression shows that the graph has amplitude $\sqrt{2}$, a period of 2π, and a phase shift of $\pi/4$ to the left, as shown in Figure 7.   ●

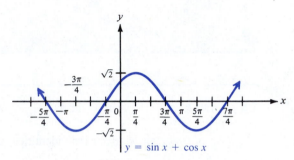

$$y = \sin x + \cos x$$

Figure 7

7.5 Exercises

Rewrite each of the following as a sum or difference of trigonometric functions.

1. $\cos 35° \sin 25°$ **2.** $2 \sin 2x \sin 4x$ **3.** $3 \cos 5x \cos 3x$ **4.** $2 \sin 74° \cos 114°$

5. $\sin(-\theta)\sin(-3\theta)$ **6.** $4\cos(-32°)\sin 15°$ **7.** $-8\cos 4y \cos 5y$ **8.** $2 \sin 3k \sin 14k$

Rewrite each of the following as a product of trigonometric functions.

9. $\sin 60° - \sin 30°$ **10.** $\sin 28° + \sin(-18°)$ **11.** $\cos 42° + \cos 148°$ **12.** $\cos 2x - \cos 8x$

13. $\sin 12\beta - \sin 3\beta$ **14.** $\cos 5x + \cos 10x$ **15.** $-3 \sin 2x + 3 \sin 5x$ **16.** $-\cos 8s + \cos 14s$

Verify that each equation in Exercises 17–26 is an identity.

17. $\tan x = \dfrac{\sin 3x - \sin x}{\cos 3x + \cos x}$ **18.** $\dfrac{\sin 5t + \sin 3t}{\cos 3t - \cos 5t} = \cot t$ **19.** $\dfrac{\cot 2\theta}{\tan 3\theta} = \dfrac{\cos 5\theta + \cos \theta}{\cos \theta - \cos 5\theta}$

20. $\dfrac{\cos \alpha + \cos \beta}{\cos \alpha - \cos \beta} = -\cot \left(\dfrac{\alpha + \beta}{2} \right) \cot \left(\dfrac{\alpha - \beta}{2} \right)$ **21.** $\dfrac{1}{\tan 2s} = \dfrac{\sin 3s - \sin s}{\cos s - \cos 3s}$

22. $\dfrac{\sin^2 5\alpha - 2 \sin 5\alpha \sin 3\alpha + \sin^2 3\alpha}{\sin^2 5\alpha - \sin^2 3\alpha} = \dfrac{\tan \alpha}{\tan 4\alpha}$ **23.** $\sin 6\theta \cos 4\theta - \sin 3\theta \cos 7\theta = \sin 3\theta \cos \theta$

24. $\sin 8\beta \sin 4\beta + \cos 10\beta \cos 2\beta = \cos 6\beta \cos 2\beta$ **25.** $\sin^2 u - \sin^2 v = \sin(u + v)\sin(u - v)$

26. $\cos^2 u - \cos^2 v = -\sin(u + v)\sin(u - v)$

27. Show that the double-angle identity for sine can be considered a special case of the identity $\sin s \cos t = (1/2)[\sin(s + t) + \sin(s - t)]$.

28. Show that the double-angle identity $\cos 2s = 2 \cos^2 s - 1$ is a special case of the identity $\cos s \cos t = (1/2) [\cos (s + t) + \cos (s - t)]$.

Use the reduction identity to simplify each of the following for angles between 0° and 360°. Use a calculator or Table 3 to find angles to the nearest degree. Choose the smallest possible positive value of α.

29. $-\sin x + \cos x$ **30.** $\sqrt{3} \sin x - \cos x$ **31.** $5 \sin \theta - 12 \cos \theta$ **32.** $12 \sin A + 5 \cos A$

33. $-15 \sin x + 8 \cos x$ **34.** $15 \sin B - 8 \cos B$ **35.** $-7 \sin \theta - 24 \cos \theta$ **36.** $24 \cos t - 7 \sin t$

37. $3 \sin x + 4 \cos x$ **38.** $-4 \sin x + 3 \cos x$

Graph each of the following by first changing to the form $y = a \sin (x + \alpha)$.

39. $y = \sqrt{3} \sin x + \cos x$ **40.** $y = \sin x - \sqrt{3} \cos x$ **41.** $y = -\sin x + \cos x$ **42.** $y = -\sin x - \cos x$

7.6 TRIGONOMETRIC EQUATIONS

So far in this chapter trigonometric identities, statements that are true for every value in the domain of the variable, have been discussed. In this section we discuss conditional equations that involve trigonometric functions. As mentioned in the introduction to this chapter, a **conditional equation** is an equation in which some replacements for the variable make the statement true, while others make it false. For example, $2x + 3 = 5$, $x^2 - 5x = 10$, and $2^x = 8$ are conditional equations. Conditional equations with trigonometric functions can usually be solved by using algebraic methods and trigonometric identities to simplify the equations.

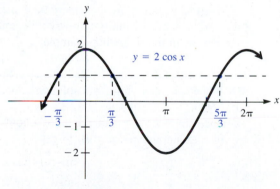

Figure 8

Trigonometric equations usually have infinitely many solutions due to the periodic nature of trigonometric functions. For example, Figure 8 illustrates the solutions of the equation $2 \cos x = 1$. The figure shows the graph of $y = 2 \cos x$ and the location of points on the graph where $2 \cos x = 1$. The corresponding x-values are shown on the x-axis. Solving the equation algebraically for $\cos x$ and then for x in the interval $[0, 2\pi)$ gives

$$2 \cos x = 1$$

$$\cos x = \frac{1}{2}$$

$$x = \frac{\pi}{3} \quad \text{or} \quad \frac{5\pi}{3}.$$

The infinite solution set is written

$$\left\{ x \,\middle|\, x = \frac{\pi}{3} + 2\pi n \text{ or } x = \frac{5\pi}{3} + 2\pi n, \, n \text{ any integer} \right\}.$$

∙ ∙ ∙ ∙ ∙ ∙ ∙ ∙ ∙

EXAMPLE 1   Solve $3 \sin x = \sqrt{2} + \sin x$ for x.

First, solve for $\sin x$. Collect all terms with $\sin x$ on one side of the equation.

$$3 \sin x = \sqrt{2} + \sin x$$

$$2 \sin x = \sqrt{2}$$

$$\sin x = \frac{\sqrt{2}}{2}$$

Between 0 and 2π, $x = \pi/4$ or $3\pi/4$. The infinite solution set is

$$\left\{ x \,\middle|\, x = \frac{\pi}{4} + 2\pi n \text{ or } x = \frac{3\pi}{4} + 2\pi n, \, n \text{ any integer} \right\}. \quad \bullet$$

When an equation involves more than one trigonometric function, it is often help-ful to use a suitable identity to rewrite the equation in terms of just one trigonometric function, as in the following example.

∙ ∙ ∙ ∙ ∙ ∙ ∙ ∙ ∙

EXAMPLE 2   Find all solutions for $\sin x + \cos x = 0$ in the interval $[0°, 360°)$.

Since $\sin x/\cos x = \tan x$ and $\cos x/\cos x = 1$, divide both sides of the equation by $\cos x$ (assuming $\cos x \neq 0$) to get

$$\sin x + \cos x = 0$$

$$\frac{\sin x}{\cos x} + \frac{\cos x}{\cos x} = \frac{0}{\cos x}$$

$$\tan x + 1 = 0$$

$$\tan x = -1.$$

For the last equation the solutions in the given interval are

$$x = 135° \quad \text{and} \quad x = 315°.$$

It was assumed here that $\cos x \neq 0$. If $\cos x = 0$, the given equation becomes $\sin x + 0 = 0$, or $\sin x = 0$. If $\cos x = 0$, it is not possible for $\sin x = 0$, so that no solutions were missed by assuming $\cos x \neq 0$.   ●

The next example shows how factoring may be used to solve a trigonometric equation.

• • • • • • • • •

EXAMPLE 3   Find all solutions for $\sin x \tan x = \sin x$ in the interval $[0°, 360°)$.

Subtract $\sin x$ from both sides, then factor on the left.

$$\sin x \tan x = \sin x$$
$$\sin x \tan x - \sin x = 0$$
$$\sin x (\tan x - 1) = 0$$

Now set each factor equal to 0.

$$\sin x = 0 \qquad\qquad \tan x - 1 = 0$$
$$\qquad\qquad\qquad\qquad \tan x = 1$$
$$x = 0° \quad \text{or} \quad x = 180° \qquad x = 45° \quad \text{or} \quad x = 225°$$

The solutions in the given interval are $0°$, $45°$, $180°$, and $225°$.   ●

There are four solutions for Example 3. Trying to solve the equation by dividing both sides by $\sin x$ would result in $\tan x = 1$, which would give $x = 45°$ or $x = 225°$. The other two solutions would not appear. The missing solutions are the ones that make the divisor, $\sin x$, equal 0. For this reason, it is best to avoid dividing by a variable expression. However, in an equation like that in Example 2, dividing both sides by a variable simplified the solution considerably.

 CAUTION   If you divide by a variable expression, *you need to check* whether the numbers that make that expression equal to 0 are solutions of the original equation.

Sometimes a trigonometric equation can be solved by first squaring both sides, and then using a trigonometric identity. This works for those identities that involve squares, like $\sin^2 x + \cos^2 x = 1$ or $\tan^2 x + 1 = \sec^2 x$. If you square both sides of an equation, *be sure to check* for any numbers that do not satisfy the original equation (even though they do satisfy the squared equation). Do this by substituting potential solutions into the original equation.

EXAMPLE 4   Find all solutions for $\tan x + \sqrt{3} = \sec x$ in the interval $[0, 2\pi)$.

Square both sides, then express $\sec^2 x$ in terms of $\tan^2 x$.

$$\tan x + \sqrt{3} = \sec x$$
$$\tan^2 x + 2\sqrt{3}\, \tan x + 3 = \sec^2 x$$
$$\tan^2 x + 2\sqrt{3}\, \tan x + 3 = 1 + \tan^2 x$$
$$2\sqrt{3}\, \tan x = -2$$
$$\tan x = -\frac{1}{\sqrt{3}}$$

The possible solutions in the given interval are $5\pi/6$ and $11\pi/6$. Now check the possible solutions in the original equation. Try $5\pi/6$ first.

$$\tan x + \sqrt{3} = \tan\frac{5\pi}{6} + \sqrt{3} = \frac{-\sqrt{3}}{3} + \sqrt{3} = \frac{2\sqrt{3}}{3}$$
$$\sec x = \sec\frac{5\pi}{6} = \frac{-2\sqrt{3}}{3}$$

By this check, $5\pi/6$ is not a solution. Now try $11\pi/6$.

$$\tan\frac{11\pi}{6} + \sqrt{3} = \frac{-\sqrt{3}}{3} + \sqrt{3} = \frac{2\sqrt{3}}{3}$$
$$\sec\frac{11\pi}{6} = \frac{2\sqrt{3}}{3}$$

Thus, $11\pi/6$ is the only solution of the original equation in the given interval.   ●

Some trigonometric equations are quadratic in form and can be solved by the methods used to solve quadratic equations.

EXAMPLE 5   Find all solutions of $\tan^2 x + \tan x - 2 = 0$ in the interval $[0, 2\pi)$.

Let $y = \tan x$, so that the equation becomes $y^2 + y - 2 = 0$, with the left side factorable as $(y - 1)(y + 2) = 0$. Substituting $\tan x$ back for y gives

$$(\mathbf{tan}\, x - 1)(\mathbf{tan}\, x + 2) = 0.$$

Set each factor equal to 0.

$$\tan x - 1 = 0 \quad \text{or} \quad \tan x + 2 = 0$$
$$\tan x = 1 \quad \text{or} \quad \tan x = -2$$

If $\tan x = 1$, then for $[0, 2\pi)$, $x = \pi/4$ or $x = 5\pi/4$. If $x = -2$, then a calculator or Table 3 gives $x = 2.0344$ or 5.1760 (approximately). The solutions in $[0, 2\pi)$ are $\pi/4$, $5\pi/4$, 2.0344, and 5.1760.   ●

When a trigonometric equation that is quadratic in form cannot be factored, the quadratic formula can be used to solve the equation.

. . . .

EXAMPLE 6   Find all solutions for $\cot^2 x + 3 \cot x = 1$ in the interval $[0°, 360°)$.

Write the equation with 0 on one side.

$$\cot^2 x + 3 \cot x - 1 = 0$$

Since the expression on the left cannot be readily factored, use the quadratic formula with $a = 1$, $b = 3$, $c = -1$, and $\cot x$ as the variable.

$$\cot x = \frac{-3 \pm \sqrt{9 + 4}}{2} = \frac{-3 \pm \sqrt{13}}{2} = \frac{-3 \pm 3.6056}{2}$$

$$\cot x = .3028 \quad \text{or} \quad \cot x = -3.3028$$

From $\cot x = .3028$, use a calculator or Table 3 to find $x = 73.2°$ or $253.2°$ (since cotangent has period 180°). Also, $\cot x = -3.3028$ leads to $x = 163.2°$ or $343.2°$ giving the solutions 73.2°, 163.2°, 253.2°, and 343.2°.   ●

The methods for solving trigonometric equations illustrated in the examples are summarized as follows.

. . . .

Solving Trigonometric Equations

. . . .

1. If only one trigonometric function is present, first solve the equation for that function.

2. If more than one trigonometric function is present, rearrange the equation so that one side equals 0. Then try to factor.

3. If method 2 does not work, try using identities to change the form of the equation. It may be helpful to square both sides of the equation first. Be sure to check all proposed solutions.

4. If the equation is quadratic in form, but not easily factorable, use the quadratic formula.

Conditional trigonometric equations where a half-angle or multiple angle is given, such as $2 \sin (x/2) = 1$, often require an additional step to solve. This extra step is shown in the following example.

. . . .

EXAMPLE 7   Find all solutions for $2 \sin x/2 = 1$ in the interval $[0°, 360°)$.

Dividing the inequality $0° \le x < 360°$ through by 2 gives

$$0° \le \frac{x}{2} < 180°,$$

so that the first step is to find all values of $x/2$ in the interval $[0°, 180°)$. Begin as before by solving for the trigonometric function.

$$2 \sin \frac{x}{2} = 1$$

$$\sin \frac{x}{2} = \frac{1}{2}$$

Both $\sin 30° = 1/2$ and $\sin 150° = 1/2$, and both $30°$ and $150°$ are in the interval $[0°, 180°)$, so

$$\frac{x}{2} = 30° \quad \text{or} \quad \frac{x}{2} = 150°,$$

from which $x = 60° \quad \text{or} \quad x = 300°.$ ●

• • • • • • • • •

EXAMPLE 8   Find all solutions for $4 \sin x \cos x = \sqrt{3}$ in the interval $[0°, 360°)$.

The identity $2 \sin x \cos x = \sin 2x$ is useful here.

$$4 \sin x \cos x = \sqrt{3}$$

$$2(2 \sin x \cos x) = \sqrt{3}$$

$$2 \sin 2x = \sqrt{3}$$

$$\sin 2x = \frac{\sqrt{3}}{2}$$

The interval $0° \leq x < 360°$ implies $0° \leq 2x < 720°$, so

$$2x = 60°, 120°, 420°, \text{ or } 480°$$

$$x = 30°, 60°, 210°, \text{ or } 240°,$$

and there are four solutions. ●

• • • • • • • • •

EXAMPLE 9   Find all solutions of $\cos 6x - \cos 2x = -\sin 4x$ in the interval $[0, 2\pi)$.

Use the identity

$$\cos x - \cos y = -2 \sin \left(\frac{x + y}{2} \right) \sin \left(\frac{x - y}{2} \right)$$

to rewrite the given equation as

$$-2 \sin \left(\frac{6x + 2x}{2} \right) \sin \left(\frac{6x - 2x}{2} \right) = -\sin 4x$$

$$-2 \sin 4x \sin 2x = -\sin 4x$$

or $2 \sin 4x \sin 2x - \sin 4x = 0.$

Factor to get $\sin 4x\,(2\sin 2x - 1) = 0$

$$\sin 4x = 0 \quad \text{or} \quad 2\sin 2x - 1 = 0$$

$$\sin 4x = 0 \quad \text{or} \quad \sin 2x = \frac{1}{2}.$$

The solutions of $\sin 4x = 0$ are given by $4x = 0 + n \cdot \pi$, so

$$x = 0 + n \cdot \frac{\pi}{4}.$$

Letting $n = 0, 1, 2, 3, 4, 5, 6, 7$ gives the solutions

$$0, \ \pi/4, \ \pi/2, \ 3\pi/4, \ \pi, \ 5\pi/4, \ 3\pi/2, \ \text{and} \ 7\pi/4.$$

If $\sin 2x = 1/2$, then $2x = \pi/6 + n \cdot 2\pi$, and $2x = 5\pi/6 + n \cdot 2\pi$. The first of these produces the solutions $\pi/12$ and $13\pi/12$, while the second produces $5\pi/12$ and $17\pi/12$. The solutions of the original equation are

$$0, \ \pi/4, \ \pi/2, \ 3\pi/4, \ \pi, \ 5\pi/4, \ 3\pi/2, \ 7\pi/4, \ \pi/12, \ 13\pi/12, \ 5\pi/12, \ \text{and} \ 17\pi/12. \quad \bullet$$

The next example shows how to solve equations involving inverse trigonometric functions.

· · · · · · · · · ·

EXAMPLE 10 Solve $\mathrm{Sin}^{-1} x - \mathrm{Cos}^{-1} x = \pi/6$.
Begin by adding $\mathrm{Cos}^{-1} x$ to both sides of the equation to get

$$\mathrm{Sin}^{-1} x = \mathrm{Cos}^{-1} x + \frac{\pi}{6}.$$

Since $\sin(\mathrm{Sin}^{-1}x) = x$, by substitution

$$\sin\left(\mathrm{Cos}^{-1} x + \frac{\pi}{6}\right) = x.$$

Let $u = \mathrm{Cos}^{-1} x$. Then

$$\sin\left(u + \frac{\pi}{6}\right) = x.$$

Use the identity for $\sin(s + t)$, which gives

$$\sin u \cos \frac{\pi}{6} + \cos u \sin \frac{\pi}{6} = x. \qquad (*)$$

Since $u = \mathrm{Cos}^{-1} x$, we have $\cos u = x$. Because of the range of $\mathrm{Cos}^{-1} x$, u can be in either quadrant I or II. Sketch triangles in each of these quadrants and label them as shown in Figure 9 on the next page. In either quadrant I or II, $\sin u$ is positive, with

$$\sin u = \sqrt{1 - x^2}.$$

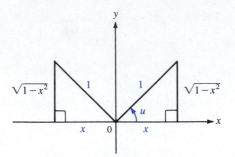

u can be in either quadrant I or II

Figure 9

Replace $\sin u$ with $\sqrt{1 - x^2}$, $\sin \pi/6$ with $1/2$, $\cos \pi/6$ with $\sqrt{3}/2$, and $\cos u$ with x, so that equation (∗) becomes

$$\sqrt{1 - x^2} \cdot \frac{\sqrt{3}}{2} + x \cdot \frac{1}{2} = x$$

$$\sqrt{1 - x^2} \cdot \sqrt{3} + x = 2x$$

$$\sqrt{3} \cdot \sqrt{1 - x^2} = x.$$

Squaring both sides gives $\qquad 3(1 - x^2) = x^2,$

from which $\qquad\qquad\qquad 3 - 3x^2 = x^2$

$$3 = 4x^2$$

$$x = \pm\sqrt{\frac{3}{4}} = \pm\frac{\sqrt{3}}{2}.$$

Check by substitution in the original equation that $\sqrt{3}/2$ is a solution, while $-\sqrt{3}/2$ is not. ●

The next example illustrates how a trigonometric equation may occur in a practical problem.

· · · · · · · · · ·

EXAMPLE 11   The altitude of a projectile in feet (neglecting air resistance) is given by

$$y = (\tan\theta)x - \frac{16}{v_0^2 \cos^2\theta}x^2,$$

where x is the range (horizontal distance covered) in feet and v_0 is the initial velocity of the projectile at an angle θ from the horizontal. See Figure 10. A projectile is fired with an initial velocity of 100 feet per second. Find the firing angle of the projectile so that it strikes the ground 312.5 feet from the firing point.

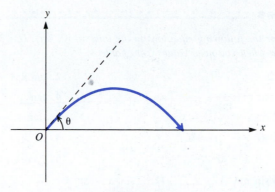

Figure 10

We want to find the value of θ so that $y = 0$ when $x = 312.5$ and $v_0 = 100$. Substitute these values into the given equation.

$$y = (\tan \theta)x - \frac{16}{v_0^2 \cos^2 \theta} x^2$$

$$0 = (\tan \theta)(312.5) - \frac{16}{100^2 \cos^2 \theta} (312.5)^2$$

$$0 = \tan \theta - \frac{16}{10{,}000 \cos^2 \theta} (312.5) \qquad \text{Divide both sides by 312.5.}$$

$$0 = \tan \theta - \frac{1}{2 \cos^2 \theta} \qquad\qquad\qquad \text{Simplify.}$$

$$0 = 2 \cos^2 \theta \tan \theta - 1 \qquad\qquad \text{Multiply both sides by } 2 \cos^2 \theta.$$

$$0 = 2 \cos^2 \theta \left(\frac{\sin \theta}{\cos \theta} \right) - 1 \qquad\qquad \tan \theta = \frac{\sin \theta}{\cos \theta}$$

$$0 = 2 \cos \theta \sin \theta - 1 \qquad\qquad \frac{\cos^2 \theta}{\cos \theta} = \cos \theta$$

$$0 = \sin 2\theta - 1 \qquad\qquad\qquad 2 \cos \theta \sin \theta = \sin 2\theta$$

$$\sin 2\theta = 1$$

$$2\theta = 90°$$

$$\theta = 45°$$

The projectile should be fired at an angle of 45° to meet the requirements of the problem. Note that θ must be in the interval $(0, 90°)$ in this situation. ●

7.6 Exercises ·

Find all solutions for the following equations in the interval $[0, 2\pi)$. Use 3.1416 as an approximation for π when you need values from Table 3.

1. $3 \tan x + 5 = 2$

2. $\tan x + 1 = 2$

3. $2 \sec x + 1 = \sec x + 3$

4. $\tan^2 x - 1 = 0$

5. $(\cot x - \sqrt{3})(2 \sin x + \sqrt{3}) = 0$

6. $(\tan x - 1)(\cos x - 1) = 0$

7. $(\sec x - 2)(\sqrt{3} \sec x - 2) = 0$

8. $(2 \sin x + 1)(\sqrt{2} \cos x + 1) = 0$

9. $\cos^2 x + 2 \cos x + 1 = 0$

10. $2 \cos^2 x - \sqrt{3} \cos x = 0$

11. $-2 \sin^2 x = 3 \sin x + 1$

12. $3 \sin^2 x - \sin x = 2$

13. $\cos^2 x - \sin^2 x = 0$

14. $\dfrac{2 \tan x}{3 - \tan^2 x} = 1$

15. $\sin 2x = 0$

16. $\cos 2x = 1$

17. $3 \tan 2x = \sqrt{3}$

18. $\cot 2x = \sqrt{3}$

19. $\sqrt{2} \cos 2x = -1$

20. $2\sqrt{3} \sin 2x = -3$

21. $\sin \dfrac{x}{2} = \sqrt{2} - \sin \dfrac{x}{2}$

22. $\cos 2x - \cos x = 0$

Find all solutions for the following equations in the interval $[0°, 360°)$. Find θ to the nearest tenth of a degree.

23. $\tan \theta + 6 \cot \theta = 5$

24. $\csc \theta = 2 \sin \theta + 1$

25. $\sec^2 \theta = 2 \tan \theta + 4$

26. $2 \tan^2 \theta \sin \theta - \tan^2 \theta = 0$

27. $5 \sec^2 \theta = 6 \sec \theta$

28. $\cos^2 \theta = \sin^2 \theta + 1$

29. $\csc^2 \theta - 2 \cot \theta = 0$

30. $3 \cot^3 \theta = \cot \theta$

31. $\sin^2 \theta \cos^2 \theta = 0$

32. $\sec^2 \theta \tan \theta = 2 \tan \theta$

33. $2 \sin 2\theta = \sqrt{3}$

34. $2 \cos 2\theta = \sqrt{2}$

35. $2\sqrt{3} \sin \dfrac{\theta}{2} = 3$

36. $2\sqrt{3} \cos \dfrac{\theta}{2} = -3$

37. $2 \sin \theta = 2 \cos 2\theta$

38. $\cos \theta - 1 = \cos 2\theta$

39. $\sin 2\theta = 2 \cos^2 \theta$

40. $\csc^2 \dfrac{\theta}{2} = 2 \sec \theta$

41. $\cos \theta = \sin^2 \dfrac{\theta}{2}$

42. $4 \cos 2\theta = 8 \sin \theta \cos \theta$

43. $2 \cos^2 2\theta = 1 - \cos 2\theta$

44. $\sin \theta = \cos \dfrac{\theta}{2}$

Give all solutions (to the nearest tenth of a degree) for each of the following. Write the solutions in the form used in Example 1.

45. $\sin x - \cos x = 1$

46. $\cos 2x = 1$

47. $\cos x \left(\sin x - \dfrac{1}{2} \right) = 0$

48. $(2 \sin x - \sqrt{3})(\cos x + 1) = 0$

49. $\tan^2 x + 2 \tan x = 3$

50. $\cot^2 x - 4 \cot x - 5 = 0$

51. $\sin 4x = \cos 4x$

52. $\tan 3x = \cot 3x$

To solve the following equations, you will need the quadratic formula. Find all solutions in the interval $[0°, 360°)$. Give solutions to the nearest tenth of a degree.

53. $9 \sin^2 x - 6 \sin x = 1$

54. $4 \cos^2 x + 4 \cos x = 1$

55. $\tan^2 x + 4 \tan x + 2 = 0$

56. $3 \cot^2 x - 3 \cot x - 1 = 0$

57. $\cot x + 2 \csc x = 3$

58. $2 \sin x = 1 - 2 \cos x$

For the following equations, use the sum and product identities of Section 7.5. Give all solutions in the interval $[0, 2\pi)$.

59. $\sin x + \sin 3x = \cos x$

60. $\cos 4x - \cos 2x = \sin x$

61. $\sin 3x - \sin x = 0$

62. $\cos 2x + \cos x = 0$

63. $\sin 4x + \sin 2x = 2 \cos x$

64. $\cos 5x + \cos 3x = 2 \cos 4x$

In an electrical circuit, let V represent the electromotive force in volts at t seconds. Assume $V = \cos 2\pi t$. Find the smallest positive value of t where $0 \le t \le 1/2$ for the values of V in Exercises 65 and 66.

65. $V = 0$

66. $V = .5$

An equation for the curve describing the altitude of a projectile is

$$y = (\tan \theta)x - \frac{16}{v_0^2 \cos^2 \theta} x^2$$

where v_0 is the initial velocity of the projectile and θ is the angle from the horizontal at which it is fired. (See Example 11.)

67. A projectile is fired with an initial velocity of 400 ft per sec at an angle of 45° with the horizontal. Find each of the following: **(a)** the range (horizontal distance covered),   **(b)** the maximum altitude.

68. Repeat Exercise 67 if the projectile is fired at 800 ft per sec at an angle of 30° with the horizontal.

A coil of wire rotating in a magnetic field induces a voltage given by

$$e = 20 \sin \left(\frac{\pi t}{4} - \frac{\pi}{2} \right),$$

where t is time in seconds. Find the smallest positive time to produce the voltages in Exercises 69 and 70.

69. 0

70. $10\sqrt{3}$

71. The equation $.342D \cos \theta + h \cos^2 \theta = 16D^2/V_0^2$ is used in reconstructing accidents in which a vehicle vaults into the air after hitting an obstruction. V_0 is the velocity in feet per second of the vehicle when it hits, D is the distance (in feet) from the obstruction to the landing point, and h is the difference in height (in feet) between the landing point and the takeoff point. Angle θ is the takeoff angle, the angle between the horizontal and the path of the vehicle. Find θ to the nearest degree if $V_0 = 60$, $D = 80$, and $h = 2$.

72. The seasonal variation in the length of daylight can be represented by a sine function. For example, the daily number of hours of daylight in New Orleans is given by

$$h = \frac{35}{3} + \frac{7}{3} \sin \frac{2\pi x}{365},$$

where x is the number of days after March 21 (disregarding leap year).
(a) On what date will there be about 14 hours of daylight?
(b) What date has the least number of hours of daylight?
(c) When will there be about 10 hours of daylight?

73. The British nautical mile is defined as the length of a minute of arc of a meridian. Since the earth is flat at its poles, the nautical mile, in feet, is given by $L = 6,077 - 31 \cos 2\theta$, where θ is the latitude in degrees. See the figure.

(a) Find the latitude(s) at which the nautical mile is 6,074 feet.

(b) At what latitude(s) is the nautical mile 6,108 feet?

(c) In the United States the nautical mile is defined everywhere as 6,080.2 feet. At what latitude(s) does this agree with the British nautical mile?*

A nautical mile is the length on any of these meridians cut by a central angle of measure 1 minute.

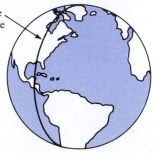

74. When a large view camera is used to take a picture of an object that is not parallel to the film, the lens board should be tilted so that the planes containing the subject, the lens board, and the film intersect in a line (see the figure). This gives the best "depth of field."*

(a) Write two equations, one relating α, x, and z, and the other relating β, x, y, and z.

(b) Eliminate z from the equations in part (a) to get one equation relating α, β, x, and y.

(c) Solve the equation from part (b) for α.

(d) Solve the equation from part (b) for β.

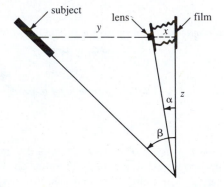

Solve each of the following equations.

75. $\operatorname{Sin}^{-1} x = \operatorname{Tan}^{-1} 3/4$

76. $\operatorname{Tan}^{-1} x = \operatorname{Cos}^{-1} 5/13$

77. $\arccos x = \arcsin 3/5$

78. $\arctan x = \arcsin(-4/5)$

79. $\operatorname{Cos}^{-1} x + 2 \operatorname{Sin}^{-1} \sqrt{3}/2 = \pi$

80. $\operatorname{Sin}^{-1} x + \operatorname{Tan}^{-1} \sqrt{3} = 2\pi/3$

81. $\operatorname{Sin}^{-1} 2x + \operatorname{Cos}^{-1} x = \pi/6$

82. $\operatorname{Sin}^{-1} 2x + \operatorname{Sin}^{-1} x = \pi/2$

83. $\operatorname{Cos}^{-1} x + \operatorname{Tan}^{-1} x = \pi/2$

84. $\operatorname{Tan}^{-1} x + \operatorname{Cos}^{-1} x = \pi/4$

*Exercises 72, 73, and 74 are from *A Sourcebook of Applications of School Mathematics* by Donald Bushaw et al. Copyright © 1980 by The Mathematical Association of America. Reprinted by permission. The material was prepared with the support of National Science Foundation Grant No. SED72-01123 A05. However, any opinions, findings, conclusions, or recommendations expressed herein are those of the authors and do not necessarily reflect the views of NSF.

7 CHAPTER SUMMARY

Key Words

identity
7.1 fundamental identities
negative angle identities
7.3 sum and difference
identities
cofunction identities

7.4 double-angle identities
half-angle identities
7.5 sum and product identities
reduction identity
7.6 conditional equation

Review Exercises

1. Use the trigonometric identities to find the remaining five trigonometric function values of x, given that $\cos x = 3/5$ and x is in quadrant IV.

2. Given $\tan x = -5/4$, where x is in the interval $(\pi/2, \pi)$, use trigonometric identities to find the other trigonometric function values of x.

3. Given $\sin x = -1/4$, $\cos y = -4/5$, and both x and y are in quadrant III, find $\sin (x + y)$ and $\cos (x - y)$.

4. Given $\sin 2\theta = \sqrt{3}/2$ and 2θ terminates in quadrant II, use trigonometric identities to find $\tan \theta$.

5. Given $x = \pi/8$, use trigonometric identities to find $\sin x$, $\cos x$, and $\tan x$.

For each item in List I, give the letter of the item in List II that completes an identity.

List I		List II
6. $\sin 35°$	**(a)** $\sin (-35°)$	**(e)** $\cos 150° \cos 60° - \sin 150° \sin 60°$
7. $\tan (-35°)$	**(b)** $\cos 55°$	**(f)** $\cot (-35°)$
8. $\cos 35°$		**(g)** $\cos^2 150° - \sin^2 150°$
9. $\cos 75°$	**(c)** $\sqrt{\dfrac{1 + \cos 150°}{2}}$	**(h)** $\sin 15° \cos 60° + \cos 15° \sin 60°$
10. $\sin 75°$		**(i)** $\cos (-35°)$
11. $\sin 300°$	**(d)** $2 \sin 150° \cos 150°$	**(j)** $\cot 125°$
12. $\cos 300°$		

For each item in List I give the letter of the item in List II that completes an identity.

List I		List II	
13. $\csc x$	**(a)** $\dfrac{1}{\sin x}$	**(d)** $\dfrac{1}{\cot^2 x}$	**(g)** $\dfrac{1}{\sin^2 x}$
14. $\tan x$			
15. $\cot x$	**(b)** $\dfrac{1}{\cos x}$	**(e)** $\dfrac{1}{\cos^2 x}$	**(h)** $1 - \cos^2 x$
16. $\sin^2 x$			
17. $\tan^2 x + 1$	**(c)** $\dfrac{\sin x}{\cos x}$	**(f)** $\dfrac{\cos x}{\sin x}$	
18. $\tan^2 x$			

Use identities to express each of the following in terms of sin θ and cos θ and simplify.

19. $\sec^2 \theta - \tan^2 \theta$

20. $\dfrac{\cot \theta}{\sec \theta}$

21. $\tan^2 \theta (1 + \cot^2 \theta)$

22. $\csc \theta + \cot \theta$

23. $\csc^2 \theta + \sec^2 \theta$

24. $\tan \theta - \sec \theta \csc \theta$

Show that each of the following is an identity.

25. $\dfrac{\sin 2x}{\sin x} = \dfrac{2}{\sec x}$

26. $2 \cos A - \sec A = \cos A - \dfrac{\tan A}{\csc A}$

27. $\dfrac{2 \tan B}{\sin 2B} = \sec^2 B$

28. $\tan \beta = \dfrac{1 - \cos 2\beta}{\sin 2\beta}$

29. $1 + \tan^2 \alpha = 2 \tan \alpha \csc 2\alpha$

30. $-\dfrac{\sin (A - B)}{\sin (A + B)} = \dfrac{\cot A - \cot B}{\cot A + \cot B}$

31. $\dfrac{\sin t}{1 - \cos t} = \cot \dfrac{t}{2}$

32. $2 \cos (A + B) \sin (A + B) = \sin 2A \cos 2B + \sin 2B \cos 2A$

33. $\dfrac{2 \cot x}{\tan 2x} = \csc^2 x - 2$

34. $\sin t = \dfrac{\cos t \sin 2t}{1 + \cos 2t}$

35. $\tan \theta \sin 2\theta = 2 - 2 \cos^2 \theta$

36. $\csc A \sin 2A - \sec A = \cos 2A \sec A$

37. $2 \tan x \csc 2x - \tan^2 x = 1$

38. $2 \cos^2 \theta - 1 = \dfrac{1 - \tan^2 \theta}{1 + \tan^2 \theta}$

39. $\sin^3 \theta = \sin \theta - \cos^2 \theta \sin \theta$

40. $\dfrac{\sin^2 x}{2 - 2 \cos x} = \cos^2 \dfrac{x}{2}$

41. $\cos^4 \theta = \dfrac{3}{8} + \dfrac{1}{2} \cos 2\theta + \dfrac{1}{8} \cos 4\theta$

42. $8 \sin^2 \dfrac{\gamma}{2} \cos^2 \dfrac{\gamma}{2} = 1 - \cos 2\gamma$

43. $\cos^2 \dfrac{x}{2} = \dfrac{1 + \sec x}{2 \sec x}$

44. $\tan \theta \cos^2 \theta = \dfrac{2 \tan \theta \cos^2 \theta - \tan \theta}{1 - \tan^2 \theta}$

45. $\tan 8k - \tan 8k \cdot \tan^2 4k = 2 \tan 4k$

46. $\sec^2 \alpha - 1 = \dfrac{\sec 2\alpha - 1}{\sec 2\alpha + 1}$

47. $\dfrac{\sin 3t + \sin 2t}{\sin 3t - \sin 2t} = \dfrac{\tan \dfrac{5t}{2}}{\tan \dfrac{t}{2}}$

48. $\sin 2\alpha = \dfrac{2(\sin \alpha - \sin^3 \alpha)}{\cos \alpha}$

49. $\tan 2\beta - \sec 2\beta = \dfrac{\tan \beta - 1}{\tan \beta + 1}$

50. $\dfrac{\sin^3 t - \cos^3 t}{\sin t - \cos t} = \dfrac{2 + \sin 2t}{2}$

51. $-\cot \dfrac{x}{2} = \dfrac{\sin 2x + \sin x}{\cos 2x - \cos x}$

52. $2 \cos^3 x - \cos x = \dfrac{\cos^2 x - \sin^2 x}{\sec x}$

Find all solutions for the following equations in the interval $[0, 2\pi)$.

53. $\sin^2 x = 1$

54. $2 \tan x - 1 = 0$

55. $2 \sin^2 x - 5 \sin x + 2 = 0$

56. $\tan x = \cot x$

57. $\sec^4 2x = 4$

58. $\tan^2 2x - 1 = 0$

59. $\sin \dfrac{x}{2} = \cos \dfrac{x}{2}$

60. $\sec \dfrac{x}{2} = \cos \dfrac{x}{2}$

61. $\cos 2x + \cos x = 0$

62. $\sin x \cos x = \dfrac{1}{4}$

Find all solutions for the following equations in the interval $[0°, 360°)$.

63. $2 \cos \theta = 1$

64. $(\tan \theta + 1)\left(\sec \theta - \dfrac{1}{2}\right) = 0$

65. $\sin^2 \theta + 3 \sin \theta + 2 = 0$

66. $\sin 2\theta = \cos 2\theta + 1$

67. $\dfrac{\sin \theta}{\cos \theta} = \tan^2 \theta$

68. $2 \sin 2\theta = 1$

69. Recall Snell's law from Exercises 95–98 of Section 6.4: $c_1/c_2 = \sin \theta_1/\sin \theta_2$, where c_1 is the speed of light in one medium, c_2 is the speed of light in a second medium, and θ_1 and θ_2 are the angles shown in the figure. Suppose that a light is shining up through water into the air as in the figure. As θ_1 increases, θ_2 approaches 90°, at which point no light will emerge from the water. Assume the ratio c_1/c_2 in this case is .752.
   (a) For what value of θ_1 does $\theta_2 = 90°$? This value of θ_1 is the *critical angle* for water.
   (b) What happens when θ_1 is greater than the critical angle?

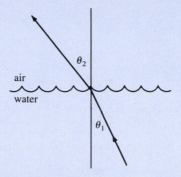

70. The angle between the downward vertical position and another position of a rhythmically moving arm is given by $y = (1/3) \sin (4\pi t/3)$, where t is time in seconds. See the figure below.
   (a) Solve the equation for t.
   (b) At what time(s) does the arm form an angle of .3 radians?

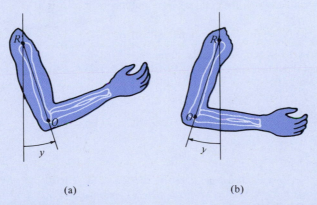

(a)   (b)

Schematic diagrams of a rhythmically moving arm. The upper arm RO rotates back and forth about the point R; the position of the arm is measured by the angle y between the actual position and the downward vertical position.

Exact values of the trigonometric functions of 15° can be found by the following method, an alternative to the use of the half-angle formulas. Start with a right triangle ABC having a 60° angle at A and a 30° angle at B. Let the hypotenuse of this triangle have length 2. Extend side BC and draw a semicircle with diameter along BC extended, center at B, and radius AB. Draw segment AE. (See the figure.) Since any angle inscribed in a semicircle is a right angle, triangle AED is a right triangle.

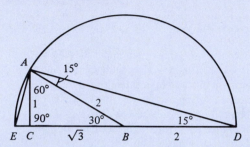

71. Show that triangle *ABD* is isosceles.

72. Show that angle *ABD* is 150°.

73. Show that angle *DAB* is 15°, as is angle *ADB*.

74. Show that *DC* has length $2 + \sqrt{3}$.

75. Since *AC* has length 1, the length of *AD* is given by $(AD)^2 = 1^2 + (2 + \sqrt{3})^2$.
Reduce this to $\sqrt{8 + 4\sqrt{3}}$, and show that this result equals $\sqrt{6} + \sqrt{2}$.

76. Use angle *ADB* of triangle *ADE* and find cos 15°.

77. Show that *AE* has length $\sqrt{6} - \sqrt{2}$. **78.** Find sin 15°.

79. Use triangle *ACE* and find tan 15°. **80.** Find cot 15°.

The following exercises are taken with permission from a standard calculus book.*

81. Let α and β be two given numbers.
  (a) Prove that if $\sin(\alpha + \beta) = \sin(\alpha - \beta)$, then either α is an odd multiple of $\pi/2$ or β is a multiple of π, or both.
  (b) Prove that if $\cos(\alpha + \beta) = \cos(\alpha - \beta)$, then either α is a multiple of π or β is a multiple of π, or both.
  (c) Prove that if $\tan(\alpha + \beta) = \tan(\alpha - \beta)$, then β is a multiple of π.

82. Note that for $\alpha = 18°$, we have $\cos 3\alpha = \sin 2\alpha$. Use the formula
$\cos 3\alpha = 4\cos^3 \alpha - 3\cos \alpha$ and show that $\sin 18° = (\sqrt{5} - 1)/4$.

*Reproduced from *Calculus*, 2nd edition, by Leonard Gillman and Robert H. McDowell, by permission of W. W. Norton & Company, Inc. Copyright © 1978, 1973 by W. W. Norton & Company, Inc.

8

APPLICATIONS OF TRIGONOMETRY; DE MOIVRE'S THEOREM

In this chapter we discuss some applications of trigonometry in surveying, navigation, electronics, and other engineering areas. The ancient Egyptians, Babylonians, and Greeks developed trigonometry to find the lengths of the sides of triangles and the measures of their angles.

In Egypt, trigonometry was used to establish land boundaries after the annual flooding of the Nile River. In Babylonia it was used in astronomy. When Greek astronomers made physical models to study the motions of the planets, they computed sides and angles of new triangles from other triangles. Indian astronomers used the Greek models to make up handbooks with tables containing the various computations. In this way, trigonometry and other branches of mathematics originated.

8.1 RIGHT TRIANGLE APPLICATIONS

Every triangle has three sides and three angles. The first few sections of this chapter show that if any three of these measures are known (if at least one measure is a side), then the other three measures can be found. The process of finding the unknown parts of a triangle is called **solving the triangle.**

Recall that if θ is an acute angle of a right triangle, then $\sin \theta$, $\cos \theta$, and $\tan \theta$ can be found from the lengths of the sides of the triangle.

$$\sin \theta = \frac{\text{opposite}}{\text{hypotenuse}} \qquad \cos \theta = \frac{\text{adjacent}}{\text{hypotenuse}} \qquad \tan \theta = \frac{\text{opposite}}{\text{adjacent}}$$

The other three values, $\cot \theta$, $\sec \theta$, and $\csc \theta$, can be found in a similar way. See Figure 1 on the next page.

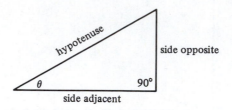

Figure 1

• • • • • • • • •

EXAMPLE 1   The right triangle in Figure 2 has sides of lengths 3, 4, and 5. Find each of the following.

(a) $\sin \theta$, $\cos \theta$, and $\tan \theta$
Use the quotients given above.

$$\sin \theta = \frac{\text{side opposite}}{\text{hypotenuse}} = \frac{3}{5}$$

$$\cos \theta = \frac{\text{side adjacent}}{\text{hypotenuse}} = \frac{4}{5}$$

$$\tan \theta = \frac{\text{side opposite}}{\text{side adjacent}} = \frac{3}{4}$$

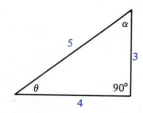

Figure 2

(b) The degree measure of angle θ, to the nearest tenth of a degree
Since $\sin \theta = 3/5 = .6000$, using a calculator or Table 3 gives

$$\theta \approx 36.9°.$$

(c) $\sin \alpha$, $\cos \alpha$, and $\tan \alpha$
Use the appropriate values from Figure 2.

$$\sin \alpha = \frac{4}{5}, \quad \cos \alpha = \frac{3}{5}, \quad \text{and} \quad \tan \alpha = \frac{4}{3}$$

(d) The degree measure of angle α
The sum of the two acute angles of a right triangle is 90°, and $\theta = 36.9°$ from part (b), so

$$\alpha \approx 90° - 36.9°$$

$$= 53.1°. \quad \bullet$$

When trigonometry is used to solve triangles or to find the measures of all sides and all angles, it is convenient to use a to represent the length of the side opposite angle A, b for the length of the side opposite angle B, and so on. The letter c is used for the hypotenuse in a right triangle.

. . . .

EXAMPLE 2   Solve right triangle *ABC*, with $A = 34.5°$ and $c = 12.7$. See Figure 3.

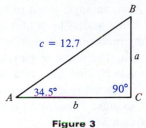

Figure 3

This triangle can be solved by finding the measures of the remaining sides and angles. By the definitions given above, $\sin A = a/c$, where $A = 34.5°$ and $c = 12.7$.

$$\sin A = \frac{a}{c}$$

$$\sin \mathbf{34.5°} = \frac{a}{\mathbf{12.7}}$$

Upon multiplying both sides by 12.7,

$$a = 12.7 \sin 34.5° \approx 12.7(.5664) \approx 7.19.$$

Now, $\cos A$ can be used to find *b*.

$$\cos A = \frac{\text{side adjacent}}{\text{hypotenuse}} = \frac{b}{c}$$

$$\cos \mathbf{34.5°} = \frac{b}{\mathbf{12.7}}$$

$$b = 12.7 \cos 34.5° \approx 12.7 \,(.8241) \approx 10.5$$

Once *b* has been found, the Pythagorean theorem could be used as a check. All that is still needed for solving triangle *ABC* is to find *B*. Since $A + B = 90°$, so that $B = 90° - A$, and $A = 34.5°$,

$$B = 90° - 34.5° = 55.5°.$$

Triangle *ABC* is now solved—the lengths of all sides and the measures of all angles are known.   ●

CAUTION   In Example 2, the value of *b* could have been found with the Pythagorean theorem. It is better, however, to use the information given in the problem rather than a result just calculated. If a mistake were to be made in finding *a*, then *b* would also be incorrect.

Recall the earlier discussion of significant digits. The number of significant digits in an answer cannot exceed the least number of significant digits in any number used to find the answer. Thus, in Example 2 the answers were given to three significant digits. (90° is an exact measure here.) Although most of the results of our work in this chapter are rounded, in the rest of the chapter, we will use the symbol "=" rather than "≈" with the understanding that these answers are actually approximations. Also, an angle given to the nearest degree is assumed to have *two* significant digits; to the nearest tenth of a degree, *three* significant digits, and so on.

Many problems with right triangles involve the angle of elevation or the angle of depression. The **angle of elevation** from point X to point Y (above X) is the acute angle made by line XY and a horizontal line through X. The angle of elevation is always measured from the horizontal. See Figure 4. The **angle of depression** from point X to point Y (below X) is the acute angle made by the line XY and a horizontal line through X.

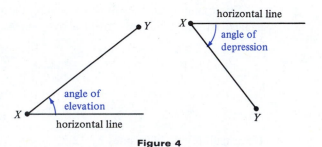

Figure 4

· · · · · · · · ·

EXAMPLE 3   Judy Bezzone knows that when she stands 123 feet from the base of a flagpole, the angle of elevation to the top is 26.7°. If her eyes are 5.30 feet above the ground, find the height of the flagpole.

The length of the side adjacent to Bezzone is known (see Figure 5) and the length of the side opposite her must be found. The ratio that involves these two values is the tangent.

$$\tan A = \frac{\text{side opposite}}{\text{side adjacent}}$$

$$\tan 26.7° = \frac{a}{123}$$

$$a = 123 \tan 26.7° = 61.9 \text{ feet}$$

Since Bezzone's eyes are 5.30 feet above the ground, the height of the flagpole is

$$61.9 + 5.30 = 67.2 \text{ feet.} \quad \bullet$$

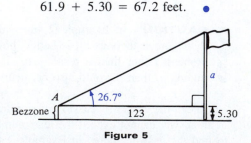

Figure 5

. . . .

EXAMPLE 4   Wang Wei needs to know the height of a tree. From a given point on the ground he finds that the angle of elevation to the top of the tree is 36.7°. He then moves back 50.0 ft. From the second point, the angle of elevation to the top of the tree is 22.2°. See Figure 6. Find the height of the tree.

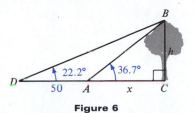

Figure 6

The figure shows two unknowns, x, the distance from the center of the trunk of the tree to a point where the first observation was made, and h, the height of the tree. Since nothing is given about the length of the hypotenuse of either triangle ABC or triangle BCD, use a ratio that does not involve the hypotenuse, the tangent.

In triangle ABC, $\tan 36.7° = \dfrac{h}{x}$ or $h = x \tan 36.7°$.

In triangle BCD, $\tan 22.2° = \dfrac{h}{50 + x}$ or $h = (50 + x) \tan 22.2°$.

Since each of these two expressions equals h, these expressions must be equal.

$$x \tan 36.7° = (50 + x) \tan 22.2°$$

Now use algebra to solve for x.

$$x \tan 36.7° = 50 \tan 22.2° + x \tan 22.2°$$
$$x \tan 36.7° - x \tan 22.2° = 50 \tan 22.2°$$
$$x(\tan 36.7° - \tan 22.2°) = 50 \tan 22.2°$$
$$x = \frac{50 \tan 22.2°}{\tan 36.7° - \tan 22.2°}$$

It was shown above that $h = x \tan 36.7°$. Substituting for x gives

$$h = \left(\frac{50 \tan 22.2°}{\tan 36.7° - \tan 22.2°}\right)(\tan 36.7°).$$

From a calculator or Table 3,

$$\tan 36.7° = .7454$$

and

$$\tan 22.2° = .4081$$

so

$$\tan 36.7° - \tan 22.2° = .7454 - .4081 = .3373,$$

with

$$h = \left(\frac{50(.4081)}{.3373}\right)(.7454) = 45 \text{ ft.} \quad \bullet$$

Many applications of trigonometry involve **bearing,** an important idea in navigation and surveying. Bearing is used to give directions. There are two common systems used to express bearing. When a single angle is given, such as 164°, bearing is measured in a clockwise direction from due north. Several sample bearings using this system are shown in Figure 7.

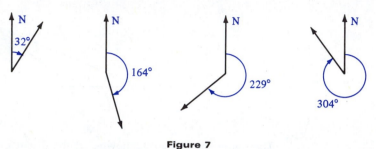

Figure 7

• • • • • • • • •

EXAMPLE 5   Radar stations A and B are on an east-west line, 3.7 kilometers apart. Station A detects a plane at C on a bearing of 61°. Station B detects the same plane on a bearing of 331°. Find the distance from A to C.

Draw a sketch showing the given information, as in Figure 8. Angle C is a right angle, since angles CAB and CBA have a sum of 90°. The necessary distance, b, can be found by using cosine.

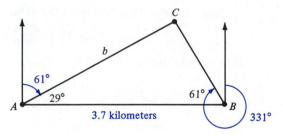

Figure 8

As the figure suggests,

$$\cos 29° = \frac{b}{3.7}$$

$$3.7 \cos 29° = b$$

$$3.7(.8746) = b$$

$$b = 3.2 \text{ kilometers.} \quad \bullet$$

The other common system for expressing bearing starts with a north-south line and uses an acute angle to show the direction, either east or west, from this line. Figure 9 shows several sample bearings using this system. The letter N or S always comes first, followed by an acute angle, and then E or W.

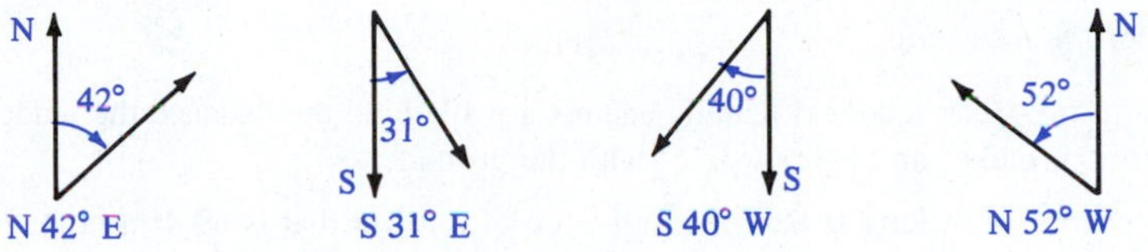

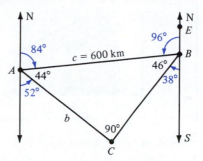

Figure 9

EXAMPLE 6   The bearing from A to C is S 52° E. The bearing from A to B is N 84° E. The bearing from B to C is S 38° W. A plane flying at 250 kilometers per hour takes 2.4 hours to go from A to B. Find the distance from A to C.

Figure 10

Figure 10 shows a sketch of the given information. Since the bearing from A to B is 84°, angle ABE is 180° − 84° = 96°, with angle ABC equal to 46°. Also, angle BAC is 180° − (84° + 52°) = 44°. Angle C is 180° − (44° + 46°) = 90°. Since a plane flying at 250 kilometers per hour takes 2.4 hours to go from A to B, the distance from A to B is 2.4(250) = 600 kilometers. To find b, the distance from A to C, use sine. (Cosine could also have been used.)

$$\sin 46° = \frac{b}{c} = \frac{b}{600}$$

$$600 \sin 46° = b$$

$$b = 430 \text{ kilometers} \quad \bullet$$

8.1 Exercises

Solve each of the following right triangles. Angle C is the right angle.

1. $B = 73.0°$, $b = 128$ in

2. $A = 61.0°$, $b = 39.2$ cm

3. $a = 76.4$ yd, $b = 39.3$ yd

4. $a = 958$ m, $b = 489$ m

5. $a = 18.9$ cm, $c = 46.3$ cm

6. $b = 219$ m, $c = 647$ m

7. $A = 53.24°$, $c = 387.1$ ft

8. $A = 13.47°$, $c = 1285$ m

9. $B = 39.10°$, $c = .6231$ m

10. $B = 82.83°$, $c = 4.825$ cm

11. $c = 7.813$ m, $b = 2.467$ m

12. $c = 44.91$ mm, $a = 32.71$ mm

13. $B = 42.432°$, $a = 157.49$ m

14. $A = 36.704°$, $c = 1461.3$ cm

15. $A = 57.209°$, $c = 186.49$ cm

16. $B = 12.099°$, $b = 7.0463$ m

17. $b = 173.921$ m, $c = 208.543$ m

18. $a = 864.003$ cm, $c = 1092.84$ cm

Solve each of the following.

19. A 39.4-m fire-truck ladder is leaning against a wall. Find the distance the ladder goes up the wall if it makes an angle of 42.5° with the ground.

20. A guy wire 87.4 m long is attached to the top of a tower that is 69.4 m high. Find the angle that the wire makes with the ground.

21. A representation of an aerial photograph of a complex of buildings is shown on the left below.* If the sun was at an angle 26.5° when the photograph was taken, how high is the building diagramed on the right? Use .48 cm as the length of the shadow.

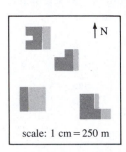

scale: 1 cm = 250 m

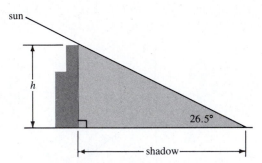

22. The figure at left below represents an aerial photograph of a cliff in a remote region of Antarctica casting a shadow. Compute the height of the cliff if the angle of elevation of the sun was 19.0° when the photograph was taken.

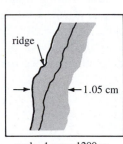

scale: 1 cm = 1200 m

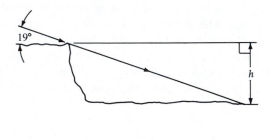

Work the following problems involving angles of elevation or depression.

23. Suppose the angle of elevation of the sun is 28.2°. Find the length of the shadow cast by a person 6.0 ft tall.

24. The shadow of a vertical tower is 58.2 m long when the angle of elevation of the sun is 36.3°. Find the height of the tower.

25. Find the angle of elevation of the sun if a 53.9 ft flagpole casts a shadow 74.6 ft long.

26. The angle of depression from the top of a building to a point on the ground is 34.8°. How far is the point on the ground from the top of the building if the building is 368 m high?

*Exercises 21 and 22 from *Plane Trigonometry*, fourth edition, by Bernard J. Rice and Jerry D. Strange. Copyright © 1986 by Prindle, Weber & Schmidt. Reprinted by permission.

27. A television camera is to be mounted on a bank wall so as to have a good view of the head teller (see the figure). Find the angle of depression that the lens should make with the horizontal.

28. A company safety committee has recommended that a floodlight be mounted in a parking lot so as to illuminate the employee exit (see the figure). Find the angle of depression of the light.

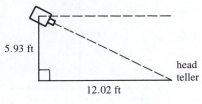

5.93 ft

12.02 ft

head teller

Exercise 27

29. Marge drives her Mercedes up a straight road inclined at an angle of 4.20° with the horizontal. She starts at an elevation of 680 ft above sea level and drives 12,400 ft along the road. Find her final altitude.

30. The road into Death Valley is straight; it makes an angle of 4.20° with the horizontal. Starting at sea level, the road descends to −121 ft. Find the distance it is necessary to travel along the road to reach bottom.

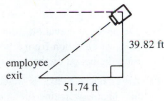

39.82 ft

employee exit

51.74 ft

Exercise 28

31. A sailboat is sailing due east at 10 mph. A powerboat is several miles south of the sailboat and capable of traveling at 20 mph. (See the figure.) What bearing will allow the powerboat to reach the sailboat in the shortest time? (*Hint:* Let the distance traveled by the two boats be 10x and 20x, where x is the number of hours until they meet.)

32. Air-traffic controllers must know the *cloud ceiling,* the altitude of the lowest point of the clouds. To determine the cloud ceiling, a spotlight located a known distance from the airport is pointed straight up and the angle of elevation from the airport to the circle of light on the bottom of the cloud is determined. (See the figure.) Find the cloud ceiling if the spotlight is located 2.0 mi from the airport and the angle of elevation is 30°.

33. A tower stands on top of a hill. From a point on the ground 148 m from a point directly under the tower, the angle of elevation to the *bottom* of the tower is 18.3°. From the same point, the angle of elevation to the *top* of the tower is 34.2°. Find the height of the tower.

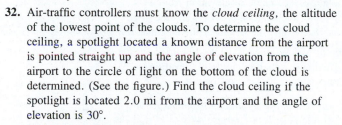

Exercise 31

34. The angle of elevation from the top of an office building in New York City to the top of the World Trade Center is 68°, while the angle of depression from the top of the office building to the bottom of the Trade Center is 63°. The office building is 290 ft from the World Trade Center. Find the height of the World Trade Center.

35. Mt. Rogers, with an altitude of 5700 ft, is the highest point in Virginia. The angle of elevation from the top of Mt. Rogers to a plane flying overhead is 33°. The straight line distance from the mountaintop to the plane is 4600 ft. Find the altitude of the plane.

36. The highest point in Texas is Guadalupe Peak. The angle of depression from the top of this peak to a small miner's cabin at elevation 2000 ft is 26°. The cabin is 14,000 ft horizontally from a point directly under the top of the mountain. Find the altitude of the top of the mountain.

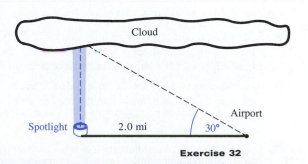

Cloud

Spotlight   2.0 mi   30°   Airport

Exercise 32

Find h in each of the following.

37.

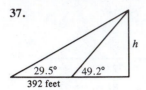

38.

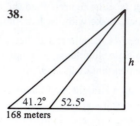

39. The angle of elevation from a point on the ground to the top of a pyramid is 35.5°. The angle of elevation from a point 135 ft further back to the top of the pyramid is 21.1°. Find the height of the pyramid.

40. A lighthouse keeper is watching a boat approach directly to the lighthouse. When she first begins watching the boat, the angle of depression of the boat is 15.8°. Just as the boat turns away from the lighthouse, the angle of depression is 35.6°. If the height of the lighthouse is 68.7 m, find the distance traveled by the boat as it approaches the lighthouse.

41. A television antenna is on top of the center of a house. The angle of elevation from a point 28.0 m from the center of the house to the top of the antenna is 27.1°, and the angle of elevation to the bottom of the antenna is 18.2°. Find the height of the antenna.

42. The angle of elevation from Lone Pine to the top of Mt. Whitney is 10.9°. If I drive 7.00 km along a straight level road toward Mt. Whitney, I find the angle of elevation to be 22.7°. Find the height of the top of Mt. Whitney above the level of the road.

In each of the following exercises, generalize problems from above by finding formulas for h in terms of k, A, and B. Assume A < B in Exercise 43 and A > B in Exercise 44.

43.

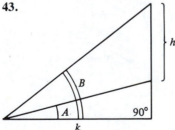

44.

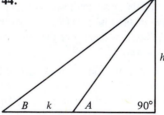

Solve each of the following problems involving bearing.

45. A ship leaves port and sails on a bearing of 28.2°. Another ship leaves the same port at the same time and sails on a bearing of 118.2°. If the first ship sails at 20.0 mph and the second sails at 24.0 mph, find the distance between the two ships after five hr.

46. Radio direction finders are set up at points *A* and *B,* which are 2.00 mi apart on an east-west line. From *A* it is found that the bearing of the signal from a radio transmitter is 36.3°, while from *B* the bearing of the same signal is 306.3°. Find the distance between the transmitter and *B.*

47. The bearing from Winston-Salem, North Carolina, to Danville, Virginia, is N 42° E. The bearing from Danville to Goldsboro, North Carolina is S 48° E. A small plane traveling at 60 mph takes 1 hr to go from Winston-Salem to Danville and 1.8 hr to go from Danville to Goldsboro. Find the distance from Winston-Salem to Goldsboro.

48. The bearing from Atlanta to Macon is S 27° E, while the bearing from Macon to Augusta is N 63° E. A plane traveling at 60 mph needs 1¼ hr to go from Atlanta to Macon and 1¾ hr to go from Macon to Augusta. Find the distance from Atlanta to Augusta.

49. The airline distance from Philadelphia to Syracuse is 260 mi, on a bearing of 335°. The distance from Philadelphia to Cincinnati is 510 mi, on a bearing of 245°. Find the bearing from Cincinnati to Syracuse.

50. A ship travels 70 km on a bearing of 27.0° and then turns on a bearing of 117.0° for 180 km. Find the distance of the end of the trip from the starting point.

51. Atoms in metals can be arranged in patterns called **unit cells.** One such unit cell, called a **primitive cell,** is a cube with an atom at each corner. A right triangle can be formed from one edge of the cell, a face diagonal and a cube diagonal as in the figure below. If each cell edge is 3.00×10^{-8} cm and the face diagonal is 4.24×10^{-8} cm, what is the angle between the cell edge and a cube diagonal?

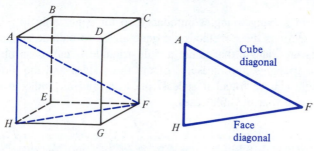

Exercise 51

52. To determine the diameter of the sun, an astronomer might sight with a transit (a device used by surveyors for measuring angles) first to one edge of the sun and then to the other, finding that the included angle equals 1.1°. Assuming that the distance from the earth to the sun is about 92,900,000 mi, calculate the diameter of the sun. See the figure.

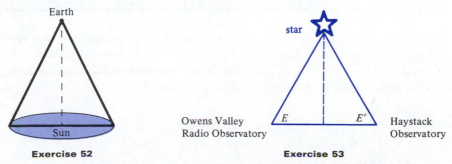

Exercise 52   **Exercise 53**

53. Very accurate measurements have shown that the distance between California's Owens Valley Radio Observatory and the Haystack Observatory in Massachusetts is 2441.2938 mi. Suppose the two observatories focus on a distant star and find that angles E and E' in the figure are both 89.99999°. Find the distance to the star from Haystack. (Assume the earth is flat.)

54. The figure shows a magnified view of the threads of a bolt.
Find x if d is 2.894 mm.

Use a right triangle to find each of the following.

55. $\sin (\text{Cos}^{-1} 1/4)$

56. $\tan (\text{Sin}^{-1} 4/7)$

57. $\cos (\text{Sin}^{-1} 2/3)$

58. $\sin (\text{Tan}^{-1} 1/5)$

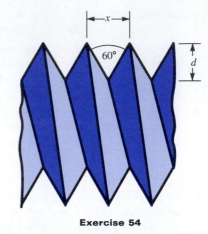

Exercise 54

8.2 OBLIQUE TRIANGLES; THE LAW OF SINES

The methods of solving triangles introduced so far apply only to right triangles. In the next few sections these methods are generalized to include all triangles, not just right triangles. A triangle that is not a right triangle is called an **oblique triangle.** As with right triangles, the measures of the three sides and the three angles of an oblique triangle can be found if at least one side and any other two measures are known. There are four possible cases.

**Solving
Oblique
Triangles**

1. One side and two angles are known.

2. Two sides and one angle not included between the two sides are known. This case may lead to more than one triangle.

3. Two sides and the angle included between the two sides are known.

4. Three sides are known.

The first two cases require the *law of sines*, which is discussed in this section. The last two cases require the *law of cosines,* discussed in the next section.

To derive the law of sines, start with an oblique triangle, such as the acute triangle in Figure 11(a) or the obtuse triangle in Figure 11(b).

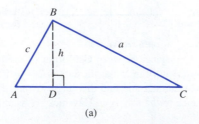

(a)

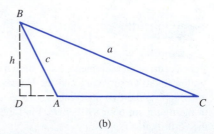

(b)

Figure 11

The following discussion applies to both triangles. First, construct the perpendicular from B to side AC. Let h be the length of this perpendicular. Then c is the hypotenuse of right triangle ADB, and a is the hypotenuse of right triangle BDC. By results given earlier,

$$\text{in triangle } ADB, \quad \sin A = \frac{h}{c} \quad \text{or} \quad h = c \sin A,$$

and $$\text{in triangle } BDC, \quad \sin C = \frac{h}{a} \quad \text{or} \quad h = a \sin C.$$

Since $h = c \sin A$ and $h = a \sin C$,

$$a \sin C = c \sin A,$$

or, upon dividing both sides by $\sin A \sin C$,

$$\frac{a}{\sin A} = \frac{c}{\sin C}.$$

In a similar way, by constructing the perpendiculars from other vertices,

$$\frac{a}{\sin A} = \frac{b}{\sin B} \quad \text{and} \quad \frac{b}{\sin B} = \frac{c}{\sin C},$$

proving the following theorem, called the **law of sines.**

Law of Sines

In any triangle ABC, with sides a, b, and c,

$$\frac{a}{\sin A} = \frac{b}{\sin B} = \frac{c}{\sin C} \quad \text{or} \quad \frac{\sin A}{a} = \frac{\sin B}{b} = \frac{\sin C}{c}.$$

In some cases, the second form is easier to use.

If two angles and one side of a triangle are known, the law of sines can be used to solve the triangle.

EXAMPLE 1   Solve triangle ABC if $A = 32.0°$, $B = 81.8°$, and $a = 42.9$ cm. See Figure 12.

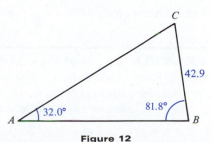

Figure 12

Start by drawing a triangle, roughly to scale, and labeling the given parts as in Figure 12. Since the values of A, B, and a are known, use the part of the law of sines that involves these variables.

$$\frac{a}{\sin A} = \frac{b}{\sin B}$$

Substituting the known values gives

$$\frac{\textbf{42.9}}{\sin \textbf{32.0°}} = \frac{b}{\sin \textbf{81.8°}}.$$

Multiply both sides of the equation by $\sin 81.8°$.

$$b = \frac{42.9 \sin 81.8°}{\sin 32.0°}$$

When using a calculator to find b, keep intermediate answers in the calculator until the final result is found. Then round to the proper number of significant digits. In this case, find $\sin 81.8°$, and then multiply that number by 42.9. Keep the result in the calculator while you find $\sin 32.0°$, and then divide. This final answer should be rounded to 3 significant figures.

$$b \approx 80.1 \text{ cm}$$

Find C from the fact that the sum of the angles of any triangle is $180°$.

$$A + B + C = 180°$$
$$C = 180° - A - B$$
$$C = 180° - \textbf{32.0°} - \textbf{81.8°} = 66.2°$$

Now use the law of sines again to find c. (Why should you not use the Pythagorean theorem?)

$$\frac{a}{\sin A} = \frac{c}{\sin C}$$

$$\frac{\textbf{42.9}}{\sin \textbf{32.0°}} = \frac{c}{\sin \textbf{66.2°}}$$

$$c = \frac{42.9 \sin 66.2°}{\sin 32.0°} \approx \frac{42.9(.9150)}{.5299} \approx 74.1$$

The length of side c is 74.1 cm.   ●

· · · · · · · · ·

EXAMPLE 2   Shirley Cicero wishes to measure the distance across the Big Muddy River. See Figure 13. She finds that $C = 112.88°$, $A = 31.10°$, and $b = 347.6$ ft. Find the required distance.

Before the law of sines can be used to find a, angle B must be found.

$$B = 180° - A - C$$
$$= 180° - 31.10° - 112.88° = 36.02°$$

Use the part of the law of sines involving A, B, and b.

$$\frac{a}{\sin A} = \frac{b}{\sin B}$$

Substitute the known values.

$$\frac{a}{\sin 31.10°} = \frac{347.6}{\sin 36.02°}$$

$$a = \frac{347.6 \sin 31.10°}{\sin 36.02°}$$

$$a = 305.3 \text{ ft} \quad \bullet$$

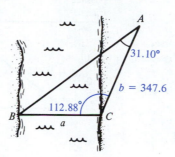

Figure 13

· · · · · · · · ·

EXAMPLE 3   Solve triangle ABC if $C = 55.6°$, $c = 8.94$ m, and $b = 25.1$ m.

Let us first look for angle B. The work is easier if the unknown is in the numerator, so start with

$$\frac{\sin B}{b} = \frac{\sin C}{c}.$$

Substitute the given values.

$$\frac{\sin B}{\mathbf{25.1}} = \frac{\sin \mathbf{55.6°}}{\mathbf{8.94}}$$

$$\sin B = \frac{25.1 \sin 55.6°}{8.94} = 2.3166$$

By this result, $\sin B$ is greater than 1. This is impossible, since $-1 \le \sin B \le 1$, for any angle B. For this reason, triangle ABC does not exist. See Figure 14. $\bullet$

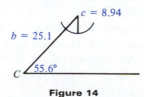

Figure 14

When any two angles and the length of a side of a triangle are given, the law of sines can be applied directly to solve the triangle. However, if only one angle and two sides are given, the triangle may not exist, as in Example 3, or there may be more than one triangle satisfying the given conditions. For example, suppose the measure of acute angle A of triangle ABC is known, along with the length of side a and the length of side b. To show this information, draw angle A and measure off a length b along its terminal side. Now draw a side of length a opposite angle A. As the chart in Figure 15 shows, there might be more than one possible outcome, so this is called the **ambiguous case of the law of sines.**

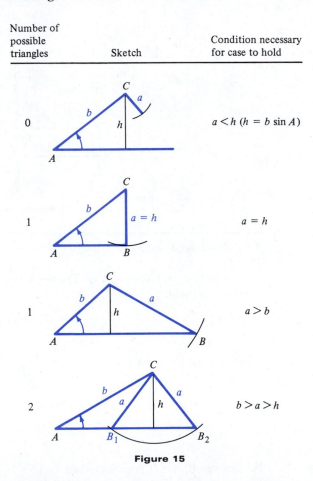

Number of possible triangles	Sketch	Condition necessary for case to hold
0		$a < h$ ($h = b \sin A$)
1		$a = h$
1		$a > b$
2		$b > a > h$

Figure 15

If angle A is obtuse, there are two possible outcomes as shown in the chart in Figure 16.

It is possible to derive formulas that show which of the various cases exist for a particular set of numerical data. However, this work is unnecessary with the law of sines. For example, if the law of sines is used, and gives $\sin B > 1$, there is no triangle at all. (Why?) A case producing two different triangles is illustrated in Example 4.

Number of possible triangles	Sketch	Condition necessary for case to hold
0	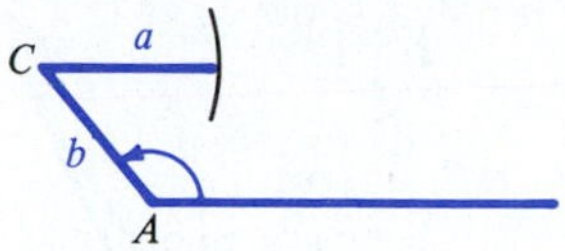	$a \leq b$
1	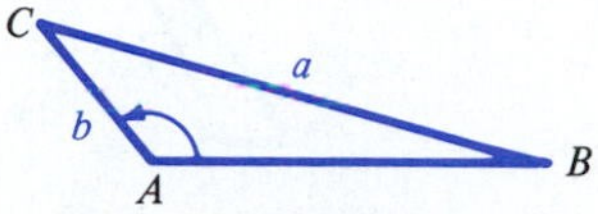	$a > b$

Figure 16

· · · · · · · · · ·

EXAMPLE 4   Solve triangle ABC if $A = 55.3°$, $a = 22.8$, and $b = 24.9$.
To begin, use the law of sines to find angle B.

$$\frac{a}{\sin A} = \frac{b}{\sin B}$$

$$\frac{22.8}{\sin 55.3°} = \frac{24.9}{\sin B}$$

$$\sin B = \frac{24.9 \sin 55.3°}{22.8}$$

$$\sin B = .8979$$

Since $\sin B < 1$, there is at least one triangle. Figure 17 shows the case if there are two triangles. Assume there are two triangles and find two possible values of B.

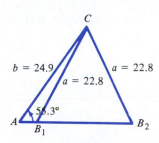

Figure 17

Since $\sin B = .8979$, one value of B is

$$B = 63.9°.$$

From the identity $\sin (180° - B) = \sin B$, another possible value of B is

$$B = 180° - 63.9°$$

$$B = 116.1°.$$

To keep track of these two different values of B, let

$$B_1 = 116.1° \quad \text{and} \quad B_2 = 63.9°.$$

Now separately solve triangles AB_1C_1 and AB_2C_2 as shown in Figure 18.

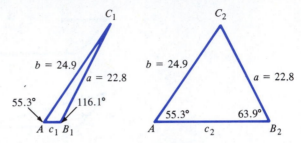

Figure 18

Since B_1 is the larger of the two values of B, find C_1 next.

$$C_1 = 180° - A - B_1$$
$$C_1 = 8.6°$$

Had this answer been negative, there would have been only one triangle. This is why the larger angle was used first. Now, use the law of sines to find c_1.

$$\frac{a}{\sin A} = \frac{c_1}{\sin C_1}$$

$$\frac{22.8}{\sin 55.3°} = \frac{c_1}{\sin 8.6°}$$

$$c_1 = \frac{22.8 \sin 8.6°}{\sin 55.3°}$$

$$c_1 = 4.15$$

Solve triangle AB_2C_2 by first finding C_2.

$$C_2 = 180° - A - B_2$$
$$C_2 = 60.8°$$

By the law of sines,

$$\frac{22.8}{\sin 55.3°} = \frac{c_2}{\sin 60.8°}$$

$$c_2 = \frac{22.8 \sin 60.8°}{\sin 55.3°}$$

$$c_2 = 24.2. \quad \bullet$$

Area   The method used to derive the law of sines can also be used to derive a useful formula for the area of a triangle. A familiar formula for the area of a triangle is

$K = (1/2)bh$, where K represents the area, b the base, and h the height. This formula cannot always be used, since in practice h is often unknown. We can derive a more useful formula from $K = (1/2)bh$ as follows. Refer to acute triangle ABC in Figure 19(a) or obtuse triangle ABC in Figure 19(b).

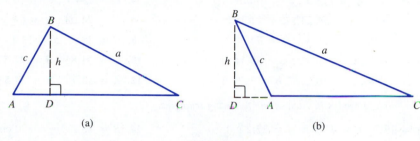

(a)

(b)

Figure 19

A perpendicular has been drawn from B to the base of the triangle. This perpendicular forms two right triangles. Using triangle ABD,

$$\sin A = \frac{h}{c},$$

or

$$h = c \sin A.$$

Substituting into the formula $K = \frac{1}{2}bh$,

$$K = \frac{1}{2} b(c \sin A)$$

$$K = \frac{1}{2} bc \sin A.$$

Any other pair of sides and the angle between them could have been used, as in the next theorem.

Area of a Triangle

> The area of a triangle is given by half the product of the lengths of two sides and the sine of the angle between the two sides.

EXAMPLE 5 Find the area of triangle MNP if $m = 29.7$ m, $n = 53.9$ m, and $P = 28.7°$.

By the last result, the area of the triangle is

$$\frac{1}{2} (29.7)(53.9) \sin 28.7° = 384 \text{ m}^2. \quad \bullet$$

8.2 Exercises ···

Solve each of the following triangles that exist.

1. $A = 46.5°$, $B = 52.8°$, $b = 87.3$ mm

2. $A = 59.5°$, $B = 48.3°$, $b = 32.9$ m

3. $A = 27.2°$, $C = 115.5°$, $c = 76.0$ ft

4. $B = 124.1°$, $C = 18.7°$, $c = 94.6$ m

5. $A = 68.41°$, $B = 54.23°$, $a = 12.75$ ft

6. $C = 74.08°$, $B = 69.38°$, $c = 45.38$ m

7. $A = 87.2°$, $b = 75.9$ yd, $C = 74.3°$

8. $B = 38.6°$, $a = 19.7$ cm, $C = 91.6°$

9. $A = 39.70°$, $C = 30.35°$, $b = 39.74$ m

10. $C = 71.83°$, $B = 42.57°$, $a = 2.614$ cm

11. $B = 42.88°$, $C = 102.40°$, $b = 3974$ ft

12. $A = 18.75°$, $B = 51.53°$, $c = 2798$ yd

Find the missing angles in each of the following triangles.

13. $A = 29.7°$, $b = 41.5$ ft, $a = 27.2$ ft

14. $B = 48.2°$, $a = 890$ cm, $b = 697$ cm

15. $C = 41.3°$, $b = 25.9$ m, $c = 38.4$ m

16. $B = 48.8°$, $a = 3850$ in., $b = 4730$ in.

17. $B = 74.3°$, $a = 859$ m, $b = 783$ m

18. $C = 82.2°$, $a = 10.9$ km, $c = 7.62$ km

19. $A = 142.13°$, $b = 5.432$ ft, $a = 7.297$ ft

20. $B = 113.72°$, $a = 189.6$ yd, $b = 243.8$ yd

In Exercises 21–28, solve each triangle that exists.

21. $A = 42.5°$, $a = 15.6$ ft, $b = 8.14$ ft

22. $C = 52.3°$, $a = 32.5$ yd, $c = 59.8$ yd

23. $B = 72.2°$, $b = 78.3$ m, $c = 145$ m

24. $C = 68.5°$, $c = 258$ cm, $b = 386$ cm

25. $A = 38.7°$, $a = 9.72$ km, $b = 11.8$ km

26. $C = 29.5°$, $a = 8.61$ m, $c = 5.21$ m

27. $B = 39.68°$, $a = 29.81$ m, $b = 23.76$ m

28. $A = 51.20°$, $c = 7986$ cm, $a = 7208$ cm

29. In the figure, a line of length c is to be drawn from point A to the dashed line in order to form a triangle. For what value(s) of c can you draw the following? **(a)** Exactly one triangle **(b)** Two triangles **(c)** No triangles

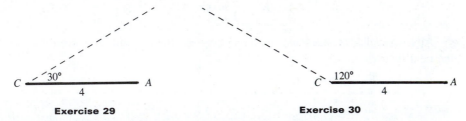

Exercise 29   **Exercise 30**

30. In the figure, a line of length c is to be drawn from point A to the dashed line in order to form a triangle. For what value(s) of c can you draw the following? **(a)** Exactly one triangle **(b)** Two triangles **(c)** No triangles

Solve each of the following exercises. Recall that bearing was discussed with right triangle applications.

31. To find the distance AB across a river, a distance $BC = 354$ m is laid off on one side of the river. In triangle ABC, it is found that $B = 112.0°$ and $C = 15.3°$. Find AB.

32. To determine the distance RS across a deep canyon, Maria lays off a distance $TR = 582$ yd. She then finds that in triangle RST, $T = 32.8°$ and $R = 102.0°$. Find RS.

33. Radio direction finders are placed at points *A* and *B,* which are 3.46 mi apart on an east-west line, with *A* west of *B*. From *A* the bearing of a certain radio transmitter is 47.7°, while from *B* the bearing is 303.0°. Find the distance between the transmitter and *A*.

34. A ship is sailing due north. Captain Odjakjian notices that the bearing of a lighthouse 12.5 km distant is 38.8°. Later on, the captain notices that the bearing of the lighthouse has become 136.0°. How far did the ship travel between the two observations of the lighthouse?

35. A folding chair is to have a seat 12.0 in. deep with angles as shown in the figure. How far down from the seat should the crossing legs be joined? (Find *x* in the figure.)

36. Pedro notices that the bearing of a tree on the opposite bank of a river is 115.0°. Kim is on the same bank as Pedro but 428 m away. She notices that the bearing of the tree is 45.3°. The river is flowing north between parallel banks. What is the distance across the river?

37. Three gears are arranged as shown in the figure. Find angle *θ*.

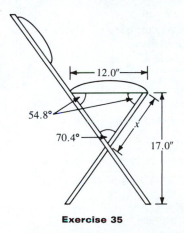

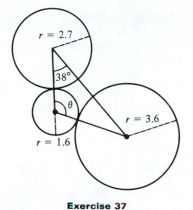

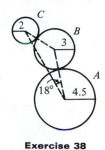

Exercise 35   Exercise 37   Exercise 38

38. Three atoms with atomic radii of 2, 3, and 4.5 are arranged as in the figure. Find the distance between the centers of atoms *A* and *C*.

39. A surveyor reported the following data about a piece of property: "The property is triangular in shape, with dimensions as shown in the figure." Use the law of sines to see if such a piece of property could exist.

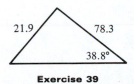

Exercise 39

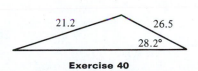

Exercise 40

40. The surveyor tries again: "A second triangular piece of property has dimensions as shown." This time it turns out that the surveyor did not consider every possible case. Use the law of sines to show why.

41. A ship sailing parallel to shore sights a lighthouse at an angle of 30° from its direction of travel. See the figure. After traveling 2 mi farther, the angle has increased to 60°. At that time, how far is the ship from the lighthouse?

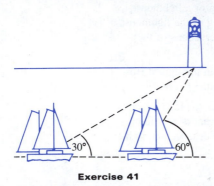

Exercise 41   **Exercise 42**

42. Several of the exercises on right triangle applications involved a figure similar to the one above, in which the angles α and β and the length of the line segment AB are known, and the length of the side CD is to be determined. Use the law of sines to obtain a formula for x in terms of α, β, and d.

Find the area of the triangles in Exercises 43–50.

43. $A = 42.5°$, $b = 13.6$ m, $c = 10.1$ m

44. $C = 72.2°$, $b = 43.8$ ft, $a = 35.1$ ft

45. $B = 124.5°$, $a = 30.4$ cm, $c = 28.4$ cm

46. $C = 142.7°$, $a = 21.9$ km, $b = 24.6$ km

47. $A = 56.80°$, $b = 32.67$ in, $c = 52.89$ in

48. $A = 34.97°$, $b = 35.29$ m, $c = 28.67$ m

49. $A = 24.42°$, $B = 56.33°$, $c = 78.40$ cm

50. $B = 48.5°$, $C = 74.3°$, $a = 462$ km

51. A painter is going to apply a special coating to a triangular metal plate on a new building. Two sides measure 16.1 m and 15.2 m. She knows that the angle between these sides is 125.0°. How many square meters should she plan to cover with the coating?

52. A realtor wants to find the area of a triangular lot. A surveyor takes measurements and finds that two sides are 52.1 m and 21.3 m, and the angle between them is 42.2°. What is the area of the lot?

Prove that each statement in Exercises 53–56 is true for any triangle ABC, with corresponding sides a, b, and c.

53. $\dfrac{a + b}{b} = \dfrac{\sin A + \sin B}{\sin B}$

54. $\dfrac{a - b}{a + b} = \dfrac{\sin A - \sin B}{\sin A + \sin B}$

55. $\dfrac{a + b}{c} = \dfrac{\cos \frac{1}{2}(A - B)}{\sin \frac{1}{2} C}$

56. $\dfrac{a - b}{c} = \dfrac{\sin \frac{1}{2}(A - B)}{\cos \frac{1}{2} C}$

57. In any triangle having sides a, b, and c, it must be true that $a + b > c$. Use this fact and the law of sines to show that $\sin A + \sin B > \sin (A + B)$ for any two angles A and B of a triangle.

58. Show that the area of a triangle having sides a, b, and c and corresponding angles A, B, and C is given by $(a^2 \sin B \sin C)/(2 \sin A)$.

59. Prove the law of sines if angle A is a right angle.

60. Derive the formula for the area of a triangle, $K = (1/2) bc \sin A$, if angle A is a right angle.

61. Use the law of sines to prove that in the triangle ABC, if angles A and B are acute and $A > B$, then the side opposite A is longer than the side opposite B.

8.3 THE LAW OF COSINES

If two sides and the angle between the two sides are given, the law of sines cannot be used to solve the triangle. Also, if all three of the sides of a triangle are given, the law of sines cannot be used to find the unknown angles. Both of these cases require the law of cosines.

We know that the Pythagorean theorem relates the three sides of a right triangle: $a^2 + b^2 = c^2$, where c is the length of the hypotenuse. The law of cosines is a generalization of this theorem for *any* triangle. To derive this law, let ABC be any oblique triangle. Choose a coordinate system with vertex B at the origin and side BC along the positive x-axis. See Figure 20.

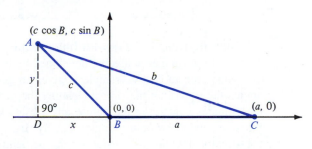

Figure 20

Let (x, y) be the coordinates of vertex A of the triangle. Verify that for angle B, whether obtuse or acute,

$$\sin B = \frac{y}{c} \quad \text{and} \quad \cos B = \frac{x}{c}.$$

(Here x is negative if B is obtuse.) From these results

$$y = c \sin B \quad \text{and} \quad x = c \cos B,$$

so that the coordinates of point A become

$$(c \cos B, c \sin B).$$

Point C has coordinates $(a, 0)$ and AC has length b. By the distance formula,

$$b = \sqrt{(c \cos B - a)^2 + (c \sin B)^2}.$$

Squaring both sides and simplifying gives

$$
\begin{aligned}
b^2 &= (c \cos B - a)^2 + (c \sin B)^2 \\
&= c^2 \cos^2 B - 2ac \cos B + a^2 + c^2 \sin^2 B \\
&= a^2 + c^2 (\cos^2 B + \sin^2 B) - 2ac \cos B \\
&= a^2 + c^2 (1) - 2ac \cos B \\
&= a^2 + c^2 - 2ac \cos B.
\end{aligned}
$$

This result is one form of the **law of cosines.** In the work above, A or C could just as easily have been placed at the origin, giving the same result but with the variables rearranged. These various forms of the law of cosines are summarized in the following theorem.

The Law of Cosines

In any triangle ABC, with sides a, b, and c,

$$
\begin{aligned}
a^2 &= b^2 + c^2 - 2bc \cos A \\
b^2 &= a^2 + c^2 - 2ac \cos B \\
c^2 &= a^2 + b^2 - 2ab \cos C.
\end{aligned}
$$

Book II of Euclid's *Elements* (third century B.C.) concerns the transformation of areas and the geometric algebra developed by the Pythagorean school. There, the law of cosines for an obtuse triangle is given in geometric form: "In obtuse-angled triangles the square on the side subtending the obtuse angle is greater than the [sum of the] squares containing the obtuse angle by twice the rectangle contained by one of the sides about the obtuse angle, namely that on which the perpendicular falls, and the straight line cut off outside by the perpendicular toward the acute angle." This geometric language is difficult to understand, but it simply means that for the obtuse triangle in Figure 20

$$b^2 = a^2 + c^2 - 2ax. \tag{$*$}$$

As shown in Figure 20, the value of x is $c \cos B$, giving

$$b^2 = a^2 + c^2 - 2ac \cos B.$$

The useful algebraic notation in $(*)$ was introduced by the great French cryptographer and amateur mathematician François Viète (1540–1603) in his *Introduction to the Analytic Art*. There he brought Euclid's ancient geometrical statement together with numerical algebra. Because of this important contribution, he is called the father of symbolic algebra.

· · · · · · · · ·

EXAMPLE 1   Solve triangle ABC if $A = 42.3°$, $b = 12.9$, and $c = 15.4$. See Figure 21.

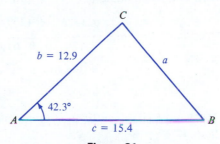

Figure 21

Find a with the law of cosines.

$$a^2 = b^2 + c^2 - 2bc \cos A$$

Substitute the given values.

$$a^2 = (12.9)^2 + (15.4)^2 - 2(12.9)(15.4) \cos 42.3°$$
$$a^2 = 166.41 + 237.16 - (397.32)(.7396)$$
$$a^2 = 403.57 - 293.86 = 109.71$$
$$a = 10.5$$

Now that a, b, c, and A are known, the law of sines can be used to find either angle B or angle C. If there is an obtuse angle in the triangle, it will be the larger of B and C. Since we cannot tell from the sine of the angle whether it is acute or obtuse, it is a good idea to find the smaller angle (which will be acute) first. In this triangle, $B < C$ because $b < c$, so use the law of sines to find B.

$$\frac{\sin 42.3°}{10.5} = \frac{\sin B}{12.9}$$

$$\sin B = \frac{12.9 \sin 42.3°}{10.5}$$

$$\sin B = .8268$$
$$B = 55.8°$$

Now find C.

$$C = 180° - A - B = 81.9° \quad •$$

· · · · · · · · ·

EXAMPLE 2   Solve triangle ABC if $C = 132.7°$, $b = 259$, and $a = 423$.

Since C is given, but c is unknown, use the law of cosines in the form $c^2 = a^2 + b^2 - 2ab \cos C$. Inserting the given data gives

$$c^2 = (423)^2 + (259)^2 - 2(423)(259) \cos 132.7°.$$

The value of cos 132.7° can be found with a calculator or Table 3.

$$\cos 132.7° = -.6782$$

Now continue finding c.

$$c^2 = (423)^2 + (259)^2 - 2(423)(259)(-.6782)$$

$$c^2 = 178{,}929 + 67{,}081 + 148{,}603$$

$$= 394{,}613$$

$$c = 628$$

The law of sines can be used to complete the solution. Check that $A = 29.7°$ and $B = 17.6°$.   ●

• • • • • • • •

EXAMPLE 3   Solve triangle ABC if $a = 9.47$, $b = 15.9$, and $c = 21.1$.

Again the law of cosines must be used. It is a good idea to find the largest angle first in case it is obtuse. Since c has the greatest length, angle C will be the largest angle. Start with

$$c^2 = a^2 + b^2 - 2ab \cos C,$$

or

$$\cos C = \frac{a^2 + b^2 - c^2}{2ab}.$$

Substituting the given values leads to

$$\cos C = \frac{(9.47)^2 + (15.9)^2 - (21.1)^2}{2(9.47)(15.9)}$$

$$= \frac{-102.7191}{301.146}$$

$$= -.341094,$$

and

$$C = 109.9°.$$

(Angle C is obtuse since cos C is negative.) Use the law of sines to find B. Verify that $B = 45.1°$. Since $A = 180° - B - C$,

$$A = 25.0°.   ●$$

As mentioned above, there are four possible cases that can arise in solving an oblique triangle. These cases are summarized as follows. (The first two cases require the law of sines, while the second two require the law of cosines. In all four cases, assume that the given information actually produces a triangle.)

Solving Triangles

Case	Abbreviation	Example
One side and two angles are known.	SAA	a, B, A known, find b $b = \dfrac{a \sin B}{\sin A}$
Two sides and one angle (not included between the two sides) are known (watch for the ambiguous case—there may be two triangles).	SSA	b, c, B known, find C $\sin C = \dfrac{c \sin B}{b}$
Three sides are known.	SSS	a, b, c, known, find A $\cos A = \dfrac{b^2 + c^2 - a^2}{2bc}$
Two sides and the angle included between the two sides are known.	SAS	a, B, c known, find b $b^2 = a^2 + c^2 - 2ac \cos B$

Area   The law of cosines can be used to find a formula for the area of a triangle when only the lengths of the three sides of the triangle are known. This formula, called Heron's area formula, is given as the next theorem. For a proof see Exercise 48.

Heron's Area Formula

If a triangle has sides of lengths a, b, and c, and if the semiperimeter s is

$$s = \frac{1}{2}(a + b + c),$$

then the area of the triangle is

$$K = \sqrt{s(s - a)(s - b)(s - c)}.$$

Heron of Alexandria (flourished A.D. 75), an applied mathematician and practicing surveyor who wrote more than fourteen works on mathematics and physics, laid out the scientific foundation for engineering and land surveying. His *Pneumatica* contains descriptions of inventions such as a simple steam engine and a device for opening temple doors by means of a fire on the altar. His *Metrica,* which is concerned with calculating the volumes of various figures, contains the formula above for the area of a triangle in terms of its sides. Since Heron's work *Metrica* was so practical, this formula was widely used.

EXAMPLE 4 Find the area of the triangle having sides of lengths $a = 29.7$ ft, $b = 42.3$ ft, and $c = 38.4$ ft.

To use Heron's area formula, first find s.

$$s = \frac{1}{2}(a + b + c)$$

$$s = \frac{1}{2}(29.7 + 42.3 + 38.4)$$

$$s = 55.2$$

The area is then given by

$$K = \sqrt{s(s - a)(s - b)(s - c)}$$

$$= \sqrt{55.2(55.2 - 29.7)(55.2 - 42.3)(55.2 - 38.4)}$$

$$K = \sqrt{55.2(25.5)(12.9)(16.8)}$$

$$K = 552 \text{ ft}^2. \quad \bullet$$

8.3 Exercises

Solve each of the following triangles.

1. $C = 28.3°$, $b = 5.71$ in, $a = 4.21$ in

2. $A = 41.4°$, $b = 2.78$ yd, $c = 3.92$ yd

3. $C = 45.6°$, $b = 8.94$ m, $a = 7.23$ m

4. $A = 67.3°$, $b = 37.9$ km, $c = 40.8$ km

5. $B = 74.80°$, $a = 8.919$ in, $c = 6.427$ in

6. $C = 59.70°$, $a = 3.725$ mi, $b = 4.698$ mi

7. $A = 112.8°$, $b = 6.28$ m, $c = 12.2$ m

8. $B = 168.2°$, $a = 15.1$ cm, $c = 19.2$ cm

Find all the angles in each of the following triangles.

9. $a = 3.00$ ft, $b = 5.00$ ft, $c = 6.00$ ft

10. $a = 4.00$ ft, $b = 5.00$ ft, $c = 8.00$ ft

11. $a = 9.31$ cm, $b = 5.73$ cm, $c = 8.24$ cm

12. $a = 28.3$ in, $b = 47.1$ in, $c = 57.9$ in

13. $a = 42.9$ m, $b = 37.6$ m, $c = 62.7$ m

14. $a = 189$ yd, $b = 214$ yd, $c = 325$ yd

15. $AB = 1240$ ft, $AC = 876$ ft, $BC = 918$ ft

16. $AB = 298$ m, $AC = 421$ m, $BC = 324$ m

17. $a = 12.54$ in, $b = 16.83$ in, $c = 21.62$ in

18. $a = 250.8$ ft, $b = 212.7$ ft, $c = 324.1$ ft

Solve each of the following problems. Use the laws in this chapter as necessary.

19. Points *A* and *B* are on opposite sides of Lake Folsom. From a third point, *C*, the angle between the lines of sight to *A* and *B* is 46.3°. If *AC* is 350 m long and *BC* is 286 m long, find *AB*.

20. The sides of a parallelogram are 4.0 cm and 6.0 cm. One angle is 58° while another is 122°. Find the lengths of the diagonals of the figure.

21. Airports *A* and *B* are 450 km apart, on an east-west line. Mei flies in a northeast direction from *A* to airport *C*. From *C* she flies 359 km on a bearing of 128° to *B*. How far is *C* from *A*?

22. Two ships leave a harbor together, traveling on courses that have an angle of 135° between them. If they each travel 402 mi, how far apart are they?

23. The plans for a mountain cabin show the dimensions given in the figure. Find *x*.

24. A hill slopes at an angle of 12.47° with the horizontal. As shown in the figure, a 459-ft tower stands 100 ft from the bottom of the hill. How much rope would be required to reach from the top of the tower to the bottom of the hill?

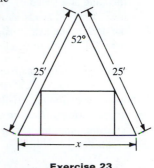

Exercise 23

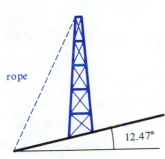

Exercise 24

Exercise 25

25. A satellite traveling in a circular orbit 1600 km above Earth is due to pass directly over a tracking station at noon.* (See the figure.) Assume that the satellite takes 2 hr to make an orbit and that the radius of the Earth is 6400 km. Find the distance between the satellite and the tracking station at 12:03 P.M.

26. A weight is supported by cables attached to both ends of a balance beam. See the figure. What angles are formed between the beam and the cables?

27. Two factories blow their whistles at 5 o'clock exactly. A man hears the two blasts at 3 seconds and 6 seconds after 5, respectively. The angle between his lines of sight to the two factories is 42.2°. If sound travels 344 m per sec., how far apart are the factories?

28. A parallelogram has sides of length 25.9 cm and 32.5 cm. The longer diagonal has a length of 57.8 cm. Find the angle opposite the diagonal.

29. A person in a plane flying a straight course observes a mountain at a bearing 24.1° to the right of its course. At that time the plane is 7.92 km from the mountain. A short time later, the bearing to the mountain becomes 32.7°. How far is the airplane from the mountain when the second bearing is taken?

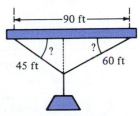

Exercise 26

*Bernice Kastner, Ph.D., SPACEMATHEMATICS. National Aeronautics and Space Administration (NASA), 1985.

30. The aircraft carrier *Tallahassee* is traveling at sea on a steady course with a bearing of 30° at 32 mph. Patrol planes on the carrier have enough fuel for 2.6 hr of flight when traveling at a speed of 520 mph. One of the pilots takes off on a bearing of 338° and then turns and heads in a straight line, so as to be able to catch the carrier, landing on the deck at the exact instant that his fuel runs out. If the pilot left at 2 P.M., at what time did he turn to head for the carrier? See the figure.

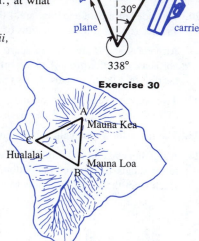

Exercise 30

To help predict eruptions from the volcano Mauna Loa on the island of Hawaii, scientists keep track of the volcano's movement by using a "super triangle" with vertices on the three volcanos shown on the map. (For example, in a recent year, Mauna Loa moved six inches north and northwest—a result of increasing internal pressure.) The data in the following exercises have been rounded.

31. $AB = 22.47928$ mi, $AC = 28.14276$ mi, $A = 58.56989°$; find BC.

32. $AB = 22.47928$ mi, $BC = 25.24983$ mi, $A = 58.56989°$; find B.

Exercises 31, 32

Find the area of each triangle in Exercises 33–40.

33. $a = 15$ in, $b = 19$ in, $c = 24$ in

34. $a = 27$ m, $b = 40$ m, $c = 34$ m

35. $a = 154$ cm, $b = 179$ cm, $c = 183$ cm

36. $a = 25.4$ yd, $b = 38.2$ yd, $c = 19.8$ yd

37. $a = 76.3$ ft, $b = 109$ ft, $c = 98.8$ ft

38. $a = 15.89$ in, $b = 21.74$ in, $c = 10.92$ in

39. $a = 74.14$ ft, $b = 89.99$ ft, $c = 51.82$ ft

40. $a = 1.096$ km, $b = 1.142$ km, $c = 1.253$ km

41. Sun wants to paint a triangular region 75 by 68 by 85 m. A can of paint covers 75 m² of area. How many cans (rounded to the larger number of cans) will be needed?

42. How many cans would be needed if the region were 8.2 by 9.4 by 3.8 m?

Use the information that $\cos A = (b^2 + c^2 - a^2)/(2bc)$ and $s = (1/2)(a + b + c)$ to show that each equation in Exercises 43–46 is true.

43. $1 + \cos A = \dfrac{(b + c + a)(b + c - a)}{2bc}$

44. $1 - \cos A = \dfrac{(a - b + c)(a + b - c)}{2bc}$

45. $\cos \dfrac{A}{2} = \sqrt{\dfrac{s(s - a)}{bc}}$ $\left(\text{Recall: } \cos \dfrac{A}{2} = \sqrt{\dfrac{1 + \cos A}{2}}\right)$

46. $\sin \dfrac{A}{2} = \sqrt{\dfrac{(s - b)(s - c)}{bc}}$ $\left(\text{Recall: } \sin \dfrac{A}{2} = \sqrt{\dfrac{1 - \cos A}{2}}\right)$

47. The area of a triangle having sides b and c and angle A is given by $\dfrac{1}{2} bc \sin A$. Show that this result can be written as

$$\sqrt{\dfrac{1}{2} bc(1 + \cos A) \cdot \dfrac{1}{2} bc(1 - \cos A)}.$$

48. Use the results of Exercises 43–47 to prove Heron's area formula.

49. Let a and b be the equal sides of an isosceles triangle. Prove that $c^2 = 2a^2(1 - \cos C)$.

50. Let point D on side AB of triangle ABC be such that CD bisects angle C. Show that $AD/DB = b/a$.

51. Use the law of cosines to prove that if one angle of a triangle is obtuse, then the side opposite the obtuse angle is longer than either of the other two sides.

52. Show that the Pythagorean theorem is a special case of the law of cosines.

8.4 VECTORS AND THEIR APPLICATIONS

As shown in earlier sections, the measure of all six parts of a triangle can be found, given at least one side and any other two measures. In this section, applications of this work to *vectors* are discussed.

Many quantities in mathematics involve magnitudes, such as 45 lb or 60 mph. These quantities are called **scalars.** Other quantities, called **vector quantities,** involve both magnitude and direction. Typical vector quantities are velocity, acceleration, and force.

A vector quantity is often represented with a directed line segment, called a **vector.** The length of the vector represents the magnitude of the vector quantity. The direction of the vector, indicated with an arrowhead, represents the direction of the quantity. For example, the vector in Figure 22 represents a force of 10 lb applied at an angle of 30° from the horizontal.

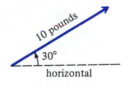

Figure 22

The term *vector* was introduced in 1846 by the Irish mathematician William Rowan Hamilton (1805–85) in the same paper in which he introduced the term *scalar*. *Vector* inherited its name from the earlier use of "radius vector." *Scalar* came from "scale of progression of numbers from negative to positive infinity." Hamilton was extraordinarily accomplished in languages—he knew fourteen of them by age 14—and was a close friend of the poet William Wordsworth.

The symbol for a vector is often printed in boldface type. To write vectors by hand, it is customary to use an arrow over the letter or letters. Thus **OP** and $\overrightarrow{OP}$ both represent vector OP. Vectors may be named with either one lowercase or uppercase letter, or two uppercase letters. When two letters are used, the first indicates the *initial point* and the second indicates the *terminal point* of the vector. Knowing these points gives the direction of the vector. For example, vectors **OP** and **PO** in

Figure 23 are not the same vectors. They have the same magnitude, but opposite directions. The magnitude of vector **OP** is written |**OP**|.

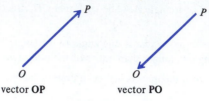

vector **OP**   vector **PO**

Figure 23

Two vectors are *equal* if and only if they have the same direction and the same magnitude. In Figure 24, vectors **A** and **B** are equal, as are vectors **C** and **D.**

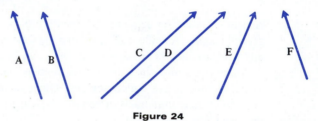

Figure 24

As Figure 24 shows, equal vectors need not coincide, but they must be parallel. Vectors **A** and **E** are unequal because they do not have the same direction, while **A** ≠ **F** because they have different magnitudes, as indicated by their different lengths.

To find the **sum** of two vectors **A** and **B,** written **A** + **B,** place the initial point of vector **B** at the terminal point of vector **A,** as shown in Figure 25. The vector with the same initial point as **A** and the same terminal point as **B** is the sum **A** + **B.** The sum of two vectors is also a vector.

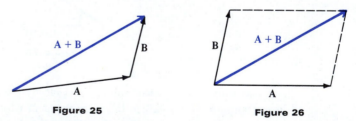

Figure 25   **Figure 26**

Another way to find the sum of two vectors is to use the **parallelogram rule.** Place vectors **A** and **B** so that their initial points coincide. Then complete a parallelogram which has **A** and **B** as two sides. The diagonal of the parallelogram with the same initial point as **A** and **B** is the same vector sum **A** + **B** found by the definition. See Figure 26.

Parallelograms can be used to show that vector **B** + **A** is the same as vector **A** + **B,** or that

$$\mathbf{A} + \mathbf{B} = \mathbf{B} + \mathbf{A},$$

so that vector addition is **commutative.**

The vector sum **A** + **B** is the **resultant** of vectors **A** and **B**. Each of the vectors **A** and **B** is a **component** of vector **A** + **B**. In many practical applications, such as surveying, it is necessary to break a vector into its **vertical** and **horizontal components**. These components are two vectors, one vertical and one horizontal, whose resultant is the original vector. As shown in Figure 27, vector **OR** is the vertical component and vector **OS** is the horizontal component of **OP**.

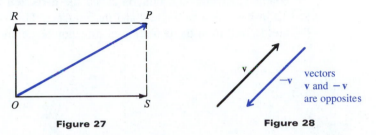

Figure 27   Figure 28

For every vector **v** there is a vector −**v** which has the same magnitude as **v** but opposite direction. Vector −**v** is called the **opposite** of **v**. See Figure 28. The sum of **v** and −**v** has magnitude 0 and is called the **zero vector**. As with real numbers, to *subtract* vector **B** from vector **A**, find the vector sum **A** + (−**B**). See Figure 29.

The **scalar product** of a real number (or scalar) k and a vector **u** is the vector k**u** which has magnitude $|k|$ times the magnitude of **u**. As suggested by Figure 30, the vector k**u** has the same direction as **u** if $k > 0$, and opposite direction if $k < 0$.

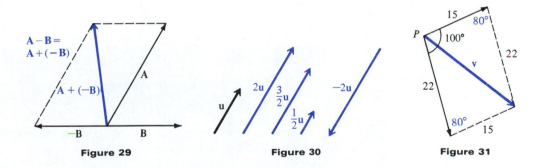

Figure 29   Figure 30   Figure 31

• • • • • • • • •

EXAMPLE 1   Two forces of 15 and 22 newtons (a *newton* is a unit of force used in physics) act on a point in the plane. If the angle between the forces is 100°, find the magnitude of the resultant force.

As shown in Figure 31, a parallelogram that has the forces as adjacent sides can be formed. The angles of the parallelogram adjacent to angle P each measure 80°, since adjacent angles of a parallelogram are supplementary. Opposite sides of the parallelogram are equal in length. The resultant force divides the parallelogram into two triangles. Use the law of cosines with either triangle to get

$$|\mathbf{v}|^2 = 15^2 + 22^2 - 2(15)(22) \cos 80°$$
$$= 225 + 484 - 115 = 594$$
$$|\mathbf{v}| = 24. \quad \bullet$$

NOTE   The diagonal of the parallelogram in Figure 31 divides angle P into two *unequal* angles. This is usually the case, unless the parallelogram has all four sides equal.

If a vector is placed so that its origin coincides with the origin of a rectangular coordinate system, then the angle between the x-axis and the vector, measured in a counterclockwise direction, is called the **direction angle** for the vector. In Figure 32, **u** has direction angle θ and magnitude r. The following basic results for vectors are derived from the definition of direction angle and earlier results.

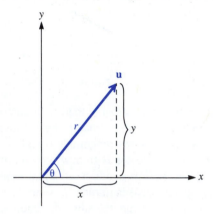

Figure 32

Basic Results for Vectors

Let a vector have direction angle θ and magnitude r. Let

$$x = r \cos \theta,$$

$$\text{and} \quad y = r \sin \theta.$$

Then the horizontal component of the vector has magnitude $|x|$ and the vertical component has magnitude $|y|$.
Also,

$$x^2 + y^2 = r^2 \quad \text{and} \quad \tan \theta = \frac{y}{x}, \quad x \neq 0.$$

EXAMPLE 2   Vector **w** has magnitude 25.0 and direction angle 41.7°. Find the magnitudes of the horizontal and vertical components of the vector.

In Figure 33, the vertical component is labeled **v** and the horizontal component is labeled **u.** Using the basic results for vectors,

$$|\mathbf{v}| = 25.0 \sin 41.7°$$
$$= 16.6,$$

and

$$|\mathbf{u}| = 25.0 \cos 41.7°$$
$$= 18.7. \quad \bullet$$

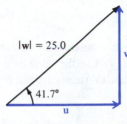

Figure 33

Let vector **u** be placed in a plane so that the initial point of the vector is at the origin, (0, 0), and the endpoint is at the point (a, b). A vector with initial point at the origin is called a **position vector** or (sometimes) a **radius vector.** A position vector having endpoint at the point (a, b) is called the **vector** (a, b)**.** To avoid confusion, the vector (a, b) is written as $\langle a, b \rangle$. The numbers a and b are called the **x-component** and **y-component,** respectively. Figure 34 shows the vector **u** = $\langle a, b \rangle$. The figure suggests the following relationship.

The magnitude of vector **u** = $\langle a, b \rangle$ is given by

$$|\mathbf{u}| = \sqrt{a^2 + b^2}.$$

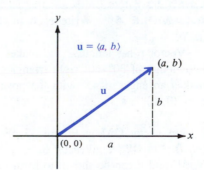

Figure 34

EXAMPLE 3   Figure 35 shows vector $\mathbf{u} = \langle 3, -2 \rangle$. Find the magnitude and direction angle for $\mathbf{u}$.

The magnitude is

$$\sqrt{3^2 + (-2)^2} = \sqrt{13}.$$

To find the direction angle θ, start with

$$\tan \theta = \frac{y}{x} = \frac{-2}{3} = -\frac{2}{3}.$$

Vector $\mathbf{u}$ has positive x-component and negative y-component, placing the vector in quadrant IV. Use the table or a calculator to show that, to the nearest tenth, an angle in quadrant IV satisfying $\tan \theta = -2/3$ is

$$\theta = -33.7°$$

or $$-33.7° + 360° = 326.3°.$$

The direction angle is 326.3°.   ●

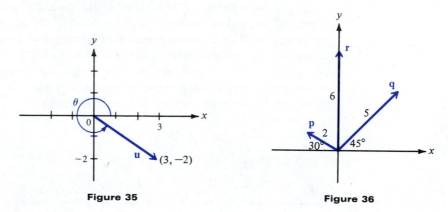

Figure 35

Figure 36

EXAMPLE 4   Write each of the position vectors of Figure 36 in the form $\langle a, b \rangle$.

Vector $\mathbf{p}$ has length 2 and makes an angle of 30° with the negative x-axis. From knowledge of 30°–60° right triangles, $\mathbf{p} = \langle -\sqrt{3}, 1 \rangle$. Vector $\mathbf{q}$ has length 5 and makes an angle of 45° with the positive x-axis, so $\mathbf{q} = \langle 5\sqrt{2}/2, 5\sqrt{2}/2 \rangle$. Finally, $\mathbf{r} = \langle 0, 6 \rangle$.   ●

Let vector $\mathbf{OM}$ in Figure 37 be given by $\langle a, b \rangle$, and vector $\mathbf{ON}$ be given by $\langle c, d \rangle$. Let $\mathbf{OP}$ be given by $\langle a + c, b + d \rangle$. With facts from geometry, points O, N, M, and P can be shown to form the vertices of a parallelogram. Since a diagonal of this parallelogram gives the resultant of $\mathbf{OM}$ and $\mathbf{ON}$, vector $\mathbf{OP}$ is given by $\mathbf{OP} = \mathbf{OM} + \mathbf{ON}$, with the resultant of $\langle a, b \rangle$ and $\langle c, d \rangle$ given by $\langle a + c, b + d \rangle$. In the same way, $k\langle a, b \rangle = \langle ka, kb \rangle$ for any real number k. These vector operations are summarized below.

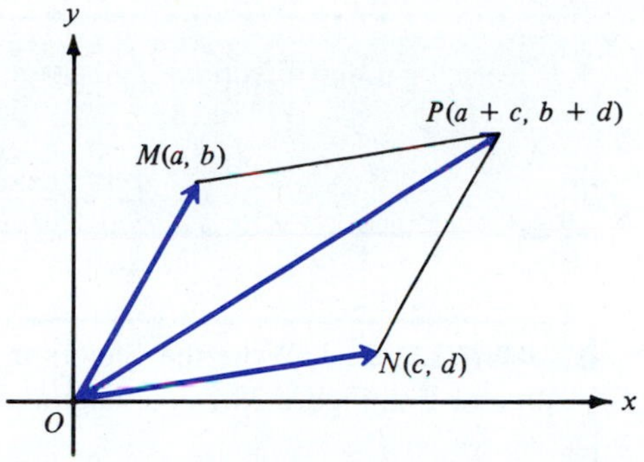

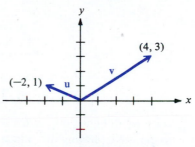

Figure 37

Vector Operations

For any real numbers a, b, c, d, and k,

$$\langle a, b \rangle + \langle c, d \rangle = \langle a + c, b + d \rangle$$
$$k \langle a, b \rangle = \langle ka, kb \rangle.$$

EXAMPLE 5   Let $\mathbf{u} = \langle -2, 1 \rangle$ and $\mathbf{v} = \langle 4, 3 \rangle$. Find each of the following. See Figure 38.

(a) $\mathbf{u} + \mathbf{v} = \langle -2, 1 \rangle + \langle 4, 3 \rangle = \langle -2 + 4, 1 + 3 \rangle = \langle 2, 4 \rangle$

(b) $-2\mathbf{u} = -2\langle -2, 1 \rangle = \langle -2(-2), -2(1) \rangle = \langle 4, -2 \rangle$

(c) $4\mathbf{u} + 3\mathbf{v} = 4\langle -2, 1 \rangle + 3\langle 4, 3 \rangle = \langle -8, 4 \rangle + \langle 12, 9 \rangle$
$\qquad = \langle -8 + 12, 4 + 9 \rangle = \langle 4, 13 \rangle$ ●

Figure 38

For a vector $\mathbf{u}$ with magnitude r and direction angle θ, it was shown above that the horizontal and vertical components of $\mathbf{u}$ are $\langle x, 0 \rangle$ and $\langle 0, y \rangle$ where

$$x = r \cos \theta \quad \text{and} \quad y = r \sin \theta.$$

Since $\mathbf{u} = \langle x, y \rangle$, this leads to the result on the following page.

• •

If a vector **u** has direction angle θ and magnitude r, then

$$\mathbf{u} = \langle r \cos \theta,\ r \sin \theta \rangle.$$

• • • • • • • • •

EXAMPLE 6 Write the vectors in Figure 39 in the form $\langle a, b \rangle$.

Vector **u** in Figure 39 has a magnitude of 5 and direction angle 60°. By the result just above,

$$\mathbf{u} = \langle 5 \cos 60°,\ 5 \sin 60° \rangle = \left\langle 5 \cdot \frac{1}{2},\ 5 \cdot \frac{\sqrt{3}}{2} \right\rangle = \left\langle \frac{5}{2},\ \frac{5\sqrt{3}}{2} \right\rangle.$$

Also,

$$\mathbf{v} = \langle 2 \cos 180°,\ 2 \sin 180° \rangle = \langle 2(-1),\ 2(0) \rangle = \langle -2,\ 0 \rangle.$$

Finally, $\mathbf{w} = \langle 6 \cos 280°,\ 6 \sin 280° \rangle$

or $\mathbf{w} \approx \langle 1.0419,\ -5.9088 \rangle.$ ●

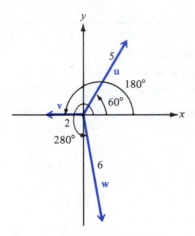

Figure 39

As shown in Figure 40, vector **u** can be thought of as the resultant of two vectors: one on the x-axis, having magnitude given by the absolute value of the x-component, and one on the y-axis, having magnitude given by the absolute value of the y-component. This idea applies to any vector $\mathbf{u} = \langle a, b \rangle$.

$$\mathbf{u} = \langle a, b \rangle = \langle a, 0 \rangle + \langle 0, b \rangle$$
$$= a \langle 1, 0 \rangle + b \langle 0, 1 \rangle$$

The vector $\langle 1, 0 \rangle$ is called the unit vector **i,** while $\langle 0, 1 \rangle$ is the unit vector **j. Unit vector** refers to the fact that the magnitude is 1. With these unit vectors, any vector $\mathbf{u} = \langle a, b \rangle$ may be written as

$$\mathbf{u} = a\mathbf{i} + b\mathbf{j}.$$

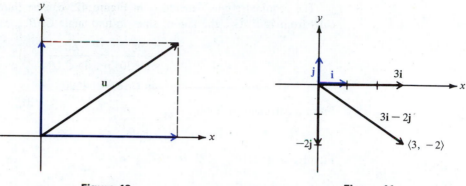

Figure 40   **Figure 41**

The vector sum $a\mathbf{i} + b\mathbf{j}$ is a **linear combination** of vectors $\mathbf{i}$ and $\mathbf{j}$. As an example, vector $\mathbf{u} = \langle 3, -2 \rangle$ in Figure 41 can be written with unit vectors as

$$\mathbf{u} = 3\mathbf{i} - 2\mathbf{j}.$$

⚠ CAUTION   Note that the *vector* $\mathbf{i}$ is not the same as the *complex number* i.

Applications of Vectors   Vectors have many applications, especially in physics. Some of these applications use the idea of an *equilibrant*: if the resultant of two forces is $\mathbf{u}$, then $-\mathbf{u}$ is the equilibrant of the two forces. The **equilibrant** is the force necessary to counterbalance the joint action of two forces.

· · · · · · · · ·

EXAMPLE 7   Find the magnitude of the equilibrant of forces of 48 and 60 newtons acting on a point A, if the angle between the forces is 50°. Then find the angle between the equilibrant and the 48-newton force.

In Figure 42, the equilibrant is $-\mathbf{v}$. The magnitude of $\mathbf{v}$, and also of $-\mathbf{v}$, is found by using triangle ABC and the law of cosines. Angle B is 130°, since adjacent angles of a parallelogram are supplementary.

$$|\mathbf{v}|^2 = 48^2 + 60^2 - 2(48)(60) \cos 130°$$
$$= 2304 + 3600 - 5760\,(-.6428)$$
$$|\mathbf{v}|^2 = 9606.5,$$

or $$|\mathbf{v}| = 98,$$

to two significant digits.

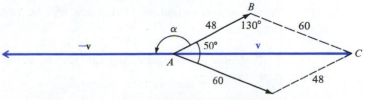

Figure 42

The required angle, labeled α in Figure 42, can be found by subtracting angle CAB from 180°. Use the law of sines to find angle CAB.

$$\frac{98}{\sin 130°} = \frac{60}{\sin CAB}$$

$$\sin CAB = .4690$$

From a calculator or table,

$$\text{angle } CAB = 28°.$$

Finally, $\alpha = 180° - 28° = 152°.$ ●

. . . .

EXAMPLE 8   Find the force required to hold a 50-lb weight on a ramp inclined at 20° to the horizontal.

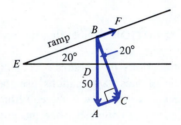

Figure 43

In Figure 43, the vertical 50-lb force represents the force due to gravity. The component **BC** represents the force with which the body pushes against the ramp, while the component **BF** represents a force that would pull the body up the ramp. Since vectors **BF** and **AC** are equal, **AC** gives the required force.

Vectors **BF** and **AC** are parallel, so angle EBD equals angle A. Since angle BDE and angle C are right angles, triangles ABC and DEB have two corresponding angles equal and so are similar triangles. Therefore, angle ABC equals angle E, which is 20°. From right triangle ABC,

$$\sin \mathbf{20°} = \frac{|\mathbf{AC}|}{\mathbf{50}}$$

$$|\mathbf{AC}| = 50 \sin 20°$$

$$|\mathbf{AC}| = 17.$$

To the nearest pound, a 17-lb force will be required to hold the weight on the ramp.   ●

Problems involving bearing can also be worked with vectors, as shown in the next example.

· · · · · · · · · ·

EXAMPLE 9   A ship leaves port on a bearing of 28° and travels 8.2 mi. The ship then turns due east and travels 4.3 mi. How far is the ship from port? What is its bearing from port?

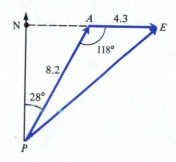

Figure 44

In Figure 44, vectors **PA** and **AE** represent the ship's lines of travel. The magnitude and bearing of the resultant **PE** must be found. Triangle *PNA* is a right triangle, so angle *NAP* = 90° − 28° = 62°. Then angle *PAE* = 180° − 62° = 118°. Use the law of cosines to find |**PE**|, the magnitude of vector **PE**.

$$|PE|^2 = 8.2^2 + 4.3^2 - 2(8.2)(4.3) \cos 118°$$
$$= 67.24 + 18.49 - 70.52(-.4695)$$
$$|PE|^2 = 118.84$$

Therefore, |**PE**| = 10.9.

To find the bearing of the ship from port, first find angle *APE*. Use the law of sines, along with the value of |**PE**| before rounding.

$$\frac{\sin APE}{4.3} = \frac{\sin 118°}{10.9}$$

$$\sin APE = \frac{4.3 \sin 118°}{10.9}$$

$$\text{angle } APE = 20.4°$$

After rounding to two significant digits, angle *APE* is 20°, and the ship is 11 mi from port on a bearing of 28° + 20° = 48°   ●

In air navigation, the air speed of a plane is its speed relative to the air, while the ground speed is its speed relative to the ground. Because of the wind, these two speeds are usually different. The ground speed of the plane is represented by the vector sum of the air speed and wind speed vectors. See Figure 45 on the next page.

· · · · · · · · · ·

EXAMPLE 10   A plane with an air speed of 192 mph is headed on a bearing of 121°. A north wind is blowing (from north to south) at 15.9 mph. Find the ground speed and the actual bearing of the plane.

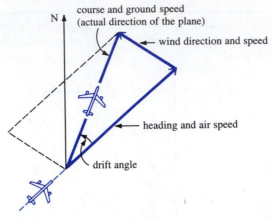

Figure 45

Figure 46

In Figure 46 the ground speed is represented by $|\mathbf{x}|$. We must find angle α to find the bearing, which will be $121° + \alpha$. From Figure 46, angle *BCO* equals angle *AOC*, which equals $121°$. Find $|\mathbf{x}|$ by the law of cosines.

$$|\mathbf{x}|^2 = 192^2 + 15.9^2 - 2(192)(15.9)\cos 121°$$
$$|\mathbf{x}|^2 = 36{,}864 + 252.81 - 6105.6(-.5150) = 40{,}261$$

Therefore, $|\mathbf{x}| = 200.7$,

or 201 mph. Now find α by using the law of sines. As before, use the value of $|\mathbf{x}|$ before rounding.

$$\frac{\sin \alpha}{15.9} = \frac{\sin 121°}{200.7}$$
$$\sin \alpha = .0679$$
$$\alpha = 3.89°$$

After rounding, α is $3.9°$. The ground speed is about 201 mph, on a bearing of $124.9°$. ●

8.4 Exercises ·

Name all pairs of vectors below that meet the conditions in Exercises 1–4.

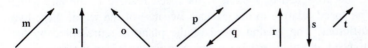

1. They appear to be equal.   **2.** They are opposites.

3. The first is a scalar multiple of the second, with the scalar positive.

4. The first is a scalar multiple of the second, with the scalar negative.

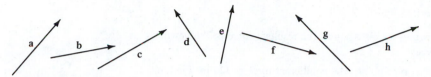

Exercises 5–16 refer to the vectors pictured above. Draw a sketch to represent each of the following vectors.

5. $-\mathbf{b}$ **6.** $-\mathbf{g}$ **7.** $3\mathbf{a}$ **8.** $2\mathbf{h}$

9. $\mathbf{a} + \mathbf{c}$ **10.** $\mathbf{h} + \mathbf{g}$ **11.** $\mathbf{b} + \mathbf{d}$ **12.** $\mathbf{a} + \mathbf{f}$

13. $\mathbf{a} - \mathbf{c}$ **14.** $\mathbf{d} - \mathbf{e}$ **15.** $\mathbf{a} + (\mathbf{b} + \mathbf{c})$ **16.** $(\mathbf{a} + \mathbf{b}) + \mathbf{c}$

Let $\mathbf{u} = \langle -2, 5 \rangle$, $\mathbf{v} = \langle 3, -2 \rangle$, and $\mathbf{w} = \langle -4, 6 \rangle$. Find each of the following vectors.

17. $\mathbf{u} + \mathbf{v}$ **18.** $\mathbf{u} - \mathbf{w}$ **19.** $6\mathbf{v} - 2\mathbf{u}$ **20.** $-3\mathbf{u} + 4\mathbf{w}$

In each of the following exercises, $\mathbf{v}$ has the given direction angle and magnitude. Find the magnitudes of the x- and y-components of $\mathbf{v}$.

21. $\theta = 45°$, $|\mathbf{v}| = 20$ **22.** $\theta = 75°$, $|\mathbf{v}| = 100$ **23.** $\theta = 60.1°$, $|\mathbf{v}| = 28.6$

24. $\theta = 35.9°$, $|\mathbf{v}| = 47.8$ **25.** $\theta = 128.0°$, $|\mathbf{v}| = 198$ **26.** $\theta = 146.0°$, $|\mathbf{v}| = 238$

Find the magnitude and direction angle for each of the following vectors.

27. $\langle 1, 1 \rangle$ **28.** $\langle -4, 4\sqrt{3} \rangle$ **29.** $\langle 8\sqrt{2}, -8\sqrt{2} \rangle$ **30.** $\langle \sqrt{3}, -1 \rangle$

31. $\langle 15, -8 \rangle$ **32.** $\langle -7, 24 \rangle$ **33.** $\langle -6, 0 \rangle$ **34.** $\langle 0, -12 \rangle$

Write each of the following vectors in the form $a\mathbf{i} + b\mathbf{j}$.

35. $\langle -5, 8 \rangle$ **36.** $\langle 6, -3 \rangle$ **37.** $\langle 2, 0 \rangle$ **38.** $\langle 0, -4 \rangle$

39. Direction angle 45°, magnitude 8 **40.** Direction angle 210°, magnitude 3

41. Direction angle 115°, magnitude .6 **42.** Direction angle 208°, magnitude .9

State a condition on $\mathbf{a}$ and $\mathbf{b}$ that implies each of the following equations or inequalities.

43. $|\mathbf{a} + \mathbf{b}| = 0$ **44.** $|\mathbf{a} + \mathbf{b}| = |\mathbf{a}| + |\mathbf{b}|$ **45.** $|\mathbf{a} + \mathbf{b}| = |\mathbf{a} - \mathbf{b}|$ **46.** $|\mathbf{a} + \mathbf{b}| > |\mathbf{a} - \mathbf{b}|$

In each of the following, two or three forces act at a point in the plane. The angle between the forces is given. Find the magnitude of the resultant force.

47. Forces of 250 and 450 newtons, forming an angle of 85°

48. Forces of 19 and 32 newtons, forming an angle of 118°

49. Forces of 17.9 and 25.8 lb, forming an angle of 105°

50. Forces of 75.6 and 98.2 lb, forming an angle of 82.8°

51. Forces of 116, 139, and 127 lb, with an angle of 141° between the 116-lb and 139-lb forces, an angle of 82° between the 127-lb and the 139-lb forces, and an angle of 137° between the 116-lb and the 127-lb forces

52. Forces of 37.8, 53.7, and 42.5 lb, with an angle of 68.5° between the 37.8-lb and 53.7-lb forces, an angle of 132° between the 53.7-lb and the 42.5-lb forces, and an angle of 159.5° between the 37.8-lb and the 42.5-lb forces

Solve each of the following problems.

53. Two forces of 692 and 423 newtons act on a point. The resultant force is 786 newtons. Find the angle between the forces.

54. Two forces of 128 and 253 lb act on a point. The equilibrant force is 320 lb. Find the angle between the forces.

55. Find the force required to hold a 100-lb box on a ramp inclined 10° with the horizontal.

56. Find the force required to keep a 3000-lb car parked on a hill which makes an angle of 15° with the horizontal.

57. A force of 25 lb is required to hold an 80-lb lawnmower on a hill. What angle does the hill make with the horizontal?

58. A force of 500 lb is required to hold a boat on a ramp inclined at 18° with the horizontal. How much does the boat weigh?

59. Niki and Suki are little dogs. Suki pulls on a rope attached to their doggie dish with a force of 3.89 lb. Niki pulls on another rope with a force of 4.72 lb. The angle between the forces is 143°. Find the direction and magnitude of the equilibrant.

60. Two people are carrying a box. One person exerts a force of 150 lb at an angle of 62.4° with the horizontal. The other person exerts a force of 114 lb at an angle of 54.9°. Find the weight of the box.

61. A crate is supported by two ropes. One rope makes an angle of 46.3° with the horizontal and has a tension of 89.6 lb on it. The other rope is horizontal. Find the weight of the crate and the tension in the horizontal rope.

62. Three forces acting at a point are in equilibrium. The forces are 980 lb, 760 lb, and 1220 lb. Find the angles between the directions of the forces. (*Hint:* Arrange the forces to form the sides of a triangle.)

63. A force of 176 lb makes an angle of 78.8° with a second force. The resultant of the two forces makes an angle of 41.1° with the first force. Find the magnitude of the second force and of the resultant.

64. A force of 28.7 lb makes an angle of 42.2° with a second force. The resultant of the two forces makes an angle of 32.6° with the first force. Find the magnitude of the second force and of the resultant.

65. A plane flies 650.0 mph on a bearing of 175.0°. A 25.0-mph wind, from a direction of 266.0°, blows against the plane. Find the resulting bearing of the plane.

66. A pilot wants to fly on a bearing of 74.9°. By flying due east, he finds that a 42.0-mph wind blowing from the south puts him on course. Find the air speed and the ground speed.

67. Starting at point *A*, a ship sails 18.5 km on a bearing of 189.0°, then turns and sails 47.8 km on a bearing of 317.0°. Find the distance of the ship from point *A*.

68. The distance between points *A* and *B* is 1.7 mi. In between *A* and *B* is a dark forest. To avoid the forest, John walks from *A* a distance of 1.1 mi on a bearing of 325°, and then turns and walks 1.4 mi to *B*. Find the bearing of *B* from *A*.

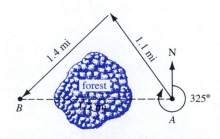

Exercise 68

69. The airline route from San Francisco to Honolulu is on a bearing of 233.0°. A jet flying at 450 mph on that bearing runs into a wind blowing at 39.0 mph from a direction 114.0°. Find the resulting bearing and ground speed of the plane.

70. A pilot is flying at 168 mph. She wants her flight path to be on a bearing of 57.6°. A wind is blowing from the south at 27.1 mph. Find the bearing the pilot should fly, and find the plane's ground speed.

71. What bearing and air speed are required for a plane to fly 400 mi due north in 2.5 hr, if the wind is blowing from a direction 328° at 11 mph?

72. Janet and Russ are pulling their daughter Lindsay on a sled. Russ pulls with a force of 18 lb at an angle of 10°. Janet pulls with a force of 12 lb at an angle of 15°. Find the resultant. See the figure.

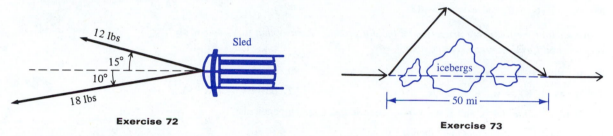

Exercise 72

Exercise 73

73. A ship sailing the North Atlantic has been warned to change course to avoid a group of icebergs. The captain turns and sails on a bearing of 62° for a while, then changes course again to a bearing of 115° until the ship reaches its original course. (See the figure.) How much farther did the ship have to travel to avoid the icebergs?

74. A car going around a banked curve is subject to the forces shown in the figure. If the radius of the curve is 100 ft, what value of θ would allow an automobile to travel around the curve at a speed of 40 ft per sec. without depending on friction?

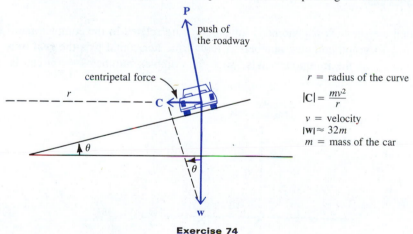

P

push of the roadway

centripetal force

C

r

θ

θ

W

r = radius of the curve

$|C| = \dfrac{mv^2}{r}$

v = velocity

$|w| \approx 32m$

m = mass of the car

Exercise 74

Let $\mathbf{u} = \langle a_1, b_1 \rangle$, $\mathbf{v} = \langle a_2, b_2 \rangle$, $\mathbf{w} = \langle a_3, b_3 \rangle$, and $\mathbf{0} = \langle 0, 0 \rangle$. Let k and m be any real numbers. Prove each of the following statements.

75. $\mathbf{u} + \mathbf{v} = \mathbf{v} + \mathbf{u}$

76. $\mathbf{u} + (\mathbf{v} + \mathbf{w}) = (\mathbf{u} + \mathbf{v}) + \mathbf{w}$

77. $-1(\mathbf{u}) = -\mathbf{u}$

78. $k(\mathbf{u} + \mathbf{v}) = k\mathbf{u} + k\mathbf{v}$

79. $\mathbf{u} + \mathbf{0} = \mathbf{u}$

80. $\mathbf{u} + (-\mathbf{u}) = \mathbf{0}$

Let $\mathbf{u} = a_1\mathbf{i} + b_1\mathbf{j}$ and $\mathbf{v} = a_2\mathbf{i} + b_2\mathbf{j}$. *The* dot product, *or* inner product, *of* $\mathbf{u}$ *and* $\mathbf{v}$, *written* $\mathbf{u} \cdot \mathbf{v}$, *is defined as* $\mathbf{u} \cdot \mathbf{v} = a_1a_2 + b_1b_2$. *Find* $\mathbf{u} \cdot \mathbf{v}$ *for each pair of vectors in Exercises 81–84.*

81. $\mathbf{u} = 6\mathbf{i} - 2\mathbf{j}, \mathbf{v} = -3\mathbf{i} + 2\mathbf{j}$

82. $\mathbf{u} = 3\mathbf{i} + 2\mathbf{j}, \mathbf{v} = -3\mathbf{i} + 7\mathbf{j}$

83. $\mathbf{u} = \langle -6, 8 \rangle$ and $\mathbf{v} = \langle 3, -4 \rangle$

84. $\mathbf{u} = \langle 0, -2 \rangle$ and $\mathbf{v} = \langle -2, 6 \rangle$

85. Let α be the angle between the vectors $\mathbf{u}$ and $\mathbf{v}$, where $0° \leq \alpha \leq 180°$. Show that
$\mathbf{u} \cdot \mathbf{v} = |\mathbf{u}|\, |\mathbf{v}| \cos \alpha$.

Use the result of Exercise 85 to find the angle between the following pairs of vectors.

86. $\mathbf{u} = \langle -2, 5 \rangle$ and $\mathbf{v} = \langle 3, -4 \rangle$

87. $\mathbf{u} = \langle 1, 8 \rangle$ and $\mathbf{v} = \langle 2, -5 \rangle$

88. $\mathbf{u} = \langle -6, -2 \rangle$ and $\mathbf{v} = \langle 3, -1 \rangle$

Prove each of the following properties of the dot product. Assume that $\mathbf{u}$, $\mathbf{v}$, *and* $\mathbf{w}$ *are vectors, and k is a nonzero real number.*

89. $\mathbf{u} \cdot \mathbf{v} = \mathbf{v} \cdot \mathbf{u}$

90. $\mathbf{u} \cdot \mathbf{u} = |\mathbf{u}|^2$

91. $\mathbf{u} \cdot (\mathbf{v} + \mathbf{w}) = \mathbf{u} \cdot \mathbf{v} + \mathbf{u} \cdot \mathbf{w}$

92. $(k\mathbf{u}) \cdot \mathbf{v} = k(\mathbf{u} \cdot \mathbf{v})$

93. If $\mathbf{u}$ and $\mathbf{v}$ are not $\mathbf{0}$, and if $\mathbf{u} \cdot \mathbf{v} = 0$, then $\mathbf{u}$ and $\mathbf{v}$ are perpendicular.

94. If $\mathbf{u}$ and $\mathbf{v}$ are perpendicular, then $\mathbf{u} \cdot \mathbf{v} = 0$.

8.5 TRIGONOMETRIC FORM OF COMPLEX NUMBERS

In this section we see how trigonometry and vectors are related to the complex numbers. One way to graph complex numbers is to call the horizontal axis the **real axis** and the vertical axis the **imaginary axis.** Then the complex number $2 - 3i$ can be graphed as shown in Figure 47.

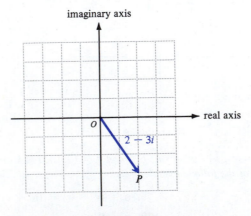

Figure 47

Each nonzero complex number graphed in this way determines a unique directed line segment, the segment from the origin to the point representing the complex number. Recall that such directed line segments (like **OP** of Figure 47) are called vectors. We have now seen three notations for a two-dimensional vector:

$$\langle a, b \rangle, \quad a\mathbf{i} + b\mathbf{j}, \quad \text{and} \quad a + bi.$$

(Remember **i** is a vector and i is a complex number.)

By definition, the sum of the two complex numbers $4 + i$ and $1 + 3i$ is

$$(4 + i) + (1 + 3i) = 5 + 4i.$$

Graphically, the sum of two complex numbers is represented by the vector which is the resultant of the vectors corresponding to the two numbers. The vectors representing the complex numbers $4 + i$ and $1 + 3i$, and the resultant vector which represents their sum, $5 + 4i$, are shown in Figure 48. As Figure 48 suggests, the result of adding two complex numbers is the resultant of the two corresponding vectors. Thus, complex numbers provide a way to perform two-dimensional vector arithmetic algebraically.

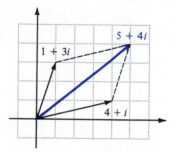

Figure 48

EXAMPLE 1   Find the resultant of $6 - 2i$ and $-4 - 3i$. Graph both complex numbers and their resultant.

The resultant is found by adding the two numbers.

$$(6 - 2i) + (-4 - 3i) = 2 - 5i$$

The graphs are shown in Figure 49. ●

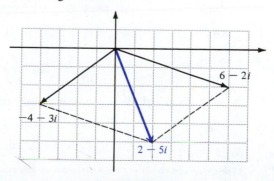

Figure 49

Figure 50 shows the complex number $x + yi$ that corresponds to a vector **OP** with direction angle θ and magnitude r. Earlier it was shown that $x = r \cos \theta$ and $y = r \sin \theta$. Substituting $x = r \cos \theta$ and $y = r \sin \theta$ into $x + yi$ gives

$$x + yi = \boldsymbol{r \cos \theta} + (\boldsymbol{r \sin \theta})i$$
$$= r(\cos \theta + i \sin \theta).$$

Trigonometric or Polar Form of a Complex Number

The expression

$$\boldsymbol{r(\cos \theta + i \sin \theta)}$$

is called the **trigonometric form** or **polar form** of the complex number $x + yi$.* The number r is called the **modulus** or **absolute value** of $x + yi$, while θ is the **argument** of $x + yi$.

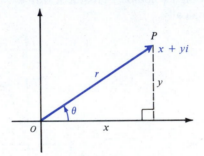

Figure 50

EXAMPLE 2   Write the following complex numbers in trigonometric form.

(a) $-\sqrt{3} + i$

Figure 51 shows the graph of the complex number $-\sqrt{3} + i$. Since $x = -\sqrt{3}$ and $y = 1$,

$$r = \sqrt{x^2 = y^2} = \sqrt{3 + 1} = 2,$$

and

$$\tan \theta = \frac{y}{x} = \frac{1}{-\sqrt{3}} = -\frac{\sqrt{3}}{3}.$$

As Figure 51 shows, θ is in quadrant II. Since

$$\tan \frac{5\pi}{6} = -\frac{\sqrt{3}}{3},$$

$\theta = 5\pi/6$ and the trigonometric form of $-\sqrt{3} + i$ is

$$x + yi = r(\cos \theta + i \sin \theta)$$
$$-\sqrt{3} + i = 2(\cos 5\pi/6 + i \sin 5\pi/6).$$

*The expression $\cos \theta + i \sin \theta$ is sometimes abbreviated as cis θ. With this notation $r(\cos \theta + i \sin \theta)$ is written as r cis θ.

(b) $-2 - 2i$

First find r and θ.

$$r = \sqrt{x^2 + y^2} = \sqrt{4 + 4} = 2\sqrt{2}$$

$$\tan \theta = \frac{y}{x} = \frac{-2}{-2} = 1$$

As shown in Figure 52, the complex number is in quadrant III, where

$$\tan 5\pi/4 = 1.$$

From these results, the trigonometric form of $-2 - 2i$ is

$$2\sqrt{2}(\cos 5\pi/4 + i \sin 5\pi/4). \quad \bullet$$

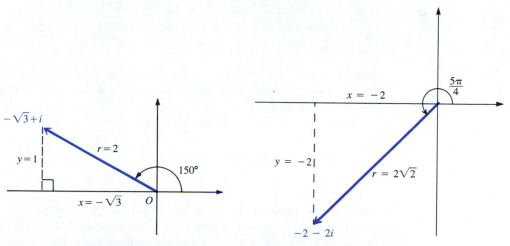

Figure 51   **Figure 52**

· · · · · · · · ·

EXAMPLE 3   Express $2(\cos 300° + i \sin 300°)$ in standard form.

Recall that $a + bi$ is the standard form for a complex number. Because $\cos 300° = 1/2$ and $\sin 300° = -\sqrt{3}/2$,

$$2(\cos 300° + i \sin 300°) = 2\left(\frac{1}{2} - i\frac{\sqrt{3}}{2}\right) = 1 - i\sqrt{3}. \quad \bullet$$

Products of Complex Numbers   The product of the two complex numbers $1 + i\sqrt{3}$ and $-2\sqrt{3} + 2i$ can be found by the methods shown earlier.

$$(1 + i\sqrt{3})(-2\sqrt{3} + 2i) = -2\sqrt{3} + 2i - 2i(3) + 2i^2\sqrt{3}$$

$$= -2\sqrt{3} + 2i - 6i - 2\sqrt{3} \qquad \text{Since } i^2 = -1$$

$$(1 + i\sqrt{3})(-2\sqrt{3} + 2i) = -4\sqrt{3} - 4i$$

This same product also can be found by first converting the complex numbers $1 + i\sqrt{3}$ and $-2\sqrt{3} + 2i$ to trigonometric form. Using the method explained above,

$$1 + i\sqrt{3} = 2(\cos 60° + i \sin 60°)$$

and
$$-2\sqrt{3} + 2i = 4(\cos 150° + i \sin 150°).$$

If the trigonometric forms are now multiplied together and if the trigonometric identities for the cosine and the sine of the sum of two angles are used, the result is

$$[2(\cos 60° + i \sin 60°)][4(\cos 150° + i \sin 150°)]$$
$$= 2 \cdot 4(\cos 60° \cdot \cos 150° + i \sin 60° \cdot \cos 150°$$
$$+ i \cos 60° \cdot \sin 150° + i^2 \sin 60° \cdot \sin 150°)$$
$$= 8[(\cos 60° \cdot \cos 150° - \sin 60° \cdot \sin 150°)$$
$$+ i(\sin 60° \cdot \cos 150° + \cos 60° \cdot \sin 150°)]$$
$$= 8[\cos (60° + 150°) + i \sin (60° + 150°)]$$
$$= 8(\cos 210° + i \sin 210°).$$

The modulus of the product, 8, is equal to the product of the moduli of the factors, $2 \cdot 4$, while the argument of the product, 210°, is the sum of the arguments of the factors, 60° + 150°.

Since $\cos 210° = -\sqrt{3}/2$ and $\sin 210° = -1/2$,

$$8(\cos 210° + i \sin 210°) = 8\left(-\frac{\sqrt{3}}{2} + i\left(-\frac{1}{2}\right)\right)$$
$$= -4\sqrt{3} - 4i,$$

which is the same result found above, suggesting an alternative way to find the product of two complex numbers using the trigonometric form of the numbers.

Generalizing, the product of the two complex numbers, $r_1(\cos \theta_1 + i \sin \theta_1)$ and $r_2(\cos \theta_2 + i \sin \theta_2)$ is

$$[r_1(\cos \theta_1 + i \sin \theta_1)] \cdot [r_2(\cos \theta_2 + i \sin \theta_2)]$$
$$= r_1 r_2(\cos \theta_1 \cos \theta_2 + i \sin \theta_1 \cos \theta_2 + i \cos \theta_1 \sin \theta_2 + i^2 \sin \theta_1 \sin \theta_2)$$
$$= r_1 r_2[(\cos \theta_1 \cos \theta_2 - \sin \theta_1 \sin \theta_2) + i(\sin \theta_1 \cos \theta_2 + \cos \theta_1 \sin \theta_2)]$$
$$= r_1 r_2[\cos (\theta_1 + \theta_2) + i \sin (\theta_1 + \theta_2)].$$

This work proves the following product theorem.

Product Theorem

If $r_1(\cos \theta_1 + i \sin \theta_1)$ and $r_2(\cos \theta_2 + i \sin \theta_2)$ are any two complex numbers, then

$$[r_1(\cos \theta_1 + i \sin \theta_1)] \cdot [r_2(\cos \theta_2 + i \sin \theta_2)]$$
$$= r_1 r_2[\cos (\theta_1 + \theta_2) + i \sin (\theta_1 + \theta_2)].$$

• • • • • • • •

EXAMPLE 4   Find the product of the two complex numbers

$$3(\cos 45° + i \sin 45°) \quad \text{and} \quad 2(\cos 135° + i \sin 135°).$$

Using the product theorem,

$$[3(\cos 45° + i \sin 45°)][2(\cos 135° + i \sin 135°)]$$
$$= \mathbf{3 \cdot 2}[\cos (\mathbf{45° + 135°}) + i \sin (\mathbf{45° + 135°})]$$
$$= 6(\cos 180° + i \sin 180°),$$

which can be expressed as $6(-1 + i \cdot 0) = 6(-1) = -6$. The two complex numbers in this example are complex factors of -6.   •

Quotients of Complex Numbers   In standard form, the quotient of the complex numbers $1 + i\sqrt{3}$ and $-2\sqrt{3} + 2i$ is

$$\frac{1 + i\sqrt{3}}{-2\sqrt{3} + 2i} = \frac{(1 + i\sqrt{3})(-2\sqrt{3} - 2i)}{(-2\sqrt{3} + 2i)(-2\sqrt{3} - 2i)}$$

$$= \frac{-2\sqrt{3} - 2i - 6i - 2i^2\sqrt{3}}{12 - 4i^2}$$

$$\frac{1 + i\sqrt{3}}{-2\sqrt{3} + 2i} = \frac{-8i}{16} = -\frac{1}{2}i.$$

Writing $1 + i\sqrt{3}$, $-2\sqrt{3} + 2i$, and $-\dfrac{1}{2}i$ in trigonometric form gives

$$1 + i\sqrt{3} = 2(\cos 60° + i \sin 60°)$$
$$-2\sqrt{3} + 2i = 4(\cos 150° + i \sin 150°)$$
$$-\frac{1}{2}i = \frac{1}{2}[\cos (-90°) + i \sin (-90°)].$$

The modulus of the quotient, 1/2, is the quotient of the two moduli, 2 and 4. The argument of the quotient, $-90°$, is the difference of the two arguments, $60° - 150° = -90°$. It would be easier to find the quotient of these two complex numbers in trigonometric form than in standard form. Generalizing from this example leads to the following theorem.

• • • • • • • • • • • •   •

Quotient
Theorem

If $r_1(\cos \theta_1 + i \sin \theta_1)$ and $r_2(\cos \theta_2 + i \sin \theta_2)$ are complex numbers, where $r_2(\cos \theta_2 + i \sin \theta_2) \neq 0$, then

$$\frac{r_1(\cos \theta_1 + i \sin \theta_1)}{r_2(\cos \theta_2 + i \sin \theta_2)} = \frac{r_1}{r_2}[\cos (\theta_1 - \theta_2) + i \sin (\theta_1 - \theta_2)].$$

· · · · · · · · ·

EXAMPLE 5   Find the quotient of

$$10[\cos(-60°) + i \sin(-60°)] \quad \text{and} \quad 5(\cos 150° + i \sin 150°).$$

Write the result in standard form.

By the quotient theorem,

$$\frac{10[\cos(-60°) + i \sin(-60°)]}{5(\cos 150° + i \sin 150°)}$$

$$= \frac{10}{5}[\cos(-60° - 150°) + i \sin(-60° - 150°)]$$

$$= \frac{10}{5}[\cos(-210°) + i \sin(-210°)].$$

Since angles of $-210°$ and $150°$ are coterminal, replace $-210°$ with $150°$ to get

$$\frac{10[\cos(-60°) + i \sin(-60°)]}{5(\cos 150° + i \sin 150°)} = 2(\cos 150° + i \sin 150°).$$

Because $\cos 150° = -\sqrt{3}/2$ and $\sin 150° = 1/2$,

$$2(\cos 150° + i \sin 150°) = 2\left(\frac{-\sqrt{3}}{2} + i \cdot \frac{1}{2}\right)$$

$$= -\sqrt{3} + i.$$

The quotient in standard form is $-\sqrt{3} + i$.   ●

8.5 Exercises ·

Graph each of the following complex numbers.

1. $-2 + 3i$ **2.** $-4 + 5i$ **3.** $2 - 2i\sqrt{3}$ **4.** $4\sqrt{2} + 4i\sqrt{2}$

5. $-4i$ **6.** $3i$ **7.** -8 **8.** 2

Find the resultant of each of the following pairs of complex numbers.

9. $2 - 3i, -1 + 4i$ **10.** $-4 - 5i, 2 + i$ **11.** $-5 + 6i, 3 - 4i$ **12.** $8 - 5i, -6 + 3i$

13. $-2, 4i$ **14.** $5, -4i$ **15.** $2 + 6i, -2i$ **16.** $4 - 2i, 5$

Write the following complex numbers in standard form.

17. $2(\cos 45° + i \sin 45°)$ **18.** $4(\cos 60° + i \sin 60°)$ **19.** $10(\cos 90° + i \sin 90°)$

20. $8(\cos 270° + i \sin 270°)$ **21.** $4(\cos 240° + i \sin 240°)$ **22.** $2(\cos 330° + i \sin 330°)$

23. $5(\cos 300° + i \sin 300°)$ **24.** $6(\cos 135° + i \sin 135°)$ **25.** $\sqrt{2}(\cos 180° + i \sin 180°)$

26. $\sqrt{3}(\cos 315° + i \sin 315°)$

Using a calculator or Table 3, complete the following chart, giving angles to the nearest tenth of a degree.

	Standard form	Trigonometric form
27.	$2 + 3i$	
28.		$(\cos 35° + i \sin 35°)$
29.		$3(\cos 250.2° + i \sin 250.2°)$
30.	$-4 + i$	
31.	$-1.8794 + .6840i$	
32.		$2(\cos 310.3° + i \sin 310.3°)$
33.	$3 + 5i$	
34.		$(\cos 110.5° + i \sin 110.5°)$

Write each of the following complex numbers in trigonometric form.

35. $3 - 3i$ **36.** $-2 + 2i\sqrt{3}$ **37.** $-3 - 3i\sqrt{3}$ **38.** $1 + i\sqrt{3}$

39. $\sqrt{3} - i$ **40.** $4\sqrt{3} + 4i$ **41.** $2 + 2i$ **42.** $-\sqrt{3} + i$

43. -4 **44.** $5i$

Find each of the following products. Write the result in standard form.

45. $[3(\cos 60° + i \sin 60°)][2(\cos 90° + i \sin 90°)]$ **46.** $[4(\cos 30° + i \sin 30°)][5(\cos 120° + i \sin 120°)]$

47. $[2(\cos 45° + i \sin 45°)][2(\cos 225° + i \sin 225°)]$ **48.** $[8(\cos 300° + i \sin 300°)][5(\cos 120° + i \sin 120°)]$

49. $[4(\cos 60° + i \sin 60°)][6(\cos 330° + i \sin 330°)]$ **50.** $[8(\cos 210° + i \sin 210°)][2(\cos 330° + i \sin 330°)]$

51. $[5(\cos 90° + i \sin 90°)][3(\cos 45° + i \sin 45°)]$

52. $[6(\cos 120° + i \sin 120°)][5(\cos (-30°) + i \sin (-30°))]$

53. $[\sqrt{3}(\cos 45° + i \sin 45°)][\sqrt{3}(\cos 225° + i \sin 225°)]$

54. $[\sqrt{2}(\cos 300° + i \sin 300°)][\sqrt{2}(\cos 270° + i \sin 270°)]$

Find the following quotients. Write the results in standard form.

55. $\dfrac{4(\cos 120° + i \sin 120°)}{2(\cos 150° + i \sin 150°)}$ **56.** $\dfrac{10(\cos 225° + i \sin 225°)}{5(\cos 45° + i \sin 45°)}$ **57.** $\dfrac{16(\cos 300° + i \sin 300°)}{8(\cos 60° + i \sin 60°)}$

58. $\dfrac{24(\cos 150° + i \sin 150°)}{2(\cos 30° + i \sin 30°)}$ **59.** $\dfrac{3(\cos 305° + i \sin 305°)}{9(\cos 65° + i \sin 65°)}$ **60.** $\dfrac{12(\cos 293° + i \sin 293°)}{6(\cos 23° + i \sin 23°)}$

61. $\dfrac{8}{\sqrt{3} + i}$ **62.** $\dfrac{2i}{-1 - i\sqrt{3}}$

63. $\dfrac{2\sqrt{6} - 2i\sqrt{2}}{\sqrt{2} - i\sqrt{6}}$ **64.** $\dfrac{4 + 4i}{2 - 2i}$

In applied work in trigonometry, it is often necessary to find the resultant of more than two vectors graphically. Find the resultant of the indicated vectors in Exercises 65 and 66. Give the resultant in standard form.

65.

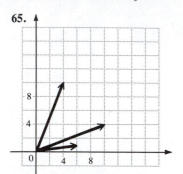

66.

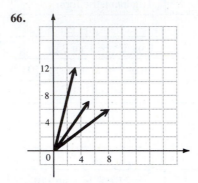

67. The alternating current in an electric inductor is $I = E/Z$ amperes, where E is the voltage and $Z = R + X_L i$ is the impedance. If $E = 8(\cos 20° + i \sin 20°)$, $R = 6$, and $X_L = 3$, find the current. Give the answer in standard form.

68. The current I in a circuit with voltage E, resistance R, capacitive reactance X_c, and inductive reactance X_L is $I = E/(R + (X_L - X_c)i)$. Find I if $E = 12(\cos 25° + i \sin 25°)$, $R = 3$, $X_L = 4$, and $X_c = 6$. Give the answer in standard form.

The complex number z, where z = x + yi, can be graphed as (x, y). Find and graph all complex numbers z satisfying the conditions in Exercises 69–74.

69. The modulus of z is 1.

70. The modulus of z is greater than 1.

71. The real part of z is 1.

72. The imaginary part of z is 1.

73. The real and imaginary parts of z are equal.

74. The real part of z equals z itself.

75. Show that the graphs of all complex numbers z with modulus 5 lie on a circle. What is the center and radius of the circle?

76. Where is the point corresponding to the complex number $z + 3i$ located relative to the point corresponding to z? What about the location of $z - 3i$ relative to the location of z?

Give a geometric condition that implies the following situation.

77. The difference between two nonreal complex numbers $a + bi$ and $c + di$ is a real number.

78. The square of a nonreal complex number $a + bi$ is a real number.

79. The modulus of the sum of two complex numbers $a + bi$ and $c + di$ is equal to the sum of their moduli.

80. The modulus of the difference of two complex numbers $a + bi$ and $c + di$ is equal to the sum of their moduli.

8.6 DE MOIVRE'S THEOREM AND *n*TH ROOTS

We have seen that in trigonometric form complex numbers are easily multiplied and divided. The most important use of trigonometric form, however, is in finding powers and roots of complex numbers. While powers of a complex number *can* be found by repeated multiplication, which may become tiresome, the only way to find roots of complex numbers is by the method to be introduced in this section. The square of $r(\cos \theta + i \sin \theta)$ can be found by multiplying the number by itself.

$$[r(\cos \theta + i \sin \theta)]^2 = [r(\cos \theta + i \sin \theta)][r(\cos \theta + i \sin \theta)]$$
$$= r \cdot r[\cos (\theta + \theta) + i \sin (\theta + \theta)]$$
$$[r(\cos \theta + i \sin \theta)]^2 = r^2(\cos 2\theta + i \sin 2\theta)$$

In the same way,

$$[r(\cos \theta + i \sin \theta)]^3 = r^3(\cos 3\theta + i \sin 3\theta).$$

These results suggest the plausibility of the following theorem, for positive integer values of *n*. Although this theorem is stated and can be proved for all *n,* we will use it only for positive integer values of *n* and their reciprocals.

De Moivre's Theorem

If $r(\cos \theta + i \sin \theta)$ is a complex number expressed in trigonometric form, and if *n* is any real number, then

$$[r(\cos \theta + i \sin \theta)]^n = r^n(\cos n\theta + i \sin n\theta).$$

For positive integer values of *n,* De Moivre's theorem can be proved by the method of mathematical induction.*

The French expatriate friend of Isaac Newton, Abraham De Moivre (1667–1754), never explicitly stated the theorem bearing his name. He did, however, revolutionize trigonometry by introducing imaginary quantities to the subject. The theorem, one of the most famous in mathematics, was named after him in 1826 by the publisher of the leading mathematics journal of its day. De Moivre had suggestions of the theorem in his writings of 1730 and might have known it as early as 1707. As with most eighteenth-century mathematics, it was Euler who explicitly stated it. He proved ''De Moivre's theorem'' in 1748.

Example 1 on the following page shows how to apply De Moivre's theorem to find a power of a complex number.

*Mathematical induction is discussed in Section 11.5.

EXAMPLE 1   Find $(1 + i\sqrt{3})^8$.

To use De Moivre's theorem, first convert $1 + i\sqrt{3}$ into trigonometric form.

$$1 + i\sqrt{3} = 2(\cos 60° + i \sin 60°)$$

Now apply De Moivre's theorem.

$$\begin{aligned}(1 + i\sqrt{3})^8 &= [2(\cos 60° + i \sin 60°)]^8 \\ &= 2^8[\cos (8 \cdot 60°) + i \sin (8 \cdot 60°)] \\ &= 256(\cos 480° + i \sin 480°) \\ &= 256(\cos 120° + i \sin 120°)\end{aligned}$$

$$(1 + i\sqrt{3})^8 = 256\left(-\frac{1}{2} + i\,\frac{\sqrt{3}}{2}\right) = -128 + 128i\sqrt{3} \quad \bullet$$

De Moivre's theorem is also used to find the nth roots of complex numbers.

Definition of nth Root

For a positive integer n, the complex number $a + bi$ is an **nth root** of the complex number $x + yi$ if

$$(a + bi)^n = x + yi.$$

To find the cube roots of the complex number $8(\cos 135° + i \sin 135°)$, for example, look for a complex number, say $r(\cos \alpha + i \sin \alpha)$, that will satisfy

$$[r(\cos \alpha + i \sin \alpha)]^3 = 8(\cos 135° + i \sin 135°).$$

By De Moivre's theorem, this equation becomes

$$r^3(\cos 3\alpha + i \sin 3\alpha) = 8(\cos 135° + i \sin 135°).$$

One way to satisfy this equation is to set $r^3 = 8$ and also set $\cos 3\alpha + i \sin 3\alpha = \cos 135° + i \sin 135°$. The first of these conditions implies that $r = 2$, and the second implies that

$$\cos 3\alpha = \cos 135° \quad \text{and} \quad \sin 3\alpha = \sin 135°.$$

These equations can be satisfied only if

$$3\alpha = 135° + 360° \cdot k, \quad k \text{ any integer,}$$

or

$$\alpha = \frac{135° + 360° \cdot k}{3}, \quad k \text{ any integer.}$$

If $k = 0$,

$$\alpha = \frac{135° + \mathbf{0°}}{3} = 45°.$$

For $k = 1$,

$$\alpha = \frac{135° + \mathbf{360°}}{3} = \frac{495°}{3} = 165°.$$

When $k = 2$,

$$\alpha = \frac{135° + \mathbf{720°}}{3} = \frac{855°}{3} = 285°.$$

In the same way, $\alpha = 405°$ when $k = 3$. However, $\sin 405° = \sin 45°$ and $\cos 405° = \cos 45°$, so all of the cube roots (three of them) can be found by letting $k = 0, 1,$ and 2.

When $k = 0$, the root is

$$2(\cos \mathbf{45°} + i \sin \mathbf{45°}).$$

When $k = 1$, the root is

$$2(\cos \mathbf{165°} + i \sin \mathbf{165°}).$$

When $k = 2$, the root is

$$2(\cos \mathbf{285°} + i \sin \mathbf{285°}).$$

In summary, the complex numbers

$$2(\cos 45° + i \sin 45°), \quad 2(\cos 165° + i \sin 165°),$$
$$\text{and} \quad 2(\cos 285° + i \sin 285°)$$

are the three cube roots of $8(\cos 135° + i \sin 135°)$. Generalizing this result leads to the following theorem.

nth Root Theorem

If n is any positive integer and r is a positive real number, then the complex number $r(\cos \theta + i \sin \theta)$ has exactly n distinct nth roots, given by

$$r^{1/n}(\cos \alpha + i \sin \alpha),$$

where
$$\alpha = \frac{\theta + 360° \cdot k}{n}, \quad k = 0, 1, 2, \ldots, n - 1.$$

Alternatively, write the formula for α as

$$\alpha = \frac{\theta}{n} + \left(\frac{360°}{n}\right) \cdot k, \quad k = 0, 1, 2, \ldots, n - 1.$$

EXAMPLE 2   Find all fourth roots of $-8 + 8i\sqrt{3}$.

First write $-8 + 8i\sqrt{3}$ in trigonometric form as

$$-8 + 8i\sqrt{3} = 16(\cos 120° + i \sin 120°).$$

Here $r = 16$ and $\theta = 120°$. The fourth roots of this number have modulus $16^{1/4} = 2$ and arguments given as follows.

If $k = 0$,
$$\frac{120° + 360° \cdot \mathbf{0}}{4} = 30°,$$

if $k = 1$,
$$\frac{120° + 360° \cdot \mathbf{1}}{4} = 120°,$$

if $k = 2$, $$\frac{120° + 360° \cdot 2}{4} = 210°,$$

if $k = 3$, $$\frac{120° + 360° \cdot 3}{4} = 300°.$$

Using these angles, the fourth roots are

$$2(\cos 30° + i \sin 30°), \quad 2(\cos 120° + i \sin 120°),$$
$$2(\cos 210° + i \sin 210°), \quad \text{and} \quad 2(\cos 300° + i \sin 300°).$$

These four roots can be written in standard form as $\sqrt{3} + i$, $-1 + i\sqrt{3}$, $-\sqrt{3} - i$, and $1 - i\sqrt{3}$. The graphs of these roots are all on a circle that has center at the origin and radius 2, as shown in Figure 53. Notice that the roots are equally spaced about the circle and are 90° apart. ●

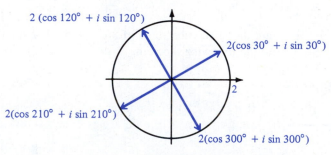

2 $(\cos 120° + i \sin 120°)$

2$(\cos 30° + i \sin 30°)$

2$(\cos 210° + i \sin 210°)$

2$(\cos 300° + i \sin 300°)$

Figure 53

EXAMPLE 3   Find all complex number solutions of $x^5 - 1 = 0$.
Write the equation as

$$x^5 - 1 = 0$$
$$x^5 = 1$$
$$x = \sqrt[5]{1}.$$

While there is only one real number solution, 1, there are five complex number solutions. To find these solutions, first write 1 in trigonometric form as

$$1 = 1 + 0i = 1(\cos 0° + i \sin 0°).$$

The modulus of the fifth roots is $1^{1/5} = 1$, and the arguments are given by

$$\frac{0° + 360° \cdot k}{5}, \quad k = 0, 1, 2, 3, \text{ or } 4.$$

By using these arguments, the fifth roots are

$$1(\cos 0° + i \sin 0°), \quad 1(\cos 72° + i \sin 72°), \quad 1(\cos 144° + i \sin 144°),$$
$$1(\cos 216° + i \sin 216°), \quad \text{and} \quad 1(\cos 288° + i \sin 288°).$$

The first of these roots equals 1; the others cannot easily be expressed in standard form. The five fifth roots all lie on a unit circle and are equally spaced around it every 72°, as shown in Figure 54. ●

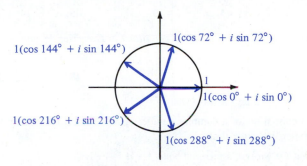

Figure 54

8.6 Exercises

Find the following powers. Write the result in standard form.

1. $[3(\cos 30° + i \sin 30°)]^3$

2. $[2(\cos 135° + i \sin 135°)]^4$

3. $\left(\cos \dfrac{\pi}{4} + i \sin \dfrac{\pi}{4}\right)^8$

4. $\left[2\left(\cos \dfrac{\pi}{3} + i \sin \dfrac{\pi}{3}\right)\right]^3$

5. $[3(\cos 100° + i \sin 100°)]^3$

6. $[3(\cos 40° + i \sin 40°)]^3$

7. $(\sqrt{3} + i)^5$

8. $(2\sqrt{2} - 2i\sqrt{2})^6$

9. $(2 - 2i\sqrt{3})^4$

10. $\left(\dfrac{\sqrt{2}}{2} - \dfrac{\sqrt{2}}{2}i\right)^8$

11. $(-2 - 2i)^5$

12. $(-1 + i)^7$

13. $(-.4283 + .5172i)^4$

14. $(1.87615 - 1.42213i)^3$

15. $\left[1.86\left(\cos \dfrac{5\pi}{9} + i \sin \dfrac{5\pi}{9}\right)\right]^{15}$

16. $\left[24.3\left(\cos \dfrac{7\pi}{12} + i \sin \dfrac{7\pi}{12}\right)\right]^3$

In Exercises 17–46, leave answers in trigonometric form.

Find and graph all cube roots of each of the following complex numbers.

17. $(\cos 0° + i \sin 0°)$

18. $(\cos 90° + i \sin 90°)$

19. $8(\cos 60° + i \sin 60°)$

20. $27(\cos 300° + i \sin 300°)$

21. $-8i$

22. $27i$

23. -64

24. 27

25. $1 + i\sqrt{3}$

26. $2 - 2i\sqrt{3}$

27. $-2\sqrt{3} + 2i$

28. $\sqrt{3} - i$

Find and graph all the following roots of 1. (Roots of 1 are sometimes called roots of unity.*)*

29. Second

30. Fourth

31. Sixth

32. Eighth

Find and graph all the following roots of i.

33. Second

34. Fourth

Find all solutions of each of the following equations.

35. $x^3 - 1 = 0$ **36.** $x^3 + 1 = 0$ **37.** $x^3 + i = 0$ **38.** $x^4 + i = 0$

39. $x^3 - 8 = 0$ **40.** $x^3 + 27 = 0$ **41.** $x^4 + 1 = 0$ **42.** $x^4 + 16 = 0$

43. $x^4 - i = 0$ **44.** $x^5 - i = 0$ **45.** $x^3 - (4 + 4i\sqrt{3}) = 0$ **46.** $x^4 - (8 + 8i\sqrt{3}) = 0$

Use a calculator to find all solutions of the equations in Exercises 47 and 48. Give answers in standard form.

47. $x^3 + 4 - 5i = 0$ **48.** $x^5 + 2 + 3i = 0$

49. Factor $f(x) = x^5 - 2$ into linear factors.

50. One of the 3 cube roots of a complex number is $2 + 2\sqrt{3}i$. Determine the standard form of its other two cube roots.

51. (a) Show that De Moivre's theorem holds for $n = -1$.
    (b) Show that for any nonzero complex number $z = r(\cos\theta + i\sin\theta)$, $z^{-1} = (1/r)(\cos\theta - i\sin\theta)$.

52. Derive the standard identities for $\cos 2\theta$ and $\sin 2\theta$ by expanding the left side of De Moivre's theorem for the case $n = 2$ and equating the real and imaginary parts of the resulting two sides of the equation.

Let $z = a + bi$. Solve the following equations for z. Give answers in trigonometric form.

53. $z^2 = 1 + i$ **54.** $z^2 = -\sqrt{2} + i\sqrt{2}$ **55.** $z^2 = 3 - 3i$ **56.** $z^2 = -\sqrt{3} - 1$

8.7 POLAR EQUATIONS

Throughout this text the rectangular Cartesian coordinate system has been used to graph equations. In 1691 the renowned Swiss mathematician and teacher Jakob Bernoulli (1654–1705) introduced **polar coordinates** to the world. Instead of using two straight lines as the frame of reference, he used one point, called the **pole,** and an infinite ray, called the **polar axis.** We draw the polar axis in the direction of the positive x-axis, as shown in Figure 55.

pole

polar axis

Figure 55

In Figure 56 the pole has been placed at the origin of a Cartesian coordinate system, so that the polar axis coincides with the positive x-axis. Point P has coordinates (x, y) in the Cartesian coordinate system. Point P can also be located by giving the directed angle θ from the positive x-axis to OP and the directed distance r from the pole to point P. The ordered pair (r, θ) gives the **polar coordinates** of point P. Although we use degree measure for θ in this section, polar coordinates may be given with θ in radians.

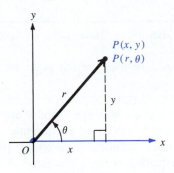

Figure 56

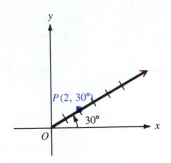

Figure 57

EXAMPLE 1   Plot each point, given its polar coordinates.

(a) $P(2, 30°)$

In this case, $r = 2$ and $\theta = 30°$, so the point P is located 2 units from the origin in the positive direction on a ray $30°$ from the polar axis, as shown in Figure 57.

(b) $Q(-4, 120°)$

Since r is negative, Q is 4 units in the negative direction from the pole on an extension of the $120°$ ray. See Figure 58.

(c) $R(5, -45°)$

Point R is shown in Figure 59. Since θ is negative, the angle is measured in the clockwise direction.   ●

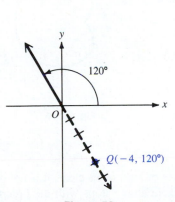

Figure 58

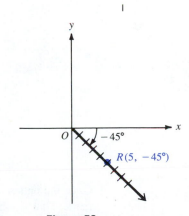

Figure 59

Parts (b) and (c) show one important difference between Cartesian coordinates and polar coordinates. While a given point in the plane can have only one pair of Cartesian coordinates, this same point can have infinitely many pairs of polar coordinates. For example, point Q in Figure 58 could also be located as $(4, 300°)$, and point R in Figure 59 is also located as $(-5, 135°)$.

. . . .

EXAMPLE 2   Give three other pairs of polar coordinates for the point (3, 140°).

Three pairs that could be used for the point are (3, −220°), (−3, 320°), and (−3, −40°). ●

An equation like $r = 3 \sin \theta$, where r and θ are the variables, is a **polar equation.** (Equations in x and y are called **rectangular** or **Cartesian equations.**) The simplest equation for many useful curves is often a polar equation.

Graphing a polar equation is much the same as graphing a Cartesian equation: Find some representative ordered pairs, (r, θ), satisfying the equation, and then sketch the graph. For example, to graph $r = 1 + \cos \theta$, first find and graph some ordered pairs (as in the table) and then connect the points in order—from (2, 0°) to (1.9, 30°) to (1.7, 45°) and so on. The graph, shown in Figure 60, is called a **cardioid** because of its heart shape.

θ	0°	30°	45°	60°	90°	120°	135°	150°	180°	270°	315°
$\cos \theta$	1	.9	.7	.5	0	−.5	−.7	−.9	−1	0	.7
$r = 1 + \cos \theta$	2	1.9	1.7	1.5	1	.5	.3	.1	0	1	1.7

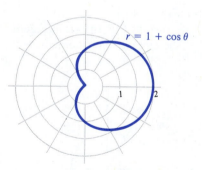

Figure 60

Once the pattern of values of r becomes clear, it is not necessary to find more ordered pairs. That is why the table above stops with the ordered pair (1.7, 315°). From the pattern, the pair (1.9, 330°) also would satisfy the equation.

. . . .

EXAMPLE 3   Graph $r^2 = \cos 2\theta$.

First complete a table of ordered pairs as shown, and then sketch the graph, as in Figure 61. The point (−1, 0°), with r negative, may be plotted as (1, 180°). Also,

(−.7, 30°) may be plotted as (.7, 210°), and so on. This curve is called a **lemniscate.**

θ	0°	30°	45°	135°	150°	180°
2θ	0°	60°	90°	270°	300°	360°
cos 2θ	1	.5	0	0	.5	1
$r = \pm\sqrt{\cos 2\theta}$	±1	±.7	0	0	±.7	±1

Values of θ for 45° < θ < 135° are not included in the table because the corresponding values of cos 2θ are negative (quadrants II and III) and so do not have real square roots. Values of θ larger than 180° give 2θ larger than 360°, and would repeat the points already found. ●

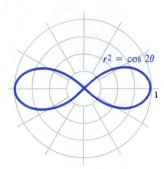

Figure 61

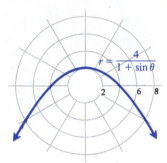

Figure 62

EXAMPLE 4 Graph $r = \dfrac{4}{1 + \sin \theta}$

Again complete a table of ordered pairs, which leads to the graph shown in Figure 62.

θ	0°	30°	45°	60°	90°	120°	135°	150°	180°	210°	225°
sin θ	0	.5	.7	.9	1	.9	.7	.5	0	−.5	−.7
$r = \dfrac{4}{1 + \sin \theta}$	4	2.7	2.3	2.1	2.0	2.1	2.3	2.7	4.0	8.0	13.3

With the points given in the table, the pattern of the graph should start to be clear. If it is not, continue to find additional points. ●

EXAMPLE 5 Graph $r = 3 \cos 2\theta$.

Because of the 2θ, the graph requires a large number of points. A few points are given on the next page. You should complete the table similarly through the first 360°.

θ	0°	15°	30°	45°	60°	75°	90°
2θ	0°	30°	60°	90°	120°	150°	180°
$\cos 2\theta$	1	.9	.5	0	−.5	−.9	−1
$r = 3 \cos 2\theta$	3	2.7	1.5	0	−1.5	−2.7	−3

Plotting these points in order gives the graph, called a **four-leaved rose.** Notice in Figure 63 how the graph is developed with a continuous curve, beginning with the upper half of the right horizontal leaf and ending with the lower half of that leaf. As the graph is traced, the curve goes through the pole four times. ●

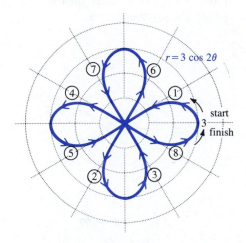

Figure 63

Generalizing from Example 5, the graphs of $r = \sin n\theta$ and $r = \cos n\theta$ are roses, with n petals if n is odd, and $2n$ petals if n is even.

· · · · · · · · ·

EXAMPLE 6   Graph $r = 2\theta$ (θ measured in radians).

Some ordered pairs are shown below. In Examples 3 to 5, it was not necessary to use negative values of θ because the trigonometric functions are periodic. Here negative values must be considered too. The radian measures have been rounded for simplicity.

θ (degrees)	−180	−90	−45	0	30	60	90	180	270	360
θ (radians)	−3.1	−1.6	−.8	0	.5	1	1.6	3.1	4.7	6.3
$r = 2\theta$	−6.2	−3.2	−1.6	0	1	2	3.2	6.2	9.4	12.6

Figure 64 shows this graph, called a **spiral of Archimedes.** ●

Sometimes an equation given in polar form is easier to graph in Cartesian form. To convert a polar equation to a Cartesian equation use the following relationships from Section 8.4, which are derived from triangle *POQ* in Figure 65.

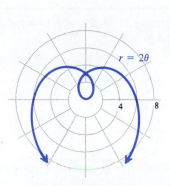

Figure 64

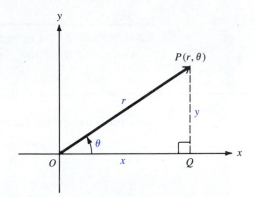

Figure 65

Converting Between Polar and Rectangular Coordinates

$$x = r \cos \theta \qquad r = \sqrt{x^2 + y^2}$$

$$y = r \sin \theta \qquad \tan \theta = \frac{y}{x}, \quad x \neq 0$$

EXAMPLE 7   Convert the equation in Example 4,

$$r = \frac{4}{1 + \sin \theta},$$

to Cartesian coordinates.

Multiply both sides of the equation by the denominator on the right, to clear the fraction.

$$r = \frac{4}{1 + \sin \theta}$$

$$r + r \sin \theta = 4$$

Now substitute $\sqrt{x^2 + y^2}$ for r and y for $r \sin \theta$.

$$\sqrt{x^2 + y^2} + y = 4$$

$$\sqrt{x^2 + y^2} = 4 - y$$

Square both sides to eliminate the radical.

$$x^2 + y^2 = (4 - y)^2$$

$$x^2 + y^2 = 16 - 8y + y^2$$

$$x^2 = -8y + 16$$

The final equation represents a parabola and can be graphed using rectangular coordinates. ●

• • • • • • • •

EXAMPLE 8   Convert the equation $3x + 2y = 4$ to a polar equation.

Use $x = r \cos \theta$ and $y = r \sin \theta$ to get

$$3x + 2y = 4$$
$$3r \cos \theta + 2r \sin \theta = 4.$$

Now solve for r. First factor out r on the left.

$$r(3 \cos \theta + 2 \sin \theta) = 4$$

$$r = \frac{4}{3 \cos \theta + 2 \sin \theta}$$

The polar equation of the line $3x + 2y = 4$ is

$$r = 4/(3 \cos \theta + 2 \sin \theta). \quad \bullet$$

Note the use of polar coordinates in the following example, taken with permission from *Calculus and Analytic Geometry,* Fifth Edition, by George Thomas and Ross Finney (Addison-Wesley, 1979).

• • • • • • • •

EXAMPLE 9   Karl von Frisch has advanced the following theory on how bees communicate information about newly discovered sources of food. A scout returning to its hive from a flower bed gives away samples of the food, and then if the bed is

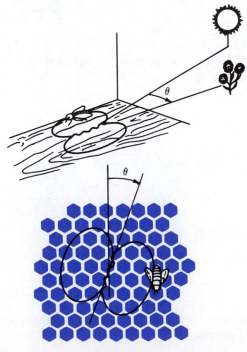

Figure 66

more than about a hundred yards away, performs a dance to show where the flowers are in relation to the position of the sun. The bee runs straight ahead for a centimeter or so, waggling from side to side, and circles back to the starting place. The bee then repeats the straight run, circling back in the opposite direction. (See Figure 66.) The dance continues this way in regular alternation. Exceptionally excited bees have been observed to dance for more than three and a half hours.

If the dance is performed inside, it is performed on the vertical wall of a honeycomb, with gravity substituting for the sun's position. A vertical straight run means that the food is in the direction of the sun. A run 30° to the right of vertical means that the food is 30° to the right of the sun, and so on. Distance (more accurately, the amount of energy required to reach the food) is communicated by the duration of the straight-run portions of the dance. Straight runs lasting three seconds each are typical for distances of about a half mile from the hive. Straight runs that last five seconds each mean about two miles.   ●

8.7 Exercises

Plot each point, given its polar coordinates.

1. $(1, 45°)$

2. $(3, 120°)$

3. $(-2, 135°)$

4. $(-4, 27°)$

5. $(5, -60°)$

6. $(2, -45°)$

7. $(-3, -210°)$

8. $(-1, -120°)$

9. $(3, 300°)$

10. $(4, 270°)$

11. $(-5, -420°)$

12. $(2, -435°)$

Graph each of the following equations for $0° \leq \theta \leq 360°$, unless other domains are specified.

13. $r = 2 + 2 \cos \theta$

14. $r = 2(4 + 3 \cos \theta)$

15. $r = 3 + \cos \theta$ (limaçon)

16. $r = 2 - \cos \theta$ (limaçon)

17. $r = \sin 2\theta$ (four-leaved rose)
(*Hint:* Use $0° \leq \theta < 360°$ every 15°.)

18. $r = 3 \cos 5\theta$ (five-leaved rose)

19. $r^2 = 4 \cos 2\theta$ (lemniscate)

20. $r^2 = 4 \sin 2\theta$ (lemniscate)

21. $r = 4(1 - \cos \theta)$ (cardioid)

22. $r = 3(2 - \cos \theta)$ (cardioid)

23. $r = 2 \sin \theta \tan \theta$ (cissoid)

24. $r = \dfrac{\cos 2\theta}{\cos \theta}$

25. $r = \dfrac{3}{2 + \sin \theta}$

26. $r = \sin \theta \cos^2 \theta$

Graph each of the following for $-\pi \leq \theta \leq \pi$, measuring θ in radians.

27. $r = 5\theta$ (spiral of Archimedes)

28. $r = \theta$ (spiral of Archimedes)

29. $r\theta = \pi$ (hyperbolic spiral)

30. $r^2 = \theta$ (parabolic spiral)

31. $\ln r = \theta$ (logarithmic spiral)

32. $\log r = \theta$ (logarithmic spiral)

Find a polar equation having the following type of graph.

33. A vertical line

34. A horizontal line

35. A straight line through the origin

36. A circle of radius k and origin as center

For each of the following equations, find an equivalent equation in Cartesian coordinates, and sketch the graph.

37. $r = 2 \sin \theta$

38. $r = 2 \cos \theta$

39. $r = \dfrac{2}{1 - \cos \theta}$

40. $r = \dfrac{3}{1 - \sin \theta}$

41. $r + 2 \cos \theta = -2 \sin \theta$

42. $r = \dfrac{3}{4 \cos \theta - \sin \theta}$

43. $r = 2 \sec \theta$

44. $r = -5 \csc \theta$

45. $r(\cos \theta + \sin \theta) = 2$

46. $r(2 \cos \theta + \sin \theta) = 2$

47. $r \sin \theta + 2 = 0$

48. $r \sec \theta = 5$

For each of the following equations, find an equivalent equation in polar coordinates.

49. $x + y = 4$

50. $2x - y = 5$

51. $x^2 + y^2 = 16$

52. $x^2 + y^2 = 9$

53. $y = 2$

54. $x = 4$

55. $y^2 = 25x$

56. $x^2 = 4y$

57. $x^2 + 9y^2 = 36$

58. $16x^2 + y^2 = 16$

For each of the conditions in Exercises 59–62 for the graph of a polar equation, determine the type of symmetry that the graph will possess.

59. Whenever (r, θ) is on the graph, then so is $(-r, -\theta)$.

60. Whenever (r, θ) is on the graph, then so is $(-r, \theta)$.

61. Whenever (r, θ) is on the graph, then so is $(r, -\theta)$.

62. Whenever (r, θ) is on the graph, then so is $(r, \pi - \theta)$.

63. Show that the distance between (r_1, θ_1) and (r_2, θ_2) is $\sqrt{r_1^2 + r_2^2 - 2r_1 r_2 \cos(\theta_1 - \theta_2)}$.

8 CHAPTER SUMMARY

Key Words

8.1 solving a triangle
angle of elevation
angle of depression
bearing
8.2 oblique triangle
law of sines
ambiguous case of the law
of sines
8.3 law of cosines
Heron's area formula
8.4 scalar

vector
resultant
scalar product
direction angle
position vector
unit vector
equilibrant
8.5 real axis
imaginary axis
trigonometric form
modulus

argument
8.6 De Moivre's theorem
nth root
8.7 polar coordinates
pole
polar axis
cardioid
lemniscate
limaçon (from Exercises)
rose
spiral of Archimedes

Review Exercises

Find the indicated parts in each right triangle in Exercises 1–6. Assume that the right angle is at C.

1. $A = 47.3°$, $b = 39.6$ cm; find B and c.
2. $A = 15.3°$, $c = 301$ m; find B.
3. $b = 68.6$ m, $c = 122.8$ m; find A and B.
4. $A = 42.2°$, $a = 689$ cm; find b and c.
5. $B = 88.3°$, $b = 402$ ft; find a and c.
6. $A = 51.74°$, $b = 29.62$ ft; find a and c.

7. When the angle of elevation of the sun is 15.8°, the shadow of a tower is 84.2 ft long. Find the height of the tower.

8. From the top of a cliff, the angle of depression to a river below is 32.2°. The river is 850 ft from a point directly below the top of the cliff. How high is the cliff?

9. From a point at the base of a mountain, the angle of elevation to the top is 21.2°. From a point 2000 ft back, the angle of elevation is 18.0°. Find the height of the mountain.

10. The figure is an illustration of one of the first steam engines.* Steam pressure forced the beam to pivot up and down at point X. The beam, in turn, moved the shaft. Gear A then revolved in a path around a gear B of equal diameter, causing gear B to rotate and turn the attached wheel. Suppose that the beam moved from Y to Z, sweeping out an angle of 40°. If YZ equals the diameter of the path of gear A around gear B, and if $XY = 203$ cm, find the diameter of gear B.

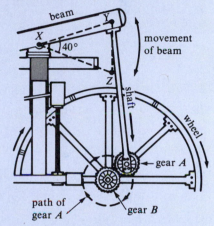

Exercise 10

Find the indicated parts of each of the following triangles.

11. $A = 101.0°$, $B = 25.0°$, $a = 165$ m; find b.
12. $A = 82.8°$, $C = 62.1°$, $b = 12.8$ cm; find a.
13. $B = 39.8°$, $b = 268$ m, $a = 430$ m; find A.
14. $C = 79.3°$, $c = 97.4$ mm, $a = 75.3$ mm; find A.
15. $A = 25.1°$, $a = 6.92$ ft, $b = 4.82$ ft; find B.
16. $C = 74.2°$, $c = 96.3$ m, $B = 39.5°$; find b.

Solve each of the following problems.

17. The angle of elevation from the top of a cliff to the top of a second cliff 290 ft away is 68°, while the angle of depression from the top of the first cliff to the bottom of the second cliff is 63°. Find the height of the second cliff.

18. A lot has the shape of the quadrilateral in the figure. What is its area?

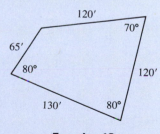

Exercise 18

*From Trigonometry with Applications by John A. Graham and Robert H. Sorgenfrey. Copyright © 1983 by Houghton Mifflin Company. Reprinted by permission.

19. A tree leans at an angle of 8.0° from the vertical. From a point 7.0 m from the bottom of the tree, the angle of elevation to the top of the tree is 68°. How tall is the tree?

20. A hill makes an angle of 14.3° with the horizontal. From the base of the hill, the angle of elevation to the top of a tree on top of the hill is 27.2°. The distance along the hill from the base to the tree is 212 ft. Find the height of the tree.

21. A person is looking out the window of a train compartment. Determine angle θ, if the relevant lengths are as shown in the figure below.

22. A baseball diamond is a square, 90 ft on a side, with home plate and the three bases as vertices. The pitcher's rubber is located 60.5 ft from home plate. Find the distance from the pitcher's rubber to each of the bases.

7 m

Exercise 19

Find the indicated parts in each of the triangles in Exercises 23–28.

23. $A = 129.6°$, $a = 127$ ft, $b = 69.8$ ft; find B.

24. $C = 51.3°$, $c = 68.3$ m, $b = 58.2$ m; find B.

25. $a = 86.1$ in., $b = 253$ in., $c = 241$ in.; find A.

26. $a = 14.8$ m, $b = 19.7$ m, $c = 31.8$ m; find B.

27. $A = 46.2°$, $b = 18.4$ m, $c = 19.2$ m; find a.

28. $B = 121.0°$, $a = 127$ ft, $c = 69.8$ ft; find b.

29. Solve the triangle having $A = 25.2°$, $a = 6.92$ yd, $b = 4.82$ yd.

30. Solve the triangle having $A = 61.7°$, $a = 78.9$ m, $b = 86.4$ m.

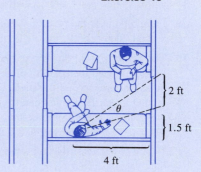

2 ft

θ

1.5 ft

4 ft

Exercise 21

Find the area of the triangles in Exercises 31–33.

31. $b = 841$ m, $c = 716$ m, $A = 149.5°$ 32. $a = 94.6$ yd, $b = 123$ yd, $c = 109$ yd

33. $a = 27.6$ cm, $b = 19.8$ cm, $C = 42.5°$

34. Raoul plans to paint a triangular wall in his A-frame cabin. Two sides measure 7.0 m each and the third side measures 6.0 m. How much paint will he need if a can of paint covers 7.5 m²?

In Exercises 35–37, use the vectors shown at right. Find each of the following.

35. **a + b**   36. **a − b**   37. **a + 3c**

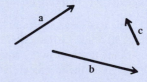

a

c

b

Find the magnitudes of the horizontal and vertical components of each of the following vectors.

38. $\alpha = 45°$, magnitude 50   39. $\alpha = 75.0°$, magnitude 69.2   40. $\alpha = 154.0°$, magnitude 964

*Suppose vector **v** has the given direction angle and magnitude. Find the magnitudes of the x- and y-components of **v**.*

41. $\theta = 45°$, $|\mathbf{v}| = 2\sqrt{2}$ 42. $\theta = 120°$, $|\mathbf{v}| = 5\sqrt{3}$ 43. $\theta = 210°$, $|\mathbf{v}| = 8$ 44. $\theta = 302°$, $|\mathbf{v}| = 25$

Find the magnitude and direction angles for each of the following vectors.

45. $\langle -6, 2 \rangle$ **46.** $\langle -6\sqrt{2}, 6\sqrt{2} \rangle$ **47.** $\langle 0, -2 \rangle$ **48.** $\langle \pi, 0 \rangle$

*Write each of the following vectors in the form a**i** + b**j**.*

49. $\langle 2, -1 \rangle$

50. $\langle -6, 3 \rangle$

51. Direction angle 30°, magnitude 20

52. Direction angle 162°, magnitude 5

Given two forces and the angle between them, find the magnitude of the resultant force.

53. Forces of 15 and 23 lb, forming an angle of 87°

54. Forces of 142 and 215 newtons, forming an angle of 112°

55. Forces of 85.2 and 69.4 newtons, forming an angle of 58.3°

56. Forces of 475 and 586 lb, forming an angle of 78.2°

Solve each of the following problems.

57. A force of 150 lb acts at a right angle to a force of 225 lb. Find the magnitude of the equilibrant and the angle it makes with the 150-lb force.

58. Forces of 320 and 294 grams act on an object. The angle between the forces is 62.5°. Find the magnitude of the resultant.

59. A box of chickens is supported above the ground to keep the foxes out. The box hangs from two ropes. One makes an angle of 52.6° with the horizontal. The tension in this rope is 89.6 lb. The second rope makes an angle of 82.5° with the first rope, and has a tension of 61.7 lb. The box weighs 10.0 lb. Find the weight of the chickens. (*Hint:* Add the *y*-components of each tension vector.)

60. A force of 186 lb just keeps a 2800-lb Toyota from rolling down a hill. What angle does the hill make with the horizontal?

61. A plane has a still-air speed of 520 mph. The pilot wishes to fly on a bearing of 310°. A wind of 37.0 mph is coming from a bearing of 212.0°. What direction should the pilot fly, and what will be her actual speed?

62. A boat travels 15 km per hr in still water. The boat is traveling across a large river, on a bearing of 130°. The current in the river, coming from the west, is at a speed of 7.0 km per hr. Find the resulting speed of the boat and its resulting direction of travel.

Find the resultant of each of the following pairs of complex numbers.

63. $7 + 3i$ and $-2 + i$

64. $2 - 4i$ and $-1 - 2i$

Graph each of the following complex numbers.

65. $5i$ **66.** $-4 + 2i$

67. $3 - 3i\sqrt{3}$ **68.** $-5 - i\sqrt{3}$

Complete the following chart.

	Standard form	Trigonometric form		Standard form	Trigonometric form
69.	$-2 + 2i$		**70.**	$1 - i$	
71.		$3(\cos 90° + i \sin 90°)$	**72.**		$4(\cos 240° + i \sin 240°)$
73.		$2(\cos 225° + i \sin 225°)$	**74.**	$-4i$	
75.	$-4 + 4i\sqrt{3}$		**76.**		$2(\cos 180° + i \sin 180°)$

Perform the indicated operations. Write answers in standard form.

77. $5(\cos 90° + i \sin 90°) \cdot 6(\cos 180° + i \sin 180°)$ **78.** $3(\cos 135° + i \sin 135°) \cdot 2(\cos 105° + i \sin 105°)$

79. $\dfrac{2(\cos 60° + i \sin 60°)}{8(\cos 300° + i \sin 300°)}$ **80.** $\dfrac{4(\cos 270° + i \sin 270°)}{2(\cos 90° + i \sin 90°)}$ **81.** $(\sqrt{3} + i)^3$

82. $(2 - 2i)^5$ **83.** $(\cos 100° + i \sin 100°)^6$ **84.** $(\cos 20° + i \sin 20°)^3$

In Exercises 85–88 give answers in trigonometric form.

85. Find the fifth roots of $-2 + 2i$.   **86.** Find the cube roots of $1 - i$.

87. Find the sixth roots of 1.   **88.** Find the fourth roots of $\sqrt{3} + i$.

Find all solutions for the following equations.

89. $x^3 + 125 = 0$ **90.** $x^4 + 16 = 0$

Graph the following using polar coordinates. Check your answer by finding an equivalent equation in Cartesian coordinates.

91. $r = \dfrac{3}{1 + \cos \theta}$ **92.** $r = \dfrac{4}{2 \sin \theta - \cos \theta}$ **93.** $r = \sin \theta + \cos \theta$ **94.** $r = 2$

Find an equivalent equation in polar coordinates.

95. $x = -3$ **96.** $y = x$ **97.** $y = x^2$ **98.** $x = y^2$

The law of sines can be used to prove the identity for sin (A + B). Let ABC be any triangle, and go through the following steps.

99. Show that $c = a \cos B + b \cos A$.

100. Multiply the terms of the result in Exercise 99 by the corresponding terms in the law of sines.

$$\frac{\sin C}{c} = \frac{\sin A}{a} = \frac{\sin B}{b}$$

101. Since $A + B + C = \pi$, we have $C = \pi - (A + B)$. Use the fact that $\sin (\pi - s) = \sin s$ to show that $\sin C = \sin (A + B)$.

102. Finally, obtain the identity for $\sin (A + B)$.

(This proof is valid only for values A and B that might be angles of a triangle. Adjustments would have to be made to generalize this result for other angles. For more details, see Mathematics Magazine, *vol. 35 (1962), p. 229.)*

103. Let $\mathbf{v}_1 = \langle a_1, b_1 \rangle$ and $\mathbf{v}_2 = \langle a_2, b_2 \rangle$ be two nonzero vectors.
   (a) Show that if $\mathbf{v}_1 = k\mathbf{v}_2$ for some scalar k, then $a_1 b_2 = a_2 b_1$.
   (b) Show that if $a_1 b_2 = a_2 b_1$, then there exists a scalar k such that $\mathbf{v}_1 = k\mathbf{v}_2$.

104. Let $\mathbf{P}$ and $\mathbf{Q}$ be nonzero vectors. Prove the following.
   (a) If $\mathbf{P}$ and $\mathbf{Q}$ are perpendicular, then $|\mathbf{P} + \mathbf{Q}|^2 = |\mathbf{P}|^2 + |\mathbf{Q}|^2$.
   (b) If $|\mathbf{P} + \mathbf{Q}|^2 = |\mathbf{P}|^2 + |\mathbf{Q}|^2$, then $\mathbf{P}$ and $\mathbf{Q}$ are perpendicular.

105. Let A, B, C be the interior angles of a triangle in which no angle is a right angle. By calculating $\tan (A + B + C)$ and observing that $A + B + C = 180°$, show that the equation $\tan A + \tan B + \tan C = \tan A \tan B \tan C$ must be satisfied.

Let $R(z)$ and $I(z)$ denote respectively the real and imaginary parts of a complex number z. Show the following in Exercises 106–108.

106. $z + \bar{z} = 2R(z)$ **107.** $z - \bar{z} = 2iI(z)$ **108.** $|R(z)| \le |z|$

109. Find all pairs (x_0, y_0) that are both polar and rectangular coordinates for the same point.

110. A regular pentagon is inscribed in a circle with center at the origin and radius 3. One vertex is on the positive x-axis. Write the polar coordinates of all the vertices.

Exercises 103, 104, 109, and 110: Reproduced from *Calculus,* by Leonard Gillman and Robert H. McDowell, 2nd Edition, by permission of W. W. Norton & Company, Inc. Copyright © 1978, 1973 by W. W. Norton & Company, Inc. Exercises 105–108: From George B. Thomas, Jr. & Ross L. Finney, *Calculus and Analytic Geometry,* seventh edition, Copyright © 1988, Addison-Wesley, Reading, Massachusetts. P. A-57. Reprinted with permission.

9

SYSTEMS OF EQUATIONS AND INEQUALITIES

Many applications of mathematics require the simultaneous solution of a large number of equations having many variables. Such a set of equations is called a **system of equations.** The **solutions** of a system of equations must satisfy every equation in the system. The definition of a linear equation given earlier can be extended to more variables: Any equation of the form

$$a_1x_1 + a_2x_2 + \cdots + a_nx_n = b$$

for real numbers $a_1, a_2, \ldots, a_n$ (not all of which are 0), and b, is a **linear equation.** If all the equations in a system are linear, the system is a **system of linear equations,** or a **linear system.** In this chapter, methods of solving systems of equations or inequalities are discussed. The matrix techniques discussed in this chapter have gained particular importance with the increasing availability of computers.

9.1 LINEAR SYSTEMS

A solution of the first-degree equation

$$a_1x_1 + a_2x_2 + \ldots + a_nx_n = b,$$

mentioned above, is an ordered set of numbers $s_1, s_2, \ldots, s_n$ such that

$$a_1s_1 + a_2s_2 + \ldots + a_ns_n = b.$$

The solution is usually written

$$(s_1, s_2, \ldots, s_n).$$

You already have had experience in solving systems of two (or three) linear equations. The methods you have studied earlier are substitution, elimination (addition or subtraction), and graphing. The following examples illustrate these methods.

. . . .

EXAMPLE 1   Solve the system

$$5x + 3y = 95 \tag{1}$$

$$2x - 7y = -3. \tag{2}$$

Any solution of this system of two equations with two variables will be an ordered pair of numbers (x, y) that satisfies both equations. In this example we will use the substitution method to solve the system.

First solve either equation for one variable. Let us solve equation (2) for x.

$$2x - 7y = -3$$

$$2x = 7y - 3$$

$$x = \frac{7y - 3}{2} \tag{3}$$

Now substitute the result for x in equation (1).

$$5x + 3y = 95$$

$$5\left(\frac{7y - 3}{2}\right) + 3y = 95$$

To solve for y, first multiply both sides of the equation by 2 to eliminate the denominator.

$$5(7y - 3) + 6y = 190$$

$$35y - 15 + 6y = 190$$

$$41y = 205$$

$$y = 5$$

Find x by substituting 5 for y in equation (3).

$$x = \frac{7y - 3}{2} = \frac{7(5) - 3}{2} = 16.$$

Check that the solution is the ordered pair $(16, 5)$ by substitution in the original system.   ●

Another way to solve a system of two equations, called the **elimination method,** involves using multiplication and addition to eliminate a variable from both equations. This elimination can be done if the coefficients of that variable in the two equations are additive inverses of each other. To achieve this, properties of algebra are used to change the system to obtain an **equivalent system,** one with the same solution set as the given system. There are three transformations that may be applied to a system to get an equivalent system. These are listed on the following page.

Transformation of a Linear System

1. Any two equations of the system may be exchanged.
2. Both sides of any equation of the system may be multiplied by any nonzero real number.
3. Any equation of the system may be replaced by the sum of that equation and a multiple of another equation in the system.

The next example illustrates the use of these transformations in the elimination method to solve a linear system.

EXAMPLE 2 Solve the system

$$3x - 4y = 1 \qquad (1)$$
$$2x + 3y = 12. \qquad (2)$$

The goal is to use the transformations to change one or both equations so the coefficients of one variable in the two equations are additive inverses of each other. Then addition of the two equations will eliminate that variable. One way to eliminate a variable in this example is to use the second transformation and multiply both sides of equation (2) by -3, giving the equivalent system

$$3x - 4y = 1$$
$$-6x - 9y = -36. \qquad (3)$$

Now multiply both sides of equation (1) by 2, and use the third transformation to add the result to equation (3), eliminating x.

$$
\begin{array}{r}
6x - 8y = 2 \\
-6x - 9y = -36 \\
\hline
-17y = -34
\end{array}
$$

The result is the system

$$3x - 4y = 1$$
$$-17y = -34. \qquad (4)$$

Multiplying both sides of equation (4) by $-1/17$ gives the equivalent system

$$3x - 4y = 1$$
$$y = 2.$$

Now substitute 2 for y in equation (1).

$$3x - 4(2) = 1$$
$$3x - 8 = 1$$
$$3x = 9$$
$$x = 3$$

The solution of the original system is $(3, 2)$. The graphs of both equations of the system in Figure 1 confirm that $(3, 2)$ satisfies both equations of the system. ●

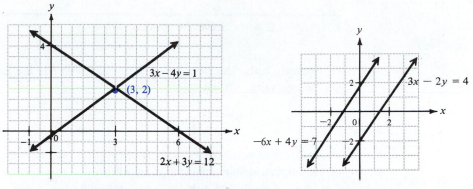

Figure 1

Figure 2

· · · · · · · · ·

EXAMPLE 3   Solve the system $3x - 2y = 4$ **(1)**

$-6x + 4y = 7.$ **(2)**

Multiply both sides of equation (1) by 2, and add to equation (2), giving the equivalent system

$$3x - 2y = 4$$

$$0 = 15.$$

Since $0 = 15$ is never true, the system has no solution. As suggested by Figure 2, this means that the graphs of the equations of the system never intersect (the lines are parallel). The solution set for this system is $\varnothing$.   ●

A system of equations with no solution is **inconsistent,** and the graphs of an inconsistent linear system are parallel lines. If the equations of a system are all equivalent, the equations are said to be **dependent equations,** and their graphs will coincide as in Figure 3. In this case the system has infinitely many solutions.

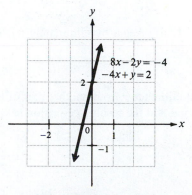

Figure 3

· · · · · · · · ·

EXAMPLE 4 Solve the system $8x - 2y = -4$ (1)
$$-4x + y = 2. \tag{2}$$

Multiply both sides of equation (1) by 1/2, and add the result to equation (2), to get the equivalent system

$$8x - 2y = -4$$
$$0 = 0.$$

The second equation, $0 = 0$, is always true, which indicates that the equations of the original system are equivalent. (With this system, the second transformation can be used to change either equation into the other.) Any ordered pair (x, y) that satisfies either equation will satisfy the system. From equation (2),

$$-4x + y = 2,$$
or
$$y = 2 + 4x.$$

The solution of the system can be written in the form of a set of ordered pairs $(x, 2 + 4x)$, for any real number x. Typical ordered pairs in the solution set are $(0, 2 + 4 \cdot 0) = (0, 2)$, $(-4, 2 + 4(-4)) = (-4, -14)$, $(3, 14)$, and $(7, 30)$. As shown in Figure 3, both equations of the original system lead to the same straight line graph. ●

Transformations can also be used to solve a system of three linear equations in three variables. In that case, first eliminate a variable from any two of the equations. Then eliminate the *same variable* from a different pair of equations. Eliminate a second variable using the resulting two equations in two variables to get an equation with just one variable whose value can now be determined. Find the values of the remaining variables by substitution.

· · · · · · · · ·

EXAMPLE 5 Solve the system $3x + 9y + 6z = 3$ (1)
$$2x + y - z = 2 \tag{2}$$
$$x + y + z = 2. \tag{3}$$

Let us begin by eliminating z from equations (2) and (3) by simply adding them to get

$$3x + 2y = 4. \tag{4}$$

To eliminate z from another pair of equations, multiply both sides of equation (2) by 6 and add the result to equation (1).

$$\begin{array}{r} 3x + 9y + 6z = 3 \\ 12x + 6y - 6z = 12 \\ \hline 15x + 15y \quad\quad = 15 \end{array} \tag{5}$$

To eliminate x from equations (4) and (5), multiply both sides of equation (4) by -5 and add the result to equation (5). Solve the new equation for y.

$$
\begin{aligned}
-15x - 10y &= -20 \\
\underline{15x + 15y} &= \underline{15} \\
5y &= -5 \\
y &= -1
\end{aligned}
$$

Using $y = -1$, find x from equation (4) by substitution.

$$
\begin{aligned}
3x + 2(-1) &= 4 \\
x &= 2
\end{aligned}
$$

Substitute 2 for x and -1 for y in equation (3) to find z.

$$
\begin{aligned}
2 + (-1) + z &= 2 \\
z &= 1
\end{aligned}
$$

Verify that the **ordered triple** $(2, -1, 1)$ satisfies all three equations.   ●

The **echelon method** is a systematic way of using the elimination method to solve a system of equations by putting the system into **triangular form.** Because the echelon method is a step-by-step procedure, it is used to solve systems of equations by computer. This procedure is given below for variables x, y, and z. It can be extended for more variables.

Putting a System of Three Equations into Triangular Form

1. Obtain 1 as the coefficient of x in the first equation.
2. Use the first equation to eliminate the x term from the second and third equations.
3. Obtain 1 as the coefficient of y in the second equation.
4. Use the second equation to eliminate the y term from the third equation.
5. Obtain 1 as the coefficient of z in the third equation.

As a preliminary step, rearrange the order of the three given equations if that will eliminate the need for some of the steps. The next example illustrates this method with the system in Example 5.

EXAMPLE 6   Solve the system

$$
\begin{aligned}
3x + 9y + 6z &= 3 \qquad &(1) \\
2x + y - z &= 2 \qquad &(2) \\
x + y + z &= 2. \qquad &(3)
\end{aligned}
$$

Multiply both sides of equation (1) by 1/3 to change the coefficient of x to 1. (Equations (1) and (3) could have been exchanged instead.)

$$x + 3y + 2z = 1 \qquad \textbf{(4)}$$
$$2x + y - z = 2$$
$$x + y + z = 2$$

To eliminate x in equation (2), multiply both sides of equation (4) by -2 and add the results to equation (2). This gives the system

$$x + 3y + 2z = 1$$
$$- 5y - 5z = 0 \qquad \textbf{(5)}$$
$$x + y + z = 2.$$

Eliminate x in equation (3) by multiplying both sides of equation (4) by -1 and adding the results to equation (3).

$$x + 3y + 2z = 1$$
$$-5y - 5z = 0$$
$$-2y - z = 1. \qquad \textbf{(6)}$$

Now multiply both sides of equation (5) by $-1/5$ to get

$$x + 3y + 2z = 1$$
$$y + z = 0 \qquad \textbf{(7)}$$
$$-2y - z = 1.$$

To eliminate y in equation (6), multiply both sides of equation (7) by 2 and add the result to equation (6). The new system is

$$x + 3y + 2z = 1 \qquad \textbf{(4)}$$
$$y + z = 0 \qquad \textbf{(7)}$$
$$z = 1. \qquad \textbf{(8)}$$

This last system is in triangular form. From the last equation, $z = 1$. Substitute $z = 1$ into equation (7) to get $y = -1$ and then substitute these values into equation (4) to find that $x = 2$. The solution of the system is $(2, -1, 1)$ as before. ●

In the rest of this section, we will use the echelon method of solving a system of equations by putting it in triangular form.

A system of equations will sometimes have more variables than equations. In this case, there is no single solution, and in fact there might be no solution at all. The next example shows how to handle these systems.

· · · · · · · · ·

EXAMPLE 7 Solve the system $\quad x + 2y + z = 4 \qquad \textbf{(1)}$
$$3x - y - 4z = -9. \qquad \textbf{(2)}$$

Eliminate x from equation (2) by multiplying both sides of equation (1) by -3 and adding the result to equation (2).

$$x + 2y + z = 4$$
$$-7y - 7z = -21 \qquad (3)$$

Multiply both sides of equation (3) by $-1/7$, giving the simplified system

$$x + 2y + z = 4$$
$$y + z = 3. \qquad (4)$$

It is not possible to eliminate more variables here—any attempt to do so would reintroduce x into the second equation. Instead, solve equation (4) for y. (We could have used z.)

$$y = 3 - z$$

Solve equation (1) for x after first substituting $3 - z$ for y.

$$x + 2(3 - z) + z = 4$$
$$x + 6 - 2z + z = 4$$
$$x - z = -2$$
$$x = -2 + z$$

For any value of z that might be chosen, y is found from the fact that $y = 3 - z$, and x from the fact that $x = -2 + z$. The solution can be written as $(-2 + z, 3 - z, z)$, where z is a real number. Typical values can be found by choosing values of z. For example, if z is 4, then $y = 3 - z = 3 - 4 = -1$, and $x = -2 + z = -2 + 4 = 2$, giving as one solution $(2, -1, 4)$. The solution set is an infinite set of ordered triples. Compare this with Example 4.

In Example 4, the solution was expressed in terms of x. Had we solved equation (4) for z instead of y, the solution would have had the form $(1 - y, y, 3 - y)$, where y is a real number. Verify that this alternative solution leads to the same set of ordered triples. It is most convenient to express solutions in terms of z when we use the echelon method. Since z can be any real number, the solution $(-2 + z, 3 - z, z)$ is said to have z **arbitrary.** ●

· · · · · · · · ·

EXAMPLE 8   Solve the system $x + 2y + z = 0$ (1)
$$4x - y + z = 0 \qquad (2)$$
$$-x - 2y - z = 0. \qquad (3)$$

Since equations (1) and (3) are equivalent, one of them can be dropped, so the system is reduced to

$$x + 2y + z = 0$$
$$4x - y + z = 0.$$

This system of two equations in three variables can be solved by the method illustrated in Example 7. Verify that the solution with z arbitrary is $(-z/3, -z/3, z)$, where z is a real number. •

In the system of equations given in Example 8, each equation has a constant term of 0. By inspection, the ordered triple $(0, 0, 0)$ is a solution of the system. However, in this case, there are infinitely many other solutions. A system of three equations in three unknowns with all constant terms equal to zero is called a **homogeneous system** of three equations in three unknowns. The ordered triple $(0, 0, 0)$ is always a solution of a homogeneous system; this solution is the **trivial solution.**

In Section 3.2 the general form of the equation of a circle was given as

$$x^2 + y^2 + Dx + Ey + F = 0.$$

This general form is used in the next example.

• • • • • • • • •

EXAMPLE 9   Find an equation for all the circles going through $(2, 1)$ and $(-4, -1)$.

A circle is completely determined if any three distinct points on the circle are given. Here only two points are known. There are infinitely many circles through these two points. The set of all the circles going through these two points is called a **family of circles.** To find the equation of this family of circles, start with the general form of the equation of a circle. ·

$$x^2 + y^2 + Dx + Ey + F = 0.$$

Substitute the values from the two given points, $(2, 1)$, and $(-4, -1)$ into this equation, producing the following system of two equations in three variables.

$$4 + 1 + 2D + E + F = 0$$
$$16 + 1 - 4D - E + F = 0$$

These equations may be rewritten as

$$F + E + 2D = -5 \qquad \qquad (1)$$
$$F - E - 4D = -17. \qquad \qquad (2)$$

Multiplying both sides of equation (1) by -1 and adding the results to equation (2), gives

$$F + E + 2D = -5$$
$$-2E - 6D = -12. \qquad \qquad (3)$$

Multiply both sides of equation (3) by $-1/2$ to get

$$F + E + 2D = -5 \qquad \qquad (1)$$
$$E + 3D = 6. \qquad \qquad (4)$$

Solve equation (4) for E: $E = -3D + 6$. Substitute this value for E into equation (1) and solve for F.

$$F + (-3D + 6) + 2D = -5$$
$$F - D = -11$$
$$F = D - 11$$

The solution is $(D, -3D + 6, D - 11)$, so that the equation of the family of circles is

$$x^2 + y^2 + Dx + (-3D + 6)y + (D - 11) = 0.$$

To find the equation for a particular circle that goes through the two given points, choose a value for D. For example, choosing $D = 3$ gives $E = -3(3) + 6 = -3$ and $F = 3 - 11 = -8$. Substituting these values into the general equation for the family of circles gives the particular equation

$$x^2 + y^2 + 3x - 3y - 8 = 0.$$

Verify that this circle goes through the two given points by showing that both ordered pairs, $(2, 1)$ and $(-4, -1)$ satisfy the equation. Figure 4 shows this circle and two others that also go through the points $(2, 1)$ and $(-4, -1)$.   ●

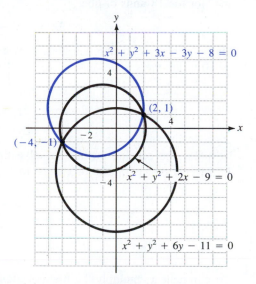

Figure 4

NOTE In the last two examples, each system turned out to have one more variable than equation. If there are two more variables than equations, there will be two arbitrary variables, and so on.

Applications of mathematics often require the solution of a system of equations, as Example 10 on the following page shows.

. . . .

EXAMPLE 10 An animal feed is made from three ingredients: corn, soybeans, and cottonseed. One unit of each ingredient provides units of protein, fat, and fiber as shown in the table below. How many units of each ingredient should be used to make a feed which contains 22 units of protein, 28 units of fat, and 18 units of fiber?

	Protein	Fat	Fiber
Corn	.25	.4	.3
Soybeans	.4	.2	.2
Cottonseed	.2	.3	.1

Let x represent the number of units of corn, y, the number of units of soybeans, and z, the number of units of cottonseed that are required. Since the total amount of protein is to be 22 units,

$$.25x + .4y + .2z = 22.$$

Also, for the 28 units of fat,

$$.4x + .2y + .3z = 28,$$

and, for the 18 units of fiber,

$$.3x + .2y + .1z = 18.$$

Multiply the first equation on both sides by 100, and the second and third equations by 10 to get the system

$$25x + 40y + 20z = 2200$$
$$4x + 2y + 3z = 280$$
$$3x + 2y + z = 180.$$

Now use the echelon method to solve this system. Multiply the first equation by 1/25 to get the coefficient of x equal to 1.

$$x + \frac{8}{5}y + \frac{4}{5}z = 88$$

$$4x + 2y + 3z = 280$$
$$3x + 2y + z = 180$$

To eliminate x, multiply the first equation by -4 and add the results to the second equation. Then multiply the first equation by -3 and add the results to the third equation.

$$x + \frac{8}{5}y + \frac{4}{5}z = 88$$

$$\frac{-22}{5}y - \frac{1}{5}z = -72 \tag{1}$$

$$\frac{-14}{5}y - \frac{7}{5}z = -84 \tag{2}$$

To get a coefficient of 1 for y, multiply equation (1) by $-5/22$ and equation (2) by $-5/14$.

$$x + \frac{8}{5}y + \frac{4}{5}z = 88$$

$$y + \frac{1}{22}z = \frac{180}{11} \qquad \text{(3)}$$

$$y + \frac{1}{2}z = 30 \qquad \text{(4)}$$

Now multiply equation (3) by -1 and add the results to equation (4), so the system is in triangular form.

$$x + \frac{8}{5}y + \frac{4}{5}z = 88$$

$$y + \frac{1}{22}z = \frac{180}{11}$$

$$\frac{5}{11}z = \frac{150}{11} \qquad \text{(5)}$$

Finally, multiply equation (5) by 11/5, getting $z = 30$. Substituting back gives $y = 15$ and $x = 40$. Verify that (40, 15, 30) is the solution of this system. The feed should contain 40 units of corn, 15 units of soybeans, and 30 units of cottonseed to meet the given requirements.   ●

9.1 Exercises

Solve each of the following systems by the substitution method.

1. $x - 5y = 8$
$\quad x = 4y$

2. $6x - y = 5$
$\quad y = 11x$

3. $4x - 5y = -11$
$\quad 2x + y = 5$

4. $-3x + 2y = 11$
$\quad 2x + 7y = 11$

5. $7x - y = -10$
$\quad 3y - x = 10$

6. $4x + 5y = 7$
$\quad 9y = 31 + 2x$

7. $5x = y - 8$
$\quad 3y = 4x + 46$

8. $-2x = 6y + 18$
$\quad -29 = 5y - 3x$

9. $\quad x = 2y - z$
$\quad 3x + y = 3z + 19$
$\quad 2x - y + 4z = -1$

10. $\quad y = 3x + z - 13$
$\quad x - y = z + 3$
$\quad x + 2y - z = -3$

11. $x - 2y + z = 3$
$\quad 3x + y - 2z = -1$
$\quad x - 4y + z = 3$

12. $-2x + 5y + 3z = -6$
$\quad 4x - y - 2z = -6$
$\quad 3x + 4y - 2z = -14$

Solve each system by elimination.

13. $3x + 2y = -6$
$\quad 5x - 2y = -10$

14. $3x - y = -4$
$\quad x + 3y = 12$

15. $2x - 3y = -7$
$\quad 5x + 4y = 17$

16. $4x + 3y = -1$
$\quad 2x + 5y = 3$

17. $5x + 7y = 6$
$\quad 10x - 3y = 46$

18. $12x - 5y = 9$
$\quad 3x - 8y = -18$

19. $6x + 7y = -2$
$\quad 7x - 6y = 26$

20. $2x + 9y = 3$
$\quad 5x + 7y = -8$

21. $\dfrac{x}{2} + \dfrac{y}{3} = 8$

$\dfrac{2x}{3} + \dfrac{3y}{2} = 17$

22. $\dfrac{x}{5} + 3y = 31$

$2x - \dfrac{y}{5} = 8$

23. $\dfrac{3x}{2} - \dfrac{y}{3} = 5$

$\dfrac{5x}{2} + \dfrac{2y}{3} = 12$

24. $\dfrac{4x}{5} + \dfrac{y}{4} = -2$

$\dfrac{x}{5} + \dfrac{y}{8} = 0$

25. $\dfrac{2x - 1}{3} + \dfrac{y + 2}{4} = 4$

$\dfrac{x + 3}{2} - \dfrac{x - y}{3} = 3$

26. $\dfrac{x + 6}{5} + \dfrac{2y - x}{10} = 1$

$\dfrac{x + 2}{4} - \dfrac{4y + 2}{5} = -3$

27. $x + y + z = 2$
$2x + y - z = 5$
$x - y + z = -2$

28. $2x + y + z = 9$
$-x - y + z = 1$
$3x - y + z = 9$

29. $x + 3y + 4z = 14$
$2x - 3y + 2z = 10$
$3x - y + z = 9$

30. $4x - y + 3z = -2$
$3x + 5y - z = 15$
$-2x + y + 4z = 14$

State the next transformation that should be carried out to solve each of the following systems by the echelon method.

31. $x + 3y + 4z = 5$
$-4x + 5y + 6z = 7$
$6x + 7y + 8z = 9$

32. $x + 2y + 3z = 4$
$y - 6z = 7$
$-\dfrac{1}{2}y + 8z = 9$

33. $x - 3y + 2z = 9$
$y - 3z = 4$
$5x + 4y + 4z = 7$

34. $-3x + 4y + 5z = 6$
$6x + 7y = 8$
$2x - 3y + z = 1$

35. $x + 2y + 3z = 4$
$4y - z = 1$
$2x + y + 5z = 6$

36. $3y + 4z = 5$
$2x + 6y + 7z = 8$
$5x + y - 9z = 9$

Solve each system in Exercises 37–44 by the echelon method.

37. $3y + 2z = 6$
$y - y = 0$
$4x + z = 8$

38. $3x - 2y - 8z = 1$
$9x - 6y - 24z = -2$
$x - y + z = 1$

39. $x + 2y + 3z = 8$
$3x - y + 2z = 5$
$-2x - 4y - 6z = 5$

40. $x + 3z = 0$
$2y - z = 9$
$4x + y = -7$

41. $5x - 2y + z = 2$
$2x + 3y - 2z = -38$
$x - 4y + 4z = 57$

42. $2x + 3y + 3z = 0$
$4x - 4y + z = -80$
$-x + 3y - 2z = 3$

43. $3x + 3y + z = -3$
$-2x - 5y + 3z = 10$
$2x + y + 2z = 1$

44. $-x + 4y + 5z = 6$
$-2x - y + 4z = -18$
$3x + y - 2z = 21$

45. Suppose that one of the three allowable transformations on a system of equations has been performed. Is there always an allowable transformation that will convert the new system back to the original system?

46. Consider the linear equation $2x + 3y = 6$. Find a second linear equation for which the system of two linear equations will have the following.   **(a)** Exactly one solution   **(b)** No solution   **(c)** Infinitely many solutions

Solve each system by elimination. (Hint: In Exercises 47–50 let $1/x = t$ and $1/y = u$.)

47. $\dfrac{2}{x} + \dfrac{1}{y} = \dfrac{3}{2}$

$\dfrac{3}{x} - \dfrac{1}{y} = 1$

48. $\dfrac{1}{x} + \dfrac{3}{y} = \dfrac{16}{5}$

$\dfrac{5}{x} + \dfrac{4}{y} = 5$

49. $\dfrac{2}{x} + \dfrac{1}{y} = 11$

$\dfrac{3}{x} - \dfrac{5}{y} = 10$

50. $\dfrac{2}{x} + \dfrac{3}{y} = 18$

$\dfrac{4}{x} - \dfrac{5}{y} = -8$

51. $\dfrac{4}{x+2} - \dfrac{3}{y-1} = 1$

$\dfrac{1}{x+2} + \dfrac{2}{y-1} = 1$

52. $\dfrac{1}{x-3} + \dfrac{5}{y+2} = -6$

$\dfrac{-3}{x-3} + \dfrac{2}{y+2} = 1$

53. $\dfrac{2}{x} + \dfrac{3}{y} - \dfrac{2}{z} = -1$

$\dfrac{8}{x} - \dfrac{12}{y} + \dfrac{5}{z} = 5$

$\dfrac{6}{x} + \dfrac{3}{y} - \dfrac{1}{z} = 1$

54. $-\dfrac{5}{x} + \dfrac{4}{y} + \dfrac{3}{z} = 2$

$\dfrac{10}{x} + \dfrac{3}{y} - \dfrac{6}{z} = 7$

$\dfrac{5}{x} + \dfrac{2}{y} - \dfrac{9}{z} = 6$

Solve each of the following systems in terms of the arbitrary variable z.

55. $x - 2y + 3z = 6$

$\quad\; 2x - \;\; y = 5$

56. $3x + 4y - \;\; z = 8$

$\quad\; x + 2z = 4$

57. $\quad 5x - 4y + \;\; z = 0$

$\quad\quad\; x + \;\; y = 0$

$\quad -10x + 8y - 2z = 0$

58. $2x + \;\; y - 3z = 0$

$4x + 2y - 6z = 0$

$\;\; x - \;\; y + \;\; z = 0$

59. $4x + y + z = 6$

$\quad\quad\; y = 2x$

60. $3x - 5y - 4z = 6$

$\quad\; 3x = z$

Write a system of linear equations for each of the following, and then use the system to solve the problem.

61. At the Sharp Ranch, 6 goats and 5 sheep sell for $305 while 2 goats and 9 sheep cost $285. Find the cost of a goat and the cost of a sheep.

62. Linda Ramirez is a building contractor. If she hires 7 day laborers and 2 concrete finishers, her payroll for the day is $608, while 1 day laborer and 5 concrete finishers cost $464. Find the daily-wage charge of each type of worker.

63. During summer vacation Hector and Ann earned a total of $1088. Hector worked 8 days fewer than Ann and earned $2 per day less. Find the number of days he worked and the daily wage he made if the total number of days worked by both was 72.

64. The perimeter of a rectangle is 42 cm. The longer side has a length of 7 cm more than the shorter side. Find the length of the longer side.

65. Thirty liters of a 50%-alcohol solution are to be made by mixing 70% solution and 20% solution. How many liters of each solution should be used?

66. A merchant wishes to make 100 lb of a coffee blend that can be sold for $4 per lb. This blend is to be made by mixing coffee worth $6 a pound with coffee worth $3 a pound. How many pounds of each will be needed?

67. Ms. Yim inherits $25,000. She deposits part at 8% annual interest and part at 10%. Her total annual income from these investments is $2300. How much is invested at each rate?

68. Chuck Sullivan earned $100,000 in a lottery. He invested part of the money at 10% and part at 12%. His total annual income from the two investments is $11,000. How much does he have invested at each rate?

69. Mr. Caminiti has some money invested at 8% and three times as much invested at 12%. His total annual income from the two investments is $2200. How much is invested at each rate?

70. A cash drawer contains only fives and twenties. There are eight more fives than twenties. The total value of the money is $215. How many of each type of bill are there?

71. How many gallons of milk (4.5% butterfat) must be mixed with 250 gal of skim milk (0% butterfat) to get lowfat milk (2% butterfat)?

72. By boat Tom can go 72 mi upstream to a fishing hole in 4 hr. Returning, he needs only 3 hr. What is the speed of the current in the stream?

73. Two cars start together and travel in opposite directions. At the end of 4 hr the cars are 656 km apart. If one car travels 20 km per hr faster than the other, find the speed of each car.

74. Two trains leave towns 192 km apart, traveling toward one another. One train travels 40 km per hr faster than the other. They pass one another two hours later. What is the speed of each train?

75. The perimeter of a triangle is 33 cm. The longest side is 3 cm longer than the medium side. The medium side is twice the shortest side. Find the length of each side of the triangle.

76. Carrie O'Sullivan invests in three ways the $30,000 she won in a lottery. With part of the money, she buys a mutual fund, paying 9% per year. The second part, $2000 more than the first, is used to buy utility bonds paying 10% per year. The rest is invested in a tax-free 5% bond. The first year her investments bring a return of $2500. How much is invested at each rate?

77. Find an expression for all of the quadratic functions whose graphs go through the points $(-1, 3)$ and $(-2, 7)$.

78. Find the values of a and b for which the line with equation $ax + by = 5$ contains the points $(1, -2)$ and $(2, 1)$.

The position of a particle moving in a straight line is given by $s = at^2 + bt + c$, where t is time in seconds and a, b, and c are real numbers.

79. If $s(0) = 5$, $s(1) = 23$, and $s(2) = 37$, find $s(8)$.

80. If $s(0) = -10$, $s(1) = 6$, and $s(2) = 30$, find $s(10)$.

9.2 NONLINEAR SYSTEMS

A system of equations in which at least one equation is *not* linear is called a **nonlinear system.** The substitution method works well for solving many such systems, as the next example shows.

· · · · · · · · ·

EXAMPLE 1   Solve the system $x^2 - y = 4$ (1)

$$x + y = -2.$$ (2)

When one of the equations in a nonlinear system is linear, it is usually best to begin by solving the linear equation for any variable. With this system, begin by solving equation (2) for y, giving

$$y = -2 - x.$$

Now substitute this result for y in equation (1) to get

$$x^2 - (-2 - x) = 4$$
$$x^2 + 2 + x = 4$$
$$x^2 + x - 2 = 0$$
$$(x + 2)(x - 1) = 0$$
$$x = -2 \quad \text{or} \quad x = 1.$$

Substituting -2 for x in equation (2) gives $y = 0$. Also, if $x = 1$, then $y = -3$. The solutions of the given system are $(-2, 0)$ and $(1, -3)$. A graph of the system is shown in Figure 5. ●

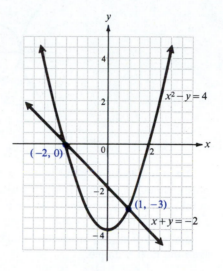

Figure 5

CAUTION If we had solved for x in equation (2) to begin the solution, we would find $y = 0$ or $y = -3$. Substituting $y = 0$ into equation (1) gives $x^2 = 4$, so $x = 2$ or $x = -2$, leading to the ordered pairs $(2, 0)$ and $(-2, 0)$. The ordered pair $(2, 0)$ does not satisfy equation (2), however. This shows the *necessity* of checking by substituting all potential solutions into each equation of the system.

Some nonlinear systems are best solved by elimination, as shown in the next example.

EXAMPLE 2   Solve the system $x^2 + y^2 = 4$ **(1)**
$$2x^2 - y^2 = 8. \qquad \textbf{(2)}$$

Adding equation (1) to equation (2) (to eliminate y) gives the new system

$$x^2 + y^2 = 4$$
$$3x^2 \qquad = 12. \qquad \textbf{(3)}$$

Solve equation (3) for x.

$$x^2 = 4$$
$$x = 2 \quad \text{or} \quad x = -2.$$

Find y by substituting back into equation (1). If $x = 2$, then $y = 0$, and if $x = -2$, then $y = 0$. The solutions of the given system are $(2, 0)$ and $(-2, 0)$. See Figure 6. ●

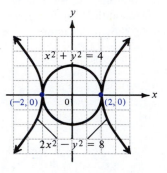

Figure 6

The elimination method works with the system in Example 2 since the system can be thought of as a system of linear equations where the variables are x^2 and y^2. In other words, the system is **linear in x^2 and y^2.** To see this, substitute u for x^2 and v for y^2. The resulting system is linear in u and v.

Some systems require a combination of the elimination method and substitution, as in the next example.

· · · · · · · · ·

EXAMPLE 3   Solve the system

$$x^2 + 3xy + y^2 = 22 \qquad \qquad \textbf{(1)}$$
$$x^2 - xy + y^2 = 6. \qquad \qquad \textbf{(2)}$$

Begin by multiplying both sides of equation (2) by -1, and adding the results to equation (1), giving the system

$$4xy = 16 \qquad \qquad \textbf{(3)}$$
$$x^2 - xy + y^2 = 6.$$

Now solve equation (3) for either x or y and substitute the result into one of the given equations. Solving for y gives

$$y = \frac{4}{x} \quad (x \neq 0). \qquad \qquad \textbf{(4)}$$

(If $x = 0$ there is no value of y which satisfies the system.) Substituting back for y in equation (2) and simplifying gives

$$x^2 - x\left(\frac{4}{x}\right) + \left(\frac{4}{x}\right)^2 = 6$$

$$x^2 - 4 + \frac{16}{x^2} = 6$$

$$x^4 - 4x^2 + 16 = 6x^2$$

$$x^4 - 10x^2 + 16 = 0$$

$$(x^2 - 2)(x^2 - 8) = 0$$

$$x^2 = 2 \quad \text{or} \quad x^2 = 8$$

$$x = \sqrt{2} \text{ or } x = -\sqrt{2} \text{ or } x = 2\sqrt{2} \text{ or } x = -2\sqrt{2}.$$

Substitute these values of x into equation (4) to find the corresponding values for y.

$$\text{If } x = \sqrt{2}, \qquad y = \frac{4}{\sqrt{2}} = 2\sqrt{2}.$$

$$\text{If } x = -\sqrt{2}, \qquad y = -\frac{4}{\sqrt{2}} = -2\sqrt{2}.$$

$$\text{If } x = 2\sqrt{2}, \qquad y = \frac{4}{2\sqrt{2}} = \sqrt{2}.$$

$$\text{If } x = -2\sqrt{2}, \quad y = -\frac{4}{2\sqrt{2}} = -\sqrt{2}.$$

The solutions of the system are $(\sqrt{2}, 2\sqrt{2}), (-\sqrt{2}, -2\sqrt{2}), (2\sqrt{2}, \sqrt{2}),$ $(-2\sqrt{2}, -\sqrt{2}).$ •

9.2 Exercises

Solve each system by the substitution method. In Exercises 1–6, graph the system.

1. $2x^2 = 3y + 23$
$y = 2x - 5$

2. $x^2 + y^2 = 5$
$-3x + 4y = 2$

3. $y = x^2$
$x + y = 2$

4. $x^2 + y^2 = 45$
$x + y = -3$

5. $y = x^2 + 6x + 9$
$x + 2y = -2$

6. $x^2 - y = -1$
$3x = y - 11$

7. $x = y - 2$
$y^2 = 2x^2 + 8$

8. $y = -x^2 + 2$
$x - y = 0$

9. $y = x^2 - 2x + 1$
$x - 3y = -1$

10. $3y = 2x + 5$
$2y^2 = 2x + 4$

11. $3x^2 + 2y^2 = 5$
$x - y = -2$

12. $x^2 - 3y^2 = 22$
$x + 3y = 2$

Solve the following systems by any method.

13. $x^2 + y^2 = 8$
$x^2 - y^2 = 0$

14. $x^2 + y^2 = 10$
$2x^2 - y^2 = 17$

15. $5x^2 - y^2 = 0$
$3x^2 + 4y^2 = 0$

16. $x^2 + y^2 = 4$
$2x^2 - 3y^2 = -12$

17. $2x^2 + 3y^2 = 5$
$3x^2 - 4y^2 = -1$

18. $3x^2 + 5y^2 = 17$
$2x^2 - 3y^2 = 5$

19. $2x^2 + 2y^2 = 20$
$3x^2 + 3y^2 = 30$

20. $x^2 + y^2 = 4$
$5x^2 + 5y^2 = 28$

21. $xy = 6$
$x + y = 5$

22. $xy = -4$
$2x + y = -7$

23. $xy = -15$
$4x + 3y = 3$

24. $xy = 8$
$3x + 2y = -16$

25. $2xy + 1 = 0$
$x + 16y = 2$

26. $-5xy + 2 = 0$
$x - 15y = 5$

27. $x^2 + 4y^2 = 25$
$xy = 6$

28. $5x^2 - 2y^2 = 6$
$xy = 2$

29. $x^2 + 2xy - y^2 = 14$
$x^2 - y^2 = -16$

30. $3x^2 + xy + 3y^2 = 7$
$x^2 + y^2 = 2$

31. $x^2 - xy + y^2 = 5$
$2x^2 + xy - y^2 = 10$

32. $3x^2 + 2xy - y^2 = 9$
$x^2 - xy + y^2 = 9$

33. $x^2 + 2xy - y^2 + y = 1$
$3x + y = 6$

34. $x^2 - 4x + y + xy = -10$
$2x - y = 10$

35. $y = 3^x$
$y = 9^{2x}$

36. $y = 5^{3x}$
$y = 125^{4x}$

37. $y = \log(x - 2)$
$y = -1 + \log(8x + 4)$

38. $y = -1 + \log(18x + 10)$
$y = \log(x + 5)$

Solve each system.

39. $y = -\sqrt{1 + x}$
$(x - 3)^2 + y^2 = 16$

40. $x^2 + y^2 = 2$
$\sqrt{x + y} = \sqrt{x - y} + 2$

41. $\dfrac{1}{x^2} - \dfrac{3}{y^2} = 14$

$\dfrac{2}{x^2} + \dfrac{1}{y^2} = 35$

42. $\dfrac{8}{\sqrt{x}} + \dfrac{9}{\sqrt{y}} = 5$

$\dfrac{12}{\sqrt{x}} - \dfrac{18}{\sqrt{y}} = -3$

Use any method to solve the following problems.

43. Find two numbers whose sum is 12 and whose product is 36.

44. Find two numbers whose sum is 17 and whose squares differ by 17.

45. Find two numbers whose ratio is $5:3$ and whose product is 135.

46. Find two numbers whose squares have a sum of 100 and a difference of 28.

47. The longest side of a right triangle is 13 m in length. One of the other sides is 7 m longer than the shortest side. Find the length of each of the two shorter sides of the triangle.

48. Does the straight line $3x - 2y = 9$ intersect the circle $x^2 + y^2 = 25$? (*Hint:* To find out, solve the system made up of these two equations.)

49. Find the equation of the straight line through $(2, 4)$ that touches the parabola $y = x^2$ at only one point. (*Note:* Recall that a quadratic equation has a unique solution when the discriminant is 0.)

50. For what value of b will the line $x + 2y = b$ touch the circle $x^2 + y^2 = 9$ in only one point?

51. For what values of a do the graphs of $x^2 + y^2 = 25$ and $x^2/a^2 + y^2/25 = 1$ have exactly two points in common?

9.3 SYSTEMS OF INEQUALITIES; LINEAR PROGRAMMING

Many mathematical descriptions of real world situations are best expressed as inequalities, rather than equalities. For example, a firm might be able to use a machine *no more* than 12 hours a day, while a production of *at least* 500 cases of a certain product might be required to meet a contract. Perhaps the simplest way to see the solution of an inequality in two variables is to draw its graph.

A line divides a plane into three sets of points—the points of the line itself and the points belonging to the two regions determined by the line. Each of these two regions is called a **half-plane.** In Figure 7 line *r* divides the plane into three different sets of points, line *r*, half-plane *P* and half-plane *Q*. The points of *r* belong to neither *P* nor *Q*. Line *r* is the boundary of each half-plane.

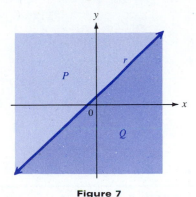

Figure 7

A **linear inequality in two variables** is an inequality of the form

$$Ax + By \leq C,$$

where A, B, and C are real numbers, with A and B not both equal to 0. (The symbol $\leq$ could be replaced with $\geq$, $<$, or $>$.) The graph of a linear inequality is a half-plane, sometimes including its boundary. For example, to graph the linear inequality $3x - 2y \leq 6$, first graph the boundary, $3x - 2y = 6$, as shown in Figure 8. Since the points of the line $3x - 2y = 6$ satisfy $3x - 2y \leq 6$, this boundary line is part

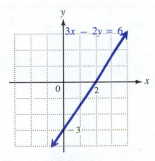

Figure 8

of the solution. To decide which half-plane—the one above the line $3x - 2y = 6$ or the one below the line—is part of the solution, solve the original inequality for y.

$$3x - 2y \leq 6$$
$$-2y \leq -3x + 6$$
$$y \geq \frac{3}{2}x - 3 \qquad \text{Change } \leq \text{ to } \geq.$$

For a particular value of x, the inequality will be satisfied by all values of y that are *greater than* or equal to $(3/2)x - 3$. This means that the solution contains the half-plane *above* the line, as shown in Figure 9.

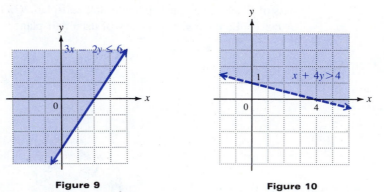

Figure 9   Figure 10

EXAMPLE 1   Graph $x + 4y > 4$.

The boundary here is the straight line $x + 4y = 4$. Since the points on this line do not satisfy $x + 4y > 4$, it is customary to make the line dashed, as in Figure 10. To decide which half-plane represents the solution, solve for y.

$$x + 4y > 4$$
$$4y > -x + 4$$
$$y > -\frac{1}{4}x + 1$$

Since y is *greater than* $(-1/4)x + 1$, the graph of the solution is the half-plane above the boundary, as shown in Figure 10.

Alternatively, or as a check, choose a point not on the boundary line and substitute into the inequality. The point $(0, 0)$ is a good choice if it does not lie on the boundary, since the substitution is easily done. Here, substitution of $(0, 0)$ into the original inequality gives

$$x + 4y > 4$$
$$0 + 4(0) > 4$$
$$0 > 4,$$

a false statement. Since the point $(0, 0)$ is below the line, the points which satisfy the inequality must be above the line, which agrees with the result above.   ●

The methods used to graph linear inequalities can be used for other inequalities of the form $y \le f(x)$ as summarized here. (Similar statements can be made for $<$, $>$, and $\ge$.)

**Graphing
Inequalities**

> **I.** For a function f, the graph of $y < f(x)$ is made up of all the points that are *below* the graph of $y = f(x)$; the graph of $y > f(x)$ is made up of all the points that are *above* the graph of $y = f(x)$.
>
> **II.** If the inequality is not or cannot be solved for y, choose a test point not on the boundary. If the test point satisfies the inequality, the graph includes all points on the same side of the boundary as the test point. Otherwise, the graph includes all points on the other side of the boundary.

The solution set of a **system of inequalities,** such as

$$x > 6 - 2y$$
$$x^2 < 2y,$$

is the intersection of the solution set of its members. This intersection is found by graphing the solution sets of both inequalities on the same coordinate axes and identifying, by shading, the region common to all graphs.

EXAMPLE 2   Graph the solution set of the system above.

Figure 11 shows the graphs of both $x > 6 - 2y$ and $x^2 < 2y$. The methods of the previous section can be used to show that the boundaries cross at the points $(2, 2)$ and $(-3, 9/2)$. The solution set of the system includes all points in the heavily shaded area. The points on the boundaries of $x > 6 - 2y$ and $x^2 < 2y$ do not belong to the graph of the solution. For this reason, the boundaries are dashed lines.   ●

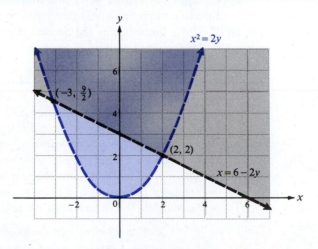

Figure 11

EXAMPLE 3   Graph the solution set of the system

$$|x| \le 3$$
$$y \le 0$$
$$y \ge |x| + 1.$$

Writing $|x| \le 3$ as $-3 \le x \le 3$ shows that it is satisfied by points in the region between -3 and 3. The set of points that satisfies $y \le 0$ includes the points below or on the x-axis. Graph $y = |x| + 1$ and use a test point to see that the solutions of $y \ge |x| + 1$ are above or on the boundary. Figure 12 shows that the solution sets of $y \le 0$ and $y \ge |x| + 1$ have no points in common; therefore the solution set for the system is $\varnothing$.   ●

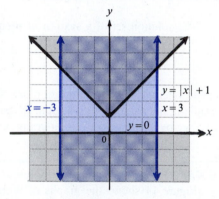

Figure 12

An important application of mathematics to business and social science is called linear programming. **Linear programming** is used to find such things as minimum cost and maximum profit. It was developed to solve problems in allocating supplies for the U.S. Air Force during the Second World War. The basic ideas of this technique will be explained with an example.

EXAMPLE 4   The Smith Company makes two products, tape decks and amplifiers. Each tape deck gives a profit of $30, while each amplifier produces $70. The company must manufacture at least ten tape decks per day to satisfy one of its customers, but no more than fifty because of production problems. Also, the number of amplifiers produced cannot exceed sixty per day. As a further requirement, the number of tape decks cannot exceed the number of amplifiers. How many of each should the company manufacture in order to obtain the maximum profit?

To begin, translate the statement of the problem into symbols.

Let x = number of tape decks to be produced daily,

                   y = number of amplifiers to be produced daily.

According to the statement of the problem given above, the company must produce at least ten tape decks (ten or more), so that

$$x \geq 10.$$

No more than 50 tape decks may be produced means that

$$x \leq 50.$$

Since not more than 60 amplifiers may be made in one day,

$$y \leq 60.$$

The number of tape decks may not exceed the number of amplifiers translates as

$$x \leq y.$$

The number of tape decks and of amplifiers cannot be negative, so that

$$x \geq 0 \quad \text{and} \quad y \geq 0.$$

Listing all the restrictions, or **constraints,** that are placed on production, gives

$$x \geq 10, \quad x \leq 50, \quad y \leq 60, \quad x \leq y, \quad x \geq 0, \quad y \geq 0.$$

To find the maximum possible profit that the company can make, subject to these constraints, begin by sketching the graph of each constraint. The only feasible values of x and y are those that satisfy all constraints; that is, the values that lie in the intersection of the graphs of the constraints. The intersection is shown in Figure 13. Any point lying inside the shaded region or on the boundary in Figure 13 satisfies the restrictions as to the number of tape decks and amplifiers that may be produced. (For practical purposes, however, only points with integer coefficients are useful.) This region is called the **region of feasible solutions.**

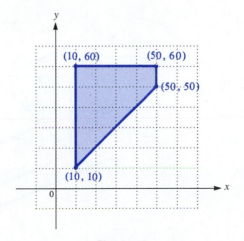

Figure 13

Since each tape deck gives a profit of $30, the daily profit from the production of x tape decks is $30x$ dollars. Also, the profit from the production of y amplifiers will be $70y$ dollars per day. The total daily profit is thus

$$\text{profit} = 30x + 70y.$$

The problem of the Smith Company may now be stated as follows: Find values of x and y in the shaded region of Figure 13 that will produce the maximum possible value of $30x + 70y$. To locate the point (x, y) that gives the maximum profit, add to the graph of Figure 13 several lines which represent some possible profit functions:

$$30x + 70y = 0,$$
$$30x + 70y = 1000,$$
$$30x + 70y = 3000,$$
$$30x + 70y = 6000.$$

Figure 14 shows the region of feasible solutions together with these lines. From the figure we see that the profit cannot be as large as 6000 because the graph for $30x + 70y = 6000$ is outside the region of feasible solutions. The point representing maximum possible profit will be found on a line, parallel to the others, between the lines $30x + 70y = 3000$ and $30x + 70y = 6000$. Profit can be maximized while still satisfying all constraints if this line just touches the feasible region. This occurs at point A, a **vertex** (or corner) point of the region of feasible solutions. Since the coordinates of this point are $(50, 60)$, the maximum profit is obtained when fifty tape decks and sixty amplifiers are produced each day. The maximum profit will be $30(50) + 70(60) = 5700$ dollars per day.

As suggested by the parallel lines of Figure 14, the optimum value will always be at a vertex point if the region of feasible solutions is bounded (enclosed). ●

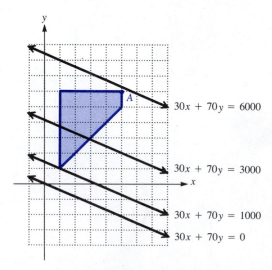

Figure 14

· · · · · · · · ·

EXAMPLE 5   Robin, who is ill, takes vitamin pills. Each day, she must have at least 16 units of Vitamin A, at least 5 units of Vitamin B_1, and at least 20 units of Vitamin C. She can choose between red pills, costing 10¢ each, which contain 8 units of A, 1 of B_1, and 2 of C; and blue pills, costing 20¢ each, which contain 2 units of A, 1 of B_1, and 7 of C. How many of each pill should she buy in order to minimize her cost and yet fulfill her daily requirements?

Let x represent the number of red pills to buy, and let y represent the number of blue pills to buy. Then the cost in pennies per day is given by

$$\text{cost} = 10x + 20y.$$

Since Robin buys x of the 10¢ pills and y of the 20¢ pills, she gets Vitamin A as follows: 8 units from each red pill and 2 units from each blue pill. Altogether, she gets $8x + 2y$ units of A per day. Since she must get at least 16 units,

$$8x + 2y \geq 16.$$

Each red pill and each blue pill supplies 1 unit of Vitamin B_1. Robin needs at least 5 units per day, so that

$$x + y \geq 5.$$

For Vitamin C the inequality is

$$2x + 7y \geq 20.$$

Also, $x \geq 0$ and $y \geq 0$, since Robin cannot buy negative numbers of the pills.

Again, total cost of the pills is minimized by the solution of the system of inequalities formed by the constraints. (See Figure 15.) The solution to this minimizing problem will also occur at a vertex point. By substituting the coordinates of the vertex points in the cost function, the lowest cost can be found.

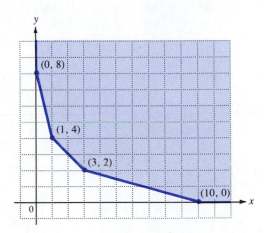

Figure 15

Point	Cost $= 10x + 20y$
$(10, 0)$	$10(10) + 20(0) = 100$
$(3, 2)$	$10(3) + 20(2) = 70 \leftarrow$ Minimum
$(1, 4)$	$10(1) + 20(4) = 90$
$(0, 8)$	$10(0) + 20(8) = 160$

Robin's best bet is to buy 3 red pills and 2 blue ones, for a total cost of 70¢ per day. She receives just the minimum amounts of Vitamins B_1 and C, but an excess of Vitamin A. Even though she has an excess of A, this is still the best buy. •

9.3 Exercises

Graph the solution of each system of inequalities in Exercises 1–24.

1. $x + y \leq 4$
$x - 2y \geq 6$

2. $2x + y > 2$
$x - 3y < 6$

3. $4x + 3y < 12$
$y + 4x > -4$

4. $3x + 5y \leq 15$
$x - 3y \geq 9$

5. $x + 2y \leq 4$
$y \geq x^2 - 1$

6. $4x - 3y \leq 12$
$y \leq x^2$

7. $y \leq -x^2$
$y \geq x^2 - 6$

8. $x + y \leq 9$
$x \leq -y^2$

9. $x - y < 1$
$-1 < y < 1$

10. $x + y \leq 36$
$-4 \leq x \leq 4$

11. $y \geq x^2 + 4x + 4$
$y < -x^2$

12. $y \geq (x - 2)^2 + 3$
$y \leq -(x - 1)^2 + 6$

13. $x + y \leq 4$
$x - y \leq 5$
$4x + y \leq -4$

14. $3x - 2y \geq 6$
$x + y \leq -5$
$y \leq 4$

15. $-2 < x < 3$
$-1 \leq y \leq 5$
$2x + y < 6$

16. $-2 < x < 2$
$y > 1$
$x - y > 0$

17. $2y + x \geq -5$
$y \leq 3 + x$
$x \leq 0$
$y \leq 0$

18. $2x + 3y \leq 12$
$2x + 3y > -6$
$3x + y < 4$
$x \geq 0$
$y \geq 0$

19. $y \geq 3^x$
$y \geq 2$

20. $y \leq \left(\dfrac{1}{2}\right)^x$
$y \geq 4$

21. $y \leq |x + 2|$
$x \geq 0$

22. $y \leq \log x$
$y \geq |x - 2|$

23. $e^{-x} - y \leq 1$
$x - 2y \geq 4$

24. $\ln x - y \geq 1$
$x^2 - 2x - y \leq 1$

25. Find a linear inequality in two variables whose graph does not intersect the solution set of $y \geq 3x + 5$.

26. Find a system of linear inequalities for which the graph is the region in the first quadrant between the pair of lines $x + 2y - 8 = 0$ and $x + 2y = 12$.

The graphs in Exercises 27–30 show regions of feasible solutions. Find the maximum and minimum values of the given expressions.

27. $3x + 5y$

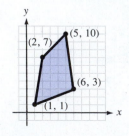

28. $6x + y$

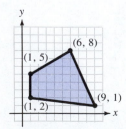

29. $40x + 75y$

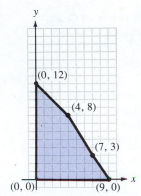

30. $35x + 125y$

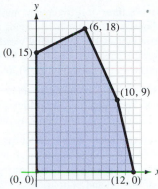

In Exercises 31–36, use graphical methods to solve each problem.

31. Find $x \geq 0$ and $y \geq 0$ such that
$$2x + 3y \leq 6$$
$$4x + y \leq 6$$
and $5x + 2y$ is maximized.

32. Find $x \geq 0$ and $y \geq 0$ such that
$$x + y \leq 10$$
$$5x + 2y \geq 20$$
$$2y \geq x$$
and $x + 3y$ is minimized.

33. Find $x \geq 2$ and $y \geq 5$ such that
$$3x - y \geq 12$$
$$x + y \leq 15$$
and $2x + y$ is minimized.

34. Find $x \geq 10$ and $y \geq 20$ such that
$$2x + 3y \leq 100$$
$$5x + 4y \leq 200$$
and $x + 3y$ is maximized.

35. Find $x \geq 0$ and $y \geq 0$ such that
$$x - y \leq 10$$
$$5x + 3y \leq 75$$
and $4x + 2y$ is maximized.

36. Find $x \geq 0$ and $y \geq 0$ such that
$$10x - 5y \leq 100$$
$$20x + 10y \geq 150$$
and $4x + 5y$ is minimized.

Write a system of inequalities for each of the following problems and then graph the region of feasible solutions of the system.

37. Ms. Oliveras was given the following advice. She should supplement her daily diet with at least 6000 USP units of vitamin A, at least 195 milligrams of vitamin C, and at least 600 USP units of vitamin D. Ms. Oliveras finds that Mason's Pharmacy carries Brand X and Brand Y vitamins. Each Brand X pill contains 3000 USP units of A, 45 milligrams of C, and 75 USP units of D, while the Brand Y pills contain 1000 USP units of A, 50 milligrams of C, and 200 USP units of D.

38. Sue is on a diet and wishes to restrict herself to no more than 1600 calories per day. Her diet consists of foods chosen from three food groups: I, meats and dairy products; II, fruits and vegetables; III, breads and other starches. These three groups contain calories per serving as shown below.

Group	I	II	III
Calories	170	50	140

Sue wishes to include four servings daily from group I and more servings from group II than from the other groups combined.

39. A manufacturer of refrigerators must ship at least 100 refrigerators to its two West Coast warehouses. Each warehouse holds a maximum of 100 refrigerators. Warehouse A holds 25 refrigerators already, while warehouse B has 20 on hand. It costs $12 to ship a refrigerator to warehouse A and $10 to ship one to warehouse B. The total shipping cost is budgeted at a maximum of $1200.

40. The California Almond Growers have 2400 boxes of almonds to be shipped from their plant in Sacramento to Des Moines and San Antonio. The Des Moines market needs at least 1000 boxes, while the San Antonio market must have at least 800 boxes.

41. The dimensions of a cylindrical can must meet several specifications. The height must not exceed 8 in. The diameter of the ends must be at least 3 in. and must not exceed the height. In addition, the lateral surface area must be no more than 4 times the combined area of the top and bottom.

Solve each of the following linear programming problems.

42. Farmer Jones raises only pigs and geese. She wants to raise no more than 16 animals with no more than 12 geese. She spends $50 to raise a pig and $20 to raise a goose. She has $500 available for this purpose. Find the maximum profit she can make if she makes a profit of $80 per goose and $40 per pig.

43. The manufacturing process requires that oil refineries must manufacture at least two barrels of gasoline for every one of fuel oil. To meet the winter demand for fuel oil, at least 3 million barrels a day must be produced. The demand for gasoline is no more than 6.4 million barrels per day. If the price of gasoline is $1.90 and the price of fuel oil is $1.50 per gallon, how much of each should be produced to maximize revenue?

44. An office manager wants to buy some filing cabinets. She knows that cabinet #1 costs $50 each, requires 6 sq ft of floor space, and holds 8 cu ft of files. Cabinet #2 costs $100 each, requires 8 sq ft of floor space, and holds 12 cu ft. She can spend no more than $700 due to budget limitations, while her office has room for no more than 72 sq ft of cabinets. She wants the maximum storage capacity within the limits imposed by funds and space. How many of each type of cabinet should she buy?

45. A small country can grow only two crops for export, coffee and cocoa. The country has 500,000 hectares* of land available for the crops. Long-term contracts require that at least 100,000 hectares be devoted to coffee and at least 200,000 hectares to cocoa. Cocoa must be processed locally, and production bottlenecks limit cocoa to 270,000 hectares. Coffee requires two workers per hectare; cocoa requires five. No more than 1,750,000 people are available for these crops. Coffee produces a profit of $220 per hectare, and cocoa a profit of $310 per hectare. How many hectares should the country devote to each crop in order to maximize profit?

46. Theo, who is dieting, requires two food supplements, I and II. He can get these supplements from two different products, A and B, as shown in the following table.

Supplement (gm/serving)	I	II
Product A	3	2
B	2	4

Theo's physician has recommended that he include at least 15 gm of each supplement in his daily diet. If product A costs 25¢ per serving and product B costs 40¢ per serving, how can he satisfy his requirements most economically?

*A hectare is a unit of land measurement; one hectare is equivalent to approximately 2.47 acres.

47. In a small town in South Carolina, zoning rules require that the window space (in sq ft) in a house be at least one-sixth of the space used up by solid walls. The monthly cost to heat the house is 20¢ per sq ft of solid walls and 80¢ per sq ft of windows. Find the maximum total area (windows plus walls) if $160 is available each month to pay for heat.

48. A machine shop manufactures two types of bolts. Each can be made on any of three groups of machines, but the time required on each group differs, as shown in the table.

Machine groups	I	II	III
Bolts Type 1	.4 hour	.5 hour	.2 hour
Type 2	.3 hour	.2 hour	.4 hour

Production schedules are made up for one week at a time. In this period there are 1200 hours of machine time available for each machine group. Type 1 bolts sell for 10¢ and type 2 bolts for 12¢. How many of each type of bolt should be manufactured per week to maximize revenue? What is the maximum revenue?

49. Seall Manufacturing Company makes color television sets. It produces a bargain set that sells for $100 profit and a deluxe set that sells for $150 profit. On the assembly line the bargain set requires 3 hr, while the deluxe set takes 5 hr. The cabinet shop spends one hour on the cabinet for the bargain set and 3 hr on the cabinet for the deluxe set. Both sets require 2 hr of time for testing and packing. On a particular production run the Seall Company has available 3900 work-hr on the assembly line, 2100 work-hr in the cabinet shop, and 2200 work-hr in the testing and packing department. How many sets of each type should it produce to make maximum profit? What is the maximum profit?

9.4 MATRIX SOLUTION OF LINEAR SYSTEMS

Since systems of linear equations occur in so many practical situations, methods of efficiently solving linear systems by computer have been developed. Computer solutions of linear systems depend on the idea of a **matrix,** a rectangular array of numbers enclosed in brackets. For example,

$$\begin{bmatrix} 2 & 3 & 7 \\ 5 & -1 & 10 \end{bmatrix}$$

is a matrix. Each number is called an **element** of the matrix.

The word *matrix* for a rectangular array of numbers was first used in 1850 by the English mathematician James Joseph Sylvester (1814–97). His friend and colleague Arthur Cayley (1821–95) developed the theory further and in 1855 wrote about multiplying and finding inverses of square matrices. At the time, he was looking for an efficient way of computing the result of substituting one linear system into another. Several of the important results in the theory of matrices can be found in the correspondence of Cayley and Sylvester. Georg Frobenius (1848–1917) showed that matrices could be used to describe other mathematical systems. By 1924, matrices with complex numbers as entries were found to be the easiest way to describe atomic systems. Today, matrices are used in nearly every branch of mathematics.

Matrices in general are discussed in more detail later in this chapter. In this section, a method is developed for solving linear systems using matrices. As an example, start with the system

$$x + 3y + 2z = 1$$
$$2x + y - z = 2$$
$$x + y + z = 2,$$

and write the coefficients of the variables and the constants as a matrix, called the **augmented matrix** of the system.

$$\begin{bmatrix} 1 & 3 & 2 & | & 1 \\ 2 & 1 & -1 & | & 2 \\ 1 & 1 & 1 & | & 2 \end{bmatrix}$$

The vertical line, which is optional, is used only to separate the coefficients from the constants. This matrix has 3 rows (horizontal) and 4 columns (vertical). To refer to a number in the matrix, use its row and column numbers. For example, the number 3 is in the first row, second column position.

The rows of this matrix can be treated just like the equations of a system of linear equations. Since the augmented matrix is nothing more than a short form of the system, any transformation of the matrix which results in an equivalent system of equations can be performed. Operations that produce such transformations are given below.

Matrix Row Transformations

For any augmented matrix of a system of linear equations, the following row transformations will result in the matrix of an equivalent system.

1. Any two rows may be interchanged.

2. The elements of any row may be multiplied by a nonzero real number.

3. Any row may be changed by adding to its elements a multiple of the elements of another row.

These transformations are just a restatement in matrix form of the transformations of systems discussed earlier. From now on, when referring to the third transformation, "a multiple of the elements of a row" will be abbreviated as "a multiple of a row."

Before using matrices to solve a linear system, the system must be arranged in the proper form, with variable terms on the left side of the equation and constant terms on the right. The variable terms must be in the same order in each of the equations. The following example illustrates the matrix method and compares it with the echelon method discussed earlier in which the system is transformed to an equivalent system in triangular form.

. . . .

EXAMPLE 1   Solve the system $3x - 4y = 1$
$$5x + 2y = 19.$$

The procedure for matrix solution is parallel to the echelon method used earlier, except for the last step. First, write the augmented matrix for the system. Here the system is already in the proper form.

Echelon method

$3x - 4y = 1$ **(1)**
$5x + 2y = 19$ **(2)**

Matrix method

$$\begin{bmatrix} 3 & -4 & | & 1 \\ 5 & 2 & | & 19 \end{bmatrix}$$

Multiply both sides of equation (1) by 1/3 so that x has a coefficient of 1.

Using row transformation (2) multiply each element of row 1 by 1/3.

$$x - \frac{4}{3}y = \frac{1}{3} \quad \textbf{(3)}$$
$$5x + 2y = 19$$

$$\begin{bmatrix} 1 & -\dfrac{4}{3} & | & \dfrac{1}{3} \\ 5 & 2 & | & 19 \end{bmatrix}$$

Eliminate x from equation (2) by adding -5 times equation (3) to equation (2).

Using row transformation (3), add -5 times the elements of row 1 to the elements of row 2.

$$x - \frac{4}{3}y = \frac{1}{3}$$
$$\frac{26}{3}y = \frac{52}{3} \quad \textbf{(4)}$$

$$\begin{bmatrix} 1 & -\dfrac{4}{3} & | & \dfrac{1}{3} \\ 0 & \dfrac{26}{3} & | & \dfrac{52}{3} \end{bmatrix}$$

Multiply both sides of equation (4) by 3/26 to get $y = 2$.

Multiply the elements of row 2 by 3/26, using row transformation (2).

$$x - \frac{4}{3}y = \frac{1}{3}$$
$$y = 2$$

$$\begin{bmatrix} 1 & -\dfrac{4}{3} & | & \dfrac{1}{3} \\ 0 & 1 & | & 2 \end{bmatrix}$$

Write the corresponding equations:

$$x - \frac{4}{3}y = \frac{1}{3}$$
$$y = 2$$

Finish the solution in either method by substituting 2 for y in the first equation to get $x = 3$. The solution of the system is $(3, 2)$.   ●

. . . .

EXAMPLE 2   Use row transformations to solve the linear system

$$2x + 6y = 28$$
$$4x - 3y = -19.$$

Begin with the augmented matrix

$$\begin{bmatrix} 2 & 6 & 28 \\ 4 & -3 & -19 \end{bmatrix}.$$

It is best to work vertically, in columns, beginning in each column with the element which is to become 1. This is similar to the order used in the first section of the chapter to arrange a system of equations in triangular form. The augmented matrix has a 2 in the first row, first column position. To get 1 in this position, use the second transformation and multiply each entry in the first row by 1/2. This is indicated below with the notation $\frac{1}{2}R_1$ next to the new first row.

$$\begin{bmatrix} 1 & 3 & 14 \\ 4 & -3 & -19 \end{bmatrix} \qquad \frac{1}{2}R_1$$

To get 0 in the second row, first column, add -4 times the first row to the second row.

$$\begin{bmatrix} 1 & 3 & 14 \\ 0 & -15 & -75 \end{bmatrix} \qquad -4R_1 + R_2$$

To get 1 in the second row, second column, multiply each element of the second row by $-1/15$, which gives

$$\begin{bmatrix} 1 & 3 & 14 \\ 0 & 1 & 5 \end{bmatrix}. \qquad -\frac{1}{15}R_2$$

This matrix corresponds to the system

$$x + 3y = 14 \qquad \text{(1)}$$
$$y = 5. \qquad \text{(2)}$$

Substitute 5 for y in equation (1) to get $x = -1$. The solution of the system is $(-1, 5)$.   ●

· · · · · · · · ·

EXAMPLE 3   Use matrix methods to solve the system

$$x - y + 5z = -6$$
$$3x + 3y - z = 10$$
$$x + 3y + 2z = 5.$$

Begin by writing the augmented matrix of the linear system.

$$\begin{bmatrix} 1 & -1 & 5 & -6 \\ 3 & 3 & -1 & 10 \\ 1 & 3 & 2 & 5 \end{bmatrix}$$

Keep in mind the triangular form that is our goal. We want the matrix to look like

$$\left[\begin{array}{ccc|c} 1 & a & b & d \\ 0 & 1 & c & e \\ 0 & 0 & 1 & f \end{array}\right].$$

Remember to work by columns, getting the required 1 in each column first, then getting the 0's below the 1. There is already a 1 in row 1, column 1. The next thing to do is get 0's in the rest of column 1. First, add to row 2 the results of multiplying row 1 by -3.

$$\left[\begin{array}{ccc|c} 1 & -1 & 5 & -6 \\ 0 & 6 & -16 & 28 \\ 1 & 3 & 2 & 5 \end{array}\right] \qquad -3R_1 + R_2$$

Now add to row 3 the results of multiplying row 1 by -1.

$$\left[\begin{array}{ccc|c} 1 & -1 & 5 & -6 \\ 0 & 6 & -16 & 28 \\ 0 & 4 & -3 & 11 \end{array}\right] \qquad -1R_1 + R_3$$

To get 1 in row 2, column 2, multiply row 2 by 1/6.

$$\left[\begin{array}{ccc|c} 1 & -1 & 5 & -6 \\ 0 & 1 & -\dfrac{8}{3} & \dfrac{14}{3} \\ 0 & 4 & -3 & 11 \end{array}\right] \qquad \dfrac{1}{6}R_2$$

Next, to get 0 in row 3, column 2, add to row 3 the results of multiplying row 2 by -4.

$$\left[\begin{array}{ccc|c} 1 & -1 & 5 & -6 \\ 0 & 1 & -\dfrac{8}{3} & \dfrac{14}{3} \\ 0 & 0 & \dfrac{23}{3} & -\dfrac{23}{3} \end{array}\right] \qquad -4R_2 + R_3$$

Finally, multiply the last row by $\dfrac{3}{23}$ to get 1 in row 3, column 3.

$$\left[\begin{array}{ccc|c} 1 & -1 & 5 & -6 \\ 0 & 1 & -\dfrac{8}{3} & \dfrac{14}{3} \\ 0 & 0 & 1 & -1 \end{array}\right] \qquad \dfrac{3}{23}R_3$$

The **main diagonal** of this matrix is the diagonal (to the left of the vertical bar) where all elements are 1. Since each element below the main diagonal is 0, the matrix is written in triangular form.

The final matrix above produces the system of equations

$$x - y + 5z = -6 \qquad (1)$$

$$y - \frac{8}{3}z = \frac{14}{3} \qquad (2)$$

$$z = -1. \qquad (3)$$

Use back-substitution into equations (1) and (2) to find the values of y and x.

$$y - \frac{8}{3}(-1) = \frac{14}{3} \qquad x - 2 + 5(-1) = -6$$

$$y = 2 \qquad\qquad x = 1$$

The solution of the system is $(1, 2, -1)$.   ●

· · · · · · · · ·

EXAMPLE 4   Solve the system $x + y = 2$
$$2x + 2y = 5.$$

Start with the augmented matrix.

$$\begin{bmatrix} 1 & 1 & 2 \\ 2 & 2 & 5 \end{bmatrix}$$

Next, add to row 2 the results of multiplying row 1 by -2.

$$\begin{bmatrix} 1 & 1 & 2 \\ 0 & 0 & 1 \end{bmatrix} \qquad -2R_1 + R_2$$

This matrix gives the system of equations

$$x + y = 2$$
$$0 = 1,$$

a system with no solution. Whenever a row of the augmented matrix is of the form $0\ 0\ 0\ .\ .\ .\ 0|a,$ (where $a \neq 0$), there will be no solution since this row corresponds to the equation $0 = a$. The solution set of the given system is $\varnothing$.   ●

· · · · · · · · ·

EXAMPLE 5   Solve the system $x + y + 3z = 5$
$$3x - 4y + z = 6.$$

The augmented matrix is

$$\begin{bmatrix} 1 & 1 & 3 & 5 \\ 3 & -4 & 1 & 6 \end{bmatrix}.$$

Get 0 in row 2, column 1.

$$\begin{bmatrix} 1 & 1 & 3 & 5 \\ 0 & -7 & -8 & -9 \end{bmatrix} \qquad -3R_1 + R_2$$

Now get 1 in row 2, column 2.

$$\begin{bmatrix} 1 & 1 & 3 & \bigm| & 5 \\ 0 & 1 & \dfrac{8}{7} & \bigm| & \dfrac{9}{7} \end{bmatrix} \qquad -\dfrac{1}{7}R_2$$

It is not possible to go further since there are just two rows, so write the system of equations

$$x + y + 3z = 5 \tag{1}$$

$$y + \frac{8}{7}z = \frac{9}{7}. \tag{2}$$

From equation (2), solve for y to get

$$y = -\frac{8}{7}z + \frac{9}{7}.$$

Substitute into equation (1) to get

$$x + \left(-\frac{8}{7}z + \frac{9}{7} \right) + 3z = 5$$

$$x = -\frac{13}{7}z + \frac{26}{7}.$$

The arbitrary variable is z; the infinitely many solutions can be written as

$$\left(\frac{-13z + 26}{7}, \frac{-8z + 9}{7}, z \right),$$

where z is any real number. ●

The cases that might occur when matrix methods are used to solve a system of linear equations are summarized below.

When the matrix is written in triangular form:

1. If the number of rows with nonzero elements to the left of the vertical line is equal to the number of variables in the system, then the system has a single solution.

2. If one of the rows has the form $[0\ 0\ .\ .\ .\ 0 \mid a]$ with $a \neq 0$, then the system has no solution.

3. If there are fewer rows in the matrix containing nonzero elements than the number of variables, then there are infinitely many solutions for the system. These solutions should be given in terms of arbitrary variables. See Example 5.

The method of using matrices to solve a system of linear equations that was developed in this section is called the **Gaussian reduction method** after the mathematician Carl F. Gauss (1777–1855). In 1811, Gauss published a paper showing how he determined the orbit of the asteroid Pallas. Over the years he had kept data on his numerous observations. Each observation produced a linear equation in six unknowns. A typical equation was

$$.79363x + 143.66y + .39493z + .95929u - .18856v + .17387w = 183.93.$$

Eventually he had twelve equations of this type. Certainly a method was needed to simplify the computation of a simultaneous solution. The method he developed was the reduction of the system from a rectangular to a triangular one, hence the Gaussian reduction. However, Gauss did not use matrices to carry out the computation.

9.4 Exercises

Write the augmented matrix for each of the following systems. Do not solve.

1. $2x + 3y = 11$
$x + 2y = 8$

2. $3x + 5y = -13$
$2x + 3y = -9$

3. $x + 5y = 6$
$y = 1$

4. $2x + 7y = 1$
$5x \quad = -15$

5. $2x + y + z = 3$
$3x - 4y + 2z = -7$
$x + y + z = 2$

6. $4x - 2y + 3z = 4$
$3x + 5y + z = 7$
$5x - y + 4z = 7$

7. $x + y \quad = 2$
$2y + z = -4$
$z = 2$

8. $x \quad = 6$
$y + 2z = 2$
$x \quad - 3z = 6$

Write the system of equations associated with each of the following augmented matrices. Do not solve.

9. $\begin{bmatrix} 2 & 1 & | & 1 \\ 3 & -2 & | & -9 \end{bmatrix}$

10. $\begin{bmatrix} 1 & -5 & | & -18 \\ 6 & 2 & | & 20 \end{bmatrix}$

11. $\begin{bmatrix} 1 & 0 & 0 & | & 2 \\ 0 & 1 & 0 & | & 3 \\ 0 & 0 & 1 & | & -2 \end{bmatrix}$

12. $\begin{bmatrix} 1 & 0 & 1 & | & 4 \\ 0 & 1 & 0 & | & 2 \\ 0 & 0 & 1 & | & 3 \end{bmatrix}$

13. $\begin{bmatrix} 3 & 2 & 1 & | & 1 \\ 0 & 2 & 4 & | & 22 \\ -1 & -2 & 3 & | & 15 \end{bmatrix}$

14. $\begin{bmatrix} 2 & 1 & 3 & | & 12 \\ 4 & -3 & 0 & | & 10 \\ 5 & 0 & -4 & | & -11 \end{bmatrix}$

Use the Gaussian reduction method to solve each of the following systems of equations. For systems with dependent equations in 3 variables, give the solution with z arbitrary; for any such equations with 4 variables, let w be the arbitrary variable.

15. $x + y = 5$
$x - y = -1$

16. $x + 2y = 5$
$2x + y = -2$

17. $x + y = -3$
$2x - 5y = -6$

18. $3x - 2y = 4$
$3x + y = -2$

19. $2x - 3y = 10$
$2x + 2y = 5$

20. $6x + y - 5 = 0$
$5x + y - 3 = 0$

21. $2x - 5y - 10 = 0$
$3x + y - 15 = 0$

22. $4x - y = 3$
$-2x + 3y = 1$

23. $x + y \quad = -1$
$y + z = 4$
$x \quad + z = 1$

24. $x \quad - z = -3$
$y + z = 9$
$x \quad + z = 7$

25. $x + y - z = 6$
$2x - y + z = -9$
$x - 2y + 3z = 1$

26. $x + 3y - 6z = 7$
$2x - y + z = 1$
$x + 2y + 2z = -1$

27. $-x + y \quad = -1$
$y - z = 6$
$x \quad + z = -1$

28. $x + y \quad = 1$
$2x \quad - z = 0$
$y + 2z = -2$

29. $2x - y + 3z = 0$
$x + 2y - z = 5$
$2y + z = 1$

30. $4x + 2y - 3z = 6$
$x - 4y + z = -4$
$-x \quad + 2z = 2$

31.
$$x - 2y + z - 5 = 0$$
$$2x + y + z + 9 = 0$$
$$-2x + 4y - 2z - 2 = 0$$

32.
$$3x + 5y - z = 0$$
$$4x - y + 2z - 1 = 0$$
$$-6x - 10y + 2z = 0$$

33.
$$x + y + z = 6$$
$$2x - y - z = 3$$

34.
$$5x - 3y + z = 1$$
$$2x + y - z = 4$$

35.
$$x - 8y + z = 4$$
$$3x - y + 2z = -1$$

36.
$$-3x + y - z = 8$$
$$2x + y + 4z = 0$$

37.
$$x + 3y - z = 0$$
$$2y - z = 4$$

38.
$$2x - y + 4z = 1$$
$$y + z = 3$$

39.
$$3x + 2y - w = 0$$
$$2x + z + 2w = 5$$
$$x + 2y - z = -2$$
$$2x - y + z + w = 2$$

40.
$$x + 3y - 2z - w = 9$$
$$4x + y + z + 2w = 2$$
$$-3x - y + z - w = -5$$
$$x - y - 3z - 2w = 2$$

41.
$$x - y + 2z + w = 4$$
$$y + z = 3$$
$$z - w = 2$$

42.
$$3x + y = 1$$
$$y - 4z + w = 0$$
$$z - 3w = -1$$

Solve each of the following word problems using matrices.

43. A working couple earned a total of $4352. The wife earned $64 per day; the husband earned $8 per day less. Find the number of days each worked if the total number of days worked by both was 72.

44. Midtown Manufacturing Company makes two products, plastic plates and plastic cups. Both require time on two machines; plates: 1 hr on machine *A* and 2 hr on machine *B*, cups: 3 hr on machine *A* and 1 hr on Machine *B*. Both machines operate 15 hr a day. How many of each product can be produced in a day under these conditions.

45. A company produces two models of bicycles, model 201 and model 301. Model 201 requires 2 hr of assembly time, and model 301 requires 3 hr of assembly time. The parts for model 201 cost $25 per bike; those for model 301 cost $30 per bike. If the company has a total of 34 hr of assembly time and $365 available per day for these two models, how many of each can be made in a day?

46. Juanita Wilson invests $10,000 in three ways. With one part, she buys mutual funds which offer a return of 8% per year. The second part, which amounts to twice the first, is used to buy government bonds at 9% per year. She puts the rest in the bank at 5% annual interest. The first year her investments bring a return of $830. How much did she invest each way?

47. To get the necessary funds for a planned expansion, a small company took out three loans totaling $25,000. The company was able to borrow some of the money at 8%. They borrowed $2000 more than one-half the amount of the 8% loan at 10%, and the rest at 9%. The total annual interest was $2220. How much did they borrow at each rate?

48. At rush hours, substantial traffic congestion is encountered at the traffic intersections shown in the figure. (All streets are one-way.) The city wishes to improve the signals at these corners so as to speed the flow of traffic. The traffic engineers first gather data. As the figure shows, 700 cars per hour came down *M* Street to intersection *A*; 300 cars per hour come to intersection *A* on 10th Street. A total of x_1 of these cars leave *A* on *M* Street, while x_4 cars leave *A* on 10th Street. The number of cars entering *A* must equal the number leaving, so that

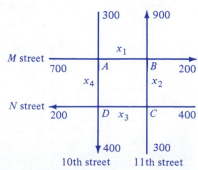

$$x_1 + x_4 = 700 + 300$$

or

$$x_1 + x_4 = 1000.$$

For intersection B, x_1 cars enter B on M Street, and x_2 cars enter B on 11th Street. The figure shows that 900 cars leave B on 11th while 200 leave on M. We have

$$x_1 + x_2 = 900 + 200$$
$$x_1 + x_2 = 1100.$$

At intersection C, 400 cars enter on N Street, 300 on 11th Street, while x_2 leave on 11th Street and x_3 leave on N Street. This gives

$$x_2 + x_3 = 400 + 300$$
$$x_2 + x_3 = 700.$$

Finally, intersection D has x_3 cars entering at N and x_4 entering on 10th. There are 400 cars leaving D on 10th and 200 leaving on N.

(a) Set up an equation for intersection D.
(b) Use the four equations to set up an augmented matrix, and then use the Gaussian method to reduce it to triangular form.
(c) Since you got a row of all zeros, the system of equations does not have a unique solution. Write three equations, corresponding to the three nonzero rows of the matrix. Solve each of the equations for x_4.
(d) One of your equations should have been $x_4 = 1000 - x_1$. What is the largest possible value of x_1 so that x_4 is not negative?
(e) Your second equation should have been $x_4 = x_2 - 100$. Find the smallest possible value of x_2 so that x_4 is not negative.
(f) Find the largest possible values of x_3 and x_4 so that neither variable is negative.
(g) Use the results of (a)–(f) to give a solution for the problem in which all the equations are satisfied and all variables are nonnegative. Is the solution unique?

Solve each of the following problems.

49. Three small boxes and 8 large boxes cost $13.30. If 6 small and 16 large boxes are purchased, the cost is $27.20. Find the cost of a small box and the cost of a large box.

50. The sum of three numbers is 20. One number is one more than another. Find the three numbers.

51. If three numbers are added, the result is -10. The second number, added to twice the third, gives 5. Find the three numbers.

52. If half of one number is added to another, the sum is twice a third number. The third number is half the second number. Find the three numbers.

For each of the following equations, determine the constants A and B that make the equation an identity. (Hint: Combine terms on the right and set coefficients of corresponding terms in the numerators equal.)

53. $\dfrac{1}{x(x-1)} = \dfrac{A}{x} + \dfrac{B}{x-1}$

54. $\dfrac{x+4}{x^2} = \dfrac{A}{x} + \dfrac{B}{x^2}$

55. $\dfrac{x}{(1-x)^2} = \dfrac{A}{1-x} + \dfrac{B}{(1-x)^2}$

56. $\dfrac{2x}{(x+2)(x-1)} = \dfrac{A}{x+2} + \dfrac{B}{x-1}$

9.5 PROPERTIES OF MATRICES

The use of matrix notation in solving a system of linear equations was shown in the previous section. In this section, the algebraic properties of matrices are discussed.

It is customary to use capital letters to name matrices. Also, subscript notation is often used to name the elements of a matrix, as in the following matrix A.

$$A = \begin{bmatrix} a_{11} & a_{12} & a_{13} & \cdots & a_{1n} \\ a_{21} & a_{22} & a_{23} & \cdots & a_{2n} \\ a_{31} & a_{32} & a_{33} & \cdots & a_{3n} \\ \cdot & \cdot & \cdot & & \cdot \\ \cdot & \cdot & \cdot & & \cdot \\ \cdot & \cdot & \cdot & & \cdot \\ a_{m1} & a_{m2} & a_{m3} & \cdots & a_{mn} \end{bmatrix}$$

With this notation, the first row, first column element is a_{11} (read "a-sub-one-one"); the second row, third column element is a_{23}; and in general, the ith row, jth column element is a_{ij}.

Matrices are classified by their size, that is, by the number of rows and columns that they contain. For example, the matrix

$$\begin{bmatrix} 2 & 7 & -5 \\ 3 & -6 & 0 \end{bmatrix}$$

has two rows (horizontal) and three columns (vertical), and is called a 2×3 matrix (read "two by three matrix"). A matrix with m rows and n columns is an $m \times n$ matrix. The number of rows is always given first.

Certain matrices have special names: an $n \times n$ matrix is a **square matrix** of order n. Also, a matrix with just one row is a **row matrix,** and a matrix with just one column is a **column matrix.**

Two matrices are **equal** if they are the same size and if corresponding elements, position by position, are equal. Using this definition, the matrices

$$\begin{bmatrix} 2 & 1 \\ 3 & -5 \end{bmatrix} \quad \text{and} \quad \begin{bmatrix} 1 & 2 \\ -5 & 3 \end{bmatrix}$$

are *not* equal (even though they contain the same elements and are the same size), since the corresponding elements differ.

· · · · · · · · ·

EXAMPLE 1

(a) From the definition of equality given above, the only way that the statement

$$\begin{bmatrix} 2 & 1 \\ p & q \end{bmatrix} = \begin{bmatrix} x & y \\ -1 & 0 \end{bmatrix}$$

can be true is if $2 = x$, $1 = y$, $p = -1$, and $q = 0$.

(b) The statement

$$\begin{bmatrix} x \\ y \end{bmatrix} = \begin{bmatrix} 1 \\ 4 \\ 0 \end{bmatrix}$$

can never be true, since the two matrices are different sizes. (One is 2×1 and the other is 3×1.) ●

Addition of matrices is defined as follows.

Definition of Addition of Matrices

To **add** two matrices of the same size, add corresponding elements. Only matrices of the same size can be added.

It can be shown that matrix addition satisfies the commutative, associative, closure, identity, and inverse properties. (See Exercises 36–40.)

EXAMPLE 2   Find each of the following sums.

(a) $\begin{bmatrix} 5 & -6 \\ 8 & 9 \end{bmatrix} + \begin{bmatrix} -4 & 6 \\ 8 & -3 \end{bmatrix} = \begin{bmatrix} 5 + (-4) & -6 + 6 \\ 8 + 8 & 9 + (-3) \end{bmatrix} = \begin{bmatrix} 1 & 0 \\ 16 & 6 \end{bmatrix}$

(b) $\begin{bmatrix} 2 \\ 5 \\ 8 \end{bmatrix} + \begin{bmatrix} -6 \\ 3 \\ 12 \end{bmatrix} = \begin{bmatrix} -4 \\ 8 \\ 20 \end{bmatrix}$

(c) The matrices $A = \begin{bmatrix} 5 & 8 \\ 6 & 2 \end{bmatrix}$ and $B = \begin{bmatrix} 3 & 9 & 1 \\ 4 & 2 & 5 \end{bmatrix}$

are different sizes, so the sum $A + B$ does not exist. ●

A matrix containing only zero elements is called a **zero matrix.** For example, $O = [0 \ \ 0 \ \ 0]$ is the 1×3 zero matrix, while

$$O = \begin{bmatrix} 0 & 0 & 0 \\ 0 & 0 & 0 \end{bmatrix}$$

is the 2×3 zero matrix.

By the additive inverse property in Chapter 1, each real number has an additive inverse: if a is a real number, there is a real number $-a$ such that

$$a + (-a) = 0 \quad \text{and} \quad -a + a = 0.$$

What about matrices? Given the matrix

$$A = \begin{bmatrix} -5 & 2 & -1 \\ 3 & 4 & -6 \end{bmatrix},$$

is there a matrix $-A$ such that

$$A + (-A) = O$$

where O is the 2×3 zero matrix? The answer is yes: the matrix $-A$ has as elements the additive inverses of the elements of A. (Remember, each element of A is a real number and, therefore, has an additive inverse.)

$$-A = \begin{bmatrix} 5 & -2 & 1 \\ -3 & -4 & 6 \end{bmatrix}$$

To check, test that $A + (-A)$ equals the zero matrix, O.

$$A + (-A) = \begin{bmatrix} -5 & 2 & -1 \\ 3 & 4 & -6 \end{bmatrix} + \begin{bmatrix} 5 & -2 & 1 \\ -3 & -4 & 6 \end{bmatrix} = \begin{bmatrix} 0 & 0 & 0 \\ 0 & 0 & 0 \end{bmatrix} = O$$

Matrix $-A$ is called the **additive inverse,** or **negative,** of matrix A. Every matrix has an additive inverse.

The real number b is subtracted from the real number a, written $a - b$, by adding a and the additive inverse of b. That is,

$$a - b = a + (-b).$$

The same definition works for **subtraction** of matrices.

Definition of Subtraction of Matrices

If A and B are two matrices of the same size, then

$$A - B = A + (-B).$$

In practice, the difference of two matrices of the same size is found by subtracting corresponding elements.

EXAMPLE 3   Find each of the following differences.

(a) $\begin{bmatrix} -5 & 6 \\ 2 & 4 \end{bmatrix} - \begin{bmatrix} -3 & 2 \\ 5 & -8 \end{bmatrix} = \begin{bmatrix} -5 - (-3) & 6 - 2 \\ 2 - 5 & 4 - (-8) \end{bmatrix} = \begin{bmatrix} -2 & 4 \\ -3 & 12 \end{bmatrix}$

(b) $[8 \quad 6 \quad -4] - [3 \quad 5 \quad -8] = [5 \quad 1 \quad 4]$

(c) The matrices

$$\begin{bmatrix} -2 & 5 \\ 0 & 1 \end{bmatrix} \quad \text{and} \quad \begin{bmatrix} 3 \\ 5 \end{bmatrix}$$

are different sizes and cannot be subtracted.   ●

In work with matrices, a real number is called a **scalar** to distinguish it from a matrix. The product of a scalar k and a matrix X is the matrix kX, each of whose elements is k times the corresponding element of X.

. . . .

EXAMPLE 4

(a) $5\begin{bmatrix} 2 & -3 \\ 0 & 4 \end{bmatrix} = \begin{bmatrix} 10 & -15 \\ 0 & 20 \end{bmatrix}$ (b) $\dfrac{3}{4}\begin{bmatrix} 20 & 36 \\ 12 & -16 \end{bmatrix} = \begin{bmatrix} 15 & 27 \\ 9 & -12 \end{bmatrix}$ •

The proofs of the following properties of scalar multiplication are left for Exercises 44–47.

. . . .

Properties of Scalar Multiplication

If A and B are matrices of the same size and c and d are real numbers.

$$(c + d)A = cA + dA$$
$$c(A + B) = cA + cB$$
$$c(A)d = cd(A)$$
$$(cd)A = c(dA).$$

We have seen how to multiply a real number (scalar) and a matrix. Now we define the product of two matrices. The procedure developed below for finding the product of two matrices may seem artificial, but it is useful in applications. The method will be illustrated before a formal rule is given. To find the product of

$$A = \begin{bmatrix} -3 & 4 & 2 \\ 5 & 0 & 4 \end{bmatrix} \quad \text{and} \quad B = \begin{bmatrix} -6 & 4 \\ 2 & 3 \\ 3 & -2 \end{bmatrix},$$

first, locate *row* 1 of A and *column* 1 of B, shown shaded below.

$$A = \begin{bmatrix} -3 & 4 & 2 \\ 5 & 0 & 4 \end{bmatrix} \quad B = \begin{bmatrix} -6 & 4 \\ 2 & 3 \\ 3 & -2 \end{bmatrix}$$

Multiply corresponding elements, and find the sum of the products.

$$(-3)(-6) + (4)(2) + (2)(3) = 32$$

This result is the element for row 1, column 1 of the product matrix.

Now use *row* 1 of A and *column* 2 of B (shown shaded below) to determine the element in row 1 and column 2 of the product matrix.

$$\begin{bmatrix} -3 & 4 & 2 \\ 5 & 0 & 4 \end{bmatrix} \begin{bmatrix} -6 & 4 \\ 2 & 3 \\ 3 & -2 \end{bmatrix}$$

Multiply corresponding elements and add the products:

$$(-3)(4) + (4)(3) + (2)(-2) = -4,$$

which is the row 1, column 2 element of the product matrix.

Next, use *row* 2 of *A* and *column* 1 of *B*; this will give the row 2, column 1 entry of the product matrix.

$$(5)(-6) + (0)(2) + (4)(3) = -18$$

Finally, use *row* 2 of *A* and *column* 2 of *B* to find the entry for row 2, column 2 of the product matrix.

$$(5)(4) + (0)(3) + (4)(-2) = 12$$

The product matrix can now be written.

$$\begin{bmatrix} -3 & 4 & 2 \\ 5 & 0 & 4 \end{bmatrix} \begin{bmatrix} -6 & 4 \\ 2 & 3 \\ 3 & -2 \end{bmatrix} = \begin{bmatrix} 32 & -4 \\ -18 & 12 \end{bmatrix}$$

As in this example, the product of a 2 × 3 matrix and a 3 × 2 matrix is a 2 × 2 matrix.

By definition, the **product** *AB* of an *m* × *n* matrix *A* and an *n* × *p* matrix *B* is found as follows. Multiply each element of the first row of *A* by the corresponding element of the first column of *B*. The sum of these *n* products is the first row, first column element of *AB*.

Also, the sum of the products found by multiplying the elements of the first row of *A* times the corresponding elements of the second column of *B* gives the first row, second column element of *AB*, and so on.

To find the *i*th row, *j*th column element of *AB*, multiply each element in the *i*th row of *A* by the corresponding element in the *j*th column of *B* (note the colored areas in the matrices below.) The sum of these products will give the element of row *i*, column *j* of *AB*.

$$A = \begin{bmatrix} a_{11} & a_{12} & a_{13} & \cdots & a_{1n} \\ a_{21} & a_{22} & a_{23} & \cdots & a_{2n} \\ \cdot & & & & \\ \cdot & & & & \\ \cdot & & & & \\ a_{i1} & a_{i2} & a_{i3} & \cdots & a_{in} \\ \cdot & & & & \\ \cdot & & & & \\ \cdot & & & & \\ a_{m1} & a_{m2} & a_{m3} & \cdots & a_{mn} \end{bmatrix} \qquad B = \begin{bmatrix} b_{11} & b_{12} & \cdots & b_{1j} & \cdots & b_{1p} \\ b_{21} & b_{22} & \cdots & b_{2j} & \cdots & b_{2p} \\ \cdot & & & & & \\ \cdot & & & & & \\ b_{n1} & b_{n2} & \cdots & b_{nj} & \cdots & b_{np} \end{bmatrix}$$

Definition of Matrix Multiplication

If the number of columns of matrix *A* is the same as the number of rows of matrix *B*, then entry c_{ij} of the product matrix $C = AB$ is found as follows:

$$c_{ij} = a_{i1}b_{1j} + a_{i2}b_{2j} + \cdots + a_{in}b_{nj}.$$

The final product will have as many rows as *A* and as many columns as *B*.

• • • • • • • • • •

EXAMPLE 5   Suppose matrix A is 3×2, while matrix B is 2×4. Can the product AB be calculated? What is the size of the product? Can the product BA be calculated? What is the size of BA?

The following diagram helps answer the questions about the product AB.

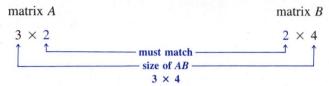

matrix A matrix B

3×2 2×4

must match
size of AB
3×4

The product AB exists, since the number of columns of A equals the number of rows of B (both are 2). The product is a 3×4 matrix. Make a similar diagram for BA.

matrix B matrix A

2×4 3×2

different

The product BA is not defined since B has 4 columns and A has only 3 rows.   ●

• • • • • • • • • •

EXAMPLE 6   Find AB and BA, if possible, where

$$A = \begin{bmatrix} 1 & -3 \\ 7 & 2 \end{bmatrix} \quad \text{and} \quad B = \begin{bmatrix} 1 & 0 & -1 & 2 \\ 3 & 1 & 4 & -1 \end{bmatrix}.$$

First decide whether AB can be found. Since A is 2×2 and B is 2×4, the product can be found and will be a 2×4 matrix. Now use the definition of matrix multiplication.

$$AB = \begin{bmatrix} 1 & -3 \\ 7 & 2 \end{bmatrix} \begin{bmatrix} 1 & 0 & -1 & 2 \\ 3 & 1 & 4 & -1 \end{bmatrix}$$

$$= \begin{bmatrix} 1(1) + (-3)(3) & 1(0) + (-3)1 & 1(-1) + (-3)4 & 1(2) + (-3)(-1) \\ 7(1) + 2(3) & 7(0) + 2(1) & 7(-1) + 2(4) & 7(2) + 2(-1) \end{bmatrix}$$

$$= \begin{bmatrix} -8 & -3 & -13 & 5 \\ 13 & 2 & 1 & 12 \end{bmatrix}$$

Since B is a 2×4 matrix, and A is a 2×2 matrix, the product BA cannot be found.   ●

• • • • • • • • • •

EXAMPLE 7   If $A = \begin{bmatrix} 1 & 3 \\ -2 & 5 \end{bmatrix}$ and $B = \begin{bmatrix} -2 & 7 \\ 0 & 2 \end{bmatrix}$, then the definition of

matrix multiplication can be used to show that

$$AB = \begin{bmatrix} -2 & 13 \\ 4 & -4 \end{bmatrix} \quad \text{and} \quad BA = \begin{bmatrix} -16 & 29 \\ -4 & 10 \end{bmatrix}. \quad ●$$

> **NOTE**   Examples 5 and 6 showed that the order in which two matrices are to be multiplied may determine whether their product can be found. Example 7 shows that even when both AB and BA can be found, they may not be equal. In general, for matrices A and B, $AB \neq BA$, so that matrix multiplication is not commutative.

Matrix multiplication does, however, satisfy the associative and distributive properties.

Properties of Matrix Multiplication

If A, B, and C are matrices such that all the following products and sums exist, then

$$(AB)C = A(BC)$$
$$A(B + C) = AB + AC$$
$$(B + C)A = BA + CA.$$

For proofs of these results for the special cases when A, B, and C are square matrices, see Exercises 41 and 42. The identity and inverse properties for matrix multiplication are discussed in a later section of this chapter.

9.5 Exercises

Find the values of the variables in each of the following.

1. $\begin{bmatrix} 2 & 1 \\ 4 & 8 \end{bmatrix} = \begin{bmatrix} x & 1 \\ y & z \end{bmatrix}$

2. $\begin{bmatrix} 2 & 5 & 6 \\ 1 & m & n \end{bmatrix} = \begin{bmatrix} z & y & w \\ 1 & 8 & -2 \end{bmatrix}$

3. $\begin{bmatrix} -7 + z & 4r & 8s \\ 6p & 2 & 5 \end{bmatrix} + \begin{bmatrix} -9 & 8r & 3 \\ 2 & 5 & 4 \end{bmatrix} = \begin{bmatrix} 2 & 36 & 27 \\ 20 & 7 & 12a \end{bmatrix}$

4. $\begin{bmatrix} a + 2 & 3z + 1 & 5m \\ 4k & 0 & 3 \end{bmatrix} + \begin{bmatrix} 3a & 2z & 5m \\ 2k & 5 & 6 \end{bmatrix} = \begin{bmatrix} 10 & -14 & 80 \\ 10 & 5 & 9 \end{bmatrix}$

In each of the following, determine whether the two matrices are equal.

5. $[1 \quad 2], \begin{bmatrix} 1 \\ 2 \end{bmatrix}$

6. $\begin{bmatrix} 1 & 2 \\ 3 & 4 \end{bmatrix}, \begin{bmatrix} 1 & 2 & 0 \\ 3 & 4 & 0 \end{bmatrix}$

Perform each of the operations in Exercises 7–14, whenever possible.

7. $\begin{bmatrix} 6 & -9 & 2 \\ 4 & 1 & 3 \end{bmatrix} - \begin{bmatrix} -8 & 2 & 5 \\ 6 & -3 & 4 \end{bmatrix}$

8. $\begin{bmatrix} 9 & 4 \\ -8 & 2 \end{bmatrix} + \begin{bmatrix} -3 & 2 \\ -4 & 7 \end{bmatrix}$

9. $\begin{bmatrix} -6 & 8 \\ 0 & 0 \end{bmatrix} - \begin{bmatrix} 0 & 0 \\ -4 & -2 \end{bmatrix}$

10. $\begin{bmatrix} 1 & -4 \\ 2 & -3 \\ -8 & 4 \end{bmatrix} - \begin{bmatrix} -6 & 9 \\ -2 & 5 \\ -7 & -12 \end{bmatrix}$

11. $\begin{bmatrix} -8 & 4 & 0 \\ 2 & 5 & 0 \end{bmatrix} + \begin{bmatrix} 6 & 3 \\ 8 & 9 \end{bmatrix}$

12. $\begin{bmatrix} 2 \\ 3 \end{bmatrix} - \begin{bmatrix} 8 & 1 \\ 9 & 4 \end{bmatrix}$

13. $\begin{bmatrix} -4x + 2y & -3x + y \\ 6x - 3y & 2x - 5y \end{bmatrix} + \begin{bmatrix} -8x + 6y & 2x \\ 3y - 5x & 6x + 4y \end{bmatrix}$

14. $\begin{bmatrix} 4k - 8y \\ 6z - 3x \\ 2k + 5a \\ -4m + 2n \end{bmatrix} - \begin{bmatrix} 5k + 6y \\ 2z + 5x \\ 4k + 6a \\ 4m - 2n \end{bmatrix}$

15. When John inventoried his bolt collection, he found that he had 7 flat-head long bolts, 9 flat-head medium, 8 flat-head short, 2 round-head long, no round-head medium, and 6 round-head short. Write this information first as a 3×2 matrix and then as a 2×3 matrix.

16. At the grocery store, Miguel bought 4 quarts of milk, 2 loaves of bread, 4 chickens, and an apple. Mary bought 2 quarts of milk, a loaf of bread, 5 chickens, and 4 apples. Write this information first as a 2×4 matrix and then as a 4×2 matrix.

Let $A = \begin{bmatrix} -2 & 4 \\ 0 & 3 \end{bmatrix}$ *and* $B = \begin{bmatrix} -6 & 2 \\ 4 & 0 \end{bmatrix}$. *Find each of the following.*

17. $2A$

18. $-3B$

19. $2A - B$

20. $-2A + 4B$

21. $-A + \dfrac{1}{2}B$

22. $\dfrac{3}{4}A - B$

Find the matrix products in Exercises 23–34, whenever possible.

23. $\begin{bmatrix} 1 & 2 \\ 3 & 4 \end{bmatrix}\begin{bmatrix} -1 \\ 7 \end{bmatrix}$

24. $\begin{bmatrix} 3 & -4 & 1 \\ 5 & 0 & 2 \end{bmatrix}\begin{bmatrix} -1 \\ 4 \\ 2 \end{bmatrix}$

25. $\begin{bmatrix} -1 & 2 & 0 \\ 0 & 3 & 2 \\ 0 & 1 & 4 \end{bmatrix}\begin{bmatrix} 2 & -1 & 2 \\ 0 & 2 & 1 \\ 3 & 0 & -1 \end{bmatrix}$

26. $\begin{bmatrix} -2 & -3 & -4 \\ 2 & -1 & 0 \\ 4 & -2 & 3 \end{bmatrix}\begin{bmatrix} 0 & 1 & 4 \\ 1 & 2 & -1 \\ 3 & 2 & -2 \end{bmatrix}$

27. $\begin{bmatrix} -2 & 1 & 4 \\ 0 & 1 & 2 \end{bmatrix}\begin{bmatrix} -2 & 1 & 0 \\ 0 & -2 & 0 \\ 4 & 1 & 2 \end{bmatrix}$

28. $\begin{bmatrix} -1 & 0 & 0 \\ 2 & 1 & 4 \end{bmatrix}\begin{bmatrix} 4 & -2 & 5 \\ 0 & 1 & 4 \\ 2 & -9 & 0 \end{bmatrix}$

29. $\begin{bmatrix} -3 & 0 & 2 & 1 \\ 4 & 0 & 2 & 6 \end{bmatrix}\begin{bmatrix} -4 & 2 \\ 0 & 1 \end{bmatrix}$

30. $\begin{bmatrix} -1 & 2 & 4 & 1 \\ 0 & 2 & -3 & 5 \end{bmatrix}\begin{bmatrix} 1 & 2 & 4 \\ -2 & 5 & 1 \end{bmatrix}$

31. $\begin{bmatrix} -2 & 4 & 6 \end{bmatrix}\begin{bmatrix} 3 \\ -2 \\ 1 \end{bmatrix}$

32. $\begin{bmatrix} 4 & 0 & 2 \end{bmatrix}\begin{bmatrix} -5 \\ 1 \\ 6 \end{bmatrix}$

33. $\begin{bmatrix} 3 \\ -2 \\ 1 \end{bmatrix}\begin{bmatrix} -2 & 4 & 6 \end{bmatrix}$

34. $\begin{bmatrix} -5 \\ 1 \\ 6 \end{bmatrix}\begin{bmatrix} 4 & 0 & 2 \end{bmatrix}$

35. The Bread Box, a small neighborhood bakery, sells four main items: sweet rolls, bread, cake, and pie. The amount of certain major ingredients (measured in cups except for eggs) required to make these items is given in matrix A.

	Eggs	Flour	Sugar	Shortening	Milk	
	1	4	1/4	1/4	1	Dozen rolls
$A =$	0	3	0	1/4	0	Loaf of bread
	4	3	2	1	1	Cake (1)
	0	1	0	1/3	0	Pie (1)

The cost per cup or per egg (in cents) for each ingredient when purchased in large lots and in small lots is given by matrix B.

$$B = \begin{bmatrix} 5 & 5 \\ 8 & 10 \\ 10 & 12 \\ 12 & 15 \\ 5 & 6 \end{bmatrix} \begin{matrix} \text{Eggs} \\ \text{Flour} \\ \text{Sugar} \\ \text{Shortening} \\ \text{Milk} \end{matrix}$$

Cost: Large lot, Small lot

(a) Use matrix multiplication to find a matrix representing the comparative costs per item under the two purchase options.

Suppose a day's orders consist of 20 dozen sweet rolls, 200 loaves of bread, 50 cakes, and 60 pies.

(b) Represent these orders as a 1×4 matrix and use matrix multiplication to write as a matrix the amount of each ingredient required to fill the day's orders.

(c) Use matrix multiplication to find a matrix representing the costs under the two purchase options to fill the day's orders.

For Exercises 36–50, let

$$A = \begin{bmatrix} a_{11} & a_{12} \\ a_{21} & a_{22} \end{bmatrix}, B = \begin{bmatrix} b_{11} & b_{12} \\ b_{21} & b_{22} \end{bmatrix}, \text{ and } C = \begin{bmatrix} c_{11} & c_{12} \\ c_{21} & c_{22} \end{bmatrix},$$

where all the elements are real numbers. Decide which of the following statements are true for these three matrices. If a statement is true, prove that it is true. If it is false, give a numerical example to show it false.

36. $A + B = B + A$ (commutative property)

37. $A + (B + C) = (A + B) + C$
(associative property)

38. $A + B$ is a 2×2 matrix.   (closure property)

39. There exists a matrix O such that $A + O = A$ and $O + A = A$.   (identity property)

40. There exists a matrix $-A$ such that $A + (-A) = O$ and $-A + A = O$.
(inverse property)

41. $(AB)C = A(BC)$ (associative property)

42. $A(B + C) = AB + AC$ (distributive property)

43. AB is a 2×2 matrix.   (closure property)

44. $c(A + B) = cA + cB$ for any real number c

45. $(c + d)A = cA + dA$ for any real numbers c and d

46. $c(A)d = cd(A)$

47. $(cd)A = c(dA)$

48. $(A + B)(A - B) = A^2 - B^2$ (where $A^2 = AA$)

49. $(A + B)^2 = A^2 + 2AB + B^2$

50. If $AB = O$, then $A = O$ or $B = O$.

The transpose, A^T, of a matrix A is found by exchanging the rows and columns of A. That is, if

$$A = \begin{bmatrix} a & b \\ c & d \end{bmatrix}, \quad \text{then} \quad A^T = \begin{bmatrix} a & c \\ b & d \end{bmatrix}.$$

Show that each of the following equations are true for matrices A and B, where

$$B = \begin{bmatrix} m & n \\ p & q \end{bmatrix}.$$

51. $(A^T)^T = A$

52. $(A + B)^T = A^T + B^T$

53. $(AB)^T = B^T A^T$

9.6 DETERMINANTS

Given any square matrix A, there is a unique real number associated with A—the *determinant* of A. In this section determinants are defined, and in the next section determinants are used to solve systems of linear equations. This technique was first published by the Swiss mathematician Gabriel Cramer (1704–52) in 1750. Although Leibniz knew the technique in 1693 and Seki Kowa of Japan knew it in 1683, their work did not become known to others. By 1812, Cauchy (1789–1857), who introduced the word *determinant,* had proved that the determinant of the product of two matrices is the product of their determinants. Carl Gustave Jacobi (1804–51) gave the convincing arguments that made determinants acceptable.

The determinant of a matrix A is written $|A|$. The determinant of a 2×2 matrix is defined as follows.

Definition of Determinant of a 2 × 2 Matrix

$$\text{If } A = \begin{bmatrix} a_{11} & a_{12} \\ a_{21} & a_{22} \end{bmatrix}, \text{ then } |A| = \begin{vmatrix} a_{11} & a_{12} \\ a_{21} & a_{22} \end{vmatrix} = a_{11}a_{22} - a_{21}a_{12}.$$

NOTE Notice that matrices are enclosed with square brackets, while determinants are denoted with vertical bars.

EXAMPLE 1 Let $A = \begin{bmatrix} -3 & 4 \\ 6 & 8 \end{bmatrix}$. Find $|A|$.

Use the definition above.

$$|A| = \begin{vmatrix} -3 & 4 \\ 6 & 8 \end{vmatrix} = -3(8) - 6(4) = -48. \quad \bullet$$

The determinant of a 3×3 matrix A is defined as follows.

Definition of Determinant of a 3 × 3 Matrix

$$\text{If } A = \begin{bmatrix} a_{11} & a_{12} & a_{13} \\ a_{21} & a_{22} & a_{23} \\ a_{31} & a_{32} & a_{33} \end{bmatrix}, \text{ then }$$

$$|A| = \begin{vmatrix} a_{11} & a_{12} & a_{13} \\ a_{21} & a_{22} & a_{23} \\ a_{31} & a_{32} & a_{33} \end{vmatrix} = (a_{11}a_{22}a_{33} + a_{12}a_{23}a_{31} + a_{13}a_{21}a_{32}) \\ - (a_{31}a_{22}a_{13} + a_{32}a_{23}a_{11} + a_{33}a_{21}a_{12}).$$

The terms on the right side of the equation in the definition of $|A|$ can be rearranged to get

$$\begin{vmatrix} a_{11} & a_{12} & a_{13} \\ a_{21} & a_{22} & a_{23} \\ a_{31} & a_{32} & a_{33} \end{vmatrix} = \begin{aligned} a_{11}(a_{22}a_{33} - a_{32}a_{23}) - a_{21}(a_{12}a_{33} - a_{32}a_{13}) \\ + a_{31}(a_{12}a_{23} - a_{22}a_{13}). \end{aligned}$$

Each of the quantities in parentheses above represents a determinant of a 2×2 matrix which is the part of the 3×3 matrix left when the row and column of the multiplier are eliminated as shown below.

$$a_{11}(a_{22}a_{33} - a_{32}a_{23}) \qquad \begin{bmatrix} a_{11} & a_{12} & a_{13} \\ a_{21} & a_{22} & a_{23} \\ a_{31} & a_{32} & a_{33} \end{bmatrix}$$

$$a_{21}(a_{12}a_{33} - a_{32}a_{13}) \qquad \begin{bmatrix} a_{11} & a_{12} & a_{13} \\ a_{21} & a_{22} & a_{23} \\ a_{31} & a_{32} & a_{33} \end{bmatrix}$$

$$a_{31}(a_{12}a_{23} - a_{22}a_{13}) \qquad \begin{bmatrix} a_{11} & a_{12} & a_{13} \\ a_{21} & a_{22} & a_{23} \\ a_{31} & a_{32} & a_{33} \end{bmatrix}$$

These determinants of 2×2 matrices are called **minors** of an element in the 3×3 matrix. The symbol M_{ij} represents the determinant of the matrix that results when row i and column j are eliminated. The following list gives some of the minors from the matrix above.

Element	Minor	Element	Minor
a_{11}	$M_{11} = \begin{vmatrix} a_{22} & a_{23} \\ a_{32} & a_{33} \end{vmatrix}$	a_{22}	$M_{22} = \begin{vmatrix} a_{11} & a_{13} \\ a_{31} & a_{33} \end{vmatrix}$
a_{21}	$M_{21} = \begin{vmatrix} a_{12} & a_{13} \\ a_{32} & a_{33} \end{vmatrix}$	a_{23}	$M_{23} = \begin{vmatrix} a_{11} & a_{12} \\ a_{31} & a_{32} \end{vmatrix}$
a_{31}	$M_{31} = \begin{vmatrix} a_{12} & a_{13} \\ a_{22} & a_{23} \end{vmatrix}$	a_{33}	$M_{33} = \begin{vmatrix} a_{11} & a_{12} \\ a_{21} & a_{22} \end{vmatrix}$

In a 4×4 matrix, the minors are determinants of 3×3 matrices, and an $n \times n$ matrix has minors that are determinants of $(n - 1) \times (n - 1)$ matrices.

To find the determinant of a 3×3 or larger matrix, first choose any row or column. Then the minor of each element in that row or column must be multiplied by $+1$ or -1, depending on whether the sum of the row numbers and column numbers is even or odd. The product of a minor and the number $+1$ or -1 is called a *cofactor*.

Let M_{ij} be the minor for element a_{ij} in an $n \times n$ matrix. The **cofactor** of a_{ij}, written A_{ij}, is

$$A_{ij} = (-1)^{i+j} \cdot M_{ij}.$$

Finally, the determinant of a 3×3 or larger matrix is found as follows.

**Finding the
Determinant
of a Matrix**

Multiply each element in any row or column of the matrix by its cofactor. The sum of these products gives the value of the determinant.

The process of forming this sum of products is called **expansion by a given row or column.** (See Exercises 55–56.)

EXAMPLE 2   Evaluate $\begin{vmatrix} 2 & -3 & -2 \\ -1 & -4 & -3 \\ -1 & 0 & 2 \end{vmatrix}$. Expand by the second column.

To find this determinant, first get the minors of each element in the second column.

$$M_{12} = \begin{vmatrix} -1 & -3 \\ -1 & 2 \end{vmatrix} = -1(2) - (-1)(-3) = -5$$

$$M_{22} = \begin{vmatrix} 2 & -2 \\ -1 & 2 \end{vmatrix} = 2(2) - (-1)(-2) = 2$$

$$M_{32} = \begin{vmatrix} 2 & -2 \\ -1 & -3 \end{vmatrix} = 2(-3) - (-1)(-2) = -8$$

Now find the cofactor of each of these minors.

$$A_{12} = (-1)^{1+2} \cdot M_{12} = (-1)^3 \cdot (-5) = (-1)(-5) = 5$$
$$A_{22} = (-1)^{2+2} \cdot M_{22} = (-1)^4 \cdot (2) = 1 \cdot 2 = 2$$
$$A_{32} = (-1)^{3+2} \cdot M_{32} = (-1)^5 \cdot (-8) = (-1)(-8) = 8$$

The determinant is found by multiplying each cofactor by its corresponding element in the matrix and finding the sum of these products.

$$\begin{vmatrix} 2 & -3 & -2 \\ -1 & -4 & -3 \\ -1 & 0 & 2 \end{vmatrix} = a_{12} \cdot A_{12} + a_{22} \cdot A_{22} + a_{32} \cdot A_{32}$$

$$= -3(5) + (-4)(2) + (0)(8)$$
$$= -15 + (-8) + 0 = -23 \quad \bullet$$

Exactly the same answer would be found using any row or column of the matrix. One reason that column 2 was used here is that it contains a 0 element, so that it was not really necessary to calculate M_{32} and A_{32} above. One learns quickly that 0's are friends in work with determinants.

Instead of calculating $(-1)^{i+j}$ for a given element, the following sign checkerboards can be used.

Array of Signs

	For 3 × 3 matrices		For 4 × 4 matrices	
	+ − +		+ − + −	
	− + −		− + − +	
	+ − +		+ − + −	
			− + − +	

The signs alternate for each row and column, beginning with + in the first row, first column position. Thus, these arrays of signs can be reproduced as needed. If we expand a 3 × 3 matrix about row 3, for example, the first minor would have a + sign associated with it, the second minor a − sign, and the third minor a + sign. These arrays of signs can be extended in this way for determinants of 5 × 5, 6 × 6, and larger matrices.

EXAMPLE 3   Evaluate
$$\begin{vmatrix} -1 & -2 & 3 & 2 \\ 0 & 1 & 4 & -2 \\ 3 & -1 & 4 & 0 \\ 2 & 1 & 0 & 3 \end{vmatrix}.$$

Expand about the fourth row, and do the arithmetic that has been left out.

$$-2\begin{vmatrix} -2 & 3 & 2 \\ 1 & 4 & -2 \\ -1 & 4 & 0 \end{vmatrix} + 1\begin{vmatrix} -1 & 3 & 2 \\ 0 & 4 & -2 \\ 3 & 4 & 0 \end{vmatrix} - 0\begin{vmatrix} -1 & -2 & 2 \\ 0 & 1 & -2 \\ 3 & -1 & 0 \end{vmatrix} + 3\begin{vmatrix} -1 & -2 & 3 \\ 0 & 1 & 4 \\ 3 & -1 & 4 \end{vmatrix}$$
$$= -2(6) + 1(-50) - 0 + 3(-41) = -185. \quad \bullet$$

There are several theorems that make it easier to calculate determinants. The theorems are true for square matrices of any order, but they are proved here only for determinants of 3 × 3 matrices.

Determinant Theorem 1

If every element in a row (or column) of matrix A is 0, then $|A| = 0$.

To prove the theorem, expand the given matrix by the row (or column) of zeros. Each term of this expansion will have a zero factor, making the final determinant 0. For example,

$$\begin{vmatrix} a_{11} & a_{12} & a_{13} \\ 0 & 0 & 0 \\ a_{31} & a_{32} & a_{33} \end{vmatrix} = -0 \cdot \begin{vmatrix} a_{12} & a_{13} \\ a_{32} & a_{33} \end{vmatrix} + 0 \cdot \begin{vmatrix} a_{11} & a_{13} \\ a_{31} & a_{33} \end{vmatrix} - 0 \cdot \begin{vmatrix} a_{11} & a_{12} \\ a_{31} & a_{32} \end{vmatrix} = 0.$$

· · · · · · · · ·

EXAMPLE 4

$$\begin{vmatrix} -3 & 7 & 0 \\ 4 & 9 & 0 \\ -6 & 8 & 0 \end{vmatrix} = 0 \quad \bullet$$

· ·

Determinant Theorem 2

If the rows of matrix A are the corresponding columns of matrix B, then $|B| = |A|$.

To prove the theorem, let

$$A = \begin{bmatrix} a_{11} & a_{12} & a_{13} \\ a_{21} & a_{22} & a_{23} \\ a_{31} & a_{32} & a_{33} \end{bmatrix} \quad \text{and} \quad B = \begin{bmatrix} a_{11} & a_{21} & a_{31} \\ a_{12} & a_{22} & a_{32} \\ a_{13} & a_{23} & a_{33} \end{bmatrix},$$

where B was obtained by interchanging the corresponding rows and columns of A. Find $|A|$ by expansion about row 1. Then find $|B|$ by expansion about column 1. You should find that $|B| = |A|$.

· · · · · · · · ·

EXAMPLE 5

(a) Let $A = \begin{bmatrix} 2 & 1 \\ 3 & 4 \end{bmatrix}$. Interchange the rows and columns of A to get matrix B.

$$B = \begin{bmatrix} 2 & 3 \\ 1 & 4 \end{bmatrix}$$

Check that $|A| = 5$ and $|B| = 5$, so that $|B| = |A|$.

(b) By determinant theorem 2, $\begin{vmatrix} 2 & 1 & 6 \\ 3 & 0 & 5 \\ -4 & 6 & 9 \end{vmatrix} = \begin{vmatrix} 2 & 3 & -4 \\ 1 & 0 & 6 \\ 6 & 5 & 9 \end{vmatrix}$. $\quad \bullet$

We now consider how each of the matrix row operations (and corresponding column operations) affect the determinant of a matrix.

<table>
<tr><td>

Determinant Theorem 3
</td><td>

If any two rows (or columns) of matrix A are interchanged to form matrix B, then $|B| = -|A|$.
</td></tr>
</table>

This theorem is proved by steps very similar to those used to prove the previous theorem. (See Exercise 57.)

EXAMPLE 6

(a) Let $A = \begin{bmatrix} 2 & 5 \\ 3 & 4 \end{bmatrix}$. Exchange the two columns of A to get the matrix

$B = \begin{bmatrix} 5 & 2 \\ 4 & 3 \end{bmatrix}$. Check that $|A| = -7$ and $|B| = 7$, so that $|B| = -|A|$.

(b) By determinant theorem 3, $\begin{vmatrix} 2 & 1 & 6 \\ 3 & 0 & 5 \\ -4 & 6 & 9 \end{vmatrix} = - \begin{vmatrix} -4 & 6 & 9 \\ 3 & 0 & 5 \\ 2 & 1 & 6 \end{vmatrix}$. ●

<table>
<tr><td>

Determinant Theorem 4
</td><td>

Suppose matrix B is formed by multiplying every element of a row (or column) of matrix A by the real number k. Then $|B| = k \cdot |A|$.
</td></tr>
</table>

The proof of this theorem is left for Exercise 58.

EXAMPLE 7   Let $A = \begin{bmatrix} 2 & -3 \\ 4 & 1 \end{bmatrix}$. Form the new matrix B by multiplying each element of the second row of A by -5.

$$B = \begin{bmatrix} 2 & -3 \\ 4(-5) & 1(-5) \end{bmatrix} = \begin{bmatrix} 2 & -3 \\ -20 & -5 \end{bmatrix}$$

Check that $|B| = -5 \cdot |A|$. ●

<table>
<tr><td>

Determinant Theorem 5
</td><td>

If two rows (or columns) of a matrix A are identical, then $|A| = 0$.
</td></tr>
</table>

To prove this theorem, note that if two rows or columns of a matrix A are interchanged to form matrix B, then $|A| = -|B|$; while if two rows of matrix A are identical and are interchanged, we still have matrix A. But then $|A| = -|A|$, which can only happen if $|A| = 0$.

. . . .

EXAMPLE 8 Since two rows are identical, $\begin{vmatrix} -4 & 2 & 3 \\ 0 & 1 & 6 \\ -4 & 2 & 3 \end{vmatrix} = 0.$ ●

The last theorem of this section is perhaps the most useful of all.

. . . .

**Determinant
Theorem 6**

Adding a constant times one row (or column) of a matrix to another row (or column) does not change the determinant of the matrix.

This theorem is proved in much the same way as the others in this section. (See Exercises 59–60 below.) It provides a powerful method for simplifying the work of finding the determinant of a 3 × 3 or larger matrix, as shown in the next example.

. . . .

EXAMPLE 9 Let $A = \begin{bmatrix} -2 & 4 & 1 \\ 2 & 1 & 5 \\ 4 & 0 & 2 \end{bmatrix}$. Find $|A|$.

First obtain a new matrix B (using theorem 6) by adding row 1 to row 2 and then (using theorem 6 again) adding 2 times row 1 to row 3.

$$B = \begin{bmatrix} -2 & 4 & 1 \\ 0 & 5 & 6 \\ 0 & 8 & 4 \end{bmatrix} \quad \begin{matrix} R_1 + R_2 \\ 2R_1 + R_3 \end{matrix}$$

Now find $|B|$ by expanding about the first column.

$$|B| = -2 \begin{vmatrix} 5 & 6 \\ 8 & 4 \end{vmatrix} = -2(20 - 48) = 56$$

By the theorem above, $|B| = |A|$ so $|A| = 56.$ ●

The following examples show how the properties of determinants are used to simplify the calculation of determinants.

. . . .

EXAMPLE 10 Without expanding, show that the value of the following determinant is 0.

$$\begin{vmatrix} 2 & 5 & -1 \\ 1 & -15 & 3 \\ -2 & 10 & -2 \end{vmatrix}$$

Examining the columns of the array shows that each element in the second column is -5 times the corresponding element in the third column. By theorem 6, add to the elements of the second column the results of multiplying the elements of the third column by 5 (abbreviated below as $5C_3 + C_2$), to get the determinant

$$\begin{vmatrix} 2 & 0 & -1 \\ 1 & 0 & 3 \\ -2 & 0 & -2 \end{vmatrix}. \qquad 5C_3 + C_2$$

The value of this determinant is 0, by theorem 1. ●

. . . .

EXAMPLE 11   Find $|A| = \begin{vmatrix} 4 & 2 & 1 & 0 \\ -2 & 4 & -1 & 7 \\ -5 & 2 & 3 & 1 \\ 6 & 4 & -3 & 2 \end{vmatrix}$

The goal is to change row 1 of the matrix (any row or column could be selected) to a row in which every element but one is 0. To begin, multiply the elements of column 2 of the matrix by -2 and add the results to the elements of column 1.

$$\begin{vmatrix} 0 & 2 & 1 & 0 \\ -10 & 4 & -1 & 7 \\ -9 & 2 & 3 & 1 \\ -2 & 4 & -3 & 2 \end{vmatrix} \qquad -2C_2 + C_1$$

Add to the elements of column 2 of the matrix the results of multiplying the elements of column 3 by -2.

$$\begin{vmatrix} 0 & 0 & 1 & 0 \\ -10 & 6 & -1 & 7 \\ -9 & -4 & 3 & 1 \\ -2 & 10 & -3 & 2 \end{vmatrix} \qquad -2C_3 + C_2$$

Row 1 of the matrix has only one nonzero number, so expand about the first row.

$$|A| = +1 \begin{vmatrix} -10 & 6 & 7 \\ -9 & -4 & 1 \\ -2 & 10 & 2 \end{vmatrix}$$

Now change column 3 of the matrix to a column with two zeros.

$$\begin{vmatrix} 53 & 34 & 0 \\ -9 & -4 & 1 \\ -2 & 10 & 2 \end{vmatrix} \qquad -7R_2 + R_1$$

$$\begin{vmatrix} 53 & 34 & 0 \\ -9 & -4 & 1 \\ 16 & 18 & 0 \end{vmatrix} \qquad -2R_2 + R_3$$

Finally, expand about column 3 of the matrix to find the value of $|A|$.

$$|A| = -1 \begin{vmatrix} 53 & 34 \\ 16 & 18 \end{vmatrix} = -1(954 - 544) = -410 \quad ●$$

In Example 11, working with *rows* of the matrix led to a *column* with only one nonzero number and working with *columns* of the matrix led to a *row* with one nonzero number.

9.6 Exercises

Find the value of each determinant. All variables represent real numbers.

1. $\begin{vmatrix} 3 & 4 \\ 5 & -2 \end{vmatrix}$

2. $\begin{vmatrix} -9 & 7 \\ 2 & 6 \end{vmatrix}$

3. $\begin{vmatrix} 0 & 4 \\ 4 & 0 \end{vmatrix}$

4. $\begin{vmatrix} 1 & 0 \\ 0 & 2 \end{vmatrix}$

5. $\begin{vmatrix} y & 2 \\ 8 & y \end{vmatrix}$

6. $\begin{vmatrix} 3 & 8 \\ m & n \end{vmatrix}$

7. $\begin{vmatrix} x & y \\ y & x \end{vmatrix}$

8. $\begin{vmatrix} 2m & 8n \\ 8n & 2m \end{vmatrix}$

Find the cofactor of each element in the second row for the following determinants.

9. $\begin{vmatrix} -2 & 0 & 1 \\ 3 & 2 & -1 \\ 1 & 0 & 2 \end{vmatrix}$

10. $\begin{vmatrix} 0 & -1 & 2 \\ 1 & 0 & 2 \\ 0 & -3 & 1 \end{vmatrix}$

11. $\begin{vmatrix} 1 & 2 & -1 \\ 2 & 3 & -2 \\ -1 & 4 & 1 \end{vmatrix}$

12. $\begin{vmatrix} 2 & -1 & 4 \\ 3 & 0 & 1 \\ -2 & 1 & 4 \end{vmatrix}$

Find the value of each determinant. All variables represent real numbers.

13. $\begin{vmatrix} 1 & 0 & 0 \\ 0 & -1 & 0 \\ 1 & 0 & 1 \end{vmatrix}$

14. $\begin{vmatrix} -2 & 0 & 1 \\ 0 & 1 & 0 \\ 0 & 0 & -1 \end{vmatrix}$

15. $\begin{vmatrix} 1 & -2 & 3 \\ 0 & 0 & 0 \\ 1 & 10 & -12 \end{vmatrix}$

16. $\begin{vmatrix} 2 & 3 & 0 \\ 1 & 9 & 0 \\ -1 & -2 & 0 \end{vmatrix}$

17. $\begin{vmatrix} 3 & 3 & -1 \\ 2 & 6 & 0 \\ -6 & -6 & 2 \end{vmatrix}$

18. $\begin{vmatrix} 5 & -3 & 2 \\ -5 & 3 & -2 \\ 1 & 0 & 1 \end{vmatrix}$

19. $\begin{vmatrix} 3 & 2 & 0 \\ 0 & 1 & x \\ 2 & 0 & 0 \end{vmatrix}$

20. $\begin{vmatrix} 0 & 3 & y \\ 0 & 4 & 2 \\ 1 & 0 & 1 \end{vmatrix}$

21. $\begin{vmatrix} i & j & k \\ -1 & 2 & 4 \\ 3 & 0 & 5 \end{vmatrix}$

22. $\begin{vmatrix} i & j & k \\ 0 & -4 & 2 \\ -1 & 3 & 1 \end{vmatrix}$

23. $\begin{vmatrix} 4 & 0 & 0 & 2 \\ -1 & 0 & 3 & 0 \\ 2 & 4 & 0 & 1 \\ 0 & 0 & 1 & 2 \end{vmatrix}$

24. $\begin{vmatrix} -2 & 0 & 4 & 2 \\ 3 & 6 & 0 & 4 \\ 0 & 0 & 0 & 3 \\ 9 & 0 & 2 & -1 \end{vmatrix}$

25. $\begin{vmatrix} 0.4 & -0.8 & 0.6 \\ 0.3 & 0.9 & 0.7 \\ 3.1 & 4.1 & -2.8 \end{vmatrix}$

26. $\begin{vmatrix} -0.3 & -0.1 & 0.9 \\ 2.5 & 4.9 & -3.2 \\ -0.1 & 0.4 & 0.8 \end{vmatrix}$

Tell why each determinant has a value of 0. All variables represent real numbers.

27. $\begin{vmatrix} 2 & 3 \\ 2 & 3 \end{vmatrix}$

28. $\begin{vmatrix} -5 & -5 \\ 6 & 6 \end{vmatrix}$

29. $\begin{vmatrix} 2 & 0 \\ 3 & 0 \end{vmatrix}$

30. $\begin{vmatrix} -8 & 0 \\ -6 & 0 \end{vmatrix}$

31. $\begin{vmatrix} -1 & 2 & 4 \\ 4 & -8 & -16 \\ 3 & 0 & 5 \end{vmatrix}$

32. $\begin{vmatrix} 1 & 0 & 0 \\ 1 & 0 & 1 \\ 3 & 0 & 0 \end{vmatrix}$

33. $\begin{vmatrix} m & 2 & 2m \\ 3n & 1 & 6n \\ 5p & 6 & 10p \end{vmatrix}$

34. $\begin{vmatrix} 7z & 8x & 2y \\ z & x & y \\ 7z & 7x & 7y \end{vmatrix}$

Use the appropriate theorems from this section to tell why each statement is true. Do not evaluate the determinants. All variables represent real numbers.

35. $\begin{vmatrix} 2 & 1 & 6 \\ 3 & 0 & 2 \\ 4 & 1 & 8 \end{vmatrix} = \begin{vmatrix} 2 & 3 & 4 \\ 1 & 0 & 1 \\ 6 & 2 & 8 \end{vmatrix}$

36. $\begin{vmatrix} 4 & -2 \\ 3 & 8 \end{vmatrix} = \begin{vmatrix} 4 & 3 \\ -2 & 8 \end{vmatrix}$

37. $\begin{vmatrix} 2 & 6 \\ 3 & 5 \end{vmatrix} = -\begin{vmatrix} 3 & 5 \\ 2 & 6 \end{vmatrix}$

38. $\begin{vmatrix} -1 & 8 & 9 \\ 0 & 2 & 1 \\ 3 & 2 & 0 \end{vmatrix} = -\begin{vmatrix} 8 & -1 & 9 \\ 2 & 0 & 1 \\ 2 & 3 & 0 \end{vmatrix}$

39. $3\begin{vmatrix} 6 & 0 & 2 \\ 4 & 1 & 3 \\ 2 & 8 & 6 \end{vmatrix} = \begin{vmatrix} 6 & 0 & 2 \\ 4 & 3 & 3 \\ 2 & 24 & 6 \end{vmatrix}$

40. $-\frac{1}{2}\begin{vmatrix} 5 & -8 & 2 \\ 3 & -6 & 9 \\ 2 & 4 & 4 \end{vmatrix} = \begin{vmatrix} 5 & 4 & 2 \\ 3 & 3 & 9 \\ 2 & -2 & 4 \end{vmatrix}$

41. $\begin{vmatrix} 3 & -4 \\ 2 & 5 \end{vmatrix} = \begin{vmatrix} 3 & -4 \\ 5 & 1 \end{vmatrix}$

42. $\begin{vmatrix} -1 & 6 \\ 3 & -5 \end{vmatrix} = \begin{vmatrix} -1 & 6 \\ 2 & 1 \end{vmatrix}$

43. $2\begin{vmatrix} 4 & 2 & -1 \\ m & 2n & 3p \\ 5 & 1 & 0 \end{vmatrix} = \begin{vmatrix} 4 & 2 & -1 \\ 2m & 4n & 6p \\ 5 & 1 & 0 \end{vmatrix}$

44. $\begin{vmatrix} -4 & 2 & 1 \\ 3 & 0 & 5 \\ -1 & 4 & -2 \end{vmatrix} = \begin{vmatrix} -4 & 2 & 1 + (-4)k \\ 3 & 0 & 5 + 3k \\ -1 & 4 & -2 + (-1)k \end{vmatrix}$

Use the method of Examples 10 and 11 to find the value of each determinant.

45. $\begin{vmatrix} 2 & 4 \\ 3 & 6 \end{vmatrix}$

46. $\begin{vmatrix} -5 & 10 \\ 6 & -12 \end{vmatrix}$

47. $\begin{vmatrix} 4 & 8 & 0 \\ -1 & -2 & 1 \\ 2 & 4 & 3 \end{vmatrix}$

48. $\begin{vmatrix} 6 & 8 & -12 \\ -1 & 0 & 2 \\ 4 & 0 & -8 \end{vmatrix}$

49. $\begin{vmatrix} 3 & 1 & 2 \\ 2 & 0 & 1 \\ 1 & 0 & -2 \end{vmatrix}$

50. $\begin{vmatrix} -2 & 2 & 3 \\ 0 & 2 & 1 \\ -1 & 4 & 0 \end{vmatrix}$

51. $\begin{vmatrix} -4 & 2 & 3 \\ 2 & 0 & 1 \\ 0 & 4 & 2 \end{vmatrix}$

52. $\begin{vmatrix} 6 & 3 & 2 \\ 1 & 0 & 2 \\ -1 & 4 & 1 \end{vmatrix}$

53. $\begin{vmatrix} 1 & 0 & 2 & 2 \\ 2 & 4 & 1 & -1 \\ 1 & -3 & 1 & 0 \\ 1 & 1 & 0 & 1 \end{vmatrix}$

54. $\begin{vmatrix} 2 & -1 & 1 & 0 \\ 1 & 1 & 0 & 1 \\ 0 & -1 & 1 & 1 \\ 1 & 2 & 1 & 2 \end{vmatrix}$

Let $A = \begin{bmatrix} a_{11} & a_{12} & a_{13} \\ a_{21} & a_{22} & a_{23} \\ a_{31} & a_{32} & a_{33} \end{bmatrix}$ *for Exercises 55–60.*

55. Find $|A|$ by expansion about row 3 of the matrix. Show that your result is really equal to $|A|$ as given in the definition of the determinant of a 3×3 matrix.

56. Repeat Exercise 55 for column 3.

57. Obtain matrix B by exchanging columns 1 and 3 of matrix A. Show that $|A| = -|B|$.

58. Obtain matrix B by multiplying each element of row 3 of A by the real number k. Show that $|B| = k \cdot |A|$.

59. Obtain matrix B by adding to column 1 of matrix A the result of multiplying each element of column 2 of A by the real number k. Show that $|A| = |B|$.

60. Obtain matrix B by adding to row 1 of matrix A the result of multiplying each element of row 3 of A by the real number k. Show that $|A| = |B|$.

61. Let A and B be any 2×2 matrices. Show that $|AB| = |A| \cdot |B|$, where $|AB|$ is the determinant of matrix AB.

62. Let A be an $n \times n$ matrix. Suppose matrix B is found by multiplying every element of A by the real number k. Express $|B|$ in terms of $|A|$.

Determinants can be used to find the area of a triangle, given the coordinates of its vertices. Given a triangle PQR with vertices (x_1, y_1), (x_2, y_2), and (x_3, y_3), as in the figure, it can be shown that the area of the triangle is given by A, where

$$A = \frac{1}{2} \begin{vmatrix} x_1 & y_1 & 1 \\ x_2 & y_2 & 1 \\ x_3 & y_3 & 1 \end{vmatrix}.$$

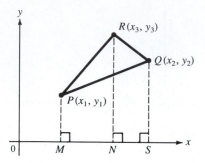

The points (x_1, y_1), (x_2, y_2), (x_3, y_3) must be taken in counterclockwise order; if this is not done then A may have the wrong sign. Alternatively, we could define A as the absolute value of 1/2 the determinant shown above. Use the formula given to find the area of the triangles in Exercises 63–66.

63. $P(0, 1)$, $Q(2, 0)$, $R(1, 3)$

64. $P(2, 5)$, $Q(-1, 3)$, $R(4, 0)$

65. $P(2, -2)$, $Q(0, 0)$, $R(-3, -4)$

66. $P(4, 7)$, $Q(5, -2)$, $R(1, 1)$

67. Prove that the straight line through the distinct points (x_1, y_1) and (x_2, y_2) has equation

$$\begin{vmatrix} x & y & 1 \\ x_1 & y_1 & 1 \\ x_2 & y_2 & 1 \end{vmatrix} = 0.$$

68. Use the result of Exercise 67 to show that three distinct points (x_1, y_1), (x_2, y_2) and (x_3, y_3) lie on a straight line if

$$\begin{vmatrix} x_1 & y_1 & 1 \\ x_2 & y_2 & 1 \\ x_3 & y_3 & 1 \end{vmatrix} = 0.$$

69. Show that the lines $a_1x + b_1y = c_1$ and $a_2x + b_2y = c_2$, when $c_1 \neq c_2$, are parallel if

$$\begin{vmatrix} a_1 & b_1 \\ a_2 & b_2 \end{vmatrix} = 0.$$

70. Prove that

$$\begin{vmatrix} 1 & 1 & 1 \\ a & b & c \\ a^2 & b^2 & c^2 \end{vmatrix} = (a - b)(b - c)(c - a).$$

9.7 CRAMER'S RULE

We have now seen how to solve a system of n linear equations with n variables using the following methods: elimination, substitution, and row transformations of matrices. Most of these systems can also be solved with determinants, as shown here.

To see how determinants arise in solving a system, write the linear system

$$a_{11}x + a_{12}y = b_1$$

$$a_{21}x + a_{22}y = b_2,$$

where each equation has at least one nonzero coefficient. Solve this system using matrix methods; begin by writing the augmented matrix

$$\left[\begin{array}{cc|c} a_{11} & a_{12} & b_1 \\ a_{21} & a_{22} & b_2 \end{array} \right].$$

Multiply each element of row 1 by $1/a_{11}$. (Here we assume $a_{11} \neq 0$.) This gives the matrix of an equivalent system:

$$\left[\begin{array}{cc|c} 1 & \dfrac{a_{12}}{a_{11}} & \dfrac{b_1}{a_{11}} \\ a_{21} & a_{22} & b_2 \end{array} \right]. \qquad \dfrac{1}{a_{11}} R_1$$

Multiply each element of row 1 by $-a_{21}$, and add the results to the corresponding element of row 2.

$$\left[\begin{array}{cc|c} 1 & \dfrac{a_{12}}{a_{11}} & \dfrac{b_1}{a_{11}} \\[2ex] 0 & a_{22} - \dfrac{a_{21}a_{12}}{a_{11}} & b_2 - \dfrac{a_{21}b_1}{a_{11}} \end{array}\right] \qquad -a_{21}R_1 + R_2$$

Multiply each element of row 2 by a_{11}.

$$\left[\begin{array}{cc|c} 1 & \dfrac{a_{12}}{a_{11}} & \dfrac{b_1}{a_{11}} \\[2ex] 0 & a_{11}a_{22} - a_{21}a_{12} & a_{11}b_2 - a_{21}b_1 \end{array}\right] \qquad a_{11}\,R_2$$

This matrix leads to the system of equations

$$x + \frac{a_{12}}{a_{11}}y = \frac{b_1}{a_{11}}$$

$$(a_{11}a_{22} - a_{21}a_{12})y = a_{11}b_2 - a_{21}b_1.$$

From the second equation of this system,

$$y = \frac{a_{11}b_2 - a_{21}b_1}{a_{11}a_{22} - a_{21}a_{12}}.$$

Both the numerator and denominator here may be written as determinants:

$$y = \frac{\begin{vmatrix} a_{11} & b_1 \\ a_{21} & b_2 \end{vmatrix}}{\begin{vmatrix} a_{11} & a_{12} \\ a_{21} & a_{22} \end{vmatrix}}. \tag{1}$$

Inserting this value of y into the first equation above shows that x can also be written with determinants as

$$x = \frac{\begin{vmatrix} b_1 & a_{12} \\ b_2 & a_{22} \end{vmatrix}}{\begin{vmatrix} a_{11} & a_{12} \\ a_{21} & a_{22} \end{vmatrix}}. \tag{2}$$

The denominator for finding both x and y is just the determinant of the matrix of coefficients of the original system. This determinant is often denoted D, so that

$$D = \begin{vmatrix} a_{11} & a_{12} \\ a_{21} & a_{22} \end{vmatrix}.$$

In equation (1), the numerator is the determinant of a matrix obtained by replacing the coefficients of y in D with the respective constants: D_y is defined as

$$D_y = \begin{vmatrix} a_{11} & b_1 \\ a_{21} & b_2 \end{vmatrix}.$$

In the same way, from equation (2), D_x is defined as

$$D_x = \begin{vmatrix} b_1 & a_{12} \\ b_2 & a_{22} \end{vmatrix}.$$

With this notation, the solution of the given system is

$$x = \frac{D_x}{D} \quad \text{and} \quad y = \frac{D_y}{D}.$$

The system has a single solution as long as $D \neq 0$. We have now proved much of the next theorem, called **Cramer's rule.**

Cramer's Rule for 2 Equations in 2 Variables

Given the system

$$a_{11}x + a_{12}y = b_1$$
$$a_{21}x + a_{22}y = b_2,$$

if $D \neq 0$, the system has the unique solution

$$x = \frac{D_x}{D} \quad \text{and} \quad y = \frac{D_y}{D}.$$

Cramer was looking for a method to determine the equation of a curve when he knew several points on the curve. In 1750 he wrote down the general equation for a curve and then substituted each point for which he had two coordinates into the equation. For this system of equations he gave ''a rule very convenient and general to solve any number of equations and unknowns which are of no more than first degree.'' This is the rule that now bears his name.

EXAMPLE 1   Use Cramer's rule to solve the system $5x + 7y = -1$
$$6x + 8y = 1.$$

To use Cramer's rule, first evaluate D, D_x, and D_y.

$$D = \begin{vmatrix} 5 & 7 \\ 6 & 8 \end{vmatrix} = 5(8) - 6(7) = -2.$$

$$D_x = \begin{vmatrix} -1 & 7 \\ 1 & 8 \end{vmatrix} = (-1)(8) - (1)(7) = -15.$$

$$D_y = \begin{vmatrix} 5 & -1 \\ 6 & 1 \end{vmatrix} = 5(1) - (6)(-1) = 11.$$

By Cramer's rule, $x = -15/(-2) = 15/2$, and $y = 11/(-2) = -11/2$. The solution is $(15/2, -11/2)$, as can be verified by substituting within the given system. ●

By much the same method as used above, Cramer's rule can be generalized to a system of n linear equations with n variables.

**General Form
of Cramer's
Rule**

Let an $n \times n$ system have linear equations of the form

$$a_{11}x_1 + a_{12}x_2 + a_{13}x_3 + \ldots + a_{1n}x_n = b_1.$$

Define D as the determinant of the $n \times n$ matrix of all coefficients of the variables. Define D_{x1} as the determinant obtained from D by replacing the entries in column 1 of D with the constants of the system. Define D_{xi} as the determinant obtained from D by replacing the entries in column i with the constants of the system. If $D \neq 0$, the unique solution of the system is

$$x_1 = \frac{D_{x1}}{D}, \; x_2 = \frac{D_{x2}}{D}, \; x_3 = \frac{D_{x3}}{D}, \; \ldots, \; x_n = \frac{D_{xn}}{D}.$$

EXAMPLE 2   Use Cramer's rule to solve the system

$$\begin{aligned}
x + y - z + 2 &= 0 \\
2x - y + z + 5 &= 0 \\
x - 2y + 3z - 4 &= 0.
\end{aligned}$$

To use Cramer's rule, the system must be rewritten in the form

$$\begin{aligned}
x + y - z &= -2 \\
2x - y + z &= -5 \\
x - 2y + 3z &= 4.
\end{aligned}$$

The determinant of coefficients, D, is

$$D = \begin{vmatrix} 1 & 1 & -1 \\ 2 & -1 & 1 \\ 1 & -2 & 3 \end{vmatrix}.$$

To find D_x, replace the elements of the first column of D with the constants of the system. Find D_y and D_z in a similar way.

$$D_x = \begin{vmatrix} -2 & 1 & -1 \\ -5 & -1 & 1 \\ 4 & -2 & 3 \end{vmatrix}, \quad D_y = \begin{vmatrix} 1 & -2 & -1 \\ 2 & -5 & 1 \\ 1 & 4 & 3 \end{vmatrix}, \quad D_z = \begin{vmatrix} 1 & 1 & -2 \\ 2 & -1 & -5 \\ 1 & -2 & 4 \end{vmatrix}$$

Verify that $D = -3$, $D_x = 7$, $D_y = -22$, and $D_z = -21$. Then, by Cramer's rule,

$$x = \frac{D_x}{D} = \frac{7}{-3} = -\frac{7}{3}, \quad y = \frac{D_y}{D} = \frac{-22}{-3} = \frac{22}{3}, \quad \text{and } z = \frac{D_z}{D} = \frac{-21}{-3} = 7.$$

The solution of the system is $(-7/3, 22/3, 7)$. ●

⚠ CAUTION   As shown in Example 2, each equation in the system must be written in the form $a_1x_1 + a_2x_2 + \ldots + a_nx_n = k$ before using Cramer's rule.

· · · · · · · · ·

EXAMPLE 3   Use Cramer's rule to solve the system

$$2x - 3y + 4z = 10$$
$$6x - 9y + 12z = 24$$
$$x + 2y - 3z = 5.$$

Verify that $D = 0$, so Cramer's rule does not apply. Use another method to determine that this system is inconsistent.   ●

Several different methods for solving systems of equations have now been shown. In general, if a small system of linear equations must be solved by pencil and paper, substitution is the best method if the various variables can easily be found in terms of each other. This happens rarely. The next choice, perhaps the best choice of all, is the elimination method. Some people like the Gaussian reduction method, which is really just a systematic way of doing the elimination method. The Gaussian reduction method is probably superior where four or more equations are involved. Cramer's rule is seldom the method of choice simply because it involves more calculations than any other method. Cramer's rule, however, like the Gaussian method, is a fast and efficient computer method for large systems.

9.7 Exercises ·

Use Cramer's rule to solve each of the following systems of linear equations.

1. $x + y = 4$
$2x - y = 2$

2. $3x + 2y = -4$
$2x - y = -5$

3. $4x + 3y = -7$
$2x + 3y = -11$

4. $3x + 2y = -4$
$5x - y = 2$

5. $2x - 3y = -5$
$x + 5y = 17$

6. $x + 9y = -15$
$3x + 2y = 5$

7. $5x + 2y = 7$
$6x + y = 8$

8. $7x + 3y = 5$
$2x + 4y = 3$

9. $10x - 8y = 1$
$-15x + 12y = 4$

10. $8x - 6y = 10$
$20x - 15y = 6$

11. $4x - y + 3z = -3$
$3x + y + z = 0$
$2x - y + 4z = 0$

12. $5x + 2y + z = 15$
$2x - y + z = 9$
$4x + 3y + 2z = 13$

13. $2x - y + 4z = -2$
$3x + 2y - z = -3$
$x + 4y + 2z = 17$

14. $x + y + z = 4$
$2x - y + 3z = 4$
$4x + 2y - z = -15$

15. $4x - 3y + z = -1$
$5x + 7y + 2z = -2$
$3x - 5y - z = 1$

16. $2x - 3y + z = 8$
$-x - 5y + z = -4$
$3x - 5y + 2z = 12$

17. $x + 2y + 3z = 4$
$4x + 3y + 2z = 1$
$-x - 2y - 3z = 0$

18. $2x - y + 3z = 1$
$-2x + y - 3z = 2$
$5x - y + z = 2$

19. $x - 2y + 3z = 4$
$5x + 7y - z = 2$
$2x + 2y - 5z = 3$

20. $-3x - 2y - z = 4$
$4x + y + z = 5$
$3x - 2y + 2z = 1$

21. $2x + 3y \quad\quad = 13$
$\quad\quad 2y - z = 5$
$\quad x \quad\quad + 2z = 4$

22. $3x \quad\quad - z = -10$
$\quad\quad y + 4z = 8$
$\quad x \quad\quad + 2z = -1$

23. $5x - y \quad\quad = -4$
$\quad 3x \quad\quad + 2z = 4$
$\quad\quad 4y + 3z = 22$

24. $3x + 5y \quad\quad = -7$
$\quad 2x \quad\quad + 7z = 2$
$\quad\quad 4y + 3z = -8$

25. $x + 2y \quad\quad = 10$
$\quad 3x \quad\quad + 4z = 7$
$\quad\quad - y - z = 1$

26. $5x - 2y \quad\quad = 3$
$\quad\quad 4y + z = 8$
$\quad x \quad\quad + 2z = 4$

27. $x \quad\quad + z \quad\quad = 0$
$\quad\quad y + 2z + w = 0$
$\quad 2x \quad\quad - w = 0$
$\quad x + 2y + 3z \quad\quad = -2$

28. $x - y + z + w = 6$
$\quad\quad 2y \quad\quad - w = -7$
$\quad x \quad\quad - z \quad\quad = -1$
$\quad\quad y \quad\quad + w = 1$

Use Cramer's rule to solve each of the following.

29. Paying $96, Mari Hall bought 5 shirts and 2 pairs of pants. Later, paying $66, she bought one more shirt and 3 more pairs of pants. Find the cost of a shirt and the cost of a pair of pants.

30. Bhiku Patel received $50,000 from the sale of a business. He invested part of the money at 16% and the rest at 18%, which resulted in his earning $8550 per year in interest. Find the amount invested at each rate.

31. A gold merchant has some 12-carat gold (12/24 pure gold), and some 22-carat gold (22/24 pure). How many grams of each should be mixed to get 25 grams of 15-carat gold?

32. A chemist has some 40% acid solution and some 60% solution. How many liters of each should be used to get 40 liters of a 45% solution?

33. How many pounds of tea worth $4.60 a pound should be mixed with tea worth $6.50 a pound to get 8 pounds of a mixture worth $5.20 a pound?

34. A solution of a drug with a strength of 5% is to be mixed with some of a 15% solution to get 15 ml of an 8% solution. How many ml of each solution should be used?

35. The cashier at an amusement park has a total of $2480, made up of fives, tens, and twenties. The total number of bills is 290, and the value of the tens is $60 more than the value of the twenties. How many of each type of bill does the cashier have?

36. Ms. Levy invests $50,000 three ways—at 8%, 8 1/2%, and 11%. In total, she receives $4436.25 per year in interest. The interest from the 11% investment is $80 more than the interest on the 8% investment. Find the amount she has invested at each rate.

For the following two exercises, use the system of equations

$$a_1x + b_1y = c_1$$
$$a_2x + b_2y = c_2.$$

37. Assume $D_x = 0$ and $D_y = 0$. Show that if $c_1c_2 \neq 0$, then $D = 0$, and the equations are dependent.

38. Assume $D = 0$, $D_x = 0$, and $b_1b_2 \neq 0$. Show that $D_y = 0$.

Use Cramer's rule to solve each system for (x, y). Assume all variables represent real numbers and all denominators are nonzero.

39. $ax + by = a^2$
$\quad\ bx + ay = b^2$

40. $ax + by = \dfrac{1}{a}$
$\quad\ \ bx + ay = \dfrac{1}{b}$

41. $\dfrac{x}{2} - \dfrac{y}{b} = 1$
$\quad\ 2x + by = 4b$

42. $a^2x + b^2y = a^2$
$\quad\ \ bx + ay = a$

9.8 MATRIX INVERSES

As shown in the exercises for an earlier section, the commutative, associative, closure, identity, and inverse properties hold for *addition* of matrices of the same size. Also, the associative and closure properties hold for *multiplication* of square matrices of the same size. There is no commutative property for multiplication, but the distributive property is valid for matrices of the proper size.

In this section two additional properties are discussed, the identity and inverse properties for multiplication of certain matrices. For the identity property to hold, there must be a matrix I such that

$$AI = A \quad \text{and} \quad IA = A$$

for any square matrix A. (Compare these products to the statement of the identity property for real numbers: $a \cdot 1 = a$ and $1 \cdot a = a$ for any real number a.)

2 × 2 Identity

If I_2 represents the 2×2 identity, then

$$I_2 = \begin{bmatrix} 1 & 0 \\ 0 & 1 \end{bmatrix}.$$

To verify that I_2 is the 2×2 identity matrix, show that $AI = A$ and $IA = A$ for any 2×2 matrix. Let

$$A = \begin{bmatrix} x & y \\ z & w \end{bmatrix}.$$

Then
$$AI = \begin{bmatrix} x & y \\ z & w \end{bmatrix}\begin{bmatrix} 1 & 0 \\ 0 & 1 \end{bmatrix} = \begin{bmatrix} x \cdot 1 + y \cdot 0 & x \cdot 0 + y \cdot 1 \\ z \cdot 1 + w \cdot 0 & z \cdot 0 + w \cdot 1 \end{bmatrix} = \begin{bmatrix} x & y \\ z & w \end{bmatrix} = A;$$

and
$$IA = \begin{bmatrix} 1 & 0 \\ 0 & 1 \end{bmatrix}\begin{bmatrix} x & y \\ z & w \end{bmatrix} = \begin{bmatrix} 1 \cdot x + 0 \cdot z & 1 \cdot y + 0 \cdot w \\ 0 \cdot x + 1 \cdot z & 0 \cdot y + 1 \cdot w \end{bmatrix} = \begin{bmatrix} x & y \\ z & w \end{bmatrix} = A.$$

Generalizing from this example, there is an $n \times n$ identity matrix having 1's on the main diagonal and 0's elsewhere.

$n \times n$ Identity Matrix

The $n \times n$ identity matrix is given by I_n where

$$I_n = \begin{bmatrix} 1 & 0 & \cdots & 0 \\ 0 & 1 & \cdots & 0 \\ \vdots & & a_{ij} & \vdots \\ 0 & 0 & \cdots & 1 \end{bmatrix}.$$

The element $a_{ij} = 1$ when $i = j$ (the diagonal elements) and $a_{ij} = 0$ otherwise.

For every nonzero real number a, there is a multiplicative inverse $1/a$ such that

$$a \cdot \frac{1}{a} = 1 \quad \text{and} \quad \frac{1}{a} \cdot a = 1.$$

(Recall: $1/a$ is also written a^{-1}.) In a similar way, if A is an $n \times n$ matrix, then its **multiplicative inverse,** written A^{-1}, must satisfy both

$$AA^{-1} = I_n \quad \text{and} \quad A^{-1}A = I_n.$$

This means that only a square matrix can have a multiplicative inverse.

CAUTION Although $a^{-1} = 1/a$ for any nonzero real number a, if A is a matrix,

$$A^{-1} \neq \frac{1}{A}.$$

In fact, $1/A$ has no meaning, since 1 is a *number* and A is a *matrix*.

The matrix A^{-1} can be found by using the row operations introduced earlier in this chapter. As an example, let us find the inverse of

$$A = \begin{bmatrix} 2 & 4 \\ 1 & -1 \end{bmatrix}.$$

Let the unknown inverse matrix be

$$A^{-1} = \begin{bmatrix} x & y \\ z & w \end{bmatrix}.$$

By the definition of matrix inverse, $AA^{-1} = I_2$, or

$$AA^{-1} = \begin{bmatrix} 2 & 4 \\ 1 & -1 \end{bmatrix} \begin{bmatrix} x & y \\ z & w \end{bmatrix} = \begin{bmatrix} 1 & 0 \\ 0 & 1 \end{bmatrix}.$$

By matrix multiplication,

$$\begin{bmatrix} 2x + 4z & 2y + 4w \\ x - z & y - w \end{bmatrix} = \begin{bmatrix} 1 & 0 \\ 0 & 1 \end{bmatrix}.$$

Setting corresponding elements equal gives the system of equations

$$2x + 4z = 1 \qquad (1)$$
$$2y + 4w = 0 \qquad (2)$$
$$x - z = 0 \qquad (3)$$
$$y - w = 1. \qquad (4)$$

Since equations (1) and (3) involve only x and z, while equations (2) and (4) involve only y and w, these four equations lead to two systems of equations,

$$\begin{matrix} 2x + 4z = 1 \\ x - z = 0 \end{matrix} \quad \text{and} \quad \begin{matrix} 2y + 4w = 0 \\ y - w = 1. \end{matrix}$$

Writing the two systems as augmented matrices gives

$$\begin{bmatrix} 2 & 4 & | & 1 \\ 1 & -1 & | & 0 \end{bmatrix} \quad \text{and} \quad \begin{bmatrix} 2 & 4 & | & 0 \\ 1 & -1 & | & 1 \end{bmatrix}.$$

Each of these systems can be solved by the Gaussian method. However, since the elements to the left of the vertical bar are identical, the two systems can be combined into one matrix

$$\begin{bmatrix} 2 & 4 & | & 1 & 0 \\ 1 & -1 & | & 0 & 1 \end{bmatrix}$$

and solved simultaneously using matrix row transformations. We need to change the numbers on the left of the vertical bar to the 2×2 identity matrix. Start by exchanging the two rows to get a 1 in the upper left corner.

$$\begin{bmatrix} 1 & -1 & | & 0 & 1 \\ 2 & 4 & | & 1 & 0 \end{bmatrix}$$

Multiply row one by -2 and add the results to row two to get

$$\begin{bmatrix} 1 & -1 & | & 0 & 1 \\ 0 & 6 & | & 1 & -2 \end{bmatrix} \qquad -2R_1 + R_2$$

Now, to get a 1 in the second row, second column position, multiply row two by 1/6.

$$\begin{bmatrix} 1 & -1 & | & 0 & 1 \\ 0 & 1 & | & 1/6 & -1/3 \end{bmatrix} \qquad (1/6)R_2$$

Finally, add row two to row one to get a 0 in the first row, second column.

$$\begin{bmatrix} 1 & 0 & | & 1/6 & 2/3 \\ 0 & 1 & | & 1/6 & -1/3 \end{bmatrix} \qquad R_2 + R_1$$

The numbers in the first column to the right of the vertical bar give the values of x and z. The second column gives the values of y and w. That is,

$$\begin{bmatrix} 1 & 0 & x & y \\ 0 & 1 & z & w \end{bmatrix} = \begin{bmatrix} 1 & 0 & 1/6 & 2/3 \\ 0 & 1 & 1/6 & -1/3 \end{bmatrix}$$

so that

$$A^{-1} = \begin{bmatrix} x & y \\ z & w \end{bmatrix} = \begin{bmatrix} 1/6 & 2/3 \\ 1/6 & -1/3 \end{bmatrix}.$$

To check, multiply A by A^{-1}. The result should be I_2.

$$AA^{-1} = \begin{bmatrix} 2 & 4 \\ 1 & -1 \end{bmatrix} \begin{bmatrix} 1/6 & 2/3 \\ 1/6 & -1/3 \end{bmatrix} = \begin{bmatrix} 1/3 + 2/3 & 4/3 - 4/3 \\ 1/6 - 1/6 & 2/3 + 1/3 \end{bmatrix} = \begin{bmatrix} 1 & 0 \\ 0 & 1 \end{bmatrix} = I_2.$$

Verify that $A^{-1}A = I_2$, also. Finally,

$$A^{-1} = \begin{bmatrix} 1/6 & 2/3 \\ 1/6 & -1/3 \end{bmatrix}.$$

In summary, to use this method of finding inverses, first form the augmented matrix $[A|I_n]$, where A is any $n \times n$ matrix and I_n is the $n \times n$ multiplication identity matrix. Find A^{-1} by performing row transformations on $[A|I_n]$ until a matrix of the form $[I_n|B]$ is found. Matrix B is the desired matrix A^{-1}. If it is not possible to get a matrix of the form $[I_n|B]$, then A^{-1} does not exist.

To decide whether two $n \times n$ matrices A and B are inverses of one another, it is only necessary to show that $AB = I_n$.

· · · · · · · · · ·

EXAMPLE 1   Find A^{-1} if $A = \begin{bmatrix} 1 & 0 & 1 \\ 2 & -2 & -1 \\ 3 & 0 & 0 \end{bmatrix}$.

Use row transformations, going through as many steps as needed.

Step 1   Write the augmented matrix $[A|I_3]$.

$$\begin{bmatrix} 1 & 0 & 1 & 1 & 0 & 0 \\ 2 & -2 & -1 & 0 & 1 & 0 \\ 3 & 0 & 0 & 0 & 0 & 1 \end{bmatrix}$$

Step 2   Since 1 is already in the upper left-hand corner as necessary, begin by selecting the row operation which will result in a 0 for the first element in the second row. Add to each element in the second row the result of multiplying the first row by -2.

$$\begin{bmatrix} 1 & 0 & 1 & 1 & 0 & 0 \\ 0 & -2 & -3 & -2 & 1 & 0 \\ 3 & 0 & 0 & 0 & 0 & 1 \end{bmatrix} \qquad -2R_1 + R_2$$

Step 3 To get 0 for the first element in the third row, add to the third row the results of multiplying each element of the first row by -3.

$$\begin{bmatrix} 1 & 0 & 1 & | & 1 & 0 & 0 \\ 0 & -2 & -3 & | & -2 & 1 & 0 \\ 0 & 0 & -3 & | & -3 & 0 & 1 \end{bmatrix} \qquad -3R_1 + R_3$$

Step 4 To get 1 for the second element in the second row, multiply the second row by $-1/2$.

$$\begin{bmatrix} 1 & 0 & 1 & | & 1 & 0 & 0 \\ 0 & 1 & 3/2 & | & 1 & -1/2 & 0 \\ 0 & 0 & -3 & | & -3 & 0 & 1 \end{bmatrix} \qquad (-1/2)R_2$$

Step 5 To get 1 for the third element in the third row, multiply the third row by $-1/3$.

$$\begin{bmatrix} 1 & 0 & 1 & | & 1 & 0 & 0 \\ 0 & 1 & 3/2 & | & 1 & -1/2 & 0 \\ 0 & 0 & 1 & | & 1 & 0 & -1/3 \end{bmatrix} \qquad (-1/3)R_3$$

Step 6 To get 0 for the third element in the first row, add to the first row the results of multiplying each element in row three by -1.

$$\begin{bmatrix} 1 & 0 & 0 & | & 0 & 0 & 1/3 \\ 0 & 1 & 3/2 & | & 1 & -1/2 & 0 \\ 0 & 0 & 1 & | & 1 & 0 & -1/3 \end{bmatrix} \qquad -1R_3 + R_1$$

Step 7 To get 0 for the third element in the second row, add to the second row the results of multiplying each element of row three by $-3/2$.

$$\begin{bmatrix} 1 & 0 & 0 & | & 0 & 0 & 1/3 \\ 0 & 1 & 0 & | & -1/2 & -1/2 & 1/2 \\ 0 & 0 & 1 & | & 1 & 0 & -1/3 \end{bmatrix} \qquad (-3/2)R_3 + R_2$$

From the last transformation, we get the desired inverse.

$$A^{-1} = \begin{bmatrix} 0 & 0 & 1/3 \\ -1/2 & -1/2 & 1/2 \\ 0 & 0 & -1/3 \end{bmatrix}$$

Confirm this by forming the product $A^{-1}A$, or AA^{-1}, each of which should be equal to I_3. ●

EXAMPLE 2 Find A^{-1} given $A = \begin{bmatrix} 2 & -4 \\ 1 & -2 \end{bmatrix}$.

Using row operations to transform the first column of the augmented matrix

$$\begin{bmatrix} 2 & -4 & | & 1 & 0 \\ 1 & -2 & | & 0 & 1 \end{bmatrix}$$

results in the following matrices:

$$\left[\begin{array}{cc|cc} 1 & -2 & 1/2 & 0 \\ 1 & -2 & 0 & 1 \end{array}\right], \qquad (1/2)R_1$$

$$\left[\begin{array}{cc|cc} 1 & -2 & 1/2 & 0 \\ 0 & 0 & -1/2 & 1 \end{array}\right]. \qquad -1\ R_1 + R_2$$

At this point, the matrix should be changed so that the second row, second column element will be 1. Since that element is now 0, there is no way to complete the desired transformation, so matrix A^{-1} does not exist. ●

If the inverse of a matrix exists, it is unique. That is, any given square matrix has no more than one inverse. The proof of this is left to Exercise 46 of this section.

Solving Systems by Inverses Matrix inverses can be used to solve square linear systems of equations. (A square system has the same number of equations as variables.) For example, given the linear system

$$a_{11}x + a_{12}y + a_{13}z = d_1$$
$$a_{21}x + a_{22}y + a_{23}z = d_2$$
$$a_{31}x + a_{32}y + a_{33}z = d_3,$$

the definition of matrix multiplication can be used to rewrite the system as

$$\left[\begin{array}{ccc} a_{11} & a_{12} & a_{13} \\ a_{21} & a_{22} & a_{23} \\ a_{31} & a_{32} & a_{33} \end{array}\right] \cdot \left[\begin{array}{c} x \\ y \\ z \end{array}\right] = \left[\begin{array}{c} d_1 \\ d_2 \\ d_3 \end{array}\right]. \qquad \textbf{(1)}$$

(To see this, multiply the matrices on the left.)

$$\text{If } A = \left[\begin{array}{ccc} a_{11} & a_{12} & a_{13} \\ a_{21} & a_{22} & a_{23} \\ a_{31} & a_{32} & a_{33} \end{array}\right], \quad X = \left[\begin{array}{c} x \\ y \\ z \end{array}\right], \quad \text{and} \quad B = \left[\begin{array}{c} d_1 \\ d_2 \\ d_3 \end{array}\right],$$

the system given in (1) becomes

$$AX = B.$$

If A^{-1} exists, then both sides of $AX = B$ can be multiplied on the left to get

$$A^{-1}(AX) = A^{-1}B$$
$$(A^{-1}A)X = A^{-1}B \qquad \text{Associative property}$$
$$I_3X = A^{-1}B \qquad \text{Inverse property}$$
$$X = A^{-1}B. \qquad \text{Identity property}$$

Matrix $A^{-1}B$ gives the solution of the system.

This method of using matrix inverses to solve systems of equations is useful when the inverse is already known or when many systems of the form $AX = B$ must be solved and only B changes.

$\cdots\cdots\cdots$

EXAMPLE 3 Use the method of matrix inverses to solve the following systems.

(a) $2x - 3y = 4$

$x + 5y = 2$

To represent the system as a matrix equation, use one matrix for the coefficients, one for the variables, and one for the constants, as follows.

$$A = \begin{bmatrix} 2 & -3 \\ 1 & 5 \end{bmatrix}, \quad X = \begin{bmatrix} x \\ y \end{bmatrix}, \quad \text{and} \quad B = \begin{bmatrix} 4 \\ 2 \end{bmatrix}.$$

As shown above, the solution is given by $X = A^{-1}B$.
To solve the system, first find A^{-1}. Verify that

$$A^{-1} = \begin{bmatrix} 5/13 & 3/13 \\ -1/13 & 2/13 \end{bmatrix}.$$

Next, find the product $A^{-1}B$.

$$A^{-1}B = \begin{bmatrix} 5/13 & 3/13 \\ -1/13 & 2/13 \end{bmatrix} \begin{bmatrix} 4 \\ 2 \end{bmatrix} = \begin{bmatrix} 2 \\ 0 \end{bmatrix}.$$

Since $X = A^{-1}B$,

$$X = \begin{bmatrix} x \\ y \end{bmatrix} = \begin{bmatrix} 2 \\ 0 \end{bmatrix}.$$

The final matrix shows that the solution of the system is $(2, 0)$.

(b) $2x - 3y = 1$

$x + 5y = 20$

This system has the same matrix of coefficients. Only matrix B is different. Use A^{-1} from part (a) and multiply by B to get

$$X = A^{-1}B = \begin{bmatrix} 5/13 & 3/13 \\ -1/13 & 2/13 \end{bmatrix} \begin{bmatrix} 1 \\ 20 \end{bmatrix} = \begin{bmatrix} 5 \\ 3 \end{bmatrix},$$

giving the solution $(5, 3)$. ●

9.8 Exercises $\cdots\cdots\cdots\cdots\cdots\cdots\cdots\cdots\cdots\cdots\cdots\cdots\cdots\cdots$

Decide whether or not the given matrices are inverses of each other. (Check to see if their product is the identity matrix I_n).

1. $\begin{bmatrix} 2 & 3 \\ 1 & 1 \end{bmatrix}$ and $\begin{bmatrix} -1 & 3 \\ 1 & -2 \end{bmatrix}$

2. $\begin{bmatrix} 5 & 7 \\ 2 & 3 \end{bmatrix}$ and $\begin{bmatrix} 3 & -7 \\ -2 & 5 \end{bmatrix}$

3. $\begin{bmatrix} 2 & 1 \\ 3 & 2 \end{bmatrix}$ and $\begin{bmatrix} 2 & 1 \\ -3 & 2 \end{bmatrix}$

4. $\begin{bmatrix} -1 & 2 \\ 3 & -5 \end{bmatrix}$ and $\begin{bmatrix} -5 & -2 \\ -3 & -1 \end{bmatrix}$

5. $\begin{bmatrix} 1 & 2 & 0 \\ 0 & 1 & 0 \\ 0 & 1 & 0 \end{bmatrix}$ and $\begin{bmatrix} 1 & -2 & 0 \\ 0 & 1 & 0 \\ 0 & -1 & 1 \end{bmatrix}$

6. $\begin{bmatrix} 0 & 1 & 0 \\ 0 & 0 & -2 \\ 1 & -1 & 0 \end{bmatrix}$ and $\begin{bmatrix} 1 & 0 & 1 \\ 1 & 0 & 0 \\ 0 & -1 & 0 \end{bmatrix}$

7. $\begin{bmatrix} 1 & 3 & 3 \\ 1 & 4 & 3 \\ 1 & 3 & 4 \end{bmatrix}$ and $\begin{bmatrix} 7 & -3 & -3 \\ -1 & 1 & 0 \\ -1 & 0 & 1 \end{bmatrix}$

8. $\begin{bmatrix} -1 & 0 & 2 \\ 3 & 1 & 0 \\ 0 & 2 & -3 \end{bmatrix}$ and $\begin{bmatrix} -1/5 & 4/15 & -2/15 \\ 3/5 & 1/5 & 2/5 \\ 2/5 & 2/15 & -1/15 \end{bmatrix}$

Find the inverse, if it exists, for each matrix.

9. $\begin{bmatrix} 1 & -1 \\ 2 & 0 \end{bmatrix}$

10. $\begin{bmatrix} -1 & 2 \\ -2 & -1 \end{bmatrix}$

11. $\begin{bmatrix} 3 & -1 \\ -5 & 2 \end{bmatrix}$

12. $\begin{bmatrix} -1 & -2 \\ 3 & 4 \end{bmatrix}$

13. $\begin{bmatrix} -6 & 4 \\ -3 & 2 \end{bmatrix}$

14. $\begin{bmatrix} 5 & 10 \\ -3 & -6 \end{bmatrix}$

15. $\begin{bmatrix} 1 & 0 & 0 \\ 0 & -1 & 0 \\ 1 & 0 & 1 \end{bmatrix}$

16. $\begin{bmatrix} 1 & 0 & 1 \\ 0 & -1 & 0 \\ 2 & 1 & 1 \end{bmatrix}$

17. $\begin{bmatrix} -1 & -1 & -1 \\ 4 & 5 & 0 \\ 0 & 1 & -3 \end{bmatrix}$

18. $\begin{bmatrix} 2 & 0 & 4 \\ 3 & 1 & 5 \\ -1 & 1 & -2 \end{bmatrix}$

19. $\begin{bmatrix} 2 & 4 & 6 \\ -1 & -4 & -3 \\ 0 & 1 & -1 \end{bmatrix}$

20. $\begin{bmatrix} 2 & 2 & -4 \\ 2 & 6 & 0 \\ -3 & -3 & 5 \end{bmatrix}$

21. $\begin{bmatrix} 1 & -2 & 3 & 0 \\ 0 & 1 & -1 & 1 \\ -2 & 2 & -2 & 4 \\ 0 & 2 & -3 & 1 \end{bmatrix}$

22. $\begin{bmatrix} 1 & 1 & 0 & 2 \\ 2 & -1 & 1 & -1 \\ 3 & 3 & 2 & -2 \\ 1 & 2 & 1 & 0 \end{bmatrix}$

Solve each system of equations by using the inverse of the coefficient matrix.

23. $2x + 3y = 10$
$x - y = -5$

24. $-x + 2y = 15$
$-2x - y = 20$

25. $2x + y = 5$
$5x + 3y = 13$

26. $-x - 2y = 8$
$3x + 4y = 24$

27. $-x + y = 1$
$2x - y = 1$

28. $3x - 6y = 1$
$-5x + 9y = -1$

Solve each system of equations by using the inverse of the coefficient matrix. The inverses for the first four problems are found in Exercises 17–20 above. Assume b is a constant in Exercises 35–38.

29. $-x - y - z = 1$
$4x + 5y = -2$
$y - 3z = 3$

30. $2x + 4z = -8$
$3x + y + 5z = 2$
$-x + y - 2z = 4$

31. $2x + 4y + 6z = 4$
$-x - 4y - 3z = 8$
$y - z = -4$

32. $2x + 2y - 4z = 12$
$2x + 6y = 16$
$-3x - 3y + 5z = -20$

33. $x + 2y + 3z = 5$
$2x + 3y + 2z = 2$
$-x - 2y - 4z = -1$

34. $x + y - 3z = 4$
$2x + 4y - 4z = 8$
$-x + y + 4z = -3$

35. $4x + 2y = 7$
$bx + 5y = 8$

36. $2x + 3y = 4$
$5x + 6y = b$

37. $2x - y = b^2$
$x + y = b$

38. $bx + 4y = 1$
$x + by = b$

Solve each system of equations by using the inverse of the coefficient matrix. The inverses were found in Exercises 21 and 22.

39. $x - 2y + 3z = 4$
$y - z + w = -8$
$-2x + 2y - 2z + 4w = 12$
$2y - 3z + w = -4$

40. $x + y + 2w = 3$
$2x - y + z - w = 3$
$3x + 3y + 2z - 2w = 5$
$x + 2y + z = 3$

Let $A = \begin{bmatrix} a & b \\ c & d \end{bmatrix}$ and let O be the 2 × 2 matrix of all zeros. Show that statements 41–44 are true.

41. $A \cdot O = O$

42. $AA^{-1} = I_2$

43. $A^{-1}A = I_2$.

44. For square matrices A and B of the same order, if $AB = O$ and if A^{-1} exists, then $B = O$.

45. The Bread Box Bakery sells three types of cakes, each requiring the amounts of the basic ingredients shown in the following matrix.

$$
\begin{array}{lccc}
\text{Types of cake} & \text{I} & \text{II} & \text{III} \\
\text{Flour (in cups)} & \begin{bmatrix} 2 & 4 & 2 \\ \text{Sugar (in cups)} & 2 & 1 & 2 \\ \text{Eggs} & 2 & 1 & 3 \end{bmatrix}
\end{array}
$$

To fill its daily orders for these three kinds of cake, the bakery uses 72 cups of flour, 48 cups of sugar, and 60 eggs.

(a) Write a 3 × 1 matrix for the amounts used daily.

(b) Let the number of daily orders for cakes be a 3 × 1 matrix X with entries x_1, x_2 and x_3. Write a matrix equation which you can solve for X, using the given matrix and the matrix from part (a).

(c) Solve the equation you wrote in part (b) to find the number of daily orders for each type of cake.

46. Prove that, if it exists, the inverse of a matrix is unique. (*Hint:* Assume there are two inverses B and C for some matrix A, so that $AB = BA = I$ and $AC = CA = I$. Multiply the first equation by C and the second by B.)

47. Give an example of two matrices A and B where $(AB)^{-1} \neq A^{-1}B^{-1}$.

48. Suppose A and B are matrices where A^{-1}, B^{-1}, and AB all exist. Show that $(AB)^{-1} = B^{-1}A^{-1}$.

49. Let $A = \begin{bmatrix} a & 0 \\ 0 & b \end{bmatrix}$. Under what conditions on a and b does A^{-1} exist?

50. Derive a formula for the inverse of $\begin{bmatrix} a & 0 \\ 0 & d \end{bmatrix}$, where $ad \neq 0$.

51. Let $A = \begin{bmatrix} a & 0 & 0 \\ 0 & b & 0 \\ 0 & 0 & c \end{bmatrix}$, where a, b and c are nonzero real numbers. Find A^{-1}.

52. Let $A = \begin{bmatrix} 1 & 0 & 0 \\ 0 & 0 & -1 \\ 0 & 1 & -1 \end{bmatrix}$. Show that $A^3 = I$ and use this result to find the inverse of A.

9.9 PARTIAL FRACTIONS

In Chapter 1 sums of rational expressions were found by combining two or more rational expressions into one rational expression. Here the reverse problem is considered: given one rational expression, express it as the sum of two or more rational expressions. A special type of sum of rational expressions is called the **partial fraction decomposition;** each term in the sum is a **partial fraction.** The technique of decomposing a rational expression into partial fractions is useful in calculus and other areas of mathematics.

To form a partial fraction decomposition of a rational expression, use the following steps.

Partial Fraction Decomposition of $\dfrac{f(x)}{g(x)}$

Step 1 If $f(x)/g(x)$ is not a proper fraction (a fraction with the numerator of lower degree than the denominator), divide $f(x)$ by $g(x)$. For example,

$$\frac{x^4 - 3x^3 + x^2 + 5x}{x^2 + 3} = x^2 - 3x - 2 + \frac{14x + 6}{x^2 + 3}.$$

Then apply the following steps to the remainder, which is a proper fraction.

Step 2 Factor $g(x)$ completely into factors of the form $(ax + b)^m$ or $(cx^2 + dx + e)^n$, where $cx^2 + dx + e$ is irreducible and m and n are integers.

Step 3 For each factor of the form $(ax + b)^m$ the decomposition must include the terms

$$\frac{A_1}{ax + b} + \frac{A_2}{(ax + b)^2} + \cdots + \frac{A_m}{(ax + b)^m}.$$

Step 4 For each factor of the form $(cx^2 + dx + e)^n$, the decomposition must include the terms

$$\frac{B_1x + C_1}{cx^2 + dx + e} + \frac{B_2x + C_2}{(cx^2 + dx + e)^2} + \cdots + \frac{B_nx + C_n}{(cx^2 + dx + e)^n}.$$

Step 5 Use algebraic techniques to solve for the constants in the numerators of the decomposition.

To find the constants in Step 5, the goal is to get a system of equations with as many equations as there are unknowns in the numerators. One method for getting these equations is to substitute values for x on both sides of the rational equation formed from Steps 3 or 4 above.

EXAMPLE 1 Find the partial fraction decomposition of

$$\frac{2x^4 - 8x^2 + 5x - 2}{x^3 - 4x}.$$

The given fraction is not a proper fraction; the numerator has higher degree than the denominator. Perform the division.

$$
\begin{array}{r}
2x \\
x^3 - 4x \overline{\smash{\big)}\, 2x^4 - 8x^2 + 5x - 2} \\
\underline{2x^4 - 8x^2 } \\
5x - 2
\end{array}
$$

The quotient is
$$\frac{2x^4 - 8x^2 + 5x - 2}{x^3 - 4x} = 2x + \frac{5x - 2}{x^3 - 4x}.$$

Now work with the remainder fraction. Factor the denominator as $x^3 - 4x = x(x + 2)(x - 2)$. Since the factors are **distinct linear factors,** use Step 3 to write the decomposition as

$$\frac{5x - 2}{x^3 - 4x} = \frac{A}{x} + \frac{B}{x + 2} + \frac{C}{x - 2}, \qquad \text{(1)}$$

where A, B, and C are constants that need to be found. Multiply both sides of equation (1) by $x(x + 2)(x - 2)$, getting

$$5x - 2 = A(x + 2)(x - 2) + Bx(x - 2) + Cx(x + 2). \qquad \text{(2)}$$

Equation (1) is an identity, since both sides represent the same rational expression. Thus, equation (2) is also an identity. Equation (1) holds for all values of x except 0, -2, and 2. However, equation (2) holds for all values of x. In particular, substituting 0 for x in equation (2) gives

$$-2 = -4A, \quad \text{so that} \quad A = \frac{1}{2}.$$

Similarly, choosing $x = -2$ gives

$$-12 = 8B, \quad \text{so that} \quad B = -\frac{3}{2}.$$

Finally, choosing $x = 2$,

$$8 = 8C, \quad \text{so that} \quad C = 1.$$

The remainder rational expression can be written as the following sum of partial fractions.

$$\frac{5x - 2}{x^3 - 4x} = \frac{1}{2x} + \frac{-3}{2(x + 2)} + \frac{1}{x - 2},$$

and the given rational expression can be written as

$$\frac{2x^4 - 8x^2 + 5x - 2}{x^3 - 4x} = 2x + \frac{1}{2x} + \frac{-3}{2(x + 2)} + \frac{1}{x - 2}.$$

Check the work by combining the terms on the right. ●

· · · · · · · · ·

EXAMPLE 2 Find the partial fraction decomposition of $\dfrac{2x}{(x - 1)^3}.$

This is a proper fraction. The denominator is already factored with **repeated linear factors.** Write the decomposition as shown, by using Step 3 above.

$$\frac{2x}{(x - 1)^3} = \frac{A}{x - 1} + \frac{B}{(x - 1)^2} + \frac{C}{(x - 1)^3}$$

Clear the denominators by multiplying both sides of this equation by $(x - 1)^3$.

$$2x = A(x - 1)^2 + B(x - 1) + C$$

Substituting 1 for x leads to $C = 2$, so that

$$2x = A(x - 1)^2 + B(x - 1) + 2. \tag{1}$$

The only root has been substituted and values for A and B still need to be found. However, *any* number can be substituted for x. For example, when we choose $x = -1$ (because it is easy to substitute), equation (1) becomes

$$-2 = 4A - 2B + 2$$
$$-4 = 4A - 2B$$
$$-2 = 2A - B. \tag{2}$$

Substituting 0 for x in equation (1) gives

$$0 = A - B + 2$$
$$2 = -A + B. \tag{3}$$

Now solve the system of equations (2) and (3) to get $A = 0$ and $B = 2$. The partial fraction decomposition is

$$\frac{2x}{(x - 1)^3} = \frac{2}{(x - 1)^2} + \frac{2}{(x - 1)^3}.$$

Three substitutions were needed because there were three constants to evaluate, A, B, and C.

To check this result, combine the terms on the right. ●

· · · · · · · · ·

EXAMPLE 3   Find the partial fraction decomposition of $\dfrac{x^2 + 3x - 1}{(x + 1)(x^2 + 2)}$.

This denominator has **distinct linear and quadratic factors** where neither is repeated. Since $x^2 + 2$ cannot be factored, it is irreducible. The partial fraction decomposition is

$$\frac{x^2 + 3x - 1}{(x + 1)(x^2 + 2)} = \frac{A}{x + 1} + \frac{Bx + C}{x^2 + 2}.$$

Multiply both sides by $(x + 1)(x^2 + 2)$ to get

$$x^2 + 3x - 1 = A(x^2 + 2) + (Bx + C)(x + 1). \tag{1}$$

First substitute -1 for x to get

$$(-1)^2 + 3(-1) - 1 = A[(-1)^2 + 2] + 0$$
$$-3 = 3A$$
$$A = -1.$$

Replace A with -1 in equation (1) and substitute any value for x. For instance, if $x = 0$,

$$0^2 + 3(0) - 1 = -1(0^2 + 2) + (B \cdot 0 + C)(0 + 1)$$
$$-1 = -2 + C$$
$$C = 1.$$

Now, letting $A = -1$ and $C = 1$, substitute again in equation (1) using another number for x. For $x = 1$,

$$3 = -3 + (B + 1)(2)$$
$$6 = 2B + 2$$
$$B = 2.$$

Using $A = -1$, $B = 2$, and $C = 1$, the partial fraction decomposition is

$$\frac{x^2 + 3x - 1}{(x + 1)(x^2 + 2)} = \frac{-1}{x + 1} + \frac{2x + 1}{x^2 + 2}.$$

Again, this work can be checked by combining terms on the right. ●

For fractions with denominators that have quadratic factors, another method is often more convenient. The system of equations is formed by equating coefficients of like terms on both sides of the partial fraction decomposition. For instance, in Example 3, after both sides were multiplied by the common denominator, the equation was

$$x^2 + 3x - 1 = A(x^2 + 2) + (Bx + C)(x + 1).$$

Multiplying on the right and collecting like terms, we have

$$x^2 + 3x - 1 = Ax^2 + 2A + Bx^2 + Bx + Cx + C$$
$$x^2 + 3x - 1 = (A + B)x^2 + (B + C)x + (C + 2A).$$

Now, equating the coefficients of like powers of x gives the three equations

$$1 = A + B$$
$$3 = B + C$$
$$-1 = C + 2A.$$

Solving this system of equations for A, B, and C would give the partial fraction decomposition. The next example uses a combination of the two methods.

· · · · · · · · ·

EXAMPLE 4   Find the partial decomposition of $\dfrac{2x}{(x^2 + 1)^2(x - 1)}$.

This expression has both a linear factor and a **repeated quadratic factor.** By Steps 3 and 4 from the box at the beginning of this section,

$$\frac{2x}{(x^2 + 1)^2(x - 1)} = \frac{Ax + B}{x^2 + 1} + \frac{Cx + D}{(x^2 + 1)^2} + \frac{E}{x - 1}.$$

Multiplication of both sides by $(x^2 + 1)^2(x - 1)$ leads to

$$2x = (Ax + B)(x^2 + 1)(x - 1) + (Cx + D)(x - 1) + E(x^2 + 1)^2. \quad \textbf{(1)}$$

If $x = 1$, equation (1) reduces to

$$2 = 4E, \quad \text{or} \quad E = \frac{1}{2}.$$

Substituting 1/2 for E in equation (1) and combining terms on the right gives

$$2x = (A + 1/2)x^4 + (-A + B)x^3 + (A - B + C + 1)x^2 +$$
$$(-A + B + D - C)x + (-B - D + 1/2). \quad \textbf{(2)}$$

To get additional equations involving the unknowns, equate the coefficients of like powers of x on the two sides of equation (2). Setting corresponding coefficients of x^4 equal,

$$0 = A + \frac{1}{2} \quad \text{or} \quad A = -\frac{1}{2}.$$

From the corresponding coefficients of x^3,

$$0 = -A + B.$$

Since $A = -1/2$, then

$$B = -1/2.$$

Using the coefficients of x^2,

$$0 = A - B + C + 1.$$

Since $A = -1/2$ and $B = -1/2$,

$$C = -1.$$

Finally, from the coefficients of x,

$$2 = -A + B + D - C.$$

Substituting for A, B, and C gives

$$D = 1.$$

With $A = -1/2$, $B = -1/2$, $C = -1$, $D = 1$, and $E = 1/2$, the given fraction has the partial fraction decomposition

$$\frac{2x}{(x^2 + 1)^2(x - 1)} = \frac{-\dfrac{1}{2}x - \dfrac{1}{2}}{x^2 + 1} + \frac{-x + 1}{(x^2 + 1)^2} + \frac{\dfrac{1}{2}}{x - 1}$$

or

$$\frac{2x}{(x^2 + 1)^2(x - 1)} = \frac{-(x + 1)}{2(x^2 + 1)} + \frac{-x + 1}{(x^2 + 1)^2} + \frac{1}{2(x - 1)}. \quad \bullet$$

The two methods discussed in this section are summarized on the following page.

To solve for the constants in the numerators of a partial fraction decomposition, use either of the following methods or a combination of the two.

Method 1   For linear factors

1. Multiply both sides of the rational expression by the common denominator.
2. Substitute the root of each factor in the resulting equation. For repeated linear factors, substitute as many other numbers as necessary to find all the constants in the numerators. The number of substitutions required will equal the number of constants A, B, ...

Method 2   For quadratic factors

1. Multiply both sides of the rational expression by the common denominator.
2. Collect terms on the right side of the resulting equation.
3. Equate the coefficients of like terms to get a system of equations.
4. Solve the system to find the constants in the numerators.

9.9 Exercises

Find the partial fraction decomposition for the following rational expressions.

1. $\dfrac{3x - 1}{x(x + 1)}$

2. $\dfrac{5}{3x(2x + 1)}$

3. $\dfrac{x + 2}{(x + 1)(x - 1)}$

4. $\dfrac{4x + 2}{(x + 2)(2x - 1)}$

5. $\dfrac{-7}{x(3x - 1)(x + 1)}$

6. $\dfrac{2x}{(x + 1)(x + 2)^2}$

7. $\dfrac{2}{x^2(x + 3)}$

8. $\dfrac{x + 2}{(x - 1)^2(x + 5)}$

9. $\dfrac{x - 1}{(x + 2)(x - 3)^2}$

10. $\dfrac{4x^2 - x - 15}{x(x + 1)(x - 1)}$

11. $\dfrac{2x + 1}{(x + 2)^3}$

12. $\dfrac{1}{x^2 + 5x + 4}$

13. $\dfrac{3}{x^2 + 4x + 3}$

14. $\dfrac{2x^5 + 3x^4 - 3x^3 - 2x^2 + x}{2x^2 + 5x + 2}$

15. $\dfrac{6x^5 + 7x^4 - x^2 + 2x}{3x^2 + 2x - 1}$

16. $\dfrac{x^3 + 4}{9x^3 - 4x}$

17. $\dfrac{x^3 + 2}{x^3 - 3x^2 + 2x}$

18. $\dfrac{-3}{x^2(x^2 + 5)}$

19. $\dfrac{2x + 1}{(x + 1)(x^2 + 2)}$

20. $\dfrac{3x - 2}{(x + 4)(3x^2 + 1)}$

21. $\dfrac{3}{x(x + 1)(x^2 + 1)}$

22. $\dfrac{1}{x(2x + 1)(3x^2 + 4)}$

23. $\dfrac{x^4 + 1}{x(x^2 + 1)^2}$

24. $\dfrac{3x - 1}{x(2x^2 + 1)^2}$

25. $\dfrac{3x^4 + x^3 + 5x^2 - x + 4}{(x - 1)(x^2 + 1)^2}$

26. $\dfrac{-x^4 - 8x^2 + 3x - 10}{(x + 2)(x^2 + 4)^2}$

27. $\dfrac{x^2}{x^4 - 1}$

28. $\dfrac{5x^5 + 10x^4 - 15x^3 + 4x^2 + 13x - 9}{x^3 + 2x^2 - 3x}$

29. $\dfrac{3x^6 + 3x^4 + 3x}{x^4 + x^2}$

30. $\dfrac{2x^2 - 12x + 4}{x^3 - 4x^2}$

9 | CHAPTER SUMMARY

Key Words

system of linear equations
9.1 elimination method
substitution method
inconsistent system
dependent equations
ordered triple
echelon method
triangular form
9.2 nonlinear system
9.3 half-plane
linear inequality

system of inequalities
linear programming
constraints
region of feasible solutions
vertex point
9.4 matrix
augmented matrix
Gaussian reduction
method
9.5 square matrix

row matrix
column matrix
scalar
9.6 determinant
minor
cofactor
9.7 Cramer's rule
9.8 identity matrix
inverse matrix
9.9 partial fractions

Review Exercises

Use the elimination or substitution method to solve each of the following linear systems. Identify any dependent or inconsistent systems.

1. $3x - 5y = -18$
$2x + 7y = 19$

2. $6x + 5y = 53$
$4x - 3y = 29$

3. $\dfrac{2}{3}x - \dfrac{3}{4}y = 13$
$\dfrac{1}{2}x + \dfrac{2}{3}y = -5$

4. $3x + 7y = 10$
$18x + 42y = 50$

5. $\dfrac{1}{x} + \dfrac{1}{y} = \dfrac{7}{10}$
$\dfrac{3}{x} - \dfrac{5}{y} = \dfrac{1}{2}$

6. $.9x - .2y = .8$
$.3x + .7y = 4.1$

7. $2x - 3y + z = -5$
$x + 4y + 2z = 13$
$5x + 5y + 3z = 14$

8. $x - 3y = 12$
$2y + 5z = 1$
$4x + z = 25$

Solve each of the following problems.

9. A student bought some candy bars, paying 25¢ each for some and 50¢ each for others. The student bought a total of 22 bars, paying a total of $8.50. How many of each kind of bar did he buy?

10. Ink worth $25 a bottle is to be mixed with ink worth $18 per bottle to get 12 bottles of ink worth $20 each. How much of each type of ink should be used?

11. Donna Sharp wins $50,000 in a lottery. She invests part of the money at 6%, twice as much as 7%, with $10,000 more than the amount invested at 6% invested at 9%. Total annual interest is $3800. How much is invested at each rate?

12. The sum of three numbers is 23. The second number is 3 more than the first. The sum of the first and twice the third is 4. Find the three numbers.

Find solutions for the following systems in terms of an arbitrary variable.

13. $3x - 4y + z = 2$
$2x + y - 4z = 1$

14. $ax + by + cz = 5$
$dx + y = 1$

Solve each system in Exercises 15–18.

15. $x^2 - 4y^2 = 19$
$x^2 + y^2 = 29$

16. $xy = 4$
$x - 6y = 2$

17. $x^2 + 2xy + y^2 = 4$
$x = 3y - 2$

18. $y = 5^{x+5}$
$y = 25^{3x}$

19. Do the circle $x^2 + y^2 = 144$ and the line $x + 2y = 8$ have any points in common? If so, what are they?

20. Find a value of b so that the straight line $3x - y = b$ touches the circle $x^2 + y^2 = 25$ in only one point.

21. Given the line $y = mx + b$ tangent to the circle $x^2 + y^2 = r^2$, find an equation for x involving m, b, and r.*

22. Find an equation of the circle having as its diameter the common chord of the two circles $x^2 + y^2 + 2x - 2y - 14 = 0$ and $x^2 + y^2 - 4x + 4y - 2 = 0$.*

Graph the solution of each system of inequalities in Exercises 23–25.

23. $x + y \le 6$
$2x - y \ge 3$

24. $x - 3y \ge 6$
$y^2 \le 16 - x^2$

25. $9x^2 + 16y^2 \ge 144$
$x^2 - y^2 \le 16$

26. A bakery makes both cakes and cookies. Each batch of cakes requires 2 hr in the oven and 3 hr in the decorating room. Each batch of cookies needs 1 1/2 hr in the oven and 2/3 hr in the decorating room. The oven is available no more than 16 hr a day, while the decorating room can be used no more than 12 hr per day. Set up a system of inequalities expressing this information, and then graph the system.

27. A candy company has 100 kg of chocolate-covered nuts and 125 kg of chocolate-covered raisins to be sold as two different mixtures. One mix will contain 1/2 nuts and 1/2 raisins, while the other mix will contain 1/3 nuts and 2/3 raisins. Set up a system of inequalities expressing this information, and then graph the system.

In Exercises 28–31, use graphical methods to find nonnegative values of x and y that meet the constraints.

28. Maximize $5x + 7y$ subject to

$2x + 4y \ge 40$

$3x + 2y \le 60.$

29. Maximize $2x + 3y$ subject to

$x + 2y \le 24$

$3x + 4y \le 60.$

30. Minimize $x + 2y$ subject to

$5x + 4y \ge 20$

$2x + 6y \ge 24$

$x + y \le 12.$

31. Minimize $4x + 3y$ subject to

$x + 2y \le 12$

$4x + 3y \ge 12$

$y \le 2x.$

32. In Exercise 26, a batch of cookies produces a profit of $20; the profit on a batch of cakes is $30. Find the number of batches of each item which will maximize profit.

33. In Exercise 27, how much of each mixture should be made to maximize revenue if the first mix sells for $6.00 per kilogram and the second mix sells for $4.80 per kilogram?

---

*Exercises 21 and 22 from *The Calculus with Analytic Geometry,* Fifth Edition, by Louis Leithold. Copyright ©1986 by Harper & Row, Publishers, Inc. Reprinted by permission.

Use the Gaussian reduction method to solve each of the following.

34. $2x + 3y = 10$
$-3x + y = 18$

35. $5x + 2y = -10$
$3x - 5y = -6$

36. $2x - y + 4z = -1$
$-3x + 5y - z = 5$
$2x + 3y + 2z = 3$

37. $5x - 8y + z = 1$
$3x - 2y + 4z = 3$
$10x - 16y + 2z = 3$

Find the values of all variables in the following.

38. $\begin{bmatrix} 5 & x + 2 \\ -6y & z \end{bmatrix} = \begin{bmatrix} a & 3x - 1 \\ 5y & 9 \end{bmatrix}$

39. $\begin{bmatrix} -6 + k & 2 & a + 3 \\ -2 + m & 3p & 2r \end{bmatrix} + \begin{bmatrix} 3 - 2k & 5 & 7 \\ 5 & 8p & 5r \end{bmatrix} = \begin{bmatrix} 5 & y & 6a \\ 2m & 11 & -35 \end{bmatrix}$

Perform each of the following operations whenever possible.

40. $\begin{bmatrix} 3 & -4 & 2 \\ 5 & -1 & 6 \end{bmatrix} + \begin{bmatrix} -3 & 2 & 5 \\ 1 & 0 & 4 \end{bmatrix}$

41. $\begin{bmatrix} 3 \\ 2 \\ 5 \end{bmatrix} - \begin{bmatrix} 8 \\ -4 \\ 6 \end{bmatrix} + \begin{bmatrix} 1 \\ 0 \\ 2 \end{bmatrix}$

42. $\begin{bmatrix} 2 & 5 & 8 \\ 1 & 9 & 2 \end{bmatrix} - \begin{bmatrix} 3 & 4 \\ 7 & 1 \end{bmatrix}$

43. $\begin{bmatrix} -3 & 4 \\ 2 & 8 \end{bmatrix}\begin{bmatrix} -1 & 0 \\ 2 & 5 \end{bmatrix}$

44. $\begin{bmatrix} 3 & 2 & -1 \\ 4 & 0 & 6 \end{bmatrix}\begin{bmatrix} -2 & 0 \\ 0 & 2 \\ 3 & 1 \end{bmatrix}$

45. $\begin{bmatrix} 1 & -2 & 4 & 2 \\ 0 & 1 & -1 & 8 \end{bmatrix}\begin{bmatrix} -1 \\ 2 \\ 0 \\ 1 \end{bmatrix}$

Find each of the following determinants.

46. $\begin{vmatrix} -1 & 8 \\ 2 & 9 \end{vmatrix}$

47. $\begin{vmatrix} -2 & 4 \\ 0 & 3 \end{vmatrix}$

48. $\begin{vmatrix} -2 & 4 & 1 \\ 3 & 0 & 2 \\ -1 & 0 & 3 \end{vmatrix}$

49. $\begin{vmatrix} -1 & 2 & 3 \\ 4 & 0 & 3 \\ 5 & -1 & 2 \end{vmatrix}$

50. $\begin{vmatrix} -1 & 0 & 2 & -3 \\ 0 & 4 & 4 & -1 \\ -6 & 0 & 3 & -5 \\ 0 & -2 & 1 & 0 \end{vmatrix}$

Explain why each of the following statements is true.

51. $\begin{vmatrix} 8 & 9 & 2 \\ 0 & 0 & 0 \\ 3 & 1 & 4 \end{vmatrix} = 0$

52. $\begin{vmatrix} 4 & 6 \\ 3 & 5 \end{vmatrix} = \begin{vmatrix} 4 & 3 \\ 6 & 5 \end{vmatrix}$

53. $\begin{vmatrix} 8 & 2 \\ 4 & 3 \end{vmatrix} = 2\begin{vmatrix} 4 & 1 \\ 4 & 3 \end{vmatrix}$

54. $\begin{vmatrix} 4 & 6 & 2 \\ -3 & 8 & -5 \\ 4 & 6 & 2 \end{vmatrix} = 0$

55. $\begin{vmatrix} 5 & -1 & 2 \\ 3 & -2 & 0 \\ -4 & 1 & 2 \end{vmatrix} = \begin{vmatrix} 5 & -1 & 2 \\ 8 & -3 & 2 \\ -4 & 1 & 2 \end{vmatrix}$

56. $\begin{vmatrix} 8 & 2 & -5 \\ -3 & 1 & 4 \\ 2 & 0 & 5 \end{vmatrix} = -\begin{vmatrix} 8 & -5 & 2 \\ -3 & 4 & 1 \\ 2 & 5 & 0 \end{vmatrix}$

Solve each of the following systems by Cramer's rule if possible. Identify any dependent or inconsistent systems.

57. $3x + y = -1$
  $5x + 4y = 10$

58. $3x + 7y = 2$
  $5x - y = -22$

59. $3x + 2y + z = 2$
  $4x - y + 3z = -16$
  $x + 3y - z = 12$

60. $5x - 2y - z = 8$
  $-5x + 2y + z = -8$
  $x - 4y - 2z = 0$

Find the inverse of each of the following matrices that has an inverse.

61. $\begin{bmatrix} 2 & 1 \\ 5 & 3 \end{bmatrix}$

62. $\begin{bmatrix} -4 & 2 \\ 0 & 3 \end{bmatrix}$

63. $\begin{bmatrix} 2 & -1 & 0 \\ 1 & 0 & 1 \\ 1 & -2 & 0 \end{bmatrix}$

64. $\begin{bmatrix} 2 & 3 & 5 \\ -2 & -3 & -5 \\ 1 & 4 & 2 \end{bmatrix}$

Use the method of matrix inverses to solve each of the following.

65. $2x + y = 5$
  $3x - 2y = 4$

66. $x + y + z = 1$
  $2x - y = -2$
  $3y + z = 2$

67. $x = -3$
  $y + z = 6$
  $2x - 3z = -9$

68. $3x - 2y + 4z = 1$
  $4x + y - 5z = 2$
  $-6x + 4y - 8z = -2$

Let $f(x) = ax + b$ and $g(x) = cx + d$. Also, let $\begin{bmatrix} a & b \\ 0 & 1 \end{bmatrix}$ correspond to f and $\begin{bmatrix} c & d \\ 0 & 1 \end{bmatrix}$ correspond to g.*

69. Show that $\begin{bmatrix} a & b \\ 0 & 1 \end{bmatrix}\begin{bmatrix} c & d \\ 0 & 1 \end{bmatrix}$ corresponds to $f \circ g$.

70. Assume that $a \neq 0$ and show that f^{-1} corresponds to $\begin{bmatrix} a & b \\ 0 & 1 \end{bmatrix}^{-1}$.

Find the partial fraction decomposition of the following rational expressions.

71. $\dfrac{x - 1}{x^3 - x^2 + 4x}$

72. $\dfrac{x + 2}{x^3 + 2x^2 + x}$

73. $\dfrac{x + 3}{x^3 + 64}$

74. $\dfrac{2x + 1}{(x + 1)(x^2 - 3x + 5)}$

*Exercises 69 and 70 from College Algebra and Trigonometry by John Schiller and Marie Wurster. Copyright © 1988 by Scott, Foresman and Company.

10

CONIC SECTIONS

Conic sections are curves that result from the intersection of a plane and a cone. Two examples are the circles and parabolas that were studied in Chapter 3. In this chapter we will look at parabolas in more detail and study the two other types of conic sections, called *ellipses* and *hyperbolas*. In Section 3, the special characteristics of the equations and graphs of the four types of conic sections are discussed. In Section 4, conic sections are rotated on their axes by means of trigonometric identities.

10.1 PARABOLAS

In Chapter 3, we showed that the graph of the quadratic function defined by the equation $y = a(x - h)^2 + k$ is a parabola with vertex at (h, k) and the vertical line $x = h$ as its axis. The equation $x = a(y - k)^2 + h$ also has a parabola as its graph. The graph of this new equation is symmetric to the graph of $y = a(x - h)^2 + k$ with respect to the line $y = x$ since each equation can be obtained by interchanging x and y in the other. Since the graph of $y = a(x - h)^2 + k$ has a vertical axis and vertex at (h, k), the graph of $x = a(y - k)^2 + h$ has a horizontal axis and vertex at (h, k).

Parabola with Horizontal Axis

The parabola with vertex at (h, k) and the horizontal line $y = k$ as axis has an equation of the form

$$x = a(y - k)^2 + h.$$

The parabola opens to the right if $a > 0$ and to the left if $a < 0$.

Parabolas are used in the design of bridges, telescopes, radar equipment, search-lights, and auto headlights. (See Section 3.6.) The ancient Maya used the reflective properties of the parabola in their ball courts.

· · · · · · · · · ·

EXAMPLE 1   Graph $x = 2y^2 + 6y + 5$.

Write this equation in the form $x = a(y - k)^2 + h$ by completing the square on y as follows.

$$x = 2(y^2 + 3y \qquad) + 5$$
$$= 2\left(y^2 + 3y + \frac{9}{4} - \frac{9}{4}\right) + 5$$
$$= 2\left(y^2 + 3y + \frac{9}{4}\right) + 2\left(-\frac{9}{4}\right) + 5$$
$$= 2\left(y + \frac{3}{2}\right)^2 + \frac{1}{2}$$

As this result shows, the vertex of the parabola is the point $(1/2, -3/2)$. The axis is the horizontal line $y = k$, or $y = -3/2$. Using the vertex and the axis and plotting a few additional points gives the graph in Figure 1. ●

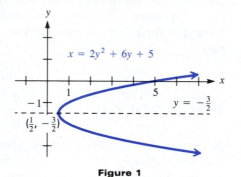

$x = 2y^2 + 6y + 5$

$y = -\frac{3}{2}$

$\left(\frac{1}{2}, -\frac{3}{2}\right)$

Figure 1

An equation of a parabola can be developed from the definition of a parabola as a set of points.

· ·

Definition of Parabola

A **parabola** is the set of points in a plane equidistant from a fixed point and a fixed line. The fixed point is called the **focus,** and the fixed line, the **directrix,** of the parabola.

An equation of a parabola can be found from the definition as follows. Let the directrix be the line $y = -c$ and the focus be the point F with coordinates $(0, c)$, as shown in Figure 2. To get the equation of the set of points that are the same

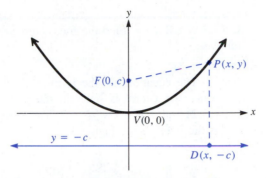

Figure 2

distance from the line $y = -c$ and the point $(0, c)$, choose one such point P and give it coordinates (x, y). Then, since $d(P, F)$ and $d(P, D)$ must have the same length, using the distance formula gives

$$d(P, F) = d(P, D)$$
$$\sqrt{(x - 0)^2 + (y - c)^2} = \sqrt{(x - x)^2 + (y - (-c))^2}$$
$$\sqrt{x^2 + (y - c)^2} = \sqrt{(y + c)^2}$$
$$x^2 + y^2 - 2yc + c^2 = y^2 + 2yc + c^2$$
$$x^2 = 4cy.$$

This discussion is summarized below.

Parabola with a Vertical Axis

> The parabola with focus at $(0, c)$ and directrix $y = -c$ has equation
>
> $$x^2 = 4cy$$
>
> The parabola has a vertical axis, opens upward if $c > 0$, and opens downward if $c < 0$.

The definition of a parabola given above has led to another form of the equation $y = ax^2$, discussed in Chapter 3, with $a = 1/(4c)$.

If the directrix is the line $x = -c$, and the focus is at $(c, 0)$, using the definition of a parabola and the distance formula leads to the equation of a parabola with a horizontal axis. (See Exercise 39.)

Parabola with a Horizontal Axis

> The parabola with focus at $(c, 0)$ and directrix $x = -c$ has equation
>
> $$y^2 = 4cx.$$
>
> The parabola opens to the right if $c > 0$, to the left if $c < 0$, and has a horizontal axis.

. . . .

EXAMPLE 2 Find the focus, directrix, vertex, and axis of the following parabolas.

(a) $x^2 = 8y$

The equation has the form $x^2 = 4cy$, so set $4c = 8$, from which $c = 2$. Since the x-term is squared, the parabola is vertical, with focus at $(0, c) = (0, 2)$ and directrix $y = -2$. The vertex is $(0, 0)$, and the axis of the parabola is the y-axis. (See Figure 3.)

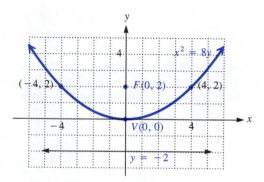

Figure 3

(b) $y^2 = -28x$

This equation has the form $y^2 = 4cx$, with $4c = -28$, so that $c = -7$. The parabola is horizontal, with focus $(-7, 0)$, directrix $x = 7$, vertex $(0, 0)$, and x-axis as axis of the parabola. Since c is negative, the graph opens to the left, as shown in Figure 4. ●

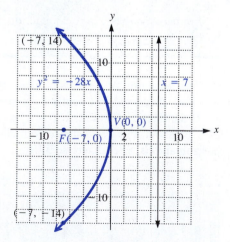

Figure 4

· · · · · · · · ·

EXAMPLE 3   Write an equation for each of the following parabolas.

(a) Focus (2/3, 0) and vertex at the origin

Since the focus (2/3, 0) is on the x-axis, the parabola is horizontal and opens to the right because $c = 2/3$ is positive. (See Figure 5.) The equation, which will have the form $y^2 = 4cx$, is

$$y^2 = 4\left(\frac{2}{3}\right)x, \quad \text{or} \quad y^2 = \frac{8}{3}x.$$

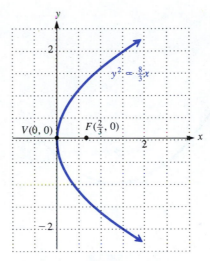

Figure 5

(b) Vertical axis, vertex at the origin, through the point $(-2, 12)$

The parabola will have an equation of the form $x^2 = 4cy$ because the axis is vertical. Since the point $(-2, 12)$ is on the graph, it must satisfy the equation. Substitute $x = -2$ and $y = 12$ into $x^2 = 4cy$ to get

$$(-2)^2 = 4c(12)$$
$$4 = 48c$$
$$c = \frac{1}{12},$$

which gives as an equation of the parabola,

$$x^2 = \frac{1}{3}y, \quad \text{or} \quad y = 3x^2. \quad \bullet$$

In Figure 6 on the following page, a horizontal line has been drawn through the focus of a parabola. Line segment AB is called the **focal chord**. As shown in Figure

6, the endpoints of the focal chord have y-values of c. The x-values of the endpoints can be found from the equation $x^2 = 4cy$.

$$x^2 = 4cy = 4c(c) = 4c^2$$
$$x = \pm 2c$$

Thus, the endpoints of the focal chord are $(-2c, c)$ and $(2c, c)$. These points are useful in graphing parabolas. Figure 7 shows the focal chord with endpoints $(c, 2c)$ and $(c, -2c)$ for a parabola with a horizontal axis.

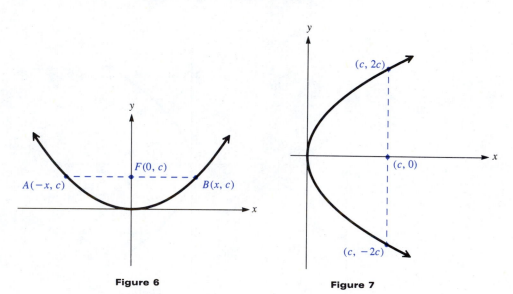

Figure 6

Figure 7

10.1 Exercises

Graph each horizontal parabola.

1. $x = y^2 + 2$ **2.** $x = -y^2$ **3.** $x = (y + 1)^2$ **4.** $x = (y - 3)^2$

5. $x = (y + 2)^2 - 1$ **6.** $x = (y - 4)^2 + 2$ **7.** $x = -2(y + 3)^2$ **8.** $x = -3(y - 1)^2 + 2$

9. $x = \frac{1}{2}(y + 1)^2 + 3$ **10.** $x = \frac{2}{3}(y - 3)^2 + 2$ **11.** $x = y^2 + 2y - 8$ **12.** $x = y^2 - 6y + 7$

13. $x = -2y^2 + 2y - 3$ **14.** $x = -4y^2 - 4y + 3$

Give the focus, directrix, axis, and endpoints of the focal chord for each of the following parabolas.

15. $x^2 = 16y$ **16.** $x^2 = 4y$ **17.** $y = -2x^2$ **18.** $y = 9x^2$

19. $x = 16y^2$ **20.** $x = -32y^2$ **21.** $x = -\frac{1}{16}y^2$ **22.** $x = -\frac{1}{4}y^2$

Write an equation for each of the following parabolas with vertex at the origin.

23. Focus $(0, -2)$ **24.** Focus $(5, 0)$ **25.** Focus $\left(-\dfrac{1}{2}, 0\right)$ **26.** Focus $\left(0, \dfrac{1}{4}\right)$

27. Through $(2, -2\sqrt{2})$, opening to the right   **28.** Through $(\sqrt{3}, 3)$, opening upward

29. Through $(\sqrt{10}, -5)$, opening downward   **30.** Through $(-3, 3)$, opening to the left

31. Through $(2, -4)$, symmetric with respect to the y-axis

32. Through $(3, 2)$, symmetric with respect to the x-axis

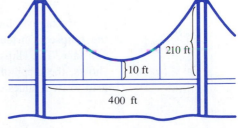

33. The cable in the center portion of a bridge is supported as shown in the figure to form a parabola. The center vertical cable is 10 ft high, the supports are 210 ft high, and the distance between the two supports is 400 ft. Find the height of the remaining vertical cables, if the vertical cables are evenly spaced. (Ignore the width of the supports and cables.)

Exercise 33

34. An arch in the shape of a parabola has the dimensions shown in the figure. How wide is the arch 9 ft up?

35. Explain why the vertex is the point on a parabola that is closest to the focus.

36. Find the equation of the parabola with vertex $(1, 2)$, axis parallel to the x-axis, and passing through the point $(13, 4)$.

37. Find the equation of the parabola having the x-axis as axis and passing through the points $(-4, 11)$ and $(2, 5)$.

38. A parabolic reflector for a car's headlight is 6 in. across and 2 in. deep. Find the distance of the bulb from the vertex. (*Note:* The bulb is located at the focus.)

Exercise 34

39. Prove that the parabola with focus $(c, 0)$ and directrix $x = -c$ has equation $y^2 = 4cx$.

40. Use the definition of a parabola to find an equation of the parabola with vertex at (h, k) and a vertical axis. Let the distance from the vertex to the focus and the distance from the vertex to the directrix be c, where $c > 0$.

10.2 ELLIPSES AND HYPERBOLAS

Ellipses   As the earth travels around the sun over a year's time, it traces out a curve called an *ellipse*. There are many applications of ellipses. In one medical application, patients with kidney stones are treated by being placed in a water bath in a tub with an elliptical cross section. Several hundred spark discharges are produced at one focus of the ellipse (see Figure 8 on the following page), with the kidney stone at the other focus. The discharges go through the water, causing the stone to break up into small pieces which can be readily excreted from the body.

An ellipse is defined as follows.

**Definition of
Ellipse**

An **ellipse** is the set of all points in a plane the sum of whose distances from two fixed points is constant. Each fixed point is called a **focus** (plural, **foci**) of the ellipse.

For example, the ellipse in Figure 8 has foci at points F and F'. By the definition, the ellipse is made up of all points P such that the sum $d(P, F) + d(P, F')$ is constant. The ellipse in Figure 8 has its center at the origin. As the vertical line test shows, the graph of Figure 8 is not the graph of a function, since one value of x can lead to two values of y.

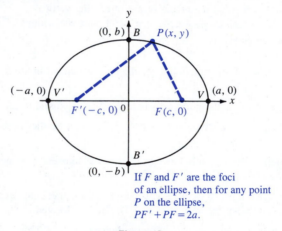

If F and F' are the foci
of an ellipse, then for any point
P on the ellipse,
$PF' + PF = 2a$.

Figure 8

To obtain an equation for an ellipse centered at the origin, let the two foci have coordinates $(-c, 0)$ and $(c, 0)$, respectively. Let the sum of the distances from any point $P(x, y)$ on the ellipse to the two foci be $2a$. By the distance formula, segment PF has length

$$d(P, F) = \sqrt{(x - c)^2 + y^2},$$

while segment PF' has length

$$d(P, F') = \sqrt{[x - (-c)]^2 + y^2} = \sqrt{(x + c)^2 + y^2}.$$

The sum of the lengths $d(P, F)$ and $d(P, F')$ must be $2a$, or

$$\sqrt{(x - c)^2 + y^2} + \sqrt{(x + c)^2 + y^2} = 2a.$$

By squaring both sides and rearranging terms, we can write this equation as

$$\frac{x^2}{a^2} + \frac{y^2}{a^2 - c^2} = 1. \tag{$*$}$$

(The algebraic details are left for Exercise 52.) Since $B(0, b)$ is on the ellipse in Figure 8,

$$d(B, F) + \quad d(B, F') = 2a$$
$$\sqrt{(-c)^2 + b^2} + \sqrt{c^2 + b^2} = 2a$$
$$2\sqrt{c^2 + b^2} = 2a$$
$$\sqrt{c^2 + b^2} = a$$
$$c^2 + b^2 = a^2$$
$$b^2 = a^2 - c^2.$$

Replacing $a^2 - c^2$ with b^2 in equation (*) gives

$$\frac{x^2}{a^2} + \frac{y^2}{b^2} = 1,$$

the **standard form** of the equation of an ellipse centered at the origin with foci on the x-axis.

Letting $y = 0$ in the standard form gives

$$\frac{x^2}{a^2} + \frac{0^2}{b^2} = 1$$
$$\frac{x^2}{a^2} = 1$$
$$x^2 = a^2$$
$$x = \pm a$$

as the x-intercepts of the ellipse. The points $V'(-a, 0)$ and $V(a, 0)$ are the **vertices** of the ellipse; the segment VV' is the **major axis.** In a similar manner, letting $x = 0$ shows that the y-intercepts are $\pm b$; the segment connecting $(0, b)$ and $(0, -b)$ is the **minor axis.** We assumed throughout the work above that the foci were on the x-axis. If the foci were on the y-axis, an almost identical proof could be used to get the standard form

$$\frac{y^2}{a^2} + \frac{x^2}{b^2} = 1.$$

CAUTION   Do not be confused by the two standard forms—in one case a^2 is associated with x^2; in the other case a^2 is associated with y^2. However, in practice it is necessary only to find the intercepts of the graph—if the positive x-intercept is larger than the positive y-intercept, the major axis is horizontal, and otherwise it is vertical. When using the relationship $a^2 - c^2 = b^2$, or $a^2 - b^2 = c^2$, choose a^2 and b^2 so that $a^2 > b^2$.

A summary of this work with ellipses follows on the next page.

Standard Equations for Ellipses

The ellipse with center at the origin and equation

$$\frac{x^2}{a^2} + \frac{y^2}{b^2} = 1 \quad (a > b)$$

has vertices $(\pm a, 0)$, endpoints of the minor axis $(0, \pm b)$, and foci $(\pm c, 0)$, where $c^2 = a^2 - b^2$.

The ellipse with center at the origin and equation

$$\frac{y^2}{a^2} + \frac{x^2}{b^2} = 1 \quad (a > b)$$

has vertices $(0, \pm a)$, endpoints of the minor axis $(\pm b, 0)$, and foci $(0, \pm c)$, where $c^2 = a^2 - b^2$.

An ellipse is symmetric with respect to its major axis, its minor axis, and its center.

EXAMPLE 1 Graph $4x^2 + 9y^2 = 36$ and find the coordinates of the foci.

To obtain the standard form for the equation of an ellipse, divide each side by 36 to get

$$\frac{x^2}{9} + \frac{y^2}{4} = 1.$$

The x-intercepts of this ellipse are ± 3, and the y-intercepts ± 2. Additional ordered pairs satisfying the equation of the ellipse may be found if desired by choosing x-values and using the equation to find the corresponding y-values. The graph of the ellipse is shown in Figure 9. Since $9 > 4$, find the foci by letting $c^2 = 9 - 4 = 5$ so that $c = \pm \sqrt{5}$. The major axis is along the x-axis so the foci are at $(-\sqrt{5}, 0)$ and $(\sqrt{5}, 0)$. ●

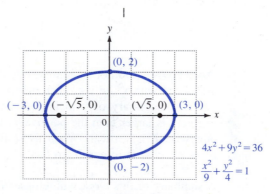

Figure 9

EXAMPLE 2   Find the equation of the ellipse having center at the origin, foci at $(0, 3)$ and $(0, -3)$, and major axis of length 8 units.

Since the major axis is 8 units long,

$$2a = 8$$

or

$$a = 4.$$

To find b^2, use the relationship $a^2 - b^2 = c^2$. Here $a = 4$ and $c = 3$. Substituting for a and c gives

$$a^2 - b^2 = c^2$$
$$4^2 - b^2 = 3^2$$
$$16 - b^2 = 9$$
$$b^2 = 7.$$

Since the foci are on the y-axis, the larger intercept, a, is used to find the denominator for y^2, giving the equation in standard form as

$$\frac{y^2}{16} + \frac{x^2}{7} = 1.$$

A graph of this ellipse is shown in Figure 10. ●

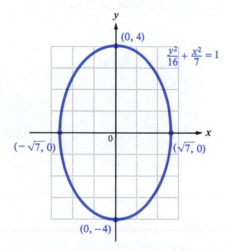

Figure 10

EXAMPLE 3   Graph $\dfrac{y}{4} = \sqrt{1 - \dfrac{x^2}{25}}$.

Square both sides to get

$$\frac{y^2}{16} = 1 - \frac{x^2}{25} \quad \text{or} \quad \frac{x^2}{25} + \frac{y^2}{16} = 1,$$

the equation of an ellipse with x-intercepts ± 5 and y-intercepts ± 4. Since $\sqrt{1 - x^2/25} \geq 0$, the only possible values of y are those making $y/4 \geq 0$, giving the half-ellipse shown in Figure 11. While the graph of the ellipse $x^2/25 + y^2/16 = 1$ is not the graph of a function, the half-ellipse in Figure 11 *is* the graph of a function. The domain of this function is the interval $[-5, 5]$ and the range is $[0, 4]$. ●

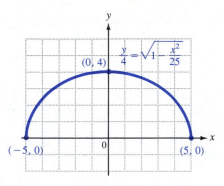

Figure 11

Just as a circle need not have its center at the origin, an ellipse may also have its center translated away from the origin.

Ellipse Centered at (h, k)

An ellipse centered at (h, k) with horizontal major axis of length $2a$ has equation

$$\frac{(x - h)^2}{a^2} + \frac{(y - k)^2}{b^2} = 1.$$

There is a similar result for ellipses having a vertical major axis.

This result can be proven from the definition of an ellipse.

EXAMPLE 4 Graph $\dfrac{(x - 2)^2}{9} + \dfrac{(y + 1)^2}{16} = 1$.

The graph of this equation is an ellipse centered at $(2, -1)$. As mentioned earlier, ellipses always have $a > b$. For this ellipse, then, $a = 4$ and $b = 3$. Since $a = 4$ is associated with y^2, the vertices of the ellipse are on the vertical line through $(2, -1)$. Find the vertices by locating two points on the vertical line through $(2, -1)$, one 4 units up from $(2, -1)$ and one 4 units down. The vertices are $(2, 3)$ and $(2, -5)$. Locate two other points on the ellipse by locating points on a horizontal line through $(2, -1)$, one 3 units to the right and one 3 units to the left. Find additional points as needed. The final graph is in Figure 12. ●

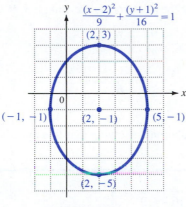

Figure 12

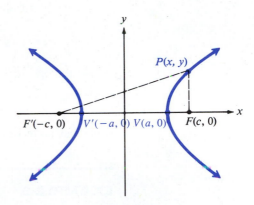

Figure 13

Hyperbolas An ellipse was defined as the set of all points in a plane the sum of whose distances from two fixed points is a constant. A *hyperbola* is defined similarly.

Definition of Hyperbola

> A **hyperbola** is the set of all points in a plane the *difference* of whose distances from two fixed points is constant. The two fixed points are called the **foci** of the hyperbola.

Suppose a hyperbola has center at the origin and foci at $F'(-c, 0)$ and $F(c, 0)$. The midpoint of the segment $F'F$ is the **center** of the hyperbola and the points $V'(-a, 0)$ and $V(a, 0)$ are the **vertices** of the hyperbola. The line segment $V'V$ is the **transverse axis** of the hyperbola. (See Figure 13.)

As with the ellipse,

$$d(V, F') - d(V, F) = (c + a) - (c - a) = 2a,$$

so the constant in the definition is $2a$, and

$$d(P, F') - d(P, F) = 2a.$$

The distance formula and algebraic manipulation similar to that used for finding an equation for an ellipse (see Exercise 51) produce the result

$$\frac{x^2}{a^2} - \frac{y^2}{c^2 - a^2} = 1.$$

Letting $b^2 = c^2 - a^2$ gives

$$\frac{x^2}{a^2} - \frac{y^2}{b^2} = 1$$

as an equation of the hyperbola in Figure 13.

Letting $y = 0$ shows that the x-intercepts are $\pm a$. If $x = 0$ the equation becomes

$$\frac{0^2}{a^2} - \frac{y^2}{b^2} = 1$$

$$-\frac{y^2}{b^2} = 1$$

$$y^2 = -b^2,$$

which has no real number solutions, showing that this hyperbola has no y-intercepts.

· · · · · · · · ·

EXAMPLE 5 Graph $\dfrac{x^2}{16} - \dfrac{y^2}{9} = 1.$

This hyperbola has x-intercepts 4 and -4 and no y-intercepts. To sketch the graph, we can find some other points that lie on the graph. For example, letting $x = 6$ gives

$$\frac{6^2}{16} - \frac{y^2}{9} = 1$$

$$-\frac{y^2}{9} = 1 - \frac{6^2}{16}$$

$$\frac{y^2}{9} = \frac{20}{16}$$

$$y^2 = \frac{180}{16} = \frac{45}{4}$$

$$y = \frac{\pm 3\sqrt{5}}{2} \approx \pm 3.4.$$

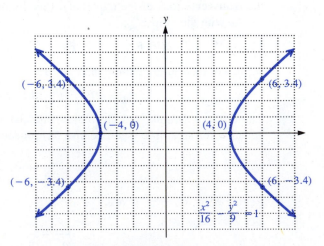

Figure 14

The graph includes the points $(6, 3.4)$ and $(6, -3.4)$. Also, letting $x = -6$ would still give $y \approx \pm 3.4$ with the points $(-6, 3.4)$ and $(-6, -3.4)$ also on the graph. These points, along with other points on the graph, were used to help sketch the final graph shown in Figure 14. ●

The basic information on hyperbolas is summarized as follows.

Standard Equations for Hyperbolas

The hyperbola with center at the origin and equation

$$\frac{x^2}{a^2} - \frac{y^2}{b^2} = 1$$

has vertices $(\pm a, 0)$ and foci $(\pm c, 0)$, where $c^2 = a^2 + b^2$.

The hyperbola with center at the origin and equation

$$\frac{y^2}{a^2} - \frac{x^2}{b^2} = 1$$

has vertices $(0, \pm a)$ and foci $(0, \pm c)$ where $c^2 = a^2 + b^2$.

Starting with the equation for a hyperbola $(x^2/a^2) - (y^2/b^2) = 1$ and solving for y, we have

$$\frac{x^2}{a^2} - 1 = \frac{y^2}{b^2}$$

$$\frac{x^2 - a^2}{a^2} = \frac{y^2}{b^2}$$

or

$$y = \pm \frac{b}{a} \sqrt{x^2 - a^2}. \qquad (*)$$

If x^2 is very large in comparison to a^2, the difference $x^2 - a^2$ would be very close to x^2. If this happens, then the points satisfying equation $(*)$ above would be very close to one of the lines

$$y = \pm \frac{b}{a} x.$$

Thus, as $|x|$ gets larger and larger, the points of the hyperbola $x^2/a^2 - y^2/b^2 = 1$ come closer and closer to the lines $y = (\pm b/a)x$. These lines, called the **asymptotes** of the hyperbola, are very helpful for graphing the hyperbola.

EXAMPLE 6   Graph $\dfrac{x^2}{25} - \dfrac{y^2}{49} = 1$ and find the coordinates of the foci.

For this hyperbola, $a = 5$ and $b = 7$. With these values, $y = (\pm b/a)x$ becomes $y = (\pm 7/5)x$. If $x = 5$, then $y = (\pm 7/5)(5) = \pm 7$, while $x = -5$ also gives

$y = \pm 7$. These four points, $(5, 7)$, $(5, -7)$, $(-5, 7)$, and $(-5, -7)$, lead to the rectangle shown in Figure 15. The extended diagonals of this rectangle are the asymptotes of the hyperbola. The hyperbola crosses the x-axis at 5 and -5, as shown in Figure 15. Find the foci by letting $c^2 = a^2 + b^2 = 25 + 49 = 74$ so that $c = \pm\sqrt{74}$. Therefore, the foci are $(\sqrt{74}, 0)$ and $(-\sqrt{74}, 0)$.   •

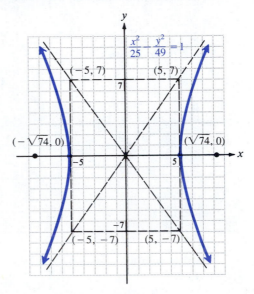

Figure 15

The rectangle used to graph the hyperbola in Example 6 is called the **fundamental rectangle.**

While $a > b$ for an ellipse, the examples above show that for hyperbolas, it is possible that $a > b$ or $a < b$; other examples would show that a might equal b, also. If the foci of a hyperbola are on the y-axis, the equation of the hyperbola is of the form

$$\frac{y^2}{a^2} - \frac{x^2}{b^2} = 1, \quad \text{with asymptotes} \quad y = \pm\frac{a}{b}x.$$

CAUTION   If the foci of the hyperbola are on the x-axis, we found that the asymptotes have equations $y = \pm(b/a)x$, while foci on the y-axis lead to asymptotes $y = \pm(a/b)x$. There is an obvious chance for confusion here; to avoid mistakes write the equation of the hyperbola in either the form

$$\frac{x^2}{a^2} - \frac{y^2}{b^2} = 1 \quad \text{or} \quad \frac{y^2}{a^2} - \frac{x^2}{b^2} = 1,$$

and replace 1 with 0. Solving the resulting equation for y produces the proper equations for the asymptotes. (The reason why this process works is explained in more advanced courses.)

EXAMPLE 7   Graph $25y^2 - 4x^2 = 100$.

Divide each side by 100 to get

$$\frac{y^2}{4} - \frac{x^2}{25} = 1.$$

This hyperbola is centered at the origin, has foci on the y-axis, and has y-intercepts 2 and -2. To find the equations of the asymptotes, replace 1 with 0.

$$\frac{y^2}{4} - \frac{x^2}{25} = 0$$

$$\frac{y^2}{4} = \frac{x^2}{25}$$

$$y^2 = \frac{4x^2}{25}$$

$$y = \pm\frac{2}{5}x$$

To graph the asymptotes, use the points $(5, 2)$, $(5, -2)$, $(-5, 2)$, and $(-5, -2)$ that determine the fundamental rectangle shown in Figure 16. The diagonals of this rectangle are the asymptotes for the graph, as shown in Figure 16.   ●

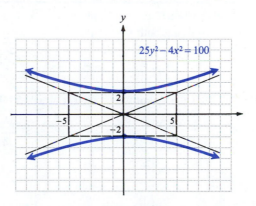

Figure 16

In each of the graphs of hyperbolas considered so far, the center is the origin and the asymptotes pass through the origin. This feature holds in general; the asymptotes of *any* hyperbola pass through the center of the hyperbola. The next example discusses a hyperbola with its center translated away from the origin.

EXAMPLE 8   Graph $\dfrac{(y + 2)^2}{9} - \dfrac{(x + 3)^2}{4} = 1$.

This equation represents a hyperbola centered at $(-3, -2)$. For this vertical hyperbola, $a = 3$ and $b = 2$. Locate the vertices by adding and subtracting 3 to the *y*-value of the center, $(-3, -2)$. Thus the vertices are at $(-3, 1)$ and $(-3, -5)$. The asymptotes have slopes $\pm 3/2$ and pass through the center $(-3, -2)$. The completed graph appears in Figure 17. ●

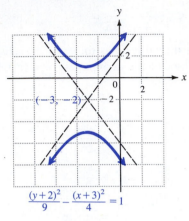

$$\frac{(y+2)^2}{9} - \frac{(x+3)^2}{4} = 1$$

Figure 17

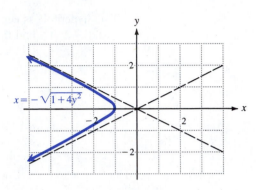

$x = -\sqrt{1 + 4y^2}$

Figure 18

Our final example shows an equation whose graph is only half of a hyperbola.

EXAMPLE 9   Graph $x = -\sqrt{1 + 4y^2}$.
   Squaring both sides gives

$$x^2 = 1 + 4y^2$$

or

$$x^2 - 4y^2 = 1.$$

To find the asymptotes, rewrite 4 as $1/(1/4)$ to change this equation into

$$x^2 - \frac{y^2}{1/4} = 0$$

or

$$\frac{1}{4}x^2 = y^2,$$

giving

$$y = \pm \frac{1}{2}x.$$

Since the given equation $x = -\sqrt{1 + 4y^2}$ restricts x to nonpositive values, the graph is the left branch of the hyperbola, as shown in Figure 18. ●

10.2 Exercises

Sketch the graph of each of the following. Give the endpoints of the major axis for ellipses and the equation of the asymptotes for hyperbolas. Give the center and foci for each figure.

1. $\dfrac{x^2}{9} + \dfrac{y^2}{4} = 1$ **2.** $\dfrac{x^2}{16} + \dfrac{y^2}{36} = 1$ **3.** $\dfrac{x^2}{9} + y^2 = 1$ **4.** $\dfrac{y^2}{16} - \dfrac{x^2}{9} = 1$

5. $\dfrac{x^2}{6} + \dfrac{y^2}{9} = 1$ **6.** $\dfrac{x^2}{8} - \dfrac{y^2}{12} = 1$ **7.** $x^2 + 4y^2 = 16$ **8.** $25x^2 + 9y^2 = 225$

9. $x^2 = 9 + y^2$ **10.** $y^2 = 16 + x^2$ **11.** $2x^2 + y^2 = 8$ **12.** $9x^2 - 25y^2 = 225$

13. $25x^2 - 4y^2 = -100$ **14.** $4x^2 - y^2 = -16$ **15.** $\dfrac{x^2}{1/9} + \dfrac{y^2}{1/16} = 1$ **16.** $\dfrac{x^2}{4/25} + \dfrac{y^2}{9/49} = 1$

17. $\dfrac{64x^2}{9} + \dfrac{25y^2}{36} = 1$ **18.** $\dfrac{121x^2}{25} + \dfrac{16y^2}{9} = 1$

Sketch the graph of each of the following.

19. $\dfrac{(x + 2)^2}{16} + \dfrac{(y + 1)^2}{9} = 1$ **20.** $\dfrac{(x - 2)^2}{25} + \dfrac{(y - 1)^2}{4} = 1$

21. $\dfrac{(x - 1)^2}{9} + \dfrac{(y + 3)^2}{25} = 1$ **22.** $\dfrac{(x + 3)^2}{16} + \dfrac{(y - 2)^2}{36} = 1$

23. $\dfrac{(x - 3)^2}{16} - \dfrac{(y + 2)^2}{49} = 1$ **24.** $\dfrac{(y - 5)^2}{4} - \dfrac{(x + 1)^2}{9} = 1$

25. $\dfrac{(y + 1)^2}{25} - \dfrac{(x - 3)^2}{36} = 1$ **26.** $\dfrac{(x + 2)^2}{16} - \dfrac{(y + 2)^2}{25} = 1$

Sketch the graph of each of the following. Identify any that are the graphs of functions.

27. $\dfrac{x}{4} = \sqrt{1 - \dfrac{y^2}{9}}$ **28.** $\dfrac{y}{2} = \sqrt{1 - \dfrac{x^2}{25}}$ **29.** $\dfrac{y}{3} = \sqrt{1 + \dfrac{x^2}{16}}$

30. $x = \sqrt{1 + \dfrac{y^2}{36}}$ **31.** $x = -\sqrt{1 - \dfrac{y^2}{64}}$ **32.** $y = -\sqrt{1 - \dfrac{x^2}{100}}$

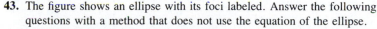

Find equations for each of the following ellipses.

33. x-intercepts ± 4; foci at $(-2, 0)$ and $(2, 0)$ **34.** y-intercepts ± 3; foci at $(0, \sqrt{3})$, $(0, -\sqrt{3})$

35. Endpoints of major axis at $(6, 0)$, $(-6, 0)$; $c = 4$ **36.** y-intercepts ± 5; $b = 2$

37. Center $(3, -2)$, $a = 5$, $c = 3$, major axis vertical

38. Center $(2, 0)$, minor axis of length 6, major axis horizontal and of length 9

Find equations for each of the following hyperbolas.

39. x-intercepts ± 3; foci at $(-4, 0)$, $(4, 0)$

40. y-intercepts ± 5; foci at $(0, 3\sqrt{3})$, $(0, -3\sqrt{3})$

41. Asymptotes $y = \pm(3/5)x$; y-intercepts $(0, 3)$, $(0, -3)$

42. Center at the origin, passing through $(5, 3)$ and $(-10, 2\sqrt{21})$, no y-intercepts

43. The figure shows an ellipse with its foci labeled. Answer the following questions with a method that does not use the equation of the ellipse.
(a) Is the point $(2, 4)$ on the ellipse?
(b) What is the length of the major axis of the ellipse?

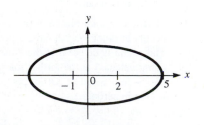

Exercise 43

44. The figure shows a hyperbola with its foci labeled. Answer the following questions with a method that does not use the equation of the hyperbola.
(a) Is the point (10, 2) on the hyperbola?
(b) What are the x-intercepts of the hyperbola?

45. Find the equation of the ellipse with foci (0, −3) and (0, 3) and passing through the point (8, 3).

46. Find the equation of the hyperbola with foci (−6, 0) and (6, 0) and passing through the point (6, 5).

47. The orbit of Mars is an ellipse, with the sun at one focus. An approximate equation for the orbit is

$$\frac{x^2}{5013} + \frac{y^2}{4970} = 1,$$

where x and y are measured in millions of miles.
(a) Find the length of the major axis.
(b) Find the length of the minor axis.

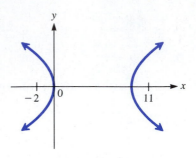

Exercise 44

48. Draftspeople often use the method shown on the sketch to draw an ellipse. Explain why the method works.

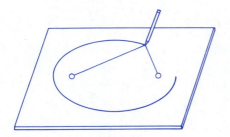

Exercise 48

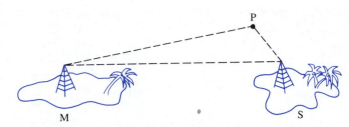

Exercise 49

49. Ships and planes often use a location finding system called LORAN. With this system, a radio transmitter at M on the figure sends out a series of pulses. When each pulse is received at transmitter S, it then sends out a pulse. A ship at P receives pulses from both M and S. A receiver on the ship measures the difference in the arrival times of the pulses. The navigator then consults a special map, showing certain curves according to the differences in arrival times. In this way, the ship can be located as lying on a portion of which curve? (This method requires three transmitters acting as two pairs.)

50. Microphones are placed at points (−c, 0) and (c, 0). An explosion occurs at point P(x, y) having positive x-coordinate. (See the figure.) The sound is detected at the closer microphone t seconds before being detected at the farther microphone. Assume that sound travels at a speed of 330 m per sec., and show that P must be on the hyperbola

$$\frac{x^2}{330^2 t^2} - \frac{y^2}{4c^2 - 330^2 t^2} = \frac{1}{4}.$$

51. Suppose a hyperbola has center at the origin, foci at $F'(-c, 0)$ and $F(c, 0)$, and the value $d(P, F') - d(P, F) = 2a$. Let $b^2 = c^2 - a^2$, and show that an equation of the hyperbola is

$$\frac{x^2}{a^2} - \frac{y^2}{b^2} = 1.$$

52. Complete the derivation of the equation of an ellipse by reducing $\sqrt{(x - c)^2 + y^2} + \sqrt{(x + c)^2 + y^2} = 2a$ to $\dfrac{x^2}{a^2} + \dfrac{y^2}{a^2 - c^2} = 1.$

10.3 CONIC SECTIONS

The graphs of parabolas, circles, hyperbolas, and ellipses are called **conic sections** since each graph can be obtained by cutting a cone with a plane as suggested by Figure 19.

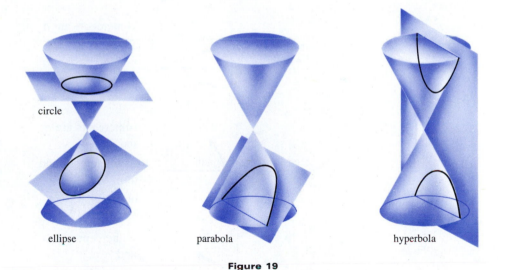

Figure 19

It turns out that all conic sections of the types presented in this chapter have equations of the form

$$Ax^2 + Cy^2 + Dx + Ey + F = 0,$$

where either A or C must be nonzero. The graphs of the conic sections are summarized in Figure 20 on the next page. Ellipses and hyperbolas having centers not at the origin can be shown in much the same way as we show circles and parabolas. Following Figure 20, the special characteristics of the equations of each of the conic sections are summarized.

Equation	Graph	Description	Identification
$y = a(x - h)^2 + k$	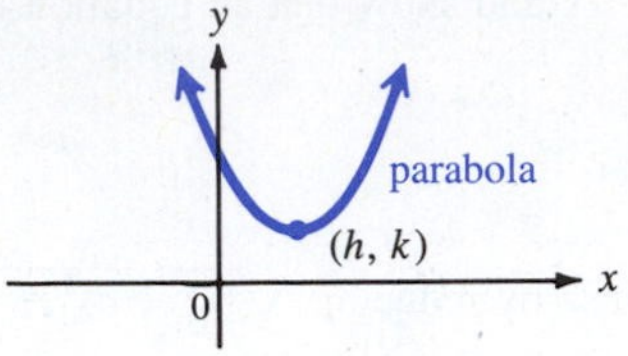 parabola	Opens upward if $a > 0$, downward if $a < 0$. Vertex is at (h, k).	x^2 term y is not squared.
$x = a(y - k)^2 + h$	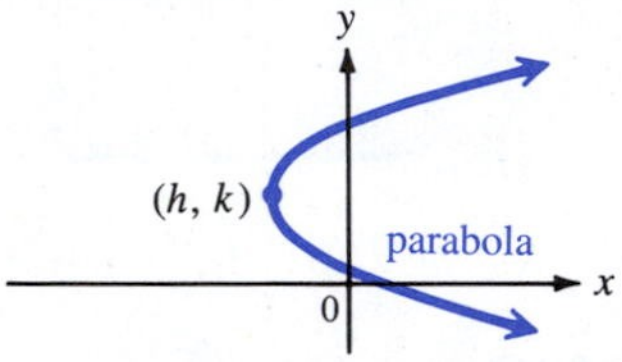 parabola	Opens to right if $a > 0$, to left if $a < 0$. Vertex is at (h, k).	y^2 term x is not squared.
$(x - h)^2 + (y - k)^2 = r^2$	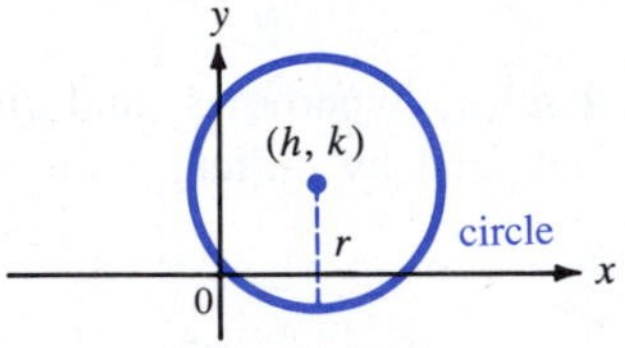 circle	Center is at (h, k), radius is r.	x^2 and y^2 terms have the same positive coefficient.
$\dfrac{x^2}{a^2} + \dfrac{y^2}{b^2} = 1$	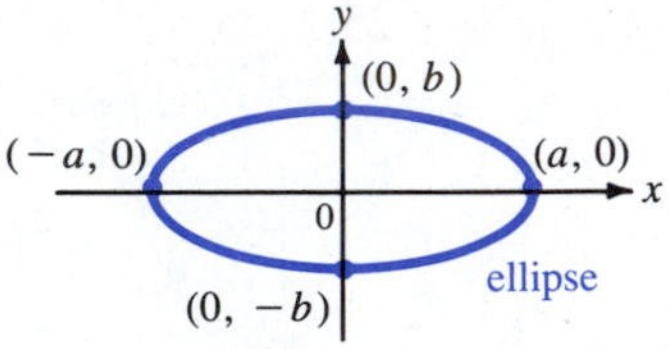 ellipse	x-intercepts are a and $-a$. y-intercepts are b and $-b$.	x^2 and y^2 terms have different positive coefficients.
$\dfrac{x^2}{a^2} - \dfrac{y^2}{b^2} = 1$	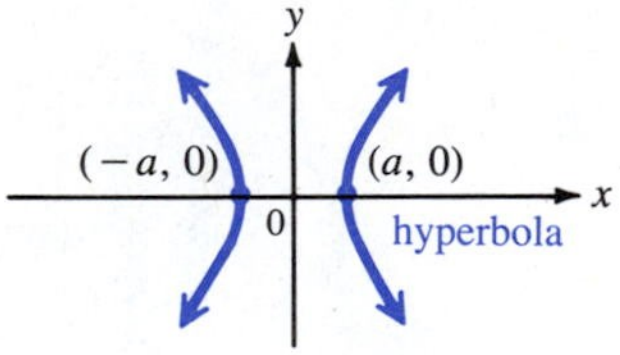 hyperbola	x-intercepts are a and $-a$. Asymptotes found from (a, b), $(a, -b)$, $(-a, -b)$, and $(-a, b)$.	x^2 has a positive coefficient. y^2 has a negative coefficient.
$\dfrac{y^2}{a^2} - \dfrac{x^2}{b^2} = 1$	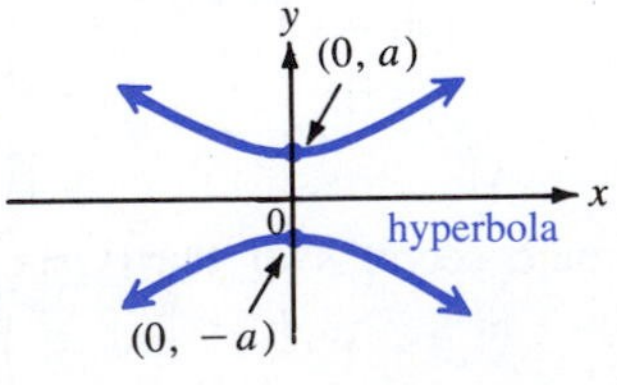 hyperbola	y-intercepts are a and $-a$. Asymptotes found from (b, a), $(b, -a)$, $(-b, a)$, and $(-b, -a)$.	y^2 has a positive coefficient. x^2 has a negative coefficient.

Figure 20

Equations of Conic Sections

Conic section	Characteristic	Example
Parabola	Either $A = 0$ or $C = 0$, but not both	$x^2 = y + 4$ $(y - 2)^2 = -(x + 3)$
Circle	$A = C \neq 0$	$x^2 + y^2 = 16$
Ellipse	$A \neq C$, $AC > 0$	$\dfrac{x^2}{16} + \dfrac{y^2}{25} = 1$
Hyperbola	$AC < 0$	$x^2 - y^2 = 1$

EXAMPLE 1   Decide on the type of conic section represented by each of the following equations, and sketch each graph.

(a) $x^2 = 25 + 5y^2$

Rewriting the equation as

$$x^2 - 5y^2 = 25$$

or

$$\frac{x^2}{25} - \frac{y^2}{5} = 1$$

shows that the equation represents a hyperbola centered at the origin, with asymptotes

$$\frac{x^2}{25} - \frac{y^2}{5} = 0,$$

or

$$y = \frac{\pm\sqrt{5}}{5}x.$$

The x-intercepts are ± 5; the graph is shown in Figure 21.

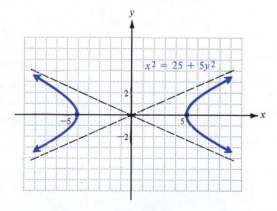

Figure 21

(b) $4x^2 - 16x + 9y^2 + 54y = -61$

Since the coefficients of the x^2 and y^2 terms are unequal and both positive, this equation might represent an ellipse. (It might also represent a single point or no points at all.) To find out, complete the square on x and y.

$$4(x^2 - 4x \quad) + 9(y^2 + 6y \quad) = -61$$
$$4(x^2 - 4x + 4 - 4) + 9(y^2 + 6y + 9 - 9) = -61$$
$$4(x^2 - 4x + 4) - 16 + 9(y^2 + 6y + 9) - 81 = -61$$
$$4(x - 2)^2 + 9(y + 3)^2 = 36$$
$$\frac{(x - 2)^2}{9} + \frac{(y + 3)^2}{4} = 1$$

This equation represents an ellipse having center at $(2, -3)$ and graph as shown in Figure 22.

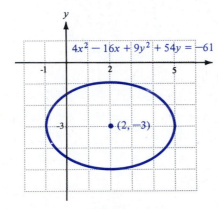

Figure 22

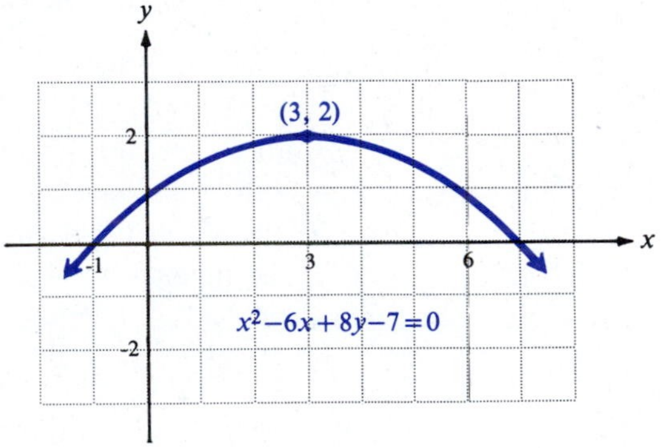

Figure 23

(c) $x^2 - 8x + y^2 + 10y = -41$

Complete the square on both x and y, as follows:

$$(x^2 - 8x + 16 - 16) + (y^2 + 10y + 25 - 25) = -41$$
$$(x - 4)^2 + (y + 5)^2 = 16 + 25 - 41$$
$$(x - 4)^2 + (y + 5)^2 = 0.$$

This result shows that the equation is that of a circle of radius 0; that is, the point $(4, -5)$. Had a negative number been obtained on the right (instead of 0), the equation would have represented no points at all, and there would be no graph.

(d) $x^2 - 6x + 8y - 7 = 0$

Since only one variable is squared (x, and not y), the equation represents a parabola. Rearrange the terms to get the term with y (the variable that is not squared) alone on one side. Then complete the square on the other side of the equation.

$$8y = -x^2 + 6x + 7$$
$$8y = -(x^2 - 6x \qquad) + 7$$
$$8y = -(x^2 - 6x + 9) + 7 + 9$$
$$8y = -(x - 3)^2 + 16$$
$$y = -\frac{1}{8}(x - 3)^2 + 2 \qquad \text{Multiply both sides by 1/8.}$$

The parabola has vertex at (3, 2), and opens downward, as shown in Figure 23. ●

▰ **CAUTION** The next example is designed to serve as a warning about a very common error.

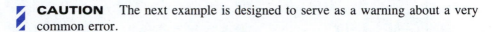

EXAMPLE 2 Graph $4y^2 - 16y - 9x^2 + 18x = -43$.
Complete the square on x and on y.

$$4(y^2 - 4y \qquad) - 9(x^2 - 2x \qquad) = -43$$
$$4(y^2 - 4y + 4) - 9(x^2 - 2x + 1) = -43 + 16 - 9$$
$$4(y - 2)^2 - 9(x - 1)^2 = -36$$

Because of the -36, it is very tempting to say that this equation does not have a graph. However, the minus sign in the middle on the left shows that the graph is that of a hyperbola. Dividing through by -36 and rearranging terms gives

$$\frac{(x - 1)^2}{4} - \frac{(y - 2)^2}{9} = 1,$$

the hyperbola centered at (1, 2), with graph as shown in Figure 24. ●

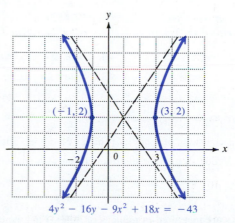

$$4y^2 - 16y - 9x^2 + 18x = -43$$

Figure 24

10.3 Exercises ···

For each of the following equations, identify the corresponding conic section. Draw a graph of each equation that has a graph.

1. $\dfrac{x^2}{4} + \dfrac{y^2}{9} = 1$

2. $\dfrac{x^2}{4} - \dfrac{y^2}{9} = 1$

3. $\dfrac{x^2}{4} - \dfrac{y^2}{4} = 1$

4. $\dfrac{x^2}{4} + \dfrac{y^2}{4} = 1$

5. $x^2 + 2x = x^2 + y - 6$

6. $y^2 - 4y = y^2 + 3 - x$

7. $x^2 = 25 + y^2$

8. $x^2 = 25 - y^2$

9. $9x^2 + 36y^2 = 36$

10. $x^2 = 4y - 8$

11. $\dfrac{(x+3)^2}{16} + \dfrac{(y-2)^2}{16} = 1$

12. $\dfrac{(x-4)}{8} - \dfrac{(y+1)^2}{2} = 0$

13. $y^2 - 4y = x + 4$

14. $11 - 3x = 2y^2 - 8y$

15. $(x+7)^2 + (y-5)^2 + 4 = 0$

16. $4(x-3)^2 + 3(y+4)^2 = 0$

17. $3x^2 + 6x + 3y^2 - 12y = 12$

18. $2x^2 - 8x + 5y^2 + 20y = 12$

19. $x^2 - 6x + y = 0$

20. $x - 4y^2 - 8y = 0$

21. $4x^2 - 8x - y^2 - 6y = 6$

22. $x^2 + 2x = y^2 - 4y - 2$

23. $4x^2 - 8x + 9y^2 + 54y = -84$

24. $3x^2 + 12x + 3y^2 = -11$

25. $6x^2 - 12x + 6y^2 - 18y + 25 = 0$

26. $4x^2 - 24x + 5y^2 + 10y + 41 = 0$

27. Identify the type of conic section consisting of the set of all points in the plane for which the distance from the point $(5, 0)$ is one-half of the distance from the line $x = 20$.

28. Identify the type of conic section consisting of the set of all points in the plane for which the distance from the point $(3, 0)$ is three-halves of the distance from the line $x = 4/3$.

The eccentricity, e, of an ellipse or a hyperbola is defined by $e = c/a$. Ellipses have eccentricity < 1 and hyperbolas have eccentricity > 1. Parabolas are said to have eccentricity $e = 1$. Find the eccentricity of each of the following ellipses and hyperbolas.

29. $12x^2 + 9y^2 = 36$

30. $8x^2 - y^2 = 16$

31. $x^2 - y^2 = 4$

32. $x^2 + 2y^2 = 8$

33. $4x^2 + 7y^2 = 28$

34. $9x^2 - y^2 = 1$

35. $x^2 - 9y^2 = 18$

36. $x^2 + 10y^2 = 10$

37. $2x^2 + y^2 = 32$

38. $5x^2 - 4y^2 = 20$

Write an equation for each conic in Exercises 39–48. Each parabola has vertex at the origin, and each ellipse or hyperbola is centered at the origin.

39. Focus at $(0, 8)$ and $e = 1$

40. Focus at $(-2, 0)$ and $e = 1$

41. Focus at $(3, 0)$ and $e = 1/2$

42. Focus at $(0, -2)$ and $e = 2/3$

43. Vertex at $(-6, 0)$ and $e = 2$

44. Vertex at $(0, 4)$ and $e = 5/3$

45. Focus at $(0, -1)$ and $e = 1$

46. Focus at $(2, 0)$ and $e = 6/5$

47. Vertical major axis of length 6 and $e = 4/5$

48. Vertical transverse axis of length 8 and $e = 7/3$

49. What is the eccentricity of a circle? (Think of a circle as an ellipse with $a = b$.)

50. The orbit of Venus around the sun is an ellipse with equation

$$\frac{x^2}{5013} + \frac{y^2}{4970} = 1,$$

where x and y are measured in millions of miles. Find the eccentricity of this ellipse.

10.4 ROTATION OF AXES

In the previous section it was shown that the equation $Ax^2 + Cy^2 + Dx + Ey + F = 0$ represents a conic section. The letter B was not used in writing the equation above; it was reserved for the one other possible second-degree term, Bxy. Every conic section has an equation of the form $Ax^2 + Bxy + Cy^2 + Dx + Ey + F = 0$ where either $A \neq 0$, $B \neq 0$, or $C \neq 0$. If $B \neq 0$ and the equation has (real) solutions, then its graph is *usually* a conic section that is rotated with respect to the xy-coordinate axes. This section shows how to use the work presented earlier to graph rotated conics.

If we begin with an xy-coordinate system having origin O and rotate the axes about O through an angle θ, the new coordinate system is called a **rotation** of the xy-system. Trigonometric identities can be used to obtain equations for converting the coordinates of a point from the xy-system to the rotated $x'y'$-system. Let P be any point other than the origin, with coordinates (x, y) in the xy-system and (x', y') in the $x'y'$-system. (See Figure 25.) Let $OP = r$, and let α represent the angle made by OP and the x'-axis. Then,

$$\cos(\theta + \alpha) = \frac{OA}{r} = \frac{x}{r}$$

$$\sin(\theta + \alpha) = \frac{AP}{r} = \frac{y}{r}$$

$$\cos \alpha = \frac{OB}{r} = \frac{x'}{r}$$

$$\sin \alpha = \frac{PB}{r} = \frac{y'}{r}.$$

These four statements can be rewritten as

$$x = r \cos(\theta + \alpha), \qquad y = r \sin(\theta + \alpha),$$
$$x' = r \cos \alpha, \qquad y' = r \sin \alpha.$$

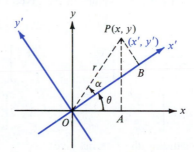

Figure 25

From the trigonometric identity for the cosine of the sum of two angles,

$$x = r \cos (\theta + \alpha)$$
$$= r(\cos \theta \cdot \cos \alpha - \sin \theta \cdot \sin \alpha)$$
$$= (r \cos \alpha) \cos \theta - (r \sin \alpha) \sin \theta$$
$$x = x' \cos \theta - y' \sin \theta.$$

Similarly, from the identity for the sine of the sum of two angles, $y = x' \sin \theta + y' \cos \theta$. This proves the following result.

Rotation Equations

If the rectangular coordinate axes are rotated about the origin through an angle θ, and if the coordinates of a point P are (x, y) and (x', y') with respect to the xy-system and the $x'y'$-system, then the **rotation equations** are

$$x = x' \cos \theta - y' \sin \theta$$

and
$$y = x' \sin \theta + y' \cos \theta.$$

EXAMPLE 1   The equation of a curve is

$$x^2 + y^2 + 2xy + 2\sqrt{2}x - 2\sqrt{2}y = 0.$$

Find the resulting equation if the axes are rotated 45°. Graph the equation.

If $\theta = 45°$, then using the rotation equations gives

$$x = \frac{\sqrt{2}}{2}x' - \frac{\sqrt{2}}{2}y'$$

and
$$y = \frac{\sqrt{2}}{2}x' + \frac{\sqrt{2}}{2}y'.$$

Substitute these values in the given equation to get

$$x^2 + y^2 + 2xy + 2\sqrt{2}x - 2\sqrt{2}y = 0$$

$$\left[\frac{\sqrt{2}}{2}x' - \frac{\sqrt{2}}{2}y'\right]^2 + \left[\frac{\sqrt{2}}{2}x' + \frac{\sqrt{2}}{2}y'\right]^2 + 2\left[\frac{\sqrt{2}}{2}x' - \frac{\sqrt{2}}{2}y'\right] \cdot \left[\frac{\sqrt{2}}{2}x' + \frac{\sqrt{2}}{2}y'\right]$$

$$+ 2\sqrt{2}\left[\frac{\sqrt{2}}{2}x' - \frac{\sqrt{2}}{2}y'\right] - 2\sqrt{2}\left[\frac{\sqrt{2}}{2}x' + \frac{\sqrt{2}}{2}y'\right] = 0.$$

Expanding these terms gives

$$\frac{1}{2}x'^2 - x'y' + \frac{1}{2}y'^2 + \frac{1}{2}x'^2 + x'y' + \frac{1}{2}y'^2 + x'^2 - y'^2$$
$$+ 2x' - 2y' - 2x' - 2y' = 0.$$

Collecting terms leads to

$$2x'^2 - 4y' = 0$$

or, finally,

$$x'^2 - 2y' = 0,$$

the equation of a parabola. The graph of this parabola is shown in Figure 26.   ●

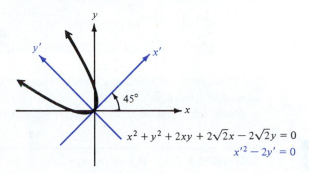

45°

$x^2 + y^2 + 2xy + 2\sqrt{2}x - 2\sqrt{2}y = 0$

$x'^2 - 2y' = 0$

Figure 26

The equation in Example 1 is a second-degree equation of the form $Ax^2 + Bxy + Cy^2 + Dx + Ey + F = 0$. Using the rotation equations eliminated the xy-term, resulting in an equation of the form $Ax^2 + Cy^2 + Dx + Ey + F = 0$, that has a conic section graph. Substituting from the rotation equations into the equation $Ax^2 + Bxy + Cy^2 + Dx + Ey + F = 0$ gives

$$A(x' \cos \theta - y' \sin \theta)^2$$
$$+ B(x' \cos \theta - y' \sin \theta)(x' \sin \theta + y' \cos \theta)$$
$$+ C(x' \sin \theta + y' \cos \theta)^2$$
$$+ D(x' \cos \theta - y' \sin \theta)$$
$$+ E(x' \sin \theta + y' \cos \theta)$$
$$+ F = 0.$$

Multiplying and collecting like terms produces the equation

$$A'x'^2 + B'x'y' + C'y'^2 + D'x' + E'y' + F' = 0,$$

where

$$B' = 2(C - A) \sin \theta \cos \theta + B(\cos^2 \theta - \sin^2 \theta).$$

To eliminate the $x'y'$ term, it is necessary to find a value of θ so that B' equals 0. That is, the equation

$$2(C - A) \sin \theta \cos \theta + B(\cos^2 \theta - \sin^2 \theta) = 0$$

must be solved for θ. This step is left for the exercises. (See Exercise 35.) Using the double-angle identities for sine and cosine leads to the solution: $B' = 0$ if

$$\cot 2\theta = \frac{A - C}{B}.$$

Angle of Rotation

The *xy*-term is removed from the general equation

$$Ax^2 + Bxy + Cy^2 + Dx + Ey + F = 0$$

by a rotation of the axes through an angle θ, $0° < \theta < 90°$, where

$$\cot 2\theta = \frac{A - C}{B}.$$

This result can be used to find the appropriate angle of rotation, θ. The rotation equations require values of $\sin \theta$ and $\cos \theta$. The following example illustrates a way to obtain $\sin \theta$ and $\cos \theta$ from $\cot 2\theta$ without first identifying the angle θ.

EXAMPLE 2   Rotate the axes and graph $52x^2 - 72xy + 73y^2 = 200$.
Here

$$\cot 2\theta = \frac{52 - 73}{-72}$$

$$= \frac{-21}{-72} = \frac{7}{24}.$$

To find $\sin \theta$ and $\cos \theta$, use the trigonometric identities

$$\sin \theta = \sqrt{\frac{1 - \cos 2\theta}{2}} \quad \text{and} \quad \cos \theta = \sqrt{\frac{1 + \cos 2\theta}{2}}.$$

Since $0° < \theta < 90°$, $0° < 2\theta < 180°$. In these two quadrants, cosine and cotangent have the same sign. Sketch a right triangle and label it as in Figure 27 to see that $\cos 2\theta = 7/25$. Then

$$\sin \theta = \sqrt{\frac{1 - 7/25}{2}} = \sqrt{\frac{9}{25}} = \frac{3}{5}$$

$$\cos \theta = \sqrt{\frac{1 + 7/25}{2}} = \sqrt{\frac{16}{25}} = \frac{4}{5}.$$

Use these values for $\sin \theta$ and $\cos \theta$ to get the rotation equations

$$x = \frac{4}{5}x' - \frac{3}{5}y' \quad \text{and} \quad y = \frac{3}{5}x' + \frac{4}{5}y'.$$

Substituting these values of x and y into the original equation gives

$$52\left(\frac{4}{5}x' - \frac{3}{5}y'\right)^2 - 72\left(\frac{4}{5}x' - \frac{3}{5}y'\right)\left(\frac{3}{5}x' + \frac{4}{5}y'\right) + 73\left(\frac{3}{5}x' + \frac{4}{5}y'\right)^2 = 200,$$

which becomes

$$52\left(\frac{16}{25}\,x'^2 - \frac{24}{25}\,x'y' + \frac{9}{25}\,y'^2\right) - 72\left(\frac{12}{25}\,x'^2 + \frac{7}{25}\,x'y' - \frac{12}{25}\,y'^2\right)$$

$$+ 73\left(\frac{9}{25}\,x'^2 + \frac{24}{25}\,x'y' + \frac{16}{25}\,y'^2\right) = 200.$$

Combining terms gives

$$25x'^2 + 100y'^2 = 200.$$

Divide both sides by 200 to get

$$\frac{x'^2}{8} + \frac{y'^2}{2} = 1,$$

an equation of an ellipse having intercepts $(0, \sqrt{2})$, $(0, -\sqrt{2})$, $(2\sqrt{2}, 0)$, and $(-2\sqrt{2}, 0)$ in the $x'y'$-system. The graph of this ellipse is shown in Figure 28. To find θ, use the fact that

$$\frac{\sin \theta}{\cos \theta} = \frac{3/5}{4/5} = \frac{3}{4} = \tan \theta.$$

from which $\theta \approx 37°$. Or, since the slope of the x'-axis is also 3/4, use that fact to identify a point on the x' axis [for instance, the point $(4, 3)$ shown in Figure 28] and then draw the axis through that point and the origin.   ●

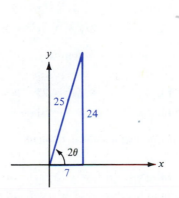

Figure 27

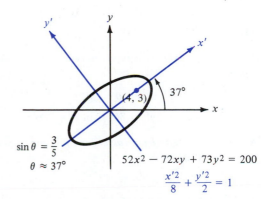

Figure 28

In addition to the conic sections discussed in this chapter, the intersection of a plane and a cone also can be a single point, a straight line, or a pair of straight lines. These atypical intersections are known as **degenerate conic sections.**

The result on the following page makes it possible to use the general equation to predict the type of graph to expect. Since the proof is lengthy, it is not included.

Equations of Conics with an *xy*-Term

If the general second-degree equation

$$Ax^2 + Bxy + Cy^2 + Dx + Ey + F = 0$$

has a graph that is a nondegenerate conic section, it will be one of the following:

(a) a circle or an ellipse if $B^2 - 4AC < 0$;

(b) a parabola if $B^2 - 4AC = 0$;

(c) a hyperbola if $B^2 - 4AC > 0$.

10.4 Exercises

Use the last result in this section to predict the type of graph of each of the following second-degree equations.

1. $4x^2 + 3y^2 + 2xy - 5x = 8$

2. $x^2 + 2xy - 3y^2 + 2y = 12$

3. $2x^2 + 3xy - 4y^2 = 0$

4. $x^2 - 2xy + y^2 + 4x - 8y = 0$

5. $4x^2 + 4xy + y^2 + 15 = 0$

6. $-x^2 + 2xy - y^2 + 16 = 0$

Find the angle of rotation θ that will remove the xy-term in each of the following equations.

7. $2x^2 + \sqrt{3}xy + y^2 + x = 5$

8. $4\sqrt{3}x^2 + xy + 3\sqrt{3}y^2 = 10$

9. $3x^2 + \sqrt{3}xy + 4y^2 + 2x - 3y = 12$

10. $4x^2 + 2xy + 2y^2 + x - 7 = 0$

11. $x^2 - 4xy + 5y^2 = 18$

12. $3\sqrt{3}x^2 - 2xy - \sqrt{3}y^2 = 25$

Use the given angle of rotation to remove the xy-term and graph each of the following equations.

13. $x^2 - xy + y^2 = 6$; $\quad \theta = 45°$

14. $2x^2 - xy + 2y^2 = 25$; $\quad \theta = 45°$

15. $8x^2 - 4xy + 5y^2 = 36$; $\quad \sin\theta = 2/\sqrt{5}$

16. $5y^2 + 12xy = 10$; $\quad \sin\theta = 3/\sqrt{13}$

Remove the xy-term from each of the following equations by performing a suitable rotation. Graph each equation.

17. $3x^2 - 2xy + 3y^2 = 8$

18. $x^2 + xy + y^2 = 3$

19. $x^2 - 4xy + y^2 = -5$

20. $x^2 + 2xy + y^2 + 4\sqrt{2}x - 4\sqrt{2}y = 0$

21. $7x^2 + 6\sqrt{3}xy + 13y^2 = 64$

22. $7x^2 + 2\sqrt{3}xy + 5y^2 = 24$

23. $3x^2 - 2\sqrt{3}xy + y^2 - 2x - 2\sqrt{3}y = 0$

24. $2x^2 + 2\sqrt{3}xy + 4y^2 = 5$

In each equation in Exercises 25–30, remove the xy-term by rotation. Then translate the axes by completing the square as shown earlier, and sketch the graph.

25. $x^2 + 3xy + y^2 - 5\sqrt{2}y = 15$

26. $x^2 - \sqrt{3}xy + 2\sqrt{3}x - 3y - 3 = 0$

27. $4x^2 + 4xy + y^2 - 24x + 38y - 19 = 0$

28. $12x^2 + 24xy + 19y^2 - 12x - 40y + 31 = 0$

29. $16x^2 + 24xy + 9y^2 - 130x + 90y = 0$

30. $9x^2 - 6xy + y^2 - 12\sqrt{10}x - 36\sqrt{10}y = 0$

31. Determine in your head the equation resulting from $\dfrac{x^2}{9} - \dfrac{y^2}{25} = 1$ when the axes are rotated by $\pi/2$.

32. Determine in your head the equation resulting from $y = -3x^2$ when the axes are rotated by $\pi/2$.

33. Show that when the axes are rotated by any angle θ, the graph of $x^2 + y^2 = 25$ remains the same. Explain why this is to be expected.

The result of substituting $x = x' \cos \theta - y' \sin \theta$ and $y = x' \sin \theta + y' \cos \theta$ into $Ax^2 + Bxy + Cy^2 + Dx + Ey + F = 0$ is $A'x'^2 + B'x'y' + C'y'^2 + D'x' + E'y' + F' = 0$.

34. Show that
$$A' = A \cos^2 \theta + B \sin \theta \cos \theta + C \sin^2 \theta$$
$$B' = 2(C - A)\sin \theta \cos \theta + B(\cos^2 \theta - \sin^2 \theta)$$
$$C' = A \sin^2 \theta - B \sin \theta \cos \theta + C \cos^2 \theta$$
$$D' = D \cos \theta + E \sin \theta$$
$$E' = E \cos \theta - D \sin \theta$$
$$F' = F.$$

35. Show that the solution of $2(C - A)\sin \theta \cos \theta + B(\cos^2 \theta - \sin^2 \theta) = 0$ is $\cot 2\theta = \dfrac{A - C}{B}$.

36. Show that $A + C = A' + C'$.

37. Show that $(A' - C')^2 + B'^2 = (A - C)^2 + B^2$.

38. Use the results of Exercises 36 and 37 to show that $B'^2 - 4A'C' = B^2 - 4AC$.

39. Use the result of Exercise 38 and the characteristics of conic sections given in Section 3 to prove the result at the end of this section.

10 | CHAPTER SUMMARY

Key Words

conic sections	**10.2 ellipse**	**hyperbola**
10.1 parabola	**foci**	**asymptotes**
focus	**vertices**	**fundamental rectangle**
directrix	**major axis**	**10.4 rotation of axes**
focal chord	**minor axis**	**rotation equations**
		degenerate conic sections

Review Exercises

Graph each of the following. Give the vertex and axis of each figure.

1. $x = -(y + 3)^2 - 9$ **2.** $x = 2(y - 7)^2 + 3$ **3.** $x = 3y^2 - 6y + 1$ **4.** $x = 4y^2 - 4y + 3$

Graph each of the following. Give the focus, directrix, and axis of each figure.

5. $y^2 = 3x$ **6.** $y^2 = -\dfrac{1}{4}x$ **7.** $x^2 - 5y = 0$ **8.** $2x^2 = y$

Write an equation for each parabola with vertex at the origin.

9. Focus $(0, -4)$

10. Focus $(3, 0)$

11. Through $(1, 7)$, opening upward

12. Through $(-2, 3)$, opening to the left

Graph each of the following and identify each graph. Give the coordinates of the vertices for each ellipse or hyperbola, and give the equations of the asymptotes for each hyperbola.

13. $\dfrac{x^2}{25} + \dfrac{y^2}{4} = 1$

14. $\dfrac{x^2}{3} + \dfrac{y^2}{16} = 1$

15. $\dfrac{x^2}{4} - \dfrac{y^2}{9} = 1$

16. $\dfrac{y^2}{100} - \dfrac{x^2}{25} = 1$

17. $(x - 3)^2 + (y + 2)^2 = 9$

18. $\dfrac{(x + 1)^2}{16} + \dfrac{(y - 1)^2}{16} = 1$

19. $x^2 = 16 + y^2$

20. $4x^2 + 9y^2 = 36$

21. $\dfrac{25x^2}{9} + \dfrac{4y^2}{25} = 1$

22. $\dfrac{100x^2}{49} + \dfrac{9y^2}{16} = 1$

23. $\dfrac{(x - 2)^2}{9} + \dfrac{(y + 3)^2}{4} = 1$

24. $\dfrac{(x - 3)^2}{4} + (y + 1)^2 = 1$

25. $\dfrac{(y + 2)^2}{4} - \dfrac{(x + 3)^2}{9} = 1$

26. $\dfrac{(x + 1)^2}{16} - \dfrac{(y - 2)^2}{4} = 1$

27. $\dfrac{x}{3} = -\sqrt{1 - \dfrac{y^2}{16}}$

28. $x = -\sqrt{1 - \dfrac{y^2}{36}}$

29. $y = -\sqrt{1 - \dfrac{x^2}{25}}$

30. $y = -\sqrt{1 + x^2}$

Write an equation for each of the following conic sections (centers at the origin).

31. Ellipse; *x*-intercept at $(4, 0)$, focus at $(-3, 0)$

32. Ellipse; vertex at $(0, 5)$, focus at $(0, 2)$

33. Hyperbola; *y*-intercept at $(0, -2)$, passing through $(2, 3)$

34. Hyperbola; focus at $(0, -7)$, transverse axis of length 10

For the equations in Exercises 35–40, name the conic and sketch the graph.

35. $4x^2 - 9y^2 = 36$

36. $y^2 + 4x^2 = 4$

37. $y^2 + x = 2$

38. $3x^2 - 4y^2 = 24$

39. $x^2 + y^2 = 16$

40. $2x^2 - y = 0$

41. Find the equation of the hyperbola consisting of all points in the plane for which the difference of the distances from $(0, 0)$ and $(0, 4)$ is 2.

42. Find the equation of the ellipse consisting of all points in the plane the sum of whose distances from $(0, 0)$ and $(4, 0)$ is 8.

Predict the type of graph of each equation.

43. $3xy - y^2 - 5 = 0$

44. $4x^2 - 2xy + y^2 + 2y - 6 = 0$

45. $x^2 - xy + 2x - 3y = 0$

46. $x^2 + 2xy - y^2 + 8 = 0$

Find the angle of rotation that will eliminate the xy-term in each equation.

47. $2\sqrt{3}x^2 + xy + \sqrt{3}y^2 + y = 2$

48. $5x^2 + 4xy + y^2 + 2x = 5$

Graph each conic after performing a suitable rotation of the axes.

49. $24xy - 7y^2 + 36 = 0$

50. $5x^2 + 8xy + 5y^2 = 9$

51. $-3xy + 9\sqrt{2}x = 15$

11

FURTHER TOPICS IN ALGEBRA

In this chapter we discuss a number of topics relating to sums of n terms, where n is a positive integer. The *binomial theorem* gives a formula for writing out the terms of $(x + y)^n$. *Sequences* (lists of numbers) are discussed with special attention to two very useful types of sequence, *arithmetic* and *geometric*. *Mathematical induction* is a method used to prove theorems about sequences and sums. The chapter ends with two related sections: the section on counting theory is used to find probabilities of events in the section on probability theory.

11.1 THE BINOMIAL THEOREM

It is often necessary to write out terms of powers of binomials of the form $(x + y)^n$. This section introduces a formula for expanding $(x + y)^n$ when n is a positive integer. Expansions of $(x + y)^n$ for a few positive integer values of n are listed below.

$$(x + y)^1 = x + y$$
$$(x + y)^2 = x^2 + 2xy + y^2$$
$$(x + y)^3 = x^3 + 3x^2y + 3xy^2 + y^3$$
$$(x + y)^4 = x^4 + 4x^3y + 6x^2y^2 + 4xy^3 + y^4$$
$$(x + y)^5 = x^5 + 5x^4y + 10x^3y^2 + 10x^2y^3 + 5xy^4 + y^5.$$

These results indicate a pattern for writing a general expression for $(x + y)^n$. First, notice that each expression begins with x raised to the same power as the binomial itself. That is, the expansion of $(x + y)^1$ has a first term of x^1, $(x + y)^2$ has a first term of x^2, $(x + y)^3$ has a first term of x^3, and so on. Also, the last term in each expansion is y to the same power as the binomial. This suggests that the expansion of $(x + y)^n$ should begin with the term x^n and end with the term y^n.

Also, the exponents on x decrease by one in each term after the first, while the exponents on y, beginning with y in the second term, increase by one in each succeeding term. It appears that the *variables* in the expansion of $(x + y)^n$ should have the following pattern.

$$x^n, \ x^{n-1}y, \ x^{n-2}y^2, \ x^{n-3}y^3, \ \ldots, \ xy^{n-1}, \ y^n$$

In this pattern, the sum of the exponents on x and y in each term is n. For example, in the third term in the list above, the variable is $x^{n-2}y^2$, and the sum of the exponents is $n - 2 + 2 = n$.

Now examine the *coefficients* in the terms of the expansions shown above. Writing the coefficients alone gives the following pattern.

Pascal's Triangle

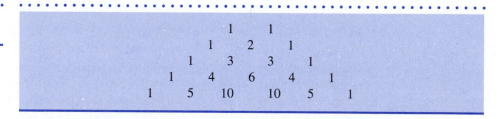

With the coefficients arranged in this way it can be seen that each number in the triangle is the sum of the two numbers directly above it (one to the right and one to the left.) For example, in the fourth row from the top, 1 is the sum of 1, the only number above it, 4 is the sum of 1 and 3, 6 is the sum of 3 and 3, and so on.

Around A.D. 1000 the Muslim scientist Abú Bakr al-Kanaji drew the attention of mathematicians in the Islamic world to this triangular array of numbers, which has many remarkable properties. It was used by Ghiyáth al-Dín Jamshid al-Káshi (*circa* 1436) in *The Calculator's Key* to develop an algorithm for finding the fifth root of a number. This triangular array was also used by the Chinese mathematicians Yang Hui (flourished 1268) and Chu Shi-kié (flourished 1303). However, the array has commonly been known as *Pascal's triangle* since the French mathematician Blaise Pascal (1623–62) published it in a work that became widely known in the West, his *Traité du Triangle Arithmétique* of 1654.

To get the coefficients for $(x + y)^6$, include a sixth row in the array of numbers given above. By adding adjacent numbers the sixth row is found to be

$$1 \quad 6 \quad 15 \quad 20 \quad 15 \quad 6 \quad 1.$$

Using these coefficients the expansion of $(x + y)^6$ is

$$(x + y)^6 = x^6 + 6x^5y + 15x^4y^2 + 20x^3y^3 + 15x^2y^4 + 6xy^5 + y^6.$$

Although it is possible to use Pascal's triangle to find the coefficients of $(x + y)^n$ for any positive integer value of n, this becomes impractical for large values of n due to the need to write out all the preceding rows. A more efficient way of finding these coefficients uses factorial notation. The number $n!$ (read "n-factorial"), is defined as follows.

Definition of n-factorial

> For any positive integer n,
>
> $$n! = n(n - 1)(n - 2) \ldots (3)(2)(1)$$
>
> and
>
> $$0! = 1.$$

For example, $5! = 5 \cdot 4 \cdot 3 \cdot 2 \cdot 1 = 120$, $7! = 7 \cdot 6 \cdot 5 \cdot 4 \cdot 3 \cdot 2 \cdot 1 = 5040$, $2! = 2 \cdot 1 = 2$, and so on.

Now look at the coefficients of the expression

$$(x + y)^5 = x^5 + 5x^4y + 10x^3y^2 + 10x^2y^3 + 5xy^4 + y^5.$$

The coefficient on the second term, $5x^4y$, is 5, and the exponents on the variables are 4 and 1. Note that

$$5 = \frac{5!}{4!1!}.$$

The coefficient on the third term is 10, with exponents of 3 and 2, and

$$10 = \frac{5!}{3!2!}.$$

The last term (the sixth term) can be written as $y^5 = 1x^0y^5$, with coefficient 1, and exponents of 0 and 5. Since $0! = 1$, check that

$$1 = \frac{5!}{0!5!}.$$

Generalizing from these examples, the coefficient for the term of the expansion of $(x + y)^n$ in which the variable part is $x^r y^{n-r}$ (where $r \le n$) will be

$$\frac{n!}{r!(n - r)!}.$$

This number, called a **binomial coefficient,** is often symbolized $\binom{n}{r}$ (read "n above r").

Definition of Binomial Coefficient

> For nonnegative integers n and r, with $r \le n$, the symbol $\binom{n}{r}$ is defined as
>
> $$\binom{n}{r} = \frac{n!}{r!(n - r)!}.$$

These binomial coefficients are just numbers from Pascal's triangle. For example, $\binom{3}{0}$ is the first number in the third row, and $\binom{7}{4}$ is the fifth number in the seventh row.

· · · · · · · · · ·

EXAMPLE 1

(a) $\dbinom{6}{2} = \dfrac{6!}{2!(6-2)!} = \dfrac{6!}{2!4!} = \dfrac{6 \cdot 5 \cdot 4 \cdot 3 \cdot 2 \cdot 1}{2 \cdot 1 \cdot 4 \cdot 3 \cdot 2 \cdot 1} = 15$

(b) $\dbinom{8}{0} = \dfrac{8!}{0!(8-0)!} = \dfrac{8!}{0!8!} = \dfrac{8!}{1 \cdot 8!} = 1$

(c) $\dbinom{10}{10} = \dfrac{10!}{10!(10-10)!} = \dfrac{10!}{10!0!} = 1$ ●

Our conjectures about the expansion of $(x + y)^n$ may be summarized as follows.

1. There are $(n + 1)$ terms in the expansion.
2. The first term is x^n and the last term is y^n.
3. The exponent on x decreases by 1 and the exponent on y increases by 1 in each succeeding term.
4. The sum of the exponents on x and y in any term is n.
5. The coefficient of the term with $x^r y^{n-r}$ is $\dbinom{n}{r}$.

These observations about the expansion of $(x + y)^n$ for any positive integer value of n suggest the **binomial theorem.**

· · · · · · · · · · · · ·
**Binomial
Theorem**

· ·

For any positive integer n and any complex numbers x and y,

$$(x + y)^n = x^n + \binom{n}{n-1} x^{n-1}y + \binom{n}{n-2} x^{n-2}y^2 +$$

$$\binom{n}{n-3} x^{n-3}y^3 + \ldots + \binom{n}{n-r} x^{n-r}y^r + \ldots$$

$$+ \binom{n}{1} xy^{n-1} + y^n.$$

As stated above, the binomial theorem is a conjecture, determined inductively by looking at $(x + y)^n$ for several values of n. A proof of the theorem using *mathematical induction* is given in Section 5 of this chapter.

· · · · · · · · · ·

EXAMPLE 2   Write out the binomial expansion of $(x + y)^9$.

Using the binomial theorem,

$$(x + y)^9 = x^9 + \binom{9}{8} x^8y + \binom{9}{7} x^7y^2 + \binom{9}{6} x^6y^3 + \binom{9}{5} x^5y^4$$

$$+ \binom{9}{4}x^4y^5 + \binom{9}{3}x^3y^6 + \binom{9}{2}x^2y^7 + \binom{9}{1}xy^8 + y^9.$$

Now evaluate each of the binomial coefficients.

$$(x + y)^9 = x^9 + \frac{9!}{8!1!} x^8 y + \frac{9!}{7!2!} x^7 y^2 + \frac{9!}{6!3!} x^6 y^3 + \frac{9!}{5!4!} x^5 y^4$$

$$+ \frac{9!}{4!5!} x^4 y^5 + \frac{9!}{3!6!} x^3 y^6 + \frac{9!}{2!7!} x^2 y^7 + \frac{9!}{1!8!} xy^8 + y^9$$

$$= x^9 + 9x^8 y + 36x^7 y^2 + 84x^6 y^3 + 126x^5 y^4 + 126x^4 y^5$$

$$+ 84x^3 y^6 + 36x^2 y^7 + 9xy^8 + y^9 \quad \bullet$$

· · · · · · · · ·

EXAMPLE 3   Expand $\left(a - \dfrac{b}{2} \right)^5$.

Write the power of the binomial as follows.

$$\left(a - \frac{b}{2} \right)^5 = \left(a + \left(-\frac{b}{2} \right) \right)^5$$

Now use the binomial theorem with $x = a$, $y = -\dfrac{b}{2}$, and $n = 5$, to get

$$\left(a - \frac{b}{2} \right)^5 = a^5 + \binom{5}{4} a^4 \left(-\frac{b}{2} \right) + \binom{5}{3} a^3 \left(-\frac{b}{2} \right)^2 + \binom{5}{2} a^2 \left(-\frac{b}{2} \right)^3 + \binom{5}{1} a \left(-\frac{b}{2} \right)^4 + \left(-\frac{b}{2} \right)^5$$

$$= a^5 + 5a^4 \left(-\frac{b}{2} \right) + 10a^3 \left(-\frac{b}{2} \right)^2 + 10a^2 \left(-\frac{b}{2} \right)^3 + 5a \left(-\frac{b}{2} \right)^4 + \left(-\frac{b}{2} \right)^5$$

$$= a^5 - \frac{5}{2} a^4 b + \frac{5}{2} a^3 b^2 - \frac{5}{4} a^2 b^3 + \frac{5}{16} ab^4 - \frac{1}{32} b^5. \quad \bullet$$

As Example 3 illustrates, any expansion of the *difference* of two terms has alternating signs.

· · · · · · · · ·

EXAMPLE 4   Expand $\left(\dfrac{3}{m^2} - 2\sqrt{m} \right)^4$. (Assume $m > 0$.)

By the binomial theorem,

$$\left(\frac{3}{m^2} - 2\sqrt{m} \right)^4 = \left(\frac{3}{m^2} \right)^4 + \binom{4}{3} \left(\frac{3}{m^2} \right)^3 (-2\sqrt{m})^1 + \binom{4}{2} \left(\frac{3}{m^2} \right)^2 (-2\sqrt{m})^2$$

$$+ \binom{4}{1} \left(\frac{3}{m^2} \right)^1 (-2\sqrt{m})^3 + (-2\sqrt{m})^4$$

$$= \frac{81}{m^8} + 4 \left(\frac{27}{m^6} \right) (-2m^{1/2}) + 6 \left(\frac{9}{m^4} \right) (4m)$$

$$+ 4 \left(\frac{3}{m^2} \right) (-8m^{3/2}) + 16m^2.$$

Here the fact that $\sqrt{m} = m^{1/2}$ was used. Finally,

$$\left(\frac{3}{m^2} - 2\sqrt{m}\right)^4 = \frac{81}{m^8} - \frac{216}{m^{11/2}} + \frac{216}{m^3} - \frac{96}{m^{1/2}} + 16m^2. \quad \bullet$$

Our observations on the binomial theorem make it possible to write any single term of a binomial expansion. For example, to find the tenth term of $(x + y)^n$, where $n \geq 9$, first notice that in the tenth term y is raised to the ninth power (since y has the power 1 in the second term, the power 2 in the third term, and so on). Because the exponents on x and y in any term must have a sum of n, the exponent on x in the tenth term is $n - 9$. Thus, the tenth term of the expansion is

$$\binom{n}{n-9}x^{n-9}y^9 = \frac{n!}{(n-9)!9!}x^{n-9}y^9.$$

This same idea can be used to obtain the result given in the following theorem.

rth Term of the Binomial Expansion

The rth term of the binomial expansion of $(x + y)^n$, where $n \geq r - 1$, is

$$\binom{n}{n-(r-1)}x^{n-(r-1)}y^{r-1}.$$

EXAMPLE 5   Find the seventh term of $(a + 2b)^{10}$.

In the seventh term $2b$ has an exponent of 6, while a has an exponent of $10 - 6$, or 4. The seventh term is

$$\binom{10}{4}a^4(2b)^6 = 210a^4(64b^6)$$

$$= 13{,}440a^4b^6. \quad \bullet$$

11.1 Exercises

Write out the binomial expansion for each of the following.

1. $(r^2 + s)^5$

2. $(m + n^2)^4$

3. $(p + 2q)^4$

4. $(3r - s)^6$

5. $\left(\frac{m}{2} - 1\right)^6$

6. $\left(3 + \frac{y}{3}\right)^5$

7. $\left(2p + \frac{q}{3}\right)^4$

8. $\left(\frac{r}{6} - \frac{m}{2}\right)^3$

9. $(m^{-2} + m^2)^4$

10. $\left(\sqrt{x} + \frac{4}{\sqrt{x}}\right)^5$

11. $(p^{2/3} - 5p^{4/3})^5$

12. $(3y^{3/2} + 5y^{1/2})^4$

Write only the first four terms in each of the following expansions.

13. $(x + 6)^{21}$

14. $(y - 8)^{17}$

15. $(3m^{-2} + 5m^{-4})^9$

16. $\left(\frac{8}{k^2} + 6k\right)^{17}$

In Exercises 17–22 write the indicated term of the binomial expansion.

17. 9th term of $(2m + n)^{10}$

18. 7th term of $(3r - 5s)^{12}$

19. 17th term of $(p^2 + q)^{20}$

20. 10th term of $(2x^2 + y)^{14}$

21. 8th term of $(x^3 + 2y)^{14}$

22. 13th term of $(a + 2b^3)^{12}$

23. Find the middle term of $(3x^7 + 2y^3)^8$.

24. Find the two middle terms of $(-2m^{-1} + 3n^{-2})^{11}$.

25. Find the value of n for which the 7th and 9th coefficients in the expansion of $(x + y)^n$ are the same.

26. Find the term in the expansion of $(3 + \sqrt{x})^{11}$ that contains x^4.

27. Find the constant term in the expansion of $\left(2x^3 + \dfrac{1}{3x^2}\right)^{10}$.

28. Find the value of n for which the coefficient of x^2 in the expansion of $(1 + x)^n$ is 21.

Use the binomial expansion to evaluate each of the following to three decimal places.

29. $(1.01)^{10}$ (*Hint:* $1.01 = 1 + .01$)

30. $(.99)^{15}$ (*Hint:* $.99 = 1 - .01$)

31. $(1.99)^8$

32. $(2.99)^3$

33. $(3.02)^6$

34. $(1.01)^9$

In later courses, it is shown that

$$(1 + x)^n = 1 + nx + \frac{n(n - 1)}{2!}x^2 + \frac{n(n - 1)(n - 2)}{3!}x^3 + \ldots$$

for any real number n (not just positive integer values) and any real number x where $|x| < 1$. This result, a generalized binomial theorem, may be used to find approximate values of powers and roots. For example,

$$\sqrt[4]{630} = (625 + 5)^{1/4} = \left[625\left(1 + \frac{5}{625}\right)\right]^{1/4} = 625^{1/4}\left(1 + \frac{5}{625}\right)^{1/4}.$$

35. Use the result above to approximate $(1 + 5/625)^{1/4}$ to the nearest thousandth. Then approximate $\sqrt[4]{630}$.

36. Approximate $\sqrt[3]{9.42}$, using the method described above.

37. Approximate $(1.02)^{-3}$.

38. Approximate $1/(1.04)^5$.

39. In the formula above, let $n = -1$ and find $(1 + x)^{-1}$.

40. Find the first four terms when $1 + x$ is divided into 1. Compare with the results of Exercise 39.

41. Work out the first four terms in the expansion of $(1 + 3)^{1/2}$ using $x = 3$ and $n = 1/2$ in the formula above. Is the result close to $(1 + 3)^{1/2} = 4^{1/2} = 2$? Why not?

42. Use the result above to show that for small values of x, $\sqrt{1 + x} \approx 1 + \dfrac{1}{2}x$.

43. When $(4x - 5)^7$ is written in the form $a_7x^7 + a_6x^6 + \ldots + a_1x + a_0$, what is the sum of the numbers $a_7, a_6, \ldots, a_1, a_0$? (*Hint:* This question can be answered without determining the values of the coefficients.)

44. Show that $\dbinom{n}{0} + \dbinom{n}{1} + \dbinom{n}{2} + \ldots + \dbinom{n}{n} = 2^n$. (*Hint:* Set $x = 1$ in the binomial expansion of $(1 + x)^n$.)

11.2 ARITHMETIC SEQUENCES

A **sequence** is a function having as domain the set of natural numbers. Instead of using $f(x)$ notation to indicate a sequence, it is customary to use a_n, where n represents an element in the domain of a sequence. The letter n is used instead of x as a reminder that n represents a *natural number*. The elements in the range of a sequence, called the **terms** of the sequence, are a_1, a_2, a_3, ... The elements of both the domain and the range of a sequence are *ordered*. The first term (range element) is found by letting $n = 1$, the second term is found by letting $n = 2$, and so on. The **general term,** or **nth term,** of the sequence is a_n.

• • • • • • • • •

EXAMPLE 1   Write the first five terms for each of the following sequences.

(a) $a_n = \dfrac{n + 1}{n + 2}$

Replacing n, in turn, with 1, 2, 3, 4, and 5 gives

$$\frac{2}{3}, \frac{3}{4}, \frac{4}{5}, \frac{5}{6}, \frac{6}{7}.$$

(b) $a_n = (-1)^n \cdot n$

Replace n, in turn, with 1, 2, 3, 4, and 5, to get

$$a_1 = (-1)^1 \cdot 1 = -1$$
$$a_2 = (-1)^2 \cdot 2 = 2$$
$$a_3 = (-1)^3 \cdot 3 = -3$$
$$a_4 = (-1)^4 \cdot 4 = 4$$
$$a_5 = (-1)^5 \cdot 5 = -5.$$

(c) $b_n = \dfrac{(-1)^n}{2^n}$

Here $b_1 = -1/2$, $b_2 = 1/4$, $b_3 = -1/8$, $b_4 = 1/16$, and $b_5 = -1/32$.   ●

A sequence is a **finite sequence** if the domain is the set $\{1, 2, 3, 4, \ldots, n\}$, where n is a natural number. An **infinite sequence** has the set of all natural numbers as its domain.

• • • • • • • • •

EXAMPLE 2   The sequence of positive even integers,

$$2, 4, 6, 8, 10, 12, 14, \ldots,$$

is infinite, but the sequence of days in June,

$$1, 2, 3, 4, \ldots, 29, 30,$$

is finite.   ●

· · · · · · · · ·

EXAMPLE 3   Find the first four terms for the sequence defined as follows:
$a_1 = 4$, and for $n > 1$, $a_n = 2 \cdot a_{n-1} + 1$.

We know $a_1 = 4$. Since $a_n = 2 \cdot a_{n-1} + 1$,

$$a_2 = 2 \cdot a_1 + 1 = 2 \cdot 4 + 1 = 9$$

$$a_3 = 2 \cdot a_2 + 1 = 2 \cdot 9 + 1 = 19$$

$$a_4 = 2 \cdot a_3 + 1 = 2 \cdot 19 + 1 = 39. \quad \bullet$$

The definition of this sequence is an example of a **recursive definition,** one in which each term is defined as an expression involving the previous term. On the other hand, the sequences in Example 1 were defined *explicitly,* with a formula for a_n that does not depend on a previous term.

A sequence in which each term after the first is obtained by adding a fixed number to the previous term is an **arithmetic sequence** (or **arithmetic progression**). The fixed number that is added is the **common difference.** The sequence

$$5, 9, 13, 17, 21, \ldots$$

is an arithmetic sequence since each term after the first is obtained by adding 4 to the previous term. That is,

$$9 = 5 + 4$$

$$13 = 9 + 4$$

$$17 = 13 + 4$$

$$21 = 17 + 4,$$

and so on. The common difference is 4.

· · · · · · · · ·

EXAMPLE 4   Find the common difference, d, for the arithmetic sequence

$$-9, -7, -5, -3, -1, \ldots .$$

Since this sequence is arithmetic, d can be found by choosing any two adjacent terms and subtracting the first from the second. If we choose -7 and -5,

$$d = -5 - (-7) = 2.$$

Choosing -9 and -7 would give $d = -7 - (-9) = 2$, the same result. $\bullet$

· · · · · · · · ·

EXAMPLE 5   Write the first five terms for each arithmetic sequence.

(a) $a_1 = 7$, $d = -3$

The first term is 7, and each succeeding term is found by adding -3 to the preceding term.

$$a_2 = 7 + d = 7 + (-3) = 4$$

$$a_3 = 4 + d = 4 + (-3) = 1$$

$$a_4 = 1 + d = 1 + (-3) = -2$$

$$a_5 = -2 + d = -2 + (-3) = -5$$

(b) $a_1 = -12$, $d = 5$

Use the definition of an arithmetic sequence.

$$a_1 = -12$$
$$a_2 = -12 + d = -12 + 5 = -7$$
$$a_3 = -7 + d = -7 + 5 = -2$$
$$a_4 = -2 + d = -2 + 5 = 3$$
$$a_5 = 3 + d = 3 + 5 = 8 \quad \bullet$$

If a_1 is the first term of an arithmetic sequence and d is the common difference, then the terms of the sequence are given by

$$a_1 = a_1$$
$$a_2 = a_1 + d$$
$$a_3 = a_2 + d = a_1 + d + d = a_1 + 2d$$
$$a_4 = a_3 + d = a_1 + 2d + d = a_1 + 3d$$
$$a_5 = a_1 + 4d$$
$$a_6 = a_1 + 5d,$$

and, by this pattern,

$$a_n = a_1 + (n - 1)d.$$

This result can be proven by mathematical induction (see Section 5 of this chapter); a summary is given below.

nth Term of an Arithmetic Sequence

In an arithmetic sequence with first term a_1 and common difference d, the nth term, a_n, is given by

$$a_n = a_1 + (n - 1)d.$$

EXAMPLE 6   Find a_{13} and a_n for the arithmetic sequence

$$-3, 1, 5, 9, \ldots$$

Here $a_1 = -3$ and $d = 1 - (-3) = 4$. To find a_{13}, substitute 13 for n in the formula for the nth term of an arithmetic sequence.

$$a_{13} = a_1 + (\mathbf{13} - 1)d$$
$$a_{13} = -3 + (12)4$$
$$a_{13} = -3 + 48$$
$$a_{13} = 45$$

To find a_n, substitute values for a_1 and d in the formula for a_n.

$$a_n = -3 + (n - 1)4$$
$$a_n = -3 + 4n - 4$$
$$a_n = 4n - 7 \quad \bullet$$

. . . .

EXAMPLE 7 Find a_{18} and a_n for the arithmetic sequence having $a_2 = 9$ and $a_3 = 15$.

Find d: $d = a_3 - a_2 = 15 - 9 = 6$.

Since
$$a_2 = a_1 + d,$$
$$9 = a_1 + 6 \quad \text{and} \quad a_1 = 3.$$

Then
$$a_{18} = 3 + (18 - 1) \cdot 6$$
$$a_{18} = 105$$

and
$$a_n = 3 + (n - 1) \cdot 6$$
$$a_n = 3 + 6n - 6$$
$$a_n = 6n - 3. \quad \bullet$$

. . . .

EXAMPLE 8 A child building a tower with blocks uses 15 for the first row. Each row has 2 blocks fewer than the previous row. If there are 8 rows in the tower, how many blocks are used for the top row?

The number of blocks in each row forms an arithmetic sequence with $a_1 = 15$ and $d = -2$. Since there are 8 rows in the tower, $n = 8$ and we want to find a_8. Using the formula

$$a_n = a_1 + (n - 1)d$$

gives
$$a_8 = 15 + (8 - 1)(-2) = 1.$$

There is just one block in the top row. $\bullet$

. . . .

EXAMPLE 9 Suppose that an arithmetic sequence has $a_8 = -16$ and $a_{16} = -40$. Find a_1.

Since $a_8 = a_1 + (8 - 1)d$, replacing a_8 with -16 gives

$$-16 = a_1 + 7d \quad \text{or} \quad a_1 = -16 - 7d.$$

Similarly,
$$-40 = a_1 + 15d \quad \text{or} \quad a_1 = -40 - 15d.$$

From these two equations, by the substitution method given earlier,

$$-16 - 7d = -40 - 15d$$
$$d = -3.$$

To find a_1, substitute -3 for d in $a_1 = -16 - 7d$.

$$a_1 = -16 - 7(-3)$$
$$a_1 = 5 \quad \bullet$$

Sum of the First n Terms   It is often necessary to add the terms of an arithmetic sequence. For example, suppose that a person borrows $3000 and agrees to pay $100 per month plus interest of 1% per month on the unpaid balance until the loan is paid off. The first month $100 is paid to reduce the loan plus interest of $(.01)3000 = 30$ dollars. The second month another $100 is paid toward the loan and $(.01)2900 = 29$ dollars) is paid for interest. Since the loan is reduced by $100 each month, interest payments decrease by $(.01)100 = 1$ dollar each month, forming the arithmetic sequence

$$30, 29, 28, \ldots , 3, 2, 1.$$

The total amount of interest paid is given by the sum of the terms of this sequence. A formula will be developed here to find this sum without adding all thirty numbers directly.

Suppose a sequence has terms $a_1, a_2, a_3, a_4, \ldots$. Then S_n is defined as the sum of the first n terms of the sequence. That is,

$$S_n = a_1 + a_2 + a_3 + \ldots + a_n.$$

If the sequence is arithmetic, the sum of the first n terms can be written as follows.

$$S_n = a_1 + [a_1 + d] + [a_1 + 2d] + \ldots + [a_1 + (n - 1)d]$$

The formula for the general term was used in the last expression. Now write the same sum in reverse order, beginning with a_n and *subtracting d*.

$$S_n = a_n + [a_n - d] + [a_n - 2d] + \ldots + [a_n - (n - 1)d]$$

Adding respective sides of these two equations term by term gives

$$S_n + S_n = (a_1 + a_n) + (a_1 + a_n) + \ldots + (a_1 + a_n)$$

or $\qquad 2S_n = n(a_1 + a_n),$

since there are n terms of $a_1 + a_n$ on the right. Now solve for S_n to get

$$S_n = \frac{n}{2}(a_1 + a_n).$$

Using the formula $a_n = a_1 + (n - 1)d$, this result for S_n can also be written as

$$S_n = \frac{n}{2}[a_1 + a_1 + (n - 1)d]$$

or $\qquad S_n = \frac{n}{2}[2a_1 + (n - 1)d],$

an alternative formula for the sum of the first n terms of an arithmetic sequence.

Sum of the First n Terms of an Arithmetic Sequence

If an arithmetic sequence has first term a_1 and common difference d, then the sum of the first n terms, S_n, is given by

$$S_n = \frac{n}{2}(a_1 + a_n)$$

or

$$S_n = \frac{n}{2}[2a_1 + (n-1)d].$$

The first formula of the theorem is used when the first and last terms are known; otherwise the second formula is used.

Either one of these formulas can be used to find the amount of interest the person above will pay on the $3000 loan. In the sequence of interest payments $a_1 = 30$, $d = -1$, $n = 30$, and $a_n = 1$. Choosing the first formula

$$S_n = \frac{n}{2}(a_1 + a_n)$$

gives

$$S_n = \frac{30}{2}(30 + 1)$$

$$= 15(31) = 465,$$

so a total of $465 interest will be paid over 30 months.

EXAMPLE 10

(a) Find S_{12} for the arithmetic sequence $-9, -5, -1, 3, 7, \ldots$.

Here $a_1 = -9$, $d = 4$, and $n = 12$. Use the second formula above.

$$S_n = \frac{n}{2}[2a_1 + (n-1)d]$$

$$S_{12} = \frac{12}{2}[2(-9) + 11(4)]$$

$$= 6(-18 + 44) = 156$$

(b) Use the formula for S_n to find the sum of the first 60 positive integers.

In this example, $n = 60$, $a_1 = 1$, and $a_{60} = 60$, so it is convenient to use the first of the two formulas:

$$S_n = \frac{n}{2}(a_1 + a_n)$$

$$S_{60} = \frac{60}{2}(1 + 60)$$

$$= 30 \cdot 61 = 1830. \quad \bullet$$

· · · · · · · · · ·

EXAMPLE 11   The sum of the first 17 terms of an arithmetic sequence is 187. If $a_{17} = -13$, find a_1 and d.

Use the first formula for S_n, with $n = 17$, to find a_1.

$$187 = \frac{17}{2}(a_1 - 13)$$
$$374 = 17(a_1 - 13)$$
$$22 = a_1 - 13$$
$$a_1 = 35$$

From the formula for a_n,

$$a_{17} = a_1 + (17 - 1)d,$$

and

$$-13 = 35 + 16d$$
$$-48 = 16d$$
$$d = -3. \quad \bullet$$

Sigma Notation   Sometimes a special shorthand notation is used for the sum of terms of a sequence. The symbol Σ, the Greek capital letter *sigma*, is used to indicate a sum. For example,

$$\sum_{i=1}^{n} a_i$$

represents the sum

$$\sum_{i=1}^{n} a_i = a_1 + a_2 + a_3 + \ldots + a_n.$$

The letter i is called the **index of summation.** Do not confuse this use of i with the use of i to represent an imaginary number. Other letters can be used instead of i. Also, the initial value of the index can be a number other than 1.

· · · · · · · · · ·

EXAMPLE 12   Find each of the following sums.

(a) $\displaystyle\sum_{i=1}^{10} (4i + 8)$

This sum can be written as

$$\sum_{i=1}^{10} (4i + 8)$$
$$= [4(1) + 8] + [4(2) + 8] + [4(3) + 8] + \ldots + [4(10) + 8]$$
$$= 12 + 16 + 20 + \ldots + 48,$$

the sum of the first ten terms of the arithmetic sequence having

$$a_1 = 4 \cdot 1 + 8 = 12,$$

$$n = 10,$$

and
$$a_n = a_{10} = 4 \cdot 10 + 8 = 48.$$

By a formula for S_n,

$$\sum_{i=1}^{10} (4i + 8) = S_{10} = \frac{10}{2}(12 + 48) = 5(60) = 300.$$

(b) $\displaystyle\sum_{j=1}^{15} (9 - j) = S_{15} = \frac{15}{2}[8 + (-6)]$

$$= \frac{15}{2}[2]$$

$$= 15$$

(c) $\displaystyle\sum_{k=3}^{9} (4 - 3k)$

The first few terms are

$$[4 - 3(3)] + [4 - 3(4)] + [4 - 3(5)] + \ldots$$
$$= -5 + (-8) + (-11) + \ldots$$

Thus $a_1 = -5$ and $d = -3$. If the sequence started with $k = 1$, there would be 9 terms. Since it starts at 3, 2 of those terms are missing, so there are 7 terms and $n = 7$.

$$\sum_{k=3}^{9} (4 - 3k) = \frac{7}{2}[2(-5) + (6)(-3)] = -98 \quad \bullet$$

11.2 Exercises

Write the first five terms of each of the following sequences.

1. $a_n = 6n + 4$ **2.** $a_n = 3n - 2$ **3.** $a_n = (-1)^n(n + 2)$ **4.** $a_n = 2^{-n+1}$

5. $a_n = \dfrac{8n - 4}{2n + 1}$ **6.** $a_n = (-2)^n(n)$ **7.** $a_n = \left(-\dfrac{1}{2}\right)^n (n^{-1})$ **8.** $a_n = \dfrac{n^2 + 1}{n^2 + 2}$

9. $a_n = \dfrac{(-1)^{n-1}(n + 1)}{n + 2}$ **10.** a_n is the nth prime number

Find the first ten terms for the sequences defined as follows.

11. $a_1 = 4$, $a_n = a_{n-1} + 5$, for $n > 1$ **12.** $a_1 = -3$, $a_n = a_{n-1} + 2$, for $n > 1$

13. $a_1 = 1$, $a_2 = 1$, $a_n = a_{n-1} + a_{n-2}$, for $n \geq 3$ (the Fibonacci sequence)

14. $a_1 = 1$, $a_n = n \cdot \cos(\pi \cdot a_{n-1})$, for $n > 1$

Write the terms of the arithmetic sequences satisfying each of the following conditions.

15. $a_1 = 4, d = 2, n = 5$

16. $a_1 = 6, d = 8, n = 4$

17. $a_2 = 9, d = -2, n = 4$

18. $a_3 = 7, d = -4, n = 4$

19. $a_1 = 4 - \sqrt{5}, a_2 = 4, n = 5$

20. $a_1 = -8, a_2 = -8 + \sqrt{7}, n = 5$

For each of the following arithmetic sequences, find d and a_n.

21. 12, 17, 22, 27, 32, 37, . . .

22. 8, 17, 26, 35, 44, 53, . . .

23. 18, 15, 12, 9, 6, . . .

24. 30, 24, 18, 12, . . .

25. $-6 + \sqrt{2}, -6 + 2\sqrt{2}, -6 + 3\sqrt{2}, \ldots$

26. $4 - \sqrt{11}, 5 - \sqrt{11}, 6 - \sqrt{11}, \ldots$

27. $x, x + m, x + 2m, x + 3m, x + 4m, \ldots$

28. $k + p, k + 2p, k + 3p, k + 4p, \ldots$

Find a_8 and a_n for each of the following arithmetic sequences.

29. $a_1 = 5, d = 2$

30. $a_1 = -3, d = -4$

31. $a_3 = 2, d = 1$

32. $a_4 = 5, d = -2$

33. $a_1 = 12, a_3 = 6$

34. $a_2 = 5, a_4 = 1$

35. $a_1 = x, a_2 = x + 3$

36. $a_2 = y + 1, d = -3$

Find a_1 for each of the following arithmetic sequences.

37. $a_9 = 47, a_{15} = 77$

38. $a_{10} = 50, a_{20} = 110$

39. $S_{20} = 1090, a_{20} = 102$

40. $S_{31} = 5580, a_{31} = 360$

Find the sum of the first ten terms for each of the following arithmetic sequences.

41. $a_3 = 5, a_4 = 8$

42. $a_2 = 9, a_4 = 13$

43. 5, 9, 13, . . .

44. 8, 6, 4, . . .

45. $a_1 = 9.428, d = -1.723$

46. $a_1 = -3.119, d = 2.422$

47. $a_4 = 2.556, a_5 = 3.004$

48. $a_7 = 11.192, a_9 = 4.812$

Evaluate each of the following sums.

49. $\sum_{i=1}^{10} (2i + 3)$

50. $\sum_{i=1}^{15} (5i - 9)$

51. $\sum_{i=1}^{12} (-5 - 8i)$

52. $\sum_{i=1}^{19} (-3 - 4i)$

53. $\sum_{i=1}^{1000} i$

54. $\sum_{i=1}^{2000} i$

55. $\sum_{i=6}^{15} (4i - 2)$

56. $\sum_{i=9}^{20} (8i + 3)$

In Exercises 57 and 58, find the value of x for which the sequence is arithmetic.

57. 3, x, 11

58. 3, x, $\dfrac{1}{2}$

59. Suppose that the sum of the first n terms in an arithmetic sequence is given by the formula $S_n = n(4n - 1)$. What is the 11th term in the sequence?

60. Determine the common difference for an arithmetic sequence in which $a_{15} - a_5$ is 50.

61. Find the sum of the first n positive integers.

62. Find the sum of the first n odd positive integers.

63. If a clock strikes the proper number of bongs each hour on the hour, how many bongs will it bong in a month of 30 days?

64. A stack of telephone poles has 30 in the bottom row, 29 in the next, and so on, with one pole in the top row. How many poles are in the stack?

65. A sky diver falls 5 m during the first second, 15 m during the second, 25 m during the third, and so on. How many meters will the diver fall during the tenth second? During the first ten seconds?

66. Deepwell Drilling Company charges a flat $500 set-up charge, plus $5 for the first foot of well drilled, $6 for the second, $7 for the third, and so on. Find the total charge for a 70-foot well.

67. An object falling under the force of gravity falls about 16 ft the first second, 48 ft during the second, 80 ft during the third, and so on. How far would the object fall during the eighth second? What is the total distance the object would fall in eight seconds?

68. The population of a city was 49,000 five years ago. Each year, the zoning commission permits an increase of 580 in the population of the city. What will the population be five years from now?

69. The sum of four terms in an arithmetic sequence is 66. The sum of the squares of the terms is 1214. Find the terms.

70. The sum of five terms of an arithmetic sequence is 5. If the product of the first and second is added to the product of the fourth and fifth, the result is 326. Find the terms.

71. A super slide of uniform slope is to be built on a level piece of land. There are to be twenty equally spaced supports, with the longest support 15 m in length, and the shortest 2 m in length. Find the total length of all the supports.

72. How much material would be needed for the rungs of a ladder of 31 rungs, if the rungs taper from 18 in. to 28 in.? Assume that the lengths of the rungs form the terms of an arithmetic sequence.

73. Find all arithmetic sequences $a_1, a_2, a_3, \ldots$, such that $a_1{}^2, a_2{}^2, a_3{}^2, \ldots$, is also an arithmetic sequence.

74. Suppose that $a_1, a_2, a_3, \ldots$ and $b_1, b_2, b_3, \ldots$ are each arithmetic sequences. Let $d_n = a_n + c \cdot b_n$, for any real number c and every positive integer n. Show that $d_1, d_2, d_3, \ldots$ is an arithmetic sequence.

75. Suppose that $a_1, a_2, a_3, a_4, a_5, \ldots$ is an arithmetic sequence. Is $a_1, a_3, a_5, \ldots$ an arithmetic sequence?

76. If $f(x) = mx + b$, show that $f(1), f(2), f(3), \ldots$ is an arithmetic sequence. What is the common difference? (*Note:* This observation shows the connection between linear functions and arithmetic sequences.)

Explain why each of the following sequences is arithmetic.

77. log 2, log 4, log 8, log 16, log 32, . . .

78. log 12, log 36, log 108, log 324, . . .

Express each of the following sums with sigma notation.

79. $5 + 7 + 9 + 11 + 13 + 15$

80. $2 + 5 + 8 + 11 + 14$

Decide which of the following statements are true for every real number x_i, real constant k, and natural number n.

81. $\displaystyle\sum_{i=1}^{n} k \cdot x_i = k^n \cdot \sum_{i=1}^{n} x_i$

82. $\displaystyle\sum_{i=1}^{n} k \cdot x_i = k \cdot \sum_{i=1}^{n} x_i$

83. $\displaystyle\sum_{i=1}^{n} k = nk$

84. $\displaystyle\sum_{i=1}^{n} k = k^n$

85. $\displaystyle\left[\sum_{i=1}^{n} x_i\right]^2 = \sum_{i=1}^{n} x_i^2$

86. $\displaystyle\sum_{i=1}^{n} (x_i - k) = \sum_{i=1}^{n} x_i - nk$

11.3 GEOMETRIC SEQUENCES

Suppose you agreed to work for 1¢ the first day, 2¢ the second day, 4¢ the third day, 8¢ the fourth day, and so on, with your wages doubling each day. How much will you earn on the 20th day, after working 5 days a week for a month? How much will you have earned altogether in 20 days? These questions will be answered in this section.

A **geometric sequence** (or **geometric progression**) is a sequence in which each term after the first is obtained by multiplying the preceding term by a constant non-zero real number, called the **common ratio.** The sequence discussed above,

$$1, \ 2, \ 4, \ 8, \ 16, \ \ldots$$

is an example of a geometric sequence in which the first term is 1 and the common ratio is 2.

If the common ratio of a geometric sequence is r, then by the definition of a geometric sequence

$$r = \frac{a_{n+1}}{a_n}$$

for every positive integer n. Therefore, the common ratio can be found by choosing any term except the first and dividing it by the preceding term.

In the geometric sequence

$$2, \ 8, \ 32, \ 128, \ \ldots$$

$r = 4$. Notice that

$$8 = 2 \cdot 4$$
$$32 = 8 \cdot 4 = (2 \cdot 4) \cdot 4 = 2 \cdot 4^2$$
$$128 = 32 \cdot 4 = (2 \cdot 4^2) \cdot 4 = 2 \cdot 4^3.$$

To generalize this, assume that a geometric sequence has first term a_1 and common ratio r. The second term can be written as $a_2 = a_1 r$, the third as $a_3 = a_2 r = (a_1 r)r = a_1 r^2$, and so on. Following this pattern, the nth term is $a_n = a_1 r^{n-1}$. Again, this result is proven by mathematical induction. (See Section 5 of this chapter.)

nth Term of a Geometric Sequence

In the geometric sequence with first term a_1 and common ratio r, the nth term is

$$a_n = a_1 r^{n-1}.$$

EXAMPLE 1 The formula for the nth term of a geometric sequence can be used to answer the first question posed at the beginning of this section. How much will be earned on the 20th day if daily wages follow the sequence

$$1, 2, 4, 8, 16, \ldots,$$

with $a_1 = 1$ and $r = 2$?

To answer the question, find a_{20}.

$$a_{20} = a_1 r^{19} = 1(2)^{19} = 524{,}288 \text{ cents, or } \$5242.88. \quad \bullet$$

EXAMPLE 2 Find a_5 and a_n for each of the following geometric sequences.

(a) 4, 12, 36, 108, . . .

The first term, a_1, is 4. To find r, choose any term except the first and divide it by the preceding term. For example,

$$r = 36/12 = 3.$$

To find the fifth term, a_5, start with $a_n = a_1 r^{n-1}$ and replace n with 5, r with 3, and a_1 with 4.

$$a_5 = 4 \cdot (3)^{5-1} = 4 \cdot 3^4 = 324$$

Also, $\qquad a_n = 4 \cdot 3^{n-1}.$

(b) 64, 32, 16, 8, . . .

Here $r = 8/16 = 1/2$, and $a_1 = 64$, so

$$a_5 = 64\left(\frac{1}{2}\right)^{5-1} = 64\left(\frac{1}{16}\right) = 4.$$

Also, $\qquad a_n = 64\left(\frac{1}{2}\right)^{n-1}. \quad \bullet$

EXAMPLE 3 Find a_1 and r for the geometric sequence with third term 20 and sixth term 160.

Use the formula for the nth term of a geometric sequence.

$$\text{For } n = 3, \quad a_3 = a_1 r^2 = 20;$$

$$\text{for } n = 6, \quad a_6 = a_1 r^5 = 160.$$

Since $a_1 r^2 = 20$, then $a_1 = 20/r^2$. Substituting this in the second equation gives

$$a_1 r^5 = 160$$

$$\left(\frac{20}{r^2}\right) r^5 = 160$$

$$20 r^3 = 160$$

$$r^3 = 8$$

$$r = 2.$$

Since $a_1 r^2 = 20$ and $r = 2$, then $a_1 = 5.$ ●

. . . .

EXAMPLE 4   A population of fruit flies is growing in such a way that each generation is 1.5 times as large as the last generation. Suppose there were 100 insects in the first generation. How many would there be in the fourth generation?

The population of each generation can be written as a geometric sequence with a_1 as the first-generation population, a_2 the second-generation population, and so on. Then the fourth-generation population will be a_4. Using the formula for a_n, with $n = 4$, $r = 1.5$, and $a_1 = 100$, gives

$$a_4 = a_1 r^3 = 100(1.5)^3 = 100(3.375) = 337.5.$$

In the fourth generation, the population will number about 338 insects.   ●

Sum of First n Terms   In applications of geometric sequences, it is often necessary to know the sum of the first n terms for the sequence. For example, a scientist might want to know the total number of insects in four generations of the population discussed in Example 4.

To find a formula for the sum of the first n terms of a geometric sequence, S_n, first write the sum as

$$S_n = a_1 + a_2 + a_3 + \ldots + a_n$$

or
$$S_n = a_1 + a_1 r + a_1 r^2 + \ldots + a_1 r^{n-1}. \tag{1}$$

If $r = 1$, $S_n = n a_1$, which is a correct formula for this case. If $r \neq 1$, multiply both sides of equation (1) by r, obtaining

$$rS_n = a_1 r + a_1 r^2 + a_1 r^3 + \ldots + a_1 r^n. \tag{2}$$

If (2) is subtracted from (1),

$$S_n - rS_n = a_1 - a_1 r^n$$

or
$$S_n(1 - r) = a_1(1 - r^n),$$

which finally gives
$$S_n = \frac{a_1(1 - r^n)}{1 - r}, \quad (r \neq 1).$$

This discussion is summarized on the next page.

Sum of *n* Terms of a Geometric Sequence

If a geometric sequence has first term a_1 and common ratio r, then the sum of the first n terms, S_n, is

$$S_n = \frac{a_1(1 - r^n)}{1 - r} \quad (r \neq 1).$$

This formula can be used to find the total fruit fly population in Example 4 over the four-generation period. With $n = 4$, $a_1 = 100$, and $r = 1.5$,

$$S_4 = \frac{100(1 - 1.5^4)}{1 - 1.5} = \frac{100(1 - 5.0625)}{-.5} = 812.5,$$

so the total population for the four generations will amount to about 813 insects.

The formula for S_n also can be used to determine the total amount earned in 20 days with daily wages of

$$1, 2, 4, 8, \ldots$$

cents. Since $a_1 = 1$ and $r = 2$,

$$S_{20} = \frac{1(1 - 2^{20})}{1 - 2}$$

$$= \frac{1 - 1048576}{-1}$$

$$= 1{,}048{,}575$$

cents or $10,485.75. Not bad for 20 days' work!

EXAMPLE 5 Find S_5 for the geometric sequence 3, 6, 12, 24, 48.

Here $a_1 = 3$ and $r = 2$. From the formula above,

$$S_5 = \frac{3(1 - 2^5)}{1 - 2} = \frac{3(1 - 32)}{-1} = \frac{3(-31)}{-1} = 93. \quad \bullet$$

EXAMPLE 6 Find $\displaystyle\sum_{i=1}^{6} 2 \cdot 3^i$.

This sum is the sum of the first six terms of a geometric sequence having $a_1 = 2 \cdot 3^1 = 6$ and $r = 3$. From the formula for S_n,

$$\sum_{i=1}^{6} 2 \cdot 3^i = S_6 = \frac{6(1 - 3^6)}{1 - 3}$$

$$= \frac{6(1 - 729)}{-2} = \frac{6(-728)}{-2} = 2184. \quad \bullet$$

Sums of Infinite Geometric Sequences Now the discussion of sums of sequences will be extended to include infinite geometric sequences such as the infinite sequence

$$2, \quad 1, \quad \frac{1}{2}, \quad \frac{1}{4}, \quad \frac{1}{8}, \quad \frac{1}{16}, \dots$$

with first term 2 and common ratio 1/2. Using the formula above gives the following sequence.

$$S_1 = 2, \quad S_2 = 3, \quad S_3 = \frac{7}{2}, \quad S_4 = \frac{15}{4}, \quad S_5 = \frac{31}{8}, \quad S_6 = \frac{63}{16}$$

These sums seem to be getting closer and closer to the number 4. For no value of n is $S_n = 4$. However, if n is large enough, then S_n is as close to 4 as desired.* This is expressed as

$$\lim_{n \to \infty} S_n = 4.$$

(Read: "the limit of S_n as n increases without bound is 4.") Since

$$\lim_{n \to \infty} S_n = 4,$$

the number 4 is called the **sum** of the infinite geometric sequence

$$2, \quad 1, \quad \frac{1}{2}, \quad \frac{1}{4}, \dots$$

and

$$2 + 1 + \frac{1}{2} + \frac{1}{4} + \frac{1}{8} + \dots = 4.$$

· · · · · · · · ·

EXAMPLE 7 Find $1 + \dfrac{1}{3} + \dfrac{1}{9} + \dfrac{1}{27} + \dots$.

Use the formula for the sum of the first n terms of a geometric sequence to get

$$S_1 = 1, \quad S_2 = \frac{4}{3}, \quad S_3 = \frac{13}{9}, \quad S_4 = \frac{40}{27},$$

and in general,

$$S_n = \frac{1\left[1 - \left(\frac{1}{3}\right)^n\right]}{1 - \frac{1}{3}}.$$

The following chart shows the value of $(1/3)^n$ for larger and larger values of n.

n	1	10	100	200
$\left(\dfrac{1}{3}\right)^n$	$\dfrac{1}{3}$	.0000169	1.9×10^{-48}	3.76×10^{-96}

*The phrases "large enough" and "as close as desired" are not nearly precise enough for mathematicians; much of a standard calculus course is devoted to making them more precise.

As n gets larger and larger, $(1/3)^n$ gets closer and closer to 0. That is,

$$\lim_{n \to \infty} \left(\frac{1}{3}\right)^n = 0,$$

making it reasonable that

$$\lim_{n \to \infty} S_n = \lim_{n \to \infty} \frac{1\left[1 - \left(\frac{1}{3}\right)^n\right]}{1 - \frac{1}{3}} = \frac{1(1 - 0)}{1 - \frac{1}{3}} = \frac{1}{\frac{2}{3}} = \frac{3}{2}.$$

Hence,

$$1 + \frac{1}{3} + \frac{1}{9} + \frac{1}{27} + \ldots = \frac{3}{2}. \quad \bullet$$

If a geometric sequence has a first term a_1 and a common ratio r, then

$$S_n = \frac{a_1(1 - r^n)}{1 - r}$$

for every positive integer n. If $-1 < r < 1$, then $\lim_{n \to \infty} r^n = 0$, and

$$\lim_{n \to \infty} S_n = \frac{a_1(1 - 0)}{1 - r} = \frac{a_1}{1 - r}.$$

This quotient, $a_1/(1 - r)$ is called the **sum of the infinite geometric sequence.** The limit, $\lim_{n \to \infty} S_n$, is often expressed as S_∞ or $\displaystyle\sum_{i=1}^{\infty} a_i$.

These results lead to the following definition.

Sum of an Infinite Geometric Sequence

The sum of the infinite geometric sequence with first term a_1 and common ratio r, where $-1 < r < 1$, is given by

$$S_\infty = \frac{a_1}{1 - r}.$$

EXAMPLE 8

(a) Find the sum $-\frac{3}{4} + \frac{3}{8} - \frac{3}{16} + \frac{3}{32} - \frac{3}{64} + \ldots$

The first term is $a_1 = -3/4$. To find r, divide any term by the preceding term. For example,

$$r = \frac{-\dfrac{3}{16}}{\dfrac{3}{8}} = -\frac{1}{2}.$$

Since $-1 < r < 1$, the formula above applies, and

$$S_\infty = \frac{a_1}{1 - r} = \frac{-\dfrac{3}{4}}{1 - \left(-\dfrac{1}{2}\right)} = -\frac{1}{2}.$$

(b) $\displaystyle\sum_{i=1}^{\infty} \left(\frac{3}{5}\right)^i = \frac{\dfrac{3}{5}}{1 - \dfrac{3}{5}} = \frac{3}{2}$ ●

11.3 Exercises ···

Write the terms of the geometric sequences that satisfy each of the following conditions.

1. $a_1 = 1/2, r = 4, n = 4$ **2.** $a_1 = 2/3, r = 6, n = 3$

3. $a_3 = 6, a_4 = 12, n = 5$ **4.** $a_2 = 9, a_3 = 3, n = 4$

Find a_5 and a_n for each of the following geometric sequences.

5. $a_1 = -3, r = -5$ **6.** $a_1 = -4, r = -2$ **7.** $a_2 = 3, r = 2$

8. $a_3 = 6, r = 3$ **9.** $a_4 = 64, r = -4$ **10.** $a_4 = 81, r = -3$

For each of the following sequences that are geometric, find r and a_n.

11. 6, 12, 24, 48, . . . **12.** 4, 16, 64, 256, . . . **13.** 3/4, 3/2, 3, 6, 12, . . .

14. $-7, -5, -3, -1, 1, 3, . . .$ **15.** $a_3 = 9, r = 2$ **16.** $a_5 = 6, r = 1/2$

Find the positive value of x for which the sequence is geometric.

17. $2, x, \dfrac{9}{2}$ **18.** $\dfrac{1}{3}, x, 12$

Find a_1 and r for each of the following geometric sequences.

19. $a_2 = 6, a_6 = 486$ **20.** $a_3 = -12, a_6 = 96$ **21.** $a_2 = 64, a_8 = 1$ **22.** $a_2 = 100, a_5 = .1$

Use the formula for S_n to find the sum of the first five terms for each of the following geometric sequences.

23. 3, 6, 12, 24, . . . **24.** 5, 20, 80, 320, . . . **25.** 12, -6, 3, $-3/2$, . . .

26. 18, -3, 1/2, $-1/12$, . . . **27.** $a_1 = 8.423, r = 2.859$ **28.** $a_1 = -3.772, r = -1.553$

Find each of the following sums.

29. $\displaystyle\sum_{i=1}^{4} 2^i$ **30.** $\displaystyle\sum_{i=1}^{6} 3^i$ **31.** $\displaystyle\sum_{j=1}^{8} 64(1/2)^j$

32. $\displaystyle\sum_{j=1}^{6} 81(2/3)^j$ **33.** $\displaystyle\sum_{i=3}^{6} 2^i$ **34.** $\displaystyle\sum_{i=4}^{7} 3^i$

Express each of the following sums with sigma notation.

35. $3 \cdot 2 + 3 \cdot 2^2 + 3 \cdot 2^3 + 3 \cdot 2^4$

36. $2 \cdot 3^4 - 2 \cdot 3^5 + 2 \cdot 3^6 - 2 \cdot 3^7 + 2 \cdot 3^8$

Find r for each of the following infinite geometric sequences. Identify any whose sum would exist.

37. 9, 18, 36, 72, 144, . . .

38. 3, 9, 27, 81, . . .

39. $-8, -16, -32, -64, . . .$

40. 12, 6, 3, 3/2, . . .

In Exercises 41–50, find each sum which exists by using the formula of this section where it applies.

41. $16 + 4 + 1 + . . .$

42. $81 + 27 + 9 + 3 + 1 + . . .$

43. $100 + 10 + 1 + . . .$

44. $128 + 64 + 32 + . . .$

45. $\dfrac{4}{5} + \dfrac{2}{5} + \dfrac{1}{5} + . . .$

46. $\dfrac{1}{3} - \dfrac{2}{9} + \dfrac{4}{27} - \dfrac{8}{81} + . . .$

47. $\displaystyle\sum_{i=1}^{\infty} (1/4)^i$

48. $\displaystyle\sum_{i=1}^{\infty} (-1/4)^i$

49. $\displaystyle\sum_{k=1}^{\infty} (.3)^k$

50. $\displaystyle\sum_{k=1}^{\infty} 10^{-k}$

51. Mitzi drops a ball from a height of 10 m and notices that on each bounce the ball returns to about 3/4 of its previous height. About how far will the ball travel before it comes to rest? (*Hint:* Consider the sum of two sequences).

52. A sugar factory receives an order for 1000 units of sugar. The production manager thus orders production of 1000 units of sugar. He forgets, however, that the production of sugar requires some sugar (to prime the machines, for example), and so he ends up with only 900 units of sugar. He then orders an additional 100 units, and receives only 90 units. A further order for 10 units produces 9 units. Finally seeing he is wrong, the manager decides to try mathematics. He views the production process as an infinite geometric progression with $a_1 = 1000$ and $r = .1$. Using this, find the number of units of sugar that he should have ordered originally.

53. After a person pedaling a bicycle removes his or her feet from the pedals, the wheel rotates 400 times the first minute. As it continues to slow down, it rotates in each minute only 3/4 as many times as in the previous minute. How many times will the wheel rotate before coming to a complete stop?

54. A pendulum bob swings through an arc 40 cm long on its first swing. Each swing thereafter, it swings only 80% as far as on the previous swing. How far will it swing altogether before coming to a complete stop?

55. Suppose you could save \$1 on January 1, \$2 on January 2, \$4 on January 3, and so on. What amount would you save on January 31? What would be the total amount of your savings during January? (*Hint:* $2^{31} = 2,147,483,648$.)

56. The final step in processing a black-and-white photographic print is to immerse the print in a chemical called "fixer." The print is then washed in running water. Under certain conditions, 98% of the fixer in a print will be removed with 15 min of washing. How much of the original fixer would be left after 1 hr of washing?

57. Certain medical conditions are treated with a fixed dose of a drug administered at regular intervals. Suppose that a person is given 2 mg of a drug each day and that during each 24-hour period, the body utilizes 40% of the amount of drug that was present at the beginning of the period.

  (a) Show that the amount of the drug present in the body at the end of n days is

$$\sum_{i=1}^{n} 2(.6)^i.$$

  (b) What will be the approximate quantity of the drug in the body after the treatment has been administered for a long period of time?

  (c) What daily dosage will guarantee that the amount of the drug in the body never exceeds 2 mg?

58. A sequence of equilateral triangles is constructed. The first triangle has sides 2 m in length. To get the second triangle, midpoints of the sides of the original triangle are connected. What is the length of the side of the eighth such triangle? See the figure below.

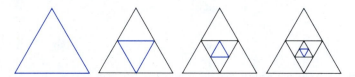

59. A sequence of equilateral triangles is constructed. The first triangle has sides 2 m in length. To get the second triangle, midpoints of the sides of the original triangle are connected. (See Exercise 58.) If this process could be continued indefinitely, what would be the total perimeter of all the triangles?

60. What would be the total area of all the triangles of Exercise 59, disregarding the overlapping?

61. A scientist has a vat containing 100 L of a pure chemical. Twenty liters is drained and replaced with water. After complete mixing, 20 L of the mixture is drained and replaced with water. What will be the strength of the mixture after 9 such drainings?

62. The half-life of a radioactive substance is the time it takes for half the substance to decay. Suppose the half-life of a substance is 3 yr, and that 10^{15} molecules of the substance are present initially. How many molecules will be present after 15 yr?

63. Each year a machine loses 20% of the value it had at the beginning of the year. Find the value of the machine at the end of 6 yr if it cost $100,000 new.

64. A bicycle wheel rotates 400 times in one minute. If the rider removes his or her feet from the pedals, the wheel will start to slow down. Each minute, it will rotate only 3/4 as many times as in the preceding minute. How many times will the wheel rotate in the fifth minute after the rider's feet are removed from the pedals? (Compare your answer to that of Exercise 53.)

65. A piece of paper is .008 inch thick. Suppose the paper is folded in half, so that its thickness doubles, for 12 times in a row. How thick would the final stack of paper be?

66. Fruit-and-vegetable dealer Greg Odjakjian paid 10¢ per lb for 10,000 lb of onions. Each week the price he charges increases by .1¢ per lb, while the onions lose 5% of their weight. If he sells the onions after six weeks, will he make or lose money? How much?

67. Find three numbers x, y, and z that are consecutive terms of both an arithmetic sequence and a geometric sequence.

68. Let $a_1, a_2, a_3, \ldots$ and $b_1, b_2, b_3, \ldots$ be geometric sequences. Let $d_n = c \cdot a_n \cdot b_n$ for any real number c and every positive integer n. Show that $d_1, d_2, d_3, \ldots$ is a geometric sequence.

Suppose that $a_1, a_2, a_3, a_4, a_5, \ldots$ is a geometric sequence with common ratio r. Show that the following sequences are geometric and give their common ratios.

69. $a_1, a_4, a_7, a_{10}, \ldots$

70. $(a_1)^2, (a_2)^2, (a_3)^2, (a_4)^2, (a_5)^2, \ldots$

71. $\sqrt{a_1}, \sqrt{a_2}, \sqrt{a_3}, \sqrt{a_4}, \sqrt{a_5}, \ldots$

72. $\dfrac{3}{a_1}, \dfrac{3}{a_2}, \dfrac{3}{a_3}, \dfrac{3}{a_4}, \dfrac{3}{a_5}, \ldots$

Explain why the following sequences are geometric.

73. $\log 6$, $\log 36$, $\log 1296$, $\log 1{,}679{,}616, \ldots$

74. $\log 2$, $\log 4$, $\log 16$, $\log 256, \ldots$

11.4 SERIES

In the previous two sections, sums of sequences were discussed. These sums are also called **series.**

Definition of Series

A **finite series** is an expression of the form

$$a_1 + a_2 + a_3 + \ldots + a_n = \sum_{i=1}^{n} a_i,$$

and an **infinite series** is an expression of the form

$$a_1 + a_2 + a_3 + \ldots + a_n + \ldots = \sum_{i=1}^{\infty} a_i,$$

Recall, in the summation notation, $\displaystyle\sum_{i=1}^{n} a_i$, i is the index of summation.

Given an expression in summation notation, it is usually not difficult to write out the first few terms of the series. For example,

$$\sum_{i=1}^{4} (2i - 5) = [2(1) - 5] + [2(2) - 5] + [2(3) - 5] + [2(4) - 5]$$
$$= -3 + (-1) + 1 + 3 = 0.$$

· · · · · · · · ·

EXAMPLE 1   Write the terms for each of the following sums. Evaluate each sum if possible.

(a) $\displaystyle\sum_{i=1}^{4} a_i = a_1 + a_2 + a_3 + a_4$

(b) $\displaystyle\sum_{i=1}^{3} (6x_i - 2)$ if $x_1 = 2$, $x_2 = 4$, $x_3 = 6$

Let $i = 1$, 2, and 3 respectively to get

$$\sum_{i=1}^{3} (6x_i - 2) = (6x_1 - 2) + (6x_2 - 2) + (6x_3 - 2).$$

Now substitute the given values for x_1, x_2, and x_3.

$$\sum_{i=1}^{3} (6x_i - 2) = (6 \cdot 2 - 2) + (6 \cdot 4 - 2) + (6 \cdot 6 - 2)$$
$$= 10 + 22 + 34 = 66$$

(c) $\displaystyle\sum_{i=1}^{4} f(x_i)\Delta x$ if $f(x) = x^2$, $x_1 = 0$, $x_2 = 2$, $x_3 = 4$, $x_4 = 6$, and $\Delta x = 2$

$$\sum_{i=1}^{4} f(x_i)\Delta x = f(x_1)\Delta x + f(x_2)\Delta x + f(x_3)\Delta x + f(x_4)\Delta x$$
$$= x_1^2\Delta x + x_2^2\Delta x + x_3^2\Delta x + x_4^2\Delta x$$
$$= 0^2(2) + 2^2(2) + 4^2(2) + 6^2(2)$$
$$= 0 + 8 + 32 + 72 = 112 \quad \bullet$$

Sometimes it is necessary to work in the opposite direction and write the terms of a series in summation notation.

· · · · · · · · ·

EXAMPLE 2   Write each series in summation notation.

(a) $1 + 4 + 7 + \ldots + 28$

To write the series in summation notation, the general term a_n must be determined. There is no standard procedure for finding an expression for a_n, and there may be more than one series that has the given terms. However, the terms of this series appear to increase by three, so it may be the sum of an arithmetic sequence with $a_1 = 1$ and $d = 3$. If so, the formula $a_n = a_1 + (n - 1)d$ can be used to find the number n of the last term. Let $a_n = 28$, $a_1 = 1$, and $d = 3$, to get

$$a_n = a_1 + (n - 1)d$$
$$28 = 1 + (n - 1)3$$
$$27 = 3n - 3$$
$$30 = 3n$$
$$n = 10.$$

Since n is a natural number, the sequence is arithmetic with general term

$$a_n = 1 + (n - 1)3 = 1 + 3n - 3 = 3n - 2.$$

The series has 10 terms and can be written in summation notation as

$$1 + 4 + 7 + \ldots + 28 = \sum_{i=1}^{10} (3i - 2).$$

(b) $5 - 15 + 45 - \ldots + 3645$

From the first three terms, successive terms appear to be those of a geometric sequence with a common ratio of -3. Use the formula for the nth term of a geometric sequence with $a_n = 3645$ to find n.

$$a_n = a_1(r)^{n-1}$$
$$3645 = 5(-3)^{n-1}$$
$$729 = (-3)^{n-1}$$

Since $729 = (3)^6 = (-3)^6$,

$$(-3)^6 = (-3)^{n-1}$$
$$6 = n - 1$$
$$n = 7,$$

and the general term is $\quad a_n = 5(-3)^{n-1}.$

In summation notation,

$$5 - 15 + 45 - \ldots + 3645 = \sum_{i=1}^{7} 5(-3)^{i-1}. \quad \bullet$$

A given series can be represented by summation notation in more than one way as shown in the next example.

EXAMPLE 3   Use summation notation to rewrite each series with the index of summation starting at the indicated number.

(a) $\displaystyle\sum_{i=1}^{8} (3i - 4); \quad 0$

Let the new index be j, where $i = j + 1$.
Then

$$\sum_{i=1}^{8}(3i - 4) = \sum_{i=1}^{i=8}(3i - 4) = \sum_{j+1=1}^{j+1=8} [3(j + 1) - 4]$$

$$= \sum_{j=0}^{j=7}(3j + 3 - 4) \quad \text{or} \quad \sum_{j=0}^{7} (3j - 1).$$

(b) $\displaystyle\sum_{i=2}^{10} i^2; \quad -1$

Here, if the new index is j, then $i = j + 3$ and

$$\sum_{i=2}^{10} i^2 = \sum_{i=2}^{i=10} i^2 = \sum_{j+3=2}^{j+3=10} (j + 3)^2$$

$$= \sum_{j=-1}^{j=7} (j + 3)^2 \quad \text{or} \quad \sum_{j=-1}^{7} (j + 3)^2. \quad \bullet$$

Polynomial functions, defined by expressions of the form

$$f(x) = a_n x^n + a_{n-1} x^{n-1} + \ldots + a_1 x + a_0,$$

can be written in compact form, using summation notation, as

$$f(x) = \sum_{i=0}^{n} a_{n-i} x^{n-i}.$$

The binomial theorem also looks much more manageable written in summation notation. The theorem can be summarized as:

$$(x + y)^n = \sum_{r=0}^{n} \binom{n}{n-r} x^{n-r} y^r.$$

Several properties of summation are given below. These provide useful shortcuts for evaluating series. The proofs of some of these properties are given in the exercises for Section 5 of this chapter.

Properties of Series

If $a_1, a_2, a_3, \ldots, a_n$ and $b_1, b_2, b_3, \ldots, b_n$ are two sequences, and c is a constant, then for every positive integer n,

(a) $\displaystyle\sum_{i=1}^{n} c = nc$

(b) $\displaystyle\sum_{i=1}^{n} ca_i = c\sum_{i=1}^{n} a_i$

(c) $\displaystyle\sum_{i=1}^{n} (a_i + b_i) = \sum_{i=1}^{n} a_i + \sum_{i=1}^{n} b_i$

(d) $\displaystyle\sum_{i=1}^{n} (a_i - b_i) = \sum_{i=1}^{n} a_i - \sum_{i=1}^{n} b_i.$

To prove property (a), expand the series to get

$$c + c + c + c + \ldots + c,$$

where there are n terms of c, so the sum is nc.

Property (c) also can be proved by first expanding the series:

$$\sum_{i=1}^{n} (a_i + b_i) = (a_1 + b_1) + (a_2 + b_2) + \ldots + (a_n + b_n).$$

Now use the commutative and associative properties to rearrange the terms.

$$\sum_{i=1}^{n} (a_i + b_i) = (a_1 + a_2 + \ldots + a_n) + (b_1 + b_2 + \ldots + b_n)$$

$$= \sum_{i=1}^{n} a_i + \sum_{i=1}^{n} b_i$$

Proofs of the other two properties are left for the exercises of Section 5.

· · · · · · · · · ·

EXAMPLE 4 Use the properties of series to evaluate $\sum_{i=1}^{6} (i^2 + 3i + 5)$.

$$\sum_{i=1}^{6} (i^2 + 3i + 5) = \sum_{i=1}^{6} i^2 + \sum_{i=1}^{6} 3i + \sum_{i=1}^{6} 5 \qquad \text{Property (c)}$$

$$= \sum_{i=1}^{6} i^2 + 3\sum_{i=1}^{6} i + \sum_{i=1}^{6} 5 \qquad \text{Property (b)}$$

$$= \sum_{i=1}^{6} i^2 + 3\sum_{i=1}^{6} i + 6(5) \qquad \text{Property (a)}$$

The following results are proved in the text and exercises of Section 5 of this chapter.

$$\sum_{i=1}^{n} i^2 = 1^2 + 2^2 + \ldots + n^2 = \frac{n(n + 1)(2n + 1)}{6}$$

and

$$\sum_{i=1}^{n} i = 1 + 2 + \ldots + n = \frac{n(n + 1)}{2}$$

For example, with $n = 6$,

$$\sum_{i=1}^{6} i^2 = \frac{6(6 + 1)(2 \cdot 6 + 1)}{6} = 7 \cdot 13 = 91$$

and

$$\sum_{i=1}^{6} i = \frac{6(6 + 1)}{2} = 3 \cdot 7 = 21.$$

Substituting these results into the work above gives

$$\sum_{i=1}^{6} (i^2 + 3i + 5) = \sum_{i=1}^{6} i^2 + 3\sum_{i=1}^{6} i + 6(5)$$

$$= 91 + 3(21) + 6(5)$$

$$= 184. \quad \bullet$$

11.4 Exercises ···

Write out the terms for each of the following sums where $x_1 = -1$, $x_2 = 0$, $x_3 = 1$, $x_4 = 2$, $x_5 = 3$.

1. $\displaystyle\sum_{i=1}^{4} (3x_i - 2)$

2. $\displaystyle\sum_{i=1}^{5} x_i^2$

3. $\displaystyle\sum_{i=1}^{3} (2x_i - x_i^2)$

4. $\displaystyle\sum_{i=1}^{4} (x_i^2 + x_i)$

5. $\displaystyle\sum_{i=1}^{5} \frac{x_i - 1}{x_i + 3}$

6. $\displaystyle\sum_{i=1}^{4} \frac{x_i}{x_i + 2}$

Write out the terms of $\displaystyle\sum_{i=1}^{4} f(x_i)\Delta x$ *with* $x_1 = 0$, $x_2 = 2$, $x_3 = 4$, $x_4 = 6$, *and* $\Delta x = .5$ *for the functions defined as follows.*

7. $f(x) = 2x - 5$

8. $f(x) = 4x + 3$

9. $f(x) = x^2 - 1$

10. $f(x) = 3 - x^2$

11. $f(x) = \dfrac{4}{x + 1}$

12. $f(x) = \dfrac{3}{2x + 1}$

Write each of the following sums using summation notation.

13. $8 + 6 + 4 + \ldots - 20$

14. $-2 + 5 + 12 + \ldots + 61$

15. $10 + 15 + 20 + \ldots + 100$

16. $5 + 2 - 1 - 4 - \ldots - 31$

17. $7 + 14 + 28 + \ldots + 1792$

18. $-4 - 12 - 36 - \ldots - 972$

19. $9 - 3 + 1 - \ldots + \dfrac{1}{81}$

20. $16 + 8 + 4 + \ldots + \dfrac{1}{16}$

21. $4 + 9 + 16 + \ldots + 169$

22. $5 + 10 + 17 + \ldots + 170$

23. $1 + \dfrac{1}{2} + \dfrac{1}{3} + \ldots + \dfrac{1}{10}$

24. $\dfrac{1}{2} + \dfrac{2}{3} + \dfrac{3}{4} + \ldots + \dfrac{14}{15}$

25. $2 + 4 + 8 + \ldots + 64$

26. $\dfrac{2}{1} + \dfrac{3}{2} + \dfrac{4}{3} + \ldots + \dfrac{12}{11}$

Use summation notation to rewrite each series with the index of summation starting at the indicated number.

27. $\displaystyle\sum_{i=1}^{5} (6 - 3i);\quad 3$

28. $\displaystyle\sum_{i=1}^{7} (5i + 2);\quad -2$

29. $\displaystyle\sum_{i=1}^{10} 2(3)^i;\quad 0$

30. $\displaystyle\sum_{i=-1}^{6} 5(2)^i;\quad 3$

31. $\displaystyle\sum_{i=-1}^{9} (i^2 - 2i);\quad 0$

32. $\displaystyle\sum_{i=3}^{11} (2i^2 + 1);\quad 0$

Use the properties of series to evaluate each summation. The following sums may be needed.

$$\sum_{i=1}^{n} i = \frac{n(n + 1)}{2} \qquad \sum_{i=1}^{n} i^2 = \frac{n(n + 1)(2n + 1)}{6} \qquad \sum_{i=1}^{n} i^3 = \frac{n^2(n + 1)^2}{4}$$

33. $\displaystyle\sum_{i=1}^{5} (5i + 3)$

34. $\displaystyle\sum_{i=1}^{5} (8i - 1)$

35. $\displaystyle\sum_{i=1}^{5} (4i^2 - 2i + 6)$

36. $\displaystyle\sum_{i=1}^{6} (2 + i - i^2)$

37. $\displaystyle\sum_{i=1}^{4} (3i^3 + 2i - 4)$

38. $\displaystyle\sum_{i=1}^{6} (i^2 + 2i^3)$

Use the series properties to rewrite each of the following summations. (Hint: Think of n as a constant.)

39. $\displaystyle\sum_{i=1}^{n}\left[4 + \left(\frac{2}{n}\right)i\right]^2\frac{2}{n}$

40. $\displaystyle\sum_{i=1}^{n}\left[\left(\frac{2}{n}\right)i + 1\right]^2\frac{2}{n}$

41. $\displaystyle\sum_{i=1}^{n}\left[3 + \left(\frac{1}{n}\right)i\right]^2\frac{1}{n}$

42. $\displaystyle\sum_{i=1}^{n}\left[5 + \left(\frac{3}{n}\right)i\right]^2\frac{3}{n}$

11.5 MATHEMATICAL INDUCTION

Many results in mathematics are claimed to be true for every positive integer. Any of these results could be checked for $n = 1$, $n = 2$, $n = 3$, and so on, but since the set of positive integers is infinite it would be impossible to check every possible case. For example, let S_n represent the statement that the sum of the first n positive integers is $n(n + 1)/2$.

$$S_n: 1 + 2 + 3 + \ldots + n = \frac{n(n + 1)}{2}.$$

The truth of this statement is easily verified for the first few values of n:

If $n = 1$, then S_1 is $\qquad\qquad 1 = \dfrac{1(1 + 1)}{2}$ which is true.

If $n = 2$, then S_2 is $\qquad 1 + 2 = \dfrac{2(2 + 1)}{2}$ which is true.

If $n = 3$, then S_3 is $\qquad 1 + 2 + 3 = \dfrac{3(3 + 1)}{2}$ which is true.

If $n = 4$, then S_4 is $\quad 1 + 2 + 3 + 4 = \dfrac{4(4 + 1)}{2}$ which is true.

Continuing in this way for any amount of time would still not prove that S_n is true for *every* positive integer value of n. To prove that such statements are true for every positive integer value of n, the following principle is often used.

Principle of Mathematical Induction

Let S_n be a statement concerning the positive integer n. Suppose that

(a) S_1 is true;

(b) for any positive integer k, $k \le n$, S_k implies S_{k+1}.

Then S_n is true for every positive integer value of n.

A proof by mathematical induction can be explained as follows. By (a) above, the statement is true when $n = 1$. By (b) above, the fact that the statement is true for $n = 1$ implies that it is true for $n = 1 + 1 = 2$. Using (b) again, it is thus true for 2 + 1 = 3, for 3 + 1 = 4, for 4 + 1 = 5, and so on. By continuing in this way, the statement must be true for *every* positive integer, no matter how large.

The situation is similar to that of a number of dominoes lined up as shown in Figure 1. If the first domino is pushed over, it pushes the next, which pushes the next, and so on until all are down.

Figure 1

Another example of the principle of mathematical induction is an infinite ladder. Suppose the rungs are spaced so that, whenever you are on a rung, you know you can move to the next rung. Then *if* you can get to the first rung, you can go as high up the ladder as you wish.

As these comments show, two separate steps are required for a proof by mathematical induction.

Proof by Induction

1. Prove that the statement is true for $n = 1$.
2. Show that, for any positive integer k, $k \le n$, S_k implies S_{k+1}.

In the next example mathematical induction is used to prove the statement S_n mentioned at the beginning of this section.

EXAMPLE 1   Let S_n represent the statement

$$1 + 2 + 3 + \ldots + n = \frac{n(n + 1)}{2}.$$

Prove that S_n is true for every positive integer n.

The proof by mathematical induction is as follows.

Step 1   Show that the statement is true when $n = 1$. If $n = 1$, S_n becomes S_1 which is

$$1 = \frac{1(1 + 1)}{2},$$

and is true.

Step 2   Show that S_k implies S_{k+1}, where S_k is the statement

$$1 + 2 + 3 + \ldots + k = \frac{k(k+1)}{2},$$

and S_{k+1} is the statement

$$1 + 2 + 3 + \ldots + k + (k+1) = \frac{(k+1)[(k+1)+1]}{2}.$$

Start with S_k.

$$1 + 2 + 3 + \ldots + k = \frac{k(k+1)}{2}$$

How can S_k be changed algebraically to match S_{k+1}? Adding $k+1$ to both sides of S_k gives

$$1 + 2 + 3 + \ldots + k + (k+1) = \frac{k(k+1)}{2} + (k+1).$$

Then, factor on the right to get

$$= (k+1)\left(\frac{k}{2} + 1\right)$$

$$= (k+1)\left(\frac{k+2}{2}\right)$$

$$1 + 2 + 3 + \ldots + k + (k+1) = \frac{(k+1)[(k+1)+1]}{2}.$$

This final result is the statement for $n = k + 1$; it has been shown that S_k implies S_{k+1}. The two steps required for a proof by mathematical induction have now been completed, so that the statement S_n is true for every positive integer value of n.   ●

• • • • • • • • •

EXAMPLE 2   Prove: if x is a real number between 0 and 1, then for every positive integer n, $0 < x^n < 1$.

Here S_1 is the statement

$$\text{if } 0 < x < 1, \text{ then } 0 < x^1 < 1,$$

which is true. S_k is the statement

$$\text{if } 0 < x < 1, \text{ then } 0 < x^k < 1.$$

Now show that this implies S_{k+1}. Multiply all members of $0 < x^k < 1$ by x to get

$$x \cdot 0 < x \cdot x^k < x \cdot 1.$$

(Here the fact that $0 < x$ is used.) Simplify to get

$$0 < x^{k+1} < x.$$

Since it is given that $x < 1$,

$$x^{k+1} < x < 1,$$

and
$$0 < x^{k+1} < 1.$$

This work shows that S_k implies S_{k+1}, and since S_1 is true, the given statement is true for every positive integer n. ●

Some statements S_n are not true for the first few values of n, but are true for all values of n that are at least equal to some fixed integer j. The following slightly generalized form of the principle of mathematical induction takes care of these cases.

Generalized Principle of Mathematical Induction

Let S_n be a statement concerning the positive integer n. Let j be a fixed positive integer. Suppose that

(a) S_j is true;

(b) for any positive integer k, $k \geq j$, S_k implies S_{k+1}.

Then S_n is true for all positive integers n, where $n \geq j$.

EXAMPLE 3 Let S_n represent the statement $2^n > 2n + 1$. Show that S_n is true for all values of n such that $n \geq 3$.

(Check that S_n is false for $n = 1$ and $n = 2$.) As before, the proof requires two steps.

Step 1 Show that S_n is true for $n = 3$. If $n = 3$, S_n is

$$2^3 > 2 \cdot 3 + 1,$$

or
$$8 > 7$$

which is true.

Step 2 Now show that S_k implies S_{k+1}, where $k \geq 3$ and

$$S_k \text{ is} \qquad 2^k > 2k + 1,$$
$$S_{k+1} \text{ is} \quad 2^{k+1} > 2(k + 1) + 1.$$

Multiply both sides of $2^k > 2k + 1$ by 2, obtaining

$$2 \cdot 2^k > 2(2k + 1),$$

or
$$2^{k+1} > 4k + 2.$$

Rewrite $4k + 2$ as $2(k + 1) + 2k$, giving

$$2^{k+1} > 2(k + 1) + 2k. \tag{1}$$

Since k is a positive integer greater than 3,

$$2k > 1. \tag{2}$$

Adding $2(k + 1)$ to both sides of inequality (2) gives

$$2(k + 1) + 2k > 2(k + 1) + 1. \tag{3}$$

From inequalities (1) and (3),

$$2^{k+1} > 2(k + 1) + 2k > 2(k + 1) + 1,$$

or $\qquad 2^{k+1} > 2(k + 1) + 1,$

as required. Thus, S_k implies S_{k+1}, and this, together with the fact that S_3 is true, shows that S_n is true for every positive integer value of n greater than or equal to 3. ●

. . . .

EXAMPLE 4 The binomial theorem can be proved by mathematical induction. That is, for any positive integer n and any complex numbers x and y,

$$(x + y)^n = x^n + \binom{n}{n - 1} x^{n-1}y + \binom{n}{n - 2} x^{n-2}y^2 +$$

$$\binom{n}{n - 3}x^{n-3}y^3 + \ldots + \binom{n}{n - r} x^{n-r}y^r + \ldots$$

$$+ \binom{n}{1} xy^{n-1} + y^n. \tag{4}$$

Let S_n be statement (4) above. Begin by verifying S_n for $n = 1$.

$$S_1: \quad (x + y)^1 = x^1 + y^1,$$

which is true.

Now assume that S_n is true for the positive integer k. Statement S_k becomes (using the definition of the binomial coefficient)

$$S_k: \quad (x + y)^k = x^k + \frac{k!}{1!(k - 1)!} x^{k-1}y + \frac{k!}{2!(k - 2)!} x^{k-2}y^2$$

$$+ \ldots + \frac{k!}{(k - 1)!1!} xy^{k-1} + y^k. \tag{5}$$

Multiply both sides of equation (5) by $x + y$.

$$(x + y)^k \cdot (x + y)$$

$$= x(x + y)^k + y(x + y)^k$$

$$= \left[x \cdot x^k + \frac{k!}{1!(k - 1)!} x^k y + \frac{k!}{2!(k - 2)!} x^{k-1} y^2 + \ldots + \frac{k!}{(k - 1)!1!} x^2 y^{k-1} + xy^k \right]$$

$$+ \left[x^k \cdot y + \frac{k!}{1!(k - 1)!} x^{k-1}y^2 + \ldots + \frac{k!}{(k - 1)!1!} xy^k + y \cdot y^k \right]$$

Rearrange terms to get

$(x + y)^{k+1}$

$$= x^{k+1} + \left[\frac{k!}{1!(k-1)!} + 1\right] x^k y + \left[\frac{k!}{2!(k-2)!} + \frac{k!}{1!(k-1)!}\right] x^{k-1} y^2 + \ldots$$

$$+ \left[1 + \frac{k!}{(k-1)!1!}\right] xy^k + y^{k+1}. \tag{6}$$

The first expression in brackets in equation (6) simplifies to $\binom{k+1}{1}$. To see this, note that

$$\binom{k+1}{1} = \frac{(k+1)(k)(k-1)(k-2) \ldots 1}{1 \cdot (k)(k-1)(k-2) \ldots 1} = k + 1.$$

Also

$$\frac{k!}{1!(k-1)!} + 1 = \frac{k!}{(k-1)!} + 1$$

$$= \frac{k! + (k-1)!}{(k-1)!} = \frac{k(k-1)! + (k-1)!}{(k-1)!}$$

$$= \frac{(k+1)(k-1)!}{(k-1)!} = k + 1.$$

The second expression becomes $\binom{k+1}{2}$, the last $\binom{k+1}{k}$, and so on. The result of equation (6) is just equation (5) with every k replaced by $k + 1$. Thus, the truth of S_n when $n = k$ implies the truth of S_n for $n = k + 1$, which completes the proof of the theorem by mathematical induction. ●

11.5 Exercises

Write out in full and verify each of the statements S_1, S_2, S_3, S_4, and S_5 for each of the following. Then use mathematical induction to prove that each of the given statements is true for every positive integer n.

1. $2 + 4 + 6 + \ldots + 2n = n(n + 1)$

2. $1 + 3 + 5 + \ldots + (2n - 1) = n^2$

Use the method of mathematical induction to prove that each of the following statements is true for every positive integer n.

3. $2 + 4 + 8 + \ldots + 2^n = 2^{n+1} - 2$

4. $1^2 + 2^2 + 3^2 + \ldots + n^2 = \dfrac{n(n+1)(2n+1)}{6}$

5. $1^3 + 2^3 + 3^3 + \ldots + n^3 = \dfrac{n^2(n+1)^2}{4}$

6. $3 + 3^2 + 3^3 + \ldots + 3^n = \dfrac{3(3^n - 1)}{2}$

7. $5 \cdot 6 + 5 \cdot 6^2 + 5 \cdot 6^3 + \ldots + 5 \cdot 6^n = 6(6^n - 1)$

8. $\dfrac{1}{1 \cdot 2} + \dfrac{1}{2 \cdot 3} + \dfrac{1}{3 \cdot 4} + \ldots + \dfrac{1}{n(n+1)} = \dfrac{n}{n+1}$

9. $\dfrac{1}{1 \cdot 4} + \dfrac{1}{4 \cdot 7} + \dfrac{1}{7 \cdot 10} + \ldots + \dfrac{1}{(3n - 2)(3n + 1)} = \dfrac{n}{3n + 1}$

10. $\dfrac{1}{2} + \dfrac{1}{2^2} + \dfrac{1}{2^3} + \ldots + \dfrac{1}{2^n} = 1 - \dfrac{1}{2^n}$

11. $1 \cdot 2 + 2 \cdot 3 + 3 \cdot 4 + \ldots + n(n + 1) = \dfrac{n(n + 1)(n + 2)}{3}$

12. $1 \cdot 4 + 2 \cdot 9 + 3 \cdot 16 + \ldots + n(n + 1)^2 = \dfrac{n(n + 1)(n + 2)(3n + 5)}{12}$

Prove each of the following results by mathematical induction. Assume $a_1, a_2, \ldots$ are the terms of an arithmetic sequence having first term a_1 and common difference d.

13. $a_n = a_1 + (n - 1)d$ **14.** $S_n = \dfrac{n}{2}(a_1 + a_n)$

Prove each result in Exercises 15–30 by mathematical induction. Assume that $a_1, a_2, a_3,$ $\ldots$ is a geometric sequence with first term a_1.

15. $a_n = a_1 \cdot r^{n - 1}$ **16.** $S_n = \dfrac{a_1(1 - r^n)}{1 - r}$

17. $(a^m)^n = a^{mn}$ (Assume a and m are constant.)   **18.** $(ab)^n = a^n b^n$ (Assume a and b are constant.)

19. $2^n > 2n$, if $n \geq 3$ **20.** $3^n > 2n + 1$, if $n \geq 2$ **21.** $3n^3 + 6n$ is divisible by 9

22. $n^2 + n$ is divisible by 2   **23.** $3^{2n} - 1$ is divisible by 8   **24.** If $a > 1$, then $a^n > 1$

25. If $a > 1$, then $a^n > a^{n-1}$ **26.** If $0 < a < 1$, then $a^n < a^{n-1}$ **27.** $2^n > n^2$, for $n > 4$

28. If $n \geq 4$, $n! > 2^n$, where $n! = n(n - 1)(n - 2) \ldots (3)(2)(1)$ **29.** $4^n > n^4$, for $n \geq 5$

30. $(1 + x)^n \geq 1 + nx$ for every $n > 1$ and fixed $x \geq -1$

31. What is wrong with the following ''proof'' by mathematical induction?

Prove: Any natural number equals the next natural number.
    To begin, we assume the statement true for some natural number k:

$$k = k + 1.$$

We must now show that the statement is true for $n = k + 1$. If we add 1 to both sides, we have

$$k + 1 = k + 1 + 1$$
$$k + 1 = k + 2.$$

Hence, if the statement is true for $n = k$, it is also true for $n = k + 1$. Thus, the theorem is proved.

32. In the country of Pango, the government prints only three-glok banknotes and five-glok notes. Prove that any purchase of 8 gloks or more can be paid for with only three-glok notes and five-glok notes. (*Hint:* Consider two cases: k gloks being paid with only three-glok notes, and k gloks requiring at least one five-glok note.)

33. Suppose that n straight lines (with $n \geq 2$) are drawn in a plane, where no two lines are parallel and no three lines pass through the same point. Show that the number of points of intersection of the lines is $(n^2 - n)/2$.

34. The series of sketches below starts with an equilateral triangle having sides of length 1. In the following steps, equilateral triangles are constructed on each side of the preceding figure. The lengths of the sides of these new triangles is 1/3 the length of the sides of the preceding triangles. Develop a formula for the number of sides of the nth figure. Use mathematical induction to prove your answer.

35. Find the perimeter of the nth figure in Exercise 34.

36. Show that the area of the nth figure in Exercise 34 is

$$\sqrt{3}\left[\frac{2}{5} - \frac{3}{20}\left(\frac{4}{9}\right)^{n-1}\right].$$

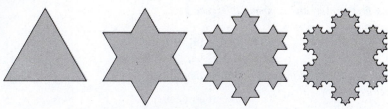

Exercises 34–36

37. A pile of n rings, each smaller than the one below it, is on a peg. Two other pegs are attached to a board with this peg. In the game called the *Tower of Hanoi* puzzle, all the rings must be moved to a different peg, with only one ring moved at a time, and with no ring ever placed on top of a smaller ring. Find the least number of moves that would be required. Prove your result with mathematical induction.

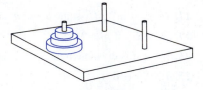

Exercise 37

38. Prove that $\displaystyle\sum_{i=1}^{n} ca_i = c\sum_{i=1}^{n} a_i$.

39. Prove that $\displaystyle\sum_{i=1}^{n} (a_i - b_i) = \sum_{i=1}^{n} a_i - \sum_{i=1}^{n} b_i$.

40. Prove De Moivre's theorem: for every positive integer n,

$$[r(\cos\theta + i\sin\theta)]^n = r^n(\cos n\theta + i\sin n\theta).$$

11.6 COUNTING PROBLEMS

If there are 3 roads from Albany to Baker and 2 roads from Baker to Creswich, in how many ways can one travel from Albany to Creswich by way of Baker? For each of the 3 roads from Albany to Baker, there are 2 different roads from Baker to Creswich. Hence, there are $3 \cdot 2 = 6$ different ways to make the trip, as shown in the **tree diagram** of Figure 2.

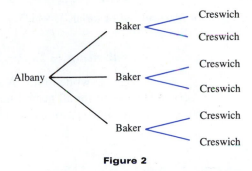

Figure 2

This example illustrates the following fundamental principle of counting.

Fundamental Principle of Counting

If one event can occur in m ways, and if for each one of these, a second event can occur in n ways, then both events can occur in mn ways.

The fundamental principle of counting can be extended to any finite number of events.

EXAMPLE 1　A restaurant offers a choice of 3 salads, 5 main dishes, and 2 desserts. Use the fundamental principle of counting to find the number of different 3-course meals that can be selected.

Three events are involved: selecting a salad, selecting a main dish, and selecting a dessert. The first event can occur in 3 ways, the second event can occur in 5 ways, and the third event can occur in 2 ways; thus there are

$$3 \cdot 5 \cdot 2 = 30 \text{ possible meals.} \quad \bullet$$

EXAMPLE 2　A teacher has 5 different books that he wishes to arrange on his desk. How many different arrangements are possible?

Five events are involved: selecting a book for the first spot, selecting a book for the second spot, and so on. For the first spot the teacher has 5 choices. After a choice has been made, the teacher has 4 choices for the second spot. Continuing in this manner, there are 3 choices for the third spot, 2 for the fourth spot, and 1 for the fifth spot. By the fundamental principle of counting, there are

$$5 \cdot 4 \cdot 3 \cdot 2 \cdot 1 \text{ or } 120 \text{ different arrangements.} \quad \bullet$$

In using the fundamental principle of counting, products such as $5 \cdot 4 \cdot 3 \cdot 2 \cdot 1$ from Example 2 occur often. For convenience in writing these products, use the symbol $n!$ (read "n factorial"), which was defined earlier for any counting number n, as follows.

$$n! = n(n - 1)(n - 2)(n - 3) \ldots (2)(1)$$

Thus $5 \cdot 4 \cdot 3 \cdot 2 \cdot 1$ is written as $5!$. Also, $3! = 3 \cdot 2 \cdot 1 = 6$. By the definition of $n!$, $n[(n - 1)!] = n!$ for all natural numbers $n \geq 2$. It is convenient to have this relation hold also for $n = 1$, so, by definition,

$$0! = 1.$$

. . . .

EXAMPLE 3   Suppose the teacher wishes to place only 3 of the 5 books on the desk. How many arrangements of 3 books are possible?

The teacher still has 5 ways to fill the first spot, 4 ways to fill the second spot, and 3 ways to fill the third. Since he wants to use only 3 books, there are only 3 spots to be filled (3 events) instead of 5, with

$$5 \cdot 4 \cdot 3 = 60 \text{ arrangements.} \quad \bullet$$

Permutations   Since each ordering of three books is considered a different arrangement, the number 60 in the example above is called the number of permutations of 5 things taken 3 at a time, written $P(5, 3) = 60$. The number of ways of arranging 5 elements from a set of 5 elements, written $P(5, 5) = 120$, was found in Example 2. A **permutation** of n elements taken r at a time is the number of ways of *arranging* r elements from a set of n elements. Generalizing from the examples above, the number of permutations of n elements, taken r at a time, denoted by $P(n, r)*$, is

$$P(n, r) = n(n - 1)(n - 2) \ldots (n - r + 1)$$
$$= \frac{n(n - 1)(n - 2) \ldots (n - r + 1)(n - r)(n - r - 1) \ldots (2)(1)}{(n - r)(n - r - 1) \ldots (2)(1)}$$
$$= \frac{n!}{(n - r)!}.$$

This proves the following result.

. . . .

Permutations of n Elements Taken r at a Time

> If $P(n, r)$ denotes the number of permutations of n elements taken r at a time, then
>
> $$P(n, r) = \frac{n!}{(n - r)!}.$$

*Alternative notations for $P(n, r)$ are P_r^n and $_nP_r$.

· · · · · · · ·

EXAMPLE 4   Find the following.

(a) The number of permutations of the letters L, M, and N
By the formula for $P(n, r)$, with $n = 3$ and $r = 3$,

$$P(3, 3) = \frac{3!}{(3 - 3)!}$$

$$= \frac{3!}{0!} = \frac{3!}{1} = 3 \cdot 2 \cdot 1 = 6.$$

The 6 permutations are

LMN, LNM, MLN, MNL, NML, NLM.

(b) The number of permutations of 2 of the 3 letters M, N, and L
Find $P(3, 2)$.

$$P(3, 2) = \frac{3!}{(3 - 2)!}$$

$$= \frac{3!}{1!} = \frac{3!}{1} = 6$$

This result is the same as the answer in part (a) because after the first 2 choices are made, the third is already determined, since only one letter is left.   ●

· · · · · · · ·

EXAMPLE 5   Suppose 8 people enter an event in a swim meet. In how many ways could the gold, silver, and bronze prizes be awarded?

Using the fundamental principle of counting, there are 3 choices to be made giving $8 \cdot 7 \cdot 6 = 336$. However, the formula for $P(n, r)$ can also be used to get the same result.

$$P(8, 3) = \frac{8!}{5!} = \frac{8 \cdot 7 \cdot 6 \cdot 5 \cdot 4 \cdot 3 \cdot 2 \cdot 1}{5 \cdot 4 \cdot 3 \cdot 2 \cdot 1}$$

$$= 8 \cdot 7 \cdot 6 = 336 \quad ●$$

· · · · · · · ·

EXAMPLE 6   In how many ways can 6 students be seated in a row of 6 desks?
Use $P(n, n)$ with $n = 6$ to get

$$P(6, 6) = 6! = 6 \cdot 5 \cdot 4 \cdot 3 \cdot 2 \cdot 1 = 720. \quad ●$$

Combinations   In the previous section we saw that there are 60 ways that a teacher can arrange 3 of 5 different books on his desk. That is, there are 60 permutations of 5 things taken 3 at a time. Suppose now that the teacher does not wish to arrange the books in a row, but rather wishes to choose, without regard to order, any 3 of the 5 books for a book sale to raise money for the school. In how many ways can this be done?

At first glance, we might say 60 again, but this is incorrect. The number 60 counts all possible *arrangements* of 3 books chosen from 5. The following 6 arrangements, however, would all lead to the same set of 3 books being given to the book sale.

mystery-biography-textbook   biography-textbook-mystery
mystery-textbook-biography   textbook-biography-mystery
biography-mystery-textbook   textbook-mystery-biography

The list shows 6 different *arrangements* of 3 books but only one *set* of 3 books. A subset of items selected *without regard to order* is called a **combination.** The number of combinations of 5 things taken 3 at a time is written $\binom{5}{3}$, or $C(5, 3)$. In this book, we will use the more common notation $\binom{5}{3}$.

To evaluate $\binom{5}{3}$, start with the $5 \cdot 4 \cdot 3$ *permutations* of 5 things taken 3 at a time. Since order doesn't matter, and each subset of 3 items from the set of 5 items can have its elements rearranged in $3 \cdot 2 \cdot 1 = 3!$ ways, $\binom{5}{3}$ can be found by dividing the number of permutations by 3!, or

$$\binom{5}{3} = \frac{5 \cdot 4 \cdot 3}{3!} = \frac{5 \cdot 4 \cdot 3}{3 \cdot 2 \cdot 1} = 10.$$

There are 10 ways that the teacher can choose 3 books for the book sale.

Generalizing this discussion gives the following formula for the number of combinations of n elements taken r at a time:

$$\binom{n}{r} = \frac{P(n, r)}{r!}.$$

A more useful version of this formula is found as follows.

$$\binom{n}{r} = \frac{P(n, r)}{r!}$$

$$= \frac{n!}{(n - r)!} \cdot \frac{1}{r!}$$

$$= \frac{n!}{(n - r)!r!}$$

This last version is the most useful for calculation. The steps above lead to the following result.

Combinations of n Elements Taken r at a Time

If $\binom{n}{r}$ represents the number of combinations of n things taken r at a time, with $r \leq n$, then

$$\binom{n}{r} = \frac{n!}{(n - r)!r!}.$$

This same result was used in work with the binomial theorem in Section 1 of this chapter. There the values of $\binom{n}{r}$ were the coefficients in the expansion of a binomial.*

. . . .

EXAMPLE 7   How many different committees of 3 people can be chosen from a group of 8 people?

Since a committee is an unordered set, use combinations to get

$$\binom{8}{3} = \frac{8!}{5!3!} = \frac{8 \cdot 7 \cdot 6 \cdot 5 \cdot 4 \cdot 3 \cdot 2 \cdot 1}{5 \cdot 4 \cdot 3 \cdot 2 \cdot 1 \cdot 3 \cdot 2 \cdot 1} = 56. \quad \bullet$$

. . . .

EXAMPLE 8   Three lawyers are to be selected from a group of 30 to work on a special project.

(a) In how many different ways can the lawyers be selected?

Here we wish to know the number of 3-element combinations that can be formed from a set of 30 elements. (We want combinations, not permutations, since order within the group of 3 doesn't matter.)

$$\binom{30}{3} = \frac{30!}{27!3!} = \frac{30 \cdot 29 \cdot 28 \cdot 27!}{27! \cdot 3 \cdot 2 \cdot 1}$$

$$= \frac{30 \cdot 29 \cdot 28}{3 \cdot 2 \cdot 1}$$

$$= 4060$$

There are 4060 ways to select the project group.

(b) In how many ways can the group of 3 be selected if a certain lawyer must work on the project?

Since 1 lawyer already has been selected for the project, the problem is reduced to selecting 2 more from the remaining 29 lawyers.

$$\binom{29}{2} = \frac{29!}{27!2!} = \frac{29 \cdot 28 \cdot 27!}{27! \cdot 2 \cdot 1} = \frac{29 \cdot 28}{2 \cdot 1} = 29 \cdot 14 = 406$$

In this case, the project group can be selected in 406 ways. $\bullet$

The formulas for permutations and combinations given in this section will be very useful in solving probability problems in the next section. Any difficulty in using these formulas usually comes from being unable to differentiate between them. Both permutations and combinations give the number of ways to choose r objects from a set of n objects. The differences between permutations and combinations are outlined on the following page.

---

*Alternative notations for the number of combinations of n elements taken r at a time are $C(n, r)$, C_r^n, and $_nC_r$.

Permutations	*Combinations*
Different orderings or arrangements of the r objects are different permutations.	Each choice or subset of r objects gives one combination. Order within the group of r objects does not matter.
$$P(n, r) = \frac{n!}{(n - r)!}$$	$$\binom{n}{r} = \frac{n!}{(n - r)!r!}$$
Clue words: arrangement, schedule, order	Clue words: group, committee, sample

In the next example, concentrate on recognizing which formula should be applied.

EXAMPLE 9   A salesman has 10 accounts in a certain city.

(a) In how many ways can he select 3 accounts to call on?

Within a selection of 3 accounts, the arrangement of the calls is not important, so there are

$$\binom{10}{3} = \frac{10!}{7!3!} = \frac{10 \cdot 9 \cdot 8}{3 \cdot 2 \cdot 1} = 120$$

ways he can select 3 accounts.

(b) In how many ways can he schedule his calls on 3 of the 10 accounts?

To schedule his calls he must *order* each selection of 3 accounts. Use permutations here, since order is important.

$$P(10, 3) = \frac{10!}{(10 - 3)!} = \frac{10!}{7!} = 10 \cdot 9 \cdot 8 = 720$$

There are 720 different orders in which he can call on 3 of the accounts.   ●

11.6 Exercises

Evaluate each expression in Exercises 1–16.

1. $P(7, 7)$ **2.** $P(5, 3)$ **3.** $P(6, 5)$ **4.** $P(4, 2)$

5. $P(8, 3)$ **6.** $P(11, 4)$ **7.** $P(7, 1)$ **8.** $P(18, 0)$

9. $\binom{6}{5}$ **10.** $\binom{4}{2}$ **11.** $\binom{15}{4}$ **12.** $\binom{9}{3}$

13. $\binom{10}{7}$ **14.** $\binom{10}{3}$ **15.** $\binom{14}{1}$ **16.** $\binom{13}{0}$

Use the multiplication principle or permutations to solve the following problems.

17. In how many ways can 4 out of 6 people be seated in a row of 4 seats?

18. In how many ways can 7 out of 10 people be assigned to 7 seats?

19. In how many ways can 5 bank tellers be assigned to 5 different windows? In how many ways can 10 tellers be assigned to the 5 windows?

20. A couple has narrowed down their choice of names for a new baby to 3 first names and 5 middle names. How many different first- and middle-name arrangements are possible?

21. How many different homes are available if a builder offers a choice of 5 basic plans, 3 roof styles, and 2 types of siding?

22. An automaker produces 7 models, each available in 6 colors, with 4 upholstery fabrics and 5 interior colors. How many varieties of the auto are available?

23. A concert is to consist of 5 works: two modern, two romantic, and one classical. In how many ways can the program be arranged?

24. If the program in Exercise 23 must be shortened to 3 works chosen from the 5, how many arrangements are possible?

25. In Exercise 23, how many different programs are possible if the two modern works are to be played first, then the two romantic, and then the classical?

26. How many 4-letter radio-station call letters can be made if the first letter must be K or W and no letter may be repeated? How many if repeats are allowed?

27. How many of the 4-letter call letters in Exercise 26 with no repeats end in K?

28. A business school gives courses in typing, shorthand, transcription, business English, technical writing, and accounting. How many ways can a student arrange his program if he takes 3 courses?

If the n objects in a permutation are not all distinguishable—that is, if there are n_1 of type 1, n_2 of type 2, and so on for r different types, then the number of distinguishable permutations *is $n!/(n_1!n_2! \ldots n_r!)$. Use this formula in the following problems.*

29. In how many ways can the letters in the word MISSISSIPPI be arranged?

30. Find the number of permutations of the letters in each of the following words: **(a)** initial **(b)** decreed.

31. Find the number of permutations of the letters in each of the following words: **(a)** little **(b)** statistics.

32. A printer has 5 A's, 4 B's, 2 C's, and 2 D's. How many different ''words'' are possible which use all these letters? (A ''word'' does not have to have any meaning here.)

33. Mike has 4 blue, 3 green, and 2 red books to arrange on a shelf.
(a) In how many ways can this be done if they can be arranged in any order?
(b) In how many distinguishable ways if books of the same color are identical and must be grouped together?
(c) In how many distinguishable ways if books of the same color are identical but need not be grouped together?

34. A child has a set of differently shaped plastic objects. There are 3 pyramids, 4 cubes, and 7 spheres.
(a) In how many ways can she arrange them in a row if they are all different colors?
(b) In how many ways if the same shapes must be grouped?
(c) In how many distinguishable ways can they be arranged in a row if blocks of the same shape are also the same color, but need not be grouped?

Use combinations to work the following problems.

35. A club has 30 members. If a committee of 4 is to be selected at random, how many different committees are possible?

36. How many different samples of 3 apples can be drawn from a crate of 25 apples?

37. A group of 3 students is to be selected randomly from a group of 12 to participate in an experimental class. In how many ways can this be done? In how many ways can the group which will not participate be selected?

38. Hal's Hamburger Heaven sells hamburgers with cheese, relish, lettuce, tomato, mustard, or ketchup. How many different hamburgers can be made using any 3 of the extras?

39. How many different 2-card hands can be dealt from a deck of 52 cards?

40. How many different 13-card bridge hands can be dealt from a deck of 52 cards?

41. Five cards are marked with the numbers 1, 2, 3, 4, and 5, shuffled, and 2 cards are then drawn. How many different 2-card combinations are possible?

42. If a bag contains 15 marbles, how many samples of 2 marbles can be drawn from it? How many samples of 4 marbles?

43. In Exercise 42, if the bag contains 3 yellow, 4 white, and 8 blue marbles, how many samples of 2 can be drawn in which both marbles are blue?

44. In Exercise 36, assume it is known that there are 5 rotten apples in the crate.
   (a) How many samples of 3 could be drawn in which all 3 are rotten?
   (b) How many samples of 3 could be drawn in which there are 1 rotten apple and 2 good apples?

45. Glendale Heights City Council is composed of 5 liberals and 4 conservatives. Three members are to be selected randomly as delegates to a convention.
   (a) How many delegations are possible?
   (b) How many delegations could have all liberals?
   (c) How many delegations could have 2 liberals and 1 conservative?
   (d) If 1 member of the council serves as mayor, how many delegations are possible which include the mayor?

46. Seven factory workers decide to send a delegation of 2 to their supervisor to discuss their grievances.
   (a) How many different delegations are possible?
   (b) If it is decided that a certain employee must be in the delegation, how many different delegations are possible?
   (c) If there are 2 women and 5 men in the group, how many delegations would include a woman?

A poker hand is made up of 5 cards drawn at random from a deck of 52 cards. Any 5 cards in one suit are called a flush. The 5 highest cards, that is, the A, K, Q, J, and 10 of any one suit are called a royal flush. Use combinations to set up each of the following. Do not evaluate.

47. Find the total number of possible poker hands.

48. How many royal flushes in hearts are possible?

49. How many royal flushes in any of the four suits are possible?

50. How many flushes in hearts are possible?

51. How many flushes in any of the four suits are possible?

52. How many combinations of 3 aces and 2 eights are possible?

Solve the following problems by using either combinations or permutations.

53. In how many ways can the letters of the word TOUGH be arranged?

54. If Matthew has 8 courses to choose from, in how many ways can he arrange his schedule if he must pick 4 of them?

55. How many samples of 3 pineapples can be drawn from a crate of 12?

56. Velma specializes in making different vegetable soups with carrots, celery, beans, peas, mushrooms, and potatoes. How many different soups can she make using any 4 ingredients?

Prove each of the following statements for positive integers n and r, with r ≤ n.

57. $P(n, n - 1) = P(n, n)$ **58.** $P(n, 1) = n$ **59.** $P(n, 0) = 1$

60. $\dbinom{n}{n} = 1$ **61.** $\dbinom{n}{0} = 1$ **62.** $\dbinom{n}{n-1} = n$ **63.** $\dbinom{n}{n-r} = \dbinom{n}{r}$

11.7 BASICS OF PROBABILITY

The study of probability theory has become increasingly popular because it has a range of practical applications. The basic ideas of probability are introduced in this section.

Consider an experiment which has one or more possible **outcomes,** each of which is equally likely to occur. For example, the experiment of tossing a coin has two equally likely possible outcomes: landing heads up (*H*) or landing tails up (*T*). Also, the experiment of rolling a die has six equally likely outcomes: landing so the face which is up shows 1, 2, 3, 4, 5, or 6 points.

The set *S* of all possible outcomes of a given experiment is called the **sample space** for the experiment. (In this text all sample spaces are finite.) A sample space for the experiment of tossing a coin consists of the outcomes *H* and *T*. This sample space can be written in set notation as

$$S = \{H, T\}.$$

Similarly, a sample space for the experiment of rolling a single die is

$$S = \{1, 2, 3, 4, 5, 6\}.$$

Any subset of the sample space is called an **event.** In the experiment with the die, for example, "the number showing is a three" is an event, say E_1, such that $E_1 = \{3\}$. "The number showing is greater than three" is also an event, say E_2, such that $E_2 = \{4, 5, 6\}$. To represent the number of outcomes which belong to event *E,* the notation $n(E)$ is used. In the experiment with the die, $n(E_1) = 1$ and $n(E_2) = 3$.

The notation $P(E)$ is used for the *probability* of an event *E.* If the outcomes in the sample space for an experiment are equally likely, then the probability of event *E* occuring is found as follows.

Definition of Probability of Event E

In a sample space with equally likely outcomes, the **probability** of an event E, written $P(E)$, is the ratio of the number of outcomes in sample space S that belong to event E, $n(E)$, to the total number of outcomes in sample space S, $n(S)$. That is,

$$P(E) = \frac{n(E)}{n(S)}.$$

To use this definition to find the probability of the event E_1 given above, start with the sample space for the experiment, $S = \{1, 2, 3, 4, 5, 6\}$, and the desired event, $E_1 = \{3\}$. Since $n(E_1) = 1$ and since there are 6 outcomes in the sample space,

$$P(E_1) = \frac{n(E_1)}{n(S)} = \frac{1}{6}.$$

EXAMPLE 1   A single die is rolled. Write the following events in set notation and give the probability for each event.

(a) E_3: the number showing is even

Use the definition above. Since $E_3 = \{2, 4, 6\}$, $n(E_3) = 3$. As shown above, $n(S) = 6$, so

$$P(E_3) = \frac{3}{6} = \frac{1}{2}.$$

(b) E_4: the number showing is greater than 4

Again $n(S) = 6$. Event $E_4 = \{5, 6\}$, with $n(E_4) = 2$. By the definition,

$$P(E_4) = \frac{2}{6} = \frac{1}{3}.$$

(c) $E_5 = \{1, 2, 3, 4, 5, 6\}$ and $P(E_5) = \frac{6}{6} = 1$

(d) $E_6 = \varnothing$ and $P(E_6) = \frac{0}{6} = 0$ ●

In part (c), $E_5 = S$. Therefore the event E_5 is a **certain event,** sure to occur every time the experiment is performed. An event which is certain to occur, such as E_5, always has a probability of 1. On the other hand, $E_6 = \varnothing$ and $P(E_6)$ is 0. The probability of an **impossible event,** such as E_6, is always 0, since none of the outcomes in the sample space satisfy the event. For any event E, $P(E)$ is between 0 and 1 inclusive.

The set of all outcomes in the sample space which do *not* belong to event E is called the **complement** of E, written E'. For example, in the experiment of drawing a single card from a standard deck of 52 cards, let E be the event ''the card is an ace.'' Then E' is the event ''the card is not an ace.'' From the definition of E', for any event E,

$$E \cup E' = S \quad \text{and} \quad E \cap E' = \emptyset.*$$

Probability concepts can be illustrated using **Venn diagrams,** as in Figure 3. The rectangle in Figure 3 represents the sample space in an experiment. The area inside the circle represents event E, while the area inside the rectangle, but outside the circle, represents event E'.

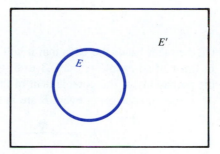

Figure 3

· · · · · · · · ·

EXAMPLE 2 In the experiment of drawing a card from a well-shuffled deck, find the probability of events E, the card is an ace, and E'.

Since there are four aces in the deck of 52 cards, $n(E) = 4$, and $n(S) = 52$. Therefore, $P(E) = \dfrac{n(E)}{n(S)} = \dfrac{4}{52} = \dfrac{1}{13}$. Of the 52 cards, 48 are not aces, so

$$P(E') = \frac{n(E')}{n(S)} = \frac{48}{52} = \frac{12}{13}. \quad \bullet$$

In Example 2, $P(E) + P(E') = (1/13) + (12/13) = 1$. This is always true for any event E and its complement E'. That is,

$$P(E) + P(E') = 1.$$

This can be restated as

$$P(E) = 1 - P(E') \quad \text{or} \quad P(E') = 1 - P(E).$$

These two equations suggest an alternate way to compute the probability of an event. For example, if it is known that $P(E) = 1/10$, then

$$P(E') = 1 - \frac{1}{10} = \frac{9}{10}.$$

*The **union** of two sets A and B is the set $A \cup B$ made up of all the elements from either A or B, or both. The **intersection** of sets A and B, written $A \cap B$, is made up of all the elements that belong to both sets at the same time.

Sometimes probability statements are expressed in terms of odds, a comparison of $P(E)$ with $P(E')$. The **odds** in favor of an event E are expressed as the ratio of $P(E)$ to $P(E')$ or as the fraction $P(E)/P(E')$. For example, if the probability of rain can be established as 1/3, the odds that it will rain are

$$P(\text{rain}) \text{ to } P(\text{no rain}) = \frac{1}{3} \text{ to } \frac{2}{3}$$

$$= \frac{1/3}{2/3}$$

$$= \frac{1}{2} \quad \text{or} \quad 1 \text{ to } 2.$$

On the other hand, the odds that it will not rain are 2 to 1 (or 2/3 to 1/3). If the odds in favor of an event are, say, 3 to 5, then the probability of the event is 3/8, while the probability of the complement of the event is 5/8.

If the odds favoring event E are m to n, then

$$P(E) = \frac{m}{m + n} \quad \text{and} \quad P(E') = \frac{n}{m + n}.$$

Now consider the probability of a **compound event** which involves an *alternative,* such as E or F. For example, in the experiment of rolling a die, suppose H is the event "the result is a 3," and K is the event "the result is an even number." What is the probability of "the result is a 3 or an even number"? We have

$$H = \{3\} \qquad \qquad P(H) = \frac{1}{6}$$

$$K = \{2, 4, 6\} \qquad \qquad P(K) = \frac{3}{6} = \frac{1}{2}.$$

therefore $\quad H \text{ or } K = \{2, 3, 4, 6\} \quad P(H \text{ or } K) = \frac{4}{6} = \frac{2}{3}.$

Notice that $P(H) + P(K) = P(H \text{ or } K)$.

Before assuming that this relationship is true in general, consider another event for this experiment, "the result is a 2," event G. Now

$$G = \{2\} \qquad \qquad P(G) = \frac{1}{6}$$

$$K = \{2, 4, 6\} \qquad \qquad P(K) = \frac{3}{6} = \frac{1}{2};$$

therefore $\quad K \text{ or } G = \{2, 4, 6\} \quad P(K \text{ or } G) = \frac{3}{6} = \frac{1}{2}.$

In this case $P(K) + P(G) \neq P(K \text{ or } G)$.

As Figure 4 shows, the difference in the two examples above comes from the fact that events H and K cannot occur simultaneously. Such events are called **mu-**

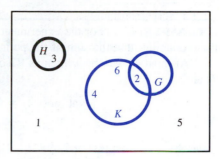

Figure 4

tually exclusive events. For these events, $H \cap K = \varnothing$, which is true in general for any two mutually exclusive events. Events K and G, however, can occur simultaneously. Both are satisfied if the result of the roll is a 2, the element in their intersection ($K \cap G = \{2\}$). This example suggests the following property.

Probability of Alternate Events

For any events E and F,

$$P(E \text{ or } F) = P(E \cup F) = P(E) + P(F) - P(E \cap F).$$

A standard deck of 52 cards, in which half are red and half are black, contains 4 *suits:* clubs (black), spades (black), hearts (red), and diamonds (red). Each suit contains 13 cards, numbered from 1 (the ace) to 10, plus a king, a queen and a jack.

EXAMPLE 3   One card is drawn from a well-shuffled deck of 52 cards. What is the probability of the following outcomes?

(a) The card is an ace or a spade.

The events "drawing an ace" and "drawing a spade" are not mutually exclusive since it is possible to draw the ace of spades, an outcome satisfying both events. The probability is

$$P(\text{ace or spade}) = P(\text{ace}) + P(\text{spade}) - P(\text{ace and spade})$$

$$= \frac{4}{52} + \frac{13}{52} - \frac{1}{52} = \frac{16}{52} = \frac{4}{13}.$$

(b) The card is a three or a king.

"Drawing a 3" and "drawing a king" are mutually exclusive events because it is impossible to draw one card that is both a 3 and a king.

$$P(3 \text{ or } K) = P(3) + P(K) - P(3 \text{ and } K)$$

$$= \frac{4}{52} + \frac{4}{52} - 0 = \frac{8}{52} = \frac{2}{13}. \quad \bullet$$

. . . .

EXAMPLE 4   For the experiment consisting of one roll of a pair of dice, find the probability that the sum of the points showing is at most 4.

"At most 4" can be written as "2 or 3 or 4." (A sum of 1 is meaningless here.) Then

$$P(\text{at most } 4) = P(2 \text{ or } 3 \text{ or } 4)$$
$$= P(2) + P(3) + P(4), \qquad (*)$$

since the events represented by "2," "3," and "4" are mutually exclusive.

The sample space for this experiment includes the 36 possible pairs of numbers from 1 to 6: (1, 1), (1, 2), (1, 3), (1, 4), (1, 5), (1, 6), (2, 1), (2, 2), and so on. The pair (1, 1) is the only one with a sum of 2, so $P(2) = 1/36$. Also $P(3) = 2/36$ since both (1, 2) and (2, 1) give a sum of 3. The pairs, (1, 3), (2, 2), and (3, 1) have a sum of 4, so $P(4) = 3/36$. Substituting into equation $(*)$ above gives

$$P(\text{at most } 4) = \frac{1}{36} + \frac{2}{36} + \frac{3}{36}$$
$$= \frac{6}{36} = \frac{1}{6}. \quad \bullet$$

11.7 Exercises ·

Write a sample space with equally likely outcomes for each of the following experiments.

1. A two-headed coin is tossed once.

2. Two ordinary coins are tossed.

3. Three ordinary coins are tossed.

4. Slips of paper marked with the numbers 1, 2, 3, 4, and 5 are placed in a box. After mixing well, two slips are drawn.

5. An unprepared student takes a three-question true/false quiz in which he guesses the answer to all three questions.

6. A die is rolled and then a coin is tossed.

Write the following events in set notation and give the probability of each event.

7. In the experiment of Exercise 2:
   (a) Both coins show the same face.
   (b) At least one coin turns up heads.

8. In Exercise 1:
   (a) The result of the toss is heads.
   (b) The result of the toss is tails.

9. In Exercise 4:
   (a) Both slips are marked with even numbers.
   (b) Both slips are marked with odd numbers.
   (c) Both slips are marked with the same number.
   (d) One slip is marked with an odd number, the other with an even number.

10. In Exercise 5:
    (a) The student gets all three answers correct.   (b) He gets all three answers wrong.
    (c) He gets exactly two answers correct.   (d) He gets at least one answer correct.

11. A marble is drawn at random from a box containing 3 yellow, 4 white, and 8 blue marbles.
    (a) A yellow marble is drawn.   (b) A blue marble is drawn.   (c) A black marble is drawn.
    (d) What are the odds in favor of drawing a yellow marble?
    (e) What are the odds against drawing a blue marble?

12. A baseball player with a batting average of .300 comes to bat. What are the odds in favor of his getting a hit?

13. In Exercise 4, what are the odds that the sum of the numbers on the two slips of paper is 5?

14. If the odds that it will rain are 4 to 5, what is the probability of rain?

15. If the odds that a candidate will win an election are 3 to 2, what is the probability that the candidate will lose?

16. A card is drawn from a well-shuffled deck of 52 cards. Find the probability that the card is each of the following: (a) a 9, (b) black, (c) a black 9, (d) a heart, (e) the 9 of hearts, (f) a face card (K, Q, J of any suit).

17. Mrs. Elliott invites 10 relatives to a party: her mother, two uncles, three brothers, and four cousins. If the chances of any one guest arriving first are equally likely, find the following probabilities.
    (a) The first guest is an uncle or a brother.
    (b) The first guest is a brother or cousin.
    (c) The first guest is a brother or her mother.

18. One card is drawn from a standard deck of 52 cards. Find the probability that the card is each of the following: (a) a 9 or a 10, (b) red or a 3, (c) a heart or black, (d) less than a 4 (Consider aces as 1's).

19. Two dice are rolled. Find the probability of the following.
    (a) The sum of the points is at least 10.
    (b) The sum of the points is either 7 or at least 10.
    (c) The sum of the points is 2 or the dice both show the same number.

20. A student estimates that his probability of getting an A in a certain course is .4; a B, .3; a C, .2; and a D, .1.
    (a) Assuming that only the grades A, B, C, D, and F are possible, what is the probability that he will fail the course?
    (b) What is the probability that he will receive a grade of C or better?
    (c) What is the probability that he will receive at least a B in the course?
    (d) What is the probability that he will get at most a C in the course?

21. If a marble is drawn from a bag containing 3 yellow, 4 white, and 8 blue marbles, what are the probabilities of the following?
    (a) The marble is either yellow or white.
    (b) It is either yellow or blue.
    (c) It is either red or white.

The table below gives a certain golfer's probabilities of scoring in various ranges on a par-70 course.

Range	Probability
below 60	.01
60–64	.08
65–69	.15
70–74	.28
75–79	.22
80–84	.08
85–89	.06
90–94	.04
95–99	.02
100 or more	.06

In a given round, find the probability that the golfer's score will be as in Exercises 22–27.

22. 90 or higher   **23.** Below par of 70   **24.** In the 70's

25. In the 90's   **26.** Not in the 60's   **27.** Not in the 60's or 70's

28. Find the odds in favor of the golfer shooting below par.

29. Find the odds against the golfer shooting in the 70s.

Francisco has set up the following table of probabilities for the number of hours it will take him to finish his homework.

Hours	1	2	3	4	5	6
Probability	.05	.10	.20	.40	.10	.15

Find the probability that the number of hours needed to finish his homework will be as follows.

30. Fewer than 3 hours   **31.** 3 hours or less   **32.** More than 2 hours

33. At least 2 hours   **34.** More than 1 hour and less than 5 hours   **35.** 8 hours

The table below shows the probability that a customer of a department store will make a purchase in the indicated range.

Cost	Probability
Below $2	.07
$2–$4.99	.18
$5–$9.99	.21
$10–$19.99	.16
$20–$39.99	.11
$40–$69.99	.09
$70–$99.99	.07
$100–$149.99	.08
$150 or over	.03

Find the probability that a customer makes a purchase which is in the following ranges.

36. $10 to $69.99   **37.** $20 or more   **38.** More than $4.99

39. Less than $100   **40.** $100 or more

In most animals and plants, it is very unusual for the number of main parts of the organism (arms, legs, toes, flower petals, etc.) to vary from generation to generation. Some species, however, have meristic variability, *in which the number of certain body parts varies from generation to generation. One researcher studied the front feet of certain guinea pigs and produced the following probabilities.**

$$P(\text{only four toes, all perfect}) = .77$$

$$P(\text{one imperfect toe and four good ones}) = .13$$

$$P(\text{exactly five good toes}) = .10$$

Find the probability of each of the following events.

41. No more than four good toes

42. Five toes, whether perfect or not

The probabilities for the outcomes of an experiment having sample space $S = \{s_1, s_2, s_3, s_4, s_5, s_6\}$ *are shown here.*

Outcomes	s_1	s_2	s_3	s_4	s_5	s_6
Probability	.17	.03	.09	.46	.21	.04

Let $E = \{s_1, s_2, s_5\}$, *and let* $F = \{s_4, s_5\}$. *Find each probability in Exercises 43–48.*

43. $P(E)$

44. $P(F)$

45. $P(E \cap F)$

46. $P(E \cup F)$

47. $P(E' \cup F')$

48. $P(E' \cap F)$

49. Let E, F, and G be events from a sample space S. Show that

$$P(E \cup F \cup G) = P(E) + P(F) + P(G) - P(E \cap F) - P(E \cap G) - P(F \cap G) + P(E \cap F \cap G).$$

(*Hint:* Let $H = E \cup F$ and use the formula for the probability of alternate events.)

11 CHAPTER SUMMARY

Key Words

11.1 **n factorial**
  binomial coefficient
  binomial theorem
11.2 **sequence**
  general term
  nth term
  arithmetic sequence
  common difference
  sigma notation

  index of summation
11.3 **geometric sequence**
  common ratio
  partial sum
  infinite geometric sequence
11.4 **series**
11.5 **mathematical induction**
11.6 **fundamental principle of counting**
  permutations

  combinations
11.7 **sample space**
  event
  probability of an event
  certain event
  impossible event
  complement
  odds
  compound event
  mutually exclusive events

---

*Excerpt from "Analysis of Variability in Number of Digits in an Inbred Strain of Guinea Pigs" by S. Wright, in *Genetics*, v. 19 (1934), 506–36. Reprinted by permission of Genetics Society of America.

Review Exercises

Use mathematical induction to prove that each of the following is true for every positive integer n.

1. $1 + 3 + 5 + 7 + \ldots + (2n - 1) = n^2$

2. $2 + 6 + 10 + 14 + \ldots + (4n - 2) = 2n^2$

3. $2 + 2^2 + 2^3 + \ldots + 2^n = 2(2^n - 1)$

4. $1^3 + 3^3 + 5^3 + \ldots + (2n - 1)^3 = n^2(2n^2 - 1)$

Use the binomial theorem to expand each of the following.

5. $(x + 2y)^4$

6. $(3z - 5w)^3$

7. $\left(3\sqrt{x} - \dfrac{1}{\sqrt{x}}\right)^5$

8. $(m^3 - m^{-2})^4$

Find the indicated term or terms for each of the following expansions.

9. Fifth term of $(3x - 2y)^6$

10. Eighth term of $(2m + n^2)^{12}$

11. First four terms of $(3 + x)^{16}$

12. Last three terms of $(2m - 3n)^{15}$

Write the first five terms for each sequence in Exercises 13–22.

13. $a_1 = 5$, $a_2 = 3$, $a_n = a_{n-1} - a_{n-2}$ for $n \geq 3$

14. $b_1 = -2$, $b_2 = 2$, $b_3 = -4$, $b_n = -2 \cdot b_{n-2}$ if n is even, and $b_n = 2 \cdot b_{n-2}$ if n is odd

15. Arithmetic, $a_2 = 6$, $d = -4$

16. Arithmetic, $a_3 = 9$, $a_4 = 7$

17. Arithmetic, $a_1 = 3 - \sqrt{5}$, $a_2 = 4$

18. Arithmetic, $a_3 = \pi$, $a_4 = 0$

19. Geometric, $a_1 = 4$, $r = 2$

20. Geometric, $a_4 = 8$, $r = 1/2$

21. Geometric, $a_1 = -3$, $a_2 = 4$

22. Geometric, $a_3 = 8$, $a_5 = 72$

23. A certain arithmetic sequence has $a_6 = -4$ and $a_{17} = 51$. Find a_1 and a_{20}.

24. For a given geometric sequence, $a_1 = 4$ and $a_5 = 324$. Find a_6.

Find a_8 for each of the following arithmetic sequences.

25. $a_1 = 6$, $d = 2$

26. $a_1 = -4$, $d = 3$

27. $a_1 = 6x - 9$, $a_2 = 5x + 1$

28. $a_3 = 11m$, $a_5 = 7m - 4$

Find S_{12} for each of the following arithmetic sequences.

29. $a_1 = 2$, $d = 3$

30. $a_2 = 6$, $d = 10$

31. $a_1 = -4k$, $d = 2k$

Find a_5 for each of the following geometric sequences.

32. $a_1 = 3$, $r = 2$

33. $a_2 = 3125$, $r = 1/5$

34. $a_1 = 5x$, $a_2 = x^2$

35. $a_2 = \sqrt{6}$, $a_4 = 6\sqrt{6}$

Find S_4 for each of the following geometric sequences.

36. $a_1 = 1$, $r = 2$

37. $a_1 = 3$, $r = 3$

38. $a_1 = 2k$, $a_2 = -4k$

Determine whether each of the following sequences is arithmetic, geometric, or neither. If the sequence is arithmetic, find the common difference. If the sequence is geometric, find the common ratio.

39. $8, -4, 2, -1, \dfrac{1}{2}, \ldots$

40. $-3, 0, 3, 6, 9, \ldots$

41. $\ln 1, \ln 2, \ln 3, \ln 4, \ldots$

42. $\ln 2, \ln 4, \ln 8, \ln 16, \ldots$

Evaluate each of the following sums which exist.

43. $18 + 9 + 9/2 + 9/4 + \ldots$

44. $20 + 15 + 45/4 + 135/16 + \ldots$

45. $-5/6 + 5/9 - 10/27 + \ldots$

46. $1/16 + 1/8 + 1/4 + 1/2 + \ldots$

47. $.9 + .09 + .009 + .0009 + \ldots$

Evaluate each of the following sums which exist.

48. $\displaystyle\sum_{i=1}^{4} \frac{2}{i}$

49. $\displaystyle\sum_{i=1}^{7} (-1)^{i+1} \cdot 6$

50. $\displaystyle\sum_{i=1}^{6} i(i + 2)$

51. $\displaystyle\sum_{i=1}^{4} \frac{i + 1}{i}$

52. $\displaystyle\sum_{i=1}^{12} (8i + 2)$

53. $\displaystyle\sum_{i=1}^{10,000} i$

54. $\displaystyle\sum_{i=1}^{6} 4 \cdot 3^{i}$

55. $\displaystyle\sum_{i=1}^{4} 8 \cdot 2^{i}$

56. $\displaystyle\sum_{i=1}^{\infty} \left(\frac{5}{8}\right)^{i}$

57. $\displaystyle\sum_{i=1}^{\infty} -10\left(\frac{5}{2}\right)^{i}$

Evaluate each of the following sums where $x_1 = 0$, $x_2 = 2$, $x_3 = 4$, $x_4 = 6$, $x_5 = 8$, $x_6 = 10$.

58. $\displaystyle\sum_{i=1}^{5} (x_i^2 - 4)$

59. $\displaystyle\sum_{i=1}^{3} (3x_i^2 + 2x_i)$

60. $\displaystyle\sum_{i=1}^{4} f(x_i) \,\Delta x; \quad f(x) = x^2 - x; \quad \Delta x = .2$

61. $\displaystyle\sum_{i=1}^{6} f(x_i)\,\Delta x; \quad f(x) = (x - 2)^3; \quad \Delta x = .1$

Write each of the following sums using summation notation.

62. $4 - 1 - 6 - \ldots - 66$

63. $10 + 14 + 18 + \ldots + 86$

64. $4 + 12 + 36 + \ldots + 972$

65. $\dfrac{5}{6} + \dfrac{6}{7} + \dfrac{7}{8} + \ldots + \dfrac{12}{13}$

66. $3(e + 1)^2 + 3(e^3 + 1)^2 + 3(e^5 + 1)^2 + 3(e^7 + 1)^2 + \ldots$

67. $\dfrac{1}{1.2} + \dfrac{1}{1.4} + \dfrac{1}{1.6} + \dfrac{1}{1.8} + \dfrac{1}{2} + \ldots$

Use summation notation to rewrite each sum with the index of summation starting at the indicated number.

68. $\displaystyle\sum_{i=1}^{8} (3 + 2i); \quad -2$

69. $\displaystyle\sum_{i=2}^{9} (4 - 6i); \quad 0$

Use the properties of summation, along with sums given earlier, to evaluate each summation.

70. $\displaystyle\sum_{i=1}^{4} (i^2 + 2i)$

71. $\displaystyle\sum_{i=1}^{6} (8 + i^3)$

Find the value of each expression in Exercises 72–75.

72. $P(9, 2)$

73. $P(6, 0)$

74. $\dbinom{8}{3}$

75. $\dbinom{10}{5}$

76. Four students are to be assigned to 4 different summer jobs. Each student is qualified for all 4 jobs. In how many ways can the jobs be assigned?

77. Nine football teams are competing for 1st, 2nd, and 3rd place titles in a statewide tournament. In how many ways can the winners be determined?

78. How many different license-plate numbers can be formed using 3 letters followed by 3 digits if no repeats are allowed? How many if there are no repeats and either letters or numbers come first?

Write sample spaces for the following.

79. A die is rolled.

80. A card is drawn from a deck containing only the thirteen spades.

81. The weight of a person is measured to the nearest half-pound; the scale will not measure more than 300 pounds.

82. A coin is tossed four times.

An urn contains five balls labeled 3, 5, 7, 9, and 11, respectively, while a second urn contains four red and two green balls. An experiment consists of pulling one ball from each urn, in turn. Write an expression for the events in Exercises 83–85.

83. The sample space

84. Event E, the first ball is greater than 5

85. Event F, the second ball is green

86. Are the outcomes in the sample space equally likely?

A company sells typewriters and copiers. Let E be the event "a customer buys a typewriter," and let F be the event "a customer buys a copier." Write each of the following using ∩, ∪, or ' as necessary.

87. A customer buys neither.

88. A customer buys at least one.

Find the odds in favor of a card drawn from an ordinary deck being as follows.

89. A club

90. A black jack

91. A red face card or a queen

A sample shipment of five swimming pool filters is chosen at random. The probability of exactly 0, 1, 2, 3, 4, or 5 filters being defective is given in the following table.

Number defective	0	1	2	3	4	5
Probability	.31	.25	.18	.12	.08	.06

Find the probability that the following number of filters is defective.

92. No more than 3

93. At least 3

A card is drawn from a standard deck of 52 cards. Find the probability that the card is as described in Exercises 94–99.

94. A black king

95. A face card or an ace

96. An ace or a diamond

97. Not a diamond

98. Not a diamond or not black

99. A diamond and black

APPENDICES

APPENDICES

APPENDIX: USING TABLES OF COMMON LOGARITHMS

As mentioned earlier, logarithms to base 10 are called **common logarithms** and are often written as log x. This convention will be used throughout this Appendix. Examples of common logarithms include

$$\log 1000 = \log 10^3 = 3$$
$$\log 100 = \log 10^2 = 2$$
$$\log 10 = \log 10^1 = 1$$
$$\log 1 = \log 10^0 = 0$$
$$\log .1 = \log 10^{-1} = -1$$
$$\log .001 = \log 10^{-3} = -3.$$

Common logarithms of numbers that are not powers of 10 are approximated with either a calculator (discussed earlier in the text) or a table of common logarithms. For example, from Table 2, a decimal approximation of log 6 is

$$\log 6 \approx .7782,$$

or, equivalently, $10^{.7782} \approx 6$. Since most logarithms are approximations anyway, it is common to replace $\approx$ with $=$ and write

$$\log 6 = .7782.$$

Table 2 gives the logarithms of numbers between 1 and 10. Since every positive number can be written in scientific notation as the product of a number between 1 and 10 and a power of 10, the logarithm of any positive number can be found by using the table and the properties of logarithms.

· · · · · · · · ·

EXAMPLE 1   Find log 6.24.

Locate the first two digits, 6.2, in the left column of the table. Then find the third digit, 4, across the top of the table. You should find

$$\log 6.24 = .7952. \quad \bullet$$

· · · · · · · · ·

EXAMPLE 2   Find log 6240.

Write 6240 using scientific notation, as

$$6240 = 6.24 \times 10^3.$$

Then use the properties of logarithms.

$$\log 6240 = \log (6.24 \times 10^3) = \log 6.24 + \log 10^3$$
$$= \log 6.24 + 3 \log 10 = \log 6.24 + 3$$

From Example 1, $\log 6.24 = .7952$, so

$$\log 6240 = .7952 + 3 = 3.7952. \quad \bullet$$

The decimal part of the logarithm, .7952 in Example 2, is called the **mantissa,** and the integer part, 3 here, is the **characteristic.** When using a table of logarithms, always make sure the mantissa is positive. The characteristic can be any integer, positive, negative, or zero.

· · · · · · · · ·

EXAMPLE 3   Find log .00587.

Use scientific notation and the properties of logarithms to get

$$\log .00587 = \log (5.87 \times 10^{-3}) = \log 5.87 + \log 10^{-3}$$
$$= .7686 + (-3) = .7686 - 3.$$

The logarithm is usually left in this form. A calculator would give the answer as -2.2314, the algebraic sum of .7686 and -3. The decimal portion in the calculator answer is a negative number. This is not the best form for the logarithm when using tables, since it is not clear which number is the mantissa.

It is possible to write the characteristic in other forms. For example, log .00587 could be written as

$$\log .00587 = 7.7686 - 10.$$

The best choice depends on the anticipated use of the logarithm. $\bullet$

· · · · · · · · ·

EXAMPLE 4   Find each of the following.

(a) $\log (2.73)^4$

Use a property of logarithms to get

$$\log (2.73)^4 = 4 \log 2.73 = 4(.4362) = 1.7448.$$

(b) $\log \sqrt[3]{.0762}$

Here

$$\log \sqrt[3]{.0762} = \log (.0762)^{1/3}$$
$$= \frac{1}{3} \log .0762 = \frac{1}{3}(.8820 - 2).$$

Since it is necessary to preserve the characteristic as an integer, change the characteristic to a multiple of 3 before multiplying by 1/3. One way to do this is to add and subtract 1 (which adds to 0) as follows.

$$.8820 - 2 = 1 - 1 + .8820 - 2$$
$$= 1.8820 - 3$$

Now complete the work above.

$$\log \sqrt[3]{.0763} = \frac{1}{3}(1.8820 - 3)$$

$$= .6273 - 1 \quad \bullet$$

Sometimes the logarithm of a number is known and the number itself must be found. The number is called the **antilogarithm,** sometimes abbreviated **antilog.** For example, .756 is the antilogarithm of .8785 − 1, since

$$\log .756 = .8785 - 1.$$

Find the antilogarithm by looking for .8785 in the body of the logarithm table. You should find 7.5 at the left and 6 at the top. Since the characteristic of .8785 − 1 is −1, the antilogarithm is

$$7.56 \times 10^{-1} = .756.$$

In exponential notation, since

$$\log .756 = .8785 - 1 = -.1215,$$

then

$$.756 = 10^{-.1215}.$$

· · · · · · · ·

EXAMPLE 5   Find each of the following antilogarithms.

(a) $\log x = 2.5340$

Find .5340 within the table; 3.4 is at the left and 2 is at the top.

$$\log x = 2 + .5340$$
$$= \log 10^2 + \log 3.42 \qquad \text{From Table 2}$$
$$= \log (3.42 \times 10^2)$$
$$= \log 342$$

Since
$$\log x = \log 342,$$
$$x = 342,$$

and 342 is the antilogarithm of 2.5340.

(b) $\log x = .7536 - 3$

Table 2 shows that the antilogarithm of .7536 − 3 is

$$5.67 \times 10^{-3} = .00567.$$

(c) $\log x = -4.0670 = -4 + (-.0670)$
$$= -4 + (-1) + [1 + (-.0670)]$$
$$= -5 + .9330$$
$$= \log .0000857$$
$$x = .0000857 \quad \bullet$$

A Appendix Exercises ··

Find the characteristic of the logarithms of each of the following.

1. 875	**2.** 9462	**3.** 2,400,000	**4.** 875,000
5. .00023	**6.** .098	**7.** .000042	**8.** .000000257

Find the common logarithms of the following numbers.

9. .000893	**10.** .00376	**11.** 68,200	**12.** 103,000
13. 7.63	**14.** 9.37	**15.** .429	**16.** 15.1
17. 235	**18.** 1.73	**19.** 48.3	**20.** 305

Find the common antilogarithms of each of the following logarithms to three significant digits.

21. 1.5366	**22.** 2.9253	**23.** .8733 − 2	**24.** .2504 − 1
25. 3.4947	**26.** 4.6863	**27.** .8039 − 3	**28.** .8078 − 2
29. 1.7938	**30.** 3.4771	**31.** − 3.7773	**32.** − 4.1325

B ⋮ APPENDIX: USING TABLES OF TRIGONOMETRIC FUNCTION VALUES

The following examples show how Table 3 is used to find values of trigonometric functions for θ in the intervals $[0°, 90°]$ or $[0, \pi/2]$. For θ in $[0°, 45°]$ or $[0, \pi/4]$, find θ by reading down one of the first two columns. Values for θ in $[45°, 90°]$ or $[\pi/4, \pi/2]$ are given by reading *up* the last two columns and referring to the names of the trigonometric functions at the *bottom* of the table. The values of θ in degrees are given to the nearest ten minutes. To use the table for decimal degrees, it is necessary to convert to degrees and minutes, to the nearest ten minutes, as in the next example.

. . . .

EXAMPLE 1   Find the following function values.

(a) sin 49° 10′

Look in the last column for 49° 10′ reading up the table and looking for sin at the bottom of the table. You should find

$$\sin 49° \ 10' = .7566.$$

(b) sec .2414

Look in the second column for .2414 radians and across the top of the table for the column headed sec.

$$\sec .2414 = 1.030$$

(c) tan 17.3°

Convert 17.3° to degrees and minutes as follows.

$$17.3° = 17° + \frac{3°}{10}$$

$$= 17° + \frac{18°}{60}$$

$$= 17° \ 18' \approx 17° \ 20'$$

to the nearest ten minutes. Now find the entry for 17° 20′ in the tan column.

$$\tan 17.3° \approx \tan 17° \ 20' = .3121$$

Interpolation could be used to get a more accurate answer. ●

The table can also be used backwards to find θ given a trigonometric function value of θ.

EXAMPLE 2 Find a value of θ in degrees that satisfies each of the following.

(a) $\cos \theta = .9063$

Look in the body of the table in a column headed cos at the top or with cos at the bottom. Since the entry is found in a column with cos at the top, read over to the first column to find $\theta = 25° \ 00'$.

(b) $\csc \theta = 1.117$

The entry 1.117 is found in a column with csc at the bottom. Read across to the last column to see that

$$\theta = 63° \ 30'.$$

Notice that the row for 63° 30' is *above* the row for 63° 00', since we read *up* the table in the last column. ●

Table 3 gives values only for angles between 0° and 90°, inclusive, or for real numbers between 0 and $\pi/2$, inclusive. For values outside this range, some of the earlier work with trigonometric functions must be used. As will be shown, the trigonometric function values for (90°, 360°) can be found by referring to the appropriate value of θ in the interval (0°, 90°) and affixing the correct sign. For any value θ in the interval (90°, 360°), the positive acute angle made by the terminal side of angle θ and the x-axis is called the **reference** or **related angle** for θ; this new angle is written θ'. See Figure 1. For example, if $\theta = 135°$, the reference angle θ' is 45°. If $\theta = 200°$, then $\theta' = 20°$.

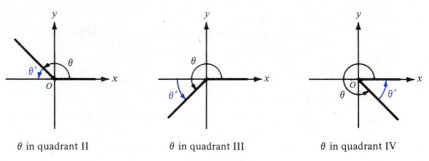

θ in quadrant II θ in quadrant III θ in quadrant IV

Figure 1

For radians, a similar definition could be given for **reference numbers**. For example, if $s = -\pi/6$, then the reference number s' is $s' = \pi/6$. Also, if $s = 5\pi/4$, then $s' = \pi/4$.

EXAMPLE 3 Find the reference angle or reference number for each of the following.

(a) 218°

As shown in Figure 2, the positive acute angle made by the terminal side of this angle and the x-axis is $218° - 180° = 38°$.

(b) 321° 10′

The positive acute angle made by the terminal side of this angle and the x-axis is $360° - 321° \, 10'$. Write $360°$ as $359° \, 60'$ so that the reference angle is

$$359° \, 60' - 321° \, 10' = 38° \, 50'.$$

By the way, an angle of $-38° \, 50'$, which has the same terminal ray as $321° \, 10'$, also has a reference angle of $38° \, 50'$.

(c) $s = 2.1031$

The reference number is found by subtracting from π. See Figure 3. Using the four-decimal-place approximation 3.1416 for π gives

$$s' = 3.1416 - 2.1031 = 1.0385. \quad \bullet$$

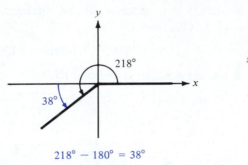

$$218° - 180° = 38°$$

Figure 2

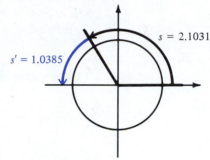

Figure 3

. . . .

EXAMPLE 4   Find each of the following values.

(a) tan 315°

Begin by finding the reference angle for $315°$, as in Figure 4. Since $315°$ is in quadrant IV, subtract $315°$ from $360°$.

$$360° - 315° = 45°$$

The value of tangent for quadrant IV angles is negative, and $\tan 45° = 1$, so

$$\tan 315° = -\tan 45° = -1.$$

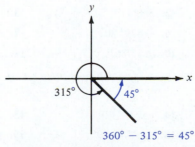

$$360° - 315° = 45°$$

Figure 4

(b) $\cos(-570°)$

To begin, find the smallest possible positive angle coterminal to $-570°$. See Figure 5. Add $2 \cdot 360°$, or $720°$, to $-570°$.

$$-570° + 720° = 150°$$

Since $150°$ is in quadrant II, find its reference angle by subtracting $150°$ from $180°$.

$$180° - 150° = 30°$$

Since cosine is negative in quadrant II,

$$\cos(-570°) = \cos 150° = -\cos 30° = -\sqrt{3}/2.$$

(c) $\sin 3.7845$

Since $s = 3.7845$ is in quadrant III, the reference number s' is found by subtracting π, or 3.1416, from s.

$$s' = 3.7845 - 3.1416 = .6429$$

The values of sine are negative in quadrant III, so

$$\sin 3.7845 = -\sin .6429 = -.5995. \quad \bullet$$

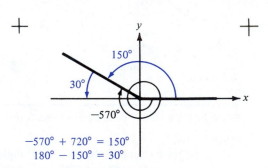

Figure 5

B Appendix Exercises

Find the reference angle or reference number for each of the following. Use 3.1416 as an approximation for π.

1. $215°$

2. $550°$

3. $-143°$

4. $-12°$

5. $-110° \ 10'$

6. $-429° \ 30'$

7. 3.21

8. 2.4

9. 5.9690

10. -1.7861

11. -4.0230

12. -2.4580

Use Table 3 to find a value for each of the following. Use 3.1416 as an approximation for π.

13. $\sin 39° \ 20'$

14. $\cos 58° \ 40'$

15. $\sin(-38° \ 40')$

16. $\csc(-168° \ 30')$

17. $\cos(-124° \ 50')$

18. $\sec 274° \ 30'$

19. $\sec 1.9024$

20. $\cot 3.1998$

21. $\sin 7.5835$

22. $\tan 6.4752$

23. $\cos(-4.0230)$

24. $\cot(-3.8426)$

Use Table 3 to find a value of θ in the interval (0°, 90°) satisfying each of the following. Give θ to the nearest ten minutes.

25. $\sin \theta = .8480$ **26.** $\tan \theta = 1.473$ **27.** $\cos \theta = .8616$ **28.** $\cot \theta = 1.257$

29. $\sin \theta = .7214$ **30.** $\sec \theta = 2.749$ **31.** $\tan \theta = 6.435$ **32.** $\sin \theta = .2784$

Find a value of s in (0, 1.5708) that makes each of the following true.

33. $\tan s = .2126$ **34.** $\cos s = .7826$ **35.** $\sin s = .9918$ **36.** $\cot s = .2994$

37. $\cot s = .0963$ **38.** $\csc s = 1.021$ **39.** $\tan s = 1.621$ **40.** $\cos s = .9272$

41. $\sin s = .4067$ **42.** $\cos s = .6626$ **43.** $\tan s = .3607$ **44.** $\sec s = 1.758$

Use Table 3 to find all values of θ to the nearest ten minutes in the interval (90°, 360°).

45. $\sin \theta = -.2560$ **46.** $\tan \theta = 1.907$ **47.** $\cos \theta = .6361$ **48.** $\sin \theta = .6777$

49. $\tan \theta = -.1257$ **50.** $\cos \theta = -.9546$ **51.** $\cot \theta = .3805$ **52.** $\csc \theta = 1.026$

53. $\sec \theta = -1.353$ **54.** $\cot \theta = -.1198$

TABLES

Table 1   **677**

TABLES

TABLE 1 Powers of *e* and Natural Logarithms

x	e^x	e^{-x}	ln x
.00	1.00000	1.00000	
.01	1.01005	.99004	−4.6052
.02	1.02020	.98019	−3.9120
.03	1.03045	.97044	−3.5066
.04	1.04081	.96078	−3.2189
.05	1.05127	.95122	−2.9957
.06	1.06183	.94176	−2.8134
.07	1.07250	.93239	−2.6593
.08	1.08328	.92311	−2.5257
.09	1.09417	.91393	−2.4079
.10	1.10517	.90483	−2.3026
.11	1.11628	.89583	−2.2073
.12	1.12750	.88692	−2.1203
.13	1.13883	.87810	−2.0402
.14	1.15027	.86936	−1.9661
.15	1.16183	.86071	−1.8971
.16	1.17351	.85214	−1.8326
.17	1.18530	.84366	−1.7720
.18	1.19722	.83527	−1.7148
.19	1.20925	.82696	−1.6607
.2	1.22140	.81873	−1.6094
.3	1.34985	.74081	−1.2040
.4	1.49182	.67032	−.9163
.5	1.64872	.60653	−.6931
.6	1.82211	.54881	−.5108
.7	2.01375	.49658	−.3567
.8	2.22554	.44932	−.2231
.9	2.45960	.40656	−.1054
1.0	2.71828	.36787	.0000
1.1	3.00416	.33287	.0953

x	e^x	e^{-x}	ln x
1.2	3.32011	.30119	.1823
1.3	3.66929	.27253	.2624
1.4	4.05519	.24659	.3365
1.5	4.48168	.22313	.4055
1.6	4.95302	.20189	.4700
1.7	5.47394	.18268	.5306
1.8	6.04964	.16529	.5878
1.9	6.68589	.14956	.6419
2.0	7.38905	.13533	.6931
2.1	8.16616	.12245	.7419
2.2	9.02500	.11080	.7885
2.3	9.97417	.10025	.8329
2.4	11.0231	.09071	.8755
2.5	12.1824	.08208	.9163
2.6	13.4637	.07427	.9555
2.7	14.8797	.06720	.9933
2.8	16.4446	.06081	1.0296
2.9	18.1741	.05502	1.0647
3.0	20.0855	.04978	1.0986
3.1	22.1979	.04505	1.1314
3.2	24.5325	.04076	1.1632
3.3	27.1126	.03688	1.1939
3.4	29.9641	.03337	1.2238
3.5	33.1154	.03020	1.2528
3.6	36.5982	.02732	1.2809
3.7	40.4473	.02472	1.3083
3.8	44.7012	.02237	1.3350
3.9	49.4024	.02024	1.3610
4.0	54.5981	.01832	1.3863
4.1	60.3402	.01657	1.4110

TABLE 1 Powers of *e* and Natural Logarithms (continued)

x	e^x	e^{-x}	ln x	x	e^x	e^{-x}	ln x
4.2	66.6863	.01500	1.4351	7.2	1339.43	.00075	1.9741
4.3	73.6997	.01357	1.4586	7.3	1480.29	.00068	1.9879
4.4	81.4508	.01228	1.4816	7.4	1635.98	.00061	2.0015
4.5	90.0170	.01111	1.5041	7.5	1808.03	.00055	2.0149
4.6	99.4842	.01005	1.5261	7.6	1998.19	.00050	2.0281
4.7	109.947	.00910	1.5476	7.7	2208.34	.00045	2.0412
4.8	121.510	.00823	1.5686	7.8	2440.59	.00041	2.0541
4.9	134.290	.00745	1.5892	7.9	2697.27	.00037	2.0669
5.0	148.413	.00674	1.6094	8.0	2980.94	.00034	2.0794
5.1	164.022	.00610	1.6292	8.1	3294.45	.00030	2.0919
5.2	181.272	.00552	1.6487	8.2	3640.94	.00027	2.1041
5.3	200.336	.00499	1.6677	8.3	4023.86	.00025	2.1163
5.4	221.406	.00452	1.6864	8.4	4447.05	.00022	2.1282
5.5	244.691	.00445	1.7047	8.5	4914.75	.00020	2.1401
5.6	270.426	.00370	1.7228	8.6	5431.65	.00018	2.1518
5.7	298.867	.00335	1.7405	8.7	6002.90	.00017	2.1633
5.8	330.299	.00303	1.7579	8.8	6634.23	.00015	2.1748
5.9	365.036	.00274	1.7750	8.9	7331.96	.00014	2.1861
6.0	403.428	.00248	1.7918	9.0	8103.08	.00012	2.1972
6.1	445.856	.00224	1.8083	9.1	8955.29	.00011	2.2083
6.2	492.748	.00203	1.8245	9.2	9897.13	.00010	2.2192
6.3	544.570	.00184	1.8405	9.3	10938.0	.00009	2.2300
6.4	601.843	.00166	1.8563	9.4	12088.4	.00008	2.2407
6.5	665.139	.00150	1.8718	9.5	13359.7	.000075	2.2513
6.6	735.093	.00136	1.8871	9.6	14764.8	.000068	2.2618
6.7	812.403	.00123	1.9021	9.7	16317.6	.000061	2.2721
6.8	897.844	.00111	1.9169	9.8	18033.8	.000055	2.2824
6.9	992.271	.00101	1.9315	9.9	19930.4	.000050	2.2925
7.0	1096.63	.00091	1.9459	10.0	22026.5	.000045	2.3026
7.1	1211.96	.00083	1.9601				

Table 2   **679**

TABLE 2   Common Logarithms

n	0	1	2	3	4	5	6	7	8	9
1.0	.0000	.0043	.0086	.0128	.0170	.0212	.0253	.0294	.0334	.0374
1.1	.0414	.0453	.0492	.0531	.0569	.0607	.0645	.0682	.0719	.0755
1.2	.0792	.0828	.0864	.0899	.0934	.0969	.1004	.1038	.1072	.1106
1.3	.1139	.1173	.1206	.1239	.1271	.1303	.1335	.1367	.1399	.1430
1.4	.1461	.1492	.1523	.1553	.1584	.1614	.1644	.1673	.1703	.1732
1.5	.1761	.1790	.1818	.1847	.1875	.1903	.1931	.1959	.1987	.2014
1.6	.2041	.2068	.2095	.2122	.2148	.2175	.2201	.2227	.2253	.2279
1.7	.2304	.2330	.2355	.2380	.2405	.2430	.2455	.2480	.2504	.2529
1.8	.2553	.2577	.2601	.2625	.2648	.2672	.2695	.2718	.2742	.2765
1.9	.2788	.2810	.2833	.2856	.2878	.2900	.2923	.2945	.2967	.2989
2.0	.3010	.3032	.3054	.3075	.3096	.3118	.3139	.3160	.3181	.3201
2.1	.3222	.3243	.3263	.3284	.3304	.3324	.3345	.3365	.3385	.3404
2.2	.3424	.3444	.3464	.3483	.3502	.3522	.3541	.3560	.3579	.3598
2.3	.3617	.3636	.3655	.3674	.3692	.3711	.3729	.3747	.3766	.3784
2.4	.3802	.3820	.3838	.3856	.3874	.3892	.3909	.3927	.3945	.3962
2.5	.3979	.3997	.4014	.4031	.4048	.4065	.4082	.4099	.4116	.4133
2.6	.4150	.4166	.4183	.4200	.4216	.4232	.4249	.4265	.4281	.4298
2.7	.4314	.4330	.4346	.4362	.4378	.4393	.4409	.4425	.4440	.4456
2.8	.4472	.4487	.4502	.4518	.4533	.4548	.4564	.4579	.4594	.4609
2.9	.4624	.4639	.4654	.4669	.4683	.4698	.4713	.4728	.4742	.4757
3.0	.4771	.4786	.4800	.4814	.4829	.4843	.4857	.4871	.4886	.4900
3.1	.4914	.4928	.4942	.4955	.4969	.4983	.4997	.5011	.5024	.5038
3.2	.5051	.5065	.5079	.5092	.5105	.5119	.5132	.5145	.5159	.5172
3.3	.5185	.5198	.5211	.5224	.5237	.5250	.5263	.5276	.5289	.5302
3.4	.5315	.5328	.5340	.5353	.5366	.5378	.5391	.5403	.5416	.5428
3.5	.5441	.5453	.5465	.5478	.5490	.5502	.5514	.5527	.5539	.5551
3.6	.5563	.5575	.5587	.5599	.5611	.5623	.5635	.5647	.5658	.5670
3.7	.5682	.5694	.5705	.5717	.5729	.5740	.5752	.5763	.5775	.5786
3.8	.5798	.5809	.5821	.5832	.5843	.5855	.5866	.5877	.5888	.5899
3.9	.5911	.5922	.5933	.5944	.5955	.5966	.5977	.5988	.5999	.6010
4.0	.6021	.6031	.6042	.6053	.6064	.6075	.6085	.6096	.6107	.6117
4.1	.6128	.6138	.6149	.6160	.6170	.6180	.6191	.6201	.6212	.6222
4.2	.6232	.6243	.6253	.6263	.6274	.6284	.6294	.6304	.6314	.6325
4.3	.6335	.6345	.6355	.6365	.6375	.6385	.6395	.6405	.6415	.6425
4.4	.6435	.6444	.6454	.6464	.6474	.6484	.6493	.6503	.6513	.6522
4.5	.6532	.6542	.6551	.6561	.6571	.6580	.6590	.6599	.6609	.6618
4.6	.6628	.6637	.6646	.6656	.6665	.6675	.6684	.6693	.6702	.6712
4.7	.6721	.6730	.6739	.6749	.6758	.6767	.6776	.6785	.6794	.6803
4.8	.6812	.6821	.6830	.6839	.6848	.6857	.6866	.6875	.6884	.6893
4.9	.6902	.6911	.6920	.6928	.6937	.6946	.6955	.6964	.6972	.6981
5.0	.6990	.6998	.7007	.7016	.7024	.7033	.7042	.7050	.7059	.7067
5.1	.7076	.7084	.7093	.7101	.7110	.7118	.7126	.7135	.7143	.7152
5.2	.7160	.7168	.7177	.7185	.7193	.7202	.7210	.7218	.7226	.7235
5.3	.7243	.7251	.7259	.7267	.7275	.7284	.7292	.7300	.7308	.7316
5.4	.7324	.7332	.7340	.7348	.7356	.7364	.7372	.7380	.7388	.7396
n	0	1	2	3	4	5	6	7	8	9

TABLE 2 Common Logarithms (continued)

n	0	1	2	3	4	5	6	7	8	9
5.5	.7404	.7412	.7419	.7427	.7435	.7443	.7451	.7459	.7466	.7474
5.6	.7482	.7490	.7497	.7505	.7513	.7520	.7528	.7536	.7543	.7551
5.7	.7559	.7566	.7574	.7582	.7589	.7597	.7604	.7612	.7619	.7627
5.8	.7634	.7642	.7649	.7657	.7664	.7672	.7679	.7686	.7694	.7701
5.9	.7709	.7716	.7723	.7731	.7738	.7745	.7752	.7760	.7767	.7774
6.0	.7782	.7789	.7796	.7803	.7810	.7818	.7825	.7832	.7839	.7846
6.1	.7853	.7860	.7868	.7875	.7882	.7889	.7896	.7903	.7910	.7917
6.2	.7924	.7931	.7938	.7945	.7952	.7959	.7966	.7973	.7980	.7987
6.3	.7993	.8000	.8007	.8014	.8021	.8028	.8035	.8041	.8048	.8055
6.4	.8062	.8069	.8075	.8082	.8089	.8096	.8102	.8109	.8116	.8122
6.5	.8129	.8136	.8142	.8149	.8156	.8162	.8169	.8176	.8182	.8189
6.6	.8195	.8202	.8209	.8215	.8222	.8228	.8235	.8241	.8248	.8254
6.7	.8261	.8267	.8274	.8280	.8287	.8293	.8299	.8306	.8312	.8319
6.8	.8325	.8331	.8338	.8344	.8351	.8357	.8363	.8370	.8376	.8382
6.9	.8388	.8395	.8401	.8407	.8414	.8420	.8426	.8432	.8439	.8445
7.0	.8451	.8457	.8463	.8470	.8476	.8482	.8488	.8494	.8500	.8506
7.1	.8513	.8519	.8525	.8531	.8537	.8543	.8549	.8555	.8561	.8567
7.2	.8573	.8579	.8585	.8591	.8597	.8603	.8609	.8615	.8621	.8627
7.3	.8633	.8639	.8645	.8651	.8657	.8663	.8669	.8675	.8681	.8686
7.4	.8692	.8698	.8704	.8710	.8716	.8722	.8727	.8733	.8739	.8745
7.5	.8751	.8756	.8762	.8768	.8774	.8779	.8785	.8791	.8797	.8802
7.6	.8808	.8814	.8820	.8825	.8831	.8837	.8842	.8848	.8854	.8859
7.7	.8865	.8871	.8876	.8882	.8887	.8893	.8899	.8904	.8910	.8915
7.8	.8921	.8927	.8932	.8938	.8943	.8949	.8954	.8960	.8965	.8971
7.9	.8976	.8982	.8987	.8993	.8998	9004	.9009	.9015	.9020	.9025
8.0	.9031	.9036	.9042	.9047	.9053	.9058	.9063	.9069	.9074	.9079
8.1	.9085	.9090	.9096	.9101	.9106	.9112	.9117	.9122	.9128	.9133
8.2	.9138	.9143	.9149	.9154	.9159	.9165	.9170	.9175	.9180	.9186
8.3	.9191	.9196	.9201	.9206	.9212	.9217	.9222	.9227	.9232	.9238
8.4	.9243	.9248	.9253	.9258	.9263	.9269	.9274	.9279	.9284	.9289
8.5	.9294	.9299	.9304	.9309	.9315	.9320	.9325	.9330	.9335	.9340
8.6	.9345	.9350	.9355	.9360	.9365	.9370	.9375	.9380	.9385	.9390
8.7	.9395	.9400	.9405	.9410	.9415	.9420	.9425	.9430	.9435	.9440
8.8	.9445	.9450	.9455	.9460	.9465	.9469	.9474	.9479	.9484	.9489
8.9	.9494	.9499	.9504	.9509	.9513	.9518	.9523	.9528	.9533	.9538
9.0	.9542	.9547	.9552	.9557	.9562	.9566	.9571	.9576	.9581	.9586
9.1	.9590	.9595	.9600	.9605	.9609	.9614	.9619	.9624	.9628	.9633
9.2	.9638	.9643	.9647	.9652	.9657	.9661	.9666	.9671	.9675	.9680
9.3	.9685	.9689	.9694	.9699	.9703	.9708	.9713	.9717	.9722	.9727
9.4	.9731	.9736	.9741	.9745	.9750	.9754	.9759	.9763	.9768	.9773
9.5	.9777	.9782	.9786	.9791	.9795	.9800	.9805	.9809	.9814	.9818
9.6	.9823	.9827	.9832	.9836	.9841	.9845	.9850	.9854	.9859	.9863
9.7	.9868	.9872	.9877	.9881	.9886	.9890	.9894	.9899	.9903	.9908
9.8	.9912	.9917	.9921	.9926	.9930	.9934	.9939	.9943	.9948	.9952
9.9	.9956	.9961	.9965	.9969	.9974	.9978	.9983	.9987	.9991	.9996
n	0	1	2	3	4	5	6	7	8	9

Table 3 **681**

TABLE 3 Trigonometric Functions in Degrees and Radians

θ (degrees)	θ (radians)	sin θ	cos θ	tan θ	cot θ	sec θ	csc θ	(radians)	(degrees)
0°00'	0000	0000	1.0000	0000	—	1.000	—	1.5708	90°00'
10	0029	0029	1.0000	0029	343.8	1.000	343.8	1.5679	50
20	0058	0058	1.0000	0058	171.9	1.000	171.9	1.5650	40
30	0087	0087	1.0000	0087	114.6	1.000	114.6	1.5621	30
40	0116	0116	9999	0116	85.94	1.000	85.95	1.5592	20
50	0145	0145	9999	0145	68.75	1.000	68.76	1.5563	10
1°00'	0175	0175	9998	0175	57.29	1.000	57.30	1.5533	89°00'
10	0204	0204	9998	0204	49.10	1.000	49.11	1.5504	50
20	0233	0233	9997	0233	42.96	1.000	42.98	1.5475	40
30	0262	0262	9997	0262	38.19	1.000	38.20	1.5446	30
40	0291	0291	9996	0291	34.37	1.000	34.38	1.5417	20
50	0320	0320	9995	0320	31.24	1.001	31.26	1.5388	10
2°00'	0349	0349	9994	0349	28.64	1.001	28.65	1.5359	88°00'
10	0378	0378	9993	0378	26.43	1.001	26.45	1.5330	50
20	0407	0407	9992	0407	24.54	1.001	24.56	1.5301	40
30	0436	0436	9990	0437	22.90	1.001	22.93	1.5272	30
40	0465	0465	9989	0466	21.47	1.001	21.49	1.5243	20
50	0495	0494	9988	0495	20.21	1.001	20.23	1.5213	10
3°00'	0524	0523	9986	0524	19.08	1.001	19.11	1.5184	87°00'
10	0553	0552	9985	0553	18.07	1.001	18.10	1.5155	50
20	0582	0581	9983	0582	17.17	1.002	17.20	1.5126	40
30	0611	0610	9981	0612	16.35	1.002	16.38	1.5097	30
40	0640	0640	9980	0641	15.60	1.002	15.64	1.5068	20
50	0669	0669	9978	0670	14.92	1.002	14.96	1.5039	10
4°00'	0698	0698	9976	0699	14.30	1.002	14.34	1.5010	86°00'
10	0727	0727	9974	0729	13.73	1.003	13.76	1.4981	50
20	0756	0756	9971	0758	13.20	1.003	13.23	1.4952	40
30	0785	0785	9969	0787	12.71	1.003	12.75	1.4923	30
40	0814	0814	9967	0816	12.25	1.003	12.29	1.4893	20
50	0844	0843	9964	0846	11.83	1.004	11.87	1.4864	10
5°00'	0873	0872	9962	0875	11.43	1.004	11.47	1.4835	85°00'
10	0902	0901	9959	0904	11.06	1.004	11.10	1.4806	50
20	0931	0929	9957	0934	10.71	1.004	10.76	1.4777	40
30	0960	0958	9954	0963	10.39	1.005	10.43	1.4748	30
40	0989	0987	9951	0992	10.08	1.005	10.13	1.4719	20
50	1018	1016	9948	1022	9.788	1.005	9.839	1.4690	10
6°00'	1047	1045	9945	1051	9.514	1.006	9.567	1.4661	84°00'
10	1076	1074	9942	1080	9.255	1.006	9.309	1.4632	50
20	1105	1103	9939	1110	9.010	1.006	9.065	1.4603	40
30	1134	1132	9936	1139	8.777	1.006	8.834	1.4573	30
40	1164	1161	9932	1169	8.556	1.007	8.614	1.4544	20
50	1193	1190	9929	1198	8.345	1.007	8.405	1.4515	10
		cos θ	sin θ	cot θ	tan θ	csc θ	sec θ	(radians)	(degrees)

θ (degrees)	θ (radians)	sin θ	cos θ	tan θ	cot θ	sec θ	csc θ	(radians)	(degrees)
7°00'	1222	1219	9925	1228	8.144	1.008	8.206	1.4486	83°00'
10	1251	1248	9922	1257	7.953	1.008	8.016	1.4457	50
20	1280	1276	9918	1287	7.770	1.008	7.834	1.4428	40
30	1309	1305	9914	1317	7.596	1.009	7.661	1.4399	30
40	1338	1334	9911	1346	7.429	1.009	7.496	1.4370	20
50	1376	1363	9907	1376	7.269	1.009	7.337	1.4341	10
8°00'	1396	1392	9903	1405	7.115	1.010	7.185	1.4312	82°00'
10	1425	1421	9899	1435	6.968	1.010	7.040	1.4283	50
20	1454	1449	9894	1465	6.827	1.011	6.900	1.4254	40
30	1484	1478	9890	1495	6.691	1.011	6.765	1.4224	30
40	1513	1507	9886	1524	6.561	1.012	6.636	1.4195	20
50	1542	1536	9881	1554	6.435	1.012	6.512	1.4166	10
9°00'	1571	1564	9877	1584	6.314	1.012	6.392	1.4137	81°00'
10	1600	1593	9872	1614	6.197	1.013	6.277	1.4108	50
20	1629	1622	9868	1644	6.084	1.013	6.166	1.4079	40
30	1658	1650	9863	1673	5.976	1.014	6.059	1.4050	30
40	1687	1679	9858	1703	5.871	1.014	5.955	1.4021	20
50	1716	1708	9853	1733	5.769	1.015	5.855	1.3992	10
10°00'	1745	1736	9848	1763	5.671	1.015	5.759	1.3963	80°00'
10	1774	1765	9843	1793	5.576	1.016	5.665	1.3934	50
20	1804	1794	9838	1823	5.485	1.016	5.575	1.3904	40
30	1833	1822	9833	1853	5.396	1.017	5.487	1.3875	30
40	1862	1851	9827	1883	5.309	1.018	5.403	1.3846	20
50	1891	1880	9822	1914	5.226	1.018	5.320	1.3817	10
11°00'	1920	1908	9816	1944	5.145	1.019	5.241	1.3788	79°00'
10	1949	1937	9811	1974	5.066	1.019	5.164	1.3759	50
20	1978	1965	9805	2004	4.989	1.020	5.089	1.3730	40
30	2007	1994	9799	2035	4.915	1.020	5.016	1.3701	30
40	2036	2022	9793	2065	4.843	1.021	4.945	1.3672	20
50	2065	2051	9787	2095	4.773	1.022	4.876	1.3643	10
12°00'	2094	2079	9781	2126	4.705	1.022	4.810	1.3614	78°00'
10	2123	2108	9775	2156	4.638	1.023	4.745	1.3584	50
20	2153	2136	9769	2186	4.574	1.024	4.682	1.3555	40
30	2182	2164	9763	2217	4.511	1.024	4.620	1.3526	30
40	2211	2193	9757	2247	4.449	1.025	4.560	1.3497	20
50	2240	2221	9750	2278	4.390	1.026	4.502	1.3468	10
13°00'	2269	2250	9744	2309	4.331	1.026	4.445	1.3439	77°00'
10	2298	2278	9737	2339	4.275	1.027	4.390	1.3410	50
20	2327	2306	9730	2370	4.219	1.028	4.336	1.3381	40
30	2356	2334	9724	2401	4.165	1.028	4.284	1.3352	30
40	2385	2363	9717	2432	4.113	1.029	4.232	1.3323	20
50	2414	2391	9710	2462	4.061	1.030	4.182	1.3294	10
		cos θ	sin θ	cot θ	tan θ	csc θ	sec θ	(radians)	(degrees)

TABLE 3 Trigonometric Functions in Degrees and Radians (continued)

θ (degrees)	θ (radians)	sin θ	cos θ	tan θ	cot θ	sec θ	csc θ	θ (radians)	θ (degrees)
14°00'	.2443	.2419	.9703	.2493	4.011	1.031	4.134	1.3265	76°00'
10	.2473	.2447	.9696	.2524	3.962	1.031	4.086	1.3235	50
20	.2502	.2476	.9689	.2555	3.914	1.032	4.039	1.3206	40
30	.2531	.2504	.9681	.2586	3.867	1.033	3.994	1.3177	30
40	.2560	.2532	.9674	.2617	3.821	1.034	3.950	1.3148	20
50	.2589	.2560	.9667	.2648	3.776	1.034	3.906	1.3119	10
15°00'	.2618	.2588	.9659	.2679	3.732	1.035	3.864	1.3090	75°00'
10	.2647	.2616	.9652	.2711	3.689	1.036	3.822	1.3061	50
20	.2676	.2644	.9644	.2742	3.647	1.037	3.782	1.3032	40
30	.2705	.2672	.9636	.2773	3.606	1.038	3.742	1.3003	30
40	.2734	.2700	.9628	.2805	3.566	1.039	3.703	1.2974	20
50	.2763	.2728	.9621	.2836	3.526	1.039	3.665	1.2945	10
16°00'	.2793	.2756	.9613	.2867	3.487	1.040	3.628	1.2915	74°00'
10	.2822	.2784	.9605	.2899	3.450	1.041	3.592	1.2886	50
20	.2851	.2812	.9596	.2931	3.412	1.042	3.556	1.2857	40
30	.2880	.2840	.9588	.2962	3.376	1.043	3.521	1.2828	30
40	.2909	.2868	.9580	.2994	3.340	1.044	3.487	1.2799	20
50	.2938	.2896	.9572	.3026	3.305	1.045	3.453	1.2770	10
17°00'	.2967	.2924	.9563	.3057	3.271	1.046	3.420	1.2741	73°00'
10	.2996	.2952	.9555	.3089	3.237	1.047	3.388	1.2712	50
20	.3025	.2979	.9546	.3121	3.204	1.048	3.356	1.2683	40
30	.3054	.3007	.9537	.3153	3.172	1.049	3.326	1.2654	30
40	.3083	.3035	.9528	.3185	3.140	1.049	3.295	1.2625	20
50	.3113	.3062	.9520	.3217	3.108	1.050	3.265	1.2595	10
18°00'	.3142	.3090	.9511	.3249	3.078	1.051	3.236	1.2566	72°00'
10	.3171	.3118	.9502	.3281	3.047	1.052	3.207	1.2537	50
20	.3200	.3145	.9492	.3314	3.018	1.053	3.179	1.2508	40
30	.3229	.3173	.9483	.3346	2.989	1.054	3.152	1.2479	30
40	.3258	.3201	.9474	.3378	2.960	1.056	3.124	1.2450	20
50	.3287	.3228	.9465	.3411	2.932	1.057	3.098	1.2421	10
19°00'	.3316	.3256	.9455	.3443	2.904	1.058	3.072	1.2392	71°00'
10	.3345	.3283	.9446	.3476	2.877	1.059	3.046	1.2363	50
20	.3374	.3311	.9436	.3508	2.850	1.060	3.021	1.2334	40
30	.3403	.3338	.9426	.3541	2.824	1.061	2.996	1.2305	30
40	.3432	.3365	.9417	.3574	2.798	1.062	2.971	1.2275	20
50	.3462	.3393	.9407	.3607	2.773	1.063	2.947	1.2246	10
20°00'	.3491	.3420	.9397	.3640	2.747	1.064	2.924	1.2217	70°00'
10	.3520	.3448	.9387	.3673	2.723	1.065	2.901	1.2188	50
20	.3549	.3475	.9377	.3706	2.699	1.066	2.878	1.2159	40
30	.3578	.3502	.9367	.3739	2.675	1.067	2.855	1.2130	30
40	.3607	.3529	.9356	.3772	2.651	1.069	2.833	1.2101	20
50	.3636	.3557	.9346	.3805	2.628	1.070	2.812	1.2072	10
		cos θ	sin θ	cot θ	tan θ	csc θ	sec θ	θ (radians)	θ (degrees)

θ (degrees)	θ (radians)	sin θ	cos θ	tan θ	cot θ	sec θ	csc θ	θ (radians)	θ (degrees)
21°00'	.3665	.3584	.9336	.3839	2.605	1.071	2.790	1.2043	69°00'
10	.3694	.3611	.9325	.3872	2.583	1.072	2.769	1.2014	50
20	.3723	.3638	.9315	.3906	2.560	1.074	2.749	1.1985	40
30	.3752	.3665	.9304	.3939	2.539	1.075	2.729	1.1956	30
40	.3782	.3692	.9293	.3973	2.517	1.076	2.709	1.1926	20
50	.3811	.3719	.9283	.4006	2.496	1.077	2.689	1.1897	10
22°00'	.3840	.3746	.9272	.4040	2.475	1.079	2.669	1.1868	68°00'
10	.3869	.3773	.9261	.4074	2.455	1.080	2.650	1.1839	50
20	.3898	.3800	.9250	.4108	2.434	1.081	2.632	1.1810	40
30	.3927	.3827	.9239	.4142	2.414	1.082	2.613	1.1781	30
40	.3956	.3854	.9228	.4176	2.394	1.084	2.595	1.1752	20
50	.3985	.3881	.9216	.4210	2.375	1.085	2.577	1.1723	10
23°00'	.4014	.3907	.9205	.4245	2.356	1.086	2.559	1.1694	67°00'
10	.4043	.3934	.9194	.4279	2.337	1.088	2.542	1.1665	50
20	.4072	.3961	.9182	.4314	2.318	1.089	2.525	1.1636	40
30	.4102	.3987	.9171	.4348	2.300	1.090	2.508	1.1606	30
40	.4131	.4014	.9159	.4383	2.282	1.092	2.491	1.1577	20
50	.4160	.4041	.9147	.4417	2.264	1.093	2.475	1.1548	10
24°00'	.4189	.4067	.9135	.4452	2.246	1.095	2.459	1.1519	66°00'
10	.4218	.4094	.9124	.4487	2.229	1.096	2.443	1.1490	50
20	.4247	.4120	.9112	.4522	2.211	1.097	2.427	1.1461	40
30	.4276	.4147	.9100	.4557	2.194	1.099	2.411	1.1432	30
40	.4305	.4173	.9088	.4592	2.177	1.100	2.396	1.1403	20
50	.4334	.4200	.9075	.4628	2.161	1.102	2.381	1.1374	10
25°00'	.4363	.4226	.9063	.4663	2.145	1.103	2.366	1.1345	65°00'
10	.4392	.4253	.9051	.4699	2.128	1.105	2.352	1.1316	50
20	.4422	.4279	.9038	.4734	2.112	1.106	2.337	1.1286	40
30	.4451	.4305	.9026	.4770	2.097	1.108	2.323	1.1257	30
40	.4480	.4331	.9013	.4806	2.081	1.109	2.309	1.1228	20
50	.4509	.4358	.9001	.4841	2.066	1.111	2.295	1.1199	10
26°00'	.4538	.4384	.8988	.4877	2.050	1.113	2.281	1.1170	64°00'
10	.4567	.4410	.8975	.4913	2.035	1.114	2.268	1.1141	50
20	.4596	.4436	.8962	.4950	2.020	1.116	2.254	1.1112	40
30	.4625	.4462	.8949	.4986	2.006	1.117	2.241	1.1083	30
40	.4654	.4488	.8936	.5022	1.991	1.119	2.228	1.1054	20
50	.4683	.4514	.8923	.5059	1.977	1.121	2.215	1.1025	10
27°00'	.4712	.4540	.8910	.5095	1.963	1.122	2.203	1.0996	63°00'
10	.4741	.4566	.8897	.5132	1.949	1.124	2.190	1.0966	50
20	.4771	.4592	.8884	.5169	1.935	1.126	2.178	1.0937	40
30	.4800	.4617	.8870	.5206	1.921	1.127	2.166	1.0908	30
40	.4829	.4643	.8857	.5243	1.907	1.129	2.154	1.0879	20
50	.4858	.4669	.8843	.5280	1.894	1.131	2.142	1.0850	10
		cos θ	sin θ	cot θ	tan θ	csc θ	sec θ	θ (radians)	θ (degrees)

Table 3 **683**

TABLE 3 Trigonometric Functions in Degrees and Radians (continued)

θ (degrees)	θ (radians)	sin θ	cos θ	tan θ	cot θ	sec θ	csc θ	θ (radians)	θ (degrees)
35°00′	6109	5736	8192	7002	1.428	1.221	1.743	9599	55°00′
10	6138	5760	8175	7046	1.419	1.223	1.736	9570	50
20	6167	5783	8158	7089	1.411	1.226	1.729	9541	40
30	6196	5807	8141	7133	1.402	1.228	1.722	9512	30
40	6225	5831	8124	7177	1.393	1.231	1.715	9483	20
50	6254	5854	8107	7221	1.385	1.233	1.708	9454	10
36°00′	6283	5878	8090	7265	1.376	1.236	1.701	9425	54°00′
10	6312	5901	8073	7310	1.368	1.239	1.695	9396	50
20	6341	5925	8056	7355	1.360	1.241	1.688	9367	40
30	6370	5948	8039	7400	1.351	1.244	1.681	9338	30
40	6400	5972	8021	7445	1.343	1.247	1.675	9308	20
50	6429	5995	8004	7490	1.335	1.249	1.668	9279	10
37°00′	6458	6018	7986	7536	1.327	1.252	1.662	9250	53°00′
10	6487	6041	7969	7581	1.319	1.255	1.655	9221	50
20	6516	6065	7951	7627	1.311	1.258	1.649	9192	40
30	6545	6088	7934	7673	1.303	1.260	1.643	9163	30
40	6574	6111	7916	7720	1.295	1.263	1.636	9134	20
50	6603	6134	7898	7766	1.288	1.266	1.630	9105	10
38°00′	6632	6157	7880	7813	1.280	1.269	1.624	9076	52°00′
10	6661	6180	7862	7860	1.272	1.272	1.618	9047	50
20	6690	6202	7844	7907	1.265	1.275	1.612	9018	40
30	6720	6225	7826	7954	1.257	1.278	1.606	8988	30
40	6749	6248	7808	8002	1.250	1.281	1.601	8959	20
50	6778	6271	7790	8050	1.242	1.284	1.595	8930	10
39°00′	6807	6293	7771	8098	1.235	1.287	1.589	8901	51°00′
10	6836	6316	7753	8146	1.228	1.290	1.583	8872	50
20	6865	6338	7735	8195	1.220	1.293	1.578	8843	40
30	6894	6361	7716	8243	1.213	1.296	1.572	8814	30
40	6923	6383	7698	8292	1.206	1.299	1.567	8785	20
50	6952	6406	7679	8342	1.199	1.302	1.561	8756	10
40°00′	6981	6428	7660	8391	1.192	1.305	1.556	8727	50°00′
10	7010	6450	7642	8441	1.185	1.309	1.550	8698	50
20	7039	6472	7623	8491	1.178	1.312	1.545	8668	40
30	7069	6494	7604	8541	1.171	1.315	1.540	8639	30
40	7098	6517	7585	8591	1.164	1.318	1.535	8610	20
50	7127	6539	7566	8642	1.157	1.322	1.529	8581	10
41°00′	7156	6561	7547	8693	1.150	1.325	1.524	8552	49°00′
10	7185	6583	7528	8744	1.144	1.328	1.519	8523	50
20	7214	6604	7509	8796	1.137	1.332	1.514	8494	40
30	7243	6626	7490	8847	1.130	1.335	1.509	8465	30
40	7272	6648	7470	8899	1.124	1.339	1.504	8436	20
50	7301	6670	7451	8952	1.117	1.342	1.499	8407	10
		cos θ	sin θ	cot θ	tan θ	csc θ	sec θ	(radians)	θ (degrees)

θ (degrees)	θ (radians)	sin θ	cos θ	tan θ	cot θ	sec θ	csc θ	θ (radians)	θ (degrees)
28°00′	4887	4695	8829	5317	1.881	1.133	2.130	1.0821	62°00′
10	4916	4720	8816	5354	1.868	1.134	2.118	1.0792	50
20	4945	4746	8802	5392	1.855	1.136	2.107	1.0763	40
30	4974	4772	8788	5430	1.842	1.138	2.096	1.0734	30
40	5003	4797	8774	5467	1.829	1.140	2.085	1.0705	20
50	5032	4823	8760	5505	1.816	1.142	2.074	1.0676	10
29°00′	5061	4848	8746	5543	1.804	1.143	2.063	1.0647	61°00′
10	5091	4874	8732	5581	1.792	1.145	2.052	1.0617	50
20	5120	4899	8718	5619	1.780	1.147	2.041	1.0588	40
30	5149	4924	8704	5658	1.767	1.149	2.031	1.0559	30
40	5178	4950	8689	5696	1.756	1.151	2.020	1.0530	20
50	5207	4975	8675	5735	1.744	1.153	2.010	1.0501	10
30°00′	5236	5000	8660	5774	1.732	1.155	2.000	1.0472	60°00′
10	5265	5025	8646	5812	1.720	1.157	1.990	1.0443	50
20	5294	5050	8631	5851	1.709	1.159	1.980	1.0414	40
30	5323	5075	8616	5890	1.698	1.161	1.970	1.0385	30
40	5352	5100	8601	5930	1.686	1.163	1.961	1.0356	20
50	5381	5125	8587	5969	1.675	1.165	1.951	1.0327	10
31°00′	5411	5150	8572	6009	1.664	1.167	1.942	1.0297	59°00′
10	5440	5175	8557	6048	1.653	1.169	1.932	1.0268	50
20	5469	5200	8542	6088	1.643	1.171	1.923	1.0239	40
30	5498	5225	8526	6128	1.632	1.173	1.914	1.0210	30
40	5527	5250	8511	6168	1.621	1.175	1.905	1.0181	20
50	5556	5275	8496	6208	1.611	1.177	1.896	1.0152	10
32°00′	5585	5299	8480	6249	1.600	1.179	1.887	1.0123	58°00′
10	5614	5324	8465	6289	1.590	1.181	1.878	1.0094	50
20	5643	5348	8450	6330	1.580	1.184	1.870	1.0065	40
30	5672	5373	8434	6371	1.570	1.186	1.861	1.0036	30
40	5701	5398	8418	6412	1.560	1.188	1.853	1.0007	20
50	5730	5422	8403	6453	1.550	1.190	1.844	9977	10
33°00′	5760	5446	8387	6494	1.540	1.192	1.836	9948	57°00′
10	5789	5471	8371	6536	1.530	1.195	1.828	9919	50
20	5818	5495	8355	6577	1.520	1.197	1.820	9890	40
30	5847	5519	8339	6619	1.511	1.199	1.812	9861	30
40	5876	5544	8323	6661	1.501	1.202	1.804	9832	20
50	5905	5568	8307	6703	1.492	1.204	1.796	9803	10
34°00′	5934	5592	8290	6745	1.483	1.206	1.788	9774	56°00′
10	5963	5616	8274	6787	1.473	1.209	1.781	9745	50
20	5992	5640	8258	6830	1.464	1.211	1.773	9716	40
30	6021	5664	8241	6873	1.455	1.213	1.766	9687	30
40	6050	5688	8225	6916	1.446	1.216	1.758	9657	20
50	6080	5712	8208	6959	1.437	1.218	1.751	9628	10
		cos θ	sin θ	cot θ	tan θ	csc θ	sec θ	(radians)	θ (degrees)

TABLE 3 Trigonometric Functions in Degrees and Radians (continued)

θ (degrees)	θ (radians)	$\sin \theta$	$\cos \theta$	$\tan \theta$	$\cot \theta$	$\sec \theta$	$\csc \theta$		
42°00′	.7330	.6691	.7431	.9004	1.111	1.346	1.494	.8378	**48°00′**
10	.7359	.6713	.7412	.9057	1.104	1.349	1.490	.8348	50
20	.7389	.6734	.7392	.9110	1.098	1.353	1.485	.8319	40
30	.7418	.6756	.7373	.9163	1.091	1.356	1.480	.8290	30
40	.7447	.6777	.7353	.9217	1.085	1.360	1.476	.8261	20
50	.7476	.6799	.7333	.9271	1.079	1.364	1.471	.8232	10
43°00′	.7505	.6820	.7314	.9325	1.072	1.367	1.466	.8203	**47°00′**
10	.7534	.6841	.7294	.9380	1.066	1.371	1.462	.8174	50
20	.7563	.6862	.7274	.9435	1.060	1.375	1.457	.8145	40
30	.7592	.6884	.7254	.9490	1.054	1.379	1.453	.8116	30
40	.7621	.6905	.7234	.9545	1.048	1.382	1.448	.8087	20
50	.7560	.6926	.7214	.9601	1.042	1.386	1.444	.8058	10
44°00′	.7679	.6947	.7193	.9657	1.036	1.390	1.440	.8029	**46°00′**
10	.7709	.6967	.7173	.9713	1.030	1.394	1.435	.7999	50
20	.7738	.6988	.7153	.9770	1.024	1.398	1.431	.7970	40
30	.7767	.7009	.7133	.9827	1.018	1.402	1.427	.7941	30
40	.7796	.7030	.7112	.9884	1.012	1.406	1.423	.7912	20
50	.7825	.7050	.7092	.9942	1.006	1.410	1.418	.7883	10
45°00′	.7854	.7071	.7071	1.000	1.000	1.414	1.414		
		$\cos \theta$	$\sin \theta$	$\cot \theta$	$\tan \theta$	$\csc \theta$	$\sec \theta$	θ (radians)	θ (degrees)

ANSWERS TO
SELECTED EXERCISES

ANSWERS TO SELECTED EXERCISES

To the Student

If you need further help with this course, you may want to obtain a copy of the *Student's Solutions Manual and Study Guide* that accompanies this textbook. This manual provides detailed step-by-step solutions to the odd-numbered exercises in the textbook as well as a practice chapter test for every chapter. Your college bookstore either has this manual or can order it for you.

Chapter 1

Section 1.1 (page 12)

1. commutative   **3.** commutative   **5.** identity   **7.** associative   **9.** identity   **11.** 24   **13.** 17   **15.** -21
17. $-6/7$ **19.** -3 **21.** natural, whole, integer, rational, real   **23.** integer, rational, real   **25.** rational, real
27. irrational, real   **29.** $(-\sqrt{36} = -6)$ integer, rational, real   **31.** undefined   **33.** 1900   **35.** 154
37. $-5, -4, -2, -\sqrt{3}, \sqrt{6}, \sqrt{8}, 3$ **39.** 3/4, 1.2, $\sqrt{2}$, 22/15, $\pi/2$, 8/5   **41.** -6 **43.** $2 - \sqrt{3}$ **45.** $3 - \sqrt{2}$
47. $x - 4$ **49.** $8 - 2k$ **51.** $56 - 7m$ **53.** $y - x$ **55.** $3 + x^2$ **57.** $1 + p^2$ **59.** $6 - \pi$ **61.** 1
63. (a) 1 (b) 1 (c) 14 (d) 2 (e) 2   **65.** (a) 2 (b) 7 (c) 0 (d) 9 (e) 9   **67.** $d(A, C) = 7$ or 1
69. first part of multiplication property   **71.** second part of multiplication property   **73.** first part of multiplication
property   **75.** triangle inequality, $|a + b| \le |a| + |b|$ **77.** property of absolute value, $|a| \cdot |b| = |ab|$
79. property of absolute value, $|a/b| = |a|/|b|$ **81.** trichotomy property   **83.** $x > 0$ and $y > 0$ or $x < 0$ and $y < 0$
85. $y < 0$ and $x > 0$ or $y > 0$ and $x < 0$ **87.** if $x = y$ or $x = -y$ **89.** if $x = 0$ **91.** -1 if $x < 0$ and 1 if $x > 0$
93. 1   **95.** x must satisfy $-9 \le x \le 9$

Section 1.2 (page 20)

1. 343   **3.** 2   **5.** 4   **7.** $-1/8$ **9.** 1/9   **11.** 1/25   **13.** 27/8   **15.** 1200   **17.** $9x^4$ **19.** .001042
21. 1,000,000   **23.** $-72m^5$ **25.** $m^{7/3}$ **27.** 1   **29.** $(1 + n)^{5/4}$ **31.** $6yz^{2/3}$ **33.** $1/d^8$ **35.** $a^{2/3}b^2$ **37.** a^3b^6
39. $x^7/5^4$ **41.** $4z^2/(x^5y)$ **43.** $30p^{3-r}$ **45.** $1/(2^3b^{y-2})$ **49.** 27,000   **51.** 27   **53.** 64 hundred dollars
55. about $64,000,000   **57.** about $10,000,000   **59.** about 86.3 mi   **61.** about 211 mi   **63.** about 29
65. about 177

Section 1.3 (page 27)

1. $3x^3 - 3x^2 + 6x + 2$ **3.** $9y^2 - 4y + 8$ **5.** $3b^2 - 2b + 7$ **7.** $3x^2 - 5x - 2$ **9.** $28r^2 + r - 2$
11. $18p^2 - 27pq - 35q^2$ **13.** $15x^2 - \frac{7}{3}x - \frac{2}{9}$ **15.** $\frac{6}{25}y^2 + \frac{11}{40}yz + \frac{1}{16}z^2$ **17.** $25r^2 - 4$ **19.** $16x^2 - 9y^2$
21. $36k^2 - 36k + 9$ **23.** $16m^2 + 16mn + 4n^2$ **25.** $8z^3 - 12z^2 + 6z - 1$ **27.** $4x^6 + 4x^3y^2 + y^4$
29. $x^3 - 6x^2y^2 + 12xy^4 - 8y^6$ **31.** $2b^5 - 8b^4 + 6b^3$ **33.** $4y^6 + 8y^5 - 24y^4 + 12y^3$ **35.** $x^3 - x^2 - x + 1$
37. $6p^3 - 11p^2 + 14p - 5$ **39.** $12k^4 + 21k^3 - 5k^2 + 3k + 2$ **41.** $a^2 - 2ab + b^2 + 4ac - 4bc + 4c^2$
43. $k^{2m} - 4$ **45.** $3p^{2x} - 5p^x - 2$ **47.** $m^{2x} - 4m^x + 4$ **49.** $27k^{3a} - 54k^{2a} + 36k^a - 8$ **51.** $y - 10y^2$
53. $-4k^{10/3} + 24k^{4/3}$ **55.** $x^2 - x$ **57.** $r - 2 + r^{-1}$ or $r - 2 + 1/r$ **59.** $5x^3 + 10x^2 + 4x - 3/x$

61. $5y^2 - 3xy + 8x^2$ **63.** $2p^3 - 2p^2 + p + 2 + \dfrac{10}{2p - 1}$ **65.** $k^2 - 5 + \dfrac{2k + 10}{k^2 + 1}$

67. $y^2 + 2y + 2/5 + \dfrac{-2y^2 + (69/5)y - (2/5)}{5y^3 - 2y + 1}$ **69.** -8 **71.** 1   **73.** -24 **79.** (a) 0, 1, 2, or 3

(b) 0, 1, 2, or 3   (c) 6   **81.** (a) m (b) m (c) $m + n$ **83.** $x^2 + 2xy + y^2 - z^2$

Section 1.4 (page 33)

1. $4m(3n - 2)$ **3.** $3r(4r^2 + 2r - 1)$ **5.** $2px(3x - 4x^2 - 6)$ **7.** $2(a + b)(1 + 2m)$ **9.** $(x - 3)(x - 8)$
11. $(4p - 1)(p + 1)$ **13.** $3(2r - 1)(2r + 5)$ **15.** $(6r - 5s)(3r + 2s)$ **17.** $x^2(3 - x)^2$ **19.** $(2m + 5)(2m - 5)$
21. $9(4r + 3s)(4r - 3s)$ **23.** $(11p^2 + 3q^2)(11p^2 - 3q^2)$ **25.** $(p^4 + 1)(p^2 + 1)(p + 1)(p - 1)$
27. $(2m - 3n)(4m^2 + 6mn + 9n^2)$ **29.** $(4 + m)(x + y)$ **31.** $(q + 3 + p)(q + 3 - p)$
33. $(a + b + x + y)(a + b - x - y)$ **35.** $(2x - y)y$ **37.** $(11m - 7)(3m - 7)$ **39.** $(x + y + 5z)(x + y - 3z)$
41. $4pq$ **43.** $r(r^2 + 18r + 108)$ **45.** $(3 - m - 2n)(9 + 3m + 6n + m^2 + 4mn + 4n^2)$ **47.** $(3k + 1)(9k^2 - 3k + 1)$
49. $(3m - n)(x - 2y)$ **51.** $(x - y)^3$ **53.** $(m^n + 4)(m^n - 4)$ **55.** $(x^n - y^{2n})(x^{2n} + x^n y^{2n} + y^{4n})$
57. $(2x^n + 3y^n)(x^n - 13y^n)$ **59.** $(5q^r - 3t^p)^2$ **61.** $[3(m + p)^k + 5][2(m + p)^k - 3]$ **63.** $(7x - 8)^2(121 - 56x)$
65. $(r - 6)/[3(r - 2)^{5/3}]$ **67.** $(3m^3 - m)/[3(m^3 + m)^{5/3}]$ **69.** $(6 + x^{-4})(3x - 2x^{-1})^2(54 + 36x^{-2} - 15x^{-4} + 22x^{-6})$
71. $(7x - 8)^{-6}(3x + 2)^{-3}(-147x - 22)$ **73.** $m^{-5}(m^4 + 3)$ **75.** $2a^{-5}(3 - 5a + 9a^3)$ **77.** $16y^{-3}(2 + 3y^4 - 4y^5)$
79. $y^{5/2}(y^2 - 3)$ **81.** $m^{-1/3}(3m - 4)$ **83.** $(3k - 2)^{-3/2}(3k + 2)$ **85.** $(r^2 - 1 + 2r)(r^2 - 1 - 2r)$
87. $(x^2 - 1 + 4x)(x^2 - 1 - 4x)$ **89.** $(m^2 - 3 + 4m)(m^2 - 3 - 4m)$

Section 1.5 (page 38)

1. $x \neq -6$ **3.** $x \neq 3/5$ and $x \neq -1$ **5.** no restrictions   **7.** $5p/2$ **9.** 8/9   **11.** $3/(t - 3)$ **13.** $(2x + 4)/x$
15. $(m - 2)/(m + 3)$ **17.** $(2m + 3)/(4m + 3)$ **19.** $(y + 2)/2$ **21.** $25p^2/9$ **23.** 2/9   **25.** 3/10   **27.** $5x/y$
29. $2(a + 4)/(a - 3)$ **31.** 1   **33.** $(n + 4)/(n - 4)$ **35.** $(x^2 - 1)/x^2$ **37.** $(x + y)/(x - y)$
39. $(x^2 - xy + y^2)/(x^2 + xy + y^2)$ **41.** $7/y$ **43.** $5/(12y)$ **45.** 1   **47.** $(6 + p)/(2p)$ **49.** $(8 - y)/(4y)$
51. $137/(30m)$ **53.** $(2y + 1)/[y(y + 1)]$ **55.** $3/[2(a + b)]$ **57.** $-2/[(a + 1)(a - 1)]$
59. $(2m^2 + 2)/[(m - 1)(m + 1)]$ **61.** $4/(a - 2)$ **63.** $5/[(a - 2)(a - 3)(a + 2)]$
65. $(6x^2 - 6x - 5)/[(x + 5)(x - 2)(x - 1)]$ **67.** $(p + 5)/[p(p + 1)]$ **69.** $-1/[x(x + h)]$ **71.** $(x + 1)/(x - 1)$
73. $-1/(x + 1)$ **75.** $(2 - b)(1 + b)/[b(1 - b)]$ **77.** 1/3   **79.** $a + b$ **81.** -1 **83.** $1/(ab)$
85. $(x^3 + 2)^3(2x^3 - 15x^2 - 2)/(x - 5)^5$ **87.** $[-5(x + 1)^4(3x^2 + 6x + 4)]/(3x^2 - 4)^6$

Section 1.6 (page 45)

1. $5\sqrt{2}$ **3.** $5\sqrt[3]{2}$ **5.** $-3\sqrt{5}/5$ **7.** $4\sqrt{5}$ **9.** $9\sqrt{3}$ **11.** $-\sqrt[3]{12}/2$ **13.** $\sqrt[4]{24}/2$ **15.** 3/2   **17.** $5\sqrt[3]{2}$
19. $2\sqrt{3}$ **21.** $13\sqrt[3]{4}/6$ **23.** $7rs^2t^5\sqrt{2r}$ **25.** $x^2yz^2\sqrt[4]{y^2z^2}$ **27.** $(p^3q - p^2q^4 + p^4)\sqrt{pq}$ **29.** 3   **31.** $3 - 2\sqrt{2}$
33. $10\sqrt[3]{2} + 16\sqrt[3]{4} + 3$ **35.** $\sqrt{15p}/(3p)$ **37.** $gh^2\sqrt{ghr}/r^2$ **39.** $-xy^2\sqrt[3]{9x^2zw}/(z^2w)$ **41.** $2x\sqrt[4]{2x^2y^3}/y^2$ **43.** $\sqrt[3]{4}/2$
45. $\sqrt[12]{2}$ **47.** $\sqrt[32]{y}$ **49.** $3\sqrt{a^2 + b^2}$ **51.** $2\sqrt{2x + 4}$ **53.** $-(1 - \sqrt{5})/2$ **55.** $3\sqrt{7} + 7$ **57.** $(\sqrt{z} + z)/(1 - z)$
59. $(-5 - 5\sqrt{3 - p})/(p - 2)$ **61.** $(a^2\sqrt{b} - b^2\sqrt{a})/(ab)$ **63.** $(p^2 + p + 2\sqrt{p(p^2 - 1)} - 1)/(-p^2 + p + 1)$
65. $(\sqrt[3]{m^2} + \sqrt[3]{mn} + \sqrt[3]{n^2})/(m - n)$ **67.** $-2(4 - \sqrt{3})(3 + \sqrt{2})/91$ or $-2(12 - 3\sqrt{3} + 4\sqrt{2} - \sqrt{6})/91$
69. $(1 - \sqrt{7})(2 - \sqrt{10})(1 + \sqrt{5})/24$ **71.** $\sqrt{17} - 4$ **73.** $(5 - x)(x - 3)^2$

Section 1.7 (page 53)

1. imaginary and complex   **3.** real and complex   **5.** imaginary and complex   **7.** complex   **9.** $0 + 10i$ **11.** $0 - 20i$
13. $5 + 2i$ **15.** $-6 - 14i$ **17.** $-5 + 0i$ **19.** $-4 + 0i$ **21.** $2 + 0i$ **23.** $7 - i$ **25.** $2 + 0i$ **27.** $1 - 10i$
29. $-14 + 2i$ **31.** $5 - 12i$ **33.** $13 + 0i$ **35.** $7 + 0i$ **37.** $0 + 25i$ **39.** $7/25 - (24/25)i$ **41.** $-1 - 2i$
43. $0 - 2i$ **45.** $[(3 - 2\sqrt{5}) + (-2 - 3\sqrt{5})i]/13$ **47.** i **49.** $-i$ **51.** 1   **53.** i **55.** $5/2 + i$
57. $-16/65 - (37/65)i$ **59.** $27/10 + (11/10)i$ **61.** $17/10 - (11/10)i$ **63.** $a = 18; b = -3$ **65.** $a = -5; b = 1$
67. $a = -3/4; b = 3$ **73.** $13 + 4i$ **75.** $a = 0$ or $b = 0$

Chapter 1 Review Exercises (page 55)
1. commutative   **3.** inverse   **5.** distributive   **7.** 20/3   **9.** 9/4   **11.** 6   **13.** $-12, -6, 0, 6$ **15.** $-\sqrt{7}$,
$\pi/4, \sqrt{11}$ **17.** $3 - \sqrt{7}$ **19.** $3 - m$ **21.** (a) 1 (b) 12   **23.** if $x \geq 0$ **25.** if A and B are the same point
27. $-8x^2 - 15x + 16$ **29.** $24k^2 - 5k - 14$ **31.** $x^2 + 4xy + 4y^2 - 2xz - 4yz + z^2$ **33.** $x^6/2$
35. $z(7z - 9z^2 + 1)$ **37.** $(3m + 1)(2m - 5)$ **39.** $5m^3(3m - 5n)(2m + n)$ **41.** $(13y^2 + 1)(13y^2 - 1)$
43. $(x + 1)(x - 3)$ **45.** $(-r^3 + 2)(7r^6 - 10r^3 + 4)$ **47.** $(m - n)(3 + 4k)$ **49.** $(2x - 1)(b + 3)$
51. $(m - 3)(2m + 3)/(5m)$ **53.** $(x + 1)/(x + 4)$ **55.** $37/(20y)$ **57.** $(x + 9)/[(x - 3)(x - 1)(x + 1)]$
59. $(3m^2 + 2m - 12)/[5(m + 2)]$ **61.** p^8 **63.** $s/(36r)$ **65.** $10\sqrt{2}$ **67.** $2^3y^4\sqrt{2m}/m^2$ **69.** $\sqrt[12]{m}$ **71.** $2q$
73. $5y^2\sqrt{3y}$ **75.** $(z\sqrt{z-1})/(z - 1)$ **77.** $\sqrt{5} - 2$ **79.** 1/216   **81.** $16r^{9/4}s^{7/3}$ **83.** $1/p^{3/2}$ **85.** $1/m^{2p}$
87. $10z^{7/3} - 4z^{1/3}$ **89.** $3p^2 + 3p^{3/2} - 5p - 5p^{1/2}$ **91.** $5 + 4i$ **93.** $29 + 37i$ **95.** $-32 + 24i$ **97.** $-2 - 2i$
99. $0 + i$ **101.** $8/5 + (6/5)i$ **103.** $-5/26 + (1/26)i$ **105.** $0 + 2i\sqrt{3}$ **107.** $x^3 + 5x$ **109.** m^6 **111.** $\dfrac{a}{2b}$

113. $\dfrac{\sqrt{a} - \sqrt{b}}{a - b}$ **115.** $4 - t - 1$ **117.** 5^2 **119.** (a) and (b)   **121.** $x \geq 0$

Chapter 2

Section 2.1 (page 62)
1. identity   **3.** conditional   **5.** identity   **7.** identity   **9.** not equivalent   **11.** equivalent   **13.** not equivalent
15. not equivalent   **17.** 4   **19.** 12   **21.** $-2/7$ **23.** $-7/8$ **25.** -1 **27.** 3   **29.** 4   **31.** 12   **33.** 3/4
35. $-12/5$ **37.** no solution   **39.** 27/7   **41.** $-59/6$ **43.** $-9/4$ **45.** no solution   **47.** -4 **49.** $-19/75$
51. $x = -3a + b$ **53.** $x = (3a + b)/(3 - a)$ **55.** $x = (3 - 3a)/(a^2 - a - 1)$ **57.** $x = 2a^2/(a^2 + 3)$
59. $x = (m + 4)/(5 + 2m)$ **61.** Dividing out $(x - 3)$ is improper.   **63.** $k = 6$

Section 2.2 (page 70)
1. $V = k/P$ **3.** $l = V/(wh)$ **5.** $g = (V - V_0)/t$ **7.** $g = 2s/t^2$ **9.** $B = 2A/h - b$ **11.** $r_1 = S/(2\pi h) - r_2$
13. $l = gt^2/(4\pi^2)$ **15.** $h = (S - 2\pi r^2)/(2\pi r)$ **17.** $f = Ab(p + 1)/24$ **19.** $R = r_1r_2/(r_2 + r_1)$ **21.** 6 cm
23. 2 liters   **25.** 6 2/3 qt   **27.** 400/3 liters   **29.** about 840 mi   **31.** 15 min   **33.** 35 km per hr   **35.** about 1486.7 mi
37. 18/5 hr   **39.** 78 hr   **41.** 40/3 hr   **43.** \$57.65   **45.** \$12,000 at 5% and \$8000 at 6%   **47.** \$54,000 at 8 1/2% and
\$8000 at 6%   **49.** \$70,000 for land that makes a profit and \$50,000 for land that produces a loss   **51.** \$267.57
53. \$20,000 at 6%, \$40,000 at 5%   **55.** $(10n - v)/5$ fives; $(v - 5n)/5$ tens   **57.** $100b/c$

Section 2.3 (page 82)
1. $-4, 2$ **3.** $-1, 4$ **5.** $-4, 0$ **7.** $-3, 2$ **9.** $\pm 3\sqrt{3}$ **11.** $3 \pm \sqrt{5}$ **13.** $(1 \pm 2\sqrt{3})/3$ **15.** $-2, 2$
17. $-7, -1$ **19.** $-3, 7$ **21.** $1 \pm \sqrt{5}$ **23.** $(-1 \pm i)/2$ **25.** $(3 \pm \sqrt{5})/2$ **27.** $(1 \pm i\sqrt{19})/10$ **29.** $-1/3$
31. $(\sqrt{2} \pm \sqrt{6})/2$ **33.** $\sqrt{2}/2, \sqrt{2}$ **35.** $(1 \pm \sqrt{13})/2$ **37.** $(3 \pm i\sqrt{7})/2$ **39.** $(1 \pm i\sqrt{7})/4$ **41.** $(1 \pm \sqrt{29})/2$
43. $1, (-1 \pm i\sqrt{3})/2$ **45.** $4, -2 \pm 2i\sqrt{3}$ **47.** $-5/2, (5 \pm 5i\sqrt{3})/4$ **49.** 0; one rational solution   **51.** 1; two
rational solutions   **53.** 8; two irrational solutions   **55.** ± 12 **57.** 121/4   **59.** 1, 9   **61.** $h = \sqrt{d^4kL/L}$ or $d^2\sqrt{kL/L}$
63. $t = \sqrt{(s - s_0 - k)g}/g$ **65.** $r = (-\pi h \pm \sqrt{\pi^2h^2 + 2\pi S})/(2\pi)$ **67.** yes   **69.** $a = 1, b = -9, c = 20$
71. $a = 1, b = -2, c = -1$ **73.** 100 yd by 400 yd   **75.** 50 m by 100 m   **77.** 9 ft by 12 ft   **79.** 50 mph
81. 16.95 cm, 20.95 cm, 26.95 cm   **85.** $k = 3$, solution is 2/3   **87.** $x = (2y \pm \sqrt{31y^2 + 9})/9$,
$y = (-2x \pm \sqrt{31x^2 - 3})/3$

Section 2.4 (page 88)
1. $\pm\sqrt{3}, \pm\sqrt{5}$ **3.** $\pm 1, \pm\sqrt{10}/2$ **5.** 4, 6   **7.** $(-6 \pm 2\sqrt{3})/3, (-4 \pm \sqrt{2})/2$ **9.** $7/2, -1/3$ **11.** $-63, 28$
13. $\pm\sqrt{5}/5$ **15.** $\pm 1, \pm 2\sqrt{6}/3$ **17.** $4, 5^{2/3}$ **19.** $\pm\sqrt{5}$ **21.** 25/16   **23.** $(7 \pm \sqrt{177})/8, (-1 \pm \sqrt{19})/3$
25. 1/2   **27.** -1 **29.** 5   **31.** -2 **33.** 2   **35.** 3/2   **37.** $-3, 1$ **39.** $-27, 3$ **41.** 9   **43.** 5/4
45. $2, -2$ **47.** $-1, 0$ **49.** 2/9, 1   **51.** $2, -2/9$ **53.** 0, 9   **55.** 1/4, 1   **57.** -1

Section 2.5 (page 95)

1. $(-1, 4)$

3. $(-\infty, 0)$

5. $(-5, -4]$

7. $-4 < x < 3$ **9.** $x \le -1$ **11.** $-2 \le x < 6$ **13.** $x \le -4$ **15.** $(-\infty, 4]$ **17.** $[-1, +\infty)$ **19.** $(-\infty, +\infty)$
21. $[-11/5, +\infty)$ **23.** $[1, 4]$ **25.** $(-6, -4)$ **27.** $(-16, 19]$ **29.** $[4, 40/3)$ **31.** $(0, 10)$ **33.** $[-2, 3]$
35. $(-\infty, 1/2) \cup (4, +\infty)$ **37.** $(-\infty, -2] \cup [0, 2]$ **39.** $(-5, 3]$ **41.** $(-\infty, -2)$ **43.** $(-\infty, 6) \cup [15/2, +\infty)$
45. $(-2, +\infty)$ **47.** $(0, 4/11) \cup (1/2, +\infty)$ **49.** $(-\infty, -2] \cup (1, 2)$ **51.** $[3/2, +\infty)$ **53.** $(5/2, +\infty)$
55. $[-8/3, 3/2] \cup (6, +\infty)$ **57.** if k is in $(-\infty, -4\sqrt{2}] \cup [4\sqrt{2}, +\infty)$ **59.** if k is in $(-\infty, 0] \cup [8, +\infty)$ **61.** (d)
63. (b) **65.** $x^2 - 7x + 10 > 0$ **67.** $x^2 + x - 12 \le 0$ **69.** $\dfrac{x+3}{x} \ge 0$ **71.** $\dfrac{x-9}{x-4} \le 0$ **73.** $[500, +\infty)$
75. $(0, 5/4), (6, +\infty)$ **77.** (a) $C = 30 + 2x$ (b) $C = 3x$ (c) $(0, 30)$ **79.** 32°F to 86°F **81.** $(9/2, 6)$ **87.** when $b > 1$

Section 2.6 (page 101)

1. $1, 3$ **3.** $-1/3, 1$ **5.** no solution **7.** $-6, 14$ **9.** $5/2, 7/2$ **11.** $-4/3, 2/9$ **13.** $-7/3, -1/7$
15. $-3/5, 11$ **17.** $[-3, 3]$ **19.** $(-\infty, -1) \cup (1, +\infty)$ **21.** no solution **23.** $[-10, 10]$ **25.** $(-4, -1)$
27. $(-\infty, -2/3) \cup (2, +\infty)$ **29.** $(-\infty, -8/3] \cup [2, +\infty)$ **31.** $(-3/2, 13/10)$ **33.** $(-3, -1)$ **35.** $(1, 5) \cup (5, +\infty)$
37. $(-\infty, 3/2] \cup [5/2, +\infty)$ **39.** 1 **41.** $3, -1/3$ **43.** $0, \pm 1$ **45.** $5/3, -4/3$ **47.** $-7/6, -3/2$ **49.** yes
51. $[-3, -\sqrt{5}] \cup [\sqrt{5}, 3]$ **53.** $(-\infty, -\sqrt{11}) \cup (-1, 1) \cup (\sqrt{11}, +\infty)$ **55.** $(-\infty, -3) \cup (-1/7, +\infty)$
57. $|x - 2| \le 4$ **59.** $|z - 12| \ge 2$ **61.** $|k - 1| = 6$ **63.** if $|x - 2| \le .0004$, then $|y - 7| \le .00001$
65. $m = 2, n = 20$

Section 2.7 (page 105)

1. $a = kb$ **3.** $x = k/y$ **5.** $r = kst$ **7.** $w = kx^2/y$ **9.** The circumference of a circle varies directly as (or is proportional to) the radius. **11.** The average speed varies directly as the distance and inversely as the time. **13.** The strength of a muscle varies directly as (or is proportional to) the cube of its length. **15.** 220/7 **17.** 18/125 **19.** m increases by $2^2 \cdot 3^4 = 324$ times **21.** 2304 ft **23.** .0444 ohms **25.** 140 kg per cm^2 **27.** \$1375 **29.** 8/9 metric tons
31. $66\pi/17$ sec. **33.** $V = 45/7$ **35.** 640 **37.** 365.24 **39.** If s is the speed and d the distance, then $s = 10\sqrt{2d}$.
41. Cost varies directly as the square of the radius; \$15.75.

Chapter 2 Review Exercises (page 109)

1. $-11/3$ **3.** $-96/7$ **5.** -2 **7.** 13 **9.** $x = -6b - a - 6$ **11.** $x = (6 - 3m)/(1 + 2k - km)$
13. $y = 6a/(4 - a)$ **15.** $P = A/(1 + i)$ **17.** $r_1 = kr_2/(r_2 - k)$ **19.** $L = V/(\pi r^2)$ **21.** $x = -4/(y^2 - 5y - 6p)$
23. \$500 **25.** 12 hours **27.** $-7 \pm \sqrt{5}$ **29.** $5/2, -3$ **31.** $7, -3/2$ **33.** $3, -1/2$ **35.** -188; two nonreal solutions **37.** 484; two rational solutions **39.** 0; one rational solution **41.** 50 m by 225 m, or 112.5 m by 100 m
43. $\pm i, \pm 1/2$ **45.** $5/2, -15$ **47.** 63/2 **49.** 3 **51.** $-2, -1$ **53.** $1, -4$ **55.** -1 **57.** 6 **59.** $(-\infty, 1]$
61. $(1, +\infty)$ **63.** $[4, 5]$ **65.** $[-5/2, 8)$ **67.** $[-4, 1]$ **69.** $(-\infty, -4] \cup [0, 4]$ **71.** $(-2, 0)$
73. $(-3, 1) \cup [7, +\infty)$ **75.** $(-\infty, a) \cup (b, +\infty)$; (a, b); at $x = a$ and $x = b$ **77.** 0 to 30 (remember, x must start at 0)
79. $-1, 5$ **81.** 11/27, 25/27 **83.** $4/3, -2/7$ **85.** $[-7, 7]$ **87.** no solution **89.** $(-2/7, 8/7)$
91. $(-\infty, -4) \cup (-2/3, +\infty)$ **93.** no solution **95.** (a) $(-\infty, +\infty)$ (b) none **97.** $(x + 2)(36/x + 3)$ or
$42 + 3x + 72/x$ **99.** $(-\infty, -7) \cup (-7, -3/2)$ **101.** $M = 1/2$ **103.** $m = kz^2$ **105.** $Y = (kMN^2)/X^3$ **107.** 27/2
109. 1372/729

Cumulative Review Exercises, Chapters 1 and 2 (page 112)

1. rational, real **3.** $(\sqrt{49} = 7)$ whole, integer, rational, real **5.** $-|-8| - |-6|, -|-2| + (-3), -|3|, -|-2|, |-8 + 2|$
7. $7 - \sqrt{5}$ **9.** $m - 3y$ **11.** $-1/125$ **13.** 1 **15.** 3^4 or 81 **17.** 1/64 **19.** $1/(a - b)$ **21.** $24a^{4/3}/b^{5/3}$
23. $rt^2/s^{1/3}$ **25.** $-5m^2 + 3m + 5$ **27.** $3k^2 - 29k + 56$ **29.** $27k^3 - 135k^2 + 225k - 125$
31. $3y^2 + yz + 11y - 2z^2 + z + 10$ **33.** $5x^3 + 10x^2 + 4x - 3/x$ **35.** $3m(m^2 + 3 + 5m^4)$ **37.** $(2q - 3)(3q + 4)$
39. $(2a + 5)(4a^2 - 10a + 25)$ **41.** $(r - p)(s + t)$ **43.** $-16z$ **45.** $4/(x - 1)$ **47.** $(3z - 2)/(3z + 2)$
49. $(3m + 11)/[2(m - 7)]$ **51.** $10\sqrt{10}$ **53.** $24\sqrt{6}$ **55.** $5p^3q^4\sqrt{2p}$ **57.** $b^4c^3\sqrt[3]{a^2c^2}$ **59.** $-(2 + \sqrt{7})/3$
61. $6 + 2i$ **63.** $17 + i$ **65.** $-12 - 5i$ **67.** $7/2 - (1/2)i$ **69.** $2/13 + (3/13)i$ **71.** $20i$ **73.** $-\sqrt{21}$
75. $-i$ **77.** $-10/13$ **79.** $-19/75$ **81.** $t = (v - v_0)/g$ **83.** 20 cm, 20 cm, 14 cm **85.** \$60,000 **87.** 10 pounds

89. 7 km per hour   **91.** $(3 \pm \sqrt{7})/5$ **93.** $-3/2, -1/4$ **95.** $(1 \pm \sqrt{6})/2$ **97.** $1 \pm i$ **99.** $(1 \pm 2i\sqrt{2})/6$
101. $-1, 1$ **103.** 7-1/2 hours   **105.** $\pm\sqrt{3}, \pm\sqrt{2}/2$ **107.** $-124, 9$ **109.** 12   **111.** 0, 18   **113.** $[6, +\infty)$
115. $(2, 9)$ **117.** $(-\infty, -6] \cup [0, +\infty)$ **119.** $(-\infty, -2] \cup [0, 3]$ **121.** $(-6, 2) \cup [6, +\infty)$
123. $(-\infty, -4) \cup (-1, +\infty)$ **125.** 1, 3   **127.** 6, $-4/5$ **129.** $b = k/a$ **131.** $g = km^2n^3$ **133.** 8
135. 1,600,000 units

Chapter 3

Section 3.1 (page 123)

1. III   **3.** IV

5.
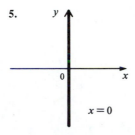

$x = 0$

7.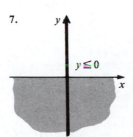

$y \le 0$

9.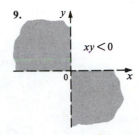

$xy < 0$

11.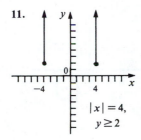

$|x| = 4,$
$y \ge 2$

13.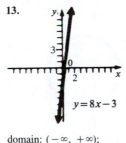

$y = 8x - 3$

domain: $(-\infty, +\infty)$;
range: $(-\infty, +\infty)$

15.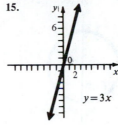

$y = 3x$

domain: $(-\infty, +\infty)$;
range: $(-\infty, +\infty)$

17.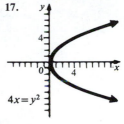

$4x = y^2$

domain: $[0, +\infty)$;
range: $(-\infty, +\infty)$

19.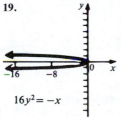

$16y^2 = -x$

domain: $(-\infty, 0]$;
range: $(-\infty, +\infty)$

21.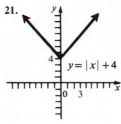

$y = |x| + 4$

domain: $(-\infty, +\infty)$;
range: $[4, +\infty)$

23.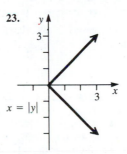

$x = |y|$

domain: $[0, +\infty)$;
range: $(-\infty, +\infty)$

25.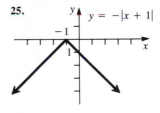

$y = -|x + 1|$

domain: $(-\infty, +\infty)$;
range: $(-\infty, 0]$

27.
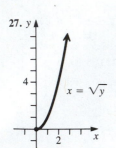

domain: $[0, +\infty)$;
range: $[0, +\infty)$

29.
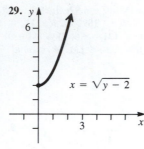

domain: $[0, +\infty)$;
range: $[2, +\infty)$

31.
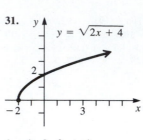

domain: $[-2, +\infty)$;
range: $[0, +\infty)$

33.
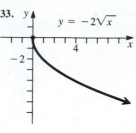

domain: $[0, +\infty)$;
range: $(-\infty, 0]$

35. $V = s^3 - 2s^2$ **37.** $A = \dfrac{1}{4}b^2$

Section 3.2 (page 131)

1. $8\sqrt{2}$; $(9, 3)$ **3.** $\sqrt{34}$; $(-11/2, -7/2)$ **5.** $\sqrt{133}$; $(2\sqrt{2}, 3\sqrt{5}/2)$ **7.** 7.616 **9.** 27.203 **11.** $(13, 10)$
13. $(-10, 11)$ **15.** yes **17.** no **19.** yes **21.** no **23.** $5, -1$ **25.** $9 + \sqrt{119}, 9 - \sqrt{119}$
27. $2\sqrt{2}, -2\sqrt{2}$ **29.**

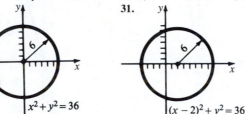

31.

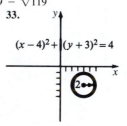

33.

35. $(x - 1)^2 + (y - 4)^2 = 9$ **37.** $(x + 8)^2 + (y - 6)^2 = 25$ **39.** $(x + 1)^2 + (y - 2)^2 = 25$
41. $(x + 3)^2 + (y + 2)^2 = 4$ **43.** $(x - 4)^2 + (y - 13/2)^2 = 45/4$ **45.** $(-3, -4)$; $r = 4$ **47.** $(-4, 7)$;
$r = 0$ (a point) **49.** $(0, 1)$; $r = 7$ **51.** $C = 10\pi, A = 25\pi$ **55.** $(2 + \sqrt{7}, 2 + \sqrt{7}), (2 - \sqrt{7}, 2 - \sqrt{7})$
57. (a) inside (b) outside (c) on (d) outside **59.** $(2, 3)$ or $(4, 1)$ **61.** yes **63.** $4 + s \geq \sqrt{(a - c)^2 + (b - d)^2}$
65. $(6, 8)$ and $(8, 11)$ **67.** yes, no

Section 3.3 (page 140)

1. yes **3.** no **5.** yes **7.** no **9.** yes **11.** no **13.** (a) 0 (b) 4 (c) 2 (d) 4 **15.** (a) -3 (b) -2 (c) 0
(d) 2 **17.** -3 **19.** undefined **21.** 4 **23.** 8 **25.** 0 **27.** 4 **29.** $(1 - 3a^2)/a^2$ **31.** $1/(a^2 - 3)$
33. $(25a^2 - 3)/\sqrt{3a + 1}$ **35.** 3 **37.** 18 **39.** 4 **41.** $(-\infty, +\infty)$; $[0, +\infty)$ **43.** $[-8, +\infty)$; $[0, +\infty)$
45. $(-\infty, 1) \cup (1, 2) \cup (2, +\infty)$ **47.** $(-\infty, -1] \cup [5, +\infty)$; $(-\infty, 0]$ **49.** $(-\infty, +\infty)$; $[0, +\infty)$ **51.** $[-5, 4]$; $[-2, 6]$
53. $(-\infty, +\infty)$; $(-\infty, 12]$ **55.** $[-3, 4]$; $[-6, 8]$ **57.** no; the domains are different **59.** (a) $8 - 3x^2 - 6xh - 3h^2$
(b) $-6xh - 3h^2$ (c) $-6x - 3h$ **61.** (a) $4x + 4h + 11$ (b) $4h$ (c) 4 **63.** $r = 2h/3$ **65.** $(0, 2000)$
67. (b) $A(r) = 440r - \pi r^2$ (c) $(0, 220/\pi)$ **69.** (b) $T(x) = [(16 + 3\pi)x^2 - 1280x + 25,600]/64\pi$ (c) $(0, 40)$
71. $\pi d^2/4$; πd **73.** $\pi r^3/6$

Section 3.4 (page 150)

1.
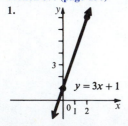

3.
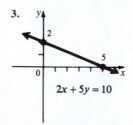

5.
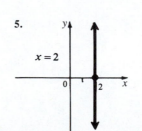

7.
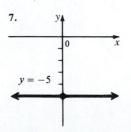

9.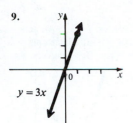
$y = 3x$

11.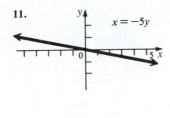
$x = -5y$

13.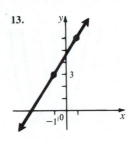

15.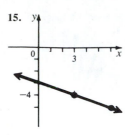

17.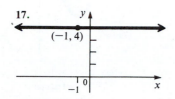
$(-1, 4)$

19.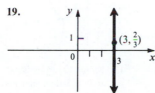
$(3, \frac{2}{3})$

21. -1 **23.** $1/3$ **25.** 2 **27.** $-3/4$ **29.** 0
31. 0 **33.** (d)   **35.** (a)   **37.** (e)   **39.** yes

41.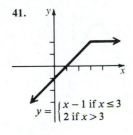
$y = \begin{cases} x - 1 & \text{if } x \le 3 \\ 2 & \text{if } x > 3 \end{cases}$

43.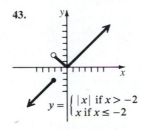
$y = \begin{cases} |x| & \text{if } x > -2 \\ x & \text{if } x \le -2 \end{cases}$

45. (a) 140 (b) 220 (c) 220 (d) 220 (e) 220
(f) 60 (g) 60 (h) $i(t)$
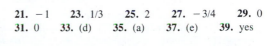

47.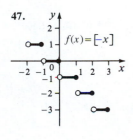
$f(x) = [-x]$

49.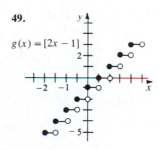
$g(x) = [2x - 1]$

51.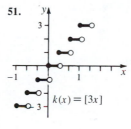
$k(x) = [3x]$

53. (a) $11 (b) $18 (c) $32 (d) $S(x)$
(e) domain: $(0, +\infty)$
(at least in theory); range:
$\{11, 18, 25, 32, 39, \ldots\}$

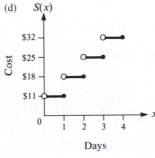
Cost
$32
$25
$18
$11
0   1   2   3   4
Days

55. $-[-x]$ ounces
57. (a) 16 (b) 11 (c) 6
(d) 8 (e) 4 (f) 0
(g) and (k) are graphed
at right (h) 0 (i) 40/3
(j) 80/3 (l) 8 (m) 6

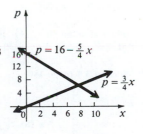
$p = 16 - \frac{5}{4}x$
$p = \frac{3}{4}x$

59. (f) -3 **61.** (b) 1

Section 3.5 (page 158)
1. $2x + y - 5 = 0$ **3.** $3x + 2y + 7 = 0$ **5.** $3x + 4y - 6 = 0$ **7.** $y - 2 = 0$ **9.** $x + 8 = 0$ **11.** $x + 3y - 10 = 0$ **13.** $x - 4y + 13 = 0$ **15.** $2x - 3y - 6 = 0$ **17.** $x + 5 = 0$ **19.** $y - 7 = 0$ **21.** $x + 3y - 11 = 0$ **23.** $2x - y - 9 = 0$ **25.** $x - y - 7 = 0$ **27.** $5x - 3y + 13 = 0$ **29.** $2x - y + 4 = 0$ **31.** $k = 2$ **33.** $k = -5$ **35.** (b)   **37.** (e)   **39.** (d)   **41.** $f(x) = x + 1; f(7) = 8$ **43.** (a) $-1/2$ (b) $-7/2$ **47.** 2   **49.** $-2x - h$ **51.** $y = 640x + 1100; m = 640$ **53.** $y = -1000x + 40,000; m = -1000$ **55.** $y = 2.5x - 70; m = 2.5$

Section 3.6 (page 169)

1.

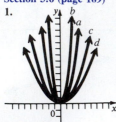

(e) It controls how broad the parabola is.

3.

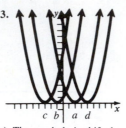

(e) The parabola is shifted horizontally.

5. vertex: (2, 0); axis: $x = 2$; x-intercept: 2; y-intercept: 4

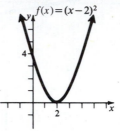

7. vertex: $(-3, -4)$; axis: $x = -3$; x-intercepts: -5, -1; y-intercept: 5

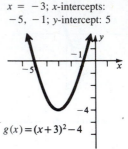

9. vertex: $(-3, 2)$; axis: $x = -3$; x-intercepts: -2, -4; y-intercept: -16

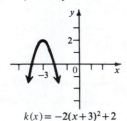

11. vertex: $(-1, -3)$; axis: $x = -1$; x-intercept: none; y-intercept: $-7/2$

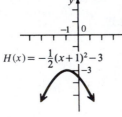

13. vertex: (1, 2); axis: $x = 1$; x-intercept: none; y-intercept: 3

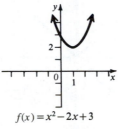

15. vertex: $(-2, 6)$; axis: $x = -2$; x-intercepts, $-2 \pm \sqrt{6}$; y-intercept: 2

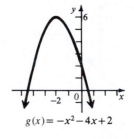

17. vertex: (1, 3); axis: $x = 1$; x-intercept: none; y-intercept: 5

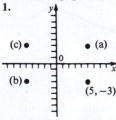

19. 3   **21.** 0   **23.** 80 by 160 ft   **25.** 5 in   **27.** 10, 10
29. (a) $x(500 - x) = 500x - x^2$ (b)

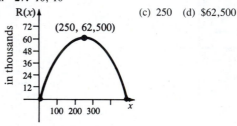

(c) 250   (d) $62,500

31. $10\sqrt{3}$ m ≈ 17.32 m   **33.** (a) B   (b) C   (c) A   **35.** (a)   **37.** (b)
39. $b = 6$ or -6 **41.** if $b^2 - 4ac < 0$ **45.** $12y = x^2$ **47.** $-8x = y^2$ **49.** $8(y - 4) = (x - 3)^2$
51. $(x - h)^2 = 4p(y - k)$ **53.** (a) $\sqrt{3}$ (b) 1/3

Section 3.7 (page 180)

1.

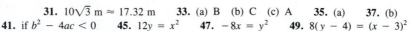

3.

5. x-axis, y-axis, and origin   **7.** y-axis   **9.** none of these symmetries   **11.** x-axis   **13.**

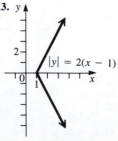

15.

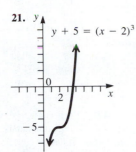

$|y| = |x + 2|$

17. The graph is a straight line with a point missing at the origin.

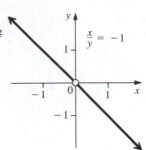

$\dfrac{x}{y} = -1$

19.

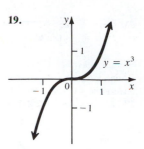

$y = x^3$

21.

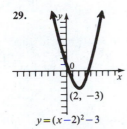

$y + 5 = (x - 2)^3$

23.

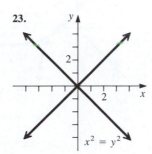

$x^2 = y^2$

25.

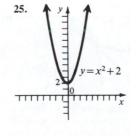

$y = x^2 + 2$

27.

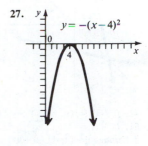

$y = -(x - 4)^2$

29.

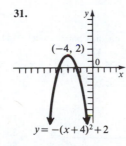

$(2, -3)$

$y = (x-2)^2 - 3$

31.

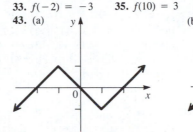

$(-4, 2)$

$y = -(x+4)^2 + 2$

33. $f(-2) = -3$ **35.** $f(10) = 3$ **37.** $f(x) = -5x + 17$

43. (a)

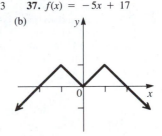

(b)

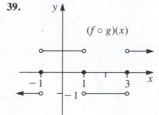

45. If $c > 0$, the graph has the same general shape; if $c < 0$, the graph is reflected about the x-axis. If $|c| > 1$, the graph is narrower; if $|c| < 1$, the graph is broader.   **47.** Many answers are possible.   **51.** true   **53.** false

Section 3.8 (page 186)
1. $10x + 2$; $-2x - 4$; $24x^2 + 6x - 3$; $(4x - 1)/(6x + 3)$; all domains are $(-\infty, +\infty)$, except for f/g, which is $(-\infty, -1/2) \cup (-1/2, +\infty)$ **3.** $4x^2 - 4x + 1$; $2x^2 - 1$; $(3x^2 - 2x)(x^2 - 2x + 1)$; $(3x^2 - 2x)/(x^2 - 2x + 1)$; all domains are $(-\infty, +\infty)$, except for f/g, which is $(-\infty, 1) \cup (1, +\infty)$ **5.** $\sqrt{2x + 5} + \sqrt{4x - 9}$; $\sqrt{2x + 5} - \sqrt{4x - 9}$; $\sqrt{(2x + 5)(4x - 9)}$; $\sqrt{(2x + 5)/(4x - 9)}$; all domains are $[9/4, +\infty)$, except f/g, which is $(9/4, +\infty)$ **7.** 55   **9.** 1848   **11.** $-6/7$ **13.** $4m^2 + 6m + 1$ **15.** 1122   **17.** 97   **19.** $256k^2 + 48k + 2$ **21.** $24x + 4$; $24x + 35$ **23.** $-5x^2 + 20x + 18$; $-25x^2 - 10x + 6$ **25.** $1/x^2$; $1/x^2$ **27.** $\sqrt{8x^2 - 4}$; $8x + 10$ **29.** $x/(2 - 5x)$; $2(x - 5)$ **31.** 4   **33.** 0   **35.** 1   **37.** 2   **39.**

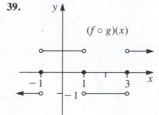

$(f \circ g)(x)$

47–51. Many answers are possible.

53. $(-c^2 + 10c - 25)/25 + 500$ **55.** $4\pi t^2$ **57.** $(f \circ g)(x) = f(-x)$; the graph is reflected about the y-axis.
$(g \circ f)(x) = -f(x)$; the graph is reflected about the x-axis.   **59.** $(-\infty, -1) \cup (-1, +\infty)$

Section 3.9 (page 195)

1. one-to-one   **3.** one-to-one   **5.** not one-to-one   **7.** one-to-one   **9.** not one-to-one   **11.** not one-to-one   **13.** not one-to-one   **15.** one-to-one   **17.** one-to-one   **19.** not one-to-one   **21.** inverses   **23.** not inverses   **25.** inverses   **27.** not inverses   **29.** not inverses   **31.** inverses   **33.** not inverses

35.
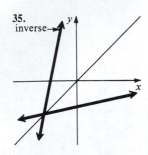

37.
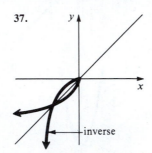

39.
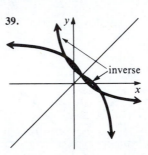

41. $f^{-1}(x) = (x + 5)/4$ **43.** $f^{-1}(x) = (-5/2)x$ **45.** $f^{-1}(x) = \sqrt[3]{-x - 2}$ **47.** not one-to-one   **49.** $f^{-1}(x) = 4/x$
51. $f^{-1}(x) = 3 - \sqrt{x}$ **53.** $f^{-1}(x) = x^2 - 6$ domain: $[0, +\infty)$ **55.** 4   **57.** 2   **59.** -2 **61.** 1.14   **63.** the number of
dollars required to build 1000 cars   **65.** $(4, 1), (-1, -4), (-4, -1)$ **67.** 2   **71.** Since f is not one-to-one, it has
no inverse f^{-1}

Chapter 3 Review Exercises (page 198)

1. $d(P, Q) = \sqrt{85}$; $(-1/2, 2)$ **3.** $(7, -13)$ **5.** $-7, -1, 8, 23$ **7.** no such points exist   **11.** $(x + 2)^2 +$
$(y - 3)^2 = 25$ **13.** $(x + 8)^2 + (y - 1)^2 = 289$ **15.** $(2, -3); r = 1$ **17.** $(-7/2, -3/2); r = 3\sqrt{6}/2$ **19.** true
21. no; $(-\infty, +\infty)$; $[0, +\infty)$ **23.** yes; $(-\infty, -2] \cup [2, +\infty)$; $[0, +\infty)$ **25.** yes; $(-\infty, +\infty)$; $(-\infty, +\infty)$ **27.** function
29. function   **31.** function   **33.** $(-\infty, +\infty)$ **35.** $(-\infty, +\infty)$ **37.** $(-\infty, +\infty)$ **39.** 6/5   **41.** 9/4
43. undefined

45.
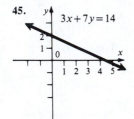
$3x + 7y = 14$

47.
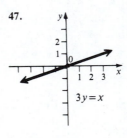
$3y = x$

49.
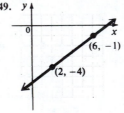
$(6, -1)$ $(2, -4)$

51. $x + 3y - 10 = 0$
53. $2x + y - 1 = 0$
55. $5x - 3y + 15 = 0$
57. $3x - y - 7 = 0$
59. $5x - 8y + 40 = 0$
61. $y = [5000/x]$

63.
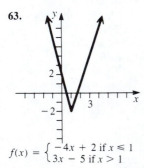
$f(x) = \begin{cases} -4x + 2 \text{ if } x \le 1 \\ 3x - 5 \text{ if } x > 1 \end{cases}$

65. $(0, 6); x = 0$
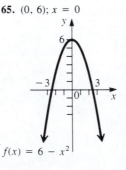
$f(x) = 6 - x^2$

67. $(2, 3); x = 2$
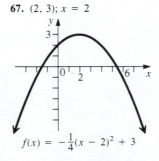
$f(x) = -\frac{1}{4}(x - 2)^2 + 3$

69. $(-7/2, 9/4); x = -7/2$
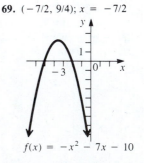
$f(x) = -x^2 - 7x - 10$

71. $(1/2, 2); x = 1/2$

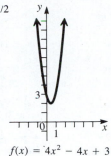

$f(x) = 4x^2 - 4x + 3$

73. h **75.** $k \le 0; x = h \pm \sqrt{-ak/a}$ **77.** 5.5 and 5.5
79. a square, 45 m on a side **81.** x-axis, y-axis, and origin **83.** none of these
85. x-axis **87.** x-axis, y-axis, and origin

89.

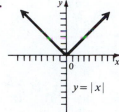

$y = |x|$

91. reflect graph in x-axis

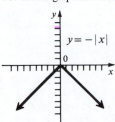

$y = -|x|$

93. move 3 units to the right, 4 down, graph is narrower

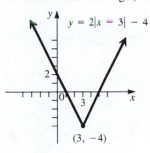

$y = 2|x - 3| - 4$

$(3, -4)$

95. $y = -2x + 3$
97. $y = (1/2)x - (3/2)$

99.

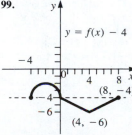

$y = f(x) - 4$

$(8, -4)$

$(4, -6)$

101.

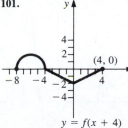

$(4, 0)$

$y = f(x + 4)$

103.

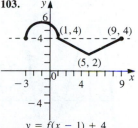

$(1, 4)$ $(9, 4)$

$(5, 2)$

$y = f(x - 1) + 4$

105. odd **107.** odd **109.** $4x^2 - 3x - 8$ **111.** 44 **113.** $16k^2 - 6k - 8$ **115.** $-23/4$ **117.** $(-\infty, +\infty)$
119. $\sqrt{x^2 - 2}$ **121.** $\sqrt{34}$ **123.** 1 **125.** not one-to-one **127.** not one-to-one **129.** not one-to-one
131. $f^{-1}(x) = (x - 3)/12$ **133.** f has no inverse **135.** the number of years after 1990 when world population
will equal 6,000,000,000 **137.** (a) $V = \pi d^3/4$ (b) $S = 3\pi d^2/2$ **139.** The conversion
of u U.S. dollars into c Canadian dollars is $c = V(u) = (1 + .15) u = 1.15u$, and the
conversion of d Canadian dollars into t U.S. dollars is $t = H(d) = (1 - .15) d = .85d$;
therefore $H(V(u)) = .85(1.15)u = .9775u \ne u$.

Chapter 4

Section 4.1 (page 208)
1. $x^2 - 3x - 2$ **3.** $m^3 - m^2 - 6m$ **5.** $3x^2 + 4x + 3/(x - 5)$ **7.** $4m^2 - 4m + 1 + [-3/(m + 1)]$
9. $x^4 + x^3 + 2x - 1 + 3/(x + 2)$ **11.** $(1/3)x^2 - (1/9)x + 1/(x - 1/3)$ **13.** $y^2 + y + 1$
15. $f(x) = (x - 1)(2x^2 + 3x + 4) - 4$ **17.** $f(x) = (x + 2)(-x^2 + 4x - 8) + 20$
19. $f(x) = (x - 3)(4x^3 + 9x^2 + 7x + 20) + 60$ **21.** -8 **23.** 2 **25.** -6 **27.** $-6 - i$ **29.** yes **31.** yes
33. no **35.** no **37.** yes **39.** -2 **41.** $k = 3$

Section 4.2 (page 215)
1. no **3.** yes **5.** no **7.** $f(x) = -(1/6)x^3 + (13/6)x + 2$ **9.** $f(x) = (-1/2)x^3 - (1/2)x^2 + x$ **11.** $f(x) = -10x^3 + 30x^2 - 10x + 30$ **13.** $-3, 2, 1 - i$; all multiplicity 1 **15.** multiplicity 1: $-i/2, i/2$; multiplicity 5: 0
17. multiplicity 2: $1 - \sqrt{3}$; multiplicity 5: $-2, 1$ **19.** $f(x) = x^2 - 10x + 26$ **21.** $f(x) = x^3 - 3x^2 + x + 1$
23. $f(x) = x^4 - 6x^3 + 10x^2 + 2x - 15$ **25.** $f(x) = x^3 - 8x^2 + 22x - 20$ **27.** $f(x) = x^4 - 6x^3 + 17x^2 - 28x + 20$
29. $-1 + i, -1 - i$ **31.** $(-1 + i\sqrt{5})/2, (-1 - i\sqrt{5})/2$ **33.** $i, 2i, -2i$ **35.** $f(x) = (x - 2)(2x - 5)(x + 3)$
37. $f(x) = (x + 4)(3x - 1)(2x + 1)$ **39.** $f(x) = (x - 3i)(x + 4)(x + 3)$ **41.** Zeros are $-2, -1, 3$;
$f(x) = (x + 2)^2(x + 1)(x - 3)$. **43.** 1, 3, or 5 **45.** at $t = 3$ and $t = -2$

Section 4.3 (page 220)
1. $\pm1, \pm1/2, \pm1/3, \pm1/6$ **3.** $\pm1, \pm1/2, \pm2, \pm4, \pm8$ **5.** $\pm1, \pm2, \pm5, \pm10, \pm25, \pm50$ **7.** $-1, -2, 5$ **9.** 2,
$-3, -5$ **11.** no rational zeros **13.** $1, -2, -3, -5$ **15.** $-4, 3/2, -1/3; f(x) = (x + 4)(2x - 3)(3x + 1)$ **17.** $1/2$;
$f(x) = 2(x - 1/2)(x^2 + 4x + 8) = (2x - 1)(x^2 + 4x + 8)$ **19.** no rational zeros; prime **21.** $-1, -2, -3, 4$;
$f(x) = (x + 1)(x + 2)(x + 3)(x - 4)$ **23.** $1; f(x) = (x - 1)(x^4 + 4x^3 - x^2 - 12x - 12)$ **25.** $-2/3, -1, 3$
27. $1, -5/4$ **29.** 1 **31.** 3 **33.** 6 in. by 7 in.; 3.0 in. by 13.8 in.

Section 4.4 (page 232)

1.
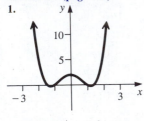
$f(x) = x^4 - 3x^2 + 2$

3.
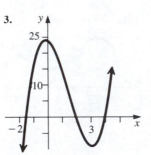
$f(x) = 2x^3 - 9x^2 - 2x + 24$

5.
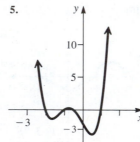
$f(x) = 2x^4 + 5x^3 - 5x - 2$

17. 2 or 0 positive and 1 negative **19.** 4 or 2 or 0 positive and 0 negative **21.** $-3,$ $-1.4, 1.4$ **23.** 1.5 **25.** no real zeros **27.** $3.24, -1.24$

29.
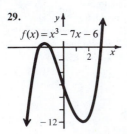
$f(x) = x^3 - 7x - 6$

31.
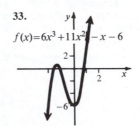
$f(x) = x^4 - 5x^2 + 6$

33.
$f(x) = 6x^3 + 11x^2 - x - 6$

35.
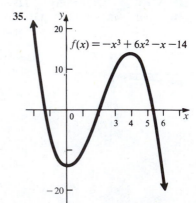
$f(x) = -x^3 + 6x^2 - x - 14$

37. $(-\infty, -1.9) \cup (.1, 3.3)$ **39.** $(-\infty, -4.2) \cup (-.5, .5) \cup (4.2, +\infty)$
41. (a)

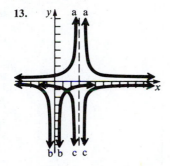

$A(x) =$
$-.015x^3 + 1.058x$

$0 \le x \le 8$

(b) between 4 and 5 hr, closer to 5 hr
(c) from about 1 hr to about 8 hr

43. (a)

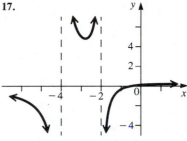

$P(t) = t^3 - 25t^2 + 200t$

(b) increasing from $t = 0$ to $t = 6\ 2/3$ and from $t = 10$ on; decreasing from $t = 6\ 2/3$ to $t = 10$

45. $f(x) = (x + 3)^2(x - 2)^2$ **47.** domain: $[-1, 0] \cup [4, +\infty)$; range $[0, +\infty)$ **49.** Neither $f(x)$ nor $f(-x)$ can have 7 sign changes (since some terms have zero coefficients), so by Descartes' rule, there cannot be 7 zeros.

Section 4.5 (page 244)
1. vertical asymptote: $x = 5$; horizontal asymptote; $y = 0$ **3.** vertical asymptote: $x = 7/3$; horizontal asymptote: $y = 0$
5. vertical asymptote: $x = -9/2$; horizontal asymptote: $y = 3/2$ **7.** vertical asymptotes: $x = 3$, $x = 1$; horizontal asymptote: $y = 0$ **9.** oblique asymptote: $y = x - 3$; vertical asymptote $x = -3$ **11.** vertical asymptotes: $x = -2$, $x = 5/2$; horizontal asymptote: $y = 1/2$

13.

15.

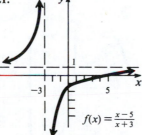

$f(x) = \dfrac{3}{(x + 4)^2}$

17.

$f(x) = \dfrac{2x + 1}{(x + 2)(x + 4)}$

19.

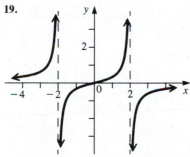

$f(x) = \dfrac{-x}{x^2 - 4}$

21.

$f(x) = \dfrac{x - 5}{x + 3}$

23.

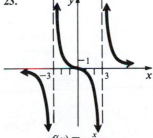

$f(x) = \dfrac{x}{x^2 - 9}$

25.

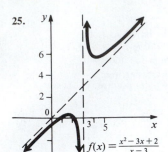

$$f(x) = \frac{x^2 - 3x + 2}{x - 3}$$

27.

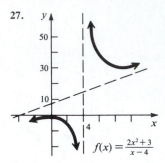

$$f(x) = \frac{2x^2 + 3}{x - 4}$$

29.

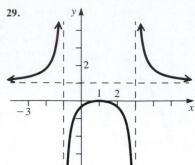

$$f(x) = \frac{(x - 1)^2}{x^2 - 2x - 3}$$

31.

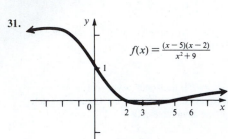

$$f(x) = \frac{(x - 5)(x - 2)}{x^2 + 9}$$

33.

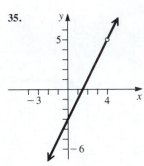

$$f(x) = \frac{-9}{x^2 + 9}$$

35.

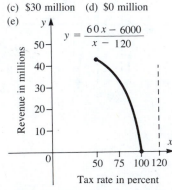

$$f(x) = \frac{(2x - 3)(x - 4)}{x - 4}$$

37.

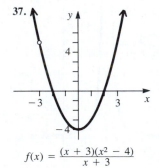

$$f(x) = \frac{(x + 3)(x^2 - 4)}{x + 3}$$

39. (a)

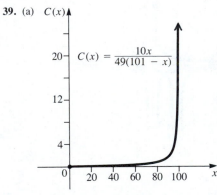

$$C(x) = \frac{10x}{49(101 - x)}$$

(b) (0, 100); winning 0 points would correspond to a cost of $0.

41. (a) $42.9 million   (b) $40 million
(c) $30 million   (d) $0 million
(e)

$$y = \frac{60x - 6000}{x - 120}$$

Tax rate in percent

43.

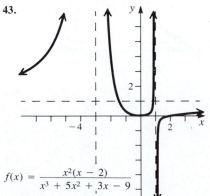

$$f(x) = \frac{x^2(x - 2)}{x^3 + 5x^2 + 3x - 9}$$

45. $f(x) = \dfrac{x - 2}{x(x - 4)}$

47. $f(x) = \dfrac{4x(x - 3)}{(x + 2)(x - 2)}$

Chapter 4 Review Exercises (page 247)
1. $q(x) = 2x^2 + 5x + 1; r = 2$ **3.** $q(x) = 2x^2 + 4x + 4; r = 14$ **5.** 11   **7.** 28   **9.** $f(x) = x^3 - 10x^2 +$
$17x + 28$ **11.** $f(x) = x^4 - x^3 - 9x^2 + 7x + 14$ **13.** no   **15.** no   **17.** $f(x) = -2x^3 + 6x^2 + 12x - 16$
19. $f(x) = x^4 - 3x^2 - 4$ **21.** Many answers are possible. Any polynomial function of the form $f(x) = a(x - b)^2(x - c)^2$ for
real numbers a, b, and c will satisfy.   **23.** $1 - i, 1 + i, 4, -3$ **25.** $1/2, -1, 5$ **27.** $4, -1/2, -2/3$ **31.** $r = 13/2$
37. $-.5, .4, 5.6$

39.

41. (a) positive, 1; negative 2
or 0   (b) -3.9
(c)

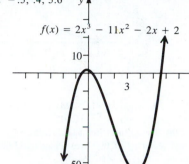

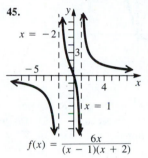

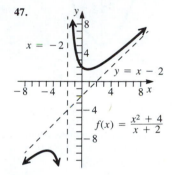

43.

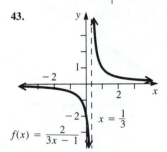

45.

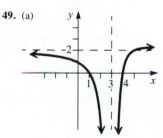

47.

49. (a)

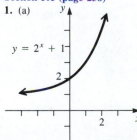

(b) $f(x) =$
$$\dfrac{2(x - 1)(x - 4)}{(x - 3)^2}$$

Chapter 5

Section 5.1 (page 258)
1. (a)

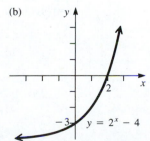

(b)

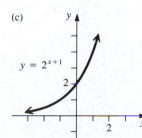

(c)

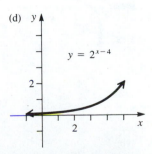

(d)

3.

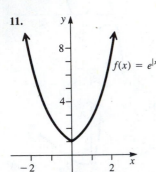

5.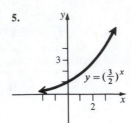

7.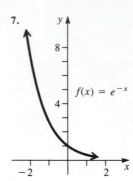

9.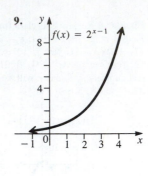

11.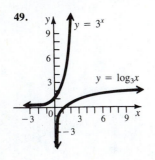

13.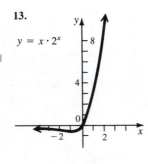

15. 1/2   **17.** -2 **19.** 0   **21.** 3   **23.** 8
25. 1/5   **27.** $-2/3$ **29.** $-3, 3$ **31.** 0
33. (a) \$14,049.28   (b) \$14,257.61   (c) \$14,332.45
35. (a) 1,000,000   (b) about 1,041,000   (c) about 1,083,000
(d) about 1,221,000   (e) $P(t)$

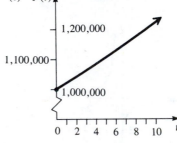

37. $f(x) = 2^x$ **39.** $f(x) = 25^x$ **41.** $f(t) = 8 \cdot 16^t$ **43.** $f(t) = 3(1/9)^t$
45. $4/(e^{-x} - e^x)^2$ **47.** 1/2   **49.** 8   **51.** 11.6 lb per sq in.   **53.** The graph
increases at a faster rate; the graph decreases at a slower rate.   **55.** 0   **57.** 2
59. neither   **61.** 2.718

Section 5.2 (page 267)
1. $\log_{10} 10,000 = 4$ **3.** $\log_{1/2} 16 = -4$ **5.** $6^2 = 36$ **7.** $10^{-4} = .0001$ **9.** 2   **11.** 1   **13.** $-1/6$ **15.** 8
17. 4   **19.** 1/2   **21.** 2   **23.** 9   **25.** 9   **27.** 1/5   **29.** 3/2   **31.** 16   **33.** $\log_2 6 + \log_2 x - \log_2 y$
35. $1 + (1/2) \log_5 7 - \log_5 3$ **37.** cannot be simplified using the properties of logarithms   **39.** $\log_k p + 2 \cdot \log_k q - \log_k m$
41. $\log_a (xy)/m$ **43.** $\log_m (a^2/b^6)$ **45.** $\log_x (a^{3/2}b^{-4})$ or $\log_x (a^{3/2}/b^4)$ **47.** $\log_b [7x(x + 2)/8]$

49.

51. (a)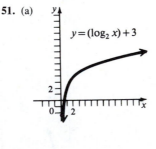

(b)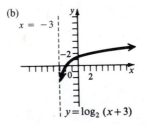

(c)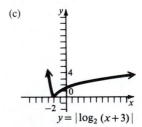

53.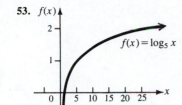
$f(x) = \log_5 x$

$x = 1$, no y-intercept

55.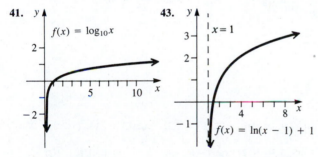
$y = \log_{1/2}(1 - x)$

$x = 0$, $y = 0$

57.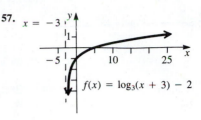
$f(x) = \log_3(x + 3) - 2$

$x = 6$, $y = -1$

59. .7781   **61.** 1.3010   **63.** $\log_6 2$ **65.** -2 **67.** 4   **69.** (a) about 239   (b) about 477   (c) about 759

(d)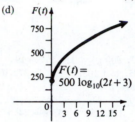
$F(t) = 500 \log_{10}(2t + 3)$

71. plus $10 \log_{10} 2 \approx 3$ **73.** $398{,}000{,}000 \, I_0$

Section 5.3 (page 274)

1. -2.1850 **3.** 2.0183   **5.** $-.0160$ **7.** 5.4330   **9.** 1   **11.** 4   **13.** -3 **15.** 1.2   **17.** 9   **19.** $-.5$
21. 6   **23.** 32   **25.** 1.3863   **27.** 2.8332   **29.** 5.8579   **31.** 11.3022   **33.** 1.43   **35.** -1.58 **37.** .96   **39.** 1.89

41.
$f(x) = \log_{10} x$

$x = 1$; no y-intercept

43.
$x = 1$
$f(x) = \ln(x - 1) + 1$

$x \approx 1.4$; no y-intercept

45. $\ln b = c$ **47.** translated up 2 units   **49.** 1
51. (a) about 350 years   (b) about 4000 years
(c) about 2300 years   **53.** 8.4   **55.** 13.5
57. 4×10^{-4} **59.** 3.2×10^{-7} **61.** (a) 961,000
(b) about 670,000

Section 5.4 (page 281)

1. 1.631   **3.** $-.080$ **5.** 3.797   **7.** 2.386   **9.** $-.123$ **11.** no solution   **13.** 17.475   **15.** -2.141,
2.141   **17.** 1.386   **19.** 11   **21.** 5   **23.** 10   **25.** 4   **27.** no solution   **29.** $-2, 2$ **31.** 2   **33.** 3
35. 1, 10   **37.** 15 seconds   **39.** 46 days   **41.** 0   **43.** 0, $\ln 5$ **45.** $x = \ln(y + \sqrt{y^2 + 1})$

47. $t = (-2/R) \ln(1 - RI/E)$ **49.** $t = \dfrac{\ln[A/(A - Pr)]}{\ln(1 + r)}$ **51.** $x = (1 - y)/2$ **53.** $f^{-1}(x) = $

$(1/2)10^x - 3$; domain: $(-\infty, +\infty)$; range $(-3, +\infty)$ **55.** $f^{-1}(x) = 7 - \log_{10}(x/3)$; domain: $(0, +\infty)$; range; $(-\infty, +\infty)$

57. $(1/3, +\infty)$ **59.** $(1, 10)$ **61.** $(-\infty, 1)$ **63.** $t = \dfrac{\ln(A/P)}{n \ln(1 + r/n)}$ **65.** (a) 611 million   (b) 1007 million

(c) 2028 million   (d) about 18 days   (e) January 19

Section 5.5 (page 287)

1. about 14 months   **3.** about 283 grams   **5.** about 11.3°   **7.** 5600 years   **9.** about 15,000 years   **11.** about 173 days   **13.** $2,166.57   **15.** $17,143.21   **17.** $12,652.54   **19.** about 13.9 years   **21.** $r = [\ln (A/P)]/t$
23. 81.25°C   **25.** 117.5°C   **27.** about 1611   **29.** in about 1.8 decades or 18 years

Chapter 5 Review Exercises (page 289)

1. 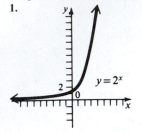   **3.** 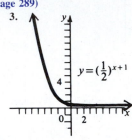   **5.** 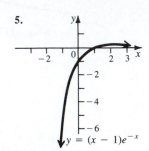   **7.**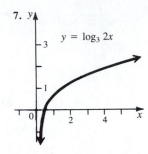

9. 5/3   **11.** 3/2   **13.** 800 g   **15.** $1790.19   **17.** about 12 years   **19.** $\log_2 32 = 5$ **21.** $\log_{1/16} (1/2) = 1/4$
23. $\log 3 = .4771$ **25.** $\ln 1.1052 = .1$ **27.** $10^{-3} = .001$ **29.** $10^{.537819} = 3.45$ **31.** 3
33. $\log_3 m + \log_3 n - \log_3 p$ **35.** $2 \ln x + 4 \ln y + (3/5)\ln m + (1/5)\ln p$ **37.** .9279   **39.** 3.0212   **41.** -5.3
43. 4.4886   **45.** 2.0794   **47.** -3.2403 **49.** 1.490   **51.** 1.303   **53.** .747   **55.** 4   **57.** 2   **59.** 2
61. $x = [\ln(y + \sqrt{y^2 + 1})]/\ln 5$ **63.** $x = (4 + a \pm \sqrt{a^2 + 12a})/2$ **65.** $c = de^{(N - a)/b}$ **67.** .8 m   **69.** 1.5 m
71. (a) 207 (b) 235 (c) 249 (d)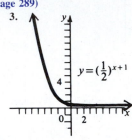
73. (a) B (b) A (c) C   **75.** by a factor of $2^5 = 32$
77. domain: $(-\infty, -1) \cup (2, +\infty)$; range: $(-\infty, +\infty)$
79. $16,095.74   **81.** $580,792.63   **83.** (a) about 286 grams (b) about 28.9 days   **85.** The correct statement is $\log_5 125 - \log_5 25 = \log_5 \dfrac{125}{25} = \log_5 5 = 1$.

Chapter 6

Section 6.1 (page 299)

1. $(-\sqrt{2}/2, -\sqrt{2}/2)$ **3.** $(-1, 0)$ **5.** $(1, 0)$ **7.** $(1, 0)$ **9.** $(0, 1)$ **11.** $(-\sqrt{2}/2, -\sqrt{2}/2)$ **13.** $(\sqrt{2}/2, \sqrt{2}/2)$
15. $(-\sqrt{2}/2, \sqrt{2}/2)$ **17.** (a) $(2/3, -\sqrt{5}/3)$ (b) $(2/3, \sqrt{5}/3)$ (c) $(-2/3, -\sqrt{5}/3)$ (d) $(-2/3, \sqrt{5}/3)$
19. (a) $(4/5, -3/5)$ (b) $(4/5, 3/5)$ (c) $(-4/5, -3/5)$ (d) $(-4/5, 3/5)$ **21.** (a) $(-1/2, -\sqrt{3}/2)$ (b) $(-1/2, \sqrt{3}/2)$
(c) $(1/2, -\sqrt{3}/2)$ (d) $(1/2, \sqrt{3}/2)$ **23.** (a) $(-2/5, \sqrt{21}/5)$ (b) $(-2/5, -\sqrt{21}/5)$ (c) $(2/5, \sqrt{21}/5)$ (d) $(2/5, -\sqrt{21}/5)$
25. 0; 1   **27.** $\sqrt{2}/2; \sqrt{2}/2$ **29.** $-1; 0$ **31.** $0; -1$ **33.** $-\sqrt{2}/2; \sqrt{2}/2$ **35.** $-\sqrt{2}/2; -\sqrt{2}/2$ **37.** IV
39. III   **41.** III   **43.** II   **45.** $(\sqrt{3}/2, 1/2)$ **47.** $\sqrt{3}/2$ **49.** $-1/2$ **51.** $-\sqrt{3}/2$ **53.** 1/2   **55.** $\sin s = 4/5$;
$\cos s = 3/5$ **57.** $\sin s = 5/8; \cos s = -\sqrt{39}/8$ **59.** $\sin s = 2\sqrt{3}/\sqrt{13}$ or $2\sqrt{39}/13; \cos s = -1/\sqrt{13}$ or $-\sqrt{13}/13$
61. $\sin s = -\sqrt{2}/\sqrt{11}$ or $-\sqrt{22}/11; \cos s = 3/\sqrt{11}$ or $3\sqrt{11}/11$ **63.** $\sin s = b/\sqrt{a^2 + b^2}; \cos s = a/\sqrt{a^2 + b^2}$
65. $-.4$ **67.** because $\sin (-x) = -\sin x$ **69.** $-$ **71.** $+$ **73.** $\pi/4 + n\pi$, where n is any integer

Section 6.2 (page 305)

1. $\tan s = \sqrt{3}/3; \cot s = \sqrt{3}; \sec s = 2\sqrt{3}/3; \csc s = 2$ **3.** $\tan s = -4/3; \cot s = -3/4; \sec s = -5/3; \csc s = 5/4$ **5.** $\tan s = -\sqrt{3}; \cot s = -\sqrt{3}/3; \sec s = 2; \csc s = -2\sqrt{3}/3$
In Exercises 7–17 and 23–33 answers are given in the order sine, cosine, tangent, cotangent, secant, cosecant.
7. $0, -1, 0$, undefined, -1, undefined   **9.** $\sqrt{2}/2, -\sqrt{2}/2, -1, -1, -\sqrt{2}, \sqrt{2}$ **11.** $-\sqrt{2}/2, \sqrt{2}/2, -1, -1,$
$\sqrt{2}, -\sqrt{2}$ **13.** $1/2, \sqrt{3}/2, \sqrt{3}/3, \sqrt{3}, 2\sqrt{3}/3, 2$ **15.** $\sqrt{3}/2, -1/2, -\sqrt{3}, -\sqrt{3}/3, -2, 2\sqrt{3}/3$ **17.** $-1/2, -\sqrt{3}/2,$
$\sqrt{3}/3, \sqrt{3}, -2\sqrt{3}/3, -2$ **19.** $+, +$ **21.** $-, -, +, +, -, -$ **23.** $1/4, \sqrt{15}/4, 1/\sqrt{15}$ or $\sqrt{15}/15, \sqrt{15}, 4/\sqrt{15}$ or
$4\sqrt{15}/15, 4$ **25.** $2\sqrt{10}/7, -3/7, -2\sqrt{10}/3, -3/(2\sqrt{10})$ or $-3\sqrt{10}/20, -7/3, 7/(2\sqrt{10})$ or $7\sqrt{10}/20$ **27.** $-\sqrt{3}/2, 1/2,$

$-\sqrt{3}$, $-1/\sqrt{3}$ or $-\sqrt{3}/3$, 2, $-2/\sqrt{3}$ or $-2\sqrt{3}/3$ **29.** $\sqrt{6}/\sqrt{7}$ or $\sqrt{42}/7$, $-1/\sqrt{7}$ or $-\sqrt{7}/7$, $-\sqrt{6}$, $-1/\sqrt{6}$ or $-\sqrt{6}/6$, $-\sqrt{7}$, $\sqrt{7}/\sqrt{6}$ or $\sqrt{42}/6$ **31.** $a/\sqrt{a^2+b^2}$, $b/\sqrt{a^2+b^2}$, a/b, b/a, $\sqrt{a^2+b^2}/b$, $\sqrt{a^2+b^2}/a$ **33.** q, $-\sqrt{1-q^2}$, $-q/\sqrt{1-q^2}$, $-\sqrt{1-q^2}/q$, $-\sqrt{1-q^2}$, $-1/\sqrt{1-q^2}$, $1/q$ **35.** impossible   **37.** possible   **39.** possible   **41.** possible   **43.** impossible   **45.** II   **47.** III   **49.** II or III   **51.** III or IV   **53.** $\sin s = \sqrt{65}/9$, $\tan s = -\sqrt{65}/4$, $\cot s = -4\sqrt{65}/65$, $\sec s = -9/4$, $\csc s = 9\sqrt{65}/65$ **55.** $\sin s = 3\sqrt{13}/13$, $\cos s = 2\sqrt{13}/13$, $\cot s = 2/3$, $\sec s = \sqrt{13}/2$ **57.** $\cos t = -2\sqrt{5}/5$, $\tan t = -1/2$, $\cot t = -2$, $\sec t = -\sqrt{5}/2$, $\csc t = \sqrt{5}$ **59.** $\sin s = -\sqrt{21}/7$, $\cos s = -2\sqrt{7}/7$, $\cot s = 2\sqrt{3}/3$, $\csc s = -\sqrt{21}/3$ **61.** $\cos s = \sqrt{1-a^2}$, $\tan s = a\sqrt{1-a^2}/(1-a^2)$, $\cot s = \sqrt{1-a^2}/a$, $\sec s = \sqrt{1-a^2}/(1-a^2)$, $\csc s = 1/a$ **63.** $4\pi/5$ **65.** $\pi/4$ **67.** +   **69.** −   **75.** (a) $\cos s$ (b) $\sin s$ (c) $\csc s$ (d) $\tan s$ (e) s (f) $\pi/2 - s$ **77.** odd   **79.** odd

Section 6.3 (page 316)
1. 320°   **3.** 235°   **5.** 90°   **7.** 179°   **9.** $\pi/3$ **11.** $\pi/2$ **13.** $3\pi/4$ **15.** $5\pi/3$ **17.** $9\pi/4$ **19.** $7\pi/9$ **21.** 60°   **23.** 315°   **25.** 330°   **27.** −30°   **29.** 900°   **31.** 63°   **33.** 2.4289   **35.** 1.1254   **37.** −.5185   **39.** −3.6612   **41.** 114.5916°   **43.** 99.6947°   **45.** 5.2254°   **47.** $5\pi/6$ **49.** at $(-3/5, 4/5)$ **51.** 450°, $5\pi/2$ **53.** −900°, -5π **55.** 3/2   **57.** 1.23   **59.** 3.09309   **61.** 980 mi   **63.** 3700 mi   **65.** 25.1 in.   **67.** 25.8 cm   **69.** 5.05 m   **71.** (a) 11.6 in. (b) 37.1°   **73.** $5\pi/4$ radians   **75.** $\pi/25$ radians per sec.   **77.** 9 min   **79.** $72\pi/5$ cm per sec.   **81.** 18π cm   **83.** 12 sec.   **85.** $3\pi/32$ radians per sec.   **87.** (a) $\pi/3$ radians per sec.   (b) 1800π ft ≈ 5650 ft (c) 10π ≈ 31.4 ft per sec.   **89.** 400π ≈ 1260 in. per min; faster   **91.** 2π radians per min or $\pi/30$ radians per sec.   **93.** $\pi/18$ radians per sec.   **95.** (a) 2π radians per day; $\pi/12$ radians per hr (b) 0 (c) $12{,}800\pi$ km per day or 533π km per hr (d) 9050π km per day or 377π km per hr

Section 6.4 (page 327)
In Exercises 1–15, answers are given in the order sine, cosine, tangent, cotangent, secant, and cosecant.
1. $\sqrt{3}/2$, $-1/2$, $-\sqrt{3}$, $-\sqrt{3}/3$, -2, $2\sqrt{3}/3$ **3.** $1/2$, $-\sqrt{3}/2$, $-\sqrt{3}/3$, $-\sqrt{3}$, $-2\sqrt{3}/3$, 2   **5.** $-\sqrt{3}/2$, $-1/2$, $\sqrt{3}$, $\sqrt{3}/3$, -2, $-2\sqrt{3}/3$ **7.** $-1/2$, $\sqrt{3}/2$, $-\sqrt{3}/3$, $-\sqrt{3}$, $2\sqrt{3}/3$, -2 **9.** $\sqrt{3}/2$, $1/2$, $\sqrt{3}$, $\sqrt{3}/3$, 2, $2\sqrt{3}/3$ **11.** $1/2$, $-\sqrt{3}/2$, $-\sqrt{3}/3$, $-\sqrt{3}$, $-2\sqrt{3}/3$, 2   **13.** 0, -1, 0, undefined, -1, undefined   **15.** -1, 0, undefined, 0, undefined, -1 **17.** $\sqrt{3}/3$, $\sqrt{3}$ **19.** $\sqrt{3}/2$, $\sqrt{3}/3$, $2\sqrt{3}/3$ **21.** -1, -1 **23.** $-\sqrt{3}/2$, $-2\sqrt{3}/3$
In Exercises 25–39 answers are given in the order sine, cosine, tangent, cotangent, secant, and cosecant.
25. $4/5$, $-3/5$, $-4/3$, $-3/4$, $-5/3$, $5/4$ **27.** $7/25$, $24/25$, $7/24$, $24/7$, $25/24$, $25/7$ **29.** $-4/5$, $-3/5$, $4/3$, $3/4$, $-5/3$, $-5/4$ **31.** $-\sqrt{2}/2$, $\sqrt{2}/2$, -1, -1, $\sqrt{2}$, $-\sqrt{2}$ **33.** $-2/3$, $\sqrt{5}/3$, $-2\sqrt{5}/5$, $-\sqrt{5}/2$, $3\sqrt{5}/5$, $-3/2$ **35.** $\sqrt{3}/4$, $-\sqrt{13}/4$, $-\sqrt{39}/13$, $-\sqrt{39}/3$, $-4\sqrt{13}/13$, $4\sqrt{3}/3$ **37.** $-5/\sqrt{26}$ or $-5\sqrt{26}/26$, $-1/\sqrt{26}$ or $-\sqrt{26}/26$, 5, $1/5$, $-\sqrt{26}$, $-\sqrt{26}/5$ **39.** $3/5$, $4/5$, $3/4$, $4/3$, $5/4$, $5/3$ **41.** 1   **43.** $1/2 + \sqrt{3}$ **45.** $-29/12$ **47.** false   **49.** true   **51.** false   **53.** true   **55.** true   **57.** 30°, 150°   **59.** 60°, 240°   **61.** 120°, 240°   **63.** 0°, 180°   **65.** .6338   **67.** −.6266   **69.** −.5711   **71.** −3.0716   **73.** .9636   **75.** −.6361   **77.** .70710678; the calculator gives an approximate answer.   **79.** $\tan 50°$ **81.** $\tan 1$ **83.** $\cos 5°$ **85.** yes   **87.** −   **89.** −   **91.** $a = 12$, $b = 12\sqrt{3}$, $d = 12\sqrt{3}$, $c = 12\sqrt{6}$ **93.** $m = 7\sqrt{3}/3$, $a = 14\sqrt{3}/3$, $n = 14\sqrt{3}/3$, $q = 14\sqrt{6}/3$ **95.** 2×10^8 m per sec.   **97.** 19°   **99.** 48.7°

Section 6.5 (page 338)
1. 2

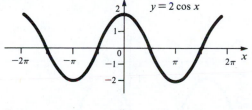
$y = 2\cos x$

3. 2/3

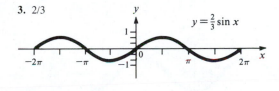
$y = \frac{2}{3}\sin x$

5. 1

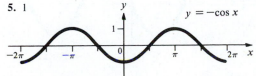
$y = -\cos x$

7. 2

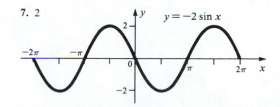
$y = -2\sin x$

9. 4π, 1, none

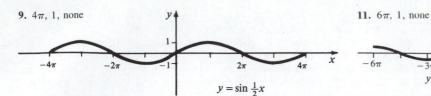

$y = \sin \frac{1}{2}x$

11. 6π, 1, none

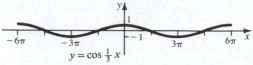

$y = \cos \frac{1}{3}x$

13. $2\pi/3$, 1, none

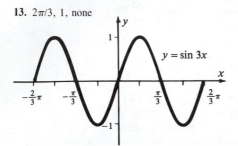

$y = \sin 3x$

15. $\pi/2$, 1, none

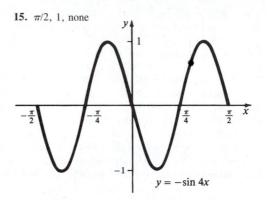

$y = -\sin 4x$

17. 8π, 2, none

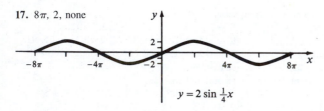

$y = 2 \sin \frac{1}{4}x$

19. $2\pi/3$, 2, none

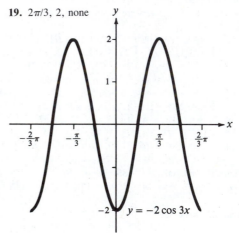

$y = -2 \cos 3x$

21. $2\pi/3$, 1/2, none

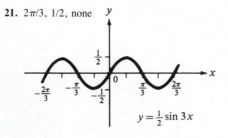

$y = \frac{1}{2} \sin 3x$

23. 2π, 1, 2

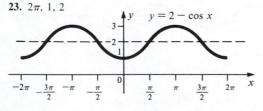

$y = 2 - \cos x$

25. 2π, 2, -3

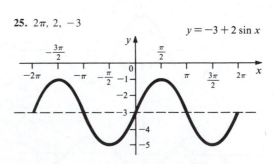

$y = -3 + 2 \sin x$

27. 2, 1, none

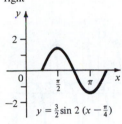

$y = \cos \pi x$

29. 1, 2π, none, $\pi/2$ to the right

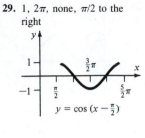

$y = \cos(x - \frac{\pi}{2})$

31. 3, 2π, none, $3\pi/2$ to the left

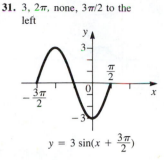

$y = 3 \sin(x + \frac{3\pi}{2})$

33. 3/2, π, none, $\pi/4$ to the right

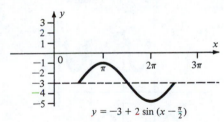

$y = \frac{3}{2} \sin 2\,(x - \frac{\pi}{4})$

35. 3, $\pi/2$, none, $\pi/4$ to the left

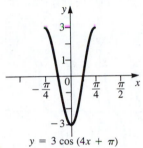

$y = 3 \cos(4x + \pi)$

37. 1/2, 4π, none, $\pi/2$ to the right

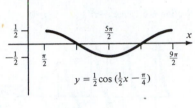

$y = \frac{1}{2} \cos(\frac{1}{2}x - \frac{\pi}{4})$

39. 2, 2π, 3 down, $\pi/2$ to the right

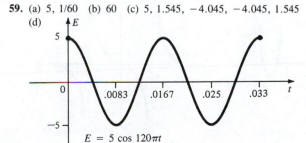

$y = -3 + 2 \sin(x - \frac{\pi}{2})$

41. $y = 4 \sin (1/2)x$ **43.** $y = -\pi \cos \pi x$ **45.** 9   **47.** odd

49.

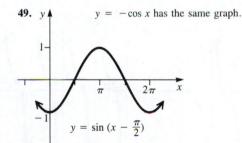

$y = -\cos x$ has the same graph.

$y = \sin(x - \frac{\pi}{2})$

51. 1, $4\pi/3$ **53.** about 35,000 years   **55.** almost 2 hours   **57.** 20

59. (a) 5, 1/60   (b) 60   (c) 5, 1.545, -4.045, -4.045, 1.545
(d)

$E = 5 \cos 120\pi t$

65.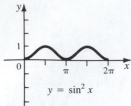

$$y = \sin^2 x$$

67.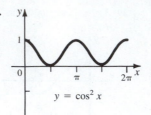

$$y = \cos^2 x$$

69.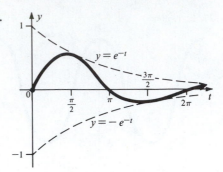

71. $(f \circ g)(x)$ where $f(x) = 5 \cos x$ and $g(x) = 3x - \pi$ **73.** $(f \circ g)(x)$ where $f(x) = \sin^2 x$ and $g(x) = 2x + 3$

Section 6.6 (page 349)

1.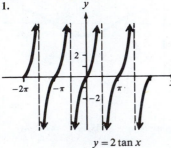

$$y = 2 \tan x$$

3.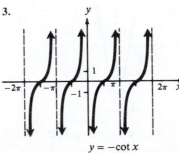

$$y = -\cot x$$

5.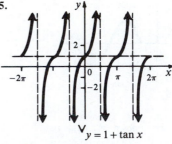

$$y = 1 + \tan x$$

7.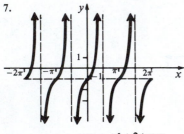

$$y = -1 + 2 \tan x$$

9.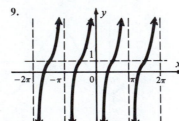

$$y = 1 - \cot x$$

11.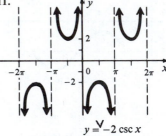

$$y = -2 \csc x$$

13.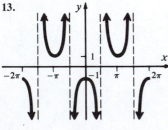

$$y = -\sec x$$

15.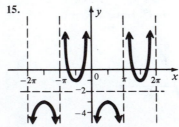

$$y = -2 - \csc x$$

17. $\pi/3$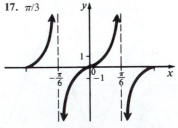

$$y = \tan 3x$$

19. $\pi/3$

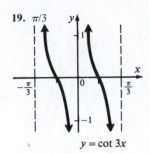

$y = \cot 3x$

21. $\pi/2$

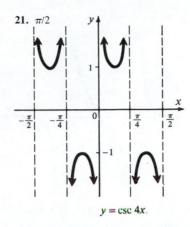

$y = \csc 4x$.

23. 4π

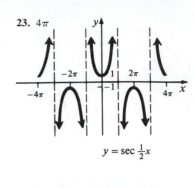

$y = \sec \frac{1}{2}x$

25. π

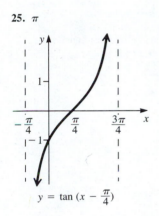

$y = \tan\left(x - \frac{\pi}{4}\right)$

27. 2π

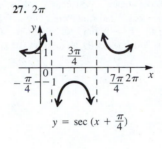

$y = \sec\left(x + \frac{\pi}{4}\right)$

29. 2π

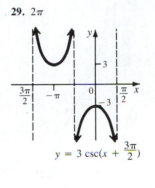

$y = 3\csc\left(x + \frac{3\pi}{2}\right)$

31. π

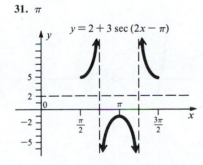

$y = 2 + 3\sec(2x - \pi)$

33. 2π

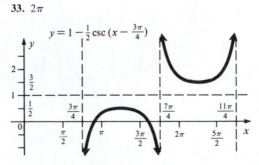

$y = 1 - \frac{1}{2}\csc\left(x - \frac{3\pi}{4}\right)$

35. $4\pi/3$

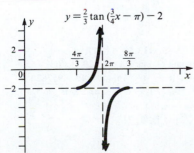

$y = \frac{2}{3}\tan\left(\frac{3}{4}x - \pi\right) - 2$

37.

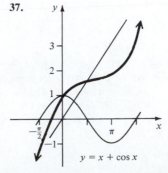

$y = x + \cos x$

39.

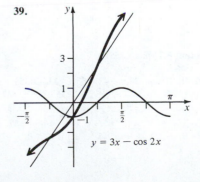

$y = 3x - \cos 2x$

41.

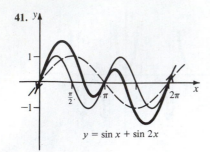

$y = \sin x + \sin 2x$

43.

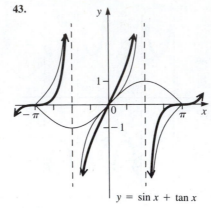

$y = \sin x + \tan x$

45.

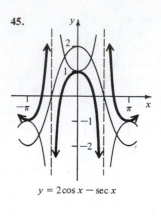

$y = 2\cos x - \sec x$

47.

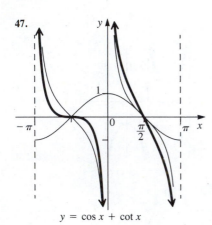

$y = \cos x + \cot x$

49. $\tan x$ and $\cot x$ **51.** $\cos x$ and $\sec x$ **53.** $b = 5$
55. $\sin x$, $\cos x$, $\tan x$, and $\cot x$
57. (a) 0   (b) -2.9 m   (c) -12.3 m   (d) 12.3 m   (e) It leads to $\tan \pi/2$, which is undefined.   (f) all t except $.25\,k$, where k is an odd integer

59.

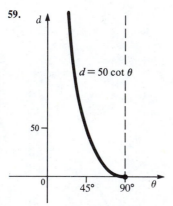

$d = 50 \cot \theta$

61.

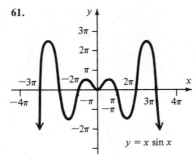

$y = x \sin x$

63.

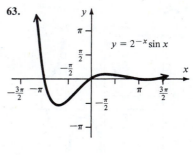

$y = 2^{-x} \sin x$

Section 6.7 (page 355)
1. $-\pi/3$ **3.** $\pi/4$ **5.** $-\pi/2$ **7.** $\pi/3$ **9.** $3\pi/4$ **11.** $-\pi/3$ **13.** $-7.67°$ **15.** $113.50°$ **17.** $22.00°$
19. $48.01°$ **21.** $-42.67°$ **23.** $.8058$ **25.** -1.332 **27.** 1.003 is not in the domain of $\mathrm{Sin}^{-1} x$.   **29.** $\sqrt{7}/3$
31. $\sqrt{5}/5$ **33.** $-\sqrt{5}/2$ **35.** $1/2$ **37.** -1 **39.** 2 **41.** $\pi/4$ **43.** 1 **45.** 0 **47.** $3\pi/4$ **49.** $.957826$
51. $.123430$ **53.** $x = \sin(y/4)$ **55.** $x = (1/2)\tan 2y$ **57.** $x = -2 + \sin y$ **59.** $\sqrt{1 - u^2}$ **61.** $\sqrt{u^2 + 1}/u$
63. $\sqrt{1 - u^2}/u$ **65.** $\sqrt{u^2 - 4}/u$ **67.** $u/\sqrt{2}$ or $\sqrt{2}u/2$ **69.** $2/\sqrt{4 - u^2}$ or $2\sqrt{4 - u^2}/(4 - u^2)$ if
$u > 0$, $-2/\sqrt{4 - u^2}/(4 - u^2)$ if $u < 0$ **71.** (a) $[0, \pi]$ (b) $[-1, 1]$ **73.** $[-2, -\sqrt{2}] \cup [\sqrt{2}, 2]$

75. $(-\infty, +\infty)$, $(0, \pi)$

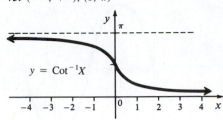

$y = \text{Cot}^{-1}X$

77. $(-\infty, -1/2] \cup [1/2, +\infty)$, $[0, \pi/2) \cup (\pi/2, \pi]$

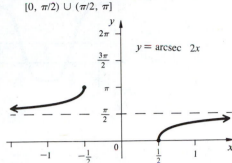

$y = \text{arcsec } 2x$

79. $[-1, 1]$, $[0, 2\pi]$

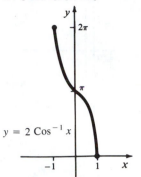

$y = 2 \text{ Cos}^{-1} x$

81. The calculator returns inverse sine values in the correct range, quadrants I or IV, only **83.** false **85.** true **87.** false **89.** false **91.** (a) 18° (b) 18° (c) 15°

Chapter 6 Review Exercises (page 357)
1. $(\sqrt{2}/2, -\sqrt{2}/2)$ **3.** $(-1, 0)$ **5.** $-\sqrt{2}/2, -\sqrt{2}/2, 1$ **7.** $-\sqrt{3}/2, 1/2, \sqrt{3}$ **9.** $(-2/3, -\sqrt{5}/3)$
11. $(2/3, \sqrt{5}/3)$ **13.** $\pi - .5, 2\pi - .5$ **15.** II **17.** III **19.** $\cos s = \sqrt{5}/3$, $\tan s = 2\sqrt{5}/5$, $\cot s = \sqrt{5}/2$,
$\sec s = 3\sqrt{5}/5$, $\csc s = 3/2$ **21.** $\sin s = -2\sqrt{5}/5$, $\cos s = \sqrt{5}/5$, $\tan s = -2$, $\csc s = -\sqrt{5}/2$ **23.** 1280° **25.** 135°
27. 1116° **29.** $3\pi/2$ **31.** $17\pi/3$ **33.** $\sqrt{3}/2$ **35.** $\sqrt{3}$ **37.** -1 **39.** Sine is $-\sqrt{2}/2$, cosine is $-\sqrt{2}/2$,
tangent is 1. **41.** 1.428 **43.** $-.9212$ **45.** -2.1176 **47.** 55.67° **49.** 12.74° **51.** .3898 **53.** .5148
55. $a = 638$, $b = 391$ **57.** $a = 32.38$ m, $c = 50.66$ m **59.** 15/32 sec. **61.** $\pi/20$ radians per sec. **63.** 285.3 cm
65. 1/2, $2\pi/3$, none, none **67.** 2, 8π, 1, none **69.** 3, 2π, none, $\pi/2$ to the left **71.** none, π, none, $\pi/8$ to the right

73.

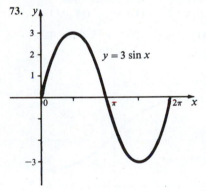

$y = 3 \sin x$

75.

$y = -\tan x$

77.

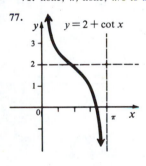

$y = 2 + \cot x$

79.

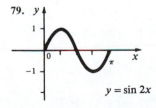

$y = \sin 2x$

81.

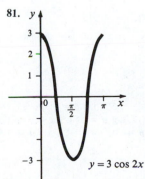

$y = 3 \cos 2x$

83.

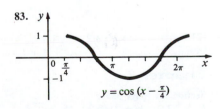

$y = \cos \left(x - \frac{\pi}{4}\right)$

85.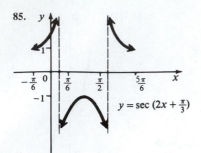

$y = \sec(2x + \frac{\pi}{3})$

87.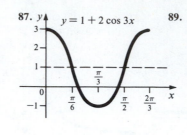

$y = 1 + 2\cos 3x$

89.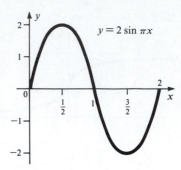

$y = 2\sin \pi x$

91.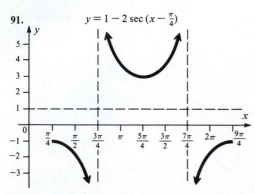

$y = 1 - 2\sec(x - \frac{\pi}{4})$

93. 3 radians

95. Since the period is π, $f(x + n\pi) = f(x)$ for integer values of n. Here $f(6\pi/5) = f(6\pi/5 + (-2)\pi) = f(-4\pi/5)$.

97.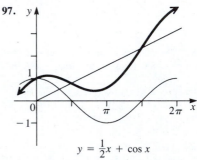

$y = \frac{1}{2}x + \cos x$

99.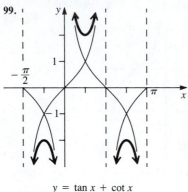

$y = \tan x + \cot x$

101. (a) 100   (b) 258   (c) 122   (d) 296   **103.** $2\pi/3$ **105.** 1.0589
107. $.6370$ **109.** $2/3$ **111.** $-\sqrt{3}/2$ **113.** π **115.** $\sqrt{1 - u^2}$
117. domain: $[-1, 1]$; range: $[-\pi/2, \pi/2]$

119. domain: $(-\infty, +\infty)$; range: $(0, \pi)$

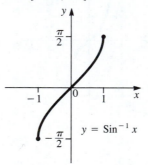

$y = \text{Sin}^{-1} x$

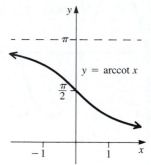

$y = \text{arccot } x$

121. (a) $x = \sin u$, $-\pi/2 \leq u \leq \pi/2$ (c) $\tan u = x/\sqrt{1 - x^2}$ (d) $u = \arctan(x/\sqrt{1 - x^2})$ (e) From part (d), $\arcsin 1/2 = \arcsin .5 = \arctan(.5/\sqrt{1 - .5^2}) = \arctan(1/\sqrt{3}) = \pi/6$.

Chapter 7

Section 7.1 (page 366)
1. $\sqrt{7}/4$ **3.** $-2\sqrt{5}/5$ **5.** $\sqrt{21}/2$ **7.** $\cos \theta = -\sqrt{5}/3$, $\tan \theta = -2\sqrt{5}/5$, $\cot \theta = -\sqrt{5}/2$, $\sec \theta = -3\sqrt{5}/5$, $\csc \theta = 3/2$ **9.** $\sin \theta = -\sqrt{17}/17$, $\cos \theta = 4\sqrt{17}/17$, $\cot \theta = -4$, $\sec \theta = \sqrt{17}/4$, $\csc \theta = -\sqrt{17}$ **11.** $\sin \theta = 2\sqrt{2}/3$, $\cos \theta = -1/3$, $\tan \theta = -2\sqrt{2}$, $\cot \theta = -\sqrt{2}/4$, $\csc \theta = 3\sqrt{2}/4$ **13.** $\sin \theta = 3/5$, $\cos \theta = 4/5$, $\tan \theta = 3/4$, $\sec \theta = 5/4$,

csc θ = 5/3   **15.** sin θ = $-\sqrt{7}/4$, cos θ = 3/4, tan θ = $-\sqrt{7}/3$, cot θ = $-3\sqrt{7}/7$, csc θ = $-4\sqrt{7}/7$ **17.** (b)   **19.** (c)
21. (a)   **23.** (a)   **25.** (d)   **27.** 1   **29.** $-\sin\alpha$ **31.** 0   **33.** $(1 + \sin\theta)/\cos\theta$ **35.** 1   **37.** -1
39. $\sin^2\theta/\cos^4\theta$ **41.** 1   **43.** $(\cos^2\alpha + 1)/(\sin^2\alpha\cos^2\alpha)$ **45.** $(\sin^2 s - \cos^2 s)/\sin^4 s$
47. $\pm\sqrt{1 + \cot^2\theta}/(1 + \cot^2\theta)$, $\pm\sqrt{\sec^2\theta - 1}/\sec\theta$ **49.** $\pm\sin\theta\sqrt{1 - \sin^2\theta}/(1 - \sin^2\theta)$, $\pm\sqrt{1 - \cos^2\theta}/\cos\theta$,
$\pm\sqrt{\sec^2\theta - 1}$, $\pm\sqrt{\csc^2\theta - 1}/(\csc^2\theta - 1)$ **51.** $\pm\sqrt{1 - \sin^2\theta}/(1 - \sin^2\theta)$, $\pm\sqrt{\tan^2\theta + 1}$, $\pm\sqrt{1 + \cot^2\theta}/\cot\theta$,
$\pm\csc\theta\sqrt{\csc^2\theta - 1}/(\csc^2\theta - 1)$ **53.** sin θ = $\pm\sqrt{2x + 1}/(x + 1)$ **61.** 4 sec θ, $3x/\sqrt{16 + 9x^2}$, $4/\sqrt{16 + 9x^2}$
63. $\sin^3\theta$, $\sqrt{1 - x^2}$, $\sqrt{1 - x^2}/x$ **65.** $(\tan^2\theta \sec\theta)/16$, $4x/\sqrt{1 + 16x^2}$, $1/\sqrt{1 + 16x^2}$ **67.** $(25\sqrt{6} - 60)/12$,
$-(25\sqrt{6} + 60)/12$

Section 7.2 (page 372)
1. $1/(\sin\theta\cos\theta)$ **3.** $1 + \cos s$ **5.** 1   **7.** 1   **9.** $2 + 2\sin t$ **11.** $(\sin\gamma + 1)(\sin\gamma - 1)$ **13.** 4 sin x
15. $(2\sin x + 1)(\sin x + 1)$ **17.** $(4\sec x - 1)(\sec x + 1)$ **19.** $(\cos^2 x + 1)^2$ **21.** sin θ **23.** 1   **25.** $\tan^2\beta$
27. $\tan^2 x$ **29.** $\sec^2 x$ **65.** identity   **67.** not an identity   **69.** not an identity

Section 7.3 (page 381)
1. cot 3°   **3.** sin 5π/12   **5.** sec 104.4°   **7.** cos $(-\pi/8)$ **9.** csc (-10.43) **11.** tan $(-86.9814°)$ **13.** 15°
15. (140/3)°   **17.** 20°   **19.** $(\sqrt{6} - \sqrt{2})/4$ **21.** $(-\sqrt{2} - \sqrt{6})/4$ **23.** $-2 + \sqrt{3}$ **25.** 0   **27.** $-\sqrt{2}/2$
29. -1 **31.** 0   **33.** $(\sqrt{3}/2)\cos\theta - (1/2)\sin\theta$ **35.** $(\sqrt{3}/2)\cos\theta + (1/2)\sin\theta$ **37.** cot x **39.** $(4 - 6\sqrt{6})/25$,
$(3 - 8\sqrt{6})/25$, $(6 + \sqrt{6})/4$ **41.** 16/65, 33/65, $-63/16$ **43.** $-77/85$, 84/85, 36/77   **45.** $(2\sqrt{638} - \sqrt{30})/56$,
$(\sqrt{290} - 2\sqrt{66})/56$, $(64\sqrt{55} + 49\sqrt{87})/1261$ **57.** You get an error message because the expression equals tan 90° which is
undefined.   **63.** 18.4°   **65.** 36/85   **67.** 63/65   **69.** $s(t) = (3\sqrt{2}/2)\cos t - (3\sqrt{2}/2)\sin t$

Section 7.4 (page 388)
1. sin 20°   **3.** tan 73.5°   **5.** tan 29.87°   **7.** cos 9x **9.** tan 4θ **11.** cos x/8   **13.** cos 30° or $\sqrt{3}/2$ **15.** sin 2π/3
or $\sqrt{3}/2$ **17.** $(1/2)\sin\pi/4$ or $\sqrt{2}/4$ **19.** $(1/2)\tan 102°$ **21.** $(1/16)\sin 59°$ **23.** cos 4α **25.** $-$ **27.** $+$
29. sin 22.5° = $\sqrt{2 - \sqrt{2}}/2$, cos 22.5° = $\sqrt{2 + \sqrt{2}}/2$, tan 22.5° = $\sqrt{3 - 2\sqrt{2}}$ or $\sqrt{2} - 1$ **31.** sin 195° =
$-\sqrt{2 - \sqrt{3}}/2$, cos 195° = $-\sqrt{2 + \sqrt{3}}/2$, tan 195° = $\sqrt{7 - 4\sqrt{3}}$ or $2 - \sqrt{3}$ **33.** sin 5π/2 = 1, cos 5π/2 = 0, tan 5π/2
is undefined   **35.** $k = 4, t = 6$ **37.** $k = 6, t = 2$ **39.** $-\sqrt{42}/12$ **41.** $-\sqrt{6}/4$ **43.** cos x = $-\sqrt{42}/12$, sin x =
$\sqrt{102}/12$, tan x = $-\sqrt{119}/7$, sec x = $-2\sqrt{42}/7$, csc x = $2\sqrt{102}/17$, cot x = $-\sqrt{119}/17$ **45.** cos 2θ = 17/25, sin 2θ =
$-4\sqrt{21}/25$, tan 2θ = $-4\sqrt{21}/17$, sec 2θ = 25/17, csc 2θ = $-25\sqrt{21}/84$, cot 2θ = $-17\sqrt{21}/84$ **47.** tan 2x = $-4/3$,
sec 2x = $-5/3$, cos 2x = $-3/5$, cot 2x = $-3/4$, sin 2x = 4/5, csc 2x = 5/4   **49.** sin α/2 = $\sqrt{3}/3$, cos α/2 = $-\sqrt{6}/3$,
tan α/2 = $-\sqrt{2}/2$, cot α/2 = $-\sqrt{2}$, sec α/2 = $-\sqrt{6}/2$, csc α/2 = $\sqrt{3}$ **69.** 4 tan^2 $x/(1 - 2\tan^2 x + \tan^4 x)$
71. 4 cos^3 $x - 3\cos x$ **73.** 84°   **75.** 60°   **77.** 3.9   **79.** $-.843580$ **81.** .537003   **83.** -1.570905
85. (a) $\sqrt{3 - 2\sqrt{2}}$ **87.** 1   **89.** $\sec^2 4x$

Section 7.5 (page 396)
1. $(1/2)[\sin 60° - \sin 10°] = (1/2)[\sqrt{3}/2 - \sin 10°]$ **3.** $(3/2)[\cos 8x + \cos 2x]$ **5.** $(1/2)[\cos 2\theta - \cos(-4\theta)] =$
$(1/2)[\cos 2\theta - \cos 4\theta]$ **7.** $-4[\cos 9y + \cos(-y)] = -4(\cos 9y + \cos y)$ **9.** 2 cos 45° sin 15°
11. 2 cos 95° cos $(-53°) = 2\cos 95°\cos 53°$ **13.** 2 cos $(15\beta/2)\sin(9\beta/2)$ **15.** $-6\cos(7x/2)\sin(-3x/2) =$
6 cos $(7x/2)\sin(3x/2)$ **29.** $\sqrt{2}\sin(x + 135°)$ **31.** 13 sin $(\theta + 293°)$ **33.** 17 sin $(x + 152°)$ **35.** 25 sin $(\theta + 254°)$
37. 5 sin $(x + 53°)$
39. $y = 2\sin(x + \pi/6)$ **41.** $y = \sqrt{2}\sin(x + 3\pi/4)$

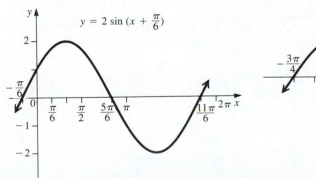

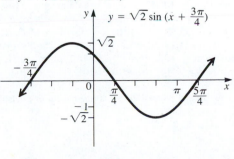

Section 7.6 (page 406)

1. $3\pi/4$, $7\pi/4$ **3.** $\pi/3$, $5\pi/3$ **5.** $\pi/6$, $7\pi/6$, $4\pi/3$, $5\pi/3$ **7.** $\pi/3$, $5\pi/3$, $\pi/6$, $11\pi/6$ **9.** π **11.** $3\pi/2$, $7\pi/6$, $11\pi/6$
13. $\pi/4$, $3\pi/4$, $5\pi/4$, $7\pi/4$ **15.** 0, $\pi/2$, π, $3\pi/2$ **17.** $\pi/12$, $7\pi/12$, $13\pi/12$, $19\pi/12$ **19.** $3\pi/8$, $5\pi/8$, $11\pi/8$, $13\pi/8$
21. $\pi/2$, $3\pi/2$ **23.** $71.6°$, $251.6°$, $63.4°$, $243.4°$ **25.** $135°$, $315°$, $71.6°$, $251.6°$ **27.** $33.6°$, $326.4°$ **29.** $45°$, $225°$
31. $0°$, $90°$, $180°$, $270°$ **33.** $30°$, $60°$, $210°$, $240°$ **35.** $120°$, $240°$ **37.** $270°$, $30°$, $150°$ **39.** $90°$, $270°$, $45°$, $225°$
41. $70.5°$, $289.5°$ **43.** $90°$, $270°$, $30°$, $150°$, $210°$, $330°$ **45.** $\{x \mid x = 90° + 360° \cdot n, \ 180° + 360° \cdot n, \ n \text{ any integer}\}$
47. $\{x \mid x = 30° + 360° \cdot n, \ 150° + 360° \cdot n, \ 90° + 360° \cdot n, \ 270° + 360° \cdot n, \ n \text{ any integer}\}$ **49.** $\{x \mid x = 45° + 180° \cdot n$,
$108.4° + 180° \cdot n, \ n \text{ any integer}\}$ **51.** $\{x \mid x = 11.3° + 45° \cdot n, \ n \text{ any integer}\}$ **53.** $53.6°$, $126.4°$, $187.9°$,
$352.1°$ **55.** $149.6°$, $329.6°$, $106.3°$, $286.3°$ **57.** $57.7°$, $159.2°$ **59.** $\pi/12$, $\pi/2$, $5\pi/12$, $3\pi/2$, $13\pi/12$, $17\pi/12$
61. 0, $\pi/4$, $3\pi/4$, π, $5\pi/4$, $7\pi/4$ **63.** $\pi/6$, $\pi/2$, $3\pi/2$, $5\pi/6$ **65.** $1/4$ sec **67.** (a) 5000 ft (b) 1250 ft **69.** 2 sec
71. $14°$ **73.** (a) $42.2°$ (b) $90°$ (c) $48.0°$ **75.** $3/5$ **77.** $4/5$ **79.** $1/2$ **81.** $-1/2$ **83.** 0

Chapter 7 Review Exercises (page 409)

1. $\sin x = -4/5$, $\tan x = -4/3$, $\sec x = 5/3$, $\csc x = -5/4$, $\cot x = -3/4$ **3.** $\sin(x + y) = (4 + 3\sqrt{15})/20$,
$\cos(x - y) = (4\sqrt{15} + 3)/20$ **5.** $\sin x = \sqrt{2 - \sqrt{2}}/2$, $\cos x = \sqrt{2 + \sqrt{2}}/2$, $\tan x = \sqrt{3 - 2\sqrt{2}}$ or $\sqrt{2} - 1$ **7.** (j)
9. (c) **11.** (d) **13.** (a) **15.** (f) **17.** (e) **19.** 1 **21.** $1/\cos^2 \theta$ **23.** $1/\sin^2 \theta \cos^2 \theta$ **53.** $\pi/2$, $3\pi/2$
55. $\pi/6$, $5\pi/6$ **57.** $\pi/8$, $3\pi/8$, $5\pi/8$, $7\pi/8$, $9\pi/8$, $11\pi/8$, $13\pi/8$, $15\pi/8$ **59.** $\pi/2$ **61.** $\pi/3$, $5\pi/3$, π **63.** $60°$, $300°$
65. $270°$ **67.** $0°$, $45°$, $180°$, $225°$ **69.** (a) $48.8°$ (b) The light beam is completely under water. **79.** $2 - \sqrt{3}$

Chapter 8

Section 8.1 (page 419)

1. $A = 17.0°$, $a = 39.1$ in., $c = 134$ in. **3.** $c = 85.9$ yd, $A = 62.8°$, $B = 27.2°$ **5.** $b = 42.3$ cm, $A = 24.1°$,
$B = 65.9°$ **7.** $B = 36.76°$, $a = 310.1$ ft, $b = 231.7$ ft **9.** $A = 50.90°$, $a = .4836$ m, $b = .3930$ m **11.** $A = 71.59°$,
$B = 18.41°$, $a = 7.413$ m **13.** $A = 47.568°$, $b = 143.97$ m, $c = 213.38$ m **15.** $B = 32.791°$, $a = 156.77$ cm,
$b = 101.00$ cm **17.** $a = 115.072$ m, $A = 33.4901°$, $B = 56.5099°$ **19.** 26.6 m **21.** 59.8 m **23.** 11 ft **25.** 35.8°
27. $26.3°$ **29.** 1590 ft **31.** $30°$ **33.** 51.6 m **35.** 8200 ft **37.** 433 ft **39.** 113 ft **41.** 5.12 m
43. $h = k(\tan B - \tan A)$ **45.** 156 mi **47.** 120 mi **49.** $38°$ **51.** $54.7°$ **53.** 6.993752×10^9 mi **55.** $\sqrt{15}/4$
57. $\sqrt{5}/3$

Section 8.2 (page 432)

In this chapter, the value of each missing angle has been calculated individually using the given information. Because of rounding, in
some cases the sum of the three angles may not be exactly $180°$.
1. $C = 80.7°$, $a = 79.5$ mm, $c = 108$ mm **3.** $B = 37.3°$, $a = 38.5$ ft, $b = 51.0$ ft **5.** $C = 57.36°$, $b = 11.13$ ft,
$c = 11.55$ ft **7.** $B = 18.5°$, $a = 239$ yd, $c = 230$ yd **9.** $B = 110.0°$, $a = 27.01$ m, $c = 21.36$ m **11.** $A = 34.72°$,
$a = 3326$ ft, $c = 5704$ ft **13.** $B_1 = 49.1°$, $C_1 = 101.2°$, $B_2 = 130.9°$, $C_2 = 19.4°$ **15.** $A = 112.3°$, $B = 26.4°$
17. no such triangle **19.** $B = 27.19°$, $C = 10.68°$ **21.** $B = 20.6°$, $C = 116.9°$, $c = 20.6$ ft **23.** no such triangle
25. $B_1 = 49.4°$, $C_1 = 91.9°$, $c_1 = 15.5$ km; $B_2 = 130.6°$, $C_2 = 10.7°$, $c_2 = 2.89$ km **27.** $A_1 = 53.23°$, $C_1 = 87.09°$, $c_1 = $
37.16 m; $A_2 = 126.77°$, $C_2 = 13.55°$, $c_2 = 8.721$ m **29.** (a) $c = 2$ or $c \geq 4$ (b) $2 < c < 4$ (c) $c < 2$ **31.** 117 m
33. 1.95 mi **35.** 10.4 in. **37.** $111°$ **39.** does not exist **41.** 2 mi **43.** 46.4 m^2 **45.** 356 cm^2 **47.** 722.9 in^2
49. 1071 sq cm **51.** 100

Section 8.3 (page 440)

1. $c = 2.83$ in., $A = 44.9°$, $B = 106.8°$ **3.** $c = 6.46$ m, $A = 53.1°$, $B = 81.3°$ **5.** $b = 9.529$ in., $A = 64.59°$, $C = $
$40.61°$ **7.** $a = 15.7$ m, $B = 21.6°$, $C = 45.6°$ **9.** $A = 29.9°$, $B = 56.3°$, $C = 93.8°$ **11.** $A = 81.4°$, $B = 37.5°$, $C = $
$61.1°$ **13.** $A = 42.0°$, $B = 35.9°$, $C = 102.1°$ **15.** $A = 47.7°$, $B = 44.9°$, $C = 87.4°$ **17.** $A = 35.37°$, $B = 50.97°$,
$C = 93.66°$ **19.** 257 m **21.** 277 km **23.** 22 ft **25.** 2000 km **27.** 1470 m **29.** 5.99 km **31.** 25.24983 mi
33. 140 in^2 **35.** 12,600 cm^2 **37.** 3650 ft^2 **39.** 1921 ft^2 **41.** 33 cans

Section 8.4 (page 454)

1. **m** and **p**, **n** and **r** **3.** **m** and **p** equal $2\mathbf{t}$ or **t** is one-half **m** or **p**; also **m** = $1\mathbf{p}$ and **n** = $1\mathbf{r}$

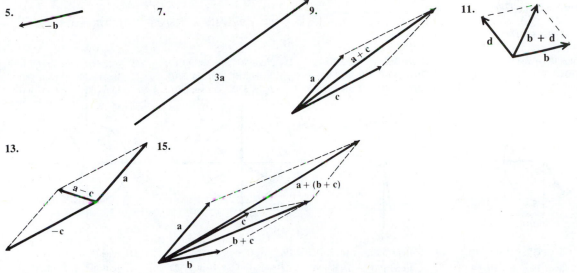

5. −b   **7.** 3a   **9.** a + c, a, c   **11.** d, b + d, b

13. a − c, a, −c   **15.** a + (b + c), a, c, b + c, b

17. <1,3>   **19.** <22, −22>   **21.** $10\sqrt{2}, 10\sqrt{2}$ **23.** 14.3, 24.8   **25.** −122, 156   **27.** $\sqrt{2}, 45°$ **29.** 16, 315°
31. 17, 332°   **33.** 6, 180°   **35.** −5i + 8j   **37.** 2i   **39.** $4\sqrt{2}i + 4\sqrt{2}j$ **41.** −.2536i + .5438j   **43.** a = −b
45. a and b must be perpendicular   **47.** 530 newtons   **49.** 27.3 lb   **51.** 84.9 lb   **53.** 94°   **55.** 17 lb
57. 18°   **59.** magnitude 2.84; angle of 55.4° with the 4.72 vector   **61.** weight: 64.8 lb; tension: 61.9 lb   **63.** 189 lb,
282 lb respectively   **65.** 172.8°   **67.** 39.2 km   **69.** 237°, 470 mph   **71.** 358°, 170 mph   **73.** The ship traveled 55.9
mi on its modified course, for additional travel of 55.9 − 50 = 5.9 mi.   **81.** −22   **83.** −50   **87.** 151°

Section 8.5 (page 464)

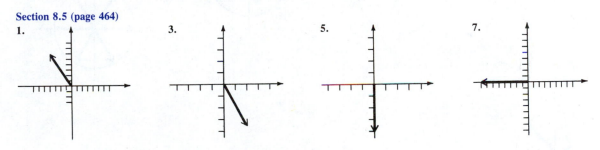

1.   **3.**   **5.**   **7.**

9. 1 + i   **11.** −2 + 2i   **13.** −2 + 4i   **15.** 2 + 4i   **17.** $\sqrt{2} + i\sqrt{2}$ **19.** 0 + 10i   **21.** $−2 − 2i\sqrt{3}$
23. $5/2 − 5i\sqrt{3}/2$ **25.** $−\sqrt{2} + 0i$ **27.** $\sqrt{13}(\cos 56.3° + i \sin 56.3°)$ **29.** −1.0162 − 2.8226i
31. 2(cos 160.0° + i sin 160.0°)   **33.** $\sqrt{34}(\cos 59.0° + i \sin 59.0°)$ **35.** $3\sqrt{2}(\cos 315° + i \sin 315°)$
37. 6(cos 240° + i sin 240°)   **39.** 2(cos 330° + i sin 330°)   **41.** $2\sqrt{2}(\cos 45° + i \sin 45°)$ **43.** 4(cos 180° + i sin 180°)
45. $−3\sqrt{3} + 3i$ **47.** 0 − 4i   **49.** $12\sqrt{3} + 12i$ **51.** $−15\sqrt{2}/2 + 15i\sqrt{2}/2$ **53.** 0 − 3i   **55.** $\sqrt{3} − i$
57. $−1 − i\sqrt{3}$ **59.** $−1/6 − i\sqrt{3}/6$ **61.** $2\sqrt{3} − 2i$ **63.** $\sqrt{3} + i$ **65.** 20 + 15i   **67.** 1.2 − .14i
69.   **71.**   **73.**   **75.** (0, 0), r = 5

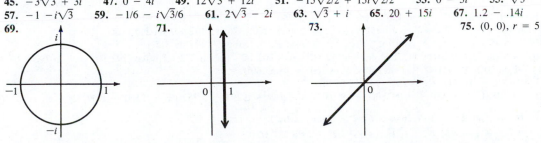

77. The terminal points of the vectors corresponding to $a + bi$ and $c + di$ must lie on a horizontal line.   **79.** The corresponding
vectors have the same direction.

Section 8.6 (page 471)

1. $0 + 27i$ **3.** $1 + 0i$ **5.** $27/2 - 27i\sqrt{3}/2$ **7.** $-16\sqrt{3} + 16i$ **9.** $-128 + 128i\sqrt{3}$ **11.** $128 + 128i$
13. $-.1892 + .0745i$ **15.** $5520 + 9550i$

17. $(\cos 0° + i \sin 0°)$,
$(\cos 120° + i \sin 120°)$,
$(\cos 240° + i \sin 240°)$

19. $2(\cos 20° + i \sin 20°)$,
$2(\cos 140° + i \sin 140°)$,
$2(\cos 260° + i \sin 260°)$

21. $2(\cos 90° + i \sin 90°)$
$2(\cos 210° + i \sin 210°)$,
$2(\cos 330° + i \sin 330°)$

23. $4(\cos 60° + i \sin 60°)$,
$4(\cos 180° + i \sin 180°)$,
$4(\cos 300° + i \sin 300°)$

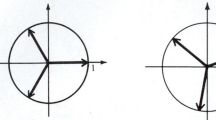
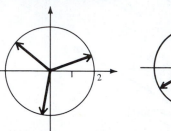
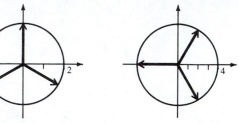

25. $\sqrt[3]{2}(\cos 20° + i \sin 20°)$,
$\sqrt[3]{2}(\cos 140° + i \sin 140°)$,
$\sqrt[3]{2}(\cos 260° + i \sin 260°)$

27. $\sqrt[3]{4}(\cos 50° + i \sin 50°)$,
$\sqrt[3]{4}(\cos 170° + i \sin 170°)$,
$\sqrt[3]{4}(\cos 290° + i \sin 290°)$

29. $(\cos 0° + i \sin 0°)$,
$(\cos 180° + i \sin 180°)$

31. $(\cos 0° + i \sin 0°)$,
$(\cos 60° + i \sin 60°)$,
$(\cos 120° + i \sin 120°)$,
$(\cos 180° + i \sin 180°)$,
$(\cos 240° + i \sin 240°)$,
$(\cos 300° + i \sin 300°)$

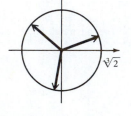
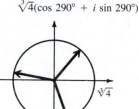

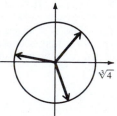

33. $(\cos 45° + i \sin 45°)$,
$(\cos 225° + i \sin 225°)$

35. $(\cos 0° + i \sin 0°)$, $(\cos 120° + i \sin 120°)$, $(\cos 240° + i \sin 240°)$ **37.** $(\cos 90° + i \sin 90°)$, $(\cos 210° + i \sin 210°)$,
$(\cos 330° + i \sin 330°)$ **39.** $2(\cos 0° + i \sin 0°)$, $2(\cos 120° + i \sin 120°)$, $2(\cos 240° + i \sin 240°)$
41. $(\cos 45° + i \sin 45°)$, $(\cos 135° + i \sin 135°)$, $(\cos 225° + i \sin 225°)$, $(\cos 315° + i \sin 315°)$
43. $(\cos 22 \ 1/2° + i \sin 22 \ 1/2°)$, $(\cos 112 \ 1/2° + i \sin 112 \ 1/2°)$, $(\cos 202 \ 1/2° + i \sin 202 \ 1/2°)$, $(\cos 292 \ 1/2° + i \sin 292 \ 1/2°)$
45. $2(\cos 20° + i \sin 20°)$, $2(\cos 140° + i \sin 140°)$, $2(\cos 260° + i \sin 260°)$
47. $1.3606 + 1.2637i$, $-1.7747 + .5464i$, $.4141 - 1.8102i$
49. $f(x) = (x - 1.1488)(x - .3550 - 1.0925i)(x + .9293 - .6752i)(x + .9293 + .6752i)(x - .3550 + 1.0925i)$

53. $\sqrt[4]{2}(\cos 22 \ 1/2° + i \sin 22 \ 1/2°)$, $\sqrt[4]{2}(\cos 202 \ 1/2° + i \sin 202 \ 1/2°)$
55. $\sqrt[4]{18}(\cos 157 \ 1/2° + i \sin 157 \ 1/2°)$, $\sqrt[4]{18}(\cos 337 \ 1/2° + i \sin 337 \ 1/2°)$

Section 8.7 (page 479)

1–11.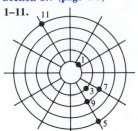

13.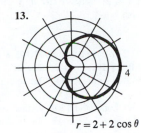
$r = 2 + 2\cos\theta$

15.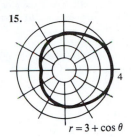
$r = 3 + \cos\theta$

17.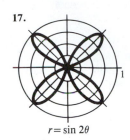
$r = \sin 2\theta$

19.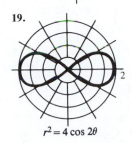
$r^2 = 4\cos 2\theta$

21.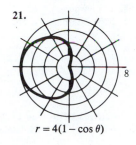
$r = 4(1 - \cos\theta)$

23.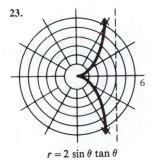
$r = 2\sin\theta\tan\theta$

25.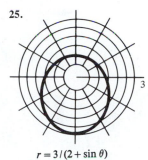
$r = 3/(2 + \sin\theta)$

27.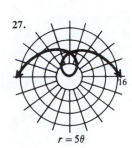
$r = 5\theta$

29.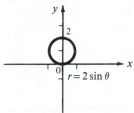
$r\theta = \pi$

31.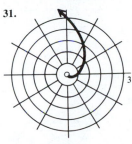
$\ln r = \theta$

33. $r = k\sec\theta$
(k any constant)

35. $\theta = k$ (k any constant)

37. $x^2 + (y - 1)^2 = 1$

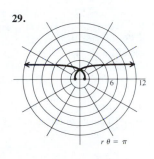
$r = 2\sin\theta$

39. $y^2 = 4(x + 1)$

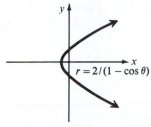
$r = 2/(1 - \cos\theta)$

41. $x^2 + y^2 + 2x + 2y = 0$ or $(x + 1)^2 + (y + 1)^2 = 2$

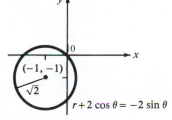
$(-1, -1)$
$\sqrt{2}$
$r + 2\cos\theta = -2\sin\theta$

43. $x = 2$

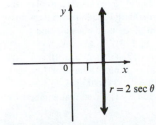
$r = 2\sec\theta$

45. $x + y = 2$

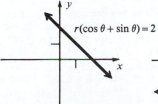
$r(\cos\theta + \sin\theta) = 2$

47. $y = -2$

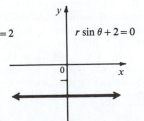
$r\sin\theta + 2 = 0$

49. $r(\cos \theta + \sin \theta) = 4$ **51.** $r = 4$ **53.** $r = 2 \csc \theta$ or $r \sin \theta = 2$ **55.** $r \sin^2 \theta = 25 \cos \theta$
57. $r^2(\cos^2 \theta + 9 \sin^2 \theta) = 36$ **59.** to the y-axis   **61.** to the x-axis

Chapter 8 Review Exercises (page 481)

1. $B = 42.7°$, $c = 58.4$ cm   **3.** $A = 56°$, $B = 34°$ **5.** $c = 402$ ft, $a = 11.9$ ft   **7.** 23.8 ft   **9.** 4000 ft
11. 71.0 m   **13.** triangle does not exist   **15.** 17.2°   **17.** 1300 ft   **19.** 13 m   **21.** 21°   **23.** 25.1°   **25.** 19.9°
27. 14.8 m   **29.** $B = 17.3°$, $C = 137.5°$, $c = 11.0$ yd   **31.** 153,000 sq m   **33.** 185 sq cm

35.

37.

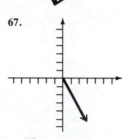

39. horizontal 17.9, vertical 66.8   **41.** 2, 2
43. $4\sqrt{3}$, 4   **45.** $2\sqrt{10}$, 161.6°   **47.** 2, 270°
49. $2\mathbf{i} - \mathbf{j}$ **51.** $10\mathbf{i}\sqrt{3} + 10\mathbf{j}$ **53.** 28 lb
55. 135 newtons   **57.** 270 lb, 56.3°   **59.** 105 lb
61. 306°, 524 mph   **63.** $5 + 4i$

65.

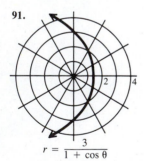

$\cos 300° + i \sin 300°$

67.

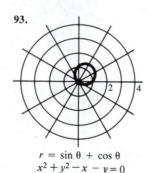

69. $2\sqrt{2}(\cos 135° + i \sin 135°)$ **71.** $0 + 3i$
73. $-\sqrt{2} - i\sqrt{2}$ **75.** $8(\cos 120° + i \sin 120°)$
77. $0 - 30i$ **79.** $-1/8 + i\sqrt{3}/8$ **81.** $0 + 8i$
83. $-1/2 - i\sqrt{3}/2$
85. $8^{1/10}(\cos 27° + i \sin 27°)$, $8^{1/10}(\cos 99° + i \sin 99°)$,
$8^{1/10}(\cos 171° + i \sin 171°)$, $8^{1/10}(\cos 243° + i \sin 243°)$,
$8^{1/10}(\cos 315° + i \sin 315°)$ **87.** $\cos 0° + i \sin 0°$,
$\cos 60° + i \sin 60°$, $\cos 120° + i \sin 120°$,
$\cos 180° + i \sin 180°$, $\cos 240° + i \sin 240°$,
89. $5(\cos 60° + i \sin 60°)$, $5(\cos 180° + i \sin 180°)$, $5(\cos 300° + i \sin 300°)$

91.

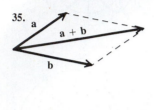

$r = \dfrac{3}{1 + \cos \theta}$

93.

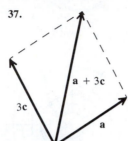

$r = \sin \theta + \cos \theta$
$x^2 + y^2 - x - y = 0$

95. $r \cos \theta = -3$ **97.** $r = \tan \theta \sec \theta$
109. all points of the form $(t, 0)$

Chapter 9

Section 9.1 (page 497)

1. $(-32, -8)$ **3.** $(1, 3)$ **5.** $(-1, 3)$ **7.** $(2, 18)$ **9.** $(4, 1, -2)$ **11.** $(1, 0, 2)$ **13.** $(-2, 0)$ **15.** $(1, 3)$
17. $(4, -2)$ **19.** $(2, -2)$ **21.** $(12, 6)$ **23.** $(4, 3)$ **25.** $(5, 2)$ **27.** $(1, 2, -1)$ **29.** $(2, 0, 3)$ **31.** Replace the
second equation with the sum of the second equation and 4 times the first equation.   **33.** Replace the third equation with the sum
of the third equation and -2 times the first equation.   **35.** Replace the third equation with the sum of the third equation and
-2 times the first equation.   **37.** $(2, 2, 0)$ **39.** no solution   **41.** $(-3, -2, 13)$ **43.** $(-3, 1, 3)$ **45.** yes
47. $(2, 2)$ **49.** $(1/5, 1)$ **51.** $(1/5, 14/3)$ **53.** $(4, 6, 1)$ **55.** $(z + (4/3), 2z - (7/3), z)$ **57.** $(-(1/9)z, (1/9)z, z)$
59. $((-1/6)z + 1, (-1/3)z + 2, z)$ **61.** goats cost \$30 and sheep cost \$25   **63.** 32 days at \$14   **65.** 18 liters of 70% and
12 liters of 20%   **67.** \$10,000 at 8% and \$15,000 at 10%   **69.** \$5000 at 8% and \$15,000 at 12%   **71.** 200 gal
73. 72 km per hr and 92 km per hr   **75.** 15 cm, 12 cm, 6 cm   **77.** $y = ax^2 + (3a - 4)x + (2a - 1)$ for all real numbers a
79. 37

Section 9.2 (page 503)

1. $(-1, -7), (4, 3)$

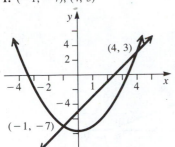

3. $(1, 1), (-2, 4)$

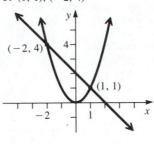

5. $(-4, 1), (-5/2, 1/4)$

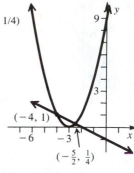

7. $(2, 4)$ **9.** $(2, 1), (1/3, 4/9)$ **11.** $(-3/5, 7/5), (-1, 1)$ **13.** $(2, 2), (2, -2), (-2, 2), (-2, -2)$ **15.** $(0, 0)$
17. $(1, 1), (1, -1), (-1, 1), (-1, -1)$ **19.** same circles, or $\{(x, y)|x^2 + y^2 = 10\}$ **21.** $(2, 3), (3, 2)$ **23.** $(-3, 5),$
$(15/4, -4)$ **25.** $(4, -1/8), (-2, 1/4)$ **27.** $(3, 2), (-3, -2), (4, 3/2), (-4, -3/2)$ **29.** $(3, 5), (-3, -5)$
31. $(\sqrt{5}, 0), (-\sqrt{5}, 0), (\sqrt{5}, \sqrt{5}), (-\sqrt{5}, -\sqrt{5})$ **33.** $(1, 3), (31/14, -9/14)$ **35.** $(0, 1)$ **37.** $(12, 1)$ **39.** $(-1, 0),$
$(6, -\sqrt{7})$ **41.** $(\sqrt{17}/17, 1), (-\sqrt{17}/17, 1), (\sqrt{17}/17, -1), (-\sqrt{17}/17, -1)$ **43.** 6 and 6   **45.** 9 and 15, or -15 and
-9 **47.** 5 m and 12 m   **49.** $y = 4x - 4$ **51.** $a \neq 5, a \neq -5$

Section 9.3 (page 512)

1.

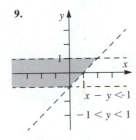

$x + y \leq 4$
$x - 2y \geq 6$

3.

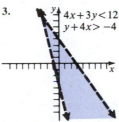

$4x + 3y < 12$
$y + 4x > -4$

5.

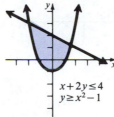

$x + 2y \leq 4$
$y \geq x^2 - 1$

7.

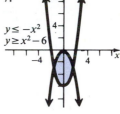

$y \leq -x^2$
$y \geq x^2 - 6$

9.

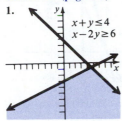

$x - y < 1$
$-1 < y < 1$

11.

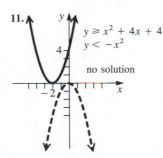

$y \geq x^2 + 4x + 4$
$y < -x^2$

no solution

13.

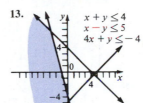

$x + y \leq 4$
$x - y \leq 5$
$4x + y \leq -4$

15.

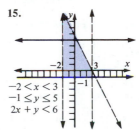

$-2 < x < 3$
$-1 \leq y \leq 5$
$2x + y < 6$

17.

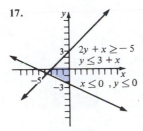

$2y + x \geq -5$
$y \leq 3 + x$
$x \leq 0, y \leq 0$

19.

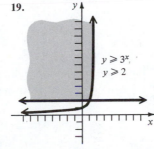

$y \geq 3^x$
$y \geq 2$

21.

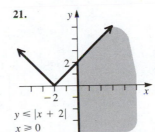

$y \leq |x + 2|$
$x \geq 0$

23.

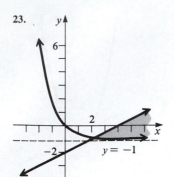

25. $y < 3x + 5$ **27.** maximum of 65 at $(5, 10)$, minimum of 8 at $(1, 1)$
29. maximum of 900 at $(0, 12)$, minimum of 0 at $(0, 0)$ **31.** $(6/5, 6/5)$
33. $(17/3, 5)$ **35.** $(105/8, 25/8)$
37. Let x = the number of Brand X pills, and let y = the number of Brand Y pills. Then $3000x + 1000y \geq 6000$, $45x + 50y \geq 195$, $75x + 200y \geq 600$, $x \geq 0$, $y \geq 0$.

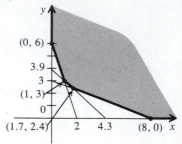

39. Let x = number shipped to warehouse A and y = number shipped to B; $0 \leq x \leq 75$, $0 \leq y \leq 80$, $x + y \geq 100$, $12x + 10y \leq 1200$.

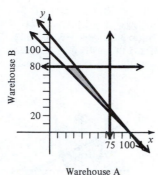

Warehouse A

41. Let h = height of the cylinder and r = radius of the cylinder; $h \leq 8$, $2r \geq 3$, $2r \leq h$, $h \leq 4r$.

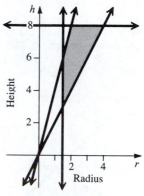

Radius

43. 6.4 million gal of gasoline and 3.2 million gal of fuel oil, for maximum revenue of $16,960,000   **45.** 250,000 hectares of each for a maximum profit of $132,500,000   **47.** a maximum of 560 sq ft with 80 sq ft of window and 480 sq ft of wall   **49.** 800 bargain sets and 300 deluxe sets for a maximum profit of $125,000

Section 9.4 (page 522)

1. $\begin{bmatrix} 2 & 3 & | & 11 \\ 1 & 2 & | & 8 \end{bmatrix}$ **3.** $\begin{bmatrix} 1 & 5 & | & 6 \\ 0 & 1 & | & 1 \end{bmatrix}$ **5.** $\begin{bmatrix} 2 & 1 & 1 & | & 3 \\ 3 & -4 & 2 & | & -7 \\ 1 & 1 & 1 & | & 2 \end{bmatrix}$ **7.** $\begin{bmatrix} 1 & 1 & 0 & | & 2 \\ 0 & 2 & 1 & | & -4 \\ 0 & 0 & 1 & | & 2 \end{bmatrix}$ **9.** $2x + y = 1$, $3x - 2y = -9$

11. $x = 2$, $y = 3$, $z = -2$ **13.** $3x + 2y + z = 1$, $2y + 4z = 22$, $-x - 2y + 3z = 15$ **15.** $(2, 3)$ **17.** $(-3, 0)$
19. $(7/2, -1)$ **21.** $(5, 0)$ **23.** $(-2, 1, 3)$ **25.** $(-1, 23, 16)$ **27.** $(3, 2, -4)$ **29.** $(2, 1, -1)$ **31.** no solution

33. $(3, 3 - z, z)$ **35.** $\left(\dfrac{-12}{23} - \dfrac{15}{23}z, \dfrac{1}{23}z - \dfrac{13}{23}, z \right)$ **37.** $\left(-6 - \dfrac{1}{2}z, \dfrac{1}{2}z + 2, z \right)$ **39.** $(-1, 2, 5, 1)$

41. $(1 - 4w, 1 - w, 2 + w, w)$ **43.** wife 40 days, husband 32 days   **45.** 5 model 201, 8 model 301   **47.** $10,000 at 8%, $7000 at 10%, $8000 at 9%   **49.** inconsistent system, no solution   **51.** $(-15 + z, 5 - 2z, z)$ **53.** $A = -1$, $B = 1$
55. $A = -1$, $B = 1$

Section 9.5 (page 531)

1. $x = 2$, $y = 4$, $z = 8$ **3.** $z = 18$, $r = 3$, $s = 3$, $p = 3$, $a = 3/4$ **5.** no   **7.** $\begin{bmatrix} 14 & -11 & -3 \\ -2 & 4 & -1 \end{bmatrix}$ **9.** $\begin{bmatrix} -6 & 8 \\ 4 & 2 \end{bmatrix}$

11. can't be done   **13.** $\begin{bmatrix} -12x + 8y & -x + y \\ x & 8x - y \end{bmatrix}$ **15.** $\begin{bmatrix} 7 & 2 \\ 9 & 0 \\ 8 & 6 \end{bmatrix}$; $\begin{bmatrix} 7 & 9 & 8 \\ 2 & 0 & 6 \end{bmatrix}$ **17.** $\begin{bmatrix} -4 & 8 \\ 0 & 6 \end{bmatrix}$ **19.** $\begin{bmatrix} 2 & 6 \\ -4 & 6 \end{bmatrix}$

21. $\begin{bmatrix} -1 & -3 \\ 2 & -3 \end{bmatrix}$ **23.** $\begin{bmatrix} 13 \\ 25 \end{bmatrix}$ **25.** $\begin{bmatrix} -2 & 5 & 0 \\ 6 & 6 & 1 \\ 12 & 2 & -3 \end{bmatrix}$ **27.** $\begin{bmatrix} 20 & 0 & 8 \\ 8 & 0 & 4 \end{bmatrix}$ **29.** can't be multiplied   **31.** $[-8]$

33. $\begin{bmatrix} -6 & 12 & 18 \\ 4 & -8 & -12 \\ -2 & 4 & 6 \end{bmatrix}$ **35.** (a) $\begin{bmatrix} 47.5 & 57.75 \\ 27 & 33.75 \\ 81 & 95 \\ 12 & 15 \end{bmatrix}$ (b) $[20\ \ 200\ \ 50\ \ 60]$, $[220\ \ 890\ \ 105\ \ 125\ \ 70]$

(c) $[11,120\ \ 13,555]$ **37.** true   **39.** true   **41.** true   **43.** true   **45.** true   **47.** true   **49.** not always true, since $AB + BA \neq 2AB$

Section 9.6 (page 542)

1. -26 **3.** -16 **5.** $y^2 - 16$ **7.** $x^2 - y^2$ **9.** $0, -5, 0$ **11.** $-6, 0, -6$ **13.** -1 **15.** 0 **17.** 0
19. $4x$ **21.** $10i + 17j - 6k$ **23.** -88 **25.** -5.5 **27.** two rows identical   **29.** one column all zeros
31. multiply each element of second row by $-1/4$; then two rows are identical   **33.** multiply each element of third column by $1/2$; then two columns are identical   **35.** rows and columns exchanged   **37.** two rows exchanged   **39.** multiply each element of second column by 3   **41.** multiply elements of first row by 1; add products to elements of second row
43. multiply elements of second row by 2   **45.** 0 **47.** 0 **49.** 5 **51.** 32 **53.** -32 **63.** $5/2$ **65.** 7

Section 9.7 (page 548)

1. $(2, 2)$ **3.** $(2, -5)$ **5.** $(2, 3)$ **7.** $(9/7, 2/7)$ **9.** $D = 0$, no solution   **11.** $(-1, 2, 1)$ **13.** $(-3, 4, 2)$
15. $(0, 0, -1)$ **17.** $D = 0$, no solution   **19.** $(197/91, -118/91, -23/91)$ **21.** $(2, 3, 1)$ **23.** $(0, 4, 2)$
25. $(31/5, 19/10, -29/10)$ **27.** $(1, 0, -1, 2)$ **29.** shirts are \$12, pants are \$18   **31.** 17.5 g of 12-carat, 7.5 g of 22-carat
33. $104/19$ lb   **35.** 164 fives, 86 tens, 40 twenties   **39.** $[(a^2 + ab + b^2)/(a + b), -ab/(a + b)]$
41. $([(2b)(b + 4)]/(b^2 + 4), [(4b)(b - 1)]/(b^2 + 4))$

Section 9.8 (page 556)

1. yes   **3.** no   **5.** no   **7.** yes   **9.** $\begin{bmatrix} 0 & 1/2 \\ -1 & 1/2 \end{bmatrix}$ **11.** $\begin{bmatrix} 2 & 1 \\ 5 & 3 \end{bmatrix}$ **13.** none   **15.** $\begin{bmatrix} 1 & 0 & 0 \\ 0 & -1 & 0 \\ -1 & 0 & 1 \end{bmatrix}$

17. $\begin{bmatrix} 15 & 4 & -5 \\ -12 & -3 & 4 \\ -4 & -1 & 1 \end{bmatrix}$ **19.** $\begin{bmatrix} 7/4 & 5/2 & 3 \\ -1/4 & -1/2 & 0 \\ -1/4 & -1/2 & -1 \end{bmatrix}$ **21.** $\begin{bmatrix} 1/2 & 1/2 & -1/4 & 1/2 \\ -1 & 4 & -1/2 & -2 \\ -1/2 & 5/2 & -1/4 & -3/2 \\ 1/2 & -1/2 & 1/4 & 1/2 \end{bmatrix}$ **23.** $(-1, 4)$ **25.** $(2, 1)$

27. $(2, 3)$ **29.** $(-8, 6, 1)$ **31.** $(15, -5, -1)$ **33.** $(-31, 24, -4)$ **35.** $\left(\dfrac{19}{20 - 2b}, \dfrac{-7b + 32}{20 - 2b} \right)$

37. $\left(\dfrac{b^2 + b}{3}, \dfrac{-b^2 + 2b}{3} \right)$ **39.** $(-7, -34, -19, 7)$ **45.** (a) $\begin{bmatrix} 72 \\ 48 \\ 60 \end{bmatrix}$ (b) $\begin{bmatrix} 2 & 4 & 2 \\ 2 & 1 & 2 \\ 2 & 1 & 3 \end{bmatrix} \begin{bmatrix} x_1 \\ x_2 \\ x_3 \end{bmatrix} = \begin{bmatrix} 72 \\ 48 \\ 60 \end{bmatrix}$ (c) 8, 8, 12

49. $ab \neq 0$ **51.** $\begin{bmatrix} 1/a & 0 & 0 \\ 0 & 1/b & 0 \\ 0 & 0 & 1/c \end{bmatrix}$

Section 9.9 (page 564)

1. $-1/x + 4/(x + 1)$ **3.** $-1/[2(x + 1)] + 3/[2(x - 1)]$ **5.** $7/x - 63/[4(3x - 1)] - 7/[4(x + 1)]$
7. $-2/(9x) + 2/(3x^2) + 2/[9(x + 3)]$ **9.** $-3/[25(x + 2)] + 3/[25(x - 3)] + 2/[5(x - 3)^2]$ **11.** $2/(x + 2)^2 - 3/(x + 2)^3$
13. $3/[2(x + 1)] - 3/[2(x + 3)]$ **15.** $2x^3 + x^2 + 1/[2(3x - 1)] + 1/[2(x + 1)]$ **17.** $1 + 1/x + 5/(x - 2) - 3/(x - 1)$
19. $-1/[3(x + 1)] + (x + 5)/[3(x^2 + 2)]$ **21.** $3/x - 3/[2(x + 1)] - 3(x + 1)/[2(x^2 + 1)]$ **23.** $1/x - 2x/(x^2 + 1)^2$
25. $3/(x - 1) + 1/(x^2 + 1) - 2/(x^2 + 1)^2$ **27.** $-1/[4(x + 1)] + 1/[4(x - 1)] + 1/[2(x^2 + 1)]$ **29.** $3x^2 + 3/x - 3x/(x^2 + 1)$

Chapter 9 Review Exercises (page 565)

1. $(-1, 3)$ **3.** $(6, -12)$ **5.** $(2, 5)$ **7.** $(-1, 2, 3)$ **9.** 10 at 25¢, 12 at 50¢   **11.** \$10,000 at 6%, \$20,000 at 7%,
\$20,000 at 9%   **13.** $[(15z + 6)/11, (14z - 1)/11, z]$
15. $(3\sqrt{3}, \sqrt{2}), (-3\sqrt{3}, \sqrt{2}), (3\sqrt{3}, -\sqrt{2}), (-3\sqrt{3}, -\sqrt{2})$
17. $(-2, 0), (1, 1)$ **19.** yes, $((8 - 8\sqrt{41})/5, (16 + 4\sqrt{41})/5), ((8 + 8\sqrt{41})/5, (16 - 4\sqrt{41})/5)$
21. $x = (-mb + \sqrt{r^2 + r^2m^2 - b^2})/(1 + m^2)$ or $(-mb - \sqrt{r^2 + r^2m^2 - b^2})/(1 + m^2)$

23.

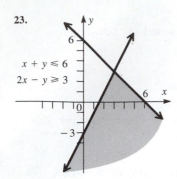

$x + y \leq 6$
$2x - y \geq 3$

25.

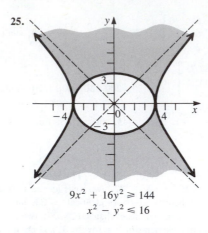

$9x^2 + 16y^2 \geq 144$
$x^2 - y^2 \leq 16$

27. 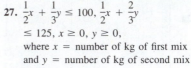 $\frac{1}{2}x + \frac{1}{3}y \leq 100, \frac{1}{2}x + \frac{2}{3}y$
$\leq 125, x \geq 0, y \geq 0,$
where $x =$ number of kg of first mix
and $y =$ number of kg of second mix

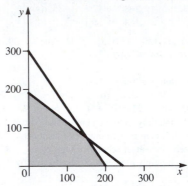

29. There is a maximum of 42 at $(12, 6)$.   **31.** There is a minimum of 12 at $(3, 0)$ or at $(6/5, 12/5)$.   **33.** 150 kg of the half-and-half, and 75 kg of the other   **35.** $(-2, 0)$ **37.** no solution   **39.** $k = -8, y = 7, a = 2, m = 3, p = 1, r = -5$

41. $\begin{bmatrix} -4 \\ 6 \\ 1 \end{bmatrix}$ **43.** $\begin{bmatrix} 11 & 20 \\ 14 & 40 \end{bmatrix}$ **45.** $\begin{bmatrix} -3 \\ 10 \end{bmatrix}$ **47.** -6 **49.** -1 **51.** One row is all zeros.   **53.** Multiply elements of

first row by 2.   **55.** New row 2 is the sum of old rows 1 and 2.   **57.** $(-2, 5)$ **59.** $(-4, 6, 2)$ **61.** $\begin{bmatrix} 3 & -1 \\ -5 & 2 \end{bmatrix}$

63. $\begin{bmatrix} 2/3 & 0 & -1/3 \\ 1/3 & 0 & -2/3 \\ -2/3 & 1 & 1/3 \end{bmatrix}$ **65.** $(2, 1)$ **67.** $(-3, 5, 1)$ **71.** $\frac{-1}{4x} + \frac{x + 3}{4(x^2 - x + 4)}$ **73.** $\frac{-1}{48(x + 4)} + \frac{x + 40}{48(x^2 - 4x + 16)}$

Chapter 10

Section 10.1 (page 574)

1.

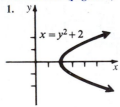

$x = y^2 + 2$

3.

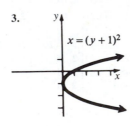

$x = (y + 1)^2$

5.

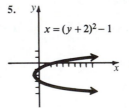

$x = (y + 2)^2 - 1$

7.

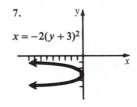

$x = -2(y + 3)^2$

9.

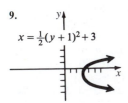

$x = \frac{1}{2}(y + 1)^2 + 3$

11.

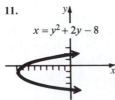

$x = y^2 + 2y - 8$

13.

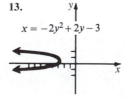

$x = -2y^2 + 2y - 3$

15. $(0, 4)$, $y = -4$, y-axis, $(-8, 4)$, $(8, 4)$ **17.** $(0, -1/8)$, $y = 1/8$, y-axis, $(-1/4, -1/8)$, $(1/4, -1/8)$ **19.** $(1/64, 0)$, $x = -1/64$, x-axis, $(1/64, -1/32)$, $(1/64, 1/32)$ **21.** $(-4, 0)$, $x = 4$, x-axis, $(-4, -8)$, $(-4, 8)$ **23.** $x^2 = -8y$ **25.** $y^2 = -2x$ **27.** $y^2 = 4x$ **29.** $x^2 = -2y$ **31.** $x^2 = -y$ **33.** 60 ft   **35.** $d(F_1, V) = \sqrt{c^2} = |c|$. For all other points $d(F_1, P) = \sqrt{x^2 + (y - c)^2} > |c|$.   **37.** $y^2 = -16(x - 57/16)$

Section 10.2 (page 586)

1. $(3, 0)$, $(-3, 0)$; $(0, 0)$; $(-\sqrt{5}, 0)$, $(\sqrt{5}, 0)$

3. $(3, 0)$, $(-3, 0)$; $(0, 0)$; $(-2\sqrt{2}, 0)$, $(2\sqrt{2}, 0)$

5. $(0, 3)$, $(0, -3)$; $(0, 0)$; $(0, -\sqrt{3})$, $(0, \sqrt{3})$

7. $(4, 0)$, $(-4, 0)$; $(0, 0)$; $(-2\sqrt{3}, 0)$, $(2\sqrt{3}, 0)$

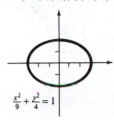

$\frac{x^2}{9} + \frac{y^2}{4} = 1$

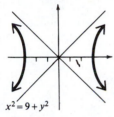

$\frac{x^2}{9} + y^2 = 1$

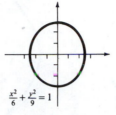

$\frac{x^2}{6} + \frac{y^2}{9} = 1$

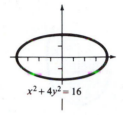

$x^2 + 4y^2 = 16$

9. $(0, 0)$; $y = \pm x$; $(-3\sqrt{2}, 0)$, $(3\sqrt{2}, 0)$

11. $(0, 2\sqrt{2})$, $(0, -2\sqrt{2})$; $(0, 0)$; $(0, -2)$, $(0, 2)$

13. $(0, 0)$; $y = \pm(5/2)x$; $(0, -\sqrt{29})$, $(0, \sqrt{29})$

15. $(1/3, 0)$, $(-1/3, 0)$; $(0, 0)$; $(-\sqrt{7}/12, 0)$, $(\sqrt{7}/12, 0)$

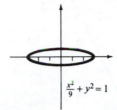

$x^2 = 9 + y^2$

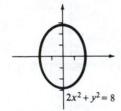

$2x^2 + y^2 = 8$

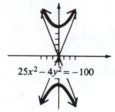

$25x^2 - 4y^2 = -100$

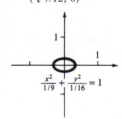

$\frac{x^2}{1/9} + \frac{y^2}{1/16} = 1$

17. $(0, 6/5)$, $(0, -6/5)$; $(0, 0)$; $(0, -3\sqrt{231}/40)$, $(0, 3\sqrt{231}/40)$

19.

21.

23.

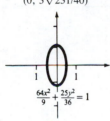

$\frac{64x^2}{9} + \frac{25y^2}{36} = 1$

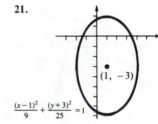

$\frac{(x + 2)^2}{16} + \frac{(y + 1)^2}{9} = 1$

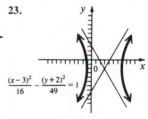

$\frac{(x-1)^2}{9} + \frac{(y+3)^2}{25} = 1$

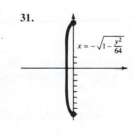

$\frac{(x-3)^2}{16} - \frac{(y+2)^2}{49} = 1$

25.

27.

29. function

31.

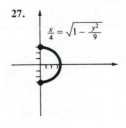

$\frac{(y+1)^2}{25} - \frac{(x-3)^2}{36} = 1$

$\frac{x}{4} = \sqrt{1 - \frac{y^2}{9}}$

$\frac{y}{3} = \sqrt{1 + \frac{x^2}{16}}$

$x = -\sqrt{1 - \frac{y^2}{64}}$

33. $x^2/16 + y^2/12 = 1$ **35.** $x^2/36 + y^2/20 = 1$ **37.** $\dfrac{(x-3)^2}{16} + \dfrac{(y+2)^2}{25} = 1$ **39.** $x^2/9 - y^2/7 = 1$

41. $y^2/9 - x^2/25 = 1$ **43.** (a) yes (b) 9   **45.** $y^2/81 + x^2/72 = 1$ **47.** (a) about 141.6 million mi (b) about 141.0 million mi   **49.** hyperbola

Section 10.3 (page 594)

1. ellipse

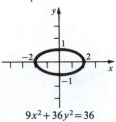

3. hyperbola

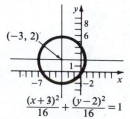

5. line

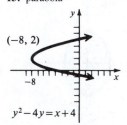

7. hyperbola

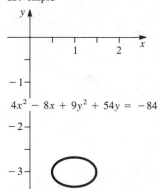

9. ellipse

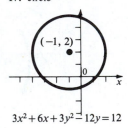

11. circle

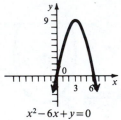

13. parabola

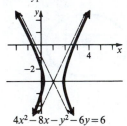

15. empty set (no graph)

17. circle

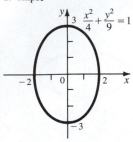

19. parabola

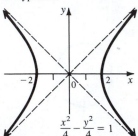

21. hyperbola

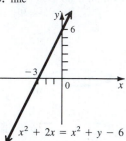

23. ellipse

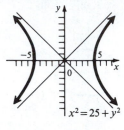

25. empty set (no graph)   **27.** ellipse   **29.** 1/2   **31.** $\sqrt{2}$ **33.** $\sqrt{21}/7$ **35.** $\sqrt{10}/3$ **37.** $\sqrt{2}/2$

39. $x^2 = 32y$ **41.** $x^2/36 + y^2/27 = 1$ **43.** $x^2/36 - y^2/108 = 1$ **45.** $x^2 = -4y$ **47.** $25x^2/81 + y^2/9 = 1$ **49.** 0

Section 10.4 (page 600)

1. circle or ellipse   **3.** hyperbola   **5.** parabola   **7.** 30°   **9.** 60°   **11.** 22.5°

13.

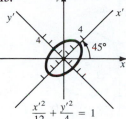

$$\frac{x'^2}{12} + \frac{y'^2}{4} = 1$$

15.

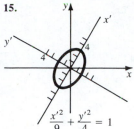

$$\frac{x'^2}{9} + \frac{y'^2}{4} = 1$$

17.

$$\frac{x'^2}{4} + \frac{y'^2}{2} = 1$$

19.

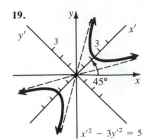

$$x'^2 - 3y'^2 = 5$$

21.

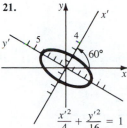

$$\frac{x'^2}{4} + \frac{y'^2}{16} = 1$$

23.

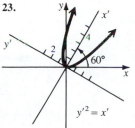

$$y'^2 = x'$$

25.

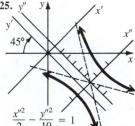

$$\frac{x''^2}{2} - \frac{y''^2}{10} = 1$$

27.

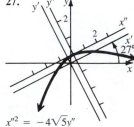

$$x''^2 = -4\sqrt{5}\,y''$$

29.

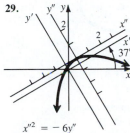

$$x''^2 = -6y''$$

31. $y'^2/9 - x'^2/25 = 1$ **33.** because the graph is a circle

Chapter 10 Review Exercises (page 601)

1.

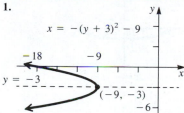

$$x = -(y + 3)^2 - 9$$

$$y = -3$$

$$(-9, -3)$$

3.

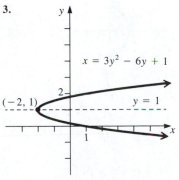

$$x = 3y^2 - 6y + 1$$

$$y = 1$$

$$(-2, 1)$$

5. $(3/4, 0),\ x = -3/4,\ y = 0$

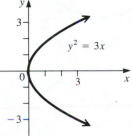

$$y^2 = 3x$$

7. $(0, 5/4),\ y = -5/4,\ x = 0$

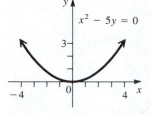

$$x^2 - 5y = 0$$

9. $x^2 = -16y$ **11.** $y = 7x^2$

13. vertices: $(-5, 0)$, $(5, 0)$

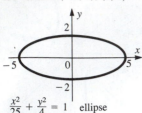

$\dfrac{x^2}{25} + \dfrac{y^2}{4} = 1$ ellipse

15. vertices: $(-2, 0)$, $(2, 0)$;
asymptotes: $y = \pm(3/2)x$

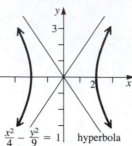

$\dfrac{x^2}{4} - \dfrac{y^2}{9} = 1$ hyperbola

17. circle

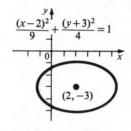

$(x - 3)^2 + (y + 2)^2 = 9$

19. vertices: $(-4, 0)$, $(4, 0)$;
asymptotes: $y = \pm x$

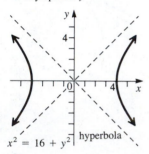

$x^2 = 16 + y^2$ hyperbola

21. vertices: $(0, 5/2)$, $(0, -5/2)$

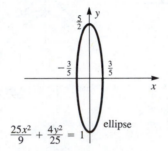

$\dfrac{25x^2}{9} + \dfrac{4y^2}{25} = 1$ ellipse

23. ellipse; vertices: $(5, -3)$, $(-1, -3)$

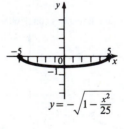

$\dfrac{(x-2)^2}{9} + \dfrac{(y+3)^2}{4} = 1$

25. hyperbola; vertices
$(-3, 0)$, $(-3, -4)$;
asymptotes: $y =$
$\pm(2/3)(x + 3) - 2$

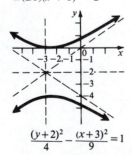

$\dfrac{(y+2)^2}{4} - \dfrac{(x+3)^2}{9} = 1$

27. semi-ellipse; vertices
$(0, 4)$, $(0, -4)$

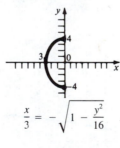

$\dfrac{x}{3} = -\sqrt{1 - \dfrac{y^2}{16}}$

29. semi-ellipse; vertices
$(-5, 0)$, $(5, 0)$

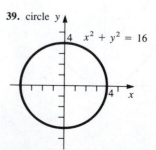

$y = -\sqrt{1 - \dfrac{x^2}{25}}$

31. $x^2/16 + y^2/7 = 1$ **33.** $y^2/4 - 5x^2/16 = 1$

35. hyperbola $4x^2 - 9y^2 = 36$

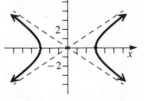

37. parabola $y^2 + x = 2$

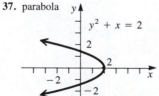

39. circle $x^2 + y^2 = 16$

41. $(y - 2)^2 - \dfrac{x^2}{3} = 1$ **43.** hyperbola   **45.** hyperbola   **47.** 15°

49.

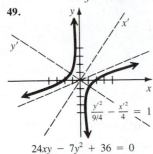

$\dfrac{y'^2}{9/4} - \dfrac{x'^2}{4} = 1$

$24xy - 7y^2 + 36 = 0$

51.

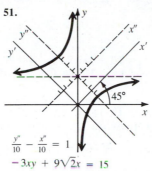

45°

$\dfrac{y''}{10} - \dfrac{x''}{10} = 1$

$-3xy + 9\sqrt{2}x = 15$

Chapter 11

Section 11.1 (page 608)

1. $r^{10} + 5r^8s + 10r^6s^2 + 10r^4s^3 + 5r^2s^4 + s^5$ **3.** $p^4 + 8p^3q + 24p^2q^2 + 32pq^3 + 16q^4$
5. $m^6/64 - 3m^5/16 + 15m^4/16 - 5m^3/2 + 15m^2/4 - 3m + 1$ **7.** $16p^4 + 32p^3q/3 + 8p^2q^2/3 + 8pq^3/27 + q^4/81$
9. $m^{-8} + 4m^{-4} + 6 + 4m^4 + m^8$ **11.** $p^{10/3} - 25p^4 + 250p^{14/3} - 1250p^{16/3} + 3125p^6 - 3125p^{20/3}$
13. $x^{21} + 126x^{20} + 7560x^{19} + 287{,}280x^{18}$ **15.** $3^9m^{-18} + 45 \cdot 3^8m^{-20} + 900 \cdot 3^7m^{-22} + 10{,}500 \cdot 3^6m^{-24}$
17. $180m^2n^8$ **19.** $4845p^8q^{16}$ **21.** $439{,}296x^{21}y^7$ **23.** $90{,}720x^{28}y^{12}$ **25.** 14   **27.** 1120/243
29. 1.105   **31.** 245.937   **33.** 758.650   **35.** 1.002, 5.010   **37.** .942   **39.** $1 - x + x^2 - x^3 + \ldots$
41. 3.0625; no, because $|x| \not< 1$ in $(1 + x)^n$.   **43.** -1

Section 11.2 (page 617)

1. 10, 16, 22, 28, 34   **3.** $-3, 4, -5, 6, -7$ **5.** 4/3, 12/5, 20/7, 28/9, 36/11   **7.** $-1/2, 1/8, -1/24, 1/64, -1/160$
9. $2/3, -3/4, 4/5, -5/6, 6/7$ **11.** 4, 9, 14, 19, 24, 29, 34, 39, 44, 49   **13.** 1, 1, 2, 3, 5, 8, 13, 21, 34, 55
15. 4, 6, 8, 10, 12   **17.** 11, 9, 7, 5   **19.** $4 - \sqrt{5}, 4, 4 + \sqrt{5}, 4 + 2\sqrt{5}, 4 + 3\sqrt{5}$ **21.** $d = 5, a_n = 7 + 5n$
23. $d = -3, a_n = 21 - 3n$ **25.** $d = \sqrt{2}, a_n = -6 + n\sqrt{2}$ **27.** $d = m, a_n = x + nm - m$ **29.** $a_8 = 19,$
$a_n = 3 + 2n$ **31.** $a_8 = 7, a_n = n - 1$ **33.** $a_8 = -9, a_n = 15 - 3n$ **35.** $a_8 = x + 21, a_n = x + 3n - 3$
37. $a_1 = 7$ **39.** $a_1 = 7$ **41.** 125   **43.** 230   **45.** 16.745   **47.** 32.28   **49.** 140   **51.** -684 **53.** 500,500
55. 400   **57.** 7   **59.** 83   **61.** $n(n + 1)/2$ **63.** 4680   **65.** 95 m, 500 m   **67.** 240 ft, 1024 ft   **69.** 9, 14,
19, 24   **71.** 170 m   **73.** All terms are the same constant.   **75.** yes   **79.** $\displaystyle\sum_{i=1}^{6} (3 + 2i)$ **81.** false
83. true   **85.** false

Section 11.3 (page 626)

1. 1/2, 2, 8, 32   **3.** 3/2, 3, 6, 12, 24   **5.** $a_5 = -1875, a_n = -3(-5)^{n-1}$ **7.** $a_5 = 24, a_n = (3/2) \cdot 2^{n-1}$ or $3 \cdot 2^{n-2}$
9. $a_5 = -256, a_n = -1 \cdot (-4)^{n-1}$ or $-(-4)^{n-1}$ **11.** $r = 2, a_n = 6 \cdot 2^{n-1}$ **13.** $r = 2, a_n = (3/4) \cdot 2^{n-1}$ or $3 \cdot 2^{n-3}$
15. $a_n = (9/4) \cdot 2^{n-1}$ or $9 \cdot 2^{n-3}$ **17.** 3   **19.** $a_1 = 2, r = 3,$ or $a_1 = -2, r = -3$ **21.** $a_1 = 128, r = 1/2,$ or $a_1 =$
$-128, r = -1/2$ **23.** 93   **25.** 33/4   **27.** 860.95   **29.** 30   **31.** 255/4   **33.** 120   **35.** $\displaystyle\sum_{i=1}^{4} 3 \cdot 2^i$ **37.** $r = 2$
39. $r = 2$ **41.** 64/3   **43.** 1000/9   **45.** 8/5   **47.** 1/3   **49.** 3/7   **51.** 70 m   **53.** 1600 rotations   **55.** $\$2^{30}$, or
$\$1{,}073{,}741{,}824; \$2^{31} - \$1$ **57.** (b) 3 mg (c) 4/5 mg   **59.** 12 m   **61.** $\approx 13.4\%$ **63.** \$26,214.40   **65.** 32.768 in.
67. any sequence of the form a, a, a **69.** r^3 **71.** $\sqrt{r}$

Section 11.4 (page 634)

1. $-5 - 2 + 1 + 4$ **3.** $-3 + 0 + 1$ **5.** $-1 - 1/3 + 0 + 1/5 + 1/3$ **7.** $-5(.5) + (-1)(.5) + 3(.5) +$
$7(.5) = -2.5 - .5 + 1.5 + 3.5$ **9.** $-1(.5) + 3(.5) + 15(.5) + 35(.5) = -.5 + 1.5 + 7.5 + 17.5$
11. $4(.5) + (4/3)(.5) + (4/5)(.5) + (4/7)(.5) = 2 + 2/3 + 2/5 + 2/7$ **13.** $\displaystyle\sum_{i=1}^{15} (10 - 2i)$ **15.** $\displaystyle\sum_{i=1}^{19} (5 + 5i)$ **17.** $\displaystyle\sum_{i=1}^{9} 7(2)^{i-1}$
19. $\displaystyle\sum_{i=1}^{7} 9\left(-\dfrac{1}{3}\right)^{i-1}$ **21.** $\displaystyle\sum_{i=2}^{13} i^2$ or $\displaystyle\sum_{i=1}^{12} (i + 1)^2$ **23.** $\displaystyle\sum_{i=1}^{10} (1/i)$ **25.** $\displaystyle\sum_{i=1}^{6} 2^i$ **27.** $\displaystyle\sum_{i=3}^{7} (12 - 3i)$ **29.** $\displaystyle\sum_{i=0}^{9} 2(3)^{i+1}$

31. $\displaystyle\sum_{i=0}^{10} [(i-1)^2 - 2(i-1)]$ or $\displaystyle\sum_{i=0}^{10} (i^2 - 4i + 3)$ **33.** 90 **35.** 220 **37.** 304 **39.** $8 + [2(n+1)]/n$
41. $9 + [3(n+1)]/n + [(n+1)(2n+1)]/(6n^2)$

Section 11.5 (page 640)
1. S_1: $2 = 1(1+1)$ S_2: $2 + 4 = 2(2+1)$ S_3: $2 + 4 + 6 = 3(3+1)$ S_4: $2 + 4 + 6 + 8 = 4(4+1)$
S_5: $2 + 4 + 6 + 8 + 10 = 5(5+1)$ **31.** It is false for $n = 1$. **35.** $3(4/3)^{n-1}$ **37.** $2^n - 1$

Section 11.6 (page 648)
1. 5040 **3.** 720 **5.** 336 **7.** 7 **9.** 6 **11.** 1365 **13.** 120 **15.** 14 **17.** 360 **19.** 120; 30,240
21. 30 **23.** 120 **25.** 4 **27.** 552 **29.** 34,650 **31.** (a) 180 (b) 50,400 **33.** (a) 362,880 (b) 6 (c) 1260
35. 27,405 **37.** 220; 220 **39.** 1326 **41.** 10 **43.** 28 **45.** (a) 84 (b) 10 (c) 40 (d) 28 **47.** $\dbinom{52}{5}$

49. $4 \cdot 1$ **51.** $4 \cdot \dbinom{13}{5}$ **53.** $P(5, 5) = 120$ **55.** $\dbinom{12}{3} = 220$

Section 11.7 (page 656)
1. $S = \{H\}$ **3.** $S = \{HHH, HHT, HTH, THH, HTT, THT, TTH, TTT\}$ **5.** Let c = correct, w = wrong. $S = $
$\{ccc, ccw, cwc, wcc, wwc, wcw, cww, www\}$ **7.** (a) $\{HH, TT\}$, 1/2 (b) $\{HH, HT, TH\}$, 3/4 **9.** (a) $\{2$ and $4\}$, 1/10
(b) $\{1$ and 3, 1 and 5, 3 and $5\}$, 3/10 (c) $\varnothing$, 0 (d) $\{1$ and 2, 1 and 4, 2 and 3, 2 and 5, 3 and 4, 4 and $5\}$, 3/5 **11.** (a) 1/5
(b) 8/15 (c) 0 (d) 1 to 4 (e) 7 to 8 **13.** 1 to 4 **15.** 2/5 **17.** (a) 1/2 (b) 7/10 (c) 2/5 **19.** (a) 1/6 (b) 1/3
(c) 1/6 **21.** (a) 7/15 (b) 11/15 (c) 4/15 **23.** .24 **25.** .06 **27.** .27 **29.** 1 to 1 **31.** .35 **33.** .95
35. 0 **37.** .38 **39.** .89 **41.** .90 **43.** .41 **45.** .21 **47.** .79

Chapter 11 Review Exercises (page 660)
5. $x^4 + 8x^3y + 24x^2y^2 + 32xy^3 + 16y^4$ **7.** $243x^{5/2} - 405x^{3/2} + 270x^{1/2} - 90x^{-1/2} + 15x^{-3/2} - x^{-5/2}$ **9.** $2160x^2y^4$
11. $3^{16} + 16 \cdot 3^{15}x + 120 \cdot 3^{14}x^2 + 560 \cdot 3^{13}x^3$ **13.** 5, 3, -2, -5, -3 **15.** 10, 6, 2, -2, -6 **17.** $3 - \sqrt{5}$, 4,
$5 + \sqrt{5}$, $6 + 2\sqrt{5}$, $7 + 3\sqrt{5}$ **19.** 4, 8, 16, 32, 64 **21.** -3, 4, $-16/3$, 64/9, $-256/27$ **23.** $a_1 = -29$, $a_{20} = 66$
25. 20 **27.** $-x + 61$ **29.** 222 **31.** $84k$ **33.** 25 **35.** 36 or -36 **37.** 120 **39.** geometric; $r = -1/2$
41. neither **43.** 36 **45.** $-1/2$ **47.** 1 **49.** 6 **51.** 73/12 **53.** 50,005,000 **55.** 240 **57.** $r > 1$ so sum
does not exist **59.** $0 + 16 + 56 = 72$ **61.** $-.8 + 0 + .8 + 6.4 + 21.6 + 51.2 = 79.2$ **63.** $\displaystyle\sum_{i=1}^{20} (6 + 4i)$

65. $\displaystyle\sum_{i=5}^{12} \frac{i}{i+1}$ **67.** $\displaystyle\sum_{i=1}^{\infty} \left(\frac{1}{1 + .2i}\right)$ **69.** $\displaystyle\sum_{i=0}^{7} (-8 - 6i)$ **71.** 489 **73.** 1 **75.** 252 **77.** 504 **79.** $S = $
$\{1, 2, 3, 4, 5, 6\}$ **81.** $S = \{0, .5, 1, 1.5, 2, \ldots, 299.5, 300\}$ **83.** $S = \{(3, R), (3, G), (5, R), (5, G), (7, R), (7, G),$
$(9, R), (9, G), (11, R), (11, G)\}$ **85.** $F = \{(3, G), (5, G), (7, G), (9, G), (11, G)\}$ **87.** $E' \cap F'$ **89.** 1 to 3
91. 2 to 11 **93.** .26 **95.** 4/13 **97.** 3/4 **99.** 0

Appendix A (page 668)
1. 2 **3.** 6 **5.** -4 **7.** -5 **9.** $.9509 - 4$ or -3.0491 **11.** 4.8338 **13.** .8825 **15.** $-.3675$
17. 2.3711 **19.** 1.6839 **21.** 34.4 **23.** .0747 **25.** 3120 **27.** .00637 **29.** 62.2 **31.** .000167

Appendix B (page 672)
1. $35°$ **3.** $37°$ **5.** $69°50'$ **7.** .0684 **9.** .3142 **11.** .8814 **13.** .6338 **15.** $-.6248$ **17.** $-.5712$
19. -3.072 **21.** .9636 **23.** $-.6361$ **25.** $58°00'$ **27.** $30°30'$ **29.** $46°10'$ **31.** $81°10'$ **33.** .2094
35. 1.4428 **37.** 1.4748 **39.** 1.0181 **41.** .4188 **43.** .3462 **45.** $194°50'$; $345°10'$ **47.** $309°30'$
49. $172°50'$; $352°50'$ **51.** $249°10'$ **53.** $137°40'$; $222°20'$

INDEX

3.1 Absolute Value Function

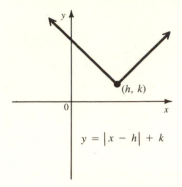

$$y = |x - h| + k$$

3.6 Quadratic Function

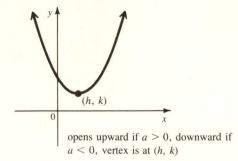

opens upward if $a > 0$, downward if $a < 0$, vertex is at (h, k)

4.4 Polynomial Functions

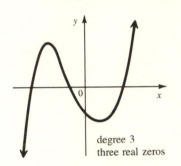

degree 3
three real zeros

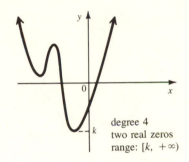

degree 4
two real zeros
range: $[k, +\infty)$

degree 5
three real zeros

4.5 Rational Function

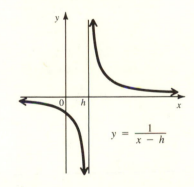

$$y = \frac{1}{x - h}$$

5.1 Exponential Functions

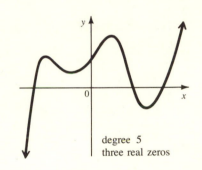

$y = a^x, 0 < a < 1$

$y = a^x, a > 1$

6.1, 6.2, 6.4 Values of Trigonometric Functions for Special Angles

Angle θ		sin θ	cos θ	tan θ
Degrees	Radians			
0°	0	0	1	0
30°	$\pi/6$	1/2	$\sqrt{3}/2$	$\sqrt{3}/3$
45°	$\pi/4$	$\sqrt{2}/2$	$\sqrt{2}/2$	1
60°	$\pi3$	$\sqrt{3}/2$	1/2	$\sqrt{3}$
90°	$\pi/2$	1	0	Undefined
180°	π	0	-1	0
270°	$3\pi/2$	-1	0	Undefined

5.2 Logarithmic Functions

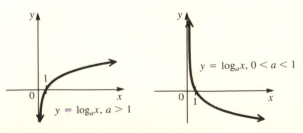

$y = \log_a x, 0 < a < 1$

$y = \log_a x, a > 1$

6.3 Conversion of Angular Measure

π radians = 180 degrees

6.4 Special Triangles

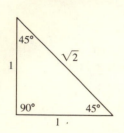
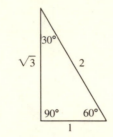

6.4 Trigonometric Functions

$$\sin \theta = \frac{y}{r} \quad \csc \theta = \frac{r}{y}$$

$$\cos \theta = \frac{x}{r} \quad \sec \theta = \frac{r}{x}$$

$$\tan \theta = \frac{y}{x} \quad \cot \theta = \frac{x}{y}$$

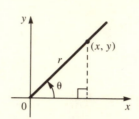

6.5, 6.6 Basic Trigonometric Graphs

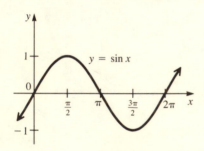
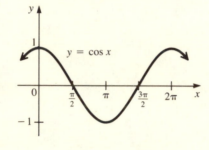
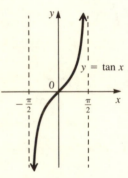

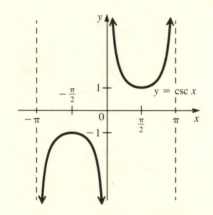
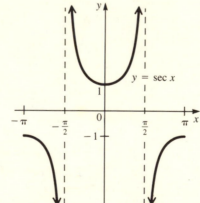
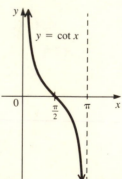

7.1 Fundamental Identities

$$\tan A = \frac{\sin A}{\cos A} \qquad \cot A = \frac{1}{\tan A}$$

$$\cot A = \frac{\cos A}{\sin A} \qquad \csc A = \frac{1}{\sin A}$$

$$\sec A = \frac{1}{\cos A}$$

$$\sin^2 A + \cos^2 A = 1$$
$$\sin^2 A = 1 - \cos^2 A$$
$$\cos^2 A = 1 - \sin^2 A$$
$$\tan^2 A + 1 = \sec^2 A$$

$$1 + \cot^2 A = \csc^2 A$$
$$\sin (-A) = -\sin A$$
$$\cos (-A) = \cos A$$
$$\tan (-A) = -\tan A$$